DEVELOPMENTS IN CORROSION PROTECTION

Edited by **M. Aliofkhazraei**

Developments in Corrosion Protection
http://dx.doi.org/10.5772/57010
Edited by M. Aliofkhazraei

Contributors

Amin. A. El-Meligi, Ion Patrascu, Eduard Gatin, R. Cara-Ilici, V. Vasilescu, S. Milicescu, Lucia Camelia Pirvu, Francisco Casanova Del Angel, Derek Northwood, Jorge Uruchurtu, Junaidah Jai, Sunday Onyebuchi Nwaubani, Gustavo Lopez, Peter Kusch, Nivin M. Mohamed Ahmed, Ime Obot, Florina Branzoi, Hui Guo, Francisco Bonnin-Pascual, Alberto Ortiz, Zhenhua Dan, Patricia Popoola, Rafael Martínez Palou, Moises Meza Pariona, M. Federica De Riccardis, Omar Fernandes Aly, Samuel Zelinka, Alicia Ares, Camila G. Dariva, Alexandre Ferreira Galio, Zahiraniza Mustaffa, Said Sayyah, Rosinei Ribeiro

Published by InTech

Janeza Trdine 9, 51000 Rijeka, Croatia

Publishing Process Manager Sandra Bakic
Technical Editor InTech DTP team
Cover InTech Design team

First published February, 2014
Printed in Croatia

Additional hard copies can be obtained from orders@intechopen.com

Developments in Corrosion Protection, Edited by M. Aliofkhazraei
p. cm.
ISBN 978-953-51-1223-5

For EU product safety concerns:
IN TECH d.o.o., Prolaz Marije Krucifikse Kozulić 3, 51000 Rijeka, Croatia,
info@intechopen.com or visit our website at intechopen.com.

Meet the editor

Dr. Mahmood Aliofkhazraei works in the corrosion and surface engineering group at the Tarbiat Modares University. He is the head of Aliofkhazraei research group. Dr. Aliofkhazraei has received several honors, including the Khwarizmi award and the best young nanotechnologist award of Iran. He is a member of the National Association of Surface Sciences, Iranian Corrosion Association, and National Elite Foundation of Iran. His research focuses on nanotechnology and its use in surface and corrosion science.

Contents

Preface

Corrosion is defined as the chemical or electrochemical reaction of material, generally a metal, with its environment which leads to change in its properties. Although corrosion occurs in all main groups of materials including metals, ceramics, polymers, and composites, it is more prevalent in metals and has detrimental effects. The first thing that comes to your mind after hearing the term "corrosion" is corrosion of a metal. Corrosion is a basically harmful phenomenon, but it can be useful in some cases. For instance, environment's pollution with corrosion products and damage to the performance of a system are among its harmful effects, whereas electric energy generation in a battery and cathodic protection of many structures are among its advantages. However, these advantages are almost nothing as compared to the costs and effects imposed by its detrimental influences.

The enormous costs of this phenomenon can be better understand through studying the published statistics on direct and indirect corrosion damages on economy of governments. A 2-year study in USA indicated that the direct cost of corrosion was 276 billion dollars which is almost 3.1% of the gross domestic product (GDP) of USA. A big share of this cost (137.9 billion dollars) is imposed to the industrial sections which cover a large number of subsets. Considering this huge cost, it is necessary to develop and expand the corrosion science and its protection technologies. This book collects new developments in corrosion science and protection. I like to appreciate all of contributors to this book and thank them for their high quality manuscripts. I wish open access publishing of this book help all researchers to benefit from this collection.

Dr. Mahmood Aliofkhazraei
Tarbiat Modares Universit
Iran

Section 1

Analysis of Corrosion

Chapter 1

Nanostructure of Materials and Corrosion Resistance

A. A. El-Meligi

Additional information is available at the end of the chapter

http://dx.doi.org/10.5772/57274

1. Introduction

The application of nanotechnology in the field of corrosion protection of metals and alloys attracts the attention of researchers. Many of these applications require good understanding of the corrosion behavior of the materials as a function of microstructure. Significant progress has been made in various aspects of synthesis of nano-scale materials. In addition, nanostructures promote selective oxidation, forming a protective oxide scale with superior adhesion to the substrate. Nanostructured materials of 1–100 nm are known for their outstanding mechanical and physical properties due to their extremely fine grain size and high grain boundary volume fraction [1]. They are important due to their unique properties that may lead to new and exciting applications [2]. Nanocomposite of polymer coating can effectively combine the benefits of organic polymers, such as elasticity and water resistance to that of advanced inorganic materials, such as hardness and permeability. The nanostructured silica coating showed comparable or better performance than hexavalent chrome passivation [3, 4]. Such behavior of nanostrucured materials, which relates to corrosion resistance, relies on materials microstructure. In fact, most properties of solids depend on the materials microstructure. The microstructure includes a number of parameters such as the chemical composition, the arrangement of the atoms (the atomic structure) and the size of a solid in one, two or three dimensions [5]. Comparable variations have been noted if the atomic structure of a solid deviates far from equilibrium or if its size is reduced to a few interatomic spacing in one, two or three dimensions. Nanoporous metals (NPMs) made by dealloying represent a class of functional materials with the unique structural properties of mechanical rigidity, electrical conductivity, and high corrosion resistance [6]. It is stated by Weissmüller et. al, that the prospect of using alloy corrosion as a means of making nanomaterials for fundamental studies and functional applications has led to a revived interest in the process. The quite distinct mechanical properties of nanoporous metals are one of the focus points of this interest, as relevant studies probe the deformation behavior of crystals at the lower end of the size scale. Furthermore, the coupling of bulk stress and strain to the forces acting along the surface of nanoporous metals provide unique opportunities for controlling the mechanical behavior through external variables such as the electrical or chemical potentials [7]. The relation between chemical architecture of thin (10 and 50 nm) chromium and tantalum oxide coatings grown by filtered cathodic arc deposition (FCAD) on carbon steel and their corrosion protection properties is reported [8]. In fact, the coatings of oxides improve sealing properties of the alloy, but there is an initiation for the localized corrosion. This may be due to the imperfectness of the coats and improper finishing for the alloy surface. The corrosion performance in chloride media of chromium coatings obtained by direct current (DC) and pulse current (PC1 and PC2) was studied [9]. The obtained results show similar hardness values and a superior corrosion performance of PC2 coatings, which is related to their roughness, morphology, wettability and mainly, their crack absence. Silver/silicon oxide core–shell nanoparticles (60 nm) were used in marine antimicrobial corrosion coatings [10]. The results show that the nanoparticles coat could enhance long-term corrosion protection in comparison with the copper biocides. This composite has a strong potential use in environmentally friendly antimicrobial coatings.

2. Properties of nanostructured materials and controlling procedures

Nanostructured materials can be produced by using different synthesis and processing methods. This will lead to variation of their grain sizes or phase dimensions in the nanometer size regime. The interest in these new ultrafine-grained materials results primarily from the special nature of their various physical, chemical, and mechanical properties and the possibilities of controlling these properties during the synthesis and subsequent processing procedures. Since it is now becoming increasingly apparent that their properties can be engineered effectively during synthesis and processing, and that they can also be produced in quantity, nanophase materials should have considerable potential for technological development in a variety of applications [11]. As the properties of solids depend on size, atomic structure and chemical composition, the nanostructured materials exhibit new properties due to one or several of the following effects [12]:

2.1. Effects of size

The effects of size result if the characteristic size of the building blocks of the microstructure is reduced to the point where critical length scales of physical phenomena (e.g. the mean free paths of electrons or phonons, a coherency length, a screening length, etc.) become comparable with the characteristic size of the building blocks of the microstructure [12].

2.2. Dimensionality change

If nanostructured materials consist of thin needle-shaped or flat, two-dimensional crystallites, they become a two- or one-dimensional system with respect to this phenomenon [13]. For example, layered materials MPS_3, where M is the transition metals, are ordered in layers with

http://dx.doi.org/10.5772/57274

interspacing of 6.4 Å, as shown in Fig. 1 [14]. These materials are crystalline and could be used for many applications, such as, a cathode for the rechargeable batteries, sensors, optical applications, and hydrogen storage [15]. Processing map for hot working of hot extruded AZ31–1.5 vol% nano-alumina magnesium composite (AZ31–NAL) prepared by disintegrated metal deposition (DMD) technique has been developed in the temperature range of 250–500 °C and strain rate range of 0.0003–10 S_{-1} [16]. The nano-alumina composite helps in reducing the preferred orientation in AZ31 alloy. This may increase its corrosion resistance.

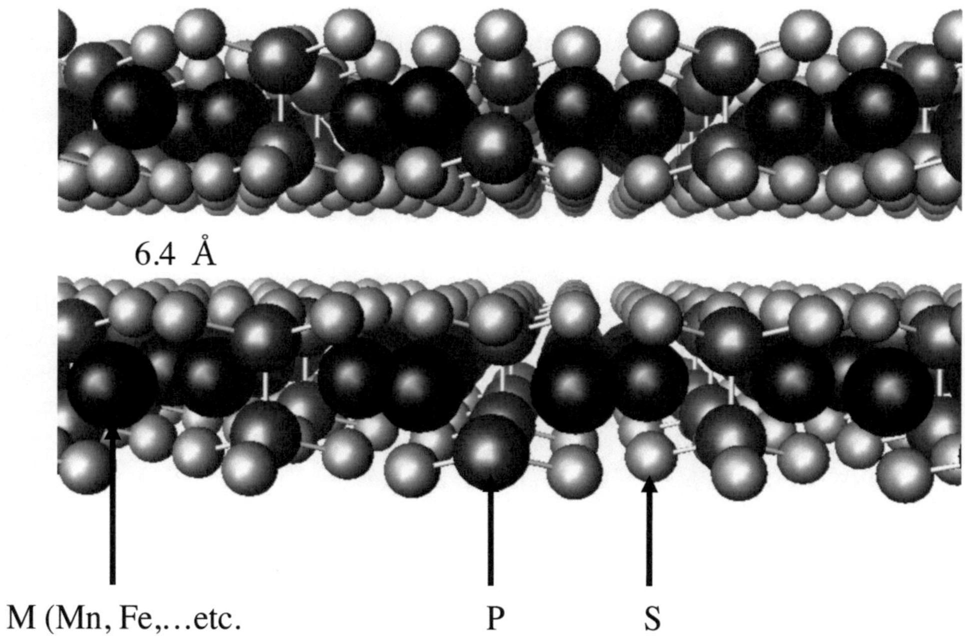

Figure 1. Layered structure of MPS3 [14].

2.3. Atomic structure

A high density of incoherent interfaces or other lattice defects such as dislocations, vacancies, will change the atomic structure [12]. The cores of lattice defects represent a constrained state of solid matter differing structurally from (unconstrained) crystals and/or glasses. As a consequence, a solid containing a high density of defect cores differs structurally from a defect-free solid with the same (average) chemical composition. The boundaries represent an example of this effect: the misfit between adjacent crystallites changes the atomic structure (e.g. the average atomic density, and the nearest-neighbour coordination, etc.) in the boundary regions relative to the perfect crystal. At high defect densities the volume fraction of defect cores becomes comparable with the volume fraction of the crystalline regions. In fact, this is the case if the crystal diameter becomes comparable with the thickness of the interfaces, i.e. for crystal sizes on the order of one or a few nanometers as is the case in nanostructured materials.

2.4. Alloying elements

The alloying elements have great influence on the alloys structure and properties. For example, the addition of alloying elements less noble than iron increases the corrosion resistance in the active region improving the protective quality of the corrosion product film in which the cations of alloying elements are significantly concentrated [17]. The influence of the alloying elements on the interface reactions of zinc coatings during the galvanization process was examined [18]. These reactions affect the crystallization and the structure and properties of the outer layer of the coatings. Depending on the type and concentration of the alloying additions in the galvanizing bath differences were induced in the crystallization process of the Fe-Zn phases. It was found that both the concentration and the distribution of the alloying elements played an important role in the growth of the phases. It has been pointed out for alloys with elements, which are capable of forming passive films, that the atoms of these elements can diffuse easily through the grain boundaries to the surface of the alloy to form the protective passive layer [19, 20].

The following cases of this type of immiscible components in nanostructured may be distinguished: solute atoms with little solubility in the lattice of the crystallites frequently segregate to the boundary cores (e.g. the free energy of the system in several alloys is reduced if large solute atoms segregate to the boundary core). The second case of nanostructured alloys results if the crystallites of nanostructured materials have different chemical compositions. Even if the constituents are immiscible in the crystalline and/or molten state (e.g. Fe and Ag), the formation of solid solutions in the boundary regions of the nanostructured has been noticed [12].

3. Corrosion of materials

Corrosion is the chemical or electrochemical reaction between a material and its environment. It produces deterioration of the material or of its properties. Also, corrosion has different forms of attack [21-23]. The basic corrosion theory is that the corrosion is an electrochemical reaction composed of two half cell reactions, an anodic reaction and a cathodic reaction. The anodic reaction releases electrons (equation 1), while the cathodic reaction consumes electrons (equations: 2-5). Each half-cell reaction has an electrical potential, known as the half-cell electrode potential. The anodic reaction potential, E_a, plus the cathodic reaction potential, E_c, adds up to E, the cell potential. If the overall cell potential is positive, the reaction will proceed spontaneously. The corrosion reactions can be represented as follows:

Anodic reaction:

$$M \rightarrow M^{n+} + ne- \tag{1}$$

Where M stands for a metal and n stands for the number of electrons that an atom of the metal will easily release, i.e. for iron and steel:

$Fe \rightarrow Fe^{2+} + 2e-$

Cathodic reactions:

- Oxygen reduction in acidic solution

$$O_2 + 4\,H^+ + 4e- \longrightarrow 2H_2O \tag{2}$$

- Oxygen reduction in neutral or basic solution

$$1/2O_2 + H_2O + 2e- \longrightarrow 2OH^- \tag{3}$$

- Hydrogen evolution from acidic solution

$$2H^+ + 2e- \longrightarrow H_2 \tag{4}$$

- Hydrogen evolution from neutral water

$$2H_2O + 2e- \longrightarrow H_2 + 2OH \tag{5}$$

The chemical attack or dry corrosion occurs under dry conditions, such as high temperatures in gaseous environments, molten salts and liquid metals. Dry corrosion process is a direct reaction between a metal and the corrosive environment. Dry corrosion is of great importance in a number of petroleum refining processes. It includes the attack of hydrogen sulfide and other sulfur compounds on steel and various alloys at elevated temperatures. Solutions to this type of corrosion generally depend on metallurgical approaches, e.g. variations in composition, heat treatment of the selected metal or alloy. In fact, high temperature corrosion of metals and alloys always involves oxidation process. For example, a metal reacts with oxygen at high temperature by initial adsorption of oxygen, chemical reaction to form the surface oxide, oxide nucleation and lateral growth into a continuous film. Such film may protect the underlying metal [24]. One of the well known oxidation is the iron oxidation in oxygen (as pure metal or in carbon steel alloys) at high temperature [25].

Wet corrosion is an electrochemical process; in practice it limits to nearly 232 ºC as an upper temperature. The electrochemical corrosion results from reaction between a metal surface and an ion-conducting environment. This process occurs by metal contact with an electrolyte for transport of electric current. Most cases of electrochemical corrosion proceed in aqueous media such as natural water, atmospheric moisture, rain, and wet soil. Also, other environments enhances the corrosion process e.g. acids, petroleum products, cooling water, chemical solutions,…etc. For example, water presents in refinery by different sources, such as the crude itself, through injection of water of steam to aid in the steam distillation of various petroleum fractions, water washing or aqueous solution contacting various intermediate and product streams in refining and petrochemical processes [21].

In fact, metal surfaces have corrosion cells in contact with an electrolyte. Different areas on the metal surface could also have different potentials with respect to the electrolyte [26, 27]. These variations could be due to a number of factors: i) metallurgical behaviors, i.e., differences in their composition, microstructure, fabrication, and field installations, and ii) environmental factors. Carbon and low alloy steels are the most widely used material in different industries in corrosive environments and the metallurgical factors have great influence on the corrosion process [28, 29].

3.1. Corrosion resistance of alloys

A number of ways have developed to increase the corrosion resistance of metals such as applying inhibitors, adding passive alloy element, putting protective coating and so on. The application of inhibitors has to follow some regulations such as environmental impact, low toxicity, and structure of inhibitors and alloys. Nanostructure inhibitors are very important in protection metals and alloys. Changing micro-structure of materials will bring surprised results to promote the corrosion resistance of metals/alloys [30-33]. This change occurs by using new micro-structures such as single crystalline, nanocrystalline or microcrystalline and amorphous to replace the traditional cast alloy [34]. Nanostrucred materials characterize by their small grain sizes (1-100 nm) and high volume fraction of grain boundaries. These properties improve unique physical, chemical and mechanical properties compared with those of their cast counterparts [34]. The properties are particularly useful in corrosion protection applications.

4. Nanosize and effect on corrosion resistance

Electrodeposition is a versatile technique for producing nanostructured materials. It is a technologically and economically viable production route to metals, alloys and metal matrix composites, both in bulk form and as coatings. Properties of nano-structured electrodeposits such as hardness, wear resistance and electrical resistivity are strongly grain size dependent. Corrosion behaviour of nanostructured-alloys has been assessed by several techniques in various environments [35–39]. Thorpe et al. reported an enhanced corrosion resistance of nanostructured- Fe32-Ni36-Cr14-P12-B6 than that of its amorphous counterpart [35]. The authors attributed this improved corrosion resistance to the observed greater Cr-enrichment of the electrochemical surface film via rapid interphase boundary diffusion. Bragagnolo et al, report improved corrosion resistance with nanostructured-Fe72Si10B15Cr3 metallic glass wires. In their study, the beneficial effects of nanostructured processing for corrosion resistance were not evident with a non-passivating alloy composition [36]. Zeiger et al, report an enhanced corrosion resistance of nanostructured-Fe-8 wt % Al in Na_2SO_4 solution [37]. The study shows that defect density (density of grain boundaries) promotes metal dissolution. As the diffusion of aluminum is fast enough in the grain boundaries, it is possible that the oxide film on nanostructured and crystalline-FeAl8 alloy provides better protection than on the same polycrystalline alloy. Weak acid and weak base ease passive film formation of nanostructured-FeA18 where Al forms stable passive film. The newly developed Ti60Cu14Ni12Sn4Nb10 nanostructured dendrite composite exhibits a particular microstructure that confers outstanding mechanical properties and improves corrosion resistance [40]. It has been reported that the presence of a ductile second phase can significantly improve the mechanical properties of nanocrystalline alloys. Newly developed Ti-base bulk alloy with a dendritic phase dispersed in a nanostructured matrix greatly enhances the usual low ductility of homogeneous nanocrystalline alloy. Polarization tests of Ti60Cu14Ni12Sn4Nb10 copper mould cast specimens indicate good corrosion resistance in acidic media (0.5N H_2SO_4), medium resistance in alkaline (0.5N NaOH) but the alloy undergoes pitting in chloride media. The behaviour of the alloy in H_2SO_4 is similar as the typical behaviour of pure titanium and titanium alloys [41]. It exhibits a wide anodic passive region between 90 mV and 1.5V (SCE), approximately, and passive current densities in the order of 10_{-3} mA due to the formation of highly protective surface films. The voltage range of passivation in NaOH electrolyte is shorter than in H_2SO_4. It finishes when the current increases to 470mV (SCE), approximately. In chloride media, sharp increase in current density interrupts the passivation during anodic polarization. The interruption occurs by a sudden passive film breakdown followed by an excessive pit growth and the subsequent dissolution of the surface of the alloy. The increase of NaCl concentration shifts pitting potentials to more negative values. This behaviour differs from general findings on polycrystalline pure titanium and dilute titanium alloys. The pitting corrosion and localized attack on the alloys don't observe in marine environments [42].

4.1. Corrosion resistance of stainless steel

Stainless steel uses as a structural material in many applications. Corrosion and wear limits its service life. Effective measures are taken to improve its corrosion and wear resistance properties [42–47]. A passive film precludes contact between corrosive environments and the metallic surface. However, weakness of the passive layer can trigger corrosion. The use of inhibitive species constitutes active corrosion protection to decrease corrosion intensity. The use of both strategies jointly could adequately protect the metallic substrate. The conversion coatings such as chromating and phosphating treatments have historically played important role in different industries. It provids corrosion protection to both ferrous and non-ferrous metals. Also in many cases the anticorrosive property of zinc phosphate treatment is insufficient for outdoor exposure. Electroplated Ni and Cr are relatively expensive and Cr(VI) is environmentally unacceptable [48]. As an alternative, siloxane based sol–gel coatings have been suggested as passive corrosion protection. It has the ability to form dense Si–O–Si network adhering to substrate through Me–O–Si bonds. The bonds form good corrosion impeding properties [48–54]. Adding specific organic functional groups such as siloxane improve adhesions on alloys surface. Clusters of metal oxides reinforces silica network during sol–gel synthesis from the respective metal alkoxides [55, 56].

Additionally, sol–gel coatings do not need high process temperature and vacuum conditions are crucial in the case of many of thin layer coating techniques [57–60]. Furthermore, sol–gel is relatively simple, low-cost and applicable to materials having complex geometry. It is considered to have low negative environmental impact. Nevertheless, defects or pores in the sol-gel coating let in the corrosive species and initiate the corrosion processes [61]. The defects indicate the need for additional active corrosion protection. Incorporation of inhibitive species into inorganic fillers enhances corrosion resistance [62, 63]. Many kinds of inhibitors could be used to protect various metallic materials [64-73]; however, very often this negatively influences the stability of the sol–gel matrix [74, 75]. Acetylenic alcohols are considered excellent inhibitors of corrosion [76]. Nanostructure sol–gel impregnated with propargyl alcohol can protect defects of sol-gel [77].

Anticorrosive primers and coatings based on conducting polymers gain momentum Owing to the stringent environmental regulations on the usage of toxic heavy metals in the formulation of corrosion protective coatings [78, 79]. The recent advancements in nanotechnology have hastened the development of high performance nanostructured coatings. They have a broad spectrum of anticorrosive activity under a wider range of hostile environments. Inhibitor coatings based on conducting polymers are either chemically or electrochemically deposited on the metal substrate [80–87].

4.2. Surface modification

ZrO_2, a bioceramic, coats stainless steel 316L to insulate the needles and improve their biocompatibility. Three types of stainless steel 316L specimens were used in the experiment: an untreated specimen, an electropolished specimen and an electropolished and plasma treated specimen. The effects of electropolishing and plasma treatment on coating strength and the relationship between coating thickness and electrical conductivity were investigated. The result showed that the coating strength of the plasma treated sample was 130% greater than that of the untreated sample and 40% greater than that of the electropolished one. The electrical insulation characteristic of ZrO_2 coating on stainless steel 316L was superior to that of uncoated stainless steel 316L. The appropriate thickness of ZrO_2 coating is 500 nm for the electrical insulation. Compared to untreated specimens, the surface roughness of treated specimens decreased by as much as 80% after 5 min, and thus, this was the most effective time to carry out the electropolishing, as shown in Fig. 2 [88].

The high-current pulsed electron beam (HCPEB) technique has been proven to be an efficient method for surface modification of alloys [89–91]. Modifications improve the overall performance of alloys such as microhardness, corrosion resistance, and wear resistance. The structure and phase transformations in the near-surface layers of a low carbon steel (0.2 wt.% C) subjected to a high-current pulsed electron beam (HCPEB) treatments have been investigated by using electron microscopy and X-ray diffraction (XRD) [92]. A nanostructure consisting of cementite and C-supersaturated austenite is formed in the near-surface region after multiple bombardments. This is confirmed by the XRD patterns shown in Fig. 3. According to the peak profile analysis, the average grain sizes of the γ phase (austenite) in the surface layer after the 5- and 10-pulse bombardments by the HCPEB are about 6.5 and 4.2 nm, respectively.

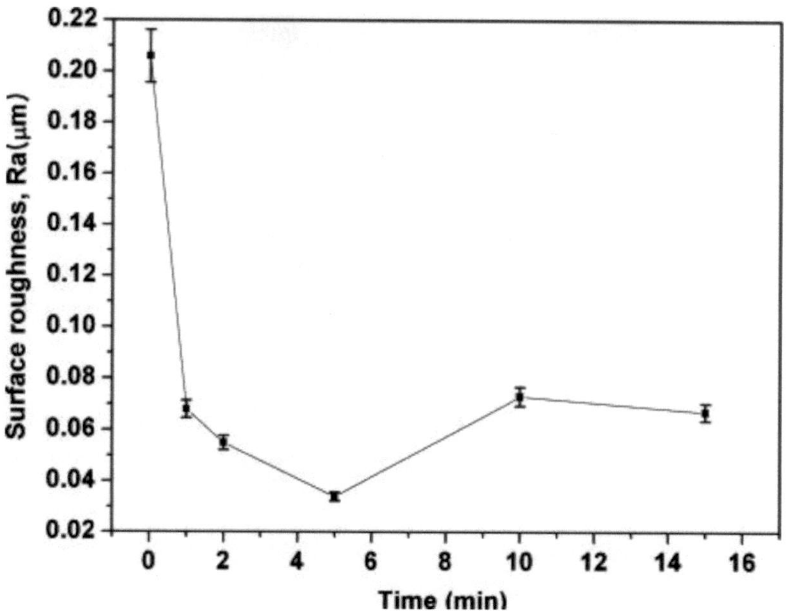

Figure 2. Variation of surface roughness as a function electropolishing time [88].

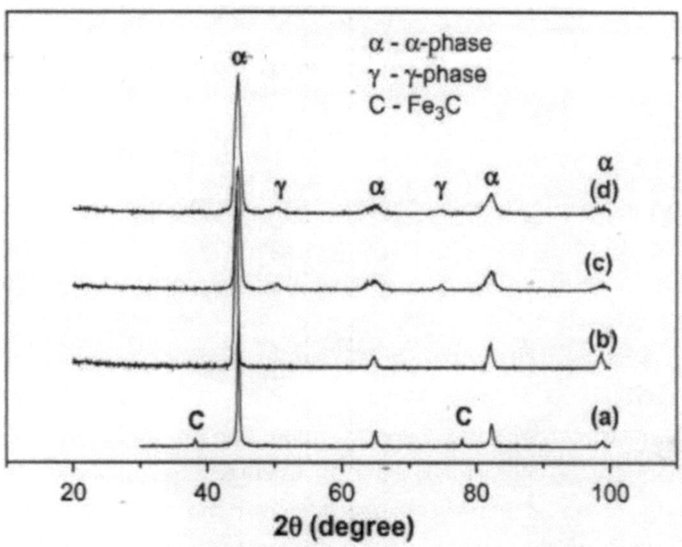

Figure 3. XRD patterns collected from the surface of the initial no. 20 steel (a), irradiated with 1 pulse (b), 5 pulses (c), and 10 pulses (d). The energy density of HCPEM is 4 J cm^2 [92].

4.3. Stainless steel corrosion and nanostructure

The family of stainless steels (SS) is large and varied; in fact, there are more than 100 registered types. Each of these was originally designated for some specific use. For example, SS316 is used for severe environments. Of course, there are many industrial processes that require a higher level of resistance to corrosion. Carburizing has been applied to austenitic stainless steel to promote increased wear and corrosion resistance [93]. Nanometric-sized carbide-based crystallites have been synthesized successfully through the pulsed nanocrystalline plasma electrolytic carburizing method on the surface of 316 austenitic stainless steel [94]. It was found that the corrosion resistances of coated samples have a direct relationship with the average size of complex carbide-based nanocrystallites. The synthesized carbide-based crystallites exhibit average sizes at around 37–80 nm. The minimum average size of nanocrystals was found to be 37.7 nm in 600 V of peak of applied cathodic voltage, 10 kHz of frequency of pulsed current, 40 °C of temperature of electrolyte, 10 min of treatment time.

Polarization resistances of different samples found through the potentiodynamic corrosion tests performed after surface hardening treatment. The polarization resistances due to corrosion in specimen groups treated at different conditions varied between 153.4 $k\Omega/cm^2$ and 635 $k\Omega/cm^2$. The figure explains which specimen has maximum or minimum corrosion resistance. From the above results, it can easily be said that a high corrosion resistance has been obtained with higher applied voltages and treatment times of carburizing. These results indicate that the corrosion resistances of the obtained layers are a direct function of the average size of the nanocrystallites [94].

Improvement in the surface layers' properties by changing their structure remains actual due to the fact that the destruction of the materials, including fatigue failure, wear and corrosion depends usually on the surface structural state [95]. The surface nanostructure, which is formed in the surface layers with thickness of about of 30–50 nm, plays a main role in the enhancement of properties [96-99]. At the same time, only a few recent studies have paid attention to the corrosion behavior of stainless steels with nanograined surface structure [100–102]. These studies have shown that the corrosion resistance can also be markedly improved, particularly by shot peening [103] or sandblasting with subsequent low-temperature annealing [104, 105]. For example, a nanocrystalline surface layer was produced on an AISI-321 stainless steel by severe plastic deformation via ultrasonic peening (UP) [106]. The nanostructured surface layer formed after straining already contains mainly the martensite nanograins characterized by an average size of about 10 nm. Grain size increased gradually up to 60 nm within the layer containing both austenite and martensite phases at a depth of about 30 nm from the treated surface. Both the microhardness behavior of the stainless steel surface and its corrosion performance in 3.5% NaCl solution can be enhanced by the UP. They are shown to be in correlation with: (i) the grain refinement process and (ii) the increase in the volume fraction of strain-induced martensite.

4.3.1. Effect of nanostructure on passive film

Nanostructure changes the thickness of passive film. XPS Cr spectra from the passive films in 3.5% NaCl solution on nanostructure (NS) coating and the cast alloy are shown in Fig. (4) [107]. After 40 s bombardment using the same bombardment parameters, the Cr^{3+} peak disappeared in the passive film on the NS coating but was still present in the film on the cast alloy. This indicates that the passive film on the NS coating was thinner than that on the cast alloy in normal NaCl solution. However, nanostructure increased the thickness of passive film on the magnesium alloy with rare earth elements[108]. It is well known that the passive film on the magnesium alloy is mainly the corrosion product layer on the sample, which blocks the dissolution of the material and inhibits the corrosion. Therefore, nanostructure promotes the dissolution of alloy and forms more products on the sample. Finally, the film layer was much thicker than that of the cast alloy. Nanostructure also changes the compact property of the passive film. After cathodic reduction, the variation of current with time could be measured at a fixed potential. If the contribution of the double layer charge is neglected, the initial drop of current density should be related to the growth of a protective film on the electrode surface. The current decreases with time as follows[109]:

$$I = 10 - \left(A + kg\ t\right) \tag{6}$$

where I represents current density, t is time, A is constant and k represents the slope of the double-log plot for potentiostatic polarization. k = -1 indicates the formation of a compact, highly protective passive film, while k = -0.5 indicates the presence of a porous film, growing as a result of a dissolution and precipitation process. A porous passive film presents on the 309 stainless steel and nanostructure improves formation of a compact film [110].

4.4. Patents of corrosion inhibition

It has been found that low levels of molybdate in combination with low levels of various phosphorous containing compounds controls corrosion in ferrous metals [111, 112]. In addition, foaming, sludge formation, and/or focused or localized corrosion has not been observed in connection with the use of the molybdate with certain phosphorous containing compounds. Additionally, the use of molybdate in combination with phosphorous containing compounds prevents or reduces the formation of precipitates, including iron and other potential precipitates [113]. A composition and method is disclosed for inhibiting the corrosion of metals in contact with an aqueous system capable of corroding the metal. The inventive composition contains a substantially water-soluble polymer of an acidic amino acid and at least one water-soluble salt of molybdenum or zinc. The composition is substantially non-toxic and environmentally acceptable and is supplied in a corrosion-inhibition amount to the otherwise metal-corrosive aqueous system [114]. A corrosion inhibitor for use with Urea Ammonium Nitrate solutions is disclosed, comprising a blend of molybdate and one or more of the inorganic phosphates (including phosphates,

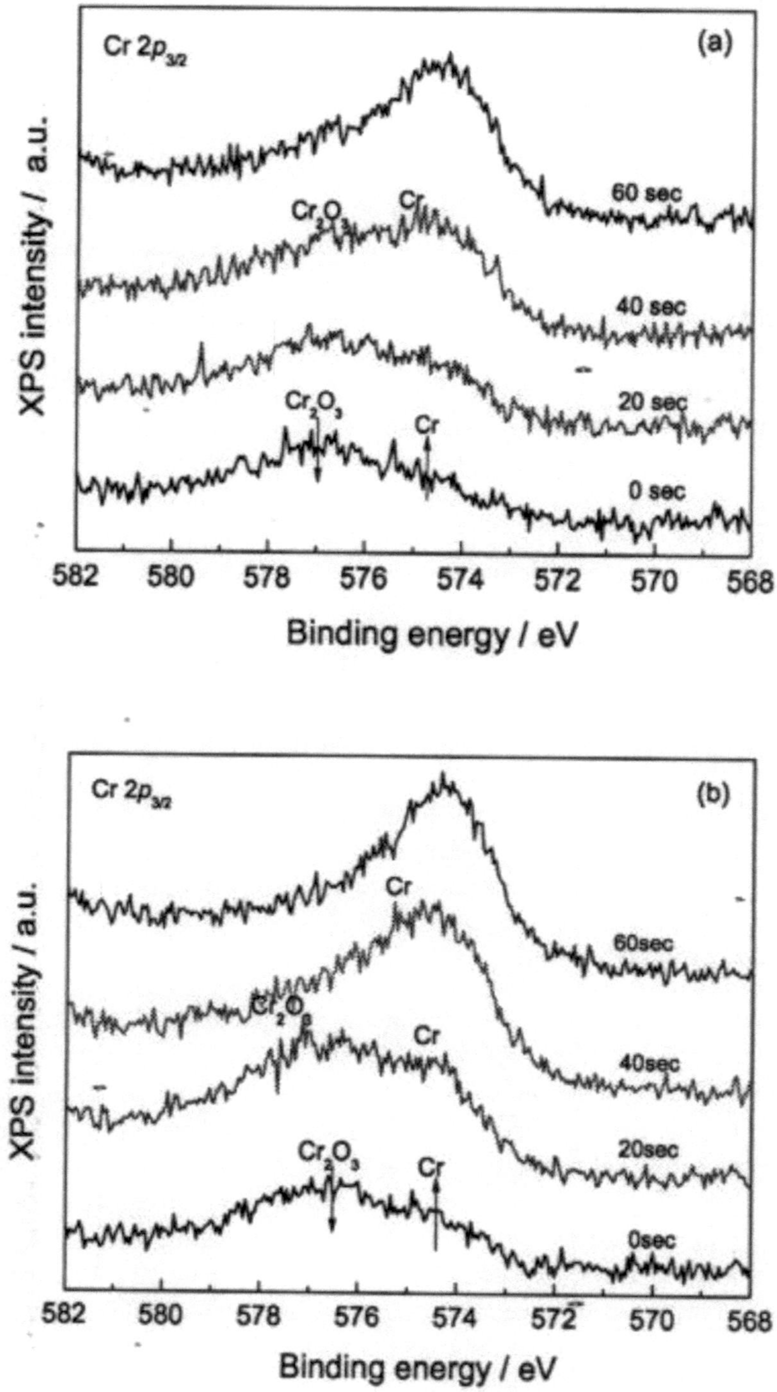

Figure 4. Cr 2p XPS depth profile spectra from the cast alloy (a) and the NC coating (b) after passivation 30 min at 0 V in 3.5% NaCl solution. The depth is indicated in spectrum[107]

polyphosphates, and pyrophosphates) and organic phosphates or phosphonates. Inorganic phosphates include, but are not limited to, SHMP (Sodium Hexametaphosphate) and TKPP (Tetra-Potassium Pyrophosphate) [115]. A corrosion inhibitor composition for a fuel, comprising a plurality of nanoparticles formed of an inorganic composition having an average longest dimension of 1 nanometer to 100 nanometers, wherein the inorganic active composition is insoluble in the fuel and is adapted to react with a corrosion causing contaminant [116]. Catalytic layers for use in the electrodes of fuel cells including a non-noble metal substrate layer coated with one or a few monolayers of noble metal, such as Pt. These thin, highly porous structures with large catalytically active surface areas, should exhibit a significantly higher power output per mg of Pt and per cm^2 of the membrane than the current Polymer Electrolyte Fuel Cells catalytic layers. [117]. Nanoparticles comprising molybdenum, methods of manufacturing nanoparticles comprising molybdenum, and nanotechnology applications of nanoparticles comprising molybdenum, such as electronics, optical devices, photonics, reagents for fine chemical synthesis, pigments and catalysts, are provided [118]. The present invention relates to methods and compositions for reducing damaging oxidation of metals. In particular, the present invention relates to nanoparticle surface treatments and use of nanoparticle surface treatments to reduce the damaging oxidation and corrosion of stainless steel and other alloy components in oxidizing and corrosive conditions [119-121].

5. Conclusion

Properties of nano-structured electrodeposits such as wear resistance, electrical resistivity and hardness depend strongly on grain sizes of metals and alloys. Nanocrystallization decreases grain size and changes surface condition. The surface condition significantly influences the electrochemical corrosion behaviors of metals/alloys in liquid system. There are a number of advanced techniques have been developed to attain the nanostructure. Many applications of the nanomaterials made to improve the devices and industries. The application of nanotechnology in the field of corrosion prevention of metals is very important. Nanometric-sized carbide-based crystallites have been synthesized successfully through the pulsed nanocrystalline plasma electrolytic carburizing method on the surface of 316 austenitic stainless steel. The corrosion resistance of coated samples has a direct relationship with the average size of complex carbide-based nanocrystallites. Significant corrosion resistance attained with higher voltages and treatment times of carburizing. These results indicate that the corrosion resistances of the obtained layers are a direct function of the average size of the nanocrystallites. In local corrosion, nanocrystallization increases the unstable points on the surface of the materials, which increases the possibility of local corrosion. However, the excellent ability of element diffusion helps to heal the local corrosion points. These points inhibit the growth of the local corrosion. The research continues to develop nanostructure of inhibitors and alloys. This will reduce the corrosion of metals and alloys.

Study (reference number)	Main Findings
[16]	The nano-alumina composite helps in reducing the preferred orientation in AZ31 alloy. This may increase its corrosion resistance.
[34]	Nanostrucred materials characterize by their small grain sizes (1-100 nm) and high volume fraction of grain boundaries. These properties improve unique physical, chemical and mechanical properties compared with those of their cast counterparts. The properties are particularly useful in corrosion protection applications.
[88]	The appropriate thickness of ZrO_2 coating is 500 nm for the electrical insulation. The electrical insulation characteristic of ZrO2 coating on stainless steel 316L was superior to that of uncoated stainless steel 316L.
[92]	A nanostructure consisting of cementite and C-supersaturated austenite improve corrosion and wear resistance for low carbon steel.
[93]	Carburizing has been applied to austenitic stainless steel to promote increased wear and corrosion resistance.
[94]	Nanometric-sized carbide-based crystallites have been synthesized successfully through the pulsed nanocrystalline plasma electrolytic carburizing method on the surface of 316 austenitic stainless steel. It was found that the corrosion resistances of coated samples have a direct relationship with the average size of complex carbide-based nanocrystallites.
[100-105]	Only a few recent studies have paid attention to the corrosion behavior of stainless steels with nanograined surface structure. These studies have shown that the corrosion resistance can also be markedly improved, particularly by shot penning or sandblasting with subsequent low-temperature annealing.
[108]	Passive film on the nanostructure increases the thickness of passive film on the magnesium alloy with rare earth elements. The passive film on the magnesium alloy is mainly the corrosion product layer on the sample, which blocks the dissolution of the material and inhibits the corrosion.
[110]	A porous passive film presents on the 309 stainless steel and nanostructure improves formation of a compact film.
[119]	Nanoparticle surface treatments use to reduce the damaging oxidation and corrosion of stainless steel and other alloy components in oxidizing and corrosive conditions.

Table 1. Main findings of some studies presented in the text.

Author details

A. A. El-Meligi*

Address all correspondence to: aemeligi@amaiu.edu.bh, ael_meligi10@hotmail.com

Head of Research and Publication, AMA International University-Bahrain, Bahrain

References

[1] Nalwa H. S. Handbook of Nanostructured Materials and Nanotechnology, Vol. 1, Academic Press, San Diego, 2000

[2] Saji V. S. and Joice Thomas, Nanomaterials for corrosion control, Current Science, Vol. 92, No. 1, 2007, 51.

[3] Honggang, J., Maggy, L., Victoria, L. T. and Enrique, J. L., Synthesis of nanostructured coatings by high velocity oxygen fuel thermal spraying. In Handbook of Nanostructured Materials and Nanotechnology, Vol. 1 (ed. Nalwa, H. S.), Academic Press, San Diego, 2000, pp. 159–209.

[4] Kumaraguru, Swaminatha P. Development of novel nanostructured materials with superior electrocatalytic and corrosion properties, PhD, University Of South Carolina, 2006.

[5] Gleiter H., Nanostructured materials: basic concepts and microstructure, Acta Materialia 2000; 48(1), 1-29.

[6] Yi Ding and Mingwei Chen, Nanoporous Metals for Catalytic and Optical Applications, Materials Research Society Bulletin, Vol. 34(08) 2009, 569-576.

[7] Jörg Weissmüller, Roger C. Newman, Hai-Jun Jin, Andrea M. Hodge and Jeffrey W. Kysar Nanoporous Metals by Alloy Corrosion: Formation and Mechanical Properties, Materials Research Society Bulletin, Vol. 34(08) 2009, 577-586.

[8] Belén Díaz, Jolant Światowsk, Vincent Maurice,,Marcin Pisarek, Antoine Seyeux, Sandrine Zanna, Sanna Tervakangas, Jukka Kolehmainen, Philippe Marcus, Chromium and tantalum oxide nanocoatings prepared by filtered cathodic arc deposition for corrosion protection of carbon steel, Surface and Coatings Technology, 206(19–20) 2012, 3903–3910.

[9] N. Imaz, M. Ostra, M. Vidal, J.A. Díez, M. Sarret, E. García-Lecina, Corrosion behaviour of chromium coatings obtained by direct and reverse pulse plating electrodeposition in NaCl aqueous solution, Corrosion Science, Accepted Manuscript, Available online 12 October 2013

[10] Yuan Le, Pengtao Hou, Jiexin Wang, Jian-Feng Chen, Controlled release active antimicrobial corrosion coatings with Ag/SiO2 core–shell nanoparticles, Materials Chemistry and Physics, 120(2–3) 2010, 351–355

[11] Richard W. Siegel, Synthesis and properties of nanophase materials, Materials Science and Engineering: A, 1993; 168(2), 189.

[12] Gleiter, H., Prog. Mater. Sci., 1998, 33, 223.

[13] Gleiter, H., Nanostruct. Mater., 1995, 6, 3.

[14] El-Meligi AA. Synthesising Pure Material of FePS3 and its Intercalation with Pyridinum, Mater. Chem. and Phys. 2005; 89(2-3), 253.

[15] Ismail N, El-Meligi A. A., Temerk Y. M., Badr M. A, and Madian M., Synthesis and characterization of MnPS3 for hydrogen sorption, J. of Solid State Chemistry, 2010; 183, 98.

[16] T. Zhong, K.P. Rao, Y.V.R.K. Prasad, F. Zhao, M. Guptad, Hot deformation mechanisms, microstructure and texture evolution in extruded AZ31–nano-alumina composite Materials Science and Engineering: A, Vol 589(1) January 2014, Pages 41–49

[17] Herr, U., Jing, J., Gonser, U. and Gleiter, H., Alloy Effects in Consolidated Binary Mixtures of Nanometer-sized Crystals, Solid St. Commun., 1990, 76, 192.

[18] Vourlias G, Pistofidis N, Stergioudis G., Tsipas D. The effect of alloying elements on the crystallization behaviour and on the properties of galvanized coatings, Crystal Research and Technology, 2004; 39(1), 23–29.

[19] Piehl L. Naphthenic Acid Corrosion in Crude Distillation Units Materials Performance, 1988; 44(1), 37.

[20] Craig H.L. CORROSION/95, NACE, Houston, Texas, 1995, p. 333.

[21] Nathan C. C. Corrosion Inhibitors, NACE, Houston, Texas, 1981.

[22] Trethewey K. R. and Chamberlain J. Corrosion for Science and Engineering, 2nd edt., Longman Group Limited, 1995.

[23] During E. D. D. Corrosion Atlas, 3rd edt., Elsevier Science, 1997.

[24] Kofstad P. High Temperature Corrosion, Elsevier Applied Science Publishers Ltd, 1988.

[25] Fontana M. J. Corrosion Engineering, "High Temperature Corrosion", Chap. 11, 3rd ed, McGraw-Hill, London, 1988..

[26] Shaikh H, Anita T, Dayal R. K, Khatak H.S. Effect of metallurgical variables on the stress corrosion crack growth behaviour of AISI type 316LN stainless steel, Corrosion Science, 2010; 52(4), 1146-1154.

[27] Pardo A, Merino, M.C, Coy A.E, Viejo F, Arrabal R, Matykina E. Effect of Mo and Mn additions on the corrosion behaviour of AISI 304 and 316 stainless steels in H2SO4, Corrosion Science,2008; 50(3), 780.

[28] Kekkonen T, Aaltonen P, Hänninen H. Metallurgical effects on the corrosion resistance of a low temperature sensitized welded aisi type 304 stainless steel Corrosion Science, 1985; 25(8-9), 821.

[29] G. Salvago, G. Fumagalli, A stochastic analysis of potential fluctuation during passive film breakdown and repair on iron, Corrosion Science, 1992; 33(6), 985.

[30] A. Gebert, U.Wolf, A. John, J. Eckert and L. Schultz: Mater. Sci. Eng. A, 2001, 299, 125.

[31] R.V.S.Rao, U.Wolf, S. Baunack, J. Eckert and A. Gebert: J. Mater. Res., 2003, 18, 97.

[32] Z. Liu, T. Wu, K. Dahm and F. Wang: Scrip. Mater., 2002, 37, 1151.

[33] X.Y. Wang and D.Y. Li: Electrochim. Acta, 2002, 47, 3939.

[34] S. Shriram, S. Mohan, N.G. Renganathan and R. Venkatachalam: Trans. IMF, 2000, 78/5, 194.

[35] Thorpe, S. J., Ramaswami, B. and Aust, K. T., Corrosion and Augur studies of a nickel-base metal-metalloid glass. J. Electrochem. Soc., 1988, 135, 2162–2170.

[36] Bragagnolo, P., Waseda, Y., Palumbo, G. and Aust, K. T., Corrosion/ coating of advanced materials. In MRS Symposium, 1989, vol. 4, pp. 469–474.

[37] Zeiger, W., Schneider, M., Scharnwber, D. and Worch, H., Nanostruct. Mater., 1995, 6, 1013–1016.

[38] Alves, H., Ferreira, M. G. S. and Koster, U., Corrosion behavior of nanocrystalline (Ni70Mo30)90B10 alloys in 0.8 M KOH solution Original, Corros. Sci., 2003, 45, 1833–1845.

[39] Barbucci, A., Farne, G., Mattaezzi, P., Riccieri, R. and Cereisola, G., Corrosion behaviour of nanocrystalline Cu90Ni10 alloy in neutral solution containing chlorides, Corros. Sci., 1999, 41, 463–475.

[40] Mato S, Alcal´a G, Woodcock T.G, Gebert A, Eckertb J, Schultz L. Electrochimica Acta 2005; 50, 2461–2467.

[41] Jovic V.D, Barsoum M.W. J. Electrochem. Soc. 2004; 151, B71.

[42] Everhart J.L. Titanium and Titanium Alloys, Reinhold Press, 1954.

[43] H. J. Chung, K. Y. Rhee, M. T. Kim and Y. C. Jung. Surface treatment of stainless steel to improve coating strength of ZrO2 coated stainless steel needle. Surface Engineering 2011; 27(2), 145.

[44] Saxena M, Prasad B.K, Dan T.K. Plat. Surf. Finish 1992; 57.

[45] Biswas K.K, Dutta S, Das S.K, Ghose M.C, Mazumdar A, Roy N. Proceedings of the Advances in Surface Treatment of Metals ASTOM-87, BARC, Bombay, 1987.

[46] Vargin V.V, in: K. Shaw(Ed.), Technology of Enamels, Trans, MaClaren and Sons, London, 1967.

[47] Brooman E.W. Met. Finish 2002; 100 (5), 42.

[48] Brooman E.W Met. Finish 2002; 100 (6), 104.

[49] Wang H, Akid R. Encapsulated cerium nitrate inhibitors to provide high-performance anti-corrosion sol–gel coatings on mild steel, Corros. Sci. 2008; 50, 1142.

[50] Twite R.L, Bierwagen G.P. Prog. Org. Coat. 1998; 33, 91.

[51] Khobaib M, Reynolds L.B, Donley M.S. Surf. Coat. Technol. 2001; 140, 16.

[52] van Ooij W.J, Zhu D, Stacy M, Seth A, Mugada T., Gandhi J, Puomi P, Tsingua. Sci. Technol. 2005; 10, 639.

[53] Mammeri F, Le Bourhis E, Rozes L, Sanchez C. J. Mater. Chem. 2005; 15, 3787.

[54] Voevodin N.N, Grebasch N.T, Soto W.S, Kasten L.S, Grant J.T, Arnold F.E, Donley M.S. Prog. Org. Coat. 2001; 41, 287.

[55] Halvarsson M, Vuorinen S. Mater. Sci. Eng. A. 1996; 209, 337.

[56] Aries L, Roy J, Sotoul J, Pantet V, Costeseque P, Aigowy T. J. Appl. Electrochem 1996; 617, 26.

[57] Dugger M.T, Chung Y.W, Bhushan B, Rothschild W. Tribiol. Trans. 1993; 36 (1), 84.

[58] Unal O, Mitchell T.E, Hever A.H. J. Am. Ceram. Soc. 1994; 77 (4), 984.

[59] Du Y.J, Damron M, Tang G, Zheng H, C.J. Chu, Osborne J.H. Prog. Org. Coat. 2001; 41, 226.

[60] Buchheit R.G, Guan H, Mahajanam S, Wong F. Prog. Org. Coat. 2003; 47, 174.

[61] Tatematsu H, Sasaki T. Cem. Concr. Compos. 2003; 25, 123.

[62] Yang H, van Ooij W.J. Plasmas Polym. 2003; 8, 297.

[63] Shchukin D.G, Zheludkevich M.L, Yasakau K.A, Lamaka S.V, Ferreira M.G.S, Mohwald H. Adv. Mater. 2006; 18, 1672.

[64] Shchukin D.G, Möhwald H. Adv. Funct. Mater. 2007; 17, 1451.

[65] Yasakau K.A, Zheludkevich M.L, Karavai O.V, Ferreira M.G.S. Prog. Org. Coat. 2008; 63, 352.

[66] Moutarlier V, Neveu B, Gigandet M.P. Surf. Coat. Technol. 2008; 202, 2052.

[67] Lamaka S.V, Zheludkevich M.L, Yasakau K.A, Serra R, Poznyak S.K, Ferreira M.G.S. Prog. Org. Coat. 2007; 58, 127.

[68] Quinet M, Neveu B, Moutarlier V, Audebert P, Ricq L. Prog. Org. Coat. 2007; 58, 46.

[69] Khramov A.N, Voevodin N.N, Balbyshev V.N, Donley M.S. Thin Solid Films. 2004; 447, 549.

[70] Voevodin N.N, Grebasch N.T, Soto W.S, Arnold F.E, Donley M.S. Surf. Coat. Technol. 2001; 140, 24.

[71] Hosseini S.M.A, Azimi A. Mater. Corros. 2008; 59, 41.

[72] Hosseini S.M.A, Quanbari M, Salari M. Ind. J. Technol. 2007; 14, 376.

[73] Hosseini S.M.A, Tajbakhsh S. Z. Phys. Chem. 2007; 221, 775.

[74] Hosseini S.M.A, Amiri M., Momeni A. Surf. Rev. Lett. 2008; 15 (4), 1.

[75] Hosseini S.M.A, Azimi A. The inhibition of mild steel corrosion in acidic medium by 1-methyl-3-pyridin-2-yl-thiourea, Corros. Sci. 2009; 51, 728.

[76] Zheludkevich M.L, Shchukin D.G, Yasakau K.A, Mohwald H, Ferreira M.G.S. Chem. Mater. 2007; 19, 402.

[77] Rozenfeld L.L. Corrosion Inhibitors, McGraw-Hill, New York, NY, 1981.

[78] Hosseini S.M.A, Jafari A.H, Jamalizadeh E. Electrochimica Acta. 2009; 54, 7207–7213.

[79] Racicot R, Brown T, Yang S.C. Synth. Met. 1997; 85, 1263.

[80] Tang Z, Alvarez N, Yang S.C. Mater. Res. Soc. Symp. Proc. (2002) 357.

[81] Karpagam V, Sathiyanarayanan S., Venkatachari G. Curr. Appl. Phys. 2008; 8 (1), 93.

[82] Tuncay A, C'olak N.Ö, Ozyilmaz G., Kemal M., Sangün R., Prog. Org. Coat. 2007; 60, 24.

[83] Ya˘gan A., Pekmez N.Ö, Yıldız A. Prog. Org. Coat. 2007; 59 (4), 297.

[84] Chen Y, Wang X.H, Li J, Lu J.L, Wang F.S. Long-term anticorrosion behaviour of polyaniline on mild Steel, Corros. Sci. 2007; 49 (7), 3052.

[85] Oh S.G, Im S.S. Curr. Appl. Phys. 2002; 2 (4), 273.

[86] Souza S.D. Surf. Coat. Technol. 2007; 201 (16–17), 7574.

[87] Armelin E, Ocampo C, Liesa F, Iribarren J.I. Prog. Org. Coat. 2007; 58 (1–4), 316.

[88] Azim S.S, Sathiyanarayanan S, Venkatachari G. Prog. Org. Coat. 2006; 56 (2–3), 154.

[89] Riaz U, Ahmad S. A, Ashraf S.M, Ahmad S. Progress in Organic Coatings. 2009; 65, 405.

[90] Motawie M., Hassan E.A, Manieh A.A, M.E. Aboul-Fetouh, A. Fakhr El-Din. J. Appl. Polym. Sci. 2003; 55 (13), 1725.

[91] Rosa V.A, Hugo R.B, Eduardo A. J. Child Chem. Soc. 2003; 48, 1.

[92] Stark R, Christiansen J. IEEE Trans. Plasma Sci. 1995; 23, 258.

[93] Witke T, Lenk A, Schultrich B. IEEE Trans. Plasma Sci. 1996; 24, 61.

[94] Korenev S.A, Huran J. IEEE Trans. Plasma Sci. 1996; 22, 253.

[95] Guan Q.F, Zoua H, Zoua G.T, Wub A.M, Haob S.Z, Zoub J.X, Qinb Y, Dongb C, Zhang Q.Y. Surface & Coatings Technology. 2005; 196, 145–149.

[96] Aliofkhazraei M, Sabour Rouhaghdam A, Heydarzadeh A, Materials Characterization. 2009; 60, 83.

[97] Lewis DB, Leyland A, Stevenson PR, Cawley J, Matthews A. Surf Coat Technol. 1993; 60, 416–23.

[98] Sulima A.M, Yevstigneev M.I. Quality of Surface Layer and Fatigue Durability of Details Made From Heatproof and Titanim Alloys, Machine Building, Moscow, 1974.

[99] Lu K, Lu J. J. Mater. Sci. Technol. 1999; 15, 193.

[100] Liu G, Lu J, Lu K. Mater. Sci. Eng. A. 2000; 286, 91.

[101] Tao N.R, Wang Z.B, Tong W.P, Sui M.L, Lu J, Lu K. Acta Mater. 2002; 50, 4603.

[102] Zhang H.W, Hei Z.K, Liu G, Lu J, Lu K. Acta Mater. 2003; 51, 1871.

[103] Chen X.H, Lu J, Lu L, Lu K. Scripta Mater. 2005; 52, 1039.

[104] T. Roland, D. Retraint, K. Lu, J. Lu, Scripta Mater. 54 (2006) 1949.

[105] Mordyuk B.N, Prokopenko G.I. Mater. Sci. Eng. A. 2006; 437, 396.

[106] T. Wang, J. Yu, B. Dong, Surf. Coat. Technol. 200 (2006) 4777.

[107] Wang X.Y, Li D.Y. Electrochim. Acta. 2002; 47, 3939.

[108] Jiang X.P, Wang X.Y, Li J.X, Li D.Y, Manc C.S, Shepard M.J, Zhai T. Mater. Sci. Eng. A 2006; 429, 30.

[109] Mordyuka B.N, Prokopenko G.I, Vasylyev M.A, Iefimov M.O. Materials Science and Engineering A. 2007, 458, 253–261.

[110] Wang X.Y, Li D.Y. Electrochim. Acta. 2002; 47, 3939.

[111] Wang T, Yu J, Dong B. Surf. Coat. Technol. 2006; 200, 4777.

[112] Liu L, Li Y, Wang F.H. Electrochim. Acta. 2007; 52, 7193.

[113] Raymond T. Cunningham, Sugar Land, Kristen T. Moore, HoustonCorrosion inhibitor for nitrogen fertilizer solutions, US 5,376,159 (1994).

[114] Dennis J. Kalota, David C. Silverman. Process for corrosion inhibition of ferrous metals. 4971724 (1990).

[115] Lai-Duien Grace Fan, Joseph C. FanInhibition of metal corrosion, US6277302 (2001).

[116] Method of use of a coating compound for liquid-fertilizer holding vessels, and liquid fertilizers, US5731032 (1998).

[117] Scott David Trahan, Eric Paul Mistretta. Corrosion inhibitor, US6953537 (2005).

[118] Jon Conrad Schaeffer, Vinod Kumar Pareek. Nanostructured corrosion inhibitors and methods of use, US7766983 (2010).

[119] Yuriy Viacheslavovich Tolmachev. Nanostructured core-shell electrocatalysts for fuel cells, US7935655 (2010).

http://dx.doi.org/10.5772/57274

[120] Tapesh Yadav. Molybdenum comprising nanomaterials and related nanotechnology, US7968503 (2011).

[121] Susan J. Kerber. Nanoparticle surface treatment, US8197613 (2012).

Chapter 2

Morphology Characterization of Pitting Corrosion on Sensitized Austenitic Stainless Steel by Digital Image Analysis

Rosinei Batista Ribeiro, Gilbert Silva,
Antônio Henriques Jr. Araujo,
Nelson Tavares Matias,
José Wilson De Jesus Silva and
Bianca Siqueira Martins Domingos

Additional information is available at the end of the chapter

http://dx.doi.org/10.5772/57256

1. Introduction

Austenitic stainless steels are selected for corrosion resistance to the atmosphere and a vast variety of chemical products. This resistance is related to the formation of a thin and adherent oxide layer (mainly Cr_2O_3), which protects the surface against aggressive environments. The most common austenitic steels are iron-chromium-nickel alloys and they are widely known as the 300 series [1,2]. These alloys usually contain between 16 and 30 wt% chromium, from 8 to 20% nickel and from 0.03 to 0.1% carbon. With increasing chromium and nickel contents, these steels become increasingly resistant to pitting corrosion. However, higher concentrations of these alloying elements result in lower carbon solubility and carbide segregation [3,4]. Heat treatments, whether intentional or accidental, may provoke carbide precipitation at grain boundaries, often by causing steel to become susceptible to intergranular corrosion. In order to minimize carbide precipitation and prevent sensitization, austenitic steels are frequently subjected to high-temperature solution treatment, commonly termed quench-annealing [5,6]. Pitting corrosion is observed in austenitic steels exposed to aqueous media containing chloride ions. This phenomenon almost always starts in anodic MnS inclusions and chromium-depleted zone adjacent to cathodic $M_{23}C_6$ (M = Cr or Mo) precipitates, and it progresses inwards in intergranular or transgranular direction depending mainly on the steel microstructure [7,8].

Many studies have been focused on the pit initiation and propagation mechanisms of these steels in chloride media, but relatively a few authors have considered the morphologic characteristics of pitting corrosion [9,10]. It should be noted that pit growth rate depends on potential distribution inside the pit, which is determined by pit shape, especially by their width/depth aspect ratio [11,12]. For pits initiated by electrochemical experiments, the pit depth or pit radius dependence on the immersion time (t) may be expressed by the equation $k(t-t_i)^n$, where t_i is the induction period, $k > 0$ and $1/3 \leq n \leq 1$ are empirical constants which depend on chloride concentration, pits shape and experimental conditions inherent of electrochemical methods [13,14]. Pits are usually modeled as having cylindrical, hemispherical or conical shapes, but in practice, they have a tendency to assume an irregular or not well-defined geometry associated with increasing size [15]. Therefore, equations, developed with these assumptions about shape, must be considered as only rough approximations.

In this study, a method based on image analysis has been used for the morphology characterization of pitting corrosion on sensitized 310S steel. For this purpose, pits have been initiated by Salt Spray exposure, and their temporal evolution has been examined by reflected light microscopy.

2. Materials and methods

As-received 310S plates [16] have been submitted to three solution heat treatment routines: heating up to 1065°C during 1 h and air cooling (condition I); heating up to 1065°C during 1 h and air cooling followed by reheating up to 670°C during 5 h and again air cooling (condition II); heating up to 1065°C during 1 h and air cooling followed by reheating up to 620°C during 24 h and again air cooling (condition III). The main purpose for annealing has been to produce a recrystallized microstructure with a uniform grain size and for dissolving detrimental chromium carbide precipitates. Then, a slow cooling or reheating within the range 550°-800°C will lead to the rejection of carbon from solution, usually as the chromium-rich carbide, even when the carbon content of the steel is very low (<0.05 wt%).

Specimens have been mechanically polished and their microstructure have been chemically etched with the following solution: 10 mL HF (48%) + 10 mL HNO_3 (65%) + 15 mL HCl (35%). Surfaces have been examined before and after Salt Spray tests [17] using a method of image analysis based on reflected light microscopy (LM). Images have been captured using a Nikon Epiphot 200 inverted metallurgical microscope coupled to a Diagnostic Instruments Insight Color QE digital camera. The NIH freeware program Image J [18] has been used for image processing, and a macro program has been developed to execute all processing and analysis steps. Quantitative parameters such as area at pit mouths, pit density and corroded area have been systematically determinate. Additionally, SEM and EDS combined techniques have been used to identify elemental composition of grains and grain boundaries. This surface analysis has been carried out using a scanning electron microscope Leica Stereoscan 440 with a coupled energy dispersive spectrometer Oxford Link Exl II.

Pits morphological characteristics have been determined by specimens vertical sectioning, under low load and rotation in a precision saw, followed by fine grinding and polishing. To ensure low deformation of profile region, corroded surfaces have been covered with an epoxy resin before cutting and mounting with phenolic resin for mechanical polishing. Analysis of pits has been based mainly on rectangularity or area-box (AB) shape parameter, defined as the ratio between pit area and minor surrounding rectangle area which encloses pit [19]. AB parameter is an effective geometry descriptor, permitting clear separation among conical pits (0.5 < AB < 0.53), spherical or hemispherical pits (0.72 < AB < 0.86) and cylindrical pits (AB ≈ 1.0). Figure 1 summarizes the class for morphological analysis.

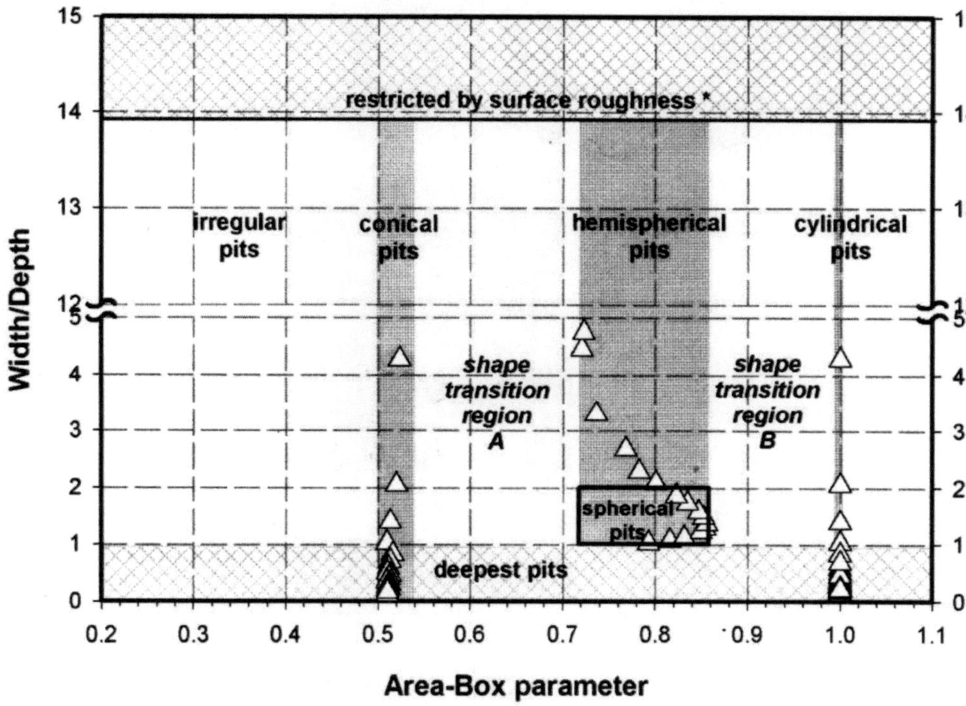

Figure 1. Diagram used for pit geometry classification. Width/depth aspect ratio versus area-box parameter.

Transition regions A (near-conical or near-hemispherical pits) and B (near-hemispherical or near-cylindrical pits) represent pits that can evolve the best-defined geometries. Irregular pits (without geometric elements) present AB values lower than 0.5. To determine pits size, both pit width and depth, the same rectangle above mentioned has been used. A representative number (60) of 1600 x 1200 x 8 bit digital images under bright-field xenon-arc illumination has been obtained for each specimen.

3. Results and discussion

Morphological and microstructural analysis has been carried out before and after Salt Spray test. Before corrosion test (Figures 2 (a) – (c), conditions I, II and III exhibit similar characteristics: a non-homogeneous microstructure with an austenite phase, different orientation grains, carbide precipitates and some holes probably provoked during the mechanical processing and metallographic preparation of steel.

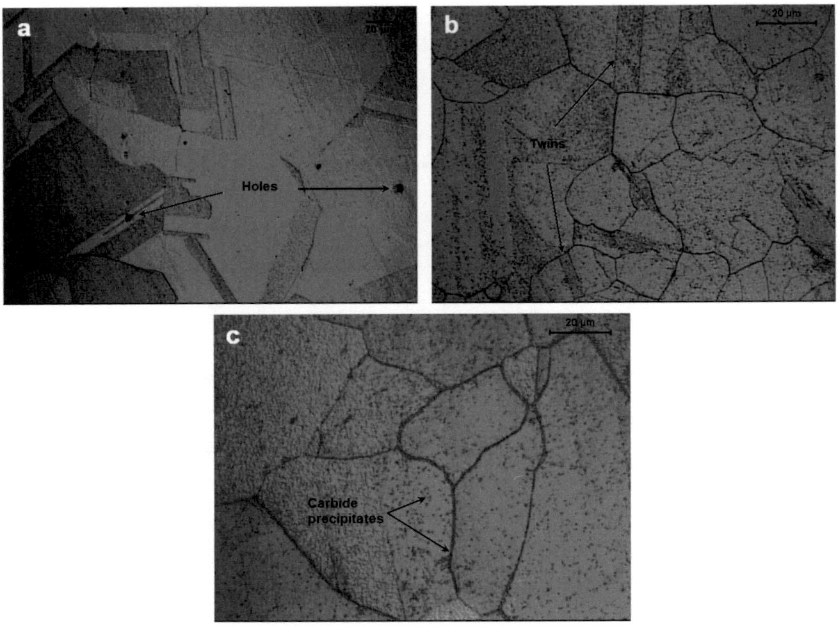

Figure 2. LM micrographs of 310S sensitized before Salt Spray test: a) condition I, b) condition II and c) condition III.

Austenite phase is present as equiaxed grains, many of which contain annealing twins. These twins are identified as bands with parallel sides and are formed when changes in the striking of atoms on close-packed (111) planes occur during recrystallization and grain growth [20]. EDS semi-quantitative analysis on grain and grain boundary of condition II is showed in Figure 3 (a) and (b).

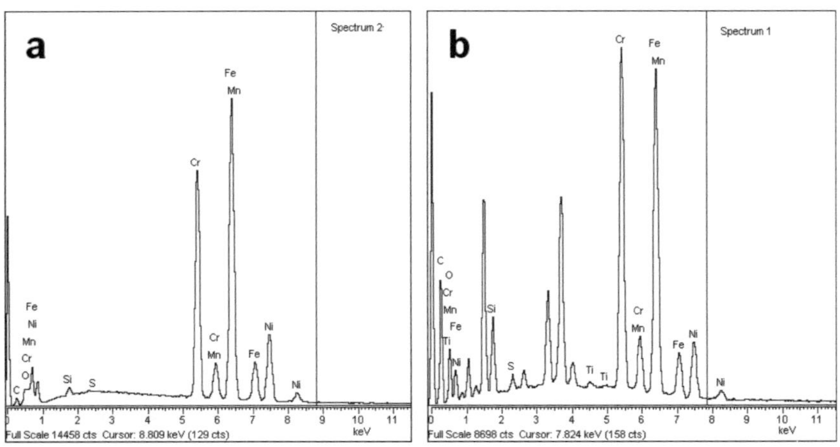

Figure 3. EDS semi-quantitative analysis upon the 310S in condition II: a) grain and b) precipitate in grain boundary.

http://dx.doi.org/10.5772/57256

The most significant results are absence of molybdenum in steel and enrichment of chromium in precipitates on grain boundary. One explanation that the Mo peak was not detected is that it probably merges with the Ti peak. Also it may be that the analysis has been punctual, and not general.

After corrosion test, surfaces have revealed pits of different shapes and sizes. Table 1 permits to compare statistical parameters measured on exposed surfaces. For all heat treatment conditions, low values of corroded areas suggest there is a low probability of pit superposition during the exposure period.

Parameter	Heat Treatment	Exposure Period (h)				
		48	120	168	216	312
Pit density	I	15.98	28.02	39.76	200.09	378.06
(mm^{-2})	II	19.03	25.15	34.12	34.16	40.20
	III	9.02	10.45	15.34	14.09	19.99
Corroded area	I	0.08	0.20	0.44	0.50	2.40
(%)	II	0.04	0.10	0.55	0.90	1.89
	III	0.09	0.08	0.80	1.20	3.90
Area at mouth	I	10.09	11.98	12.65	12.32	13.07
(μm^2)	II	8.87	9.67	20.14	25.08	29.76
	III	13.45	15.11	25.23	26.23	35.28

Table 1. Variation of pitting corrosion parameters measured on exposed surfaces

Thus, the temporal variation in pit density and pit area may be directly related to the rates of pit nucleation and two-dimensional growth, respectively. This way, nucleation rate has remained approximately constant for conditions II and III, and it has increased for condition I. In this last condition, the pits growth is slower. These facts are mainly related with a number of anodic sites and anodic/cathodic area ratio variation during exposure period.

Figure 4a-c shows cross-section images of sensitized steels after 312 h of Salt Spray test, according to ASTM B117 - 11 [17]. The choice of time interval is empirical and is based on the following relationship: with increasing exposure time, changes occur in corrosion resistance behavior, in distribution and in morphology of pits at material as received and treated.

Conditions I and II, pits are nucleated in grains and grain boundaries, while in condition III, pits are preferentially nucleated in boundaries of small grains. Generally, carbides are too fine to be resolved by reflected light microscope, but are indirectly revealed by deep etching of affected grain boundaries (Figure 4c). Pits classification and distribution have been determined from cross-section images. Table 2 indicates that for all conditions: (i) the hemispherical, near-

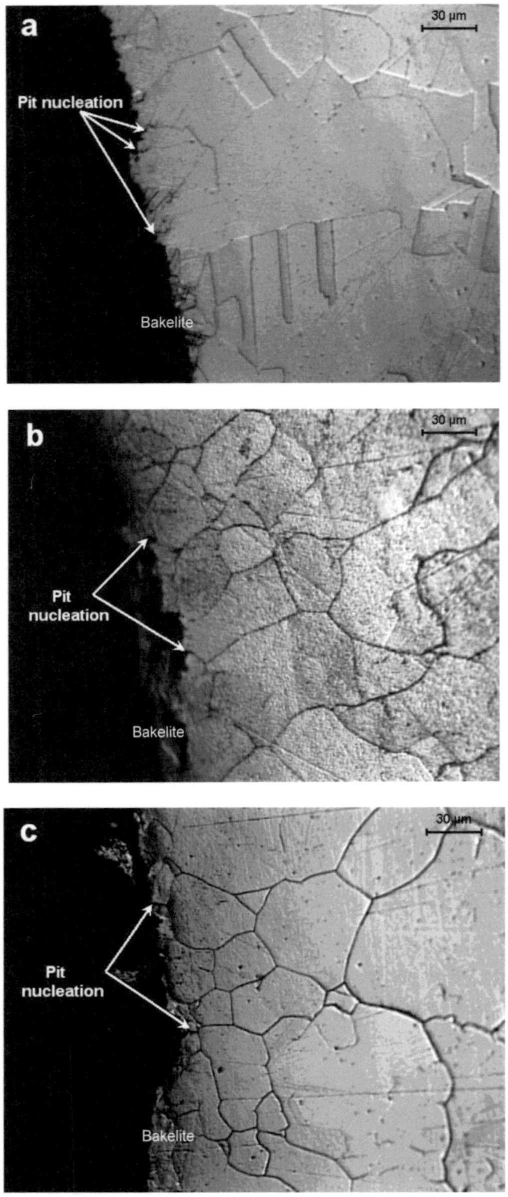

Figure 4. Cross-section images of 310S sensitized after 312 h of Salt Spray test: a) condition I, b) condition II and c) condition III.

hemispherical, near-conical and near-cylindrical pits represent more than 80% of the pit population; (ii) they appear not to undergo geometric transition associated with an increasing exposure period; (iii) pits with regular geometry represent less than 45% of pit population; (iv) cylindrical pits are absent. This geometry distribution may be attributed to similar micro-structural characteristics of different heat treatment conditions.

Pit Geometry	Heat Treatment	Exposure Period (h)				
		48	120	168	216	312
	I	6.41	0.67	0.66	4.19	0.00
Irregular (%)	II	1.71	7.43	3.75	5.36	6.67
	III	0.00	3.17	15.87	6.44	12.30
	I	1.28	4.67	1.32	2.80	1.23
Conical (%)	II	0.00	6.93	5.00	2.68	1.02
	III	0.00	2.08	2.59	13.30	6.56
Transition	I	7.70	54.66	14.47	26.57	3.47
Region A (%)	II	20.50	24.36	24.16	27.68	30.45
	III	18.14	27.09	26.98	61.80	39.33
	I	19.23	29.33	28.29	17.48	20.99
Hemispherical (%)	II	23.08	18.80	36.71	41.96	42.56
	III	24.29	28.40	29.16	15.88	24.60
Transition	I	65.38	10.67	55.26	48.96	74.31
Region B (%)	II	54.71	42.48	30.38	22.32	19.30
	III	57.57	39.26	25.40	2.58	17.21
	I	0.00	0.00	0.00	0.00	0.00
Cylindrical (%)	II	0.00	0.00	0.00	0.00	0.00
	III	0.00	0.00	0.00	0.00	0.00

Table 2. Classification and distribution of pits percentages

Table 3 shows pit width and depth distributions. To diminish influence of pre-existing holes and pits superposition on the count and statistical treatment of data, it has been replaced the mean values by the medians [21-28].

Detailed analysis in this table indicates that for all conditions: (i) for each exposure period, width inside pit is in good agreement with the width at pit mouth estimated from Table 1 (supposing circular mouth); (ii) pit depth rapidly increases during the first 48 hours of Salt Spray exposure, and then it is apparently stabilized; (iii) pits are deeper than wide, which suggests that metal dissolution rate is higher at pit bottom than at the pit wall; (iv) pit depth increases from Transition regions B to Irregular geometry; (v) irregular pits are also the widest ones.

This class of pits may have nucleated in holes, Figure 2 (a), and they have grown by an intergranular way and/or they have nucleated in boundaries of small grains and during their growth provoke grains separation, as showed in Figures 4b and 4c. Although irregular pits represent a small percent of pit population, in practice the maximum depth values are very important because the deepest pits can initiate cracks when steel is subjected to mechanical stress.

Pit Geometry	Heat Treatment	Parameter	Exposure Period (h)				
		(μm)	48	120	168	216	312
Irregular	I	Width	7.12	14.88	11.73	5.74	-
		Depth	33.17	27.19	16.31	27.47	-
	II	Width	7.29	9.52	6.01	7.77	8.15
		Depth	11.01	9.80	23.32	15.19	25.26
	III	Width	-	6.65	6.35	7.04	6.96
		Depth	-	17.15	27.79	7.15	12.84
Conical	I	Width	4.58	9.01	6.86	4.65	4.72
		Depth	21.46	14.59	6.29	11.77	18.60
	II	Width	-	6.51	4.79	4.57	5.02
		Depth	-	6.88	16.81	13.02	15.29
	III	Width	-	4.86	5.01	6.30	6.35
		Depth	-	20.62	19.60	6.46	8.78
Transition Region A	I	Width	4.74	6.89	6.64	4.38	6.34
		Depth	14.02	7.34	8.57	14.64	14.31
	II	Width	5.50	6.66	4.39	5.53	6.22
		Depth	10.50	6.88	15.00	10.68	12.99
	III	Width	4.37	5.71	4.17	6.06	5.80
		Depth	20.15	9.65	13.43	6.18	7.91
Hemispherical	I	Width	4.82	6.02	5.89	3.91	3.67
		Depth	11.19	6.85	6.99	8.66	14.02
	II	Width	5.99	5.76	5.07	5.55	6.38
		Depth	9.16	6.32	10.13	7.54	10.28
	III	Width	3.24	4.97	5.01	5.54	4.88
		Depth	13.09	8.33	9.48	5.59	7.00
Transition Region B	I	Width	5.05	5.54	5.16	4.74	4.78
		Depth	5.40	5.21	4.73	5.16	5.91
	II	Width	5.65	4.52	6.60	6.36	6.37
		Depth	6.30	5.29	7.39	6.51	7.55
	III	Width	4.79	5.10	5.12	3.19	4.72
		Depth	5.64	6.01	6.14	5.98	4.91
Cylindrical	I	Width	-	-	-	-	-
		Depth	-	-	-	-	-
	II	Width	-	-	-	-	-
		Depth	-	-	-	-	-
	III	Width	-	-	-	-	-
		Depth	-	-	-	-	-

Table 3. Variation of pit width and depth with increasing exposure period

http://dx.doi.org/10.5772/57256

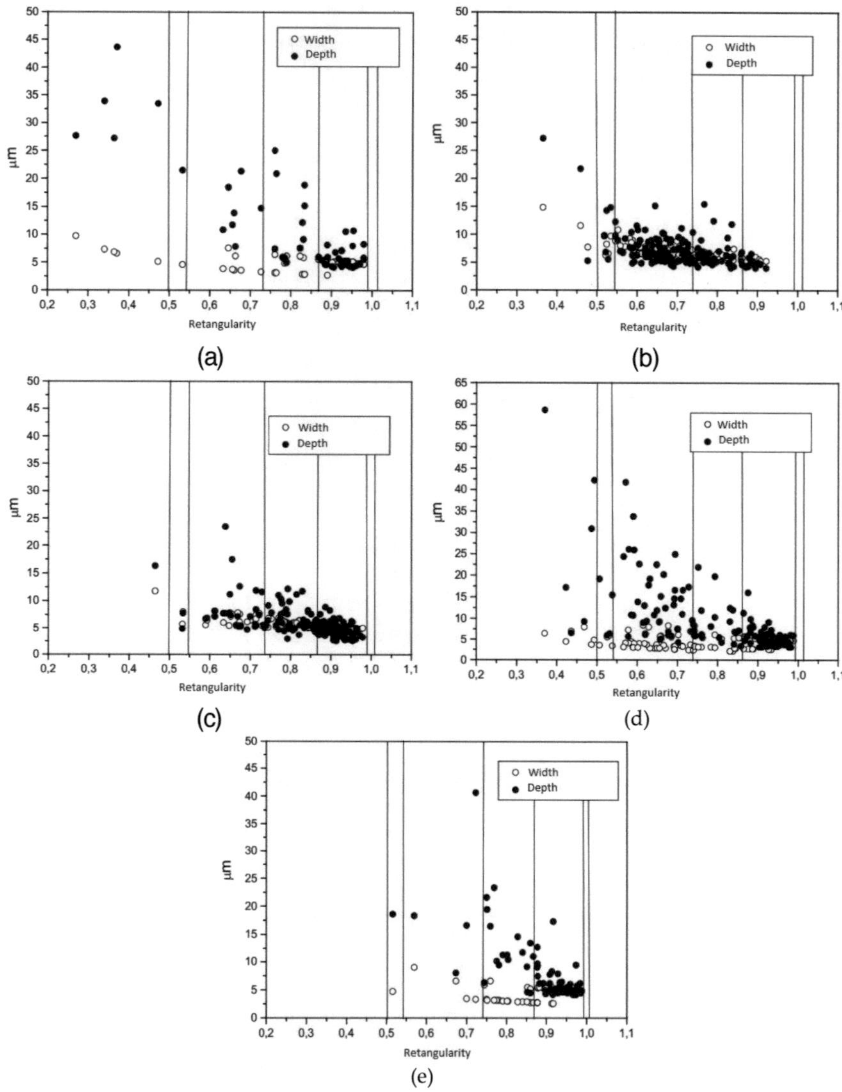

Figure 5. Size of pits formed on the surface of stainless steel 310S treated 1065 °C for 1 hour and cooled in air, after being subjected to corrosion testing via salt spray (a) 48 hours, (b) 120 hours (c) 168 hours, (d) 216 hours (e) 312 hours

The Figures 5 (a), (b), (c), (d) and (e) show the following order in the median size of the width and depth associated with the morphology of pits: irregular > conic > transition A > Hemisphere > B transition; it was not observed the existence of cylindrical pits in the system.

The pits are more profound (around 3 times) than wide and growing preferably in depth. The longer the exposure time, the greater the depth of pits between different morphologies of pits (irregular: 33.17 µm and conic: 18.60 µm in 312 hours and irregular: 27.47 µm in 48 hours 33.17 µm in 312 hours).

The formation of pitting corrosion in stainless steel 310S revealed differences, through association with the morphology, distribution, size and growth trend that govern all temporal evolution of pits and their interface (grain boundary and morphology of the precipitate).

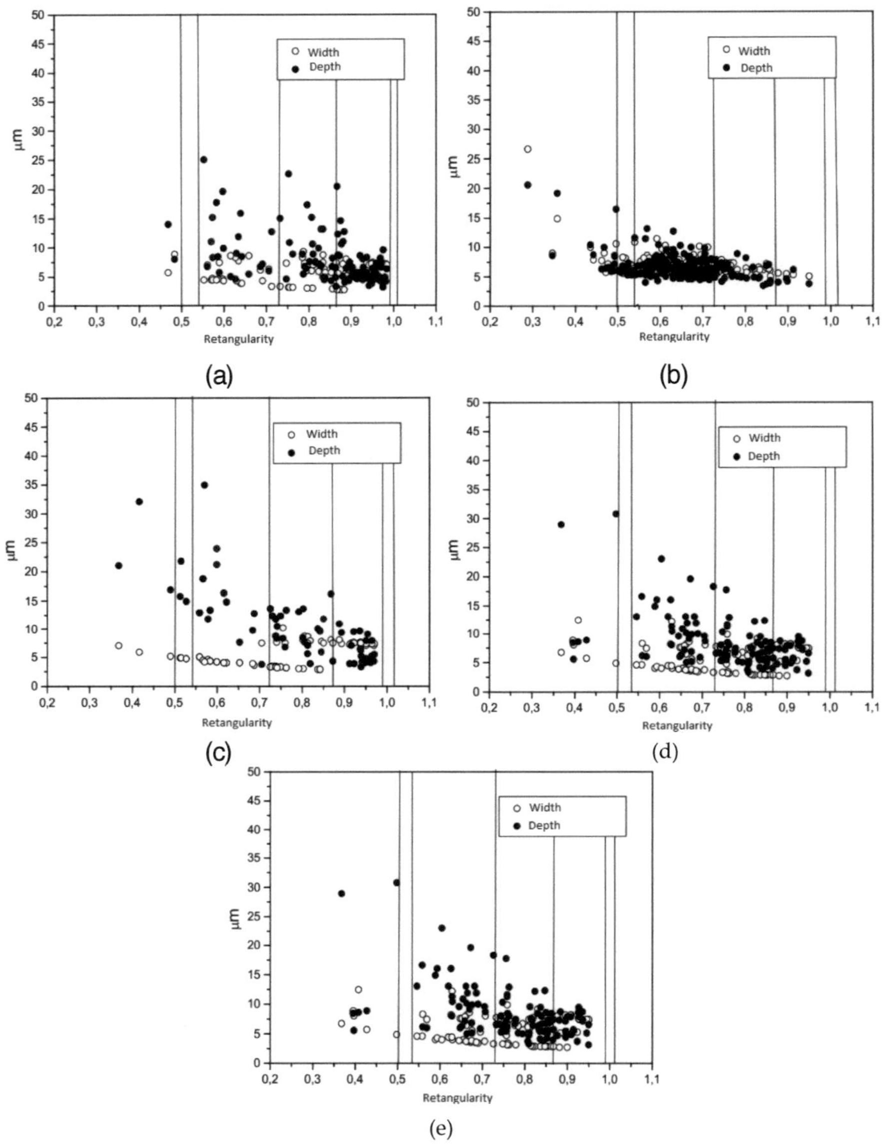

Figure 6. Size of pits formed on the surface of stainless steel 310S treated 670° C for 5 hours and cooled in air, after being subjected to corrosion testing via the salt spray (a) 48 hours, (b) 120 hours (c) 168 hours, (d) 216 hours (e) 312 hours

In Figure 6, it is shown the following distribution and classification of pits: hemispherical > transition region A > transition region B > irregular > conical. The pits are present in larger quantities in transition region B and hemispheric, there is virtually no evidence of presence and evolution of cylindrical pits in the system.

It is concluded that the distribution and classification of pits in stainless steel 310S have the following trends: hemispheric > transition region B > transition region A > conical > irregular, sometimes, transition region B > hemispheric > transition region A > conical > irregular.

There was no presence of cylindrical pits in the system. The temporal evolution of the parameters that govern the distribution of pits is similar in stainless steel 310S when treated

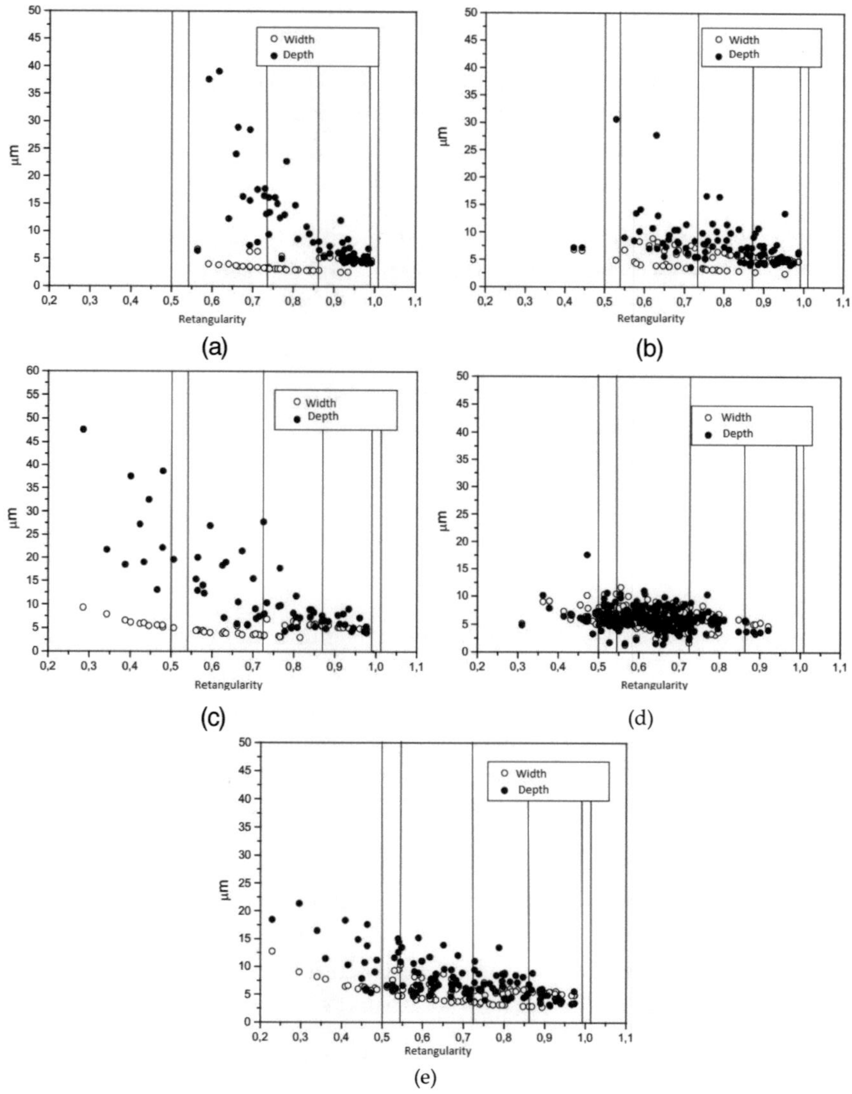

Figure 7. Size of pits formed on the surface of stainless steel 310S treated 620° C for 24 hours and cooled in air, after being subjected to corrosion testing via the salt spray (a) 48 hours, (b) 120 hours (c) 168 hours, (d) 216 hours (e) 312 hours

in the temperature range 620° C - 670° C. By increasing the exposure time, there is increased occurrence of pits and hemispherical in the transition region A, but the time course is similar in steel when treated at 1065° C for 1 hour and air-cooled to 820° C for 1 hour and air-cooled.

Changes in the distributions have been observed, namely an increase in the number of pits in the transition region and the irregular geometries and the decrease in the transition region B in the steel in the as received and treated at 1065° C for 1 hour and cold water and ice.

Figure 7 presents the following order in the median size of the width and depth associated with morphology of pits: irregular > conical > transition > hemispherical > transition B > cylindrical.

The pits are more profound (around 3 times) than wide and growing preferably in depth. It is noted that the higher exposure time, the greater the depth of pits even between different morphologies of pits (irregular: 27.79 μm and conic: 19.60 μm in 216 hours and irregular: 7.15 μm in 120 hours and 216 hours in 27.79 μm) and constant width.

The evolution kinetics of pitting corrosion has similarities to austenitic stainless steel, while the material in the state as-received, with microstructural characteristics induced by mechanical deformation, type: twinned that formed a barrier in pits formation.

However, the treated material forming the corrosion process is facilitated by making a channel sensitization induction of corrosion by extending the grain boundaries with increasing volume fraction of chromium carbides precipitated and their circular morphology.

4. Conclusions

The experimental methodology was enhanced austenitic stainless steels ABNT 310S to evaluate the classification, distribution and morphology of pits in alloy liable to vary in composition and heat treatment. The alloys after sensitization heat treatments showed formation of chromium carbides at grain boundaries and inner specifically stainless steel 310S shown in greater quantity and coarser microstructure contributing to increased pitting corrosion, as the data obtained in surface analysis and profiles.

After the corrosion tests at different conditions under salt spray, it was observed that the aspect ratio of pits for stainless 310S in the state as received and treated, the geometry corresponds to circular. The 310S stainless steel showed higher values and the same trend for the statistical parameters pits, but the alloy was lower localized corrosion resistance than stainless steel 304, apparently by the difference in the microstructure and chemical composition.

On superficial analysis in stainless steel 310S it was showed up as received the following order in the morphology of pits: hemispheric > transition region A > transition region B > irregular > conic, and the pits are present in larger amounts in hemispheric and transition region A. The pits are wider than deep, preferably grow in width. The cylindrical pits were not observed in the system.

When the alloy is treated at each exposure time, it has the same order in the amount of pits: transition region B > hemispheric > transition region A > conical > irregular, particularly the pits are deeper than wide. Significant changes in morphology and size distribution of pits on stainless steel when treated 310S are correlated to the role of chromium carbide precipitates, in particular the amount and morphology.

Through the median size of the width and depth of pits it was proved the following order: irregular > conical > transition region A > Hemispheric > transition region B. 310S stainless steel in the state as received and processed, the growth rate of pits, specifically the width and depth are different, the last of which appears to be larger and change with time of exposure, thus showing the high values in the irregular pits and distribution in large amounts of pitting in the transition region A and B, sometimes almost conical and hemispherical.

After Salt Spay test, 310S austenitic stainless steel sensitized by means of different heat treatments has exhibited similar microstructures and similar pits morphological characteristics. Nucleation rates and growth may be associated with quantity and distribution of chromium carbides. In this sense, condition I and II are the most susceptible to pitting, in particular the first one. In these two conditions, the pits are nucleated in grains and grain boundaries, while in condition III, pits are preferentially nucleated in boundaries of small grains. Then, the pits grow more rapidly in depth than in width being able to carry out a partial or total separation of grains. Pits are mainly hemispherical, near-hemispherical, near-conical and near-cylindrical. However, irregular pits are also important because they are the deepest ones, which can initiate a stress-corrosion crack, thus transforming one type of localized attack into another.

Acknowledgements

This research has been supported by FUNDUNESP (process 01258/08-DFP) and CNPq (process 305224/2004-2).

Author details

Rosinei Batista Ribeiro[1,3*], Gilbert Silva[2], Antônio Henriques Jr. Araujo[1], Nelson Tavares Matias[1,3], José Wilson De Jesus Silva[1,3] and Bianca Siqueira Martins Domingos[2]

*Address all correspondence to: rosinei.ribeiro@pq.cnpq.br

1 Universidade do Estado do Rio de Janeiro – Faculdade de Tecnologia – FAT-UERJ - Campus Regional de Resende/RJ, Brazil

2 Universidade Federal de Itajubá – UNIFEI – Itajubá/MG, Brazil

3 Faculdades Integradas Teresa D'Ávila – FATEA – Lorena/SP, Brazil

References

[1] H.K.D.H. Bhadeshia, R.W.K. Honeycombe, Steels, Microstructure and Properties, 3rd Ed., Butterworth-Heinemann Apr. 1, 2011 – Technology & Engineering 360 pages.

[2] P. Marcus, V. Maurice, H.-H. Strehblow, Localized corrosion (pitting): A model of passivity breakdown including the role of the oxide layer nanostructure, Corrosion Science, Volume 50, Issue 9, September 2008, Pages 2698–2704.

[3] A. Pardo, M.C. Merino, A.E. Coy, F. Viejo, R. Arrabal, E. Matykina, Pitting corrosion behavior of austenic stainless steels – combining effects Mn of and Mo additions, Corrosion Science, Vol. 50, Issue 6, Jun. 2008, pp. 1796-1806.

[4] L. Speckert, G.T. Burstein, Combined anodic/cathodic transient currents within nucleating pits on Al–Fe alloy surfaces, Corrosion Science Volume 53, Issue 2, Feb. 2011, Pages 534–539

[5] Z.Y. Liu, X.G. Li, Y.F. Cheng, Understand the occurrence of pitting corrosion of pipeline carbon steel under cathodic polarization, Electrochimica Acta, Volume 60, 15 January 2012, Pages 259–263

[6] Zixue Su, Georg Hähner and Wuzong Zhou, Investigation of the pore formation in anodic aluminium oxide, *J. Mater. Chem.*, 2008, 18, pp. 5787-5795.

[7] A. Pardo, M.C. Merino, A.E. Coy, R. Arrabal, F. Viejo, E. Matykina, Corrosion behaviour of magnesium/aluminium alloys in 3.5 wt.% NaCl, Corrosion Science, Vol. 50, Issue 3, Mar. 2008, pp. 823-834.

[8] Guocheng LÜ, Haidong CHENG, Chunchun XU, Zonghu HE, Effect of Strain and Chloride Concentration on Pitting Susceptibility for Type 304 Austenitic Stainless Steel, Chinese Journal of Chemical Engineering, Volume 16, Issue 2, April 2008, Pages 314–319

[9] Hosni M. Ezuber, Influence of temperature and thiosulfate on the corrosion behavior of steel in chloride solutions saturated in CO_2 Materials & Design, Volume 30, Issue 9, October 2009, Pages 3420–3427.

[10] Ghahari, S M; Krouse, D P; Laycock, N J; Rayment, T; Padovani, C; Suter, T; Mokso, R; Marone, F; Stampanoni, M; Monir, M; Davenport, A J, Pitting corrosion of stainless steel measuring and modelling pit propagation in support of damage prediction for radioactive waste containers, Corrosion Engineering, Science and Technology, Volume 46, Number 2 April 2011, pp. 205-211(7).

[11] H.W. Pickering, The role of electrode potential distribution in corrosion processes, Mater. Sci. Eng. A198, 1995, pp. 213-223.

[12] G. Engelhardt, M. Urquidi-Macdonald, D.D. Macdonald, A simplified method for estimating corrosion cavity growth rates, Corros. Sci. 39, 1997, pp. 419-441.

[13] Standard Guide for Examination and Evaluation of Pitting Corrosion, ASTM G46-94, 1999, pp. 169-175

[14] E.N. Codaro, R.Z. Nakazato, A.L. Horovistiz, L.M.F. Ribeiro, R.B. Ribeiro, L.R.O. Hein, Mater. Sci. Eng. A 341, 2003, pp. 202-210.

[15] T. Huang, G.S. Frankel, Influence of grain structure on anisotropic localized corrosion kinetics of AA7XXX-T6 alloys, Corros. Eng. Sci. & Technol. 41, 2006, pp. 192-199.

[16] Standard Specification for Chromium and Chromium-Nickel Stainless Steels plate, sheet, and strip for pressure vessels and for general applications, ASTM A240/A240M-09a, 2009, pp. 1-13.

[17] Standard Practice for Operating Salt Spray (Fog) Apparatus, ASTM B117-11, 2011, p. 1-10.

[18] W.S. Rasband, ImageJ, U. S. National Institutes of Health, Bethesda, Maryland, USA, http://rsb.info.nih.gov/ij/, 1997-2013.

[19] Marinalda C. Pereira, José W. J. Silva, Heloisa A. Acciari, Eduardo N. Codaro, Luis R. O. Hein, Morphology Characterization and Kinetics Evaluation of Pitting Corrosion of Commercially Pure Aluminium by Digital Image Analysis, *Materials Sciences and Applications*, 2012, 3, 287-293.

[20] G. Krauss, Steels, Heat treatment and processing principles, ASM international, Ohio, USA, 2000, Chap. 12, pp. 351-375.

[21] Giuseppe Carlo Marano, Rita Greco, Sara Sgobba, A comparison between different robust optimum design approaches: Application to tuned mass dampers, Probabilistic Engineering Mechanics, Volume 25, Issue 1, January 2010, Pages 108–118

[22] S.S.M. Tavares, V. Moura, V.C. da Costa, M.L.R. Ferreira, J.M. Pardal, Microstructural changes and corrosion resistance of AISI 310S steel exposed to 600–800 °C, Materials Characterization, Volume 60, Issue 6, June 2009, Pages 573-578

[23] C. García, F. Martín, Y. Blanco, M.L. Aparicio, Effect of ageing heat treatments on the microstructure and intergranular corrosion of powder metallurgy duplex stainless steels, Corrosion Science, Volume 52, Issue 11, November 2010, Pages 3725-3737

[24] A. Iversen, B. Leffler, 3.04 - Aqueous Corrosion of Stainless Steels, Shreir's Corrosion, Volume 3, 2010, Pages 1802-1878

[25] S.S.M. Tavares, V.G. Silva, J.M. Pardal, J.S. Corte, Investigation of stress corrosion cracks in a UNS S32750 superduplex stainless steel, Engineering Failure Analysis, In Press, 2013

[26] Ribeiro, R. B. ; Codaro, E. N. ; Hein, L. R. O. ; Rosa, J. L. ; Mariotto, S. F. F. ; Meraz, E. A.. Estudo da corrosão localizada em aço inoxidável austenítico ABNT 310S submetido à exposição em névoa salina. Matéria (UFRJ), v. 13, p. 1024-1030, 2009.

[27] Gravalos, Márcio Tadeu; Martins, Marcelo; Diniz, Anselmo Eduardo and MEI, Paulo Roberto. Influência da rugosidade na resistência à corrosão por pite em peças torneadas de aço inoxidável superaustenítico. Rem: Rev. Esc. Minas [online]. 2010, vol.63, n. 1, pp. 77-82.

[28] Veroli, Alyne Bernardes. Avaliação da resistência ao desgaste de aços inoxidáveis endurecíveis por precipitação nitretados. Dissertação de Mestrado. São Carlos: UFSCar, 2012. 83 p.

Chapter 3

Resistance to Corrosion and Passivity of 316L Stainless Steel Directionally Solidified Samples

Claudia Marcela Méndez,
Mónica Mariela Covinich and Alicia Esther Ares

Additional information is available at the end of the chapter

http://dx.doi.org/10.5772/57275

1. Introduction

The three main types of crystalline structure that stainless steel can be classified are Austenitic, Ferritic and Martensitic. Austenite, also known as gamma phase iron (γ-Fe), is a metallic, non-magnetic allotrope of iron or a solid solution of iron, with an alloying element.

Ferrite, also known as α-ferrite (α-Fe) or alpha iron is a materials science term for pure iron, with a body-centered cubic crystal structure. It is this crystalline structure which gives steel and cast iron their magnetic properties. Martensite most commonly refers to a very hard form of steel crystalline structure, but it can also refer to any crystal structure that is formed by displacive transformation (Lambers et. al., 2009 and Batra et al., 2003).

These are alloys containing chromium and nickel (sometimes manganese and nitrogen), structured around the type 302 composition of iron, 18%Cr, and 8%Ni. Austenitic steels are not hardenable by heat treatment. The most familiar stainless steel is probably type 304, sometimes called T304 or simply 304. Type 304 surgical stainless steel is austenitic steel containing 18-20%Cr and 8-10%Ni (Kilicli & Erdogan, 2008).

Primary stainless steel used in aviation construction. Chemical and steel industry applicable grades are 308, 308L, 316, 316L, 316LN Nitrogen bearing, 312, 309L, 310L L denotes carbon percentage less than 0.03%, mostly used for corrosion heat resistance have work hardening properties, welding primarily done by TIG and MMAW process. Another grade, 312 is used for dissimilar steel welding, also known as universal alloy steel as unknown composition steels can be welded. For high temperature application, above 600 °C, 309 and 310 grades are preferred.

According to previous studies in the literature, there are eight separate types of corrosion, with only a few having a major impact on stainless steel (DOD Technical Bulletin – Corrosion Prevention and Detection):

1. General corrosion, this type of corrosion occurs when there is an overall breakdown of the passive film. The entire surface of the metal will show a uniform sponge like appearance. Halogens penetrate the passive film of stainless and allow corrosion to occur. Fluorine, chlorine, bromine, iodine and astatine are some of the most active.

2. Crevice corrosion, this is a problem with stainless fasteners used in seawater applications, because of the low pH of salt water.

3. Pitting, Stainless that had its passivation penetrated in a small spot becomes an anodic, with the passivated part remaining cathodic, causing pit type corrosion.

4. Galvanic corrosion, the current flows from the anodic towards the cathodic metal, and in the process slowly removes material from the anodic metal. Seawater makes a good electrolyte, and thus, galvanic corrosion is a common problem in this environment.

5. Intergranular corrosion, all austentic stainless steels contain a small amount of carbon. At extremely high temperature, such as welding, the carbon forces local chrome to form chromium carbide around it, thus starving adjacent areas of the chrome it needs for its own corrosion protection.

6. Selective leaching, fluids will remove metal during a de-ionization or de-mineralization process. This usually happens inside a pipe and is rarely a fastener problem.

7. Erosion corrosion, this corrosion happens when the velocity of an abrasive fluid removes the passivation from a stainless. This is almost exclusively limited to pipe interiors.

8. Stress corrosion, also called stress corrosion cracking or chloride stress corrosion. Chlorides are probably the single biggest enemy of stainless steel. Next to water, chloride is the most common chemical found in nature. In most environments, the PPM are so small the effects on stainless are minute. But in extreme environments, such as indoor swimming pools, the effects can be extreme and potentially dangerous. If a stainless part is under tensile stress, the pitting mentioned above will deepen, and cracking may take place. If stainless steel bolts are used under tensile stress, in an environment where chlorine corrosion is likely, the potential for stress corrosion cracking should be examine carefully.

Hamdy et al. in 2006 realized the electrochemical impedance spectroscopy study of the corrosion behavior of some Niobium bearing Stainless Steels in 3.5% NaCl, concluding that according to EIS measurements, increasing Nb content results in increasing the localized corrosion resistance of austenitic stainless steels in NaCl. Determined that cold deformation has a critical rule on the corrosion resistance of stainless steel. The corrosion resistance increased by increasing CD up to 23%. The best localized corrosion resistance was obtained from the alloy containing 1.24% Nb and 23% cold deformation.

Pardo et al. in 2008 attempts to provide a further understanding of the effect of Mo and Mn additions on the corrosion resistance of two austenitic stainless steels, AISI 304 and 316, in 30 wt.% H_2SO_4, concluding that the general corrosion resistance of stainless steels in 30 wt.% H_2SO_4 increased with the addition of molybdenum as an alloy element, in such a way that the corrosion current density was one order of magnitude lower than for stainless steels with low molybdenum content.

Meanwhile, manganese addition did not have any significant effect on the behaviour of the stainless steels in acid medium. Manganese does not have a significant influence on the corrosion resistance of the stainless steels, due to the little trend to form insoluble compounds in acid medium.

Azuma et al. in 2004 analyzed the crevice corrosion behaviour of stainless steels containing 25 mass% Cr, 3 mass% Mo and various amounts of Ni in natural seawater. The results showed that ferritic steels containing nickel were more resistant to corrosion than both ferritic steels without nickel and austenitic steels. The superiority of the Ni bearing ferritic steel over the other steels was in close agreement with the depassivation pH of those steels in acidic chloride solutions. The results showed that the addition of Ni to ferritic steel was effective in decreasing the depassivation pH and the dissolution rate in acidic chloride solutions at crevices.

Passive films were grown in potentiodynamic mode, by cyclic voltammetry on AISI 316 and AISI 304 stainless steels by Freire et al, 2012. The composition of these films was investigated by X-ray photoelectron spectroscopy (XPS). The electrochemical behaviour and the chemical composition of the passive films formed by cyclic voltammetry were compared to those of films grown under natural conditions (by immersion at open circuit potential, OCP) in alkaline solutions simulating concrete. The study included the effect of pH of the electrolyte and the effect of the presence of chloride ions. The XPS results revealed important changes in the passive film composition, which becomes enriched in chromium and depleted in magnetite as the pH decreases.

On the other hand, the presence of chlorides promotes a more oxidised passive layer. The XPS results also showed relevant differences on the composition of the oxide layers for the films formed under cyclic voltammetry and/or under OCP.

Khalfallah et al. in 2011 investigated the surface modification of AISI316 stainless steel by laser melting was investigated experimentally using 2 and 4 kW laser power emitted from a continuous wave CO_2 laser at different specimen scanning speeds ranged from300 to 1500 mm/min. Also, an investigation is reported of the introduction of carbon into the same material by means of laser surface alloying, which involves pre-coating the specimen surfaces with graphite powder followed by laser melting.

The aim of these treatments was to enhance corrosion resistance by the rapid solidification associated with laser melting and also to increase surface hardness without affecting the bulk properties by increasing the carbon concentration near the surface. Different metallurgical techniques such as optical microscopy, scanning electron microscopy (SEM), and X-ray diffraction (XRD) were used to characterize the microstructure of the treated zone. The microstructures of the laser melted zones exhibited a dendritic morphology with a very fine scale with a slight increase in hardness from 200 to 230 Hv. However, the laser alloyed samples with carbon showed microstructure consisting of g dendrite surrounded by a network of eutectic structures (g + carbide). A significant increase in hardness from 200 to 500 Hv is obtained. Corrosion resistance was improved after laser melting, especially in the samples processed at high laser power (4 kW). There was shift in Icorr and Ecorr toward more noble values and a lower passive current density than that of the untreated materials. These improvements in corrosion resistance were attributed to the fine and homogeneous dendritic structure, which was found throughout the melted zones. The corrosion resistance of the carburized sample was lower than the laser melted sample.

Liou et al., 2002, studied the effects of nitrogen content and the cooling rate on the reformation of austenite in the Gleeble simulated heat-affected zone (HAZ) of 2205 duplex stainless steels (DSSs). The variation of stress corrosion cracking (SCC) behavior in the HAZ of 40 wt%CaCl2 solution at 100°C was also studied. Grain boundary austenite (GBA), Widmanstatten austenite (WA), intergranular austenite (IGA) and partially transformed austenite (PTA) were present in the HAZ. The types and amounts of these reformed austenites varied with the cooling rate and nitrogen content in the DSS. U-bend tests revealed that pitting corrosion and selective dissolution might assist the crack initiation, while the types and amounts of reformed austenite in the HAZ affected the mode of crack propagation.

The presence of GBA was found to promote the occurrence of intergranular stress corrosion cracking. WA, IGA and PTA were found to exhibit a beneficial effect on SCC resistance by deviating the crack propagation path.

Mottu et al. in 2004 studied 316LVM, coldworked austenitic stainless steel, was implanted at 49 keV with molybdenumions. Implantation doses varied between $1*10^{15}$ and $3.5*10^{16}$ ions cm^{-2}. The structure of the implanted layer was examined by grazing incidence X-ray diffraction and the chemical composition was characterized by X-ray photoelectron spectroscopy combined with argon sputtering. Pitting corrosion studies were carried out on both unimplanted and implanted stainless steels in neutral chloride medium. The relationship between the pitting corrosion resistance, the structural and chemical modifications induced by Mo implantation was discussed. As a function of molybdenum ion dose, an expansion of fcc austenite was first observed, then above $8*10^{15}$ ions cm^{-2} a new bcc structure appeared and finally the implanted layer was partially amorphized. Electrochemical studies revealed that ion implantation enhances the pitting corrosion resistance. Increase in molybdenum implantation dose was beneficial up to $8*10^{15}$ ions cm^{-2} in improving the pitting corrosion resistance, beyond which it had a detrimental effect.

Freire et al., 2009, studied iron oxide thin layers formed on mild steel substrates in alkaline media by the application of different anodic potentials were studied in order to characterize their morphology, composition and electrochemical behaviour, in particular under conditions of cathodic protection. The surface composition was evaluated by X-Ray Photoelectron Spectroscopy (XPS) and Auger Electron Spectroscopy (AES). The morphology of the surface oxides was studied via Atomic Force Microscopy (AFM). The electrochemical behaviour of the surface oxides was studied using Electrochemical Impedance Spectroscopy (EIS). The results showed that the surface film is composed by Fe^{2+} oxides and Fe^{3+} oxides and/or hydroxides. The contribution of Fe^{2+} species vanishes when the potential of film formation increases in the passive domain. Two distinct phases were differentiated in the outer layers of the surface film, which proves that film growing is topotactic in nature.

Park and Kwon examined the effects of Mn on the localized corrosion, anodic dissolution behavior, and repassivation kinetics of Fe–18Cr–xMn (x = 0, 6, 12) alloys using potentiodynamic tests with or without microelectrochemical cell, electrochemical impedance spectroscopy (EIS), and rapid scratch electrode tests. The addition of Mn to Fe–18Cr alloy significantly degrades passivity by decreasing the resistance to localized corrosion, and also by expanding the active region in the noble direction. The decrease in the resistance to localized corrosion of the alloys is due primarily to an increase in the number and size of Mn-containing oxides, acting as initiation sites for pitting corrosion. It was demonstrated using a micro-electrochemical test that the inherent protectiveness of passive film is also considerably reduced by the Mn addition, even though the deleterious influences of Mn-containing oxides are completely excluded. Mn facilitates the metal dissolution reaction by enhancing the activity of iron adsorbed intermediate or by generating second intermediate species (possibly manganese adsorbed intermediate) acting as another dissolution path. Further, the Mn addition appears to suppress the passivation process by reducing the activity of Cr-adsorbed species in an acidic solution, and hence the repassivation rate is significantly decreased with Mn content of the alloys.

Machuca et al., 2012 conducted measurements to evaluate localized corrosion on UNS S31603, UNS S31803, UNS S32750, UNS S31254 and UNS N08825 in natural seawater. Critical pitting and crevice temperatures were assessed using a potentiostatic technique and critical potentials for pitting and crevice corrosion initiation and repassivation were identified using potentiodynamic polarization at temperatures from 5 to 40 °C. Passivity breakdown always occurred through pitting and crevice growth above a transition temperature. Below this temperature, pitting corrosion was not observed on any of the alloys regardless of the applied potential, but initiation of crevice corrosion occurred after the alloys reached a transpassive potential.

The summary of main results about corrosion behaviour and passivity of stainless steels can be seen in Table 1.

Date	Authors	Summary of main results
2001	Ilevbare, G.O. ; Burstein, G.T.	The presence of Mo assists the rapid repassivation of the bare metal so that further metal dissolution and, therefore, metastable pit formation is prevented.
2002	Polo, J.L. ; Cano, E.; Bastidas, J.M.	A depressed capacitive loop at high frequencies indicating a charge transfer process; a second capacitive loop at intermediate frequencies attributed to adsorption_/desorption processes; and a third capacitive response at low frequencies associated with a diffusion process through corrosion products inside pits.
2002	Liou, H.Y.; Hsieh, R.I.; Tsai, W.T.	The presence of Widmanstatten, intergranular austenite and partially transformed austenite exhibited beneficial effect on SCC resistance by deviating the crack propagation path.

Date	Authors	Summary of main results
2004	Azuma, S. ; Kudo, T.; Miyuki, H. ; Yamashita, M.; Uchida, H.	Crevice corrosion behaviours of stainless steels containing 25% Cr, 3% Mo and various amounts of Ni were investigated by immersion tests in natural seawater and in a 10% ferric chloride solution. In the seawater immersion, 4% Ni ferritic steel showed higher resistance than 0% Ni ferritic steel and 30% Ni austenitic steel. In contrast, the CCT in 10% ferric chloride solution was steadily decreased with increasing the Ni content.
2005	Mottu, N. ; Vayer, M. ; Dudognon, J. ; Erre, R.	Improvement of the repassivation of the passive layer observed in this group has to be related to the increase of Mo concentration in the passive layer.
2006	Hamdy, A.S.; El-Shenawy, E. ; El-Bitar, T.	Increasing Nb content results in increasing the localized corrosion resistance of austenitic stainless steels in NaCl. Cold deformation (CD) has a critical rule on the corrosion resistance of stainless steel.
2008	Pisareka, M.; Kędzierzawskib, P.; Płociński, T.; Janik-Czachor, M.; Kurzydłowski, K.J.	The MnS inclusions do not undergo passivation; thus a large density of locations susceptible to pitting attack in a chloride-containing environment is available.
2008	Pardo, A. ; Merino, M.C. ; Coy, A.E. ; Viejo, F.; Arrabal, R. ; Matykina, E.	Molybdenum seems to play, therefore, a dual role in the improvement of the general corrosion resistance of the stainless steels by modification of passive film composition and modification of active dissolution by formation of insoluble oxides.
2008	Al-Fozan, S.A.; Malik, A.U.	The carbon steels 304 SS and 316L SS have been markedly affected by water line corrosion.
2009	Freire, L.; Nóvoa, X.R. ; Montemor, M.F. ; Carmezim, M.J.	The structure of the passive film is potential dependent and affects the efficiency of the cathodic protection current.
2010	Park, K.; Kwon, H.	The repassivation rate of Fe–18Cr–xMn (x = 0, 6, 12) was significantly reduced with Mn content.The susceptibility to metastable pitting corrosion was significantly increased with Mn content.
2010	Cai, B.; Liu, Y.; Tian, X.; Wang, F.; Li, H.; Ji, R.	Three regions, namely, the passive, active (consisting of severely attacked, commonly attacked and lightly attacked regions) and variable regions, can be observed on most crevice corrosion sites.
2011	Khalfallah, I.Y.; Rahoma, M.N.; Abboud, J.H. ; Benyounis, K.Y.	The microhardness was not increased after melting treatment but corrosion resistance improved especially when a high power is used. There was a shift in I_{corr} toward lower and nobler values after this treatment.
2012	Freire, L. ; Catarino, M.A. ; Godinho, M.I. ; Ferreira, M.J. ; Ferreira, M.G.S. ; Simoes, A.M.P.; Montemor, M.F.	The films formed on AISI 316 under d.c. potentiodynamic polarization are thicker, since the metallic contribution was hardly or even could not be detected. Moreover, the outer layers of the cycled films are extremely rich in Fe^{3+} oxides and depleted in Fe^{2+}, chromium oxides and nickel oxides.
2012	Machuca, L.L. ; Bailey, S.I. ; Gubner, R.	The evaluation of marine crevice corrosion mainly through the investigation of repassivation potentials is the best approach to characterize the resistance of the alloys to localized corrosion.

Table 1. Principal results about corrosion studies of stainless steels.

http://dx.doi.org/10.5772/57275

In the directional solidification process, the molten metal at the bottom of the mold (near to the cooling system) begins to cool and solidify before the rest of the mold does. As the metal on the bottom of the mold cools, this front of solidification moves steadily upward. By controlling the rate of flow for the molten metal feed and introducing thermal variations in the mold, shrink defects can be eliminated.

Directional solidification and progressive solidification describe types of solidification within castings. Directional solidification describes solidification that occurs from farthest end of the casting and works its way towards the sprue. Progressive solidification, also known as parallel solidification (Stefanescu, 2008), is solidification that starts at the walls of the casting and progresses perpendicularly from that surface (Chastain, 2004).

Directional solidification can be used as a purification process. Since most impurities will be more soluble in the liquid than in the solid phase during solidification, impurities will be "pushed" by the solidification front, causing much of the finished casting to have a lower concentration of impurities than the feedstock material, while the last solidified metal will be enriched with impurities.

The development of crystallographic texture during directional solidification has been analysed quantitatively in columnar castings of the Ni-base superalloys, CMSX4 and CM186LC, produced with a range of cooling rates (Ardakani, 2000).

Since directional solidification technique was firstly used to prepare turbine blade in 1960's, research work has provided an in-deep understanding of relationship of alloys, geometry, microstructure, and process conditions, the application of this knowledge to industry has proven to be challenging at best. Until now, directional solidification technique becomes more and more important research area and active technology in the field of materials science and engineering.

Directional solidification technology is also an important research instrumentally to study solidification theory. Because directional solidification can achieve controllable cooling rate with a broad range, solidification structure from near-equilibrium to far-from equilibrium condition can be obtained. A serious of scientific phenomena such as interface evolution, crystal growth instability, solute redistribution, and phase selection during the course have received significant attention in recent years.

Columnar-to-quiaxed transition (CET) is one possible event in casting process. This occurs during columnar growth, usually during directional solidification, when equiaxed grains begin to form, grow, and subsequently stop the columnar growth. Normally either a columnar or equiaxed grain structures desired in industry applications, so that consistent mechanical properties are achieved throughout the casting (Gäumann et al., 2001, McFadden et al., 2009, Reinhart et al., 2005, Spittle, 2006 and Ares et al., 2000, 2002, 2007, 2010).

The CET has been examined in different type of alloys because of the advantages offered by equiaxed grain solidification. In commercial practice, attempts are made to produce either wholly columnar structures (the production of directionally solidified turbine blades) or wholly equiaxed structures. Interest in the CET therefore stems from the wish to theoretically understand the solidification conditions that define the transition between these two extremes of structures.

In previous research, the authors of this work carried out experiments in which the conditions of columnar to equiaxed transition (CET) in directional solidification of dendritic alloys were determined The alloy systems in this work include Pb–Sn (Ares & Schvezov, 2000), Al–Cu (Ares et. al., 2011), Al–Mg (Ares et. al., 2003), Al–Zn and Zn-Al alloys (Ares & Schvezov, 2007).

This work aims to study the overall influence of the variation of the structure (equiaxed, columnar and columnar-to-equiaxed transition, CET) on the corrosion resistance of 316 stainless steel in aqueous 3% NaCl (pH = 5.5) using cyclic potentiodynamic polarization techniques and electrochemical impedance spectroscopy (EIS) and investigate the relationship between the corrosion resistance of material and the secondary dendritic spacing evolution.

2. Directional solidification and structures

The alloys were prepared from pure materials of different grades, and the samples were solidified directionally upwards in a furnace which was described elsewhere (Ares et al., 2000, 2007, 2010). The temperature measurements were performed using K-type thermocouples which were protected with ceramic shields and connected to a data acquisition system controlled by a computer.

The thermocouples were previously calibrated using different temperature points: demineralized water at the freezing and boiling points (corrected by atmospheric pressure) and Fe (99.999 wt%) and Stainless Steel (99.999 wt%) at their melting points. The accuracy of thermocouples was determined to be between ± 0.25 K. The samples were melted in alumina molds of 23 mm i.d. and 25 mm e.d., with a flat bottom and a cylindrical uniform section of 200 mm high.

Five thermocouples were positioned at 20 mm intervals on the centerline of the cylinder mould from the bottom surface of the mould. During a solidification experiment, the temperature measured by each thermocouple was recorded at regular intervals of time.

Different intervals were previously tested, and as a result of this exercise, an interval of 1 second was selected. For the data processing, the readings made every 0.10 seconds during 1 second were averaged, and this value was associated to the middle of the averaged interval.

This procedure was chosen as a result of a compromise between memory available in the data logger, the time taken for an entire solidification experiment, and the degree of precision to capture the main phenomena occurring during the columnar to equiaxed transition.

The experimental procedure was as follows. At first, the liquid metal into the mould in the furnace was allowed to reach the selected temperature. Once a uniform temperature was reached, less than 1 K of difference registered between the five thermocouples located at different position in the melt, the furnace power was turned off and the melt was allowed to solidify from the bottom. Heat was extracted through a cooling system, which consisted on a copper disk attached to a copper coil, both cooled by running water. The solidification velocity was adjusted by changing the temperature and flow of water and also by adding plates of materials between the copper plate and the crucible which changed the effective value of the thermal conductivity.

The crucible was also isolated on the top to reduced heat losses from the top of the furnace to a minimum. Temperature data acquisition was started approximately at the same time that the furnace power was turned off in order to record the entire unidirectional solidification process. The measurements showed that the initial temperature (superheat) ranged from 1353.2 K to 933.6 K in all the experiments. With this experimental setup unidirectional heat flow was achieved and also convection associated with the pouring of the liquid into the mould was eliminated.

After directional solidification, the cylindrical ingots were cut on a mid longitudinal plane. Then, the samples were polished with sand paper and etched with 1: 1: 1 HCl / HNO_3 / H_2O. Etching was performed at room temperature [20]. The position of the transition was located by visual observation and optical microscopy, and the distance from the bottom of the sample was measured with a ruler, see Figure 2.

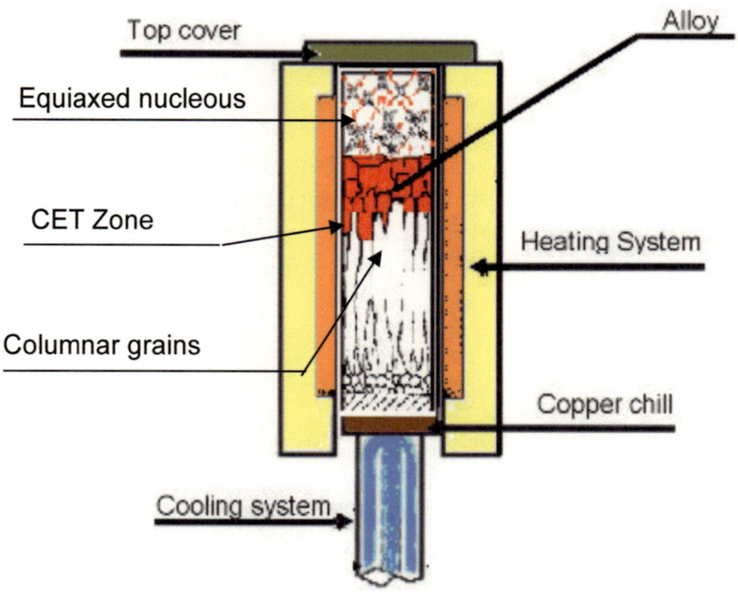

Figure 1. Schematic of experimental device for directional solidification tests.

The microstructure was analyzed using Scanning Electron Microscopy (SEM) and an inverted metallographic microscope (Arcano XTL 3400). Utilizing ASTM E112 standard norm and a typical histogram showing the frequency of the size of the equiaxed grains for each sample, the equiaxed grain size was measured, at equally spaced intervals, the counting starts at the position where the transition begun. The columnar region was divided in similar way and the width and length of the grains measured directly (Ares et al., 2007, 2010).

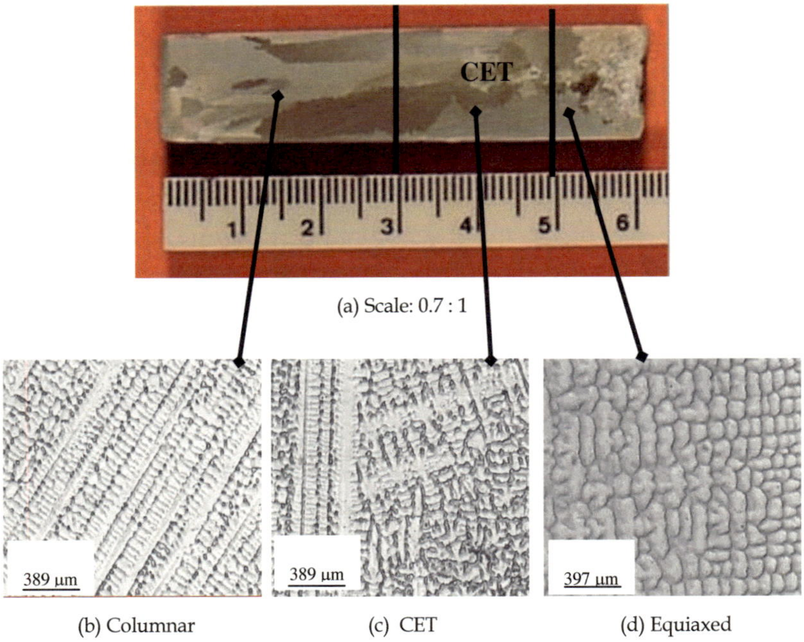

Figure 2. Obtained structures during directional solidification of 316 L stainless steel. (a) Macrostructure. (b-d) Representative microstructures in columnar, CET and equiaxed zones of the samples.

For measuring the secondary dendritic spacing, λ_2, samples were again grounded and polished until a mirror finish. Etching was performed to reveal the microstructure, which was observed by an optical microscope. With the help of TSView® image processing program, the secondary dendritic spacing value, λ_2, was obtained for each sample, averaging 15 measures in each zone of the samples.

3. Electrochemical tests

For the electrochemical tests, samples of 20 mm in length of each zone and for each concentration were prepared as test electrodes (see Figure 3), polished with sandpaper (from SiC # 80 until # 1200) and washed with distilled water and dried by natural flow of air. Table 2 shows the compositions used for analysis.

All the electrochemical tests were conducted in 3wt% NaCl solution (pH = 5.5) at room temperature using an IM6d Zahner®-Elektrik potentiostat coupled to a frequency analyzer system. Oxygen was removed from the 3% solution of NaCl in the cell by a stream of nitrogen.

A conventional three-compartment glass electrochemical cell with its compartments separated by ceramic diaphragms was used. The test electrodes consisted of sections of the ZA ingots (see Figure 3) were positioned at the glass corrosion cell kit (leaving a rectangular area in

Element	Sample					
	A (316)	**B (316L)**	**C (316L)**	**D (316L)**	**E (316L)**	**F (316)**
C	0.08	0.03	0.03	0.03	0.03	0.08
Si	1.00	1.00	1.00	1.00	1.00	1.00
Mn	2.00	2.00	2.00	0.50	2.00	2.00
P	0.045	0.045	0.045	0.045	0.045	0.045
S	0.030	0.030	0.030	0.030	0.030	0.030
Cr	18.00	18.00	18.00	18.00	18.00	18.00
Ni	10.00	14.00	14.00	14.00	22.00	10.00
Mo	2.00	3.00	8.00	3.00	3.00	8.00
Fe	Balance	Balance	Balance	Balance	Balance	Balance

Table 2. Composition of the samples (weight percent)

contact with the electrolyte). The potential of the test electrode was measured against a saturated calomel reference electrode (0.242 V vs NHE), provided with a Luggin capillary tip. The Pt sheet was used as a counter electrode.

Impedance spectra were obtained in the frequency range of 10^{-3} Hz and 10^{5} Hz at open circuit potential.

For comparison purposes, experiments with different structures were conducted under the same experimental conditions. All the corrosion tests experiments were triplicate and the average values and graphical outputs are reported.

The corrosion resistance of the material under study was investigated by performing cyclic potentiodynamic polarization curves, according to ASTM G61-86.

The samples were pre-catodized during 3 minutes at a potential of -1000 mV, lower than the open circuit potential. After that time the specimen was left open circuit for an hour, recording the open circuit potential. The scanning of the polarization curve began from the open circuit potential to more anodic potential at a rate of 0.1666 mV / s. When the current reached 5 mA/ cm^2, the effect on the potentiodynamic polarization was reversed, continuing until the line intersects the curve outward, closing the hysteresis loop, or until it reaches the corrosion potential. The above procedure was performed for each obtained sample.

After completion the polarization tests, the samples were examined using Arcano® metallographic microscope.

In corrosion studies using EIE technique, the obtained impedance spectra are usually analyzed using electrical circuits composed by components such as resistance (R), capacitance (C), inductance (L), etc., combined all to reproduce the measured impedance spectra. These circuits are called "equivalent circuit" (Ferreira & Dawson, 1985).

The equivalent circuit shown in Figure 4 was proposed, where R_{Ω} corresponds to the resistance of the electrolyte, R_1 corresponds to the resistance to charge transfer, R_2 accompanies the double layer and is referred to the resistance of the oxide layer; capacitances C come from the constant phase elements: C_1 values are attributed to the ability of the double layer and C_2 correspond to porous oxides capabilities (Polo et al., 1999).

To carry out the measurements, all samples were sanded CSi gradation # 1500, the assays were performed under constant nitrogen bubbling.

The test conditions were as follows:

- At -1000 mV during 3 minutes:
- Open circuit for 1 hour
- Potential amplitude: 10 mV/s around the open circuit potential
- Frequency range: 100 kHz^{-} to 1 mHz
- Points / decade: 10
- Stabilization of the system: 600 seconds

The measurements were performed at open circuit potential. For adjustment of the results we used the method of nonlinear least squares designed by Bouckamp (Winston Revie, 2000). We worked with a Gamry EIS Instrument, in combination with the LYP M7 potentiostat.

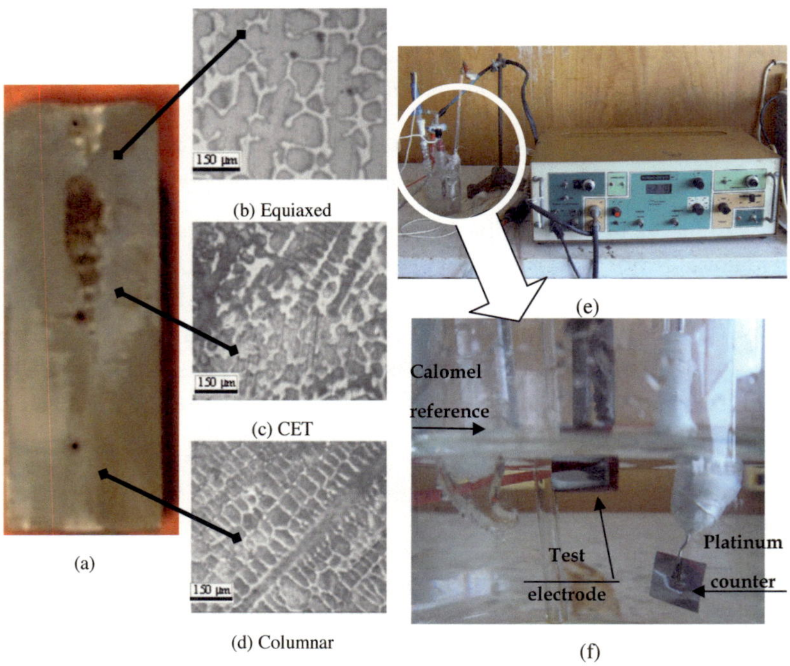

Figure 3. (a) Macrostructure. (b) to (d) Structures of work electrodes (columnar, CET and equiaxed). (e) Experimental device for electrochemical tests. (f) A glass corrosion cell kit with a platinum counter electrode and a sutured calomel reference electrode (SCE).

http://dx.doi.org/10.5772/57275

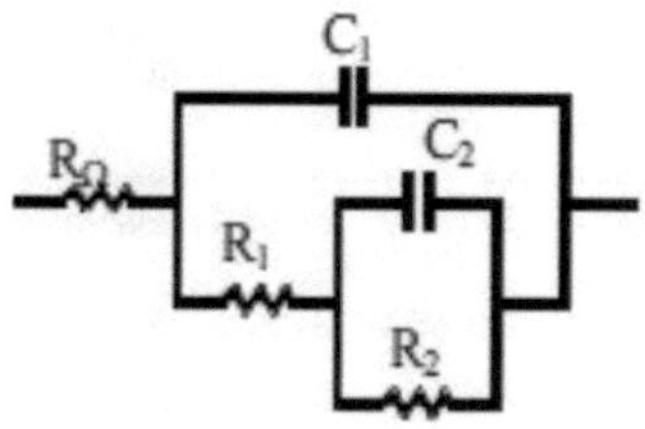

Figure 4. Proposed equivalent circuit.

4. Cyclic potentiodynamic polarization curves

Figure 3 shows the polarization curves obtained in 3% NaCl solution for one of the samples.

Localized corrosion was strongly influenced by the alloy composition, comparing the difference between the corrosion potential and the pitting potential in the same test, positioned longitudinally and then transversely to the reference electrode. It was observed that the samples for the group of 316 steels showed higher values than those found for samples of 316L stainless steel, they demonstrated that the potential start of pitting close to 1000 mV, while those rarely exceeded 500 mV at pitting potential (Figure 5).

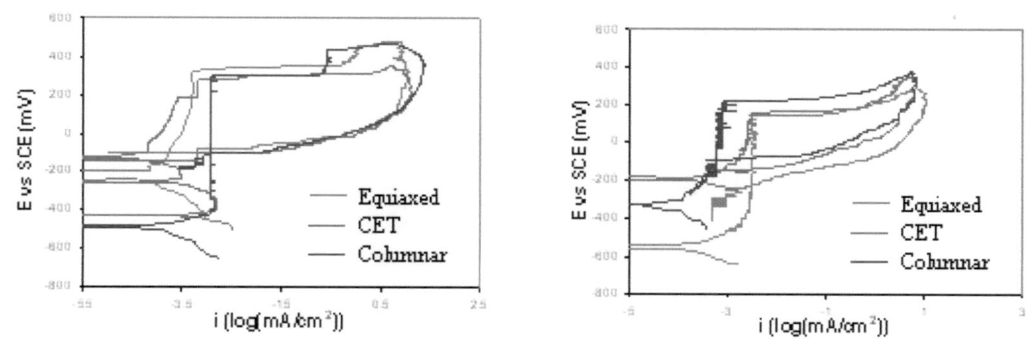

Figure 5. Cyclic potentiodynamic polarization curves of E sample, in 3% NaCl, (a) longitudinally and (b) transversally positioned, with respect to the reference electrode.

Once the process of pits formation started, evidenced by a sharp increase in the current density with respect to the potential, the specimen that turned to passivity almost immediately achieved after reversing the direction of the polarization was the specimen belonging to the group 316 stainless steels containing molybdenum in greater proportion. The other suffered bites that could not passivated again completely. As an example, in Figure 6 it is possible to observe the pits having one specimen.

The lower carbon content alloys were provided in some areas of passivity defined.

In Table 3 can be seen the various corrosion parameters of the analyzed samples.

Sample			Corrosion Potential, E_{corr}	Pitting Potential, E_{pic}	Repassivation potential , E_{rep}	E_{pic} - E_{corr}	E_{pic} - E_{rep}	E_{rep} - E_{corr}
F	Long	equiaxed	-131	932	915	1063	17	1046
		CET	-540	845	833	1385	12	1373
		columnar	-170	923	328	1093	595	498
	Trans	equiaxed	-312	891	878	1203	13	1190
		CET	-437	885	876	1322	9	1313
		columnar	-552	942	-252	1494	1194	300
B	Long	equiaxed	-580	246	-86	826	332	494
		CET	-501	443	-135	944	578	366
		columnar	-488	441	-123	929	564	365
	Trans	equiaxed	-560	121	-23	681	144	537
		CET	-563	76	-9	639	85	554
		columnar	-290	464	-348	754	812	-58
E	Long	equiaxed	-222	322	-72	544	394	150
		CET	-429	271	142	700	129	571
		columnar	-486	292	-114	778	406	372
	Trans	equiaxed	-556	140	-229	696	369	327
		CET	-196	157	-146	353	303	50
		columnar	-330	220	-113	550	333	217
C	Long	equiaxed	-504	324	-215	828	539	289
		CET	-568	250	-224	818	474	344
		columnar	-673	359	-168	1032	527	505
	Trans	equiaxed	-436	193	-206	629	399	230
		CET	-503	158	-176	661	334	327
		columnar	-583	81	-262	664	343	321
D	Long	equiaxed	-123	334	-163	457	497	-40
		CET	-208	278	-259	486	537	-51
		columnar	-184	162	-311	346	473	-127
	Trans	equiaxed	-65	495	-98	560	593	-33
		CET	-232	569	-330	801	899	-98

http://dx.doi.org/10.5772/57275

Sample			Corrosion Potential, E_{corr}	Pitting Potential, E_{pic}	Repassivation potential , E_{rep}	E_{pic} - E_{corr}	E_{pic} - E_{rep}	E_{rep} - E_{corr}
		columnar	-195	296	-244	491	540	-49
A	Long	equiaxed	-366	1006	-161	1372	1167	205
		CET	-130	904	-25	1034	929	105
		columnar	-120	911	-191	1031	1102	-71
	Trans	equiaxed	-71	991	-116	1062	1107	-45
		CET	-259	906	906	1165	0	1165
		columnar	-246	996		1242		

Table 3. Potential values found in the studied samples (mV).

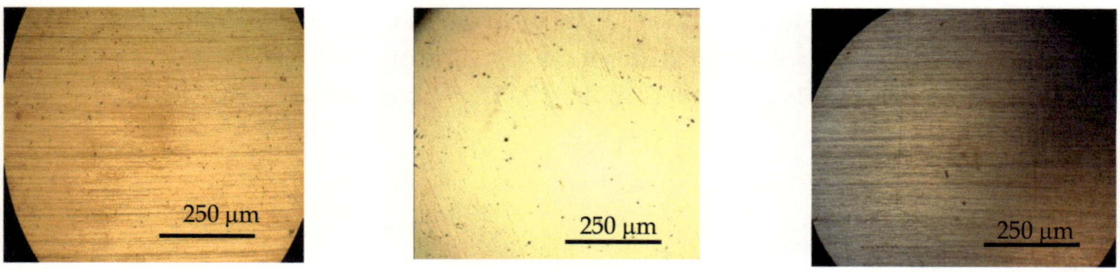

Sample F Sample B Sample E

Figure 6. Surface of the samples after being subjected to cyclic potentiodynamic polarization testing in 3% solution of NaCl (magnification 100x).

5. Electrochemical Impedance Spectroscopy (EIE)

EIE measurements were performed at open circuit potential, or the potential for corrosion, because all the samples without exception have become passives at this potential. Therefore, the obtained results, corresponds to the behaviour of the passive layer formed.

Figure 7 shows the Bode and Nyquist obtained for one of the alloys studied, it can be observed a good agreement between experimental and simulated data. Table 4 lists the values found on the parameters used in the simulation.

Impedance diagrams represented in a Nyquist diagram for all compositions showed similar shapes to each other, show a depression below the real axis. This flattening may be due to improper cell design, surface roughness or surface porosity.

From these tests it was observed that the porosity of the passive layer formed on the metal surface was variable, being more compact in areas with equiaxed structure in some alloys, and in areas with columnar structure, in others. Passive film 316L alloy with 8% Mo was, in general, the less porous the compositions studied, being therefore a protective oxide layer. This is consistent with the polarization tests conducted in which this alloy had a good performance against pitting corrosion within the group of 316L stainless steel. The other samples showed more or less compact films, depending on the type of structure in question, as noted above. The degree of porosity of the passive layer, given by the values of n_2 [9], can influence the behavior of the alloy against pitting corrosion, whose study exceeds the limits of this work.

The resistance of formed oxides related to the parameter R_2, also had a variable behaviour, depending largely on the type of solidification structure, and the alloy composition. In this type of testing could not be establishing a clear relationship between the content of alloying elements in the metal and their behaviour. However, the oxide layers formed more resistant in certain areas of the samples, mostly those with columnar grains and with CET, were not necessarily the most compact.

One example is found in the area of one of the columnar steel samples 316, which showed resistance values the oxide layer by an order of magnitude greater than the other, with a medium to high porosity. Said alloy, corresponding to the composition of 0.08% C and 2% Mo, chopped reached potentials which were around 1000 mV, but without passivation capability.

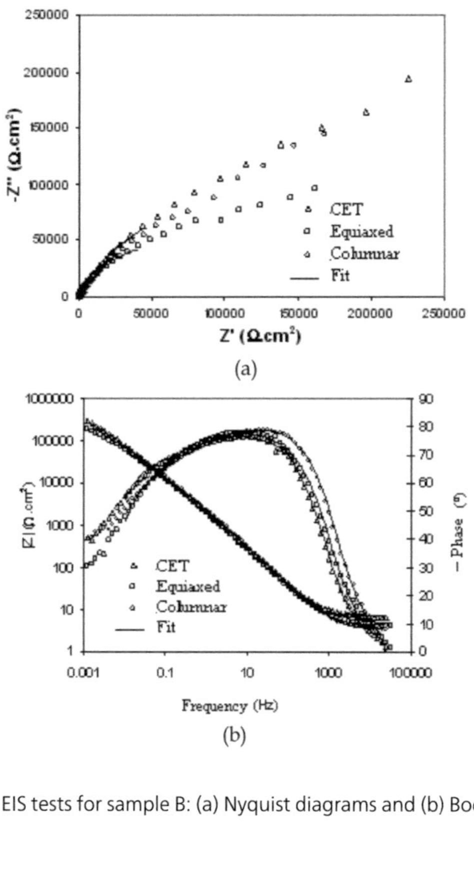

Figure 7. Diagrams obtained by EIS tests for sample B: (a) Nyquist diagrams and (b) Bode diagrams.

Parameter	Sample								
	C			D			A		
	Columnar	CET	Equiaxed	Columnar	CET	Equiaxed	Columnar	CET	Equiaxed
R_Ω (Ω cm^2)	5.46	3.30	3.84	6.00	4.17	3.56	4.83	3.45	3.28
C_1 (μFcm^{-2})	52.26	102.47	48.51	89.38	58.63	48.86	26.54	59.74	53.34
n_1	0.91	0.88	0.91	0.88	0.88	0.90	0.88	0.91	0.89
R_1 (kΩcm^2)	195.04	17.17	48.32	98.88	10.08	35.01	19.78	20.23	40.62
R_2 (kΩ cm^2)	507.74	77.11	208.51	452.16	488.99	317.10	1347.50	168.92	603.60
C_2 (μFcm^{-2})	33.77	78.48	22.12	27.50	29.39	31.08	9.11	31.63	27.90
n_2	0.79	0.64	0.70	0.67	0.58	0.61	0.58	0.63	0.82
χ^2	1.10^{-3}	1.10^{-3}	1.10^{-3}	1.10^{-3}	1.10^{-3}	1.10^{-3}	1.10^{-3}	1.10^{-3}	1.10^{-3}
Parameter	Sample								
	F			B			E		
	Columnar	CET	Equiaxed	Columnar	CET	Equiaxed	Columnar	CET	Equiaxed
R_Ω (Ω cm^2)	5.18	4.38	3.18	3.39	6.66	4.38	4.57	4.72	4.94
C_1 (μFcm^{-2})	45.67	70.47	103.37	74.00	91.70	104.07	67.27	101.88	55.20
n_1	0.87	0.90	0.86	0.90	0.87	0.87	0.87	0.88	0.90
R_1 (kΩ cm^2)	10.25	10.20	3.30	8.81	30.51	38.08	22.30	6.67	14.53
R_2 (kΩ cm^2)	326.31	145.91	35.30	15.41	189.82	84.68	403.28	116.04	348.93
C_2 (μFcm^{-2})	46.74	62.31	126.72	36.60	34.55	73.40	21.63	165.06	23.88
n_2	0.55	0.61	0.53	0.60	0.73	0.80	0.62	0.66	0.66
χ^2	1.10^{-3}	1.10^{-3}	1.10^{-3}	1.10^{-3}	1.10^{-3}	1.10^{-3}	1.10^{-3}	1.10^{-3}	1.10^{-3}

Table 4. Setting parameters of the EIS simulation at the corrosion potential.

6. Microstructure: Secondary dendritic arm spacing

In Figure 8 it is possible to observe the measurements of secondary dendritic arm spacings, λ_2, along the samples.

The values of secondary dendritic spacing, λ_2, obtained from the micrographs were plotted as a function of the distance from the base of the sample (Figure 9), following the order: columnar, CET and equiaxed zones, the average values in each zone are summarized in Table 5. The values of λ_2 obtained for 316 steels were 20 μm below the values of λ_2 for 316L steels, but in all the samples, values of λ_2 in the zones of the samples with columnar solidification structure were lower than those found in zones with equiaxed structure. The values of measurements in the CET zones were variable, being intermediate between columnar and equiaxed zones in some cases, or resulting the lower values from the three structures in others cases.

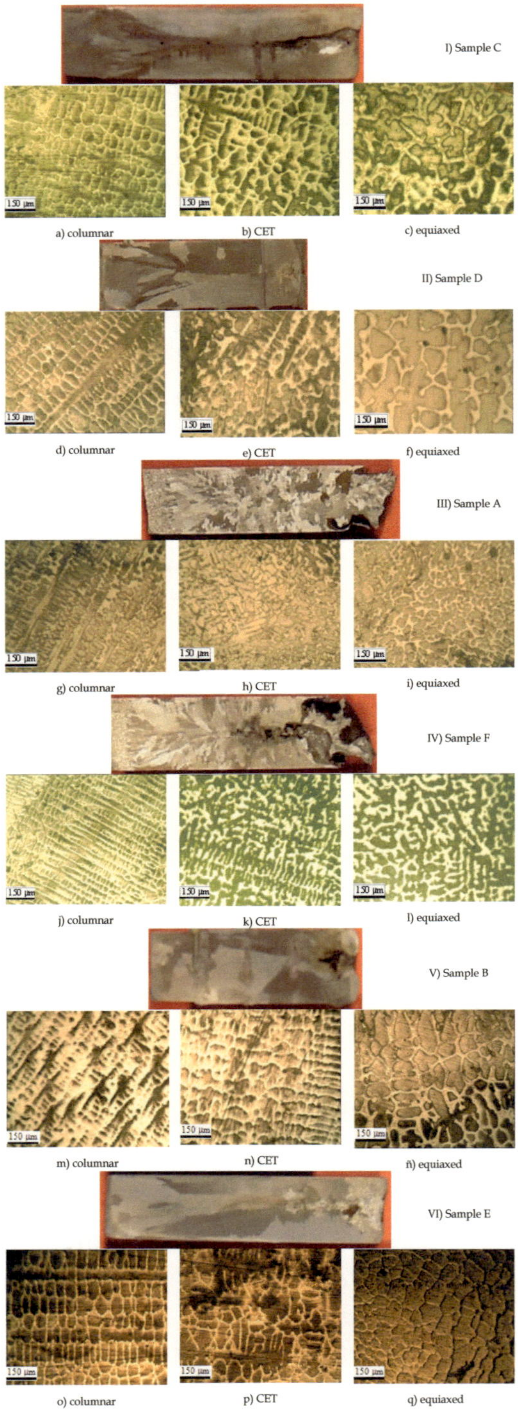

Figure 8. Solidification structures of 316 L. I-VI) Macrostructures; a-q) Microstructures.

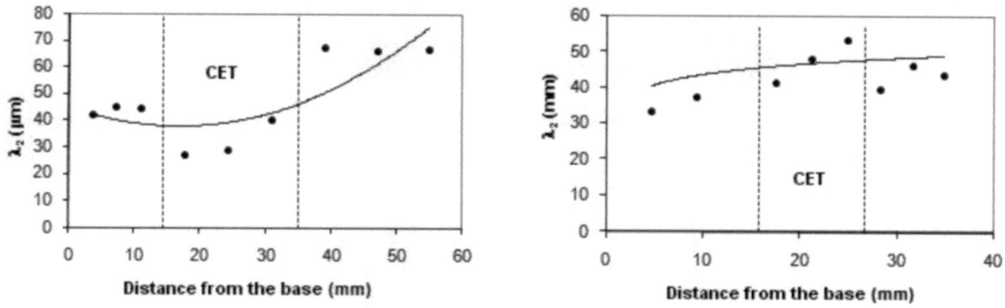

(a) (b)

Figure 9. Secondary dendritic arm spacings versus distance, for 316 L samples: (a) Sample C and (b) Sample B.

		Sample					
		C	D	A	F	B	E
Secondary dendritic arm spacing, λ_2 (µm)	Columnar	43.57	66.40	29.85	26.51	48.20	50.73
	CET	31.75	49.96	41.14	34.65	47.59	58.79
	Equiaxed	66.37	74.21	42.86	38.53	43.04	60.28

Table 5. Averages values of secondary dendritic arm spacings for each sample.

7. Summary

Present knowledge about resistance to corrosion and passivity of 316L stainless steel directionally solidified samples is limited. This is the complex problem and it is necessary to deal with it as the separate question in the analysis of microstructure evolution and electrochemical properties of the stainless steels.

Alloys corresponding to stainless steels 316 L group presented passivity in defined zones and with varying of re-passivation, while 316 stainless steels began to pitting corrosion at much higher potentials.

In general, 316 steel samples with higher content of Mo were the more resistant to pitting corrosion, in zones corresponding to equiaxed and CET solidification structures.

Molybdenum content in the alloys was beneficial, promoting re-passivation in samples that containing a higher proportion of this element.

The oxide layer formed on the stainless steel corresponding to Sample C was one of the less porous and with higher values of the transfer of charge resistance, while Sample A, formed a protective passive film in the columnar zone but with very low charge-transfer-resistance, and a high porosity, making it susceptible to localized corrosion.

Zones with equiaxed structure showed the highest values of secondary dendritic arm spacings within the same sample, and a good ability to re-passivation after pitting, while zones with columnar structure and with CET having passive zones larger than equiaxed structure.

The steel of Sample F had the best performance against located corrosion, while Sample D was the most susceptible to this type of attack.

Acknowledgements

One of the authors, Alicia Esther Ares, would like to thanks Consejo Nacional de Investigaciones Científicas y Técnicas (CONICET) from Argentina for the financial support.

Author details

Claudia Marcela Méndez[1], Mónica Mariela Covinich[1] and Alicia Esther Ares[2*]

*Address all correspondence to: aares@fceqyn.unam.edu.ar

1 Materials Laboratory, Faculty of Sciences, National University of Misiones, Posadas, Argentina

2 Researcher from National Scientific and Technical Research Council, Argentina

References

[1] Al-Fozan, S.A.; Malik, A. U. (2008) Effect of seawater level on corrosion behavior of different alloys, Desalination, Vol. 228, p.p. 61–67.

[2] Ardakani, M.G.; D'Souza, N. ; Wagner, A.; Shollock, B.A.; McLean, M. (2000) Competitive Grain Growth and Texture Evolution during Directional Solidification of Superalloys, Superalloys 2000, Edited by T.M. Pollock, R.D. Kissinger, R.R. Bowman, K.A. Green, M. McLean, S. Olson, and J.J. Schirra, TMS (The Minerals, Metals & Materials Society).

[3] Ares, A.E.; Schvezov, C.E. (2000) Solidification Parameters During the Columnar-to-Equiaxed Transition in Lead-Tin Alloys. Metallurgical and Materials Transactions A, Vol. 31, 1611-1625.

[4] Ares, A.E.; Gueijman, S.F.; Schvezov, C.E. (2010) Experimental Study of the Columnar-to-Equiaxed Transition During Directional Solidification of Zinc-Aluminum Alloys and Composites, J. Crystal Growth, Vol. 312, pp. 2154-2170.

[5] Ares, A.E.; Gueijman, S.F.; Caram, R.; C.E. Schvezov (2005) Analysis of Solidification Parameters During Solidification of Lead and Aluminum Base Alloys. J. Crystal Growth, Vol. 275, pp. 235-240.

[6] Ares, A.E.; Schvezov, C.E. (2007) Influence of Solidification Thermal Parameters on the Columnar to Equiaxed Transition of Al-Zn and Zn-Al Alloys. Metallurgical and Materials Transactions A, Vol. 38, pp. 1485-1499.

[7] Ares, A.E.; Gueijman, S.F.; Schvezov, C.E. (2002) Semi-Empirical Modeling for Columnar and Equiaxed Growth of Alloys. J. Crystal Growth, Vol. 241, pp. 235-240.

[8] Azuma, S.; Kudo, T.; Miyuki, H.; Yamashita, M.; Uchida, H. (2004) Effect of nickel alloying on crevice corrosion resistance of stainless steels, Corrosion Science, Vol. 46, p.p. 2265–2280.

[9] Batra, U., Ray, S., Prabhakar, S.R. (2003) Effect of austenitization on austempering of copper alloyed ductile iron. *J Mater Eng Perf.* , Vol. 12 (5): p.p. 597–601.

[10] Cai, B.; Liu, Y.; Tian, X.; Wang, F.; Li,H.; Ji, R. (2010) An experimental study of crevice corrosion behaviour of 316L stainless steel in artificial seawater, Corrosion Science, Vol. 52, p.p. 3235-3242.

[11] Chastain, S. (2004), Metal casting: a sand casting manual for the small foundry, Vol. II 4, Stephen Chastain, ISBN 978-0-9702203-3-2.

[12] Chupatanakul, S., Nash, P. (2006) Dilatometric measurement of carbon enrichment in austenite during bainite transformation. *J Mater Sci.,* Vol. 41 (15): p.p. 4965–9.

[13] DOD Technical Bulletin – Corrosion Prevention and Detection, Department of the Navy, United States of America.

[14] Ferreira, M.G., Dawson, L.J. (1985) Electrochemical studies of the passive film of 316 stainless steel in chloride media, UMIST, Corrosion and Protection Centre, Manchester, England M60 1QD.

[15] Freire, L.; Catarino, M.A.; Godinho, M.I.; Ferreira, M.J.; Ferreira, M.G.S.; Simões, A.M.P.; Montemor, M.F. (2012) Electrochemical and analytical investigation of passive films formed on stainless steels in alkaline media, Cement & Concrete Composites Vol. 34, p.p.1075–1081.

[16] Freire, L.; Nóvoa, X.R.; Montemor, M.F.; Carmezim, M.J. (2009) Study of passive films formed on mild steel in alkaline media by the application of anodic potentials, Materials Chemistry and Physics, Vol. 114, p.p. 962–972.

[17] Gäumann, M.; Bezençon, C.; Canalis, P. et al. (2001) Single-Crystal Laser Deposition of Superalloys: Processing-Microstructure Maps. Acta Mater., Vol. 49, pp.1051-1062.

[18] Hamdy, A.S.; El-Shenawy, V.; El-Bitar, T. (2006) Electrochemical Impedance Spectroscopy Study of the Corrosion Behavior of Some Niobium Bearing Stainless Steels in 3.5% NaCl, Int. J. Electrochem. Sci., Vol. 1, p.p. 171-180.

[19] Ilevbare, G.O. ; Burstein, G.T. (2001) The role of alloyed molybdenum in the inhibition of pitting corrosion in stainless steels, Corrosion Science, Vol. 43, p.p. 485-513.

[20] Khalfallah, I.Y.; Rahoma, M.N.; Abboud, J.H.; Benyounis, K.Y. (2011) Microstructure and corrosion behavior of austenitic stainless steel treated with laser, Optics & Laser Technology, Vol. 43, p.p. 806–813.

[21] Kilicli, V., Erdogan, M. (2008) The Strain-Hardening Behavior of Partially Austenitized and the Austempered Ductile Irons with Dual Matrix Structures. *J Mater Eng Perf.* , Vol. 17 (2), p.p. 240–249.

[22] Lambers, H.G., Tschumak, S., Maier, H.J., Canadinc, D. (2009) Role of Austenitization and Pre-Deformation on the Kinetics of the Isothermal Bainitic Transformation. *Metal Mater Trans A.* Vol. 40 (6) p.p. 1355-1366.

[23] Liou, H.Y.; Rong-Iuan Hsieh, R.I.; Tsai. W.T. (2002) Microstructure and stress corrosion cracking in simulated heat-affected zones of duplex stainless steels, Corrosion Science, Vol. 44, p.p. 2841–2856.

[24] Machuca, L.L. ; Bailey, S.I. ; Gubner, R. (2012) Systematic study of the corrosion properties of selected high-resistance alloys in natural seawater, Corrosion Science, Vol. 64, p.p. 8–16.

[25] McFadden, S.; Browne D.J.; Gandin C.A. (2009) A Comparison of Columnar-to-Equiaxed Transition Prediction Methods Using Simulation of the Growing Columnar Front. Metall Mater Trans A, Vol. 40, pp. 662-672.

[26] Mottu, N.; Vayer, M.; Dudognon, J.; Erre, R. (2004) Structure and composition effects on pitting corrosion resistance of austenitic stainless steel after molybdenum ion implantation, Surface & Coatings Technology, Vol. 200, p.p. 2131– 2136.

[27] Orazem, M. E., Tribollet, B. (2008) "Electrochemical Impedance Spectroscopy"; John Wiley and Sons, Inc.

[28] Pardo, A., Merino, M.C., Coy, A.E., Viejo, F., Arrabal, R., Matykina, E. (2008) Pitting corrosion behaviour of austenitic stainless steels – combining effects of Mn an Mo additions, *Corrosion Science*, Vol. 50 p. p. 1796-1806.

[29] Pardo, A.; Merino, M.C.; Coy, A.E.; Viejo, F.; Arrabal, R.; Matykina, E. (2008) Effect of Mo and Mn additions on the corrosion behaviour of AISI 304 and 316 stainless steels in H2SO4, Corrosion Science, Vol. 50, p.p. 780–794.

[30] Park, K.; Kwon, H.S. (2010) Effects of Mn on the localized corrosion behavior of Fe–18Cr alloys, Electrochimica Acta Vol. 55, p.p. 3421–3427.

[31] Polo, J. L., Torres, C. L., Cano, E., Bastidas, J. M. (1999) Estudio de impedancia de la corrosión del acero inoxidable AISI 316L en las regiones pasiva y de picadura, *Revista de Metalurgia*, Vol. 35, p. 368-378.

[32] Polo, J.L. ; Cano, E. , Bastidas, J.M. (2002) An impedance study on the influence of molybdenum in stainless steel pitting corrosion, Journal of Electroanalytical Chemistry, Vol. 537, p.p. 183-187.

[33] Pisareka, M.; Kędzierzawskib, P.; Płociński, T.; Janik-Czachor, M.; Kurzydłowski, K.J. (2008) Characterization of the effects of hydrostatic extrusion on grain size, surface composition and the corrosion resistance of austenitic stainless steels, Materials Characterization, Vol. 59, p.p. 1292-1300.

[34] Reinhart G.; Mangelinck-Noël N.; Nguyen-Thi H.; Schenk T.; Gastaldi J.; Billia B.; Pino P.; Härtwig J.;Baruchel J. (2005) Investigation of Columnar–Equiaxed Transition and Equiaxed Growth of Aluminium Based Alloys by X-ray Radiography. Mater Sci Eng A, Vol. 413–414, pp. 384–388.

[35] Spittle J.A. (2006) Columnar to Equiaxed Grain Transition in as Solidified Alloys, International Materials Reviews, Vol.51, No.4, pp. 247–269.

[36] Stefanescu, D.M. (2008), Science and Engineering of Casting Solidification (2nd ed.), Springer, ISBN 978-0-387-74609-8.

[37] Winston Revie, R. (2000) Uhlig's Corrosion Handbook, Austenitic and Ferritic Stainless Steels, 2° ed., EE.UU., p. 601-628.

Chapter 4

Stress Corrosion Cracking

O. F. Aly and M. Mattar Neto

Additional information is available at the end of the chapter

http://dx.doi.org/10.5772/57349

1. Introduction

Stress Corrosion Cracking (SCC) is a sudden and difficult-to-predict severe degradation mode of failure of nuclear, petrochemical, and other industries. This chapter aims to give a general view for SCC based in the authors experience on more than ten years working with this kind of failure (mainly in PWR Nuclear Plant) in the Brazilian Energy and Nuclear Research Institute.

SCC is a cause of several serious accidents due to sudden failures difficult to predict, in equipments related to industrial plants, pressure vessels, high pressure piping, ducts, and structures. One gives three following examples: a) Silver Bridge colapse in 1967, over Ohio River at Point Pleasant, West Virginia, USA with 46-killed people [1]; b) Catastrophic disk rupture of a steam turbine from nuclear power plant Hinkley Point Power Station, England in 1969 with enormous material losses, machine destruction, and financial losses due to the long period of operation impeachment [2]; c) Flixborough accident, England in 1974, due to a reactor failure, has caused 28 killed people, several injured people, and big material losses [3].

SCC may be classified as an Environmental Assisted Cracking (EAC), besides Corrosion Fatigue (CF) and Hydrogen Induced Cracking (HIC). The relationship between these three types of failures can be showed in Figure 1 where the EAC domain is the union of the three circles, each one representing the three failure modes. The SCC is caused by three main factors: a) Material susceptibility; b) Environmental condition; c) Tensile stresses (applied and residual). Sometimes CF is considered a particular case of SCC where the load is cyclical, and HIC should be considered as a mechanism of SCC [4].

The EAC scientific interest began in the late 19th century due to apparently spontaneous cracks which occurred in brass cartridges cases belonging to the British Army in India during the monsoon seasons: so they were appealed "season cracks", before when this kind of crack was just understood, and after appealed "stress corrosion crack" [5] – the reference [5] is a comprehensive article which should be read by all that want more information about the historical research of EAC.

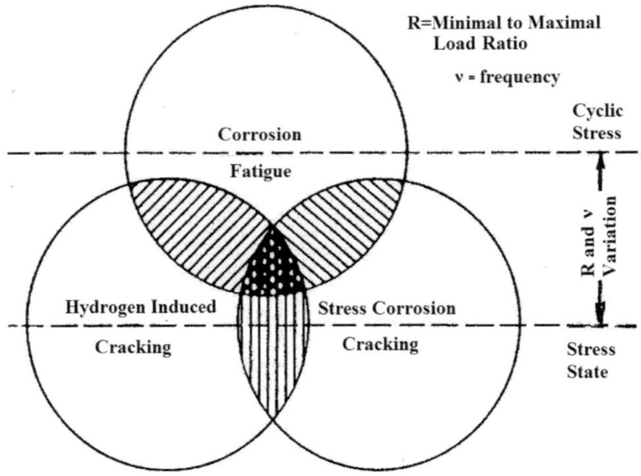

Figure 1. Diagram showing the relationship between SCC, CF, and HIC. When the frequency ν is less than 0.1Hz, SCC and HIC are possible; above this value it is CF. Adapted from [4].

2. Stress corrosion cracking

The stress corrosion cracking is brittle, practically without material loss, and visible corrosion products. It is normally "river branched" (the crack on the material is similar to a river – the primary crack, and its tributaries – the multibranched secondary ones (in the case of CF crack there are few branches). The cracks which occur just below the yield strength of the material, and could be intergranular or intergranular as showed in Fig. (2) [6].

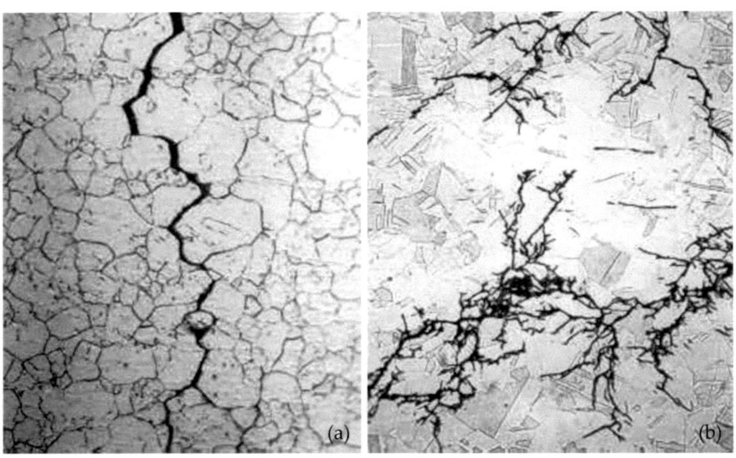

Figure 2. SCC propagation: (a) intergranular SCC of an Inconel heat exchanger tube (X500 micrography); (b) transgranular: the micrography (X300) illustrates SCC in a 316 stainless steel chemical processing piping system. Note the multi-branched transgranular crack pattern. Adapted from [6].

http://dx.doi.org/10.5772/57349

SCC normally occurs when one has a material susceptibility, involved tensile stress, and an aggressive environment. In the Table 1, it is showed some common SCC systems [7].

The stress corrosion cracking initiation and propagation is a very complex degradation process, which depends on several parameters; these can be classified in microstructural, mechanical and environmental [8], and its intricate relationship which causes the failure is showed in Fig. (3) [9].

Especific ions and substances which cause damages	Alloys susceptible to SCC	Temperature (ºC)
Halogen group		
Fluoride íons	Sensitized austenitic stainless steels	Room
Gas chlorine	Low alloys and high strength steels	Room
Gas iodides	Zirconium alloys	300
Oxygen group (systems H_2O-O_2-H_2)		
Dissolved O_2 in liquid H_2O	Sensitized stainless steels	300
Gas hydrogen in room temperature	Low alloys and high strength steels	Room
Gas hydrogen in high pressure and temperature	Medium and low strength steels	>200
Oxygen group (systems S, Se, Te)		
Politionic Acids ($H_2S_nO_6$)	Sensitized austenitic stainless steels, sensitized Inconel 600	Room
Sulphydric gas (H_2S)	Low alloys and high strength steels	Room
Sulphyde impurities in aqueous solution	High strength steels (acelerated hydrogen induced cracking)	Room
Nitrogen Group		
Liquid N_2O_4	High strength titanium alloys	50
N, P, As, Sb, Bi : metal alloy elements	Stainless steels (with Cl^- presence): accelerated cracking	Room
Carbon Group (C, Si, Ge, Sn, Pb)		
Carbonate ions on aqueous solution	Carbon steel	100
CO-CO_2-H_2O Gases	Carbon steel	....
Pb ions on aqueous solution	High Ni alloys	....

Table 1. Especific ions and substances which cause SCC in various alloys, when are present in low concentrations, and as impurities, adapted from [7].

The microstructural factors are [8]: (1) grain boundary chemistry and segregation; (2) thermal treatment which can cause intragranular and intergranular metallic carbide distribution; (3) grain size and cold work or plastic deformation which determine the yield strength: these factors can be described as A in Fig. (3) [9]. The mechanical factors are: (4) applied and residual stresses: these stresses for various geometries can be used as stress intensity K (optionally, strain and strain rate which can be also described related to stresses). The environmental factors are: (5) temperature T; (6) activity of $[H]^+$ or pH; (7) solution or water chemistry; (8) inhibitors or pollutants in solution: these two last ones can be described as [*x*] in Fig. (3); (9) electrode and corrosion potentials E and E_o; (10) partial pressure of hydrogen which reflects on potential.

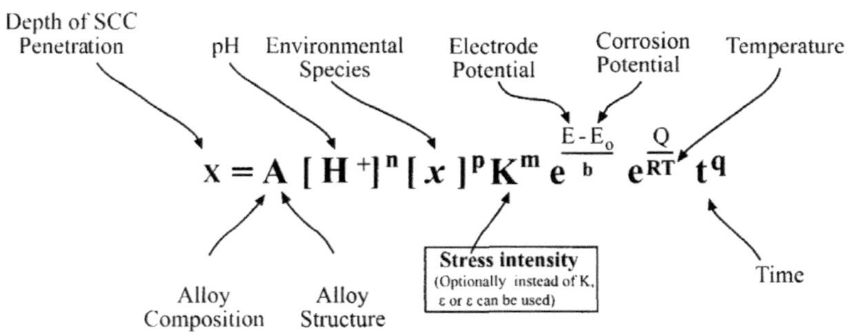

Figure 3. General relationship for SCC process depends on many parameters; where b, m, n, p, q are adjusted constants, Q is the thermal activation energy, and R is the universal gas constant. Adapted from [9].

The time evolution of a SCC could be represented by Fig. (4) [10] where is considered incubation; initiation and coalescence; when K_{ISCC} is reached, a fast increase of the propagation is observed; when the upper bound crack growth rate is reached (Faraday upper bound) an anodic dissolution limit to crack velocity is established, but it may happen more generally be any chemical or diffusion rate limiting process; when K_{IC} is reached the fracture increasing is brittle and produced by mechanical influence rather than environmental influence.

Concerning the parameters described in Fig. (3) which are influent in the SCC, one could note the following: "A" is a parameter which represents the alloy dependence in the SCC process. For example, it will be different to an Alloy whose material grain boundaries present different precipitate patterns: if one considers the nickel Alloy 600 in primary ("pure") water at high temperatures (upper than 280°C), the SCC susceptibility is greater for a material with few intergranular precipitates than for another material with more intergranular precipitates regularly spaced around the grain boundary contour. Small grains also are less susceptible than greater ones. If the material is cold worked this parameter should be different than a material which did not suffer this treatment: normally cold work material is more susceptible to SCC than the original. So, "A" is a parameter which depends on the thermomechanical treatment imposed to the material: for each heat of material there is a distinct "A". The pH ("$[H]^+$") and potential ("E" and "E_0") determine a dominium where the susceptibility to SCC is variable according to several regions (or submodes) of passivity, general corrosion, SCC, and others. Also, according to the electrode potential variation it could occur hydrogen

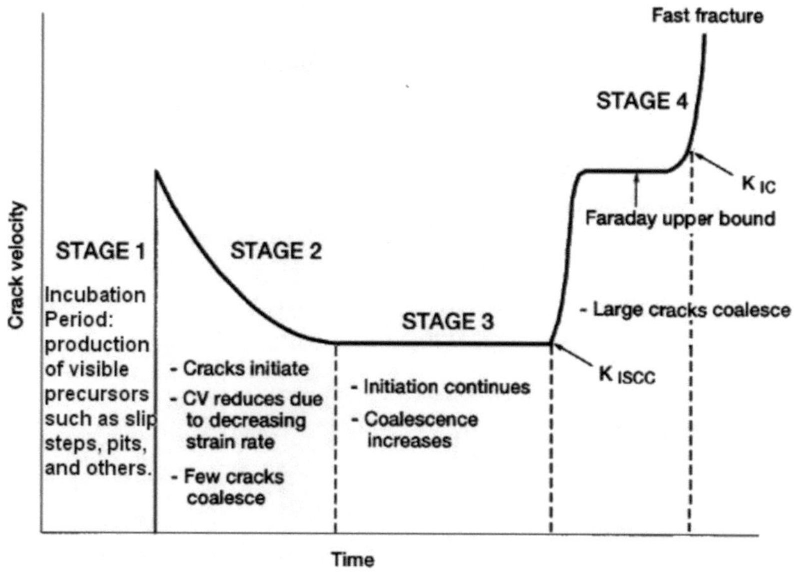

Figure 4. Time evolution of a multiple SCC: adapted from [10].

permeation variation in the material, and this could provoke HIC. Environmental species ("[x]") are influent on SCC as showed in Table 1. The tensile stress influence through "K"-parameter is reflected in the Fig. (4). Temperature ("T") influences the susceptibility to SCC according to the Arrhenius law, then this parameter is showed inlaid in the parcel "e^(Q/RT)". Finally the time "t" is fundamental in the SCC evolution as showed in the Fig. (4) [10].

3. Stress corrosion mechanisms and models

The SCC mechanisms could be classified in two families: Anodic SCC, and Cathodic SCC. The first is governed by anodic metal dissolution and the second governed by hydrogen permeated into the metal which provokes hydrogen embrittlement, and consequently HIC. In the first case are for example, the stainless steels, and in the second the high strength steels [12].

There are several models to represent these phenomena; based on anodic mechanisms: the slip dissolution / film rupture of Ford and Andresen [13], the enhanced surface mobility theory of Galvele [14], the coupled environment fracture model of Macdonald and Urquidi-Macdonald [15]; the internal oxidation mechanism of Scott and Le Calvar [16]; based in cathodic mechanisms: the hydrogen induced cracking models of Shen and Shewmon [17], Magnin and others [18], [19]; based in numeric and empiric observations: the numerical model of Rebak and Smialowska [4], the semi-empirical-probabilistic model of Staehle [9], [20]. For a comprehensive review of several of these models see mainly [8] and [20].

Two important models to be applied in stainless steel and nickel alloys respectively, for nuclear applications are here described, and are anodic: the slip-step dissolution and film rupture model [13], and the internal oxidation model [16], both applied for SCC propagation. If one substitutes the oxygen action in this last model by hydrogen action, it will be transformed in a cathodic model based on HIC.

The slip-step dissolution and film rupture model is one of the most used engineering models for SCC in nuclear applications, mainly involving stainless steel with boiling water nuclear reactors (BWR) high temperature water: it has been developed in General Electric laboratories [13]. According to the mechanism of this model, crack occurs due to metallic corrosion preferentially along an active path as a grain boundary or crystal sleep plane, in an interactive process including electrochemical dissolution and the weakening, rupture and repassivation of the metallic oxide film. The Fig. (5) [21] illustrates partially this process.

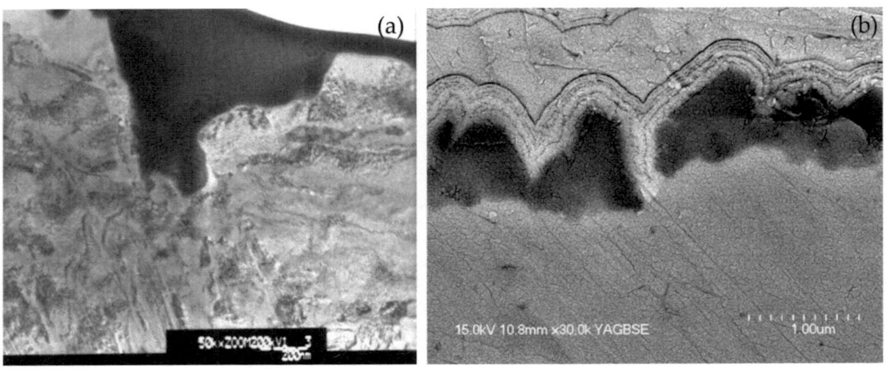

Figure 5. Slip bands details of SCC process: (a) Film structure of surface oxide on test material (austenitic stainless steel). The band width length is about 300 nm; (b) Detail shows a slip step found in a specimen section, where this SCC rupture process can be clearly seen in this step. Adapted from [21].

The crack growth rate is postulated to be sustained by periodic strain-induced rupture of the film, and the required rupture strain provided by transient creep. The mathematic expression is given through equation (1) [22].

$$V_{ct} = \frac{M}{z\rho F}\frac{Q_f}{\varepsilon_f}\left(\frac{d\varepsilon}{dt}\right)_{ct} \qquad (1)$$

where: V_{ct} is the crack tip growth rate, M is the molecular weight of the material, z is the charge of the anodic dissolved material, ϱ is the material density, F is the Faraday constant, Q_f is charge density per film rupture event, ε_f is the oxide fracture strain, and $(d\varepsilon/dt)_{ct}$ is the crack tip strain rate.

Note that the first fraction of the equation (1) is referred to the corrosion anodic dissolution of the material governed by Faraday law. The second and third fractions are referred to the mechanics of the model. Ford and Andresen, from GE laboratories, have simplified this

http://dx.doi.org/10.5772/57349

equation in a form $V_{ct}=AK^n$ where the process is governed by corrosion (Faraday law),tensile stress (through stress intensity factor K), and material interactions with the environment (constants A and n).

An applied example is shown in reference [8] through equation (2), for stress corrosion crack growth in 304 stainless steel and nickel base alloys for BWR, considering strain rate dependence of the stress intensity.

$$V_{scc} = (7.8\times10^{-3} n^{3.6})(4.1\times10^{-14} K^4)^n \tag{2}$$

where: V_{scc} is the crack growth rate in cm/s, n is an environment and material chemistry parameter related to repassivation rate, and K is the stress intensity in MPa√m.

This model has been applied mainly to stainless steel and nickel alloys in light water reactors and other structural materials from nuclear plants, and it´s included as a software element in the supervisory systems for component life prediction and evaluation *in situ* of machines.

The internal oxidation model is also used for SCC in nuclear applications, mainly involving nickel alloys with pressurized water nuclear reactors (PWR), and has been developed by P. M. Scott and M. Le Calvar [16]. This model is based on the embrittlement mechanism caused by a layer of adsorbed oxygen atoms which interacts with the material grain boundaries precipitates, and their dislocations. It depends also on oxygen diffusivity on the material. This mechanism could also produce high pressure gas bubbles which enhance the evolution of the intergranular stress corrosion cracking (IGSCC). This mechanism has good agreement to explain SCC in nickel alloys (such Alloys 600 and 690) in high temperature pressurized water of nuclear reactors (PWSCC). The model equation (3) explains the crack growth rate according to this mechanism.

$$V_{IGSCC} = \left(\frac{81kTD_0}{512\gamma^3 a^2}\right)\left(\frac{\delta N_S}{6\pi z}\right)^{1/2} \sigma_p K_I \tag{3}$$

where: V_{IGSCC} is the intergranular stress corrosion crack growth rate, k is the Boltzmann constant, T is the absolute temperature, D_0 is the grain boundary diffusion coefficient for oxygen in considered material, γ is the surface energy, a is the atomic volume, δ is the grain boundary width, N_S is the surface solubility of the oxygen, z is the number of sites explored per gas atom jump, σ_p is the stress contour of the plane strain plastic zone radius of the process crack zone, K_I is the crack tip stress intensity factor.

4. Stress corrosion cracking tests

The applicable types of tests could be classified, according to the stressing modes which are input to the test specimens, in: a) constant total strain; b) constant load; c) constant strain rate

[23] (despite of this report has more than 40 years, it is yet a valuable synthesis of the tests and their comparison). Nowadays it seems that the more usual types of tests are b) and c). This first is normally slower than the second, considered as an accelerated test. This is usually from an order of hundreds of hours, while the other could reach thousands of hours. The specimens could be cylindrical, or prismatic according to the fracture mechanics tests normalization. There are also frequently pre-cracked specimens, rather than plain specimens.

The authors experience is concerned to the slow strain rate test (SSRT), which is a dynamic test where it is imposed to the specimen a slow strain rate through external force over a monitored section, or over a notched region of this specimen, or over a fatigue pre-cracked to the evaluation of SCC material resistance. The imposed slow strain rate is normally between 10^{-4} and 10^{-7} s^{-1}: the local strain rate should be slow enough to make time to occur corrosion processes, and quick enough to cause cracks or damage in a specimen during a reasonable time [24]. The SSRT tests are carried out in accordance with ASTM G 129-95 standard [25]. The specimens are prepared according to ASTM G49-2000 and ASTM E8-2000 standards [26], [27]. The tests are performed at an open circuit potential and the specimens were exposed to the environment for at least 24 hours before applying load to stabilize the surface oxide layer [28]. In the Fig. (6) is showed some exemplified aspects of this type of test [28].

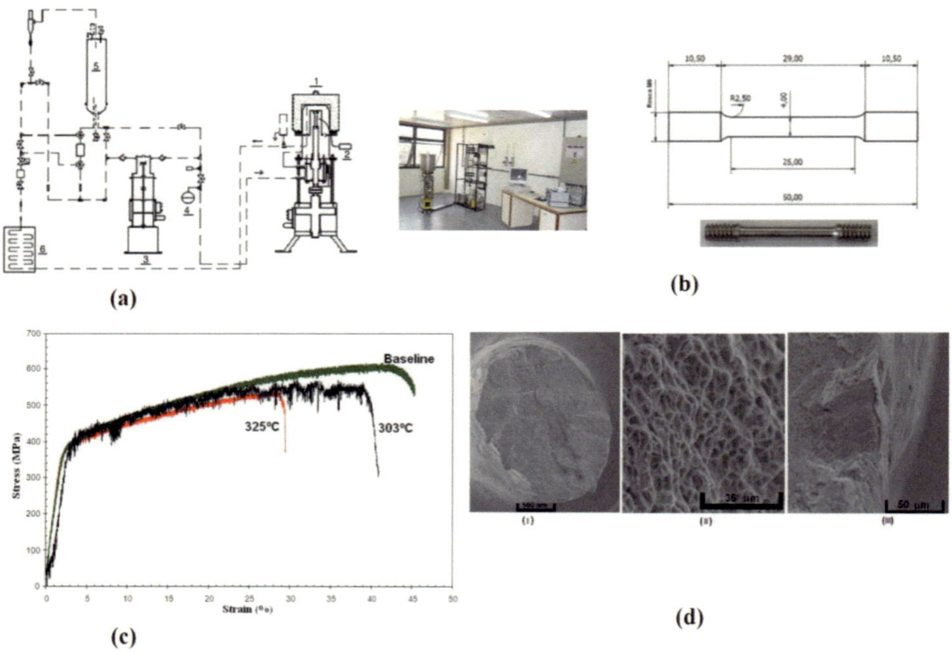

Figure 6. (a) Diagram of the installation for SCC tests: (1) autoclave, (2) Pt electrode, (3) medium circulation pump, (4) pressure accumulator, (5) medium storage tank and (6) cooler, including installation photography: adapted from [24]; (b) cylindrical specimen (mm); (c) Stress - Strain curves of Alloy182 weld of Alloy 182 obtained from SSRT at strain rate of $3x10^{-7}$ s^{-1} in PWR primary water condition at 303°C and 325°C; (d) SEM micrographs of Alloy182 weld fractured surface of the SSRT at strain rate of 3x10-7 s-1 in PWR primary water at 303°C (a) overview (b) detail of ductile fracture (c) and detail of SCC fracture failure, adapted from [28].

http://dx.doi.org/10.5772/57349

5. Proposed methodology for modeling

The integration between theoretical and experimental parts for SCC modeling is best given to an adequate methodology for modeling. This methodology allows a large overview on stress corrosion cracking phenomenology.

Staehle has proposed a three-dimensional diagram [20], which shows the conditions to occur the modes of PWSCC and other corrosion modes on Alloy 600. The base of this diagram is shown in Fig. (7), which represents the potential x pH or Pourbaix diagram for this material in primary water at high temperature (300 to 350°C). One superimposes over it, the corrosion submodes, using experimental data published on literature or from original tests. These submodes are regions of potential where the different modes of surface material-environment interactions can occur, like stress corrosion, pitting, generalized corrosion or passivation. One can put over this 2D-diagram an additional third dimension which represents the "useful strength" of the material as affected by the environment at that point, the strength fraction to SCC. If one replaces the strength fraction to SCC, which is a semi-quantitative measure to SCC susceptibility (and which could be extracted directly from the slow strain rate tests, as a comparative measurement between test in environment of SCC, and in environment neutral to SCC), by kinetic initiation and propagation models to SCC, a generalized methodology to express the SCC can be obtained [20].

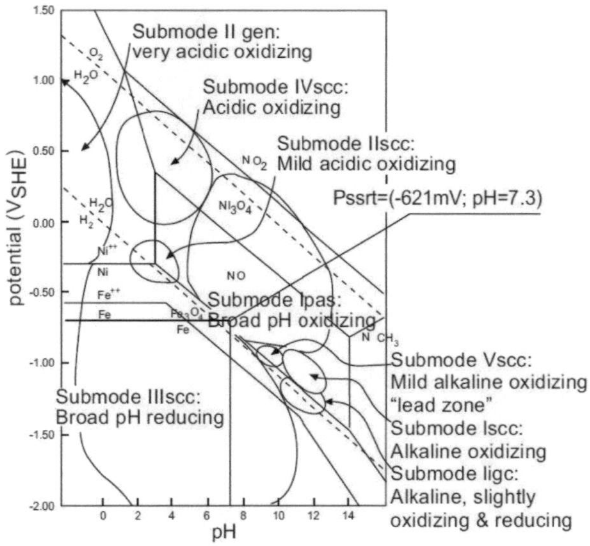

Figure 7. Pourbaix diagram V x pH for Alloy 600 in the range of 300°C used as base marked with submodes regions. The Brazilian CDTN estimated result has been plotted over it (Pssrt). Adapted from [20].

In the D Sc. Thesis [29], a preliminary usage of this methodology was proposed, departing from Fig. (7): first, the SCC susceptibility from tests realized on the slow strain rating test equipment installed at CDTN in Brazil has been marked over this diagram. Literature data already plotted over the diagram has been proved compatible. After this marked point based on realized tests, it has been supposed different kinetic models. A summary of this work, and some models obtained are in reference [30].

This methodology for modeling could be used, for example, for a Brazilian nuclear power plant, taking into account, the plant materials, and the characteristics of its design and operation, such as the heat material fabrication processes, material composition, plant thermomechanical history, primary water chemical composition, and operational temperature conditions at this plant. Tests can be realized in laboratory (such as CDTN) or from field data.

The methodology is based on empirical-experimental and/or theoretical modeling approaches, such as shown in Fig. (8) which is based in an deterministic similar equation (2) comparison with literature data [31].

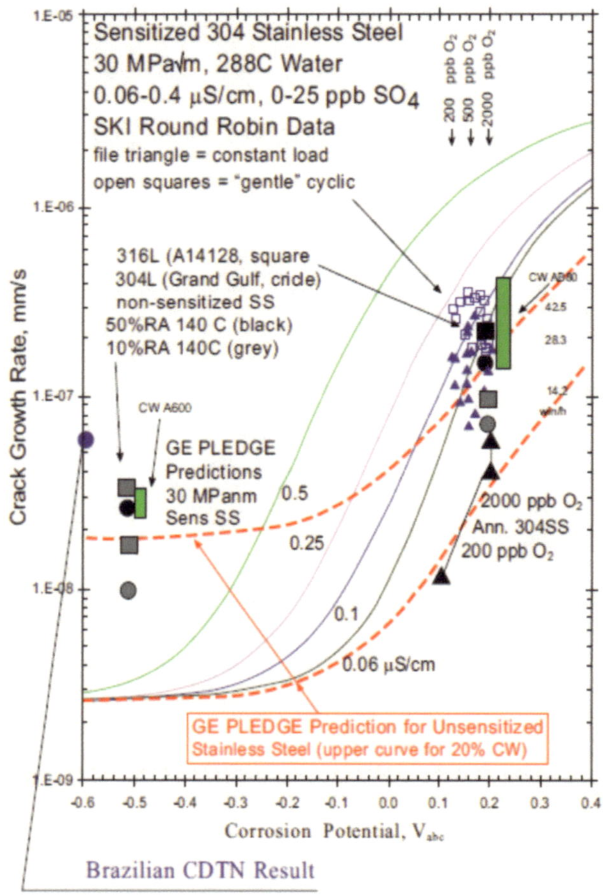

"Original Figure reprinted with permission of The Minerals, Metals & Materials Society."

Figure 8. Predicted crack growth rate for Alloy 600 and stainless steel at 288°C, based on the GE PLEDGE Code: it´s marked a Brazilian result which was tested at a temperature of 303°C. Adapted from [31].

Then, coupling the obtained modeling curves such as in the Fig. (8) with the point (potential, pH) where the specimen is, over the Pourbaix diagram marked with submodes (Fig. (7)) - this joint information can be used to evaluate the crack growth rate (*e.g.* according to equations (1) to (3)) of the component which suffers PWSCC at determined submode. If the same specimen is used in another submode (*e.g.* such as I_{SCC} – alkaline oxidizing or III_{SCC} – broad pH reducing, according to Fig. (7)), the initiation and growth kinetics may be different.

6. Current and future developments

Nowadays there are some tendencies in the SCC studies development: a better understanding of some SCC mechanism factors such as the hydrogen role (hydrogen induced cracks, embritllement, enhanced plasticity, etc.), and the precipitates (intergranular and intragranular) role in these mechanisms. Another issue is the new test methods for SCC study, such high resolution observations at nanometric scale, through synchroton, and X-ray tomography. New methods for SCC detection have been developed such as electrochemical noise, acoustic emission. The probability modeling to manage life cycle engineering considering complexes initiation modes such as pitting has been developed. The models have been improved such as the film-induced cleavage. Various of these studies are available in the last Environment-Induced Cracking of Materials (EICM-2) Proceedings [32]. Some interesting works have been developed in important scientific and technological centers such as École de Mines à Paris to a better understanding of SCC [33], and University of Oxford-Department of Materials: in this last, a new experimental approach which enables mechanical testing at micrometric scale, using micromachining cantilevers observed to a focused ion beam (FIB), have been developed. This is applied to study the status of oxidized grain boundaries in the SCC mechanism [34].

It would like adequate to develop various models according to various mechanisms available to different combinations material-environment-tensile tension: it could be done departing of various mechanisms description: an important reference is [35].

Another important issue is the SCC simulation development, which represents a modeling *pari passu* according to a global mechanism performance. Important references of this issue are representing by references [36], and [37].

7. Summary

An overview about stress corrosion cracking approaching according to the authors experience has been given in this chapter. Stress corrosion cracking is a very complex mode of degradation and theme which could be at least adequately developed in an entire book. Authors expect that this overview chapter could be better developed by the research through the essential given references.

Acknowledgements

Capes (Coordenação de Aperfeiçoamento de Pessoal de Nível Superior- Brasil) for the research fund, IPEN/CNEN-USP (Instituto de Pesquisas Energéticas e Nucleares/ Conselho Nacional de Energia Nuclear- São Paulo University– Brazil) for the research opportunity, infrastructure, and chapter publication costs sponsorship.

Author details

O. F. Aly and M. Mattar Neto

*Address all correspondence to: ofaly@ipen.br, ofaly1@gmail.com

IPEN/CNEN-USP – Energy and Nuclear Research Institute, São Paulo, Brasil

References

[1] Wikipedia. "Silver Bridge", Access on March 2013: http://en.wikipedia.org/wiki/Silver_Bridge.

[2] Japan Failure Knowledge Database. *"Burst of Steam Turbine Rotor in Nuclear Power Plant". Access on March 2013:http://www.sozogaku.com/fkd/en/cfen/CB1031029.html*

[3] Roberge PR. "Stress Corrosion Cracking of Chemical Reactor: The Flixborough explosion, UK 1974", Corrosion Doctors, Access on March 2013: http://www.corrosion-doctors.org/ProcessIndustry/flixborough.htm

[4] Hertzberg, RW. *Deformation and fracture mechanics of engineering materials*, New York, N.Y.: John Wiley & Sons, 1989.

[5] Shipilov, S.A. "Stress corrosion cracking and corrosion fatigue: a record of progress, 1873-1973" In.: Shipilov, S.A.; Jones, R.H. ;Olive, J.-M.; Rebak, R.B. (Ed.)"EICM-2 - Second International Conference on Environment-Induced Cracking of Metals, The Banff Centre, Banff, Alberta, Canada, September 19-23, 2004, Proceedings Elsevier: London, 1st Edition, 2008, Vol 1. pp. 507-557.

[6] NACE Resource Center: "Stress Corrosion Cracking", 2010. Access on June 2013: http://events.nace.org/library/corrosion/Forms/scc.asp

[7] ASM: American Society of Metals. "Stress-corrosion cracking", In.: ASM Metals Handbook of Corrosion. Materials Park, OH: 2002 ASM International, V. 13 p. 828-860.

[8] Rebak RB, Szklarska-Smialowska Z. "The mechanism of stress corrosion cracking of alloy 600 in high temperature water", Corros. Sci. 38 (1996) 971-988.

[9] Staehle RW, Bases for Predicting the Earliest Penetrations Due to SCC for Alloy 600 on the secondary Side of PWR Steam Generators, Argonne National Laboratory, 2001, NUREG/CR-6737, ANL-01/20 RWS 151, Argonne, Illinois, Sept. 2001.

[10] Scott P, Combrade P, Kilian R, Roth A, Andresen P, Kim Y. "Status Review of Initiation of Environmentally Assisted Cracking and Short Crack Growth". EPRI, Palo Alto, CA: 2005. 1011788.

[11] Aly OF. "Modeling of Primary Water Stress Corrosion Cracking at Control Rod Drive Mechanism Nozzles of Pressurized Water Reactors" (in Portuguese). São Paulo: IPEN/CNEN-USP 2006. DSc. Thesis. Access for resume and download on June 2013: http://www.teses.usp.br/teses/disponiveis/85/85134/tde-22032012-154040/pt-br.php

[12] Schmutz P. "Stress Corrosion Cracking". EMPA-Laboratory for JoiningTechnologies and Corrosion, Dübendorf, Swiss, 2013, 24 pp. Access on June 2013: http://www.surface.mat.ethz.ch/education/courses/surfaces_interfaces_and_their_applications_II/SIandAII_Ch9_Stress_Corrosion_Cracking

[13] Andresen PL, Ford FP. "Life prediction by mechanistic modeling and system monitoring of environmental cracking of iron and nickel alloys in aqueous systems", Mat.Sci. Eng. A103 (1988)167-184.

[14] Galvele JR, "A stress corrosion cracking mechanism based on surface mobility" *Corros. Sci.* v.27 n.1 (1987) 1-33.

[15] Macdonald D, Urquidi-Macdonald M, An advanced coupled environment fracture model for predicting crack growth rates, In.: Parkins Symp. on Fundamentals Aspects of Stress Corrosion Cracking. The Minerals, Metals and Materials Society, Warrendale, PA (1992) pp. 443.

[16] Scott PM, Le Calvar M, Some possible mechanisms of intergranular stress corrosion cracking of alloy 600 in PWR primary water, In.: Proc. 6th Int. Symp. On Environmental Degradation of Materials in Nuclear Power Systems Water Reactors, San Diego, CA, l-5 August 1993. The Minerals, Metals and Materials Society, Warrendale, PA (1993) p. 657.

[17] Shen CH, Shewmon PG, "IGSCC A Mechanism for Hydrogen–Induced Intergranular Stress Corrosion Cracking in Alloy 600", Met. Trans., 21A, (1990), 1261–1271.

[18] Magnin T, Boursier J-M, Noel D, Rios R, Vaillant F, Corrosion deformation interaction during stress corrosion cracking of alloy 600 in primary water, In Proc. 6th Int. Symp. on Environmental Degradation of Materials in Nuclear Power Systems - Water Reactors, San Diego, CA, l-5 August 1993. The Minerals, Metals and Materials Society, Warrendale, PA (1993) p. 669.

[19] Rios R, Magnin T, Noel D, Bouvier O de, "Critical Analysis of Alloy 600 Stress Corrosion Cracking Mechanisms in Primary Water," Metallurgical and Materials Transactions A, 26A, (1995), 925-939.

[20] Staehle RW. "Combining design and corrosion for predicting life", in: R.N. Parkins (Ed.), Life Prediction of Corrodible Structures, Vol. 1, NACE International, Houston, 1994, pp. 138-291.

[21] Wang S, Takeda Y, Sakaguchi K, Shoji T. "The Initiation of Environmentally Assisted Cracking in BWR High Temperature Water", In.: Proceedings of the 12th International Conference on Environmental Degradation of Materials in Nuclear Power System – Water Reactors –Edited by T.R. Allen, P.J. King, and L. Nelson TMS (The Minerals, Metals & Materials Society), pp.49-53 (2005).

[22] Thompson CD, Krasodomski HT, Lewis N, Makar GL, "Prediction of pure water stress corrosion cracking (PWSCC) in nickel base alloys using crack growth rate models", KAPL Atomic Power Laboratory: Schenectady, NY, 1995 (KAPL-P-000005).

[23] Parkins RN, Mazza F, Royuela JJ, Scully JC. "Report prepared for the European Federation of Corrosion Working Party on Stress Corrosion Test Methods". *Br. Corros. J.*, vol. 7, pp. 154-167. July, 1972.

[24] Matias A, Schvartzman MMAM. "Development of a methodology for evaluation of susceptibility to stress corrosion cracking in nuclear reactor environment", *Proc. of Inac 2005,* Inac, Santos, Brazil, September 2005. (in Portuguese)

[25] ASTM G 129. "Standard Test Methods for Slow Strain Rate Testing to Evaluate the Susceptibility of Metallic Materials to Environmentally Assisted Cracking" *Annual book of ASTM Standards*. West Conshohocken, PA: ASTM (1995).

[26] ASTM G 49. "Standard Test Methods for Preparation and Use of Direct Tension Stress-Corrosion Test Specimens". *Annual book of ASTM Standards*. West Conshohocken, PA: ASTM (2000).

[27] ASTM E8. "Standard Test Methods for Tension Testing of Metallic Materials *Annual book of ASTM Standards*. West Conshohocken, PA: ASTM (2000).

[28] Lima LIL, Schvartzman MMAM, Quinan MAD, Soares AEG, Piva SPT. "Stress corrosion cracking of alloy 182 weld in a PWR water environment", *Proc. of Inac 2011,* Inac, Santos, Brazil, October 2011.

[29] Aly OF. Modeling of Primary Water Stress Corrosion Cracking At Control Rod Drive Mechanism Nozzles of Pressurized Water Reactors. São Paulo: IPEN/CNEN-USP 2006. (D. Sc. Thesis in Portuguese). Access on May 2012: http://www.teses.usp.br/teses/disponiveis/85/85134/tde-22032012-154040/pt-br.php

[30] Aly OF; Andrade AHP; Mattar Neto M; Schvartzman MMAM. Results On Modeling of Primary Water Stress Corrosion Cracking at Control Rod Drive Mechanism Nozzles of Pressurized Water Reactors. In: 19th International Conference on Structural

Mechanics in Reactor Technology, 2007, Toronto. SMiRT 19 Transactions. Raleigh, NC, USA, 2007. pp. 1-8.

[31] Andresen PL, Emigh PW, Morra MM, Hickling J, Effects Of PWR Primary Water Chemistry And Deaerated Water On SCC...*Proceedings of the 12th International Conference on Environmental Degradation of Materials in Nuclear Power System – Water Reactors – Edited by T.R. Allen, P.J. King, and L. Nelson TMS (The Minerals, Metals & Materials Society), 2005* pp.994.

[32] Shipilov SA; Jones, RH; Olive J-M; Rebak, RB (Ed.)"EICM-2 - Second International Conference on Environment-Induced Cracking of Metals, The Banff Centre, Banff, Alberta, Canada, September 19-23, 2004, Proceedings Elsevier: London, 1st Edition, 2008, 2 Vols.

[33] Laghoutaris P, *Corrosion sous contrainte de l'Alliage 600 en milieu primaire des Réacteurs à Eau Pressurisée: apport à la compréhension des mécanismes.* Doctorat Sciences et génie des matériaux, 2009 MAT- Centre des matériaux PM Fourt, Paristech - ENSMP p.214.

[34] Lozano-Perez S, Dugdale H, Armstrong DEJ, Roberts SG, Micromechanical testing of oxidized grain boundaries: Understanding stress corrosion cracking mechanisms. UK: Oxford University, Department of Materials (Resumé), Access on March 2013: http://www.emc2012.org/documents/Abstracts/Abstracts/EMC2012_0545.pdf.

[35] Newman RC, "Stress Corrosion Cracking Mechanisms. Corrosion Mechanisms in Theory and Practice", eds P. Marcus and J. Oudar, 2nd ed, pp 399-450, Marcel Dekker, New York (2002).

[36] Smith RW. "Computer simulation of intergranular stress corrosion cracking via hydrogen embrittlement". Modelling Simul.Mater. Sci.Eng., 8 (2000), 629-648.

[37] Rímoli JJ. *A Computational Model for Intergranular Stress Corrosion Cracking* Pasadena, CA, USA: California Institute of Technology (Caltech). PhD. Thesis, 2009. Acesso em maio de 2013: http://thesis.library.caltech.edu/1808/3/Julian_Rimoli_Thesis.pdf.

Chapter 5

Constant Dew Point Corrosion Tests for Metals

Zhenhua Dan, Izumi Muto and Nobuyoshi Hara

Additional information is available at the end of the chapter

http://dx.doi.org/10.5772/57291

1. Introduction

Aluminium alloys are widely used in the automobile industry, in construction and in other fields where energy consumption and environmental safety are a high priority due to their high strength-to-weight ratio, their infinite recycling properties and high corrosion resistance [1-3]. Besides the prohibitively high costs of maintaining degraded aluminium products degradation in the form of localized attacks has resulted in fatalities, particularly in Cl-containing environments. The atmospheric corrosion of aluminium alloys is therefore a research field which has been extensively investigated by many researchers and groups mostly using field exposure tests [4-8]. The complex combination of environmental factors in the outdoor atmosphere, such as the amounts of airborne salt and contamination species, as well as the ever-changing weather conditions, makes it difficult to clarify the corrosion process. Conventional laboratory scale tests have been employed to evaluate the atmospheric corrosion resistance of materials. Among them, the salt spray test (ASTM B117) is widely used as an acceleration test [9-11]. Cyclic corrosion tests, such as ASTM G85-A5 and SAE J2334, and constant relative humidity tests have also been used to simulate actual atmospheric corrosion at a laboratory level [12-14]. However, in terms of corrosion morphology, these tests do not always provide good reproducibility of practical atmospheric corrosion behaviour. By measuring fluctuations in the night and day temperature and relative humidity, Muto *et al.* found that the dew point of outdoor air remains approximately constant and that humidity depends on the air temperature [15]. A testing method based on this finding has thus been proposed to simulate the actual environment of atmospheric corrosion on stainless steel [15]. The testing method, called the constant dew point corrosion test, employs a diurnal cycle of temperature and humidity at a constant dew point temperature. It has been demonstrated that this method reproduces atmospheric corrosion well in the laboratory not only in the case of stainless steel [15] and Zn alloys [16-17] but also for aluminium and its alloys [18, 19]. Basing on our experimental results, this testing methodology has been registrated to be an international standard, ISO 16539, in 2013 [20].

The atmospheric corrosion of aluminium and its alloys was investigated in the present chapter. We focused on the applicability and reproducibility of the constant dew point corrosion tests in the field of the atmospheric corrosion of the aluminium alloys. The evaluation of the atmospheric corrosion of aluminium and its alloys was conducted from the aspects of the corrosion rates, corrosion morphology and corrosion product composition in the field exposure tests and the constant dew point corrosion tests.

2. Samples preparation and method used

In our researches, high-purity aluminium (4N Al) and two kinds of aluminium alloys, O type commercial pure aluminium (AA1100) and T6 tempered Al-Mg-Si alloy (AA6061), were tested. The chemical composition of the alloys is given in Table 1. Usually, all the samples (each with a dimension of 50 mm × 50 mm × 2 mm) were polished by emery papers. The samples were then chemically etched in a 10 wt% NaOH solution at 343 K for 30 seconds, rinsed with water, immersed in a 30 wt% HNO_3 solution at 298 K for 30 seconds, and again rinsed with water.

Field exposure tests were performed in Miyakojima (E125°19`N24°44`), Choshi (E140°45`N35°43`) and Aobayama in Sendai (E140°50`N38°15`), Japan, with a map as shown in Fig. 1. The climates in Japan Weathering Testing Center (JWTC), which is 2 km far from the North Pacific Ocean, in Miyakojima are subtropical marine conditions. The climates in Choshi, which is 4 km far from the North Pacific Ocean, are typical weathering conditions as other sites in Japan. The exposure site in Aobayama campus, Tohoku University in Sendai is 6 km far from the North Pacific Ocean locating on the top of a small hill with an altitude of 100 meters.

	Si	Fe	Cu	Mn	Mg	Cr	Zn	Ti
4N Al	0.0003	0.0019	0.0004	--	0.0004	--	--	0.0003
AA1100	0.07	0.6	0.14	0.01	--	--	--	--
AA6061	0.61	0.41	0.26	0.03	1.01	0.24	0.02	0.03

Table 1. Chemical composition of 4N Al, AA1100 and AA6061 (wt%)

The tested samples were set on exposure racks facing southward at an inclination of 30° from the horizon. The two samples in each category were exposed, with one sample used for evaluating the corrosion mass losses and the other one for observing the surface and analyzing the corrosion products. Temperature, T_{air}, the relative humidity of the surrounding atmosphere, $\Phi_{\alpha ir}$, and the surface temperature of 4N Al panels, T_{Al}, were monitored by the sensors and recorded by data loggers. The accuracy of relative humidity was as high as ±2.5 % and the resolution was 0.03 %. The accuracy of temperature measurements was within ±0.2 °C, and

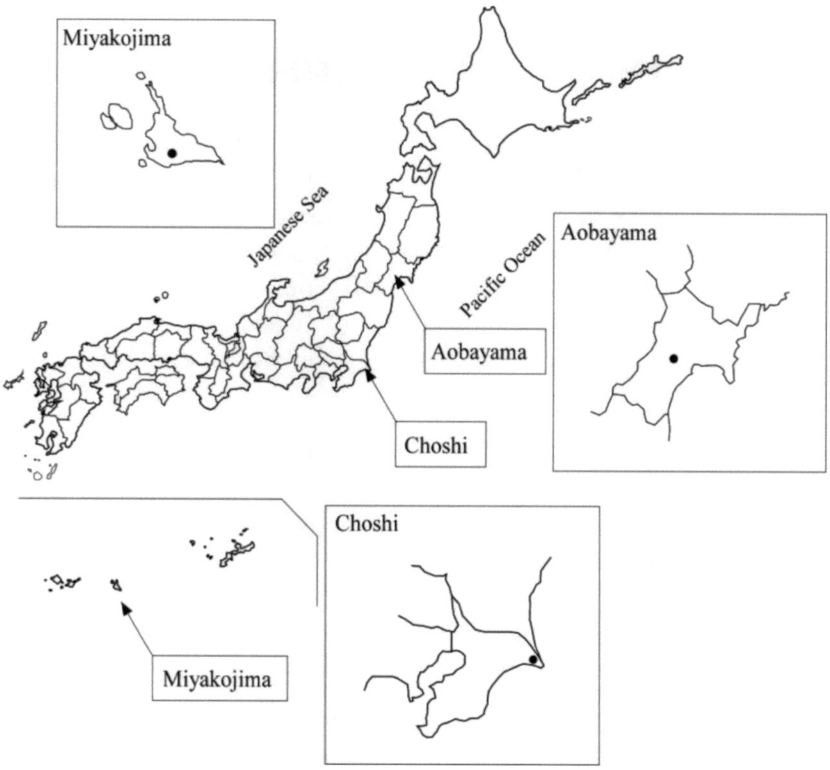

Figure 1. The exposure sites for the two-year field exposure tests performed in Japan.

the drift was less than 0.1 °C per year. The sampling interval was set at 3 minutes. The relative humidity near the 4N Al panels, Φ_{Al}, was dependent on the surface temperature of the panels, and was calculated as $\Phi_{Al} = P_{w\_air} \times \Phi_{air} / P_{w\_Al}$, where the saturated water vapour pressure, P_w (in Pa), is given as a function of temperature, T (in °C), by the following Sonntag Equation [21]:

$$ln\left(P_w\right) = -6096.9385\left(T+273.15\right)^{-1} + 21.2409642 - 2.711193\times10^{-2}\times\left(T+273.15\right) + 1.673952\times10^{-5}\times\left(T+273.15\right)^2 + 2.433502\,ln\left(T+273.15\right) \quad (1)$$

The Magnus formula [21] was then used for calculating the dew point, T_d (in °C), on the Al panel surface.

$$T_d(T,\Phi) = \frac{\lambda\cdot\left(\ln\left(\frac{\Phi}{100}\right)+\frac{\beta\cdot T}{\lambda+T}\right)}{\beta-\left(\ln\left(\frac{\Phi}{100}\right)+\frac{\beta\cdot T}{\lambda+T}\right)} \quad (2)$$

where T in °C, Φ in %, β=17.62 and λ=243.12 °C.

The time of wetness (TOW) was calculated from the collected weathering data. The chloride deposition and SO_2 deposition in Miyakojima and Choshi were monitored by the JWTC. The dry gauze method, JIS Z 2382, was used for measuring the chloride deposition. This method employs a dry gauze screen (100 cm^2) for sampling the chloride rather than the gauze wick (100 cm^2) used in the wet candle method, ASTM G 140 and ISO 9225. The sulphation cylinder method, JIS Z 2382, was used for measuring the SO_2 deposition. This method employs a cylinder with a PbO_2-coated gauze (100 cm^2) wound around it for SO_2 collection rather than the PbO_2-coated filter paper (28.26 cm^2) used in the sulphation plate method, ASTM G 91 and ISO 9225. The SO_2 deposition in Aobayama was estimated from the data of SO_2 concentration in the atmosphere reported by the Japanese government's Ministry of the Environment.

The chloride deposition in Aobayama was estimated from the amount of chloride species deposited on pure Ti panels exposed at the same time. The exposed side of the Ti sample was carefully washed by deionized water to make a 100 ml solution. This solution was filtered with a 0.45 μm membrane filter and then analysed by ion chromatography (IC). Due to the washing effect of rainwater, the amount of chloride determined is somewhat different from the chloride deposition measured by the conventional method. The meteorological parameters of the three exposure sites are summarized in Table 2.

The corrosion mass loss was evaluated after the removal of the corrosion products. The corrosion products were removed by dipping test samples into a solution of phosphoric acid (50 ml/l H_3PO_4) and chromium trioxide (20 g/l CrO_3) at 90 °C for 5-10 minutes. The samples were weighed by an electric balance with a precision of 0.01 mg.

	Atmospheric data				Precipitation		
	Temperature, °C	Relative humidity %	Time of wetness, h•year^{-1}	Rainfall, mm•year^{-1}	pH	Cl^-, mg•m^{-2}•day^{-1}	SO_2, mg•m^{-2}•day^{-1}
Miyakojima	23.9	77	4350	2176.5 [1)]	5.12-5.16 [3)]	56.5 [1)]	2.88 [1)]
Choshi	14.7	76	4500	1782.1 [1)]	4.6-5.11 [3)]	27.4 [1)]	3.73 [1)]
Aobayama	12.7	73	2190	1303.5 [2)]	4.6 [3)]	5.1 [4)]	1.52 [3)]

Table 2. Summary of climate data from the field-exposure sites

1. Average values reported by Japan Weathering Test Center for the test period (Miyakojima: July, 2007-June 2008; Choshi: Aug. 2007- July 2008)

2. Reported by Japan Meteorological Agency (Sendai: Dec. 2007- Nov. 2008)

3. Reported by Japan Ministry of the Environment

4. Measured by the Ti plate method (see text)

A climatic chamber (ESPEC, SH6610) was used to perform the constant dew point corrosion tests. Each of the following environmental factors was able to be individually manipulated: the relative humidity, the temperature, the time of wetness (TOW), and the amount of airborne salt. Diluted synthetic seawater (ASTM D 1141-90) was used as a test solution to form a thin electrolyte layer with a fixed initial thickness of about 500 μm. The surface of the specimen was covered with a polyimide adhesive tape (Permacel, P-221) with the exception of a test area of 10×10 mm. The chloride ion deposition on the specimen surface was adjusted to 1g m^{-2} by dropping a fixed volume of the test solution onto the test area. The dew point inside the chamber was controlled at a constant level during cyclic changes in temperature and relative humidity, as will be described in 3.2. The climatic chamber was equipped with an air circulation fan to ensure air flow so that uniform conditions inside the chamber could be maintained. After each wet-dry cycle (24 h) was completed, the specimen surface was gently rinsed with deionized water, dried in an N_2 stream, and covered with a fresh thin electrolyte layer. Although the SO_2 concentration of the chamber air was not controlled in the present study, the presence of SO_4^{2-} ions in the synthetic seawater (2.7 g/l) resulted in the condition where some SO_2 was deposited and oxidised on the sample surface. If we assume that the deposited SO_2 was completely oxidised to SO_4^{2-}, it can be estimated that the chloride deposition of 1 g m^{-2} of the synthetic seawater is accompanied by an SO_2 deposition of about 0.14 g m^{-2}.

The morphology of the samples after the corrosion tests was observed by a scanning electron microscope (SEM; Philips, XL-30). An energy dispersive X-ray spectrometer (EDX; Philips, XL-30) was employed to analyze the elemental distributions at local parts of the samples exposed for 3 and 12 months at the three different sites. The samples used for cross-sectional observation were embedded into epoxy resin and finely polished by a 1 μm diamond paste.

A Fourier transform infrared spectrometer (FT-IR; JASCO, FT-IR 4200) was used to identify the corrosion products. The assignment of the characteristic peaks was on the basis of the spectra measured from analytical grade $Al(OH)CO_3$ and $Al(OH)_3$ powder. Basic aluminium sulphate ($Al_x(OH)_ySO_4nH_2O$) was made in the laboratory and analyzed by FT-IR. The characteristic peaks of dawsonite ($NaAlCO_3(OH)_2$) were assigned by referring to the data in the literature.

3. Important findings

3.1. Characteristics of atmospheric corrosion in the field exposure tests

3.1.1. Real-time changes in temperature, relative humidity of the ambient atmosphere and surface temperature of specimens

The synchronous changes of four parameters, i.e. the relative humidity of the ambient air (Φ_{air}), the relative humidity of the tested 4N Al samples (Φ_{Al}), the temperature of the ambient air (T_{air}) and the temperature of the tested 4N Al samples (T_{Al}) between July 18, 2007 and July 25, 2007 in Miyakojima are shown in Fig. 2a. In the daytime, the temperature of the samples increased due to their exposure to sunshine. While the temperature decreased from radiation cooling in the night, the relative humidity increased and a dew film consequently formed on the surface. The change in relative humidity was inversely proportional to the temperature change. However, the collected relative humidity and temperature data shows that the dew point of both the atmosphere and the samples remains constant for short-term periods. This can be seen, in Fig. 1b, where experimental Φ_{Al} vs T_{Al} plots are compared with theoretical Φ-T curves for different dew point temperatures. These results are consistent with the findings of earlier studies reported by Muto *et al.* [15]. In addition, the time of wetness was 49.7% of the calendar time in Miyakojima, 51.4% in Choshi and 25.0% in Aobayama, respectively. The weathering data show that the long-term dew point varied with both the exposure time and the seasons, as shown in Fig. 3. The average dew point changed on a large scale, from 5.5 to 28.4 °C in Miyakojima, from -10.1 to 24.6 °C in Choshi and from -13.0 to 23.6 °C in Aobayama. However, the short-term dew point remains constant with a drift of 5 °C.

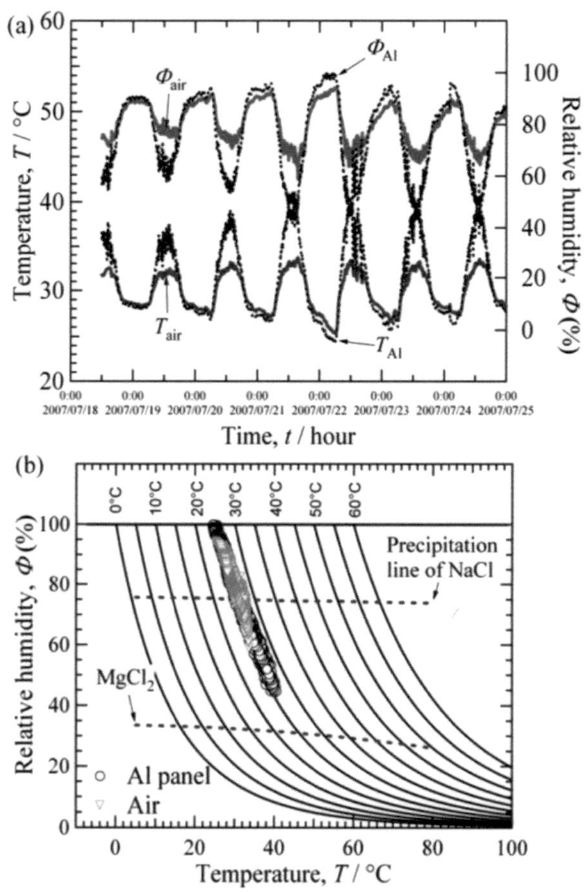

Figure 2. Changes in relative humidity, Φ and temperature, T, of the ambient atmosphere and 4N aluminium (a) and Φ vs. T plot (b) in Miyakojima from July 18th to July 25th, 2007. [18]

http://dx.doi.org/10.5772/57291

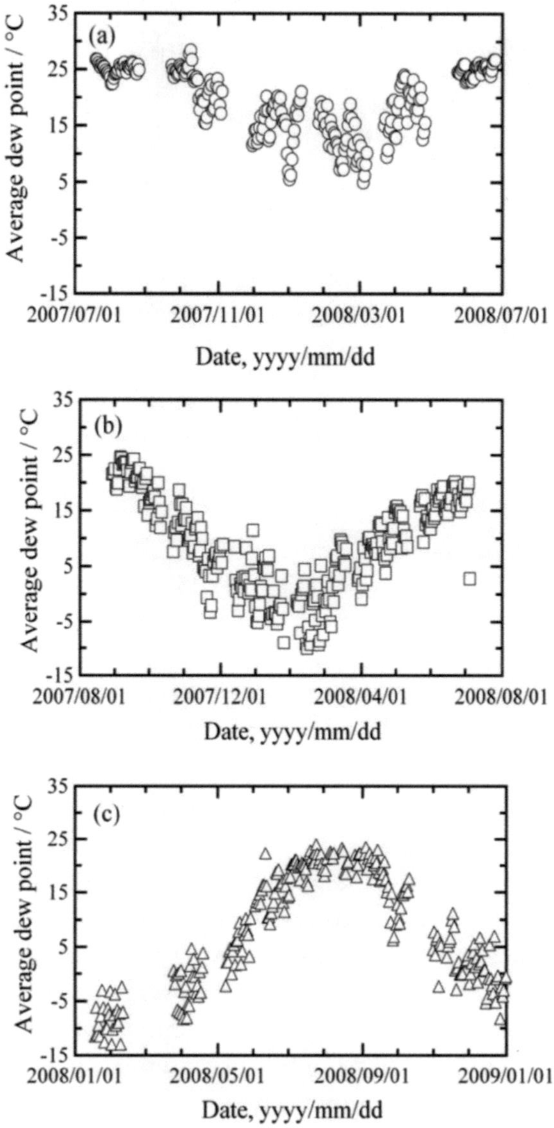

Figure 3. Changes in the dew point in Miyakojima (a), Choshi (b) and Aobayama (c) during the one year of exposure tests. [18]

3.1.2. Corrosion mass loss

The corrosion mass losses of aluminium and its alloys exposed at the three different sites are plotted against exposure time in Fig. 4. As described in the experimental section, the corrosion mass loss for each exposure period was determined by just one sample, and thus some degree of uncertainty can be expected. Though corrosion mass losses increased with exposure time, their increasing rate slowed down with the lapse of time. Corrosion mass losses, ΔW, as a

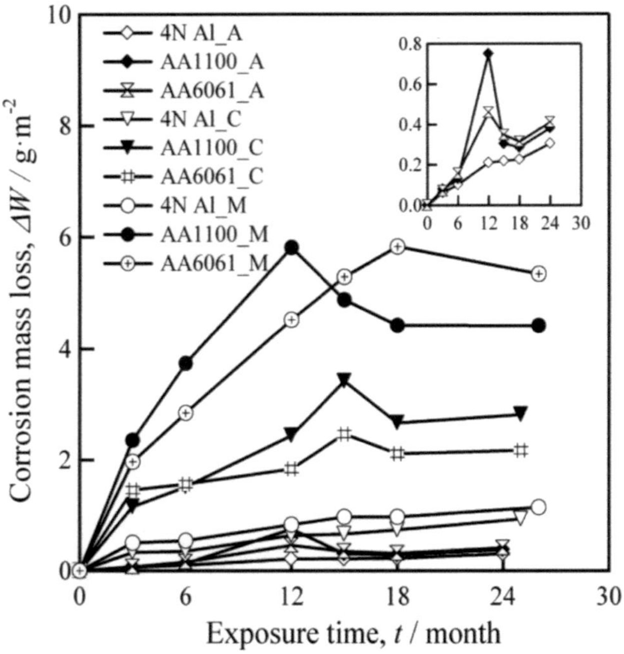

Figure 4. Corrosion mass losses of 4N Al, AA1100 and AA6061 at the three exposure sites as a function of exposure time. The magnified zone at the top right shows the corrosion mass loss in Aobayama. (A: Aobayama; C: Choshi; M: Miyakojima). [18]

function of the exposure time, *t*, were well yielded to general damage functions in the form of $\Delta W = A \times t^n$, where *A* and *n* are constants [22, 23]. Most of the *n* values of the fitted damage functions were close to 0.5, indicating that the rate of atmospheric corrosion is controlled by a diffusion process through the insoluble corrosion product which functions as an unperturbed layer. Corrosion resistance increased in the following order: 4N Al > AA6061 > AA1100. Meanwhile, the order of the corrosiveness of the exposure sites is ranked as Miyakojima > Choshi > Aobayama. The corrosion mass losses of aluminium and its alloys depended not only on the exposure time but also on the meteorological parameters of the exposure sites. To a greater or less extent, the *n* value also strongly depends on such environmental factors as the deposition rate of sulphur dioxide and sea salts, relative humidity and temperature. The corrosion rates of aluminium and its alloys were suppressed during the field exposure, particularly after the first 3 months.

3.1.3. Corrosion morphology

Both AA1100 and AA6061 suffered from tarnishing and pitting corrosion observable to the naked eye. Some parts of the surface remained stable with a complete absence of local attacks even after 12 months of exposure. The roughness of the surface and the ratio of the corroded areas increased with the exposure time. The 4N Al in three sites presented excellent corrosion resistance even though some micro-sized pits were detecte.

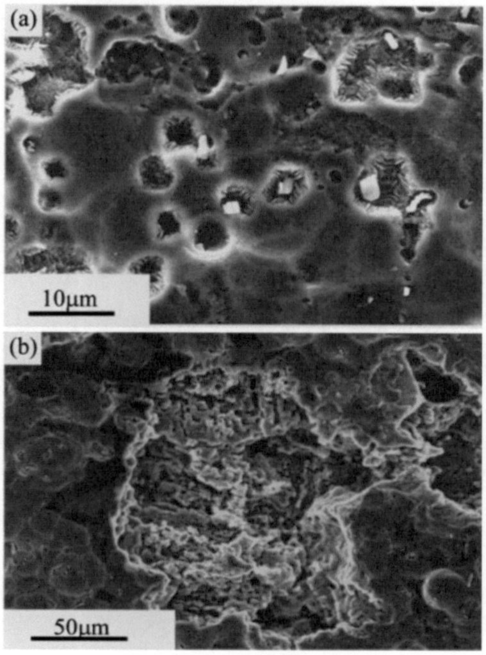

Figure 5. Corrosion morphology of AA1100 exposed in Miyakojima for 3 (a) and 12 (b) months, respectively. [18]

Figure 5 shows the corrosion morphology of AA1100 exposed in Miyakojima for 3 and 12 months. The sample was observed by SEM after the removal of the corrosion products. After 3 months of exposure, pits were formed around some particles, such as inclusions, second phases or the intermetallics, as shown in Fig. 5a. Some pits grew up to several tenths of micrometers after 12 months of exposure, as shown in Fig. 5b As exposure was extended, some more seriously corroded sites formed visible pits of up to a hundred micrometers in size.

3.1.4. Corrosion products

The corrosion products of aluminium and its alloys exposed in the fields were porous and rich in cracks. It was confirmed that corrosion primarily attacked the grain boundaries of AA6061 and a high-facet crystallographic morphology was frequently observed, which is typical for aluminium dissolution under an open circuit potential [24, 25]. In addition, the corrosion product cups were usually circular and typical of those which generate around such cathodic intermetallics as Al_3Fe and Al_2Cu in AA1100 [26, 27]. These intermetallics Al3Fe and Al2Cu usually have more noble corrosion potentials in comparison to pure Al. Al3Fe and Al2Cu particles usually act as cathodes to form micro-coupling cells with adjacent Al matrix, and trigger the dissolution of surrounding Al matrix of intermetallics to form the trenches just as shown in Fig. 5a.

The corrosion products were analyzed by energy dispersive X-ray spectrometry (EDX) and FT-IR. The EDX analysis showed that the surface regions of the corrosion products consisted of various elements, including Al, Na, O, S, and Cl, after aluminium and its alloys were exposed

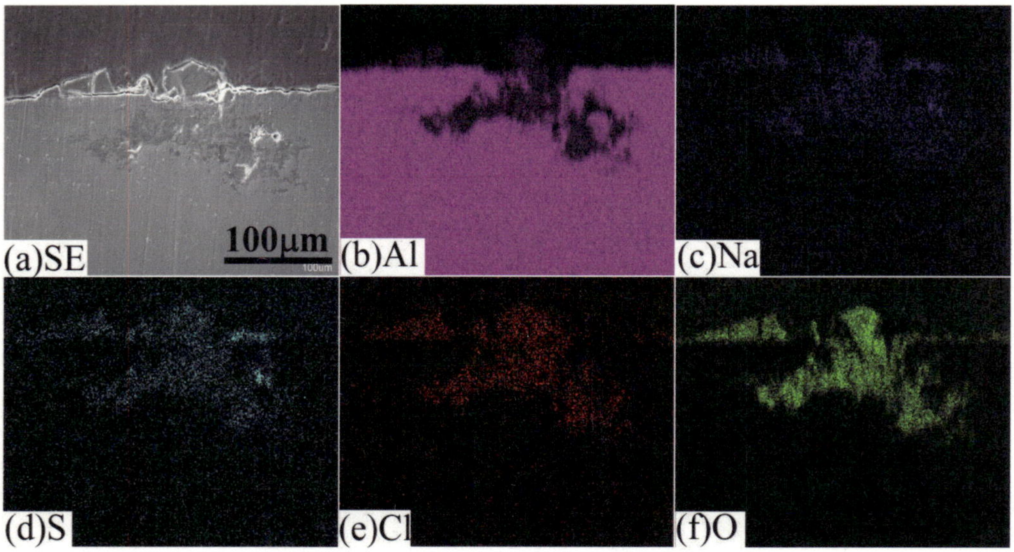

Figure 6. Cross-sectional morphology (a) and X-ray mapping of Al (b), Na (c), S (d), Cl (e) and O (f) obtained from 3-month exposed AA1100 in Miyakojima. [18]

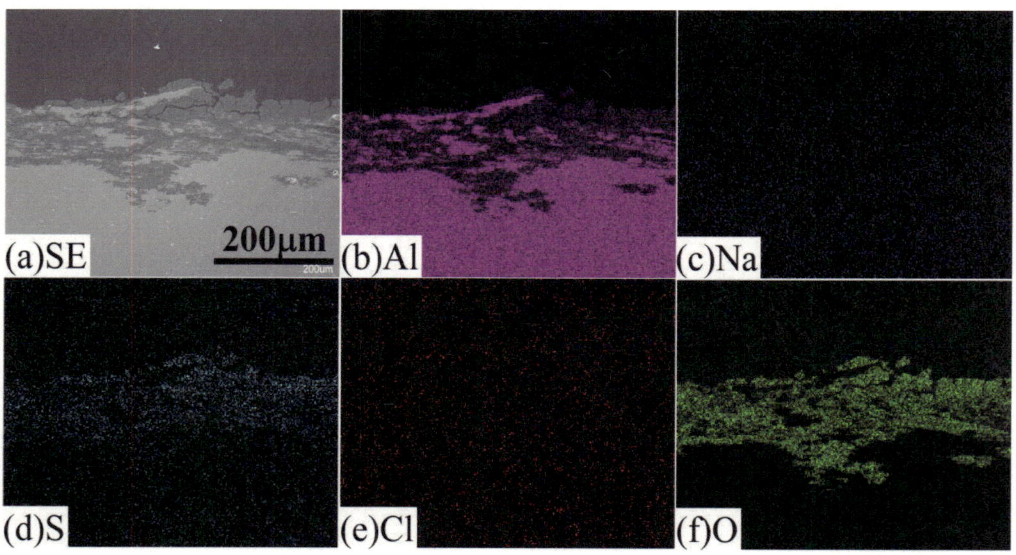

Figure 7. Cross-sectional morphology (a) and X-ray mapping of Al (b), Na (c), S (d), Cl (e) and O (f) obtained from 12-month exposed AA1100 in Miyakojima. [18]

in Miyakojima for 3 months. However, as the exposure time increased up to 12 months, the main elements distributed on the surface of the corrosion products were Al, S, and O. Further analysis was conducted to obtain cross-sectional elemental distribution maps, and the results are shown in Fig. 6 and 7 for AA1100 exposed in Miyakojima for 3 and 12 months, respectively. The data in Fig. 6 indicates that the corrosion products formed after 3 months of exposure consisted mainly of Al, Na, S, C, Cl and O. Chlorine- and sulphur-containing species were distributed inside the pits. However, the corrosion products formed after 12 months of exposure were composed of Al, O and S, as shown in Fig. 7. No signals were detected from many of the elements in the corrosion products after 3 months, notably for Na, Mg, Cl and C, indicating that none of these chemical species accumulated in the corrosion products, perhaps because the rain washes away soluble species once they are deposited.

Further analysis of the corrosion products was performed by FT-IR. Fig. 8a shows the FT-IR spectra of corrosion products formed on AA1100 after 3, 12 and 26 months of exposure in Miyakojima. Figure 8b shows the spectra of analytic grade basic aluminium carbonate, aluminium hydroxide and lab-synthesized basic aluminium sulphate. For the spectrum of dawsonite and also for that of basic aluminium sulphate, we referred to the literature [28-32]. The characteristic adsorption peaks of basic aluminium sulphate are 1135 (ν3 SO_4 asymmetric stretching), 980 (ν1 SO_4 symmetric stretching), 715 (ν4 SO_4 in-plane deformation), 610 (ν4 SO_4), 566 (ν2 SO_4 out-of-plane deformation SO_4) and 450 (ν2 SO_4) cm^{-1} [28-30, 32]. The characteristic adsorption peaks of dawsonite are 3274, 1560 and 1390 (doublet peaks, ν3 CO_3 asymmetric stretching) cm^{-1}, 854 (ν2 CO_3 out-of-plane deformation) cm^{-1} and 540-480 cm^{-1}(a broad peak) [31]. As shown in Fig. 8 (b), the spectrum of $Al(OH)CO_3$ included several peaks, such as 1560, 1390 cm^{-1} (doublet peaks), 2464 and 2603 cm^{-1} (doublet peaks), 1103 (ν1 CO_3 symmetric stretching), 980 cm^{-1} (δ-OH stretching), 854 cm^{-1} (ν2 CO_3 out-of-plane deformation) and 540-480 cm^{-1} (a broad peak). The peaks of dawsonite frequently overlapped with those of basic aluminium carbonate at different positions. After 3 months of exposure in Miyakojima, a small doublet peak around 2500 cm^{-1} appeared which can be assigned to the CO_3^{2-} species, proving that the corrosion products in the surface region consisted of either basic aluminium carbonate ($Al(OH)CO_3$) or dawsonite, and the broad peaks in the low wavenumber region were assigned to sulphate ions. However, the EDS elemental maps (Fig. 7) show the coexistence of Na in the corrosion products, suggesting that the dawsonite species were the main carrier of the carbonate ions. With increasing exposure time, the doublet peak of carbonate became weaker and the peaks in the low wavenumber region became stronger, which can be attributed to the incorporation of sulphate ions. Moreover, the stronger intensity of the ν3 SO_4 asymmetric stretching peak around 1135 cm^{-1} indicates an increase of the basic aluminium sulphate in the corrosion product layer. Generally, if field-exposed samples yield a strong and broad peak in the range of 3200 to 3600 cm^{-1}, it is assigned to the O-H stretching mode, and a medium peak at 1640 cm^{-1} was assigned to the H-O-H bending mode. The latter peak indicates the presence of free water combined with the corrosion products. The FT-IR spectra demonstrated that the existence of carbon and sulphur in the form of CO_3^{2-} and SO_4^{2-}. It was concluded that the corrosion products which formed in the first three months were aluminium hydroxide, basic aluminium sulphate, basic aluminium chloride and Na-containing basic aluminium carbonate. After 12 months of exposure, the basic aluminium carbonate and basic aluminium chloride content decreased; as a result, aluminium hydroxide and basic aluminium sulphate became the main constituents.

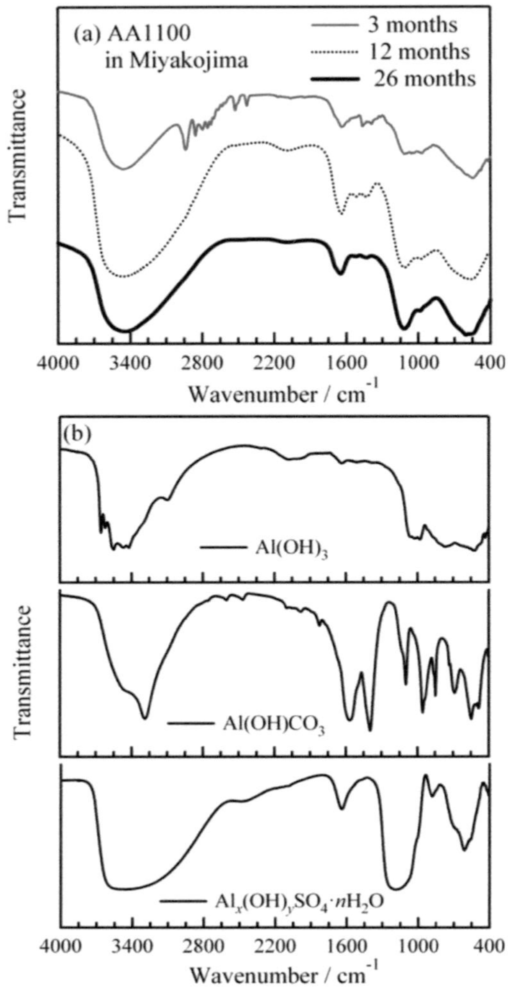

Figure 8. FT-IR spectra of corrosion products formed on AA1100 after 3, 12, and 26 months of exposure in Miyakojima (a) and reference spectra of aluminium hydroxide, basic aluminium carbonate and lab-made basic aluminium sulphate (b). [18]

3.2. Characteristics of atmospheric corrosion in the constant dew point corrosion tests

The practical dew point in Miyakojima from July 18, 2007 to July 25, 2007 in the middle of the summer season was about 26 °C in Fig. 2b. In order to simulate the practical atmospheric conditions in Miyakojima in summer season, the relative humidity and temperature were controlled according to the patterns shown in Fig. 9a, and the corresponding iso-dew point line is shown in Fig. 9b. The dew point of the pattern was maintained at 28 °C. The time of wetness was about 60%. The single cycle was 24 hours, and consisted of wetting (condensation) and drying (evaporation) stages. The chloride deposition was set at a relatively high level of 1 g·m^{-2} to simulate a severe coastal site.

http://dx.doi.org/10.5772/57291

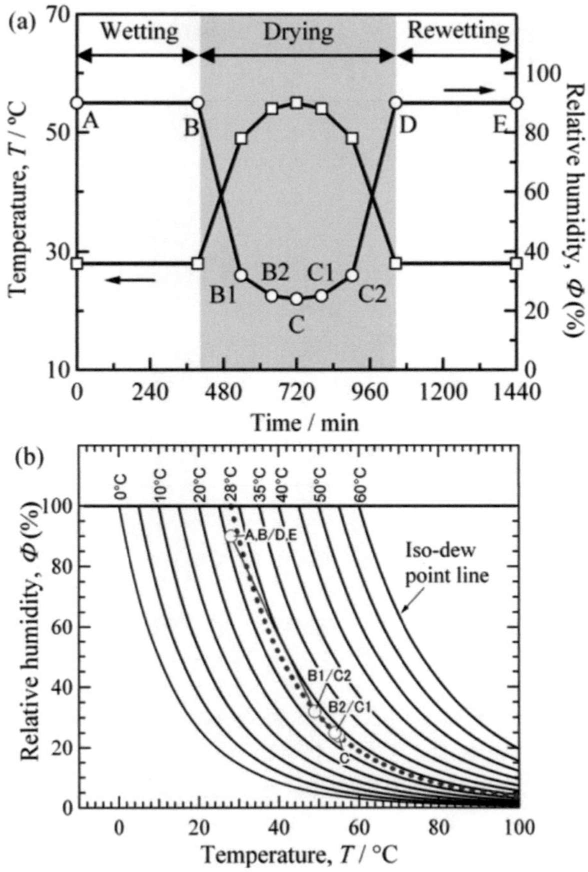

Figure 9. Real-time change pattern of relative humidity, Φ and temperature, T, during the constant dew point corrosion test with a dew point of 28 °C (a) and corresponding Φ vs. T plot (b). [18]

3.2.1. Corrosion mass loss

The corrosion mass losses of aluminium and its alloys exposed in the constant dew condition are shown in Fig. 10. The tests were repeated two or three times for each alloy. The averaged values and the maximum and minimum values of mass losses were determined and are shown in Fig. 10 as legends and error bars, respectively.

In a similar manner as the field exposure test, the mass loss increases as the number of test cycles increases, and is expressed in the form of the general damage function. The comparison of the fitted parameters of the general damage functions obtained from the constant dew point corrosion tests was conducted between the field exposure tests. The experimental data collected from the laboratory tests were in good agreement with those collected from the field experiments. A comparison of the two sets of results indicates that the corrosion mass loss of aluminium and its alloys after the constant dew point corrosion tests showed almost the same tendencies as the field exposure tests. The n value was close to 0.5, indicating that the corrosion product layer functioned as a barrier for diffusion. Nevertheless, the constant dew point

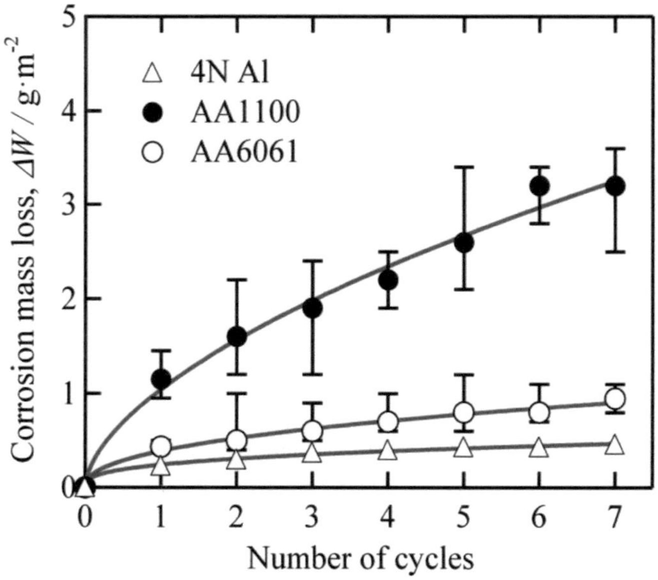

Figure 10. Corrosion mass losses of 4N Al, AA1100 and AA6061 after constant dew point tests as a function of the cycle number. [18]

corrosion test accelerated the corrosion. When the corrosion mass loss value of AA1100 after 7 cycles, 3.22 $g \cdot m^{-2}$, is put into the fitted damage function for the same alloy exposed in Miyakojima, $\Delta W = 4.35 \times t^{0.48}$, it is found that the constant dew point test of 7 cycles (7 days) corresponds to the field exposure test for 104 days in Miyakojima. This indicates that the constant dew point test accelerates the corrosion of AA1100 by a factor of about 15. In the same manner as for AA1100, the acceleration ratio of the constant dew point test with respect to the field exposure test in Miyakojima is estimated to be about 2 and 14 for AA6061 and 4N Al, respectively. Except for AA6061, the constant dew point test for a week corresponds to the field exposure test for about 3 months in Miyakojima. Atmospheric corrosion does not proceed continuously but occurs during the period when the metal surface is covered with a thin corrosive electrolyte layer. The samples do not always meet this corrosion condition in actual atmospheric corrosion environments, but a corrosive electrolyte layer is formed for in every wet-dry cycle of the constant dew point test. Therefore, the constant dew point test accelerates the corrosion.

3.2.2. Corrosion morphology

Macroscopically, the main forms of corrosion attacks seem to be pitting corrosion and tarnishing. The corrosion products exhibited a porous, cracked state. Cups of corrosion products formed on the surface as they did in some cases in the field. Trenches formed around some particles after the first cycle of exposure (Fig. 11a). The pit formed in the substrate with similar shapes as those formed in the field (Fig. 11b). The localized corrosion gradually grew deeper and wider. The corrosion behaviour was influenced by microstructure. The existence

http://dx.doi.org/10.5772/57291

of the cathodic intermetallics (Al_3Fe and Al_2Cu) in AA1100 and AA6061, and sacrificial anodic intermetallics (Mg_2Si) in AA6061 caused preferential electrochemical dissolution in the form of micro-coupling cells [33]. Pits formed on the aluminium alloys around cathodic intermetallics of several micrometers in size, which indicated that the localized corrosion initiated around intermetallics [33-35]. Some pits grew to several hundreds of micrometers in size after short exposure. The corrosion morphologies, including the structure of the corrosion products, the corrosion patterns and the shapes of the cups and pits produced, were similar with those visible in samples from the field.

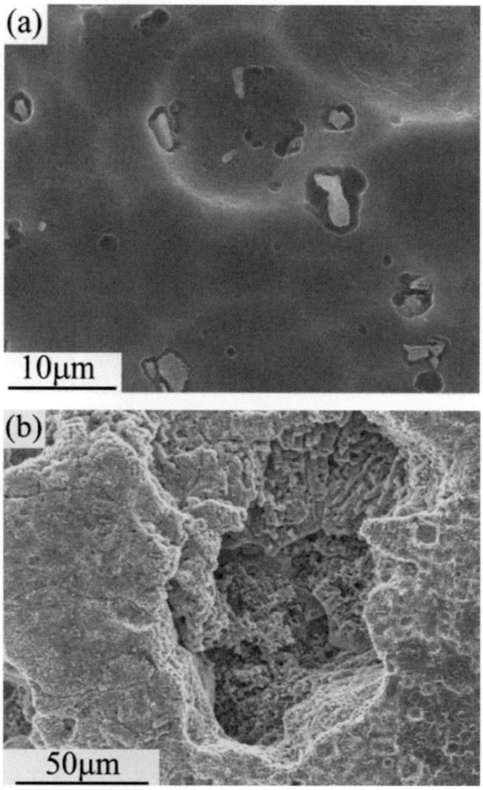

Figure 11. Corrosion morphology of AA1100 exposed for 1 (a) and 7 (b) cycles of constant dew point corrosion tests with a chloride deposition of 1 g·m^{-2} and a dew point of 28 °C. [18]

3.2.3. Corrosion products

The element distribution profiles inside the pits formed after 7 cycles of constant dew point corrosion tests are shown in Fig. 12. While there was Cl inside the pits, and enriched Cl at the bottom of the pits, sulphur was highly concentrated in the outermost layer of the corrosion products. Generally, the element distribution profiles were similar with those obtained from 3-month exposed AA1100.

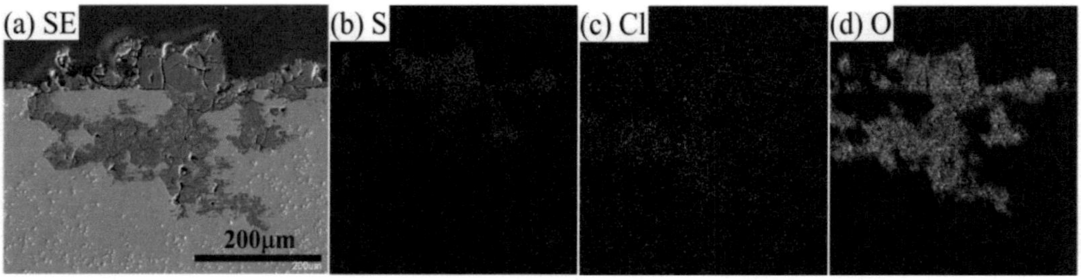

Figure 12. Cross-sectional morphology (a) and X-ray mapping of S (b), Cl (c) and O (d), obtained from AA1100 after 7 cycles of exposure in the constant dew point test with a chloride deposition of 1 g·m^{-2} and a dew point of 28 °C. [18]

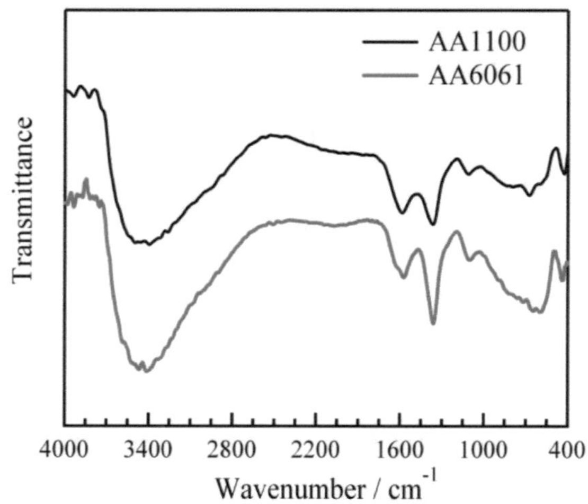

Figure 13. FT-IR spectra of corrosion products on AA1100 and AA6061 after 7 cycles of the constant dew point test with a chloride deposition of 1 g·m^{-2} and a dew point of 28 °C. [18]

Figure 13 shows the FT-IR spectrum of the corrosion products formed on AA1100 and AA6061 after 7 days of constant dew point corrosion tests in the laboratory. The corrosion products include carbonate ions (1390, 1560 cm^{-1}), sulphate ions (1085, 640, and 450 cm^{-1}) and water (1640 cm^{-1}). This is similar to the FT-IR results of field-exposed aluminium and its alloys. Both EDX and FT-IR analysis demonstrated that the corrosion products consisted mainly of basic aluminium sulphate, basic aluminium chloride, aluminium hydroxide and basic aluminium carbonate. The washing effect of rain is difficult to reproduce in a short time. Due to limited exposure time, the amount of basic aluminium sulphate and aluminium hydroxide in the corrosion product layer was considerably lower than that formed in the field samples with more than 12 months of exposure. The other corrosion products were close to those from the initial case of field exposure tests.

The observed data of the dew point change in the surrounding atmosphere and at the surface showed that the short-term dew point remained at a constant level. The dew point is associated with the heat capacity of the materials and the ratio of water vapour and dry air. Heat flux in a closed system is mainly determined by the evaporation (heating) and condensation (cooling) of the water content in the air. Though the practical open environment cannot be regarded as a closed system, at least for short periods of time, the surrounding region of the samples can be considered a closed system, where the heat flux and the flow of humid air remain stable and move freely around the samples [15, 36]. Consequently, the heat flux around the exposed samples remains constant, which results in the occurrence of the constant dew point [36-38]. The constant dew point corrosion testing model therefore has a more concrete base than other simulation methodologies in terms of its ability to model the practical conditions.

The accumulation of the sulphate ions in the localized corroded sites was observed in both the field exposure tests and the constant dew point tests indicated by the element mapping data. Due to effects of daily alternations of the wet-dry surface state and washing of raining, very few highly soluble chlorine-containing species accumulated inside the pits [5, 39]. On the other hand, ten times more sulphate ions accumulated than chloride ions [6, 40, and 41]. With increasing exposure time, such compounds as basic aluminium carbonate and basic aluminium chloride were washed away by rain while the insoluble ones, such as aluminium hydroxide and basic aluminium sulphate, remained as long-term stable corrosion products. The constant dew point test with 7 cycles is incapable of reproducing the process completely due to the limited exposure time. Consequently, strong peaks of carbonate were detected in the lab-exposed samples. Based on the surface morphology, corrosion mass loss and chemical compounds, we were able to reproduce the initial stage of the atmospheric corrosion of aluminium and its alloys by using the constant dew point test over the initial period.

4. Conclusion

In order to examine whether the constant dew point corrosion test can appropriately reproduce the atmospheric corrosion of aluminium alloys, corrosion tests were performed on AA1100, AA6061 and 4N Al under the cyclic wet-dry conditions at the constant dew point of 28 °C and with the chloride deposition of 1 $g\cdot m^{-2}$, which simulated marine atmospheric environments in the summer season in Japan. It was found that the corrosion mass loss increased as the number of cycles increased in accordance with a power-law formula. The corrosion pattern after the 7-cycle test was pitting, and the corrosion products consisted of basic aluminium sulphate, basic aluminium chloride, aluminium hydroxide, and basic aluminium carbonate. The corrosion rates, corrosion morphology and corrosion product composition were similar to those observed on the samples exposed for a few months in Miyakojima, a typical coastal site. The constant dew point corrosion test, therefore was shown capable not only of reproducing the atmospheric corrosion of aluminium, but also of accelerating it.

Acknowledgements

This research is partially supported by the Ministry of Education, Culture, Sports, Science, and Technology (MEXT) through a Grant-in-Aid for Young Scientists (B) under Grant No. 24760567 and Grant-In-Aid for Science Research in a Priority Area on "Research and Development Project on Advanced Materials Development and Integration of Novel Structured Metallic and Inorganic Materials".

Author details

Zhenhua Dan, Izumi Muto and Nobuyoshi Hara

Department of Materials Science, Tohoku University, Sendai, Japan

References

[1] Hatch JE. Aluminium: Properties and Physical Metallurgy. In: American Society of Metals, Metals Park, OH, 1984, p2-24.

[2] Davis JR. ASM Specialty Handbook: Aluminium and Aluminium Alloys. In: ASM International, Metals Park, OH, USA, 1993, p8-55.

[3] Green JAS. Aluminium Recycling and Processing for Energy Conservation and Sustainability. ASM International, Metals Park, OH, USA, 2007.

[4] McGeary FL., Summerson TJ., Ailor WH. Atmospheric Exposure of Nonferrous Metals and Alloys-Aluminium: Seven-Year Data, Metal Corrosion in the Atmosphere, ASTM STP 435, American Society for Testing and Materials, 1968. p141-174.

[5] Friel JJ. Atmospheric Corrosion Products on Al, Zn, and AlZn Metallic Coatings. Corrosion 1986;42:422-6.

[6] Graedel TE. Corrosion Mechanisms for Aluminum Exposed to the Atmosphere. Journal of Electrochemical Society 1989;136: 204C-12C.

[7] Vera R., Delgado D., Rosales BM. Effect of Atmospheric Pollutants on the Corrosion of High Power Electrical Conductors: Part 1. Aluminium and AA6201 Alloy. Corrosion Science 2006;48: 2882-900.

[8] de la Fuente D., Otero-Huerta E., Morcillo M. Studies of Long-term Weathering of Aluminium in the Atmosphere. Corrosion Science 2007;49:3134-48.

[9] Blücher DB., Lindström R., Svensson JE., Johansson LG. The Effect of CO_2 on the NaCl-Induced Atmospheric Corrosion of Aluminum. Journal of Electrochemical Society 2001;148:B127-31.

[10] Lyon SB., Thompson GE., Johnson JB. Accelerated Atmospheric Corrosion Testing Using a Cyclic Wet/Dry Exposure Test: Aluminum, Galvanized Steel, and Steel. Corrosion 1987;43: 719-26.

[11] Shi Y., Zhang Z., Su J., Cao F., Zhang J. Electrochemical Noise Study on 2024-T3 Aluminum Alloy Corrosion in Simulated Acid Rain under Cyclic Wet–dry Condition. Electrochimica Acta 2006;51:4977-86.

[12] El-Mahdy GA., Kim KB. AC Impedance Study on the Atmospheric Corrosion of Aluminum under Periodic Wet–dry Conditions. Electrochimica Acta 2004;49:1937-48.

[13] Deflorian F., Rossi S., Prosseda S. Improvement of Corrosion Protection System for Aluminium Body Bus Used in Public Transportation. Materials and Design 2006;27:758-69.

[14] Nazarov A., Thierry D. Rate-determining Reactions of Atmospheric Corrosion. Electrochimica Acta 2004;49:2717-24.

[15] Muto I., Sugimoto K. Modeling of Atmospheric Corrosion Environments and Its Application to Constant Dew-point Corrosion. Zairyo-to-Kankyo 1998;47:519-27.

[16] Muto I., Yoshida H., Ogawa H., Hara N. Effect of Alloying Elements on Atmospheric Corrosion Behavior of Zinc Die-casting Alloys. Journal of Japan Institute of Metals 2008;72: 337-46.

[17] Shinozaki J., Muto I., Ogawa H., Hara N. Effects of Third Element Addition on Atmospheric Corrosion Resistance of Zinc-aluminum Die-cast Alloys. Journal of Japan Institute of Metals 2009;73:533-41.

[18] Dan ZH., Takigawa S., Muto I., Hara N. Applicability of Constant Dew Point Corrosion Tests for Evaluating Atmospheric Corrosion of Aluminium Alloys. Corrosion Science 2011; 53: 2006-14.

[19] Dan ZH., Muto I., Hara N. Effects of Environmental Factors on Atmospheric Corrosion of Aluminium and Its Alloys under Constant Dew Point Conditions, Corrosion Science 2012; 57: 22-9.

[20] Muto I. Corrosion of Metals and Alloys- Accelerated Cyclic Corrosion Tests with Exposure to Synthetic Ocean Water Salt-Deposition Process-"Dry" and "Wet" Conditions at Constant Absolute Humidity. International standard (ISO) 2013(E);16539:1-19

[21] Sonntag D. Important New Values of the Physical Constants of 1986, Vapour Pressure Formulations Based on the ITS-90, and Psychrometer Formulae. Journal of Meteorology1990;40:340-4.

[22] Dean SW., Anthony WH. Atmospheric Corrosion of Wrought Aluminium Alloys during a Ten-year Period, Degradation of Metals in the Atmosphere, ASTM STP 965, in: S.W. Dean and T. S. Lee (Eds.), American Society for Testing and Materials, Philadelphia, 1988, pp. 191-205.

[23] Kobus J. Long-term Atmospheric Corrosion Monitoring. Materials and Corrosion 2000;51:104-8.

[24] Baumgärtner M., Kaesche H. Aluminum Pitting in Chloride Solutions: Morphology and Pit Growth Kinetics. Corrosion Science 1990;31:231-6.

[25] Guillaumin V., Mankowski G. Localized Corrosion of 6056 T6 Aluminium Alloy in Chloride Media. Corrosion Science 2000;42:105-25.

[26] Birbilis N., Buchheit RG. Investigation and Discussion of Characteristics for Intermetallic Phases Common to Aluminum Alloys as a Function of Solution pH. Journal of Electrochemical Society 2008;155:C117-26.

[27] Park JO., Paik CH., Huang YH., Alkire RC. Influence of Fe-Rich Intermetallic Inclusions on Pit Initiation on Aluminum Alloys in Aerated NaCl. Journal of Electrochemical Society 1999;146:517-23.

[28] Contreras CA., Sugita S., Ramos E. Preparation of Sodium Aluminate from Basic Aluminum Sulfate. Advances in Technology of Materials and Materiasl Processing Journal 2006;8:122-35.

[29] Kloprogge JT., Geus JW., Jansen JBH., Seykens D. Thermal Stability of Basic Aluminum Sulphate. Thermochimica Acta 1992; 209:265-76.

[30] Kloprogge JT., Frost RL. The Dehydroxylation of Basic Aluminum Sulfate: An Infrared Emission Spectroscopic Study. Thermochimica Acta 1998;320:245-52.

[31] Yalfani MS., Santiago M., Pérez-Ramírez J. In Situ Studies During Thermal Activation of Dawsonite-type Compounds to Oxide Catalysts. Journal of Material Chemistry 2007;17:1222-29.

[32] Bishop JL., Murad E. The Visible and Infrared Spectral Properties of Jarosite and Alunite. American Mineralogist 2005;90:1100-7.

[33] Wong KP., Alkire RC., Local Chemistry and Growth of Single Corrosion Pits in Aluminium, Journal of Electrochemical Society 137 (1990) 3010-5.

[34] Vermilyea DA. Concerning the Critical Pitting Potential. Journal of Electrochemical Society 1971;118:529-31.

[35] Tanem BS., Svenningsen G., Mädalen J. Relations between Sample Preparation and SKPFM Volta Potential Maps on an EN AW-6005 Aluminium Alloy. Corrosion Science 2005;47:1506-19.

[36] Dean SW., Reiser DB., Time of Wettness and Dew Formation: A Model of Atmospheric Heat Transfer, in: Kirk WW., Lawson HH. (Eds.), Atmospheric Corrosion, ASTM STP 1239, American Society for Testing and Materials, Philadelphia, PA, 1995, p3-10.

[37] Cole IS., Holgate R., Kao P., Ganther W. The Rate of Drying of Moisture from a Metal Surface and its Implication for Time of Wetness. Corrosion Science 1995;37:455-65.

[38] lncropera FP., DeWitt DP. Fundamentals of Heat and Mass Transfer (3rd Edn.), John Wiley and Sons, New York, 1990.

[39] Foley RT., Nguyen TH. Chemical Nature of Aluminium Corrosion. Journal of Electrochemical Society 1982;129:464-7.

[40] Adams F., Rawajfih Z. Basaluminite and Alunite: A Possible Cause of Sulfate Retention by Acid Soils, Soil Science Society of American Proceedings 1977;41:686-92.

[41] Nordstrom DK. The Effect of Sulfate on Aluminum Concentrations in Natural Waters: Some Stability Relations in the System $A1_2O_3$-SO_3-H_20 at 298 K, Geochimica et Cosmochimica Acta 1982;46:681-92.

Chapter 6

Structure Investigations of Rare-Earth Doped Nano-Particles – Extracted from Oxyfluoride Glass Ceramics by Thermal Induction and Corrosion Treatment

Hui Guo, Yu Hua, Lijuan Zhao and Yiming Li

Additional information is available at the end of the chapter

http://dx.doi.org/10.5772/57213

1. Introduction

Trivalent Lanthanide (Ln^{3+}) ions doped materials are well-known due to their superior luminescent properties and widely utilization in display field and fluorescent lights. Compared to organic fluorophores and semi-conducting nano-particles, Ln^{3+}-doped inorganic nano-particles have high photochemical stability, sharp emission bandwidths and large anti-Stokes shifts, therefore can be applied as excellent luminescence materials [1,2]. More recently, they also attract great interests as a new class of bioprobes, because their long-lived and intense emissions offer promising applications to biosensing, biological labeling and imaging technology [3-9]. Therefore, many researchers devote themselves to explore how to synthesize Ln^{3+}-doped inorganic nano-particles. Ever since the first report on oxyfluoride glass ceramics (GCs) [10], they have received great attention from many researchers due to the perfect mechanical stability and the efficient visible to infrared emission [11-14]. However, the luminescent nano-particles embedded in the bulk glass matrix can not be applied in mentioned fields directly, and how to free them from glass matrix is a problem to be solved. Unfortunately, there have been much research on the properties of nano-particles in GCs but no publication has been reported about how to obtain free and pure nano-particles into aqueous solution from the oxide glass matrix and how to apply it to the fields mentioned above, especially in biological field. Only Mortier and Patriarche [15] have reported an original way to synthesize single-crystal PbF_2 nano-particles used oxide GeO_2-PbO glass as the template and then the nano-particles were produced by dissolving the amorphous GeO_2-PbO phase of GCs with hydrofluoric acid. However, the corrosion reaction between the amorphous GeO_2-PbO and hydrofluoric acid can hardly be complete so that the obtained nano-particles had an amorphous phase as a shell in their experiment. Therefore, it is of great significance to develop convenient routes to fabricate inorganic nano-particles doped rare-earth (RE) ion in aqueous solution from GCs in order to meet their practical application requirements.

This chapter describes an innovative route to fabricate Ln^{3+}-doped nano-particles, with high efficient luminescent intensity and adjustable size, from oxyfluoride glass by the thermal induction and corrosion treatment [16]. Then the structure of obtained nano-particles can be investigated directly because optical properties of Ln^{3+} ions are very sensitive to the local environment [17,18] and Ln^{3+} ions are normally used as probes to survey the local structure in luminescent materials [19-23]. To determine the chemical composition of the obtained nano-particles accurately, energy dispersive X-ray spectroscopy (EDS) has been applied after getting rid off the interference of oxide glass matrix. X-ray diffraction (XRD) measurements were performed to identify the crystallization phases of nanocrystals with a power diffractometer (D/Max-2500), using CuKa as the radiation. To prove the structure of nano-particles, Rietveld analyses of XRD patterns were also carried out with the Fullprof program [24] or MAUD (material analysis using diffraction) program [25] in the range 10°-120° using the profile function of pseudo-voigt with axial divergence asymmetry. For the Rietveld analyses, the XRD data were acquired in the step-scan mode with a step of 0.02° (2θ) at a counting time of 1 s per step. High resolution transmission electron microscope (HRTEM) analysis was performed to observe the morphology of samples on a Philips TZOST TEM operating at 200kV. STEM and EDS line scan were performed on a Tecnai G2F30 field emission transmission electron microscope (FE-TEM) using HAADF as imaging mode. The emission spectra were recorded on an Edinburgh Instruments FLS920 spectrofluoremeter equipped with the continuous (450 W) xenon lamps.

2. Nano-particles preparation

The chemical compositions of the precursor glasses are shown in Table 1. About 20 g of raw materials were fully mixed and melted in a covered platinum crucible in air atmosphere at 1,000 °C for 2 h, and then cast into a steel plate [26-28]. To obtain nano-particles in glass ceramic, the glass samples were subsequently heat-treated at 470 °C for 8 h at the nucleation temperature measured by differential thermal analysis (DTA). Using DTA equipment (TA-Inst 2100), samples were held in a Pt crucible and analyzed against a calcined Al_2O_3 reference at a heating rate of 20 °C min^{-1}.

Compositions	Mole Fraction
SiO_2	50-y-x
Al_2O_3	y
PbF_2	40
CdF_2	10
RE_2O_3	x

Table 1. Compositions of the raw materials.

Figure 1(a) shows the XRD pattern of the investigated glass ceramics. After thermal induction, the XRD pattern of GCs presented intense diffraction peaks, which can be easily assigned to the β-PbF_2:Er^{3+},Yb^{3+} phase [29,30]. The lattice constant of β-PbF_2 crystal with the fluorite structure is a = 0.575nm. The slightly smaller lattice constant of the present nano-particles can be interpreted by that the nano-particles are a β-PbF_2 solid solution in which Pb^{2+} ions with ionic radius of 0.129 nm are partially substituted by RE-ion with ionic radius about 0.100nm [29]. This result is further supported by TEM of our GCs shown in Figure 1(b). The micrograph reveals that nano-particles are uniformly dispersed in the glass matrix. Furthermore, the particle size distribution has been shown in Figure 1(c). The average size of as-prepared nano-particles determined from TEM image is 42.6 nm.

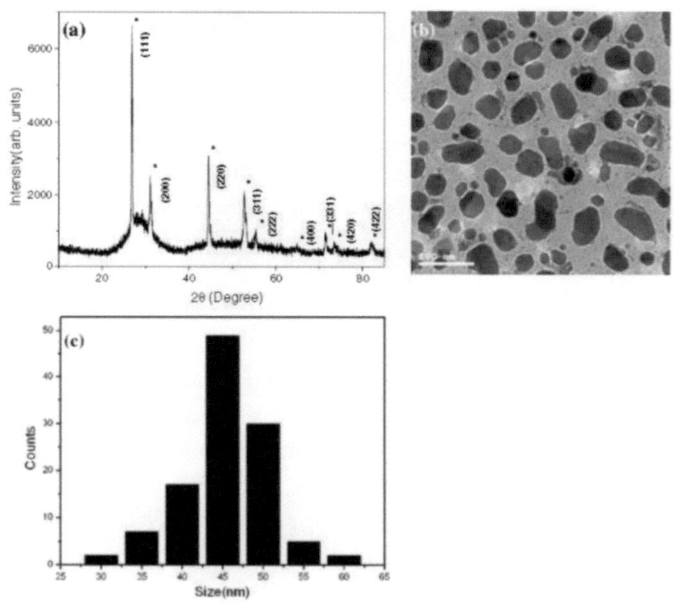

Figure 1. (a) XRD patterns of glass ceramics after thermal induction. The asterisks correspond to JCPDS file NO. 06-0251 (β-PbF_2). (b) TEM micrographs of nano-particles existing in glass matrix. (c) The size distribution of nano-particles in glass matrix [16]. Reproduced from Nanoscale Res. Lett., 2008, 3(12), 516. Copyright 2008, Springer.

To extract the nano-particles from the oxyfluoride glass ceramics, the corrosion treatment methods are as follows:

i. Firstly, the powder of GCs (about 500 mg) was immersed into 10.89 mol/L hydrofluoric acid. Then a benzene isolated layer was added to cover the hydrofluoric acid in the improved setting, as shown in Figure 2. Thus, corrosion reaction becomes more thorough and during the corrosion process, the benzene-blocking-layer separates the reactants and the air atmosphere, which is able to avoid oxidation efficiently.

ii. Secondly, the mixed solution was stirred for about 20 h to get rid off silicate glass matrix with the formula of $SiO_2 + 4HF = SiF_4\uparrow + 2H_2O$. After corrosion treatment, the

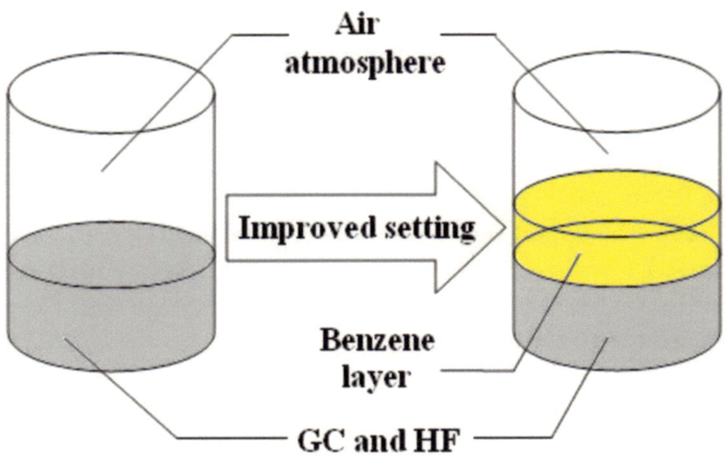

Figure 2. A general description of the corrosion process design.

nano-particles were deposited by solid-liquid separation. Then pH value was adjusted to neutrality by repeated adding distilled water.

iii. Finally, the nano-particles in aqueous solution were dispersed using sodium lauryl benzenesulfate and vacuum dried at 65 °C for 24 h to acquire the powder of nano-particles.

As shown in Figure 3, XRD spectra and TEM image of the obtained nano-particles after corrosion treatment are presented. Phase identification of the nano-particles has been studied by using XRD. The typical XRD patterns show strong diffraction peaks that can be indexed to the β-PbF_2:Er^{3+},Yb^{3+} phase due to the appreciably shift of XRD peaks position and former research results. No additional or intermediate phase is detected in the sample. The peaks position and half peak breadth of XRD are in concordance with those measured in GCs, which indicates that nano-particles are released from glass matrix to aqueous solution. Moreover, the baseline of XRD is nearly a straight line compared with that of XRD of GCs, which also indicates that silicate glass host is corroded completely. The Figure 3(b) is the TEM image of nano-particles in aqueous solution and Figure 3(c) reveals the size distribution. The average particle size is obtained as 41.7 nm, which is very close to the former calculation result of nano-particles existing in glass matrix 42.6 nm. And the comparison of Figure 1 and Figure 3 indicates that the nano-particles formed in glass matrix are released to aqueous solution and their structure, shape and size distribution in glass host are kept well. The potential reason could be the restriction of glass network structure during the thermal induction process. The nano-particles exhibit a narrow size distribution and display a good dispersibility in aqueous solution.

To illustrate the luminescent properties of the nano-particles can be kept after corrosion treatment, Figure 4 reveals the main up-conversion (UC) emission of the $^2H_{11/2}/^4S_{3/2}/^4F_{9/2}\rightarrow{}^4I_{15/2}$ transition in GCs and in aqueous solution under 980nm diode laser excitation. The weak ultraviolet UC radiation of the $^4G_{11/2}\rightarrow{}^4I_{15/2}$ transition were not shown here, since we focus on the ratio of red and green fluorescent intensities, which could show the variation of crystal

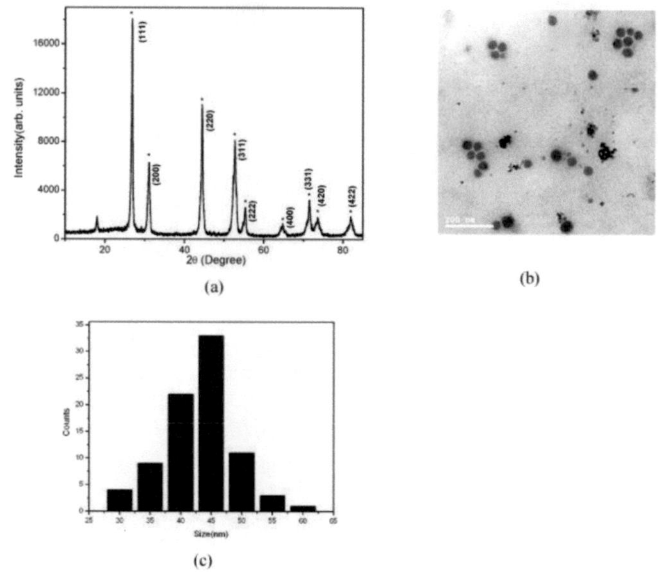

Figure 3. (a) XRD patterns of glass ceramics after thermal induction. The asterisks correspond to JCPDS file NO. 06-0251 (β-PbF_2). (b) TEM micrographs of nano-particles existing in glass matrix. (c) The size distribution of nano-particles in glass matrix [16]. Reproduced from Nanoscale Res. Lett., 2008, 3(12), 516. Copyright 2008, Springer.

lattice field around the RE-ion. Compared to GCs, the red fluorescence intensity of nano-particles in aqueous solution dramatically increases. The UC mechanisms of the green and red emissions had been researched by our groups [30]. And the Er^{3+} ion can be promoted to the $^4I_{11/2}$ state through ground state absorption (GSA) of laser photons, and then to the $^2H_{11/2}$ state by use of the excited state absorption (ESA) or energy transfer UC (ETU) processes. The $^4S_{3/2}$ state can also be populated by relaxations from the upper $^2H_{11/2}$ state. The green luminescence is emitted by the transition from $^2H_{11/2}$ and $^4S_{3/2}$ state to $^4I_{15/2}$ ground state. Hence, the green UC luminescence seldom subjects to influence of phonon energy difference of crystal field around the RE ions. The red UC luminescence, however, comes from the phonon-assisted quantum cutting (PQC) process, which primary generated in low phonon energy crystal lattice field. It is worthwhile to point out that luminescent signal include emission from RE ions in glass host and fluoride nano-particles at luminescent spectra. In GCs, most of RE ions are incorporated in the fluoride crystal phase, which mainly emit red UC luminescence. Nevertheless, a part of RE ions exists in glass host and mainly emits green UC luminescence. As Figure 4 shows, the ratio value of red and green UC luminescent intensities increase from 1:1.31 in glass ceramics to 1:0.61 in nano-particles aqueous solution. The key reason for the red UC enhancement arises from the fact that the glass host is corroded by hydrofluoric acid, the fluoride nano-particles formerly embedded in glass matrix are released to aqueous solution and eventually red UC luminescent intensities increase sharply. These results are consistent with the conclusion drawn by XRD and TEM. The nano-particles in aqueous solution have the same luminescent properties as them existing in glass host.

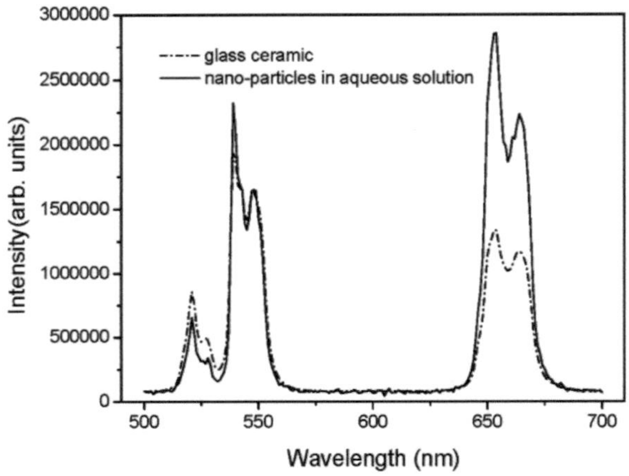

Figure 4. UC red and green emissions of GCs and nano-particles in aqueous solution under diode laser excitation of 980 nm [16]. Reproduced from Nanoscale Res. Lett., 2008, 3(12), 516. Copyright 2008, Springer.

In conclusion, the nano-particles can be extracted from glass matrix by the mentioned thermal induction and corrosion treatment. Their structure and luminescent properties can be kept as well. However, although the preparation method is based on a fundamental consideration, our proposed corrosion treatment has its limitations: (i) the matrix that need be corroded can react with the hydrofluoric acid, in general, they are some oxide compounds; (ii) the nano-particles can not react with the hydrofluoric acid, in general, they are some fluoride nanocrystals such as PbF_2, CaF_2, BaF_2, $NaYF_4$, and so on. But notably, these RE-doped fluoride nano-particles attract much more attention compared with oxide nano-particles because of their low phonon frequency. Therefore, our proposed corrosion treatment can be applied widely to any other eligible systems.

3. Structure investigations of the obtained nano-particles

In general, the optical properties of Ln^{3+} ions are very sensitive to the local environment, and especially the emission intensity of Ln^{3+}-doped materials are closely correlated with the surrounding crystal-field [18,31]. Therefore it is of primary importance to identify the structure of nano-particles in rare-earth doped GCs [32,33]. However, from $Cd_xPb_{1-x}F_2$ model to $Pb_{1-x}Er_xF_{2+x}$ model, most of structural investigations of nano-particles in GCs are indirect due to the effect of the glass matrix. The quantitative tetragonal structure has been proposed directly until the corrosion treatment method has been reported. Based on corrosion treatment, additionally, the distribution of different RE ions in GCs can also be determined clearly. Some proposed structure models in different stage are summarized in Table 2 below. And then, the distribution of different RE ions in GCs is discussed detailed in parts 3.1 and 3.2.

Glass Matrix	Doped ions	Do corrosion treatment or not	Structure of Nano-particles	Refs.
SiO_2-Al_2O_3	Pr^{3+}, Er^{3+}, Yb^{3+}	Not	$Cd_xPb_{1-x}F_2$	[10,11]
SiO_2-Al_2O_3	Er^{3+}	Not	β-PbF_2:Er^{3+}	[34]
GeO_2-PbO	Er^{3+}	Not	β-PbF_2:Er^{3+}	[35,36]
SiO_2-Al_2O_3	Er^{3+}	Not	$Pb_{1-x}Er_xF_{2+x}$	[26]
-- --	Nd^{3+}, Eu^{3+}, Er^{3+}	Not	$Pb_{1-x}Er_xF_{2+x}$	[37]
SiO_2-Al_2O_3	Eu^{3+}	Not	Tetragonal	[33]
SiO_2-Al_2O_3	Er^{3+}, Yb^{3+}	Do	Tetragonal $PbREF_5$	[24]
SiO_2-Al_2O_3	Nd^{3+}	Do	Core-shell structure	[25]
SiO_2-Al_2O_3	Yb^{3+}, Tb^{3+}, Tm^{3+}, *et al.*	Not	Cubic Pb_3REF_9 and Tetragonal $PbREF_5$	[38]

Table 2. Different proposed structure models of nano-particles in glass ceramics systems.

3.1. Distribution of heavy RE ions and the tetragonal structure proposed

Oxyfluoride glass with composition (mol%) of $30SiO_2$-$15.5Al_2O_3$-$40PbF_2$-$10CdF_2$-$0.5Er_2O_3$-$4Yb_2O_3$ was thermal treated by 460 °C, 480 °C, 500 °C, 520 °C, 540 °C, 560 °C and subsequently cooled to room temperature to be GCs (labeled as GC@460 °C, GC@480 °C, *etc.*). Afterwards, corresponding nanocrystals (NCs) (labeled as NC@460 °C, NC@480 °C, *etc.*) were obtained by corrosion treatment mentioned above [16]. For samples of Pb/Cd = 40/10, the EDS measurement of the NCs presents that at% of NCs is Pb:RE:F = 1.06:1:4.99, which is very close to the stoichiometric proportion 1:1:5. These indicate that RE ions are incorporated into fluoride nano-particles during thermal treatment. GC samples which are thermal treated at different temperatures remain essentially the same proportion. The detected trace of Cd in EDS up to only 0.09 at% indicates that the nano-particles is RE^{3+} doped β-PbF_2 (here RE^{3+} = Er^{3+} or Yb^{3+}), rather than $Cd_xPb_{1-x}F_2$. And even the trace of Cd in EDS of samples of Pb/Cd = 25/25 is only 0.19 at%. The same low atomic concentration confirms that Cd hardly congregated into the nanocrystals, which is basically independent of the initial content of Cd in raw materials. This RE^{3+} doped β-PbF_2 composition has been accepted by more and more researchers in the past decade [26,39,40].

Figure 5 shows the XRD patterns of the GCs as made. After crystallization, several diffraction peaks are exhibited. Apparently, XRD patterns are divided into two groups by the shape of peaks as: one is low temperature group (460 °C, 480 °C, 500 °C) and the other is high temperature group (520 °C, 540 °C, 560 °C). Some peaks of XRD corresponding to high temperature group split quite obviously compared with those corresponding to low temperature group. Since the chemical composition acquired from EDS of all these samples is almost the same and the sum of relative intensities of the splitting XRD peaks is basically equal to the relative intensities of the corresponding peaks which do not split, we believe that the basic structures of crystals of the two groups should be similar in general.

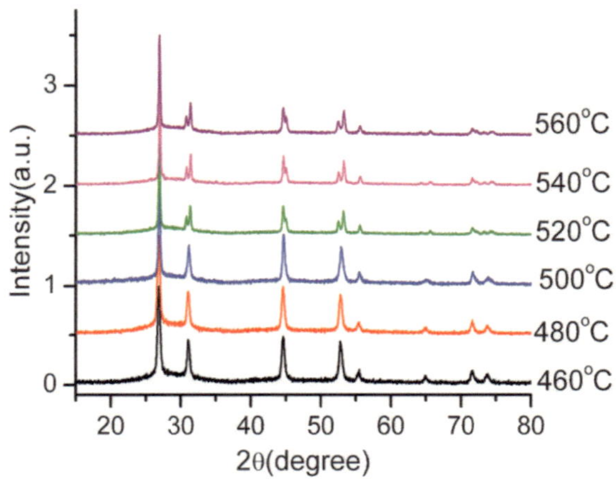

Figure 5. X-ray diffraction patterns of GCs obtained from the oxyfluoride glasses after thermal treatments at different temperatures [24]. Reproduced from Phys. Chem. Chem. Phys., 2011, 13(4), 1499. Copyright 2011, Royal Society of Chemistry.

The conclusion that the most commonly found trivalent RE site in CaF_2 is one of tetragonal C_{4v} (*4mm*) crystal-field symmetry. Since β-PbF_2 has a structure of CaF_2 (*i.e.*, fluorite structure), the formation mechanism of RE^{3+} doped crystal is clearly alike. Additionally, Beggiora *et al.* have proved that the F^- interstitial mechanism is energetically more favorable than Pb vacancy compensation as a charge compensation mechanism by computer simulation methods [26]. Mainly based on the points above, we put forward the structure displayed in Figure 6. RE^{3+} is substituted for Pb^{2+} at the center of the tetragonal lattice with four more interstitial F^- ions residing in the midpoint of the edges besides the original eight F^- in fluorite structure. This structure has a typical D_{4h} (*4/mmm*) point symmetry, which results in a higher symmetry, but includes all the symmetry operations of C_{4v} (*4mm*). The space group corresponds to *P4/mmm* (NO.123). Site symmetries and other atomic parameters are listed in Table 2 ("Ox." and "Wyck." are short for the oxidation number and the Wyckoff letter, respectively).

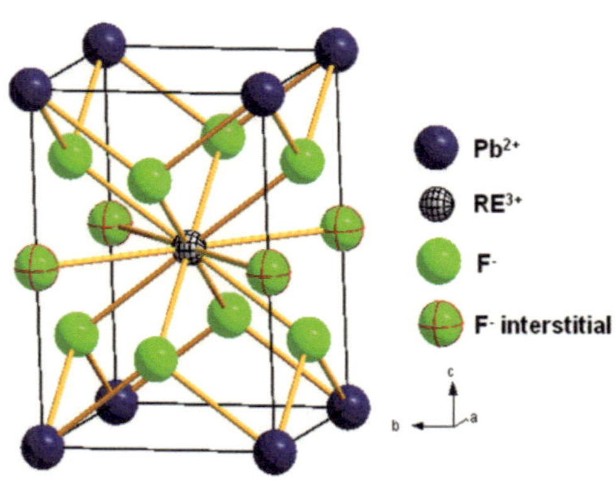

Figure 6. Stereoscopic view of structure of nano-particles embedded in glass ceramic [24]. Reproduced from Phys. Chem. Chem. Phys., 2011, 13(4), 1499. Copyright 2011, Royal Society of Chemistry.

Atom	Ox.	Wyck.	Site symmetry	x/a	y/b	z/c
Pb	2	1a	*4/mmm*	0	0	0
Yb	3	1d	*4/mmm*	1/2	1/2	1/2
F	-1	4i	*2/mm.*	0	1/2	1/4
F interstitial	-1	1b	*4/mmm*	0	0	1/2

Table 3. Atomic parameters of the structure model (Space-group:*P4/mmm* (NO.123)-tetragonal). Adapted from Ref. [24].

XRD patterns of the six GC samples as made were compared with the theoretical calculation by program Diamond 3.1d based on the structure proposed above. For high temperature group samples, take GC@540 °C for example, as shown in Figure 7(a), the lattice parameters were set as a = 4.026 Å, c = 5.793 Å. Since the calculation results can only be expressed as the form of vertical bar theoretically (see vertical bars in Figure 7(a), lower panel), the XRD trace (raw data) was converted to the form of integral intensities of peaks (*i.e.*, the peak area value, see vertical bars in Figure 7(a), upper panel) in order to compare with the theoretical results. Indexing according to the tetragonal model was also taken as exhibited in Figure 7(a), lower panel. As seen from Figure 7(a), the positions and integral intensities of peaks derived from experiment are consistent very well with those calculated based on the tetragonal model. Quantitative calculation indicates that the average relative errors between theoretical and experimental values are only 0.052 % and 2.363 % in 2θ and relative intensity for every peak, respectively.

Much the same, with GC@480 °C as the representative, Figure 7(b) shows the compared XRD data of the low temperature group samples. With the indexing of tetragonal, lattice parameters were set as a = 4.065 Å, c = 5.749 Å. Calculated average relative errors between theoretical and experimental values are only 0.068 % in 2θ and 2.199 % in relative intensity for every peak, respectively. These results quantitatively support the proposed structure.

To further confirm the validity of the proposed structural model, the XRD refinements of the NCs were made by the Rietveld method with the Fullprof program. Based on the Lattice parameters and the atomic coordinates adopted above, the scaling factors, zero corrections and peak shape parameters (including FWHM, asymmetry and pseudo-Voigt profile functions) were deduced. Figure 8(a) and Figure 8(b) show the final Rietveld plot of GC@480 °C and GC@540 °C respectively. Although the envelope lines due to the existing of glassy matrix and peaks broadening due to the small size of the nano-particles exert a serious impact on the results of fitting, the corresponding factors R_p(= 7.86%), R_{wp}(= 10.6%) for GC@480°C and R_p(= 8.76%), R_{wp}(= 12.4%) for GC@540 °C indicate a relatively good agreement between the experimental and calculated patterns.

3.2. Distribution of light RE ions and the core-shell structure

Precursor glasses (PGs) with the composition $44SiO_2$-$5Al_2O_3$-$40PbF_2$-$10CdF_2$-$1NdF_3$ (mole fraction) were prepared by traditional melting-quenching method. With the prepared precursor glasses, the corresponding transparent GCs were obtained by thermal treatment for 48

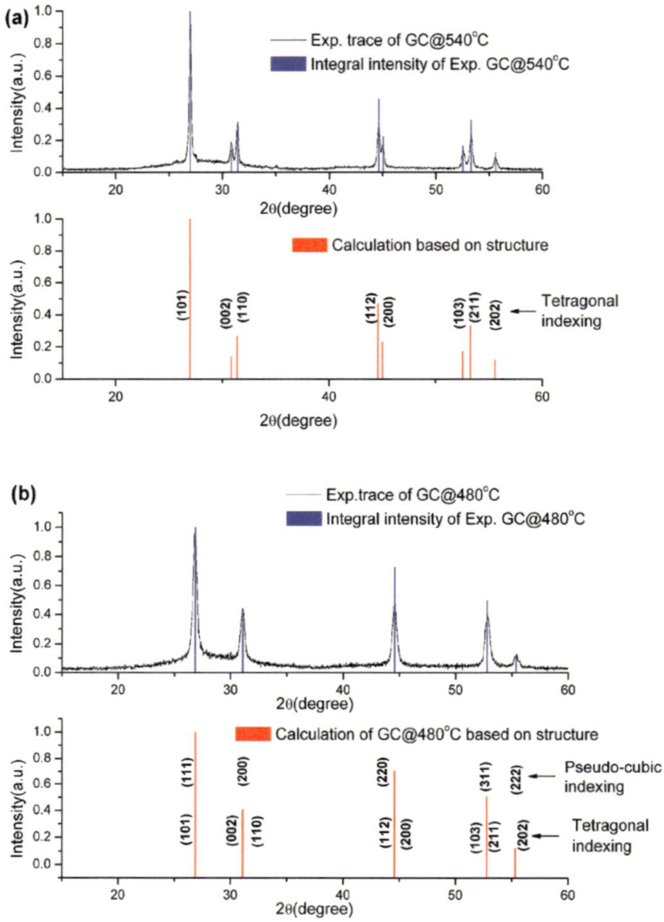

Figure 7. Quantitative comparison of theoretical and experimental X-ray diffraction patterns of glass ceramics thermal treated at (a) 540 °C and (b) 480 °C [24]. Reproduced from Phys. Chem. Chem. Phys., 2011, 13(4), 1499. Copyright 2011, Royal Society of Chemistry.

hours at 440 °C, which is determined by differential thermal analysis (DTA). With corrosion treatment, the corresponding fluoride nano-particles were obtained afterwards [16].

Figure 9(a) shows the diffraction patterns of compounds of GCs, along with the diffraction pattern of PGs without thermal treatment. Two curves show an overlap of the peaks corresponding to the amorphous matrix of Si-Al glass. The differences between PGs and GCs indicate that the crystallization process takes place in glass matrix under thermal treatment, and the characteristic diffraction peaks of pure cubic β-PbF_2 (JCPDS: 06-0251) emerge. XRD patterns of the GCs were compared with the theoretical calculation by program Diamond 3.1d based on the structure of pure cubic β-PbF_2. Since the calculation results can only be expressed in the form of a vertical bar theoretically (see vertical bars in Figure 9(c)), the XRD trace (raw data) was converted to the form of integral intensities of peaks (i.e., the peak area value, see

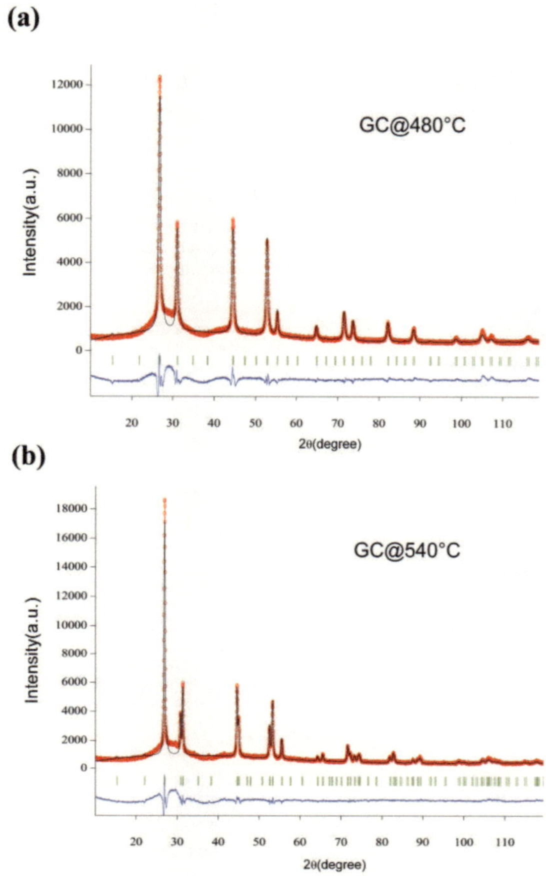

Figure 8. Rietveld analyses for X-ray diffraction patterns of (a) GC@480 °C and (b) GC@540 °C. The positions of the Bragg reflections are indicated by vertical bars (|). The difference between the experimental (small circles) and the calculated (solid line) intensities from the model is shown by the plot in the lower part of the diagram [24]. Reproduced from Phys. Chem. Chem. Phys., 2011, 13(4), 1499. Copyright 2011, Royal Society of Chemistry.

vertical bars in Figure 9(b)) in order to compare with the theoretical results. Indexing was also taken as exhibited in Figures 9(b) and (c). As seen in Figures 9(b) and (c), the positions and integral intensities of peaks derived from experiment are very consistent with the calculation based on the structure of pure cubic β-PbF_2. Quantitative calculation indicates that the average relative errors between the theoretical and experimental values are only 0.98% and 4.37% in 2θ and the relative intensity for every peak, respectively. In order to further confirm the validity of the nanocrystals structure, pure cubic β-PbF_2, X-ray data refinement was made using the program MAUD. The analysis was started assuming the structure of β-PbF_2 (F_{m3m} NO. 225) for the nanocrystal phase; and the structure of cubic amorphous silica-Al glass (P_{213} NO. 198) for the amorphous phase. The used crystalline structures can all be found in the database of the MAUD program. The corresponding factors R_{wp} (= 4.69%), R_{exp} (= 4.27%) indicate a relatively good agreement between the experimental and calculated patterns.

According to the results of obtained quantitative analysis, about 8.6 mol% to 13.5 mol% of the original Pb would contribute to form crystalline β-PbF_2 during the thermal treatment. In addition, the structure cell parameter of nanocrystal phase corresponds to that of pure cubic β-PbF_2, which further checks the structure of nanocrystal phase.

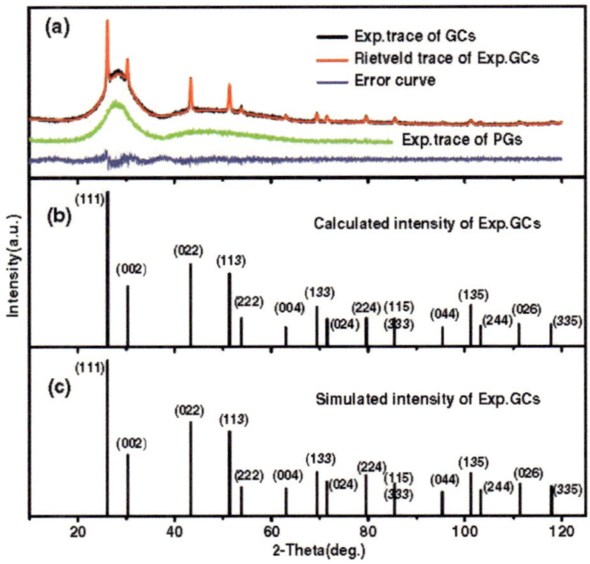

Figure 9. (a) X-ray diffraction patterns of GCs and PGs, Rietveld analyses for XRD pattern of GCs and the error curve. The calculated line spectrum from peak area of GCs (b) and the simulated line spectrum of GCs from Diamond program (c) are presented for comparison [25]. Reproduced from Nanoscale Res. Lett., 2012, 7(1), 275. Copyright 2012, Springer.

The Nd^{3+} ions could not be incorporated into nanocrystal phase by reason that the structure of nanocrystal phase is pure cubic β-PbF_2. Corrosion-treated methods were applied to acquire the NCs so as to explore the information on chemical compositions and structures of NCs and weaken the interference of glass matrix. For Nd^{3+}-doped samples, an unusual change happens when the nanocrystals embedded in glass matrix abstracted by etching process. Unlike our pervious work [24] of Er^{3+}/Yb^{3+}-codoped GCs that only β-PbF_2 phase was observed, the complex mixture were obtained after etching off the glass matrix of GCs doped with Nd^{3+}. With peak indexing, the additional intense diffraction peaks appear to be NdF_3 (JCPDS: 09-0416) and Pb_3AlF_9 H_2O phase (JCPDS: 45-1458). The new crystal phase NdF_3 generates in the course of the etching treatment, namely Nd^{3+} ions existed in glass matrix reacting with F^- ions in acid generates NdF_3. Likewise, Pb and Al ions of glass matrix reacting with F^- ions in acid generates Pb_3AlF_9 H_2O phase. MAUD program based on Rietveld method was used to calculate the relative amounts of β-PbF_2 and NdF_3 phases. The crystal structure information of the concerned phases is acquired from the Crystallography Open Database (http://www.crystallography.net/). The refinement results and difference plot for the observed and calculated patterns of NPs are shown in Figure 10. The refining factors of R_{wp} (= 8.07%) and R_{exp} (= 4.02%) indicate a good agreement between the experimental and calculated patterns.

The structure cell parameter of β-PbF_2 obtained from refinement of NCs (0.5890 nm) and GCs (0.5898 nm) is nearly the same as that of pure β-PbF_2 (0.5927 nm), which further verifies the existence of Nd^{3+} outside the NCs. According to the else relative results obtained from Rietveld analysis, the cell parameters of the other two phases are nearly the same as that of the standard structures separately, which further certifies the authenticity of the complex phases suggested above. From the results of the relative weight fraction obtained from refinement, moreover, through the sample arithmetic conversion, we can acquire that about 10.9 mol% the Pb of raw material contribute to form crystalline β-PbF_2 phase during the thermal treatment. This conclusion is in accordance with Rietveld analysis data from Figure 9.

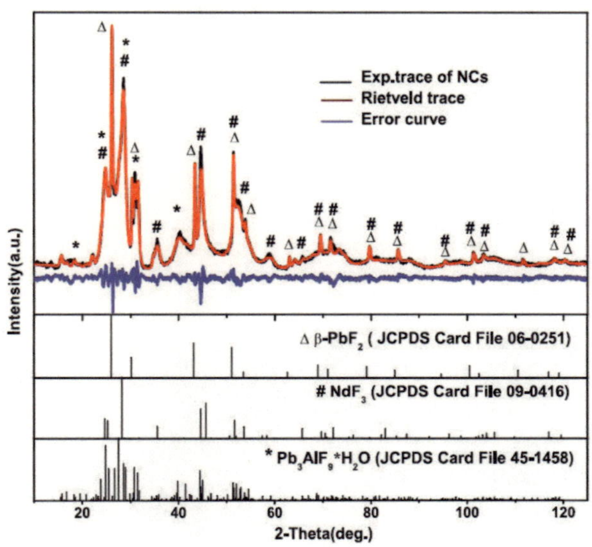

Figure 10. Rietveld analyses for XRD pattern of NCs and the error curve are presented. The line spectrum of β-PbF_2, NdF_3 and Pb_3AlF_9 H_2O are revealed for comparison [25]. Reproduced from Nanoscale Res. Lett., 2012, 7(1), 275. Copyright 2012, Springer.

High resolution TEM micrographs of GCs and NCs are displayed in Figure 11, providing a visual characterization of the samples and a directly analysis method. In GCs, as shown in Figure 11(a), only spherical β-PbF_2 nanocrystals with an average size about 22 nm distribute in glass matrix, which has a good consistency with the calculated size by using the Scherrer formula. However, in NCs, besides the 22 nm-nanocrystals already existed in GCs, as shown in Figure 11(b), a large amount of additional crystals and flake-like substrates also emerge. Particularly, most of these additional crystals assemble around the known β-PbF_2 nanocrystals, forming a core-shell-like structure. According to the HRTEM micrographs, the periodic arrayed crystal planes can be viewed clearly and part of symbolic interplanar distance (*d*) values, which measured accurately for both GC and NCs, have been labeled. The measured values of *d*, 0.294nm or 0.295nm for both Figures 11(a) and (b), are attributed to crystal plane (002) of pure cubic β-PbF_2 that as the core, while the *d* values 0.303nm and 0.316nm around β-PbF_2 nanocrystal belong to crystal planes (020) and (-121) of shell NdF_3 respectively, and the other one *d* value 1.141nm is for the symbolic crystal plane (001) of the substrates Pb_3AlF_9 H_2O. This conclusion is in accordance with the XRD refinement result of NCs, which has been displayed in Figure 10 above. On the basis of Rietveld refining results and HRTEM micrographs, the Nd^{3+} ions are reserved in glass matrix after thermal treatment and aggregated outside β-PbF_2 particles in the form of NdF_3 crystals when glass matrix are removed through the etching process. Therefore, a simplified model is proposed to describe the distribution of Nd^{3+} ions both in GCs and NCs, as shown in the insets of Figures 11(a) and (b). That is to say, with thermal treatment on PGs, the phase segregation process happens and on the one hand, as shown in the inset of Figure 11(a), Nd^{3+} ions with a inhomogeneous distribution form an aggregate structure, which has been reported on previous research [41,42], on the other hand Pb and F element get together to form high temperature phase β-PbF_2 nanocrystals, while the formation of Nd^{3+} clusters prevents the process of incorporation into β-PbF_2 phase. During the etching process for removing glass matrix, the Nd^{3+} ions meet F^- ions in the acid solution and generate NdF_3 crystals. And then the large quantities of NdF_3 crystals adsorb the surface of the β-PbF_2 nanoparticles. The model for NCs is simplified as a core-shell structure for further analysis, viz. β-PbF_2 particle as core and the covering NdF_3 crystals as shell, as shown in the inset of Figure 11(b). This rough simplification is acceptable considering the quantitative analysis result of NCs.

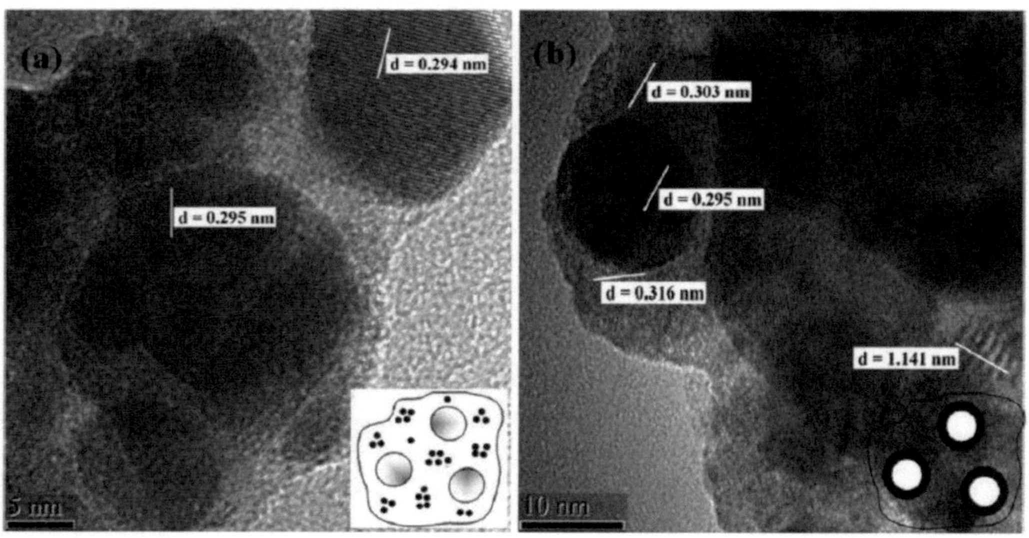

Figure 11. High resolution TEM micrographs with the marked interplanar distance *d* values of GCs (a) and NCs (b). The insets are the simplified models for GCs and NCs, where the large white spheres stand for β-PbF_2 nano-particles and small black spheres in GCs are for Nd^{3+} ions [25]. Reproduced from Nanoscale Res. Lett., 2012, 7(1), 275. Copyright 2012, Springer.

In order to verify the anticipation, EDS line scans in HAADF mode across the β-PbF_2 particles are carried out on a Tecnai G2F30 FE-TEM to offer signal intensity values for different elements. The line scanning paths are shown in Figures 12(a) and (b) and the results are shown in Figures

12(e) and (f). In GCs, the detected trace of Nd in EDS like that of Cd indicates that nanocrystalline phase is pure β-PbF_2, rather than β-PbF_2:RE, as previously suggested by many researchers. Even the trace of Nd in EDS of the other samples doped different concentration of Nd, the former researches confirmed that Cd hardly congregated in the nanocrystals, which also indicated that none of Nd ions were incorporated into β-PbF_2 crystalline phase [24,26,39]. Figure 12(c) shows EDS line scan for the ideal spheric body model of crystalline β-PbF_2 in GCs. As an ideal sphere model, dot *A* is for the center of it. Axial direction (*Z*) is the orientation of EDS line scan. Dot *O* and length *x* are the initial point and the distance of line scan separately. From Figure 12(c), we can get the relation formula the content of Pb satisfies:

$$Y_1 \propto 2\sqrt{R^2-(R-x)^2} \tag{1}$$

Where Y_1 is the relative intensity of element Pb, *R* is the half size of β-PbF_2 nanocrystal and *x* is the length of line scan. Through conventional mathematical analysis, simulated curve is shown in Figure 12(e), which is in accordance with the fitting curve of experimental measurement. In the same way, the ideal core-shell structure model of NCs, which is shown in Figure

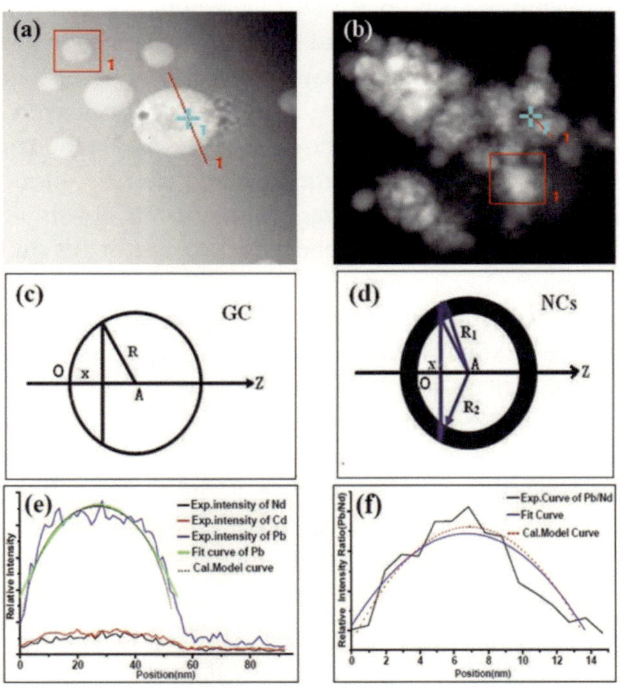

Figure 12. EDS line scan graphs in HAADF mode (a) and (b); The two pictures (c) and (d) are EDS line scan for the ideal models of crystalline β-PbF_2 and core-shell structure in NCs, respectively; The relative element signal intensity of Pb, Nd and Cd in GC, the fit curve and the calculated model curve of element Pb are presented for comparison (e); The fit curve of relative element intensity ratio Pb/Nd in NCs and the calculated core-shell model curve are also showed (f) [25]. Reproduced from Nanoscale Res. Lett., 2012, 7(1), 275. Copyright 2012, Springer.

12(d), is analyzed by conventional mathematical simulation, as shown in Figure 12(f). From Figure 12(d), the signal intensity ratio of the content of Pb and Nd can be expressed by following formula:

$$Y_2 \propto \frac{\sqrt{R_2^2-(R_2-x)^2}}{\sqrt{R_1^2-(R_2-x)^2}-\sqrt{R_2^2-(R_2-x)^2}} \tag{2}$$

Where Y_2 is the relative element intensity ratio Pb/Nd, R_1 and R_2 is the half size of core-shell model and β-PbF_2 nanocrystal respectively, and x is the length of line scan. The accordance with experimental results, as shown in Figures 12(e) and (f), proves the rationality of the model. Therefore, it is concluded that all of the Nd^{3+} ions locate in glass matrix, while few of the Nd^{3+} ions dope into crystalline phase.

4. Conclusions

An innovative route to fabricate nano-particles from oxyfluoride glass by the thermal induction and corrosion treatment was described. Although thermal induction and corrosion treatment were used to prepare β-PbF_2:RE^{3+} nano-particles in aqueous solution in this chapter, it may directly apply to any other silicate glass ceramics doped with other RE ions and embedded various composition of nano-particles, since the preparation method is based on a fundamental consideration. After obtaining the free and pure nano-particles, the structure of nano-particles can be studied directly. A tetragonal phase model with the chemical formula $PbREF_5$ has been proposed for heavy RE-doped nano-particles in this study. The corresponding space-group *P4/mmm* (NO.123) are proved by quantitative EDS and XRD analyses. Two specific crystalline phases with the same space group have been observed at 460 °C-500 °C and 520 °C-560 °C, respectively. Moreover, a super "pseudo-cubic" cell based on our tetragonal model may give a good explanation to the probable previous cubic-symmetry-misunderstanding by researchers. The RE-doping mechanism is considered to be RE^{3+} replacement of Pb^{2+} as well as interstitial F^- charge compensation. However, unlike the study of heavy RE ions doped systems, core-shell-like structure of pure β-PbF_2 surface absorbed NdF_3 crystals generated after thermal treatment for light RE ions (Nd^{3+}) doped nano-particles. This study has paved a way to more comprehensively understand the properties of the light RE ions doped oxyfluoride glass. The results here would benefit further research on the optical properties and applications of such materials. Recent years, our group has paid much attention on the research of the fabrication of Ln^{3+}-doped nano-particles from oxyfluoride glass by the thermal induction and corrosion treatment, and has done much work on the structural investigations of the obtained nano-particles. In the present and future research, to broaden the application of this material, we will pay much more efforts on the dispersion and surface modification of nano-particles during or after the corrosion process to fabricate the transparent luminescent collosol. And so far, we have made some progress on this work.

http://dx.doi.org/10.5772/57213

Acknowledgements

This work is supported by National Science Fund for Talent Training in Basic Sciences under Grant No.J1103208. The authors are also grateful for the fruitful discussions with Nan Hu, Ming Zhang and Xinxing Zhang.

Author details

Hui Guo[1], Yu Hua[1], Lijuan Zhao[1,2] and Yiming Li[1]

1 The Key Laboratory of Weak-Light Nonlinear Photonics, Ministry of Education, School of Physics, Nankai University, Tianjin, China

2 Applied Physics School of TEDA, Nankai University, Tianjin, China

References

[1] Chatterjee, D. K.; Gnanasammandhan, M. K.; and Zhang, Y. Small Upconverting Fluorescent Nanoparticles for Biomedical Applications, *Small* 2010; 6(24) 2781-2795.

[2] Chen, J.; Guo, C. R.; Wang, M.; Huang, L.; Wang, L. P.; Mi, C. C.; Li, J.; Fang, X. X.; Mao, C. B.; and Xu, S. K. Controllable Synthesis of $NaYF_4$:Yb,Er Upconversion Nanophosphors and their Application to *invivo* Imaging of *Caenorhabditis elegans*, *J. Mater. Chem.* 2011; 21(8) 2632-2638.

[3] Auzel, F. Upconversion and Anti-Stokes Processes with f and d Ions in Solids, *Chem. Rev.* 2004; 104(1) 139-174.

[4] Mai, H. X.; Zhang, Y. W.; Si, R.; Yan, Z. G.; Sun, L. D.; You, L. P.; and Yan, C. H. High-Quality Sodium Rare-Earth Fluoride Nanocrystals: Controlled Synthesis and Optical Properties, *J. Am. Chem. Soc.* 2006; 128(19) 6426-6436.

[5] Kumar, B. R.; Nyk, M.; Ohulchanskyy, T. Y.; Flask, C. A.; and Prasad, P. N. Combined Optical and MR Bioimaging Using Rare Earth Ion Doped $NaYF_4$ Nanocrystals, *Adv. Funct. Mater.* 2009; 19(6) 853-859.

[6] Lin, C. C.; Xiao, Z. R.; Guo, G. Y.; Chan, T. S.; and Liu, R. S. Versatile Phosphate Phosphors $ABPO_4$ in White Light-Emitting Diodes: Collocated Characteristic Analysis and Theoretical Calculations, *J. Am. Chem. Soc.* 2010; 132(9) 3020-3028.

[7] Wang, F.; Han, Y.; Lim, C. S.; Lu, Y. H.; Wang, J.; Xu, J.; Chen, H. Y.; Zhang, C.; Hong, M. H.; and Liu, X. G. Simultaneous Phase and Size Control of Upconversion Nanocrystals Through Lanthanide Doping, *Nature* 2010; 463 1061-1065.

[8] Wang, J. W.; Hao, J. H.; and Tanner, P. A. Upconversion Luminescence of an Insulator Involving a Band to Band Multiphoton Excitation Process, *Opt. Express* 2011; 19(12) 11753-11758.

[9] Haase, M.; and Schäfer, H. Nanopartikel für die Aufwärtskonversion, *Angew. Chem.* 2011; 123(26) 5928-5950.

[10] Wang, Y.; and Ohwaki, J. New Transparent Vitroceramics Codoped with Er^{3+} and Yb^{3+} for Efficient Frequency Upconversion, *Appl. Phys. Lett.* 1993; 63(24) 3268-3270.

[11] Tick, P. A.; Borrellia, N. F.; Cornelius, L. K.; and Newhouse, M. A. Transparent Glass Ceramics for 1300 nm Amplifier Applications, *J. Appl. Phys.* 1995; 78(11) 6367-6374.

[12] Driesen, K.; Tikhomirov, V. K.; Görller-Walrand, C.; Rodríguez, V. D.; and Seddon, A. B. Transparent Ho^{3+}-Doped Nano-Glass-Ceramics for Efficient Infrared Emission, *Appl. Phys. Lett.* 2006; 88(7) 073111.

[13] Tikhomirov, V. K.; Driesen, K.; Görller-Walrand, C.; and Mortier, M. Broadband Telecommunication Wavelength Emission in Yb^{3+}-Er^{3+}-Tm^{3+} Co-Doped Nano-Glass-Ceramics, *Opt. Express* 2007; 15(15) 9535-9540.

[14] Tikhomirov, V. K.; Driesen, K.; and Görller-Walrand, C. Low-Energy Robust Host Heavily Doped with Dy^{3+} for Emission at 1.3 to 1.4 μm, *Phys. Status Solidi A* 2007; 204(3) 839-845.

[15] Mortier, M.; and Patriarche, G. Oxide Glass Used as Inorganic Template for Fluorescent Fluoride Nanoparticles synthesis, *Opt. Mater.* 2006; 28(12) 1401-1404.

[16] Yu, H.; Hu, N.; Wang, Y. N.; Wang, Z. L.; Gan, Z. S.; and Zhao, L. J. The Fabrication of Nano-Particles in Aqueous Solution from Oxyfluoride Glass Ceramics by Thermal Induction and Corrosion Treatment, *Nanoscale Res. Lett.* 2008; 3(12) 516-520.

[17] Lehmann, O.; Kompe, K.; and Haase, M. Synthesis of Eu^{3+}-Doped Core and Core/Shell Nanoparticles and Direct Spectroscopic Identification of Dopant Sites at the Surface and in the Interior of the Particles, *J. Am. Chem. Soc.* 2004; 126(45) 14935-14942.

[18] Hao, S. W.; Sun, L.; Chen, G. Y.; Qiu, H. L.; Xu, C.; Soitah, T. N.; Sun, Y.; and Yang, C. H. Synthesis of Monoclinic Na_3ScF_6:1 mol% Er^{3+}/2 mol% Yb^{3+} Microcrystals by a Facile Hydrothermal Approach, *J. Alloys Compd.* 2012; 522 74-77.

[19] Werts, M. H. V.; Jukes, R. T. F.; and Verhoeven, J. W. The Emission Spectrum and the Radiative Lifetime of Eu^{3+} in Luminescent Lanthanide Complexes, *Phys. Chem. Chem. Phys.* 2002; 4(9) 1542-1548.

[20] Boyer, J. C.; Vetrone, F.; Capobianco, J. A.; Speghini, A.; and Bettinelli, M. Variation of Fluorescence Lifetimes and Judd-Ofelt Parameters between Eu^{3+} Doped Bulk and Nanocrystalline Cubic Lu_2O_3, *J. Phys. Chem. B* 2004; 108(52) 20137-20143.

[21] Cross, A. M.; May, P. S.; Van Veggel F. C. J. M.; and Berry, M. T. Dipicolinate Sensitization of Europium Luminescence in Dispersible 5%Eu:LaF_3 Nanoparticles, *J. Phys. Chem. C* 2010; 114(35) 14740-14747.

[22] Wang, Y. H.; Liu, Y. S.; Xiao, Q. B.; Zhu, H. M.; Li, R. F.; and Chen, X. Y. Eu^{3+} Doped KYF_4 Nanocrystals: Synthesis, Electronic Structure, and Optical Properties, *Nanoscale* 2011; 3(8) 3164-3169.

[23] Kar, A.; and Patra, A. Impacts of Core-Shell Structures on Properties of Lanthanide-Based Nanocrystals: Crystal Phase, Lattice Strain, Downconversion, Upconversion and Energy Transfer, *Nanoscale* 2012; 4(12) 3608-3619.

[24] Hu, N.; Yu, H.; Zhang, M.; Zhang, P.; Wang, Y. Z.; and Zhao, L. J. The Tetragonal Structure of Nanocrystals in Rare-Earth Doped Oxyfluoride Glass Ceramics, *Phys. Chem. Chem. Phys.* 2011; 13(4) 1499-1505.

[25] Yu, H.; Guo, H.; Zhang, M.; Liu, Y.; Liu, M.; and Zhao, L. J. Distribution of Nd^{3+} Ions in Oxyfluoride Glass Ceramics, *Nanoscale Res. Lett.* 2012; 7(1) 275.

[26] Beggiora, M.; Reaney, I. M.; and Islam, M. S. Structure of the Nanocrystals in Oxyfluoride Glass Ceramics, *Appl. Phys. Lett.* 2003; 83(3) 467-469.

[27] Méndez-Ramos, J.; Lavín, V.; Martín, I. R.; Rodríguez-Mendoza, U. R.; Rodríguez, V. D.; Lozano-Gorrín, A. D.; and Núñez, P. Site Selective Study of Eu^{3+}-Doped Transparent Oxyfluoride Glass Ceramics, *J. Appl. Phys.* 2003; 94(4) 2295-2301.

[28] Luo, W. Q.; Wang, Y. S.; Cheng, Y.; Bao, F.; and Zhou, L. H. Crystallization and Structural Evolution of YF_3-SiO_2 Xerogel, *Mater. Sci. Eng. B* 2006; 127(2-3) 218-223.

[29] Yu, H.; Zhao, L. J.; Meng, J.; Liang, Q.; Yu, X. Y.; Tang, B. Q.; and Xu, J. J. Nanocrystal Formation and Structure in Oxyfluoride Glass Ceramics, *Chin. Phys.* 2005; 14(9) 1799-1802.

[30] Yu, H.; Zhao, L. J.; Liang, Q.; Meng, J.; Yu, X. Y.; Tang, B. Q.; Tang, L. Q.; and Xu, J. J. Red Up-Conversion Luminescence Process in Oxyfluoride Glass Ceramics Doped with Er^{3+}/Yb^{3+}, *Chin. Phys. Lett.* 2005; 22(6) 1500-1503.

[31] Lehmann, O.; Kömpe, K.; and Haase, M. Synthesis of Eu^{3+}-Doped Core/Shell Nanoparticles and Direct Spectroscopic Identification of Dopant Sites at the Surface and in the Interior of the particles, *J. Am. Chem. Soc.* 2004; 126(17) 14935-14942.

[32] Méndez-Ramos, J.; Lavín, V.; Martín, I. R.; Rodríguez-Mendoza, U. R.; Rodríguez, V. D.; Lozano-Gorrín, A. D.; and Núñez, P. Role of the Eu^{3+} Ions in the Formation of Transparent Oxyfluoride Glass Ceramics, *J. Appl. Phys.* 2001; 89(10) 5307-5310.

[33] Tikhomirov, V. K.; Furniss, D.; Seddon, A. B.; Reaney, I. M.; Beggiora, M.; Ferrari, M.; Montagna, M.; and Rolli, R. Fabrication and Characterization of Nanoscale, Er^{3+}-Doped, Ultratransparent Oxy-Fluoride Glass Ceramics, *Appl. Phys. Lett.* 2002; 81(11) 1937-1939.

[34] Driesen, K.; Tikhomirov, V. K.; and Görller-Walrand, C. Eu^{3+} as a Probe for Rare-Earth Dopant Site Structure in Nano-Glass-Ceramics, *J. Appl. Phys.* 2007; 102(2) 024312.

[35] Mortier, M.; and Patriarche, G. Structural Characterisation of Transparent Oxyfluoride Glass-Ceramics, J. Mater. Sci. 2000; 35(19) 4849-4856.

[36] Mortier, M.; Goldner, P.; Chateau, C.; and Genotelle, M. Erbium Doped Glass-Ceramics: Concentration Effect on Crystal Structure and Energy Transfer between Active Ions, *J. Alloys Compd.* 2001; 323&324 245-249.

[37] Tyagi, A. K.; Patwe, S. J.; Achary, S. N.; and Mallia, M. B. Phase Relation Studies in $Pb_{1-x}M'_xF_{2+x}$ Systems (0.0≤x≤1.0; M'=Nd^{3+}, Eu^{3+} and Er^{3+}), *J. Solid State Chem.* 2004; 177(4-5) 1746-1757.

[38] Ge, J.; Zhao, L. J.; Guo, H.; Lan, Z. J.; Chang, L. F.; Li, Y. M.; and Yu, H. Structure and Distortion of Lead Fluoride Nanocrystals in Rare Earth Doped Oxyfluoride Glass Ceramics, *Phys. Chem. Chem. Phys*. 2013; 15(40) 17281-17286.

[39] Tikhomirov, V. K.; Mortier, M.; Gredin, P.; Patriarche, G.; Görller-Walrand, C.; and Moshchalkov, V. V. Preparation and Up-Conversion luminescence of 8 nm Rare-Earth Doped Fluoride Nanoparticles, *Opt. Express* 2008; 16(19) 14544-14549.

[40] Silva, M. A. P.; Dantelle, G. ; Mortier, M. ; Monteil, A.; Ribeiro, S. J. L. ; Messaddeq, Y. ; Briois, V. ; and Poulain, M. Local Order Around Rare Earth Ions During the Devitrification of Oxyfluoride Glasses, *J. Chem. Phys.* 2008; 128(24) 244516.

[41] Lavin, V.; Iparraguirre, I.; Azkargorta, J.; Mendioroz, A.; Gonzalez-Platas, J.; Balda, R.; and Fernandez, J. Stimulated and Upconverted Emissions of Nd^{3+} in a Transparent Oxyfluoride Glass-Ceramic, *Opt Mater* 2004; 25 201-208.

[42] Mendez-Ramos, J.; Abril, M.; Martin, I. R.; Rodriguez-Mendoza, U. R.; Lavin, V.; Rodriguez, V. D.; Nunez, P.; and Lozano-Gorrin, A. D. Ultraviolet and Visible Upconversion Luminescence in Nd^{3+}-Doped Oxyfluoride Glasses and Glass Ceramics Obtained by Different Preparation Methods, *J Appl Phys* 2006; 99 113511-113515.

Chapter 7

Recent Advances in Computational Design of Organic Materials for Corrosion Protection of Steel in Aqueous Media

Ime Bassey Obot

Additional information is available at the end of the chapter

http://dx.doi.org/10.5772/57245

1. Introduction

1.1. Cost of corrosion

Cost of corrosion studies have been undertaken by several countries including the United States of America, the United Kingdom, Germany, Japan, India and China. In the United States (USA) the awareness of the cost of corrosion has been maintained at a high level. Professor Uhlig; the National Bureau of Standards (NBS), now the National Institute of Standards and Technology (NIST); Battelle Columbus Laboratories (BCL); and CC Technologies Laboratories (CC), along with NACE International and the Federal Highway Administration (FHWA), have contributed to the current knowledge on the cost of corrosion. In the United Kingdom (U.K.), the Hoar Committee has played a stellar role in estimating the cost of corrosion. In Japan, three approaches (Uhlig method, Hoar method, and NBS-BCL model) have been used. In India, the Uhlig method was followed in 1958, while the NBS-BCL model was adopted in 1986. Countries have attempted to relate the cost of corrosion to their gross national product (GNP). The common findings of these studies revealed that the annual cost of corrosion ranged from approximately 1 to 5 percent of Gross National Product (GDP) of each nation. In a widely-cited study (NACE Corrosion Costs Study) by the National Association of Corrosion Engineers, NACE, the direct cost of corrosion in the U.S. was estimated to equal $276 Billion in 1998, approximately 3.1 % of GDP (Fig.1) [1]. A significant milestone in the effect of corrosion on the U.S. economy occurs in 2012 when the total cost of corrosion in the US exceeds $1 trillion annually for the first time [2]. The annual cost of corrosion worldwide is estimated at $ 2.2 Trillion (2010), which is about 3 % of the world's gross domestic product (GDP) of $ 73.33 Trillion [3]. This amount is staggering and could be use to provide food and basic social amenities to the world's poor which is part of the millennium development goals (MDGs) target by 2015.

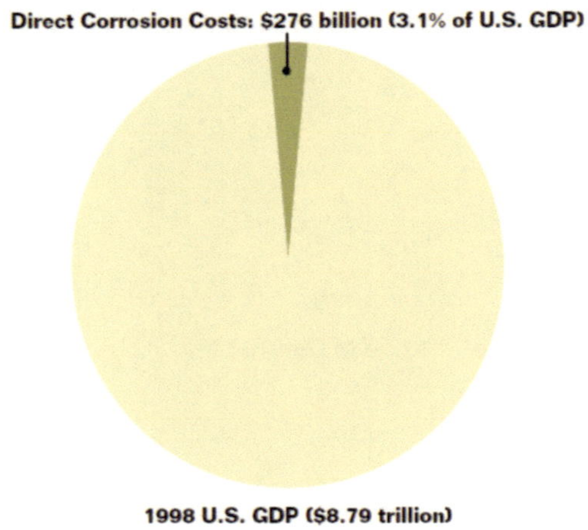

Figure 1. The impact of corrosion on the US economy.

1.2. The corrosion process

Corrosion can be defined in many ways. Some definitions are very narrow and deal with a specific form of corrosion, while others are quite broad and cover many forms of deterioration. The word corrode is derived from the Latin *corrodere,* which means "to gnaw to pieces." The general definition of corrode is to eat into or wear away gradually, as if by gnawing. For the purpose of this chapter, corrosion may be defined as the deterioration of a metal or its properties due to its reaction with its environment. The environment consists of the entire surrounding in contact with the material (the emphasis in this chapter is on aqueous corrosion or corrosion in environments where water is present). It is the process that accounts for the reverse extractive metallurgy in which metals or metal components deteriorate upon reaction with harsh environments. The reaction may be chemical, electrochemical and/or mechanical. Corrosion is an undesirable process and represents the tendency of pure metals and alloys to return to thermodynamically more stable compounds. Thus, according to Shier, et al. [4], corrosion is defined as the reaction of an engineering material with its environment leading to a consequent deterioration in properties of material. Corrosion, its control and prevention is an interdiscplinary science which is served by many disciplines including Metallurgy, Electrochemistry, Material Science, Physics, Computational Chemistry and Chemical Engineering.

1.3. Corrosion mechanism

Corrosion processes are usually electrochemical in nature, having the essential features of a battery. When metal atoms are exposed to an environment containing water molecules, they can give up electrons, becoming themselves positively charged ions provided an electrical circuit can be completed. Electrolytic corrosion consists of two partial processes: anodic (oxidation) and cathodic (reduction) reaction with consequent electron transfer between the two reactants. The two reactions occur at the same time and simultaneously upon a metal's surface. Thus, the metal is electrically charged and the rate of oxidation equals that of reduction. The loss of electron that is the usual manifestation of the corrosion process is a result of the anodic reaction (oxidation) in which the metal, M, loses electron (s) enter into ionic state as given in the equation below:

$$M \rightarrow M^{n+} + ne^- \quad (1)$$

The anodic reaction may occur uniformly over the metal surface or may be localized at a specific region. If the dissolved metal ion can react with the solution to form an insoluble compound, then a build-up of corrosion products may accumulate at the anodic site.

In the absence of applied voltage, the electron(s) generated by the anodic reaction are consumed by the cathodic reaction, which is either hydrogen evolution reaction or the oxygen-reduction reaction.

The equation for the hydrogen-evolution reaction, is written as shown below:

$$2H^+ + 2e^- \rightarrow H_2 \quad (2)$$

This is the predominant cathodic reaction when corrosion occurs in low pH. The hydrogen evolution reaction can cause a serious problem since atomic hydrogen may enter the metal and cause embrittlement. Another important cathodic reaction is usually predominant in the neutral and alkaline media. The rate of this reaction depends upon the supply of oxygen to the metal surface. The equation for these reactions are as follows:

$$O_2 + 4H^+ + 4e^- \rightarrow 2H_2O \quad (3)$$

$$O_2 + 2H_2O + 4e^- \rightarrow 4OH^- \quad (4)$$

Equs (3) and (4) represent the cathodic reaction in neutral and basic media respectively.

1.3.1. Uniform corrosion

Uniform attack or general corrosion occurs on the surface of a metal in the following way: If a metal is in contact with a conductive solution, some areas may become either anodic or cathodic sites. On the anode, the oxidation of the metal will occur, and the electrons supplied will be taken up at the cathode through the metal itself. Small currents will then flow from one to the other. The ions produced at the anode will migrate through the solution towards the cathode, where they will form an oxide film with the product of the oxygen reduction. The location of the anode and cathode changes at all times, which makes the whole surface of the metal to corrode (Fig. 2).

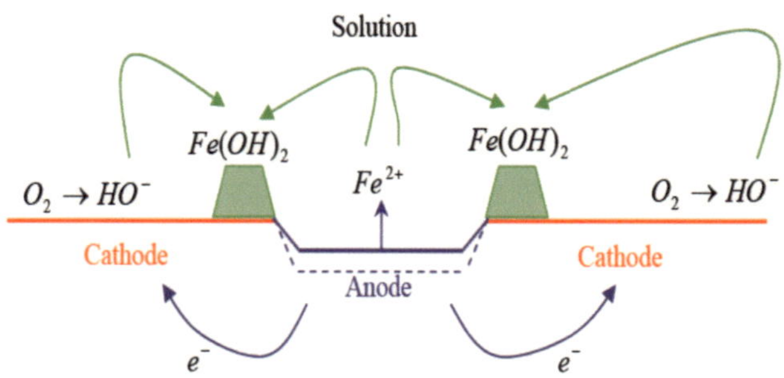

Figure 2. Uniform corrosion process of iron

1.3.2. Pitting

Pitting is a local form of corrosion (Fig. 3). This may occur at the surface of a metal due to the presence of an inhomogeneity (inclusion or break in a passive film). For this reason, aluminium alloys and stainless steel, which form passive films, are more likely to suffer from pitting. The cathode/anode area ratio as well as the pH will influence the process.

- In the pit (the anode) the oxidation of the metal takes place, releasing metal ions and electrons. The electrons then migrate to the cathode through the metal while the ions go into the solution. The presence of water then induces a further reaction known as hydrolysis:

$$Fe \rightarrow Fe^{2+} + 2e^{-} \tag{5}$$

$$3Fe^{2+} + 4H_2O \rightarrow Fe_3O_4 + 8H^{+} + 2e^{-} \tag{6}$$

- On the surface of the metal (the cathode), reduction of oxygen takes place:

$$O_2 \rightarrow OH^{-} \tag{7}$$

Due to the hydrolysis of the metal ions into the pit, the conditions becomes acidic and the walls can not repassivate. The process is autocatalytic. Since the process is localized, the anode and cathode do not change their location. The pit itself will be small compared with the surface area of the metal. The anodic site will be small compared with the cathodic site, and the metal loss in the pit will be high. For this reason, the process is said to be under cathodic control, inducing a high level of penetration.

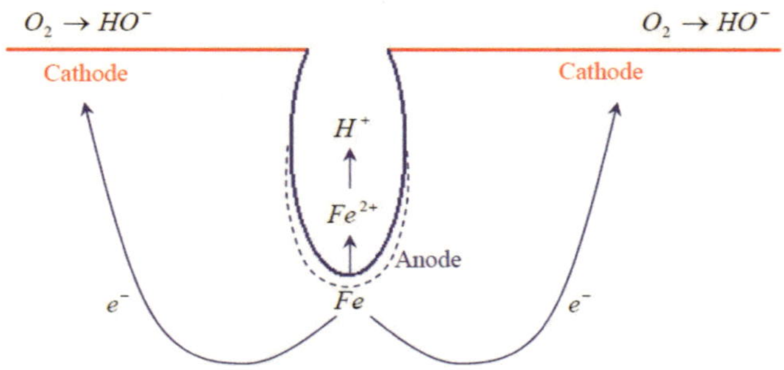

Figure 3. Pitting corrosion process of iron.

1.4. Corrosion inhibitors

A corrosion inhibitor may be defined, in general terms, as a substance that when added in a small concentration to an environment effectively reduces the corrosion rate of a metal exposed to that environment. Inhibition is used internally with carbon steel pipes and vessels as an economic corrosion control alternative to stainless steels and alloys, coatings, or non-metallic composites. A particular advantage of corrosion inhibition is that it often can be implemented or changed *in situ* without disrupting a process. The major industries using corrosion inhibitors are oil and gas exploration and production, petroleum refining, chemical manufacturing, heavy manufacturing, water treatment, and the product additive industries. The total consumption of corrosion inhibitors in the United States has doubled from approximately $600 million in 1982 to nearly $1.1 billion in 1998 [5]. US demand for corrosion inhibitors alone is forecast to rise 4.1 percent per year to $2.5 billion in 2017, with volume demand approaching 1.7 billion pounds. Growth will be driven by higher oil and natural gas output, particularly from shale formations, as well as by increasing chemical production and an expanding economy. Additionally, robust increases in construction spending will support demand for corrosion inhibitors used in cement and concrete, industrial coatings, and metal applications. The industry will continue to invest in the development of new, less costly products such as organic corrosion inhibitors with better environmental profiles and improved performance.

The use of organic adsorption type corrosion inhibitors is a widely spread and cost-effective method to control carbon steel corrosion in the oil and gas production. The chemical functionalities employed in inhibitors formulation are very diverse and commercial inhibitors usually contain active components such as amides, amines, imidazolines, quaternary ammonium salts, etc.The inhibitors efficiency depends on a wide range of factors: flow patterns, solution chemistry, temperature, pressure, etc. The mechanism of action of inhibitors is difficult to establish and there is a large amount of research done on the subject.

Selecting a corrosion inhibitor, these must be considered [6]: (i). It should be synthesized conveniently from relatively cheap raw materials, *i.e.* (bio)organic precursors, or easily and cost-effectively extracted from natural resources. (*ii*). The presence of nitrogen, oxygen, sulfur, phosphorus and multiple bonds or aromatic rings in the inhibitor molecule is preferred. This causes increased adsorption of the compound on the metal surface and to the enhancement of the inhibition efficiency. (iii). It should not cause production problems due to incompatibility with the system fluids, system conditions and other chemicals in the system. (iv). It should be non-toxic, not present a health risk to operators and maintenance personnel. (v). It should be acceptable to the environment, *i.e.* it should be environmentally friendly ("green"), if discharged with the effluent.

In the past two decades, the research in the field of ''green'' corrosion inhibitors has been addressed toward the goal of using cheap, effective molecules at low or ''zero'' environmental impact [7,8]. The following pie chart shows world consumption of corrosion inhibitors on a value basis as at 2008 (Fig. 4) [5]:

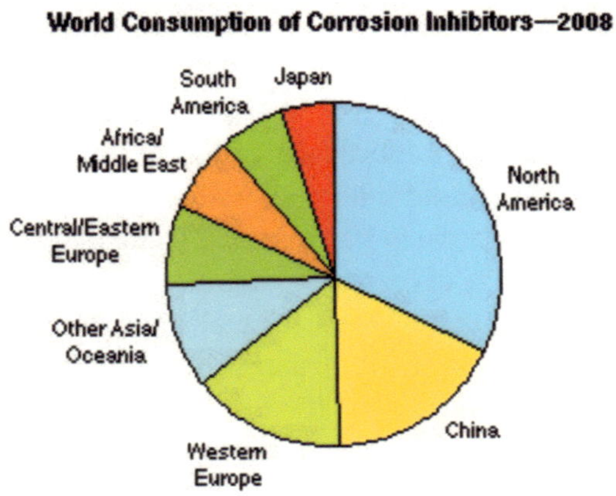

Figure 4. World consumption of corrosion inhibitors [5].

2. Brief theoretical background of Density Functional Theory (DFT)

The purpose of this section is to present a brief, but not an exhaustive, review of the basic theoretical background of the density functional theory. DFT, itself, has been described extensively elsewhere [9-11], but it is necessary to give a brief theoretical and computational concepts relevant to the application of DFT as a tool of choice for the computational design of organic corrosion inhibitors, which facilitates their adsorption onto metal and surfaces.

There are numerous quantum based approaches that have been applied to organic materials design including Hatree-Fock, Moller-Plesset, Coupled Cluster and Semi-emperical methods. However, the most popular for organic material design and calculations of chemical reactivity has been Density Functional Theory (DFT). The main reason for its popularity has been the relative accuracy with significantly lower computational cost compared to some of the other methods listed above.

The recent impact of density functional theory (DFT) in the development of quantum electrochemistry is considerable, and can be linked to achievements at the end of the 1980s when gradient-corrected and hybrid functional methods were introduced [12]. Based on the well-known Hohenberg-Kohn theorems (Kohn was awarded a Nobel Prize in physics in 1964 for his work on DFT in the same year that Pople was also awarded a Nobel Prize in the same field [13]), DFT focuses on the electron density, *ρ(r)*, itself as the carrier of all information in the molecular (or atomic) ground state, rather than on the single electron wave function, of which there is one per electon. Because the electron density arises from the collective contributions of all electrons, a considerable simplification of the many-bodied Scrhodinger equation results, because of the reduction in the complexity (number of degrees of freedom) of the system. In summary, the Hohenberg-Kohn theorem establishes that the ground state of an electronic system is just a functional of the electronic density. In principle, one only needs the knowledge of the density to calculate all of the properties of a molecular system.

In DFT, the ground state total energy for an N-electron system is expressed in terms of the three-dimensional ground-state electronic density *ρ(r)* and the external potential *v(r)* in the form [14]:

$$E[\rho] = F[\rho] + \int dr \rho(r) v(r), \tag{8}$$

where F[ρ] is the universal functional of Hohenberg-Kohn given by the sum of the electronic kinetic energy functional, and

$$N = \int dr \rho(r), \tag{9}$$

guarantees the proper normalization of the electron density.

A generalized DFT expression is [15]:

$$E_{DFT}[\rho] = T_S[\rho] + E_{ne}[\rho] + J[\rho] + E_{xc}[\rho] \tag{10}$$

where T_s is the kinetic energy functional (*S* denotes that the kinetic energy is obtained from a Slater determinant), E_{ne} is the electron-nuclear attraction functional, *J* is the Coulomb part of the electron-electron repulsion functional, and E_{xc} represents the exchange correlation functional. The dependence of each of these terms on the electron density, ρ, is represented by ρ in brackets following each term. Unfortunately, the exchange-correlation energy functional, which should be universal, is not known and practical (and approximate) solutions are obtained by the use of the so-called Kohn-Sham (KS) orbitals, which differ from other kinds of orbitals mostly by the fact that the sum of the squares of the occupied KS orbitals is the true electron density of the system, an assumption which is only approximated in other quantum chemical methods, such as Hartree-Fock.

The ground-state electronic density can be obtained from the solution of the Euler-Langrange equation corresponding to the minimization of the energy given in Equ. (8), subject to the restriction by Equ. (9), that is,

$$\mu = \left(\frac{\delta E}{\delta \rho(r)}\right)_v = v(r) + \frac{\delta F}{\delta \rho(r)}, \tag{11}$$

where μ, the undetermined Langrange multiplier, is the chemical potential that measures the escaping tendency of the electrons from a system. Electrons flow from places with higher chemical potential to places with lower chemical potential up to the point in which μ becomes constant throughout space. A brife history of DFT is presented in Table 1.

Year	Event
1926	Old DFT developed by Thomas-Fermi theory.
50's and 60's	Slater and co-workers develop Xα as crude Kohn-Sham Local density Approximations (KS-LDA).
1965	Modern DFT begins with Kohn-Sham equations. By introducing orbitals, 99% of the kinetic energy was gotten right, accurate ρ(r), and only need to approximate a small contribution, $EXC[\rho]$.
1965	Kohn-Sham (KS) also suggested local density approximation (LDA) and gradient expansion approximation.
1993	More modern functionals i.e. Generalized Gradient Approximations (GGA's) and Hybrids shown to be usefully accurate for thermochemistry.
1998	Kohn and Pople won Nobel prize in Chemistry.
2010	DFT applied in solving problems in materials science, geology, soil science, astrophysics, protein folding etc.

Table 1. Brief History and Timelines of DFT

3. DFT based chemical reactivity indicators

3.1. Global reactivity parameters

The design of organic corrosion inhibitors for metal protection (those that bind covalently to metal surfaces in order to prevent them from corroding), requires the use of chemical descriptors that encapsulate how covalently reactive a given organic molecule is. Quantum chemical descriptors are related to the electronic structure of organic materials and to the chemical mechanisms that are involved in covalent bond formation between them and the metal surfaces. Many efficient corrosion inhibitors are organic compounds rich in heteroatoms such as nitrogen, oxygen, sulfur and π-bonds. It has been reported that organic inhibitors bearing both nitrogen and sulphur atoms on the same molecular structure are more effective than inhibitors having only nitrogen atom. It is generally accepted that corrosion inhibition efficiency of organic compounds is related to their adsorption properties. The adsorption of these molecules depend mainly on certain physicochemical properties of the inhibitor molecule such as, functional groups, steric factor, aromaticity, electron density at the donor atoms and p-orbital character of donating electrons, electronic structure of the molecules and the strength of interaction between the inhibitor and the metal surface. The organic inhibitor should not only donate electrons to the unoccupied d- orbital of the metal, but can also accept electrons from the d-orbital of the metal leading to the formation of a feed-back bond

The basic relationship of the density functional theory of chemical reactivity is, precisely, the one established by Parr, Donnelly, Levy and Palke [15], that links the chemical potential (μ) of DFT with the first derivatives of the with respect to the number of electrons, and therefore with the negative of the electronegativity χ,

$$\mu = \left(\frac{\delta E}{\delta N}\right)_v = -\chi. \tag{12}$$

The next fundamental aspect comes from the identification of the concept of chemical hardness with the second derivative of the energy with respect to the number of electrons [16], that is,

$$\eta = \left(\frac{\delta^2 E}{\delta N^2}\right)_v = \left(\frac{\delta \mu}{\delta N}\right)_v. \tag{13}$$

In accordance with the earlier work of Iczkowski and Margrave [17], it should be noted that, when assuming a quadratic relationship between E and N and in a finite difference approximation, Equ. (12) may be written as:

$$\chi = -\mu = \left(\frac{I + A}{2}\right) \tag{14}$$

In terms of the energies of HOMO and LUMO molecular orbitals, hardness is given as [18]:

$$\eta = \frac{I - A}{2}, \eta = -\frac{E_{LUMO} - E_{HOMO}}{2} \tag{15}$$

where I and A are the ionization potential and electron affinity, respectively, thereby recovering the electronegativity definition of Mulliken [19]. Moreover, a theoretical justification was provided for Sanderson's principle of electronegativity equalization, which states that when two or more atoms come together to form a molecule, their electronegativities become adjusted to the same intermediate value [20].

The global softness (S), is the inverse of global hardness and is given as [21]:

$$S = \frac{1}{2\eta} = \left(\frac{\partial N}{\partial \mu}\right)_{v(r)}. \tag{16}$$

The global electrophilicity index (ω) was introduced by Parr et al. [22] and is given by:

$$\omega = \frac{\mu^2}{2\eta}. \tag{17}$$

Using the parabolic model, it was shown that global electrophilicity index (ω), can also be written as [10]:

$$\omega = \frac{(I + A)^2}{8(I + A)}. \tag{18}$$

The electron charge transfer, ΔN, from base B to acid A, and the assiociated energy change ΔE is given as [23]:

$$\Delta N = \frac{\mu_B - \mu_A}{2(\eta_A + \eta_B)} \text{ and } \Delta E = -\frac{(\mu_B - \mu_A)^2}{2(\eta_A + \eta_B)} \tag{19}$$

Substituting the subscripts A and B by metal and mol to designate a metal surface and a molecule, respectively, and replacing the μ by - χ gives:

$$\Delta N = \frac{\chi_{metal} - \chi_{mol}}{2(\eta_{metal} + \eta_{mol})} = \frac{\Phi - \chi_{mol}}{2\eta_{mol}} \tag{20}$$

$$\Delta E = -\frac{(\chi_{metal} - \chi_{mol})^2}{4(\eta_{metal} + \eta_{mol})} = \frac{(\Phi - \chi_{mol})^2}{4\eta_{mol}} \tag{21}$$

In the second equality of Eqs 20 and 21, the electronegativity of metal surface is replaced by the work function for metal surface, Φ.

3.2. Local reactivity parameters

The local reactivity of organic inhibitors can be analyzed through an evaluation of the Fukui indices [24]. These are measures of the chemical reactivity, as well as an indication of the reactive regions and the nucleophilic and electrophilic behavior of the molecule. The density functional definition of the Fukui function provides a firm foundation of the frontier-electron theory [25, 26]. Using a scheme of finite difference approximations from Mulliken population analysis of atoms in inhibitors and depending upon the direction of electron transfer we have:

$$f_k^+ = q_k(N+1) - q_k(N) \text{ (for nucleophilic attack)} \tag{22}$$

$$f_k^- = q_k(N) - q_k(N-1) \text{ (for electrophilic attack)} \tag{23}$$

$$f_k^o = \frac{q_k(N+1) - q_k(N-1)}{2} \text{ (for radical attack)} \tag{24}$$

where q_k is the gross charge of atom k in the molecule i.e. the electron density at a point r in space around the molecule. N corresponds to the number of electrons in the molecule, with $N+1$ corresponding to a singly-charged anion, with an electron added to the LUMO of the neutral molecule; and $N-1$ corresponding to singly-charged cation with an electron removed from the HOMO of the neutral molecule.

The local softness [25] is defined as:

$$s(r) = \left[\frac{\partial n(r)}{\partial \mu}\right]_{v(r)} \tag{25}$$

and this equation yields the global softness upon integration:

$$S = \int s(r)dr. \tag{26}$$

The relationship between the local softness and the Fukui function,

$$s(r) = Sf(r), \tag{27}$$

reflects that these quantities contain the same information about the relative site reactivity in a molecule.

A number of computational methods are available for the calculation of theoretical descriptors. A list of commonly used quantum chemical reactivity descriptors is given in Table 2.

Descriptors	Explanation
Global quantum chemical descriptors	
E_{HOMO}	Energy of the highest occupied molecular orbital
E_{LUMO}	Energy of the lowest unoccupied molecular orbital
$IP(\approx -E_{HOMO})$	Ionization potential: Removing an electron from a molecular system X ($X \rightarrow X^{+} + e^{-}$)
$EA(\approx -E_{LUMO})$	Electron affinity: Attaching an additional electron to a molecular system X ($X + e^{-} \rightarrow X^{-}$)
$\mu = \left(\frac{\delta E}{\delta N}\right)_v$	Chemical potential, defined as the change in electronic energy *E* upon change in total number of electrons *N*
$\chi = -\mu \approx -1/2(E_{HOMO} + E_{LUMO})$	Absolute electronegativity
$\eta = -\left(\frac{\delta \mu}{\delta N}\right)_v \approx -(E_{HOMO} - E_{LUMO})$	Molecular hardness, defined as the change in chemical potential μ upon change in total number of electrons *N*
$S = \frac{1}{2\eta}$	Molecular softness
α	Molecular polarizability; Note that molecules arrange themselves towards a state of minimum polarizability and maximum hardness
$\omega = \frac{\mu^2}{2\eta} = \frac{\chi^2}{2\eta}$	Electrophilicity index
Charge distribution	
QA(r)	Net atomic charges (at atom *r*)
PSA	Polar surface area, describing the spacial surface density distribution
μ	Molecular dipole moment
Site specific molecular descriptors (Local quantum descriptors)	
$f^{+}(r) = \left(\frac{\delta \rho(r)}{\delta N}\right)^{+}_{v(r)} \approx \rho_{N+1}(r) - \rho_N(r)$	Electrophilic Fukui function, defined as the change in electron density ρ at atom *r* upon addition of electrons to the system (N = number of electrons
$f^{-}(r) = \left(\frac{\delta \rho(r)}{\delta N}\right)^{-}_{v(r)} \approx \rho_N(r) - \rho_{N-1}(r)$	Nucleophilic Fukui function, defined as the change in electron density ρ at atom *r* upon removal of electrons from the system (N = number of electrons
$\omega(r) = \omega \times f^{+}(r)$	Local electrophilicity index
$\omega(r) = \omega \times f^{-}(r)$	Local nucleophilicity index
$\Delta f(r) = f^{+}(r) - f^{-}(r)$or $\Delta f(r) \approx \rho_{LUMO}(r) - \rho_{HOMO}(r)$	Reactivity-selectivity descriptor or dual descriptor, while $f^{+}(r)$ measures reactivity towards nucleophilic and $f(r)$ towards electrophilic attacks; therefore, electrophilic sites are identified by $\Delta f(r) > 0$. ρLUMO and ρHOMO are the electron densities of the LUMO and HOMO orbitals, respectively.

Table 2. Important molecular descriptors used to model chemical reactivity and molecular characterization/design of organic materials for metallic protection.

http://dx.doi.org/10.5772/57245

4. Commonly used DFT softwares

There are a wide variety of computational chemistry packages available to suit the needs of most researchers in the area of organic corrosion inhibitor modeling. They differ in terms of price, popularity and theoretical methods available. The list below covers some of the most popularly used in the areas of material design and characterization of organic corrosion inhibitors. It is not an endorsement of any software by the author and the decision to use any of the listed software is purely that of the reader. Commonly used softwares for quantum chemical computation is presented in Table 3.

4.1. TURBOMOLE

Turbomole is a quantum chemical program package, initially developed in the group of Prof. Dr. Reinhart Ahlrichs at the University of Karlsruhe and at the Forschungszentrum Karlsruhe. The package has an exceptional DFT subsystem.

4.2. GAUSSIAN

This is probably the most popular computational chemistry package in use today. It has over 42 years of history and has numerous publications. It was originally developed by Pople in the early 1970's. It is marketed by Gaussian Inc. The code uses Gaussian orbitals to build up the basis sets. In its original form it was primarily a Hartree- theory code. However, today it offers a range of capabilities including DFT, Semi-empirical, Hybrid QM/MM and numerous other approaches. Starting from the basic laws of quantum mechanics, Gaussian predicts the energies, molecular structures, and vibrational frequencies of molecular systems, along with numerous molecular properties derived from these basic computation types.

4.3. GAMESS

GAMESS is a program for *ab initio* molecular quantum chemistry. Briefly, GAMESS can compute SCF wavefunctions ranging from RHF, ROHF, UHF, GVB, and MCSCF. Correlation corrections to these SCF wavefunctions include Configuration Interaction, second order perturbation Theory, and Coupled-Cluster approaches, as well as the Density Functional Theory approximation.

4.4. PC GAMESS

PC GAMESS is a freely available ab initio and DFT computational chemistry program developed to offer high performance on Intel-compatible x86, AMD64, and EM64T processors. It was initially based on the free GAMESS(US) program sources but extends its functionality in some important areas. PC GAMESS has very efficient MP2 energy and gradient modules.

4.5. ORCA

ORCA is a flexible, efficient and easy-to-use *general* purpose tool for quantum chemistry written by F. Neese, with contributions from U. Becker, D. Ganiouchine, S. Koßmann, T.

Petrenko, C. Riplinger and F. Wennmohs. It offers specific emphasis on spectroscopic properties of open-shell molecules and features a wide variety of standard quantum chemical methods ranging from semiempirical methods to DFT to single- and multireference correlated ab initio methods. It can also treat environmental and relativistic effects.

4.6. CASINO

CASINO is a code for performing quantum Monte Carlo (QMC) electronic structure calculations for finite and periodic systems. The code is developed by the Theory of Condensed Matter - Quantum Monte Carlo (TCM - QMC) group of Richard Needs at the University of Cambridge largely by Mike Towler, considerably assisted from 2002 by Neil Drummond and from 2004 by Pablo López Ríos.

4.7. CRYSTAL

CRYSTAL can be used to compute the electronic structure of periodic systems within Hartree Fock and density functional theory. A variety of exchange functionals (LDA, VBH, BECKE, PWGGA, PBE) and correlation functionals (VWN, PWLSD, PZ, VBH, LYP, P86, PWGGA, PBE) are currently available, as well as the hybrid functionals, B3LYP and B3PW. Restricted closed shell, restricted open shell and unrestricted Hartree-Fock methods may be used.

4.8. NWCHEM

NWChem is a computational chemistry package that is designed to run on high-performance parallel supercomputers as well as conventional workstation clusters. It aims to be scalable both in its ability to treat large problems efficiently, and in its usage of available parallel computing resources.

4.9. ABINIT

ABINIT is a package whose main program allows one to find the total energy, charge density and electronic structure of systems made of electrons and nuclei (molecules and periodic solids) within Density Functional Theory (DFT), using pseudopotentials and a planewave basis. ABINIT also includes options to optimize the geometry according to the DFT forces and stresses, or to perform molecular dynamics simulations using these forces, or to generate dynamical matrices, Born effective charges, and dielectric tensors. Excited states can be computed within the Time-Dependent Density Functional Theory (for molecules), or within Many-Body Perturbation Theory (the GW approximation).

4.10. MOPAC

MOPAC (Molecular Orbital PACkage) is a semiempirical quantum chemistry program based on Dewar and Thiel's NDDO approximation used for the prediction of chemical properties and modeling of chemical reactions. It is used by chemists and biochemists for both research and teaching, and runs on Windows and Linux platforms.

4.11. VIEWMOL

Viewmol is an open source graphical front end for computational chemistry programs. It is able to graphically aid in the generation of molecular structures for computations and to visualize their results. It can read and write many known file types, including *Turbomole* files

4.12. ADF

Amsterdam Density Functional (ADF) This package has over a 32 year track record of development. It was originally written by Baerends (Vrije University, Amsterdam) and Zieglar (University of Calgary). It is based on Slater atomic orbitals and has the ability to work with a range of different systems; surfaces, periodic cells, and gas phase molecules.

4.13. DMol

This cose was originally developed by Delley (Paul Scherrer Institute) in late 1980's. Like ADF and Gaussian it uses localized orbitals to describe the electron density. However, unlike these codes the basis sets are numerical, making the code very fast. However, the trade off is that the code has a finite level of accuracy as the calculations cannot be made more accurate by progressively adding basis functions. It has been succesfully used for solids, surfaces and gas phase molecules. It is marketed by Accelrys as part of their Material studio software suite and has been used in over 700 publications.

SOFTWARES-QUANTUM MECHANICS	
NAME	DESCRIPTION
ADF	Quantum Mechanics (QM)
ABINIT	QM (Molecular and Periodic Systems)
AMPAC	QM (Semi-empirical)
CRYSTAL	QM for Periodic Systems
Dalton	QM Specializing in magnetic/electric properties
Discovery Studio	QM & MM/MD suite of software with a user-friendly Interface for Life Sciences
GAMESS-US	QM
Gaussian	QM
Jaguar (Qsite)	QM (QM/MM)
Materials Studio	MM/MD & QM suite of software with a user-friendly GUI for Materials Sciences
MOLPRO	QM (specializing in high-level calculations)
NWChem	QM
SIESTA	QM specializing in electron transport and Solids
VASP	QM specializing in QMD

MM= Molecular mechanics, QMD= Quantum Molecular Dynamics

Table 3. Commonly used softwares for quantum mechanics and their descriptions.

5. Some recent studies of inhibitor design using DFT reactivity indices

In designing new corrosion inhibitors from organic compounds using DFT, the energy of the highest occupied molecular orbital (E_{HOMO}), the energy of the lowest unoccupied molecular orbital (E_{LUMO}), the HOMO-LUMO energy gap, the dipole moments and molecular polarizability are the most important electronic parameters to be considered. As reported by several researchers [27-29], excellent corrosion inhibitors are usually those organic compounds which do not only offer electrons to unoccupied orbital of the metal, but also accept free electrons from the metal. It is also well established in the literature that the higher the HOMO energy of the inhibitor, the greater the trend of offering electrons to unoccupied d- orbital of the metal, and the higher the corrosion inhibition efficiency [30, 31]. In addition, the lower the LUMO energy, the easier the acceptance of electrons from metal surface, as the LUMO–HOMO energy gap decreased and the efficiency of inhibitor improved. Furthermore, the dipole moment μ is a measure of the polarity of a covalent bond, which is related to the distribution of electrons in a molecule [32]. Although literature is inconsistent with the use of μ as a predictor for the direction of a corrosion inhibition reaction, it is generally agreed that the large values of μ favour the adsorption of inhibitor onto a metal surface [33]. The polarizability $<\alpha>$ is an indicator of the linear response of the electron density in the presence of an infinitesimal electric field, which depends on the second derivative of energy with respect to the electric field. The higher values of $<\alpha>$ facilitate the strong adsorption process of corrosion inhibitors onto metallic surfaces and hence, high inhibition efficiency [34].

The work of Parr, Pearson and others [35, 36], within the framework of Density Functional Theory (DFT), has established a general rule predicting the stability of the electronic structure of molecules. The general rule of this new paradigm is that the index of chemical reactivity and stability of a molecule is its global hardness,η. Increase in hardness increases the movement of the system towards a more stable configuration and when a chemical species moves away from its equilibrium configuration its hardness value decreases. When a system evolves towards a state of greater hardness, its stability increases. When a system evolves towards the state of lower hardness, its stability decreases. η, is again used as an index of chemical reactivity. The higher the value of η, the lesser is its reactivity. The global softness, S, the inverse concept of hardness, is useful for a straightforward prediction of chemical reactivity. The soft molecules undergo changes in electron density more easily than the hard molecules and are more reactive than the hard molecules. A molecule having higher S value is more reactive than a molecule having smaller S value. In general, it can be said that the increase in softness increases chemical reactivity and increase in hardness decreases chemical reactivity. Thus in the design of new corrosion inhibitors using DFT from a family of organic molecules with related structures, a molecule with lower value of hardness and high value of softness is expected to have a high inhibition efficiency when its adsorbed onto a metal surface.

Other useful reactivity and selectivity indices to be considered when designing new corrosion inhibitors using DFT include: the fraction of electrons transferred from the inhibitor molecule to the metal surfaces, the partial atomic charges and the condensed Fukui functions. The number of electrons transferred (ΔN) indicates the tendency of a molecule to donate electrons. The higher the value of ΔN is, the greater the tendency of a molecule to donate electrons to the electron poor species. In the case of corrosion inhibitors, a higher ΔN implies a greater tendency to interact with the metal surface (i.e., a greater tendency to adsorb on the metal surface) [23]. Beside reactivity indicators, several molecular properties are good indicators of selectivity (i.e., the regions on the molecule on which certain type of reactions are likely to occur) and include the partial atomic charges and the condensed Fukui functions. The atom with the highest negative partial charge is considered a possible site for an attack by electron deficient species [37]. The condensed Fukui functions are local selectivity descriptors. These functions inform about the centers in a molecule on which nucleophilic, electrophilic and radical reactions are most likely to occur. The Fukui functions for the electron rich centers (i.e., atoms susceptible to electrophilic attack) and electron deficient centers (i.e., atoms that are susceptible to nucleophilic attack) are often estimated using the finite difference approximation approach [38]. The preferred site for nucleophilic attack is the atom (or region) in the molecule where f^+ has the highest value while the site for electrophilic attack is the atom (or region) in the molecule where the value of f^- is the highest.

Computational methods have evolved as important tools in corrosion inhibitor design during the last two decade, as calculations can provide a large amount of information of large numbers of compounds within reasonable timeframe. Results from such studies can be used as suitable starting points for further experimental studies. Computational methods can also be important tools in the development of more suitable compounds to be used for metal protection, starting from already available compounds and through structural modifications identify derivatives with improved metal protection efficacy. Recently, there are lots of studies in the literature on the computational study of organic materials useful for metal protection. The goal of these studies is to gain insights at the molecular level on the interaction of these organics with metal surfaces. This important approach is very vital in the design of new and effective corrosion inhibitors for industrial applications especially in the oil and gas sectors. Some of the recent researches are reviewed below. They are by no mean exhaustive as excellent review exist elsewhere [39].

5.1. DFT theoretical study of 7-R-3methylquinoxalin-2(1H)-thiones (R=H; CH_3; Cl) as corrosion inhibitors in hydrochloric acid

Quantum chemical approach at DFT/B3LYP/6-31G(d,p) level of theory was used to calculate some structural and electronic properties of three quinoxaline derivatives to ascertain the correlation between their experimental corrosion inhibition efficiencies on mild steel in acidic media and some of the computed parameters [40]. The global reactivity indicators show that Me-Q=S is a better inhibitor than Q=S and Cl-Q=S, respectively. Quantum chemical parameters calculated include E_{HOMO}, the dipole moment (μ), electronegativity (χ) and the amount of electrons transferred from organic inhibitor to the metal surface (ΔN) e.t.c. The high value of electron donor capacity of Me-Q=S when compared to Q=S and Cl-Q=S makes it more effective among the three compounds investigated. The substitution of the methyl group by the chlorine group in C7 position leads to a notable decrease in the adsorption capacity of the inhibitors which is as a result of the electronic effect of the substituent directly attached to the phenyl ring (Fig. 5). The Fukui functions and the electron density distribution showed that the S atom was probably the main adsorption site.

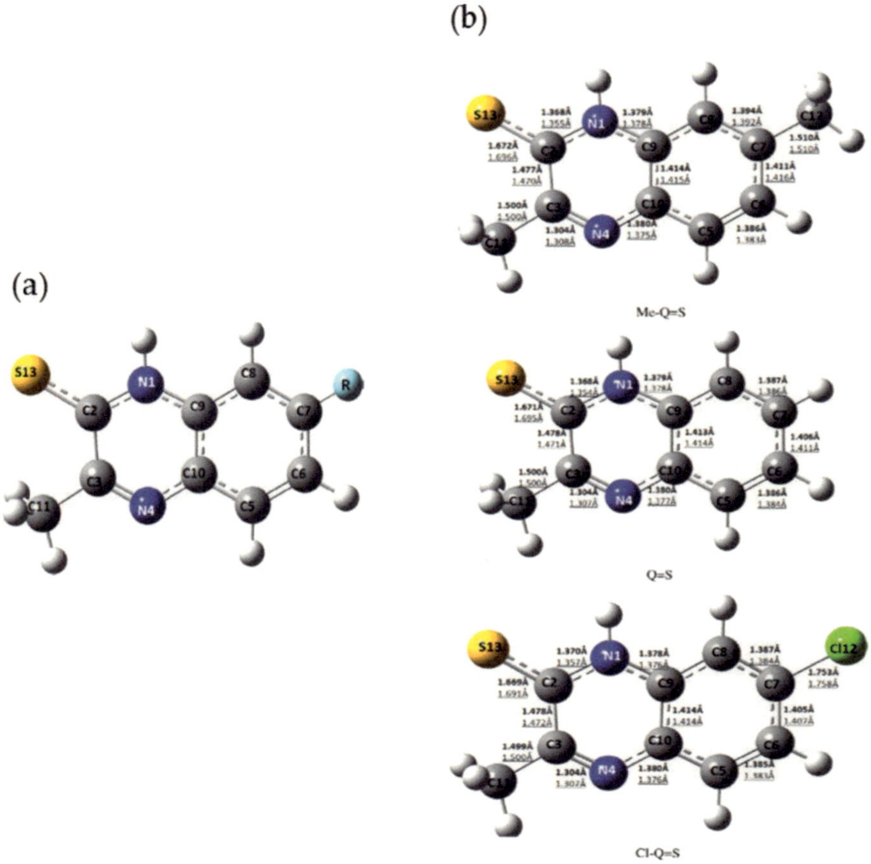

Figure 5. (a) The 7-R-3methylquinoxalin-2(1H)-thiones (R=H; CH_3; Cl) (b) Selected geometrical parameters of the compounds calculated at in gas (bold) and aqueoes phases (underlined) [Adapted from ref. [40]].

5.2. DFT theoretical studies of antipyrine Schiff bases as corrosion inhibitors

Quantum chemical calculations based on DFT/B3LYP/6-31G level were performed to find the relation between the molecular structure of the inhibitor and the inhibition efficiency of some schiff bases namely: (benzylidine amino)antipyrine (a), 4-hydroxy 3-(benzylidine amino)antipyrine (b), 2-hydroxy3-(benzylidine amino)antipyrine (c), and 2-hydroxy 3-(naphthylidine amino)antipyrine (d) [41]. The quantum chemical parameters confirmed that the replacement of phenyl group by naphthyl group increases the inhibition efficiency of the inhibitor toward the metal surface which means that the inhibitor (d) had higher inhibition efficiency than the inhibitor (c). Moreover, the substitution of the hydrogen atom of phenyl ring by a hydroxyl group increases the adsorption of the inhibitor toward the metal surface which means that the inhibitors (c) and (b) had higher inhibition efficiency than inhibitor (a). Also the substituent hydroxyl group in the ortho position in the case of inhibitor (c) favored more the adsorption of the inhibitor than that in the para position. This means that the inhibition efficiency of inhibitor (c) is higher than that of inhibitor (b). It is concluded from the calculations that the real controlling quantum factors to determine the inhibition efficiency of antipyrine inhibitors are energy gap *(ΔE)*, total negative charges (TNC),and softness (*S*) of the molecules (Fig. 6).

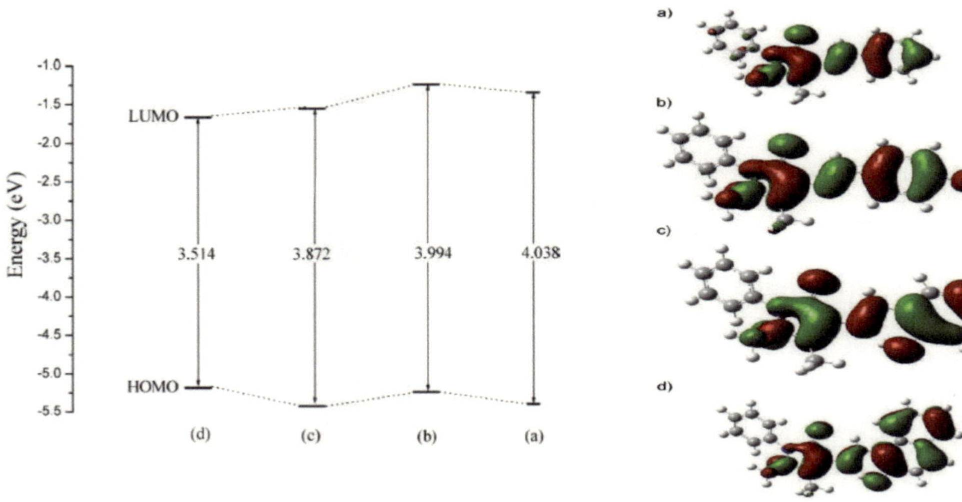

Figure 6. (a) Correlation diagram and energy gap values of inhibitors (a-d); (b) HOMO orbitals of inhibitors (a-d) [Adapted from ref. [41]].

5.3. Theoretical evaluation of corrosion inhibition performance of imidazoline compounds with different hydrophilic groups

Corrosion inhibition performance of four 1-R-2-undecyl-imidazoline compounds (R=CH_2COOH (A), CH_2CH_2OH (B), $CH_2CH_2NH_2$ (C) and H (D)) for carbon steel was evaluated by quantum chemistry and molecular mechanics methods [42]. The quantum chemical study such as E_{HOMO}, E_{LUMO}, and global reactivity parameters indicates that the inhibition efficiency of the four inhibitors followed the order of A > B > C > D. Further investigation using molecular mechanics method showed that these inhibitor molecules would form a self assembled monolayer (SAM) on the Fe surface, and related parameters such as alkyl chain distance, title angle, interaction energy, and cohesive energy were compared. These parameters implies that the SAM became less compact and stable from molecules A to D, and the interaction between the SAM and the Fe surface was also weakened. These results denote the same inhibition efficiency order of the four inhibitors revealed by quantum chemistry calculation. The results by the two simulation methods were validated by previously reported experimental results. The results were more reliable than that obtained via each single method, and could provide a model for the design of new organic corrosion inhibitors.

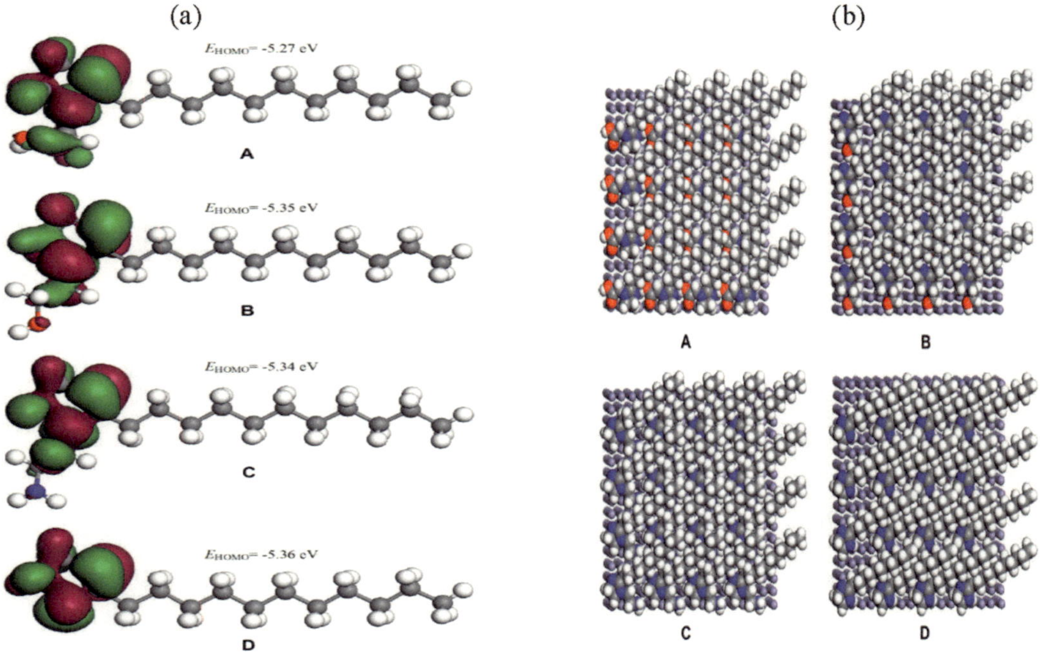

Figure 7. (a) HOMO of molecules A-D; (b) Optimized SAM for A-D molecules formed on Fe (0 0 1) surface [Adapted from ref. [42]].

5.4. Theoretical evaluation of the inhibition properties of two thiophene derivatives on corrosion of carbon steel in acidic media.

The density functional theory at the B3LYP/6-311G++(d,p) basis set level calculations were performed on two thiophene derivatives used as corrosion inhibitors, namely 2-methylthiophene (MT) and 2-(aminomethyl)thiophene (AMT) (Fig. 8), to investigate the correlation between its molecular structure and the corresponding inhibition efficiency (IE%) [43]. The inhibition efficiency is closely related to the quantum chemical parameters, E_{HOMO}, E_{LUMO}, ΔE, μ, η, and ΔN for the protonated inhibitors. It was shown that AMT molecule has the highest inhibition efficiency, which is probably due to the presence of amino group. Both inhibitors had a considerable tendency for protonation and the most probable site of protonation was the α-position of the thiophene ring. Finally, this study confirms the reliability of the quantum chemical methods to study the inhibition of corrosion of metal surface.

5.5. Theoretical study on relationship between structure of mercapto-triazole derivatives and inhibition performance

Relationships between corrosion inhibition efficiency of five kinds of mercapto-triazole inhibitors and their molecular electronic properties were theoretically studied at the level of DFT/B3LYP with 6-31+G (d, p) base sets [44]. The detailed studies revealed that corrosion inhibition efficiency and frontier orbital energy level E_{HOMO} show close correlation. The HOMO

http://dx.doi.org/10.5772/57245

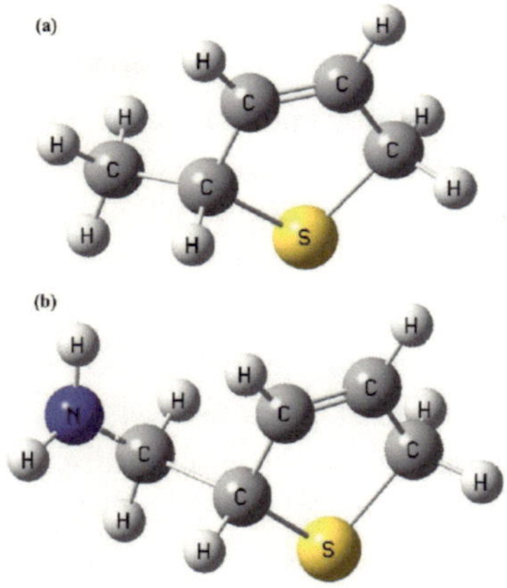

Figure 8. Optimized structures of molecules (a=MT, b=AMT) [Adapted from ref. [43]].

orbital were mainly delocalized around the triazole ring. By analyzing the molecular electrostatic potential surfaces for each molecule, it was found that most of the negative potential is concentrated on the triazole ring R2 and a small part on benzene ring R1. R2 plays a main role and R1 plays a synergistic role during the process of inhibitor adsorption on a metal surface. Further investigation indicates that the interaction energies between corrosion inhibitors and Fe(100) are positively correlated with corrosion inhibition efficiencies. It was theoretically predicted that the main structure of the inhibitor (rings R2 and R1) plays an important role for these inhibitors, as does the HOMO energy. Based on this information, attempt were made to design some superior homologous corrosion inhibitors. Example of the new designed inhibitor is given in Fig. 9.

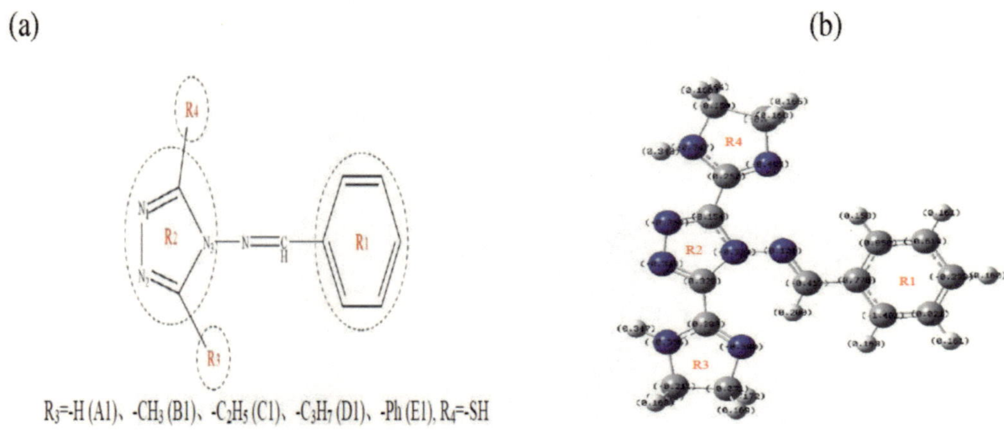

Figure 9. (a) The five mercapto-triazole inhibitors studied; (b) Optimized structure of the new designed inhibitor [Adapted from ref. [44]].

5.6. Experimental and theoretical studies for corrosion inhibition of carbon steel by imidazoline derivative in 5% NaCl saturated $Ca(OH)_2$ solution

Quantum chemical calculations and molecular dynamic (MD) simulations were applied to analyze the experimental data and elucidate the adsorption behavior and inhibition mechanism of 1-[N,N`-bis(hydroxylethylether)-aminoethyl]-2-stearicimidazoline (HASI) for carbon steel in 5% NaCl saturated $Ca(OH)_2$ solution [37]. Density functional theory (DFT) calculations suggest that the N=C-N region in imidazoline ring is the most active reaction site for the inhibitor adsorption on metal surface via the donor–acceptor interactions between the lone electron pairs on nitrogen atoms together with the π-electrons of heterocyclic and the vacant d-orbital of iron atoms. The adsorption of inhibitor on three typical surfaces (Fe (1 0 0), Fe_2O_3 (1 1 0) and Fe_3O_4 (1 0 0)) takes nearly parallel to the surface so as to maximize its contact with the surface, as shown as the MD simulations (Fig. 10). The experiments incorporating the theoretical calculation and MD simulation can provide an insight into the understanding of interactions between the inhibitor molecules and the carbon steel surface.

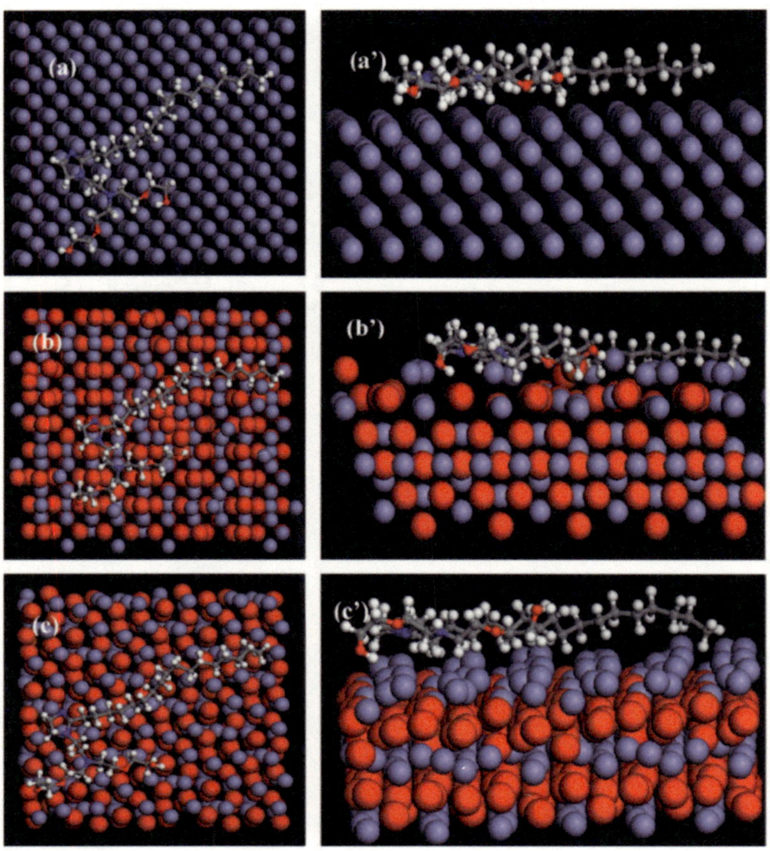

Figure 10. Equilibrium asdorption of HASI on (a) Fe, (b) Fe_3O_4 and (c) Fe_2O_3 surfaces.The left is top view and the right is side view [Adapted from ref. [37]].

6. Future developments

Density functional theory (DFT) has become an attractive theoretical method, because it gives exact, basic, and vital parameters for small and even complex molecules at lower cost. Moreover by applying DFT methodology, we can understand reactivity behaviour in terms of hard and soft acid/base (HSAB) theory that provides a systematic way for analyzing and predicting inhibitor/surface interaction. However, it has been observed that it is difficult to acquire information about the actual interaction and binding strength between organic inhibitor and metal surface which is pertinent in estimating inhibitor performance.

The above deficiency of DFT can be solved by using molecular dynamic simulations, which in recent times have been used to study the interaction between corrosion inhibitors and metal surfaces. MD simulations gives information at the molecular level on the adsorption of the inhibitor molecules on corroding metal surfaces. Furthermore, vital information on the conformation of inhibitors adsorbed onto metal surfaces and the interaction energy between them can be readily obtained [45]. Thus, MD simulations methodology can provide powerful insights into the design of inhibitor systems with superior properties and differences in inhibition efficiency among homologues inhibitors (inhibitors bearing similar structures) can be systematically explained. Furthermore, the use of molecular dynamic simulations in conjunction with DFT, will model completely the complex interaction involved in corrosion inhibition process of carbon steel (for example, the adsorption of water, electrolyte anion and the inhibitor onto the metal surface (Fig. 11) [46]. This will help in the calculation of parameters like adsorption and binding energies of the organic inhibitor materials, the electrolyte anion and water onto the metal surface.

Another important consideration in the future design of organic corrosion inhibitor will be to modeled the interaction between the inhibitor molecule and the metal oxide layer using DFT in conjunction with the molecular dynamic simulations approach. During the corrosion of mild steel in aqueous CO_2 environments, $FeCO_3$, Fe_3O_4 and Fe_2O_3 have be found in corrosion product films [47]. It should be noted that the process of metal corrosion is complex and heterogeneous due to the presence of various anodic and cathodic sites on the metal surface. For $FeCO_3$ surface, cation Fe^{2+} is present as anodic reactive site, which can bind to electrophilic attacking centers in inhibitors molecules. Anion CO_3^{2-} is present as a cathodic reactive center, which can bind to nucleophilic centers in inhibitor molecules. Therefore, inhibitor molecule can be adsorbed onto the $FeCO_3$ surface through interaction with these sites, and as a result, the transport of corrosive species is restricted from solution to metal surface. Thus, electrochemical reactions are retarded. The result of the molecular dynamic simulations calculation will enable the determination of the binding energy, which is an important factor to characterize the adsorption ability of an inhibitor molecule onto the $FeCO_3$ surface (Figs. 12 and 13) [48]. The binding energies will enable us to predict the most efficient inhibitors from a family of organic compounds with related structures.

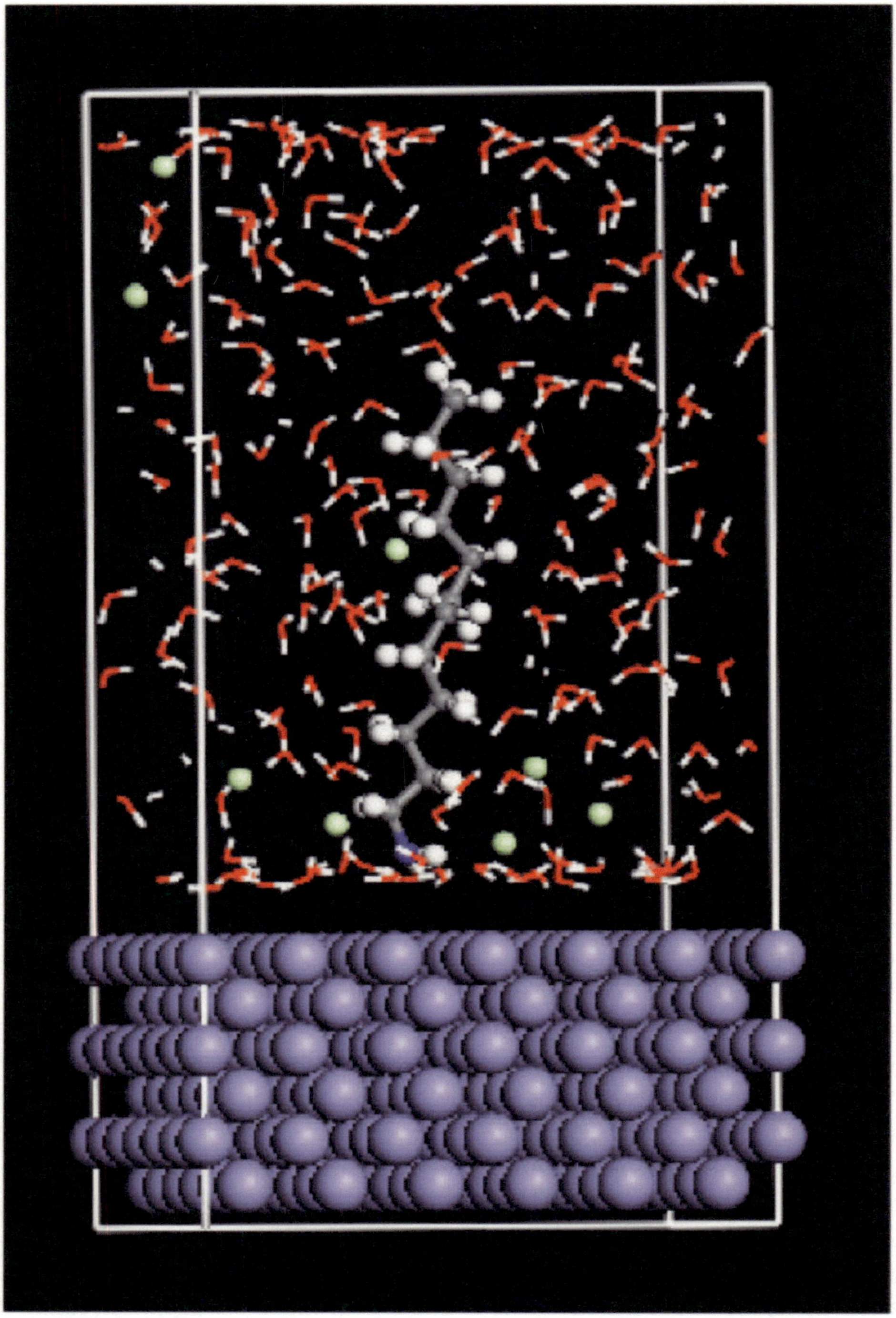

Figure 11. Equilibrium configuration of dodecylamine in aqueous solutions containing chloride ions [46].

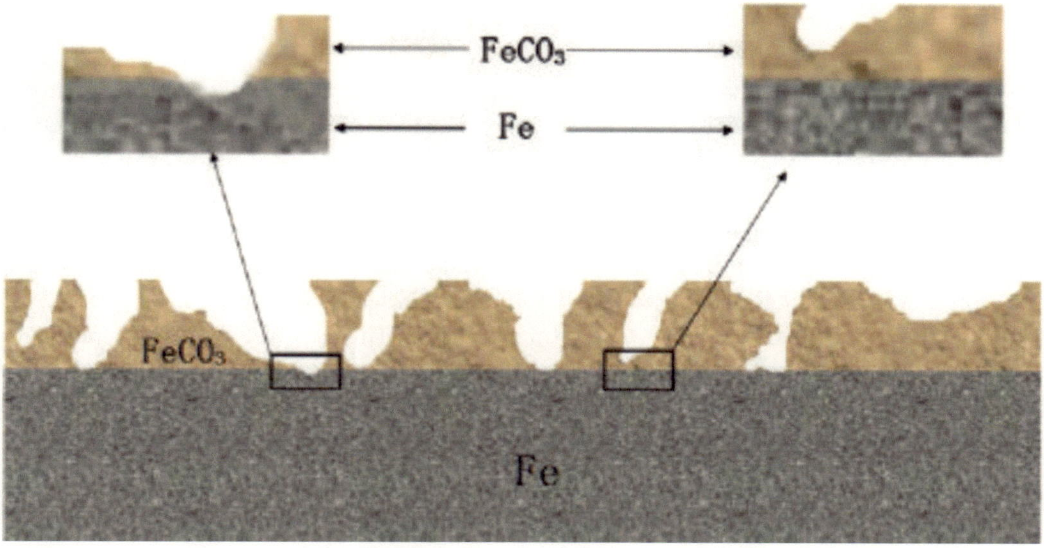

Figure 12. Sketch of corroded Fe surface in CO_2 containing environment [48].

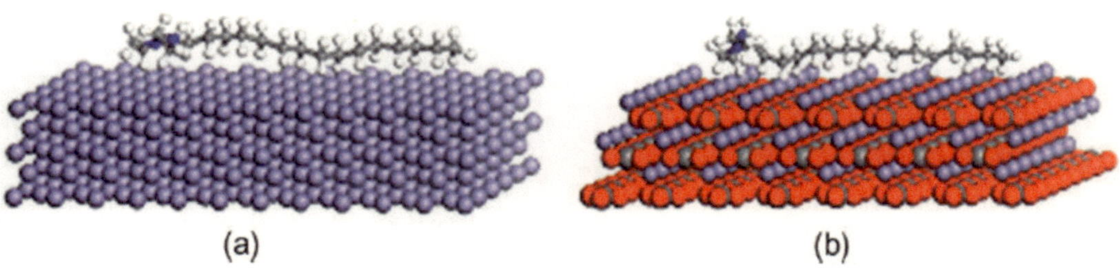

Figure 13. Equilibrium adsorption configuration of inhibitor 1-(2-aminoethyl)-2-alkyl-imidazoline derivative on (a) Fe and $FeCO_3$ (b) surfaces [48].

Finally, An important future consideration in the design of new and effective corrosion inhibitors using DFT will be to identify organic structures that have low toxicity in addition to having high corrosion protection efficiencies. In this regard, LogS, LogP, dipole moment and quantum chemical descriptors which are often used to model toxicities of therapeutic drugs could also be used for organic corrosion inhibitors since there are many similarities between the properties of drugs and the properties of corrosion inhibitors. For both applications (drugs and corrosion inhibitors), the organics need to have some solubility in water (Log S) and also a certain level of lipophilicity (Log P) (i.e. partitioning into the fatty tissues compared to partitioning from water to the metal surface).

7. Conclusion

This chapter has focused on describing the use of modern quantum chemical methods, primarily in the form of Density Functional Theory as a tool in the design of organic materials for the protection of metals in aqueous solution. It is evident that DFT is a powerful technique with a unique capability of performing practical calculations on complex, many bodied systems, such as large organic molecules of which most inhibitors comprise. DFT offers the facility to design, at the molecular level, the molecular structure, including the choice and location of various functional groups and to build in selected aromaticity, as dictated by experimental feedback from studies on existing inhibitors. The use of DFT in conjunction with molecular dynamic simulations, will model completely the complex interaction involved in corrosion inhibition process of carbon steel (for example, the adsorption of water, electrolyte anion and the inhibitor onto the metal surface and also the state of the metal surface whether it is bare or oxide covered.

Acknowledgements

The author acknowledged the support of King Fahd University of Petroleum and Minerals (KFUPM). Dr. Zuhair M. Gasem, Director, Center of Research Excellence in Corrosion at KFUPM, Kingdom of Saudi Arabia is also acknowledged for encouragement.

Author details

Ime Bassey Obot

Address all correspondence to: obot@kfupm.edu.sa

Center of Research Excellence in Corrosion, King Fahd University of Petroleum and Minerals, Dhahran, Saudi Arabia

References

[1] "Corrosion Cost and Preventive Strategies in the United States" Publication No.FHWA-RD-01-156.

[2] "Cost of Corrosion to exceed $1 Trillion in the United States in 2012-G2MT Labs-The Future of Materials Conditions Assessment," http://www.g2mtlabs.com/2011/06/nace-cost-of-corrosion-study-update/.

[3] A. Al Hashem, "Corrosion in the Gulf Cooperation Council (GCC) states: Statistics and Figures," in proceedings of the Corrosion UAE, Abu Dhabi, UAE, 2011.

[4] L. L. Shier, R. A. Jarman, G. T. Burstein, 1994. Corrosion, third ed., Butterworth-Heinemann, UK. Vol.1, p. 151.

[5] S. Muller, S. Ritzvi, K. Yokose, W. Yang, M. Jackel, Corrosion Inhibitors, *SCUP Home*, October (2009).

[6] S.J. Keny, A.G. Kumbhar, C. Thinaharan, G. Venkateswaran, Gallic acid as a corrosion inhibitor of carbon steel in chemical decontamination formulation, Corros. Sci. 50 (2008) 411–419.

[7] I. B. Obot, N. O. Obi-Egbedi, Ipomoea involcrata as an ecofriendly inhibitor for aluminium in alkaline medium, *Port. Electrochim. Acta* 27 (2009) 517–524.

[8] I. B. Obot, N. O. Obi-Egbedi, An interesting and efficient green corrosion inhibitor for aluminium from extracts of Chlomolaena odorata L. in acidic solution, *J. Appl. Electrochem.* 40 (2010) 1977–1983.

[9] R.R. Dogonadze, Theory of Molecular Electrode Kinetics, in: N.S. Hush (Ed.), Reactions of Molecules at Electrodes, Interscience Pub., London. 1971. pp. 135-227.

[10] P. Geerlings, F. De Proft, W. Langenaeker, Conceptual density functional theory, *Chem. Rev.* 103 (2003) 1793-1873.

[11] A. Nagy, Density functional theory and application to atoms and molecules, *Phys. Rev.* 298 (1998) 1-79.

[12] A. D. Becke, Density Functional Calculations of Molecular Bond Energies. J. Chem. Phys. 84 (1986) 4524-4529.

[13] P. Hohenberg, W. Kohn, Inhomogeneous electron gas, Phys. Rev. B 136 (1964) 864-871.

[14] S. Bell, T.J. Dines, B.Z. Chowdhry, R. Withnall, Computational Chemistry Using Modern Electronic Structure Methods, *J. Chem. Edu.* 84(8) 2007) 1364-1370.

[15] R. G. Parr, R. A. Donnelly, M. Levy, W. E. Palke, Electronegativity: The density functional viewpoint, *J. Chem. Phys.* 68 (1978) 3801-3807.

[16] R. G. Parr, R. G. Pearson, Absolute hardness: Companion parameter to absolute electronegativity, *J. Am. Chem. Soc.* 105 (1983) 7512-7516.

[17] R.P. Iczkowski, J.L. Margrave, Electronegativity, *J. Am. Chem. Soc.* 83 (1961) 3547-3551.

[18] J.F. Janak, Proof that $\partial E/\partial n_i = \epsilon_i$ in density functional theory, Phys. Rev. B 18 (1978) 7165-7168.

[19] R. S. Mulliken, A New Electroaffinity Scale; Together with Data on Valence States and on Valence Ionization Potentials and Electron Affinities, J. Chem. Phys. 2 (1934) 782-793.

[20] R.T. Sanderson, Chemical Bonds and Bond Energy; Academic, New York. 1976.

[21] W. Yang, R.G. Parr, Hardness, softness and the Fukui function in the electronic theory of metals and catalysis, *Proc. Natl. Acad. Sci.* 82(1985) 6723-6726.

[22] R.G. Parr, L. Sventpaly, S. Liu, Electrophilicity index, *J. Am. Chem. Soc.* 121 (1999) 1922-1924.

[23] N. Kovacevic, A. Kokalj, DFT study of interaction of Azoles with Cu(111) and Al(111) surfaces: Role of azole Nitrogen atoms and Dipole-Dipole Interactions, *J. Phys. Chem. C* 115 (2011)24189-24197.

[24] K. Fukui, Role of Frontier orbitals in chemical reactions, *Science* 218 (1982) 747-754.

[25] W. Yang, R.G. Parr, Hardness, softness and the Fukui function in the electronic theory of metals and catalysis, *Proc. Natl. Acad. Sci.* 82 (1985) 6723-6726.

[26] K.F. Khaled, Studies of iron corrosion inhibition using chemical, electrochemical and computational simulation techniques, Electrochim, Acta 55 (2010) 6523-6532.

[27] E.S. El- Ashry, S.A. Senior, QSAR of lauric hydrazide and its salts as corrosion inhibitors by using the quantum chemical and topological descriptors, Corros. Sci. 53 (2011) 1025-1034.

[28] A. Lesar, I. Milošev, Density functional study of the corrosion inhibition properties of 1,2,4 triazole and its amino derivatives, Chem. Phys. Lett. 483 (2009)198-203.

[29] G. Gece, S. Bilgiç, Quantum chemical study of some cyclic nitrogen compounds as corrosion inhibitors of steel in NaCl media, Corros. Sci. 51(2009) 1876-1878.

[30] I.B. Obot, N.O. Obi-Egbedi, Adsorption properties and inhibition of mild steel in sulphuric acid solution by ketoconazole: experimental and theoretical investigation, *Corros. Sci.* 52 (2010) 198-204.

[31] I.B. Obot, N.O. Obi-Egbedi, S.A. Umoren, The synergistic inhibitive effect and some quantum chemical parameters of 2,3-diaminonapthalene and iodide ions on the hydrochloric acid corrosion of aluminium, *Corros. Sci.* 51 (2009) 276-282.

[32] X. Li, S. Deng, H. Fu, T. Li, Adsorption and inhibition effect of 6-benzylaminopurine on cold rolled steel in 1.0 M HCl, *Electrochim. Acta* 54 (2009) 4089-4098.

[33] F. Kandemirli, S. Sagdinc, Theoretical study of corrosion inhibition of amides and thiosemicarbazones, Corros. Sci. 49 (2007) 2118-2130.

[34] M.K. Awad, M.R. Mustafa, M. M. Abo Elnga, Computational simulation of the molecular structure of some triazole as inhibitors for the corrosion of metal surface. J. Mol. Struct. (THEOCHEM) 959 (2010) 66-74.

[35] R.G. Parr, P.K. Chattaraj, Principle of maximum hardness. J. Am. Chem. Soc. 113 (1991) 1854-1855.

[36] R.G. Pearson, Hard and soft acids and bases. J. Am. Chem. Soc. 85 (1963) 3533-3543.

[37] L. Feng, H. Wang, F. Wang, Experimental and theoretical studies for corrosion inhibition of carbon steel by imidazolines derivative in 5 % NaCl saturated Ca $(OH)_2$ solution, *Electrochim. Acta.* 58 (2011) 427-436.

[38] C. Wang, J. Zhang, P. Yang, M. An, Electrochemical behaviors of Janus Green B in through-hole copper electroplating: An insight by experiment and density functional theory calculations using safranine T as a comparison, *Electrochim. Acta* 92 (2013) 356-364.

[39] G. Gece, The use of quantum chemical methods in corrosion inhibitor studies, *Corros. Sci.* 50 (2008) 2981-2992.

[40] Z. El Adnani, M. Mcharfi, M. Sfaira, M. Benzakour, A.T. Benjelloun, M. Ebn Touhami, DFT theoretical study of 7-R-3methylquinoxalin-2(1H)-thiones (R=H; CH_3; Cl) as corrosion inhibitors in hydrochloric acid, Corros. Sci. 68 (2013) 223-230.

[41] R. M. Issa, M. K. Awad, F.M. Atlam, DFT theoretical studies of antipyrine Schiff bases as corrosion inhibitors, *Materials and corrosion* 61 (2010) 709-714.

[42] J. Zhang, G. Qiao, S. Hu, Y. Yan, Z. Ren, L. Yu, Theoretical evaluation of corrosion inhibition performance of imidazoline compounds with different hydrophilic groups, 53 (2011) 147-152.

[43] G. Gece, Theoretical evaluation of the inhibition properties of two thiophene derivatives on corrosion of carbon steel in acidic media 63 (2012) 1-5.

[44] S. Chen, S. Scheiner, T. Kar, U. Adhikari, Theoretical study on relationship between structure of mercapto-triazole derivatives and inhibition performance. Int. J. Electrochem. Sci. 7 (2012) 7128-7139.

[45] S. John, M. Kuruvilla, A. Joseph, Adsorption and inhibition effect of methyl carbamate on copper metal in 1 N HNO_3: An experimental and theoretical study, RSC Adv. 3 (2013) 8929-8934.

[46] Y. Tang, L. Yao, C. Kong, W. Yang, Y. Chen, Molecular dynamic simulation of dodecylamine adsorption on iron surfaces, Corros, Sci. 53 (2011) 2046-2049.

[47] R. S. Perkins, J. D. Garber, Raman spectroscopy of iron in aqueous carbonate solutions, *J. Solut. Chem.* 32 (2003) 265-272.

[48] J. Zhang, J. Liu, W. Yu, Y. Yan, L. You, L. Liu, Molecular modeling of the inhibition mechanism of 1-(2-aminoethyl)-2-alkyl-imidazoline, *Corros. Sci.* 52 (2010) 2059.

Chapter 8

Microcorrosion Analysis and Their Effect in the Operation of Industrial Equipment of the Electronics Industry of Mexicali

Gustavo López Badilla,
María Marcela Acosta Gómez,
Elizabeth Romero Samaniego and
Sandra Luz Toledo Perea

Additional information is available at the end of the chapter

http://dx.doi.org/10.5772/57012

1. Introduction

The activities in the industrial electronic have been of great interest since the 1950 to the present days. These operations have been more important from the end of the Second World War (Lopez B. G. et al, 2007). Scientific research in MED in the design and the manufacture with more intensity revolutionized this area with the developing of great quantity of studies to improve these microelectronic components and systems used and fabricated in industrial plants and other activities (Lopez Badilla G., 2008). These investigations developed new devices such as the transistor in 1957, the microchip integrated circuits in 1966, that have been used since that time in the micro integrated control devices (MICD). These components led the develop of computer systems in 1970, the Scanning Electron Microscopy (SEM) in 1977 and some measuring instruments such as the oscilloscope in 1978 that are utilized in companies, in the researching and manufacturing process and scholar laboratories of electronic engineering (G. López Badilla, et al, 2013). Today, the MED are used in various industrial processes that manufacture microsystems, and these makes other different functions of industrial operations in the agricultural, medical, aerospace and space (Lopez G. et al, 2010). Some microelectronic devices are utilized as supply electrical power, control, indication, separation and organization in the manufacturing processes. Sometimes, for environmental factors as air pollution and climatic parameters as relative humidity (RH) and temperature, generates electrical failures originated by the corrosion process, causing economic losses by low yielding of the industrial equipments and machinery with MED. Corrosion in microelectronic devices is known as microcorrosion and is not detected at naked eye in the manufacturing processes, until the MED fails and causes uncontrolled situations in industrial plants, and preoccupy to specialized personnel, managers and owners. The microcorrosion is observed only by the SEM technique and other micro and nano procedures (Gustavo López B. et al, 2013). The indoor pollution generated by variations of climatic factors and CL of chemical agents from the outdoor sources, originated aggressive environments and the deterioration of materials very fast. For this reason, the MED are manufactured according to their functionality and corrosion resistance. At present, the materials used in the electronics industry in the world, are: aluminum, copper, chromium, tin, nickel, gold, palladium, platinum, silver, titanium, tungsten and copper alloys, tin-silver-palladium (Bella, 2002).

2. Electronics industry

The electronics industry has grown tremendously, and have a lot relation with other industries, increasing gradually in the developing countries, with a wide variety of electronics manufacturing equipments, which are used frequently (Lopez B. G, et al, 2010). In Mexico, there are a lot industrial parks with national and foreign companies located in the most important cities as Mexico City, Guadalajara, Monterrey and bordering cities of the United States (USA), as Ciudad Juarez, Mexicali, Nogales, Reynosa and Tijuana. In Mexicali, a report by the Industry Association of Mexicali (IAM, Asociación de Maquiladoras de Mexicali (AMAQ), in Spanish)-(AMAQ, 2012) indicated in 2010, the location of 150 industrial plants installed, being 100 of the electronics industry.

3. Design of electronic equipment

Sometimes, microelectronic devices with metallic materials from suppliers are affected by corrosion in the transportation process, in the reception of materials and in the storage of these MED (Lopez Badilla G. et al, 2011). Personal of warehouse of the electronics industry report some MED deteriorated each week of the year in this region of Mexico. Sometimes are reported the cases when it fails in the manufacturing processes or with the users, and is necessary pay the warranty generating economic losses (G. López-Badilla, et al, 2013). Factors as concentration levels of sulfurs mainly formed by the presence of hydrogen sulfide (H2S) and sulfurs oxide (SOX) that exceed the AQS principally in the winter season, and the variations of RH and temperature, promotes the formation of aggressive environments in indoors of the electronics industry and originates very fast the corrosive phenomena in this city, that deteriorate the MED very fast (G. L. Badilla et al, 2013). Corrosion in metallic materials used in the electronics industry are similar for other materials, but with two differences: being the first the applied voltage (105 to 106 V / cm), causing deterioration in the majority of MED by the uncontrolled flow current between conductors and ionic impurities (ASHRAE, 1999). The second difference is the size of components that are sometimes microns and require very low concentrations of pollution and film moisture to generate corrosion (Tahara A. et al, 2005). This cause electrical failures of any microelectronic device in short periods. Some electrical failures in these types of components have been reported by the migration of ions released during the formation of corrosive agents as metal oxidation, reducing to the cathodic process and generating dendrites (Lopez Badilla G. et al, 2013). This originates conductivity between the electrodes, allowing the growth of whiskers (being a detachment of metals) principally in cadmium, tin, gold, silver, zinc and copper-nickel alloy. Corrosion in welds is also considered important because it causes lack of adequate adhesion of the MED in the electronic boards (Gustavo López-Badilla, et al, 2013).

4. Industrial operations in the electronics industry

Technological advances in the electronics industry require increasingly smaller devices as the MED, with more components in their encapsulated and with major quantity of functions (Lopez Badilla G., 2008). The Micro Metal Oxide Semiconductor Field Effect Transistor (MMOSFET) and the Micro Bipolar Junction Transistor (MBJT), which are the most important basic electronic components in the electronic devices and systems, are developed as a special semiconductor with smaller dimensions than 0.1 microns (G. López-Badilla, et al, 2013). The number of functions of the MED depends on the amount of components increasing their operation capacity. The silicon semiconductor structure protects the internal elements of variations of RH and temperature, external connections to the chip and can dissipate the heat generated internally. A great advantage of use a material as copper in the MED is for their low cost and good electrical and thermic properties (G. López Badilla, et al, 2013), but is very susceptible to the corrosion phenomena. (Abdulaziz, 2003). The electronics industry which is located in the Mexicali city, have a wide variety of MED in the industrial equipments and machinery, and sometimes in indoors of these companies, are exposed to uncontrolled environments, generating the fast deterioration of electrical connections of the MED. In the manufacture of electronic devices and systems that is performed in the electronics industry, are presented four main steps, as mention in the figure 1. These stages are: (a) storage area, where are collected materials for the production process, (b) manufacturing zone, where is transformed the raw material, (c) inspection area where is evaluated the quality of products and (d) shipping zone where is installed the final product to send to customers Lopez B. G. et al, 2007).

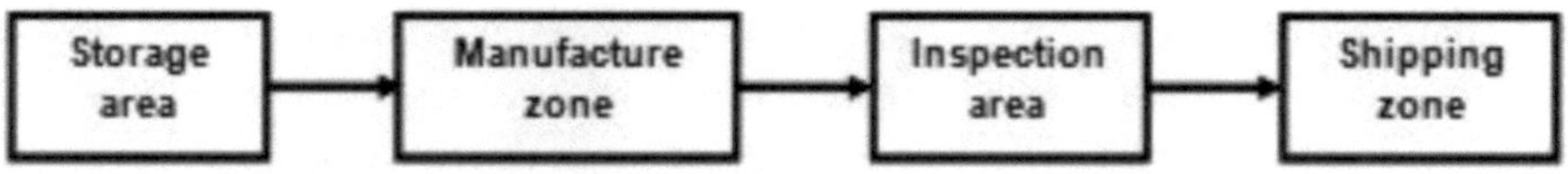

Figure 1. Steps of the manufacturing process in the electronics industry

5. Atmospheric corrosion in the electronics industry

Atmospheric corrosion is an electrochemical phenomenon generated by a wet film formed on the metallic surfaces of metals as copper used in the MED, where sometimes is visible and in other is invisible (Gustavo L. Badilla et al, 2013). This film is the electrolyte where the metal is dissolved, generating the metal ions movement, which promotes the corrosion phenomenon. The corrosion products forms dendrites and whiskers on the connections and joints of microelectronic equipment connectors. A feature of the electronic components is its design that avoids the accumulation of water, by variations of moisture. These are important factors in the formation of the electrolyte layer which cover the metal (Cerrud-Sánchez. S.M et al, 2005). The film formed on the metal surface of metals used in the MED utilized in indoors of the electronics industry, is thinner than the film formed in metals exposed at outdoor conditions. It requires a very precise analysis to detect the electrochemical process, indicating that the corrosion in indoor environments is different type that in the outdoor zones. This mean that the analysts of corrosion, specialized methods to detect the corrosion phenomena in metallic materials used in microelectronic devices and systems, with greater capacity and at micro scale to detection the corrosion, to reduce and control this electrochemical process. Air pollutants with more aggressiveness are the H2S and SO2 emitted from the principal outdoor source near of the Mexicali city, called the Cerro Prieto geothermal field, located at 25 km of this city, and these chemical agents are emitted and dispersed to the Mexicali city. The geothermic plant provides of electricity to the Mexicali city, their valley and some bordering cities of USA. The exposition of these chemical agents causes some tarnishing in the metals of MED. Other outdoor sources with a minimum effect are the traffic vehicle and the oxidation ponds composed of sewage (G.-López-Badilla, et al, 2013).

6. Corrosion in the electronics industry of Mexicali

Presence of corrosion of metals used in indoors of the electronics industry in Mexicali appears as an electrochemical process involving two aspects: wet and dry climate combined with the presence of H2S and SOX mainly. This increases the corrosion rate (CR). In the period of 2009 to 2010, the electronics industry reported an increment of 40% in the electrical failures of the MED, and specialized people was concerned and have interest for a company of the electronics industry in Mexicali for the presence of corrosion (G. López et al, 2012). A industrial plant of this city allowed the realization of a researching of the fast deterioration of the MED. For this reason, the MED of electronic industrial systems and machinery metallic were evaluated in periods of 1, 3, 6, 12, 24 and 36 months in the manufacturing and warehouse areas of the company analyzed, to determine the CL and the principal causes of the corrosion phenomena. The generation of electrical failures in industrial electronic equipment, provoked defective MED, because at certain times of the year will occur more often. An analysis reported in the operational yielding, indicated that electrical failures were increased by up to 40%. This occurred principally in the months from July to August of 2008 when the RH and temperature were higher than 75% and 40 °C, and from December of 2009 to January of 2010, and when the

RH and temperature were near of 90% and 15 °C. From December of 2009 to February of 2010, was showed the greenhouse effect where air pollutants are not dispersed easily. This information supports the hypothesis of the climatic and pollution conditions that were the principal parameters of the generation of corrosion in the electronics industry of Mexicali (Gustavo López B. et al, 2013).The topography of this city and the winds, which originates the dispersion of the air pollutants were other factors of the presence of this electrochemical phenomenon. Certain pollutants in an area of the city and its valley, reach other zones and in some times with lower levels are generated the aggressive environments and the corrosion process. It is for this reason that this research was developed to evaluate the indoor atmospheric corrosion in a company and the correlation with the production yielding of the industrial equipment and machinery. Besides the air pollutants mentioned which influence the corrosion process, there are other small particles that cause the formation of corrosive agents as microorganisms (Leidecker, 2006). Analysts of microcorrosion, mention that the absorption method is the best technique to evaluate the corrosion products of copper, nickel and zinc. This indicates a strong influence of carboxylic acids and organic species, such as acetates, oxalates formats and internal emitted by anthropogenic sources as burning wood and tires, that are small aspects in the electrochemical phenomena, but generates deterioration in metallic surfaces of electronic devices (Fontana, 1986). An example of the formation of corrosive agents is shown by a study conducted in Japan with the comparison of metals of copper and tin, after three months of exposure to sulfate occurred in the electronic industry of Mexicali is showed next in Figure 2.

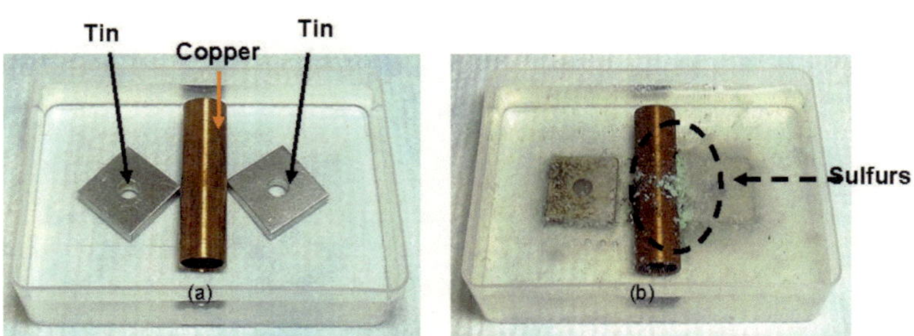

Figure 2. Analysis of copper metal: (a) before and (b) after the corrosion process

7. Description of the project

This research project was performed to determine the corrosive effect of air pollutants that adhere to the joints and connections of the MED used in the electronics industry of Mexicali. The main objective is know the kinetic of the corrosion phenomenon generated in the metallic materials of the MED and establishes the prevention and control of corrosion. With the information obtained from this study, was indicated the CL in deterioration of the MED obtained of the evaluations in indoor of the electronics industry, as based on the presence of atmospheric and climatic parameters. The analysis was made with the specifications of ASTM standards, and was determined the CL. This analysis simulates the deterioration of electrical connections of the MED in harsh environments with the presence of sulfates combined with the variations of RH and temperature. Extreme climatic conditions of outdoors of the electronics industry sometimes are uncontrollable in the indoor areas of industrial plants installed in Mexicali. The corrosion process was promoted very easy with values higher than 75% and 30 °C, that are very common in the majorly of the periods of the year in Mexicali. This promotes the electrochemical phenomena and forms small and big spots in the metallic surfaces of the MED. To control this problematic situation, it was necessary to use techniques and methods of protection, such as corrosion inhibitors and surface treatments (Ponchak, 2004).

8. Comparative studies of microcorrosion

Several studies mention about the effect of microcorrosion generation in the operation yielding of microdevices used in the electronics industry. This process of analysis indicates that at the minimum presence of corrosion in these electronic components, generates a significant impact on the deterioration of electronic microdevices. Below, shows a descriptive analysis of some studies in various areas of the world in Table 1.

Authors	Title	General aspects	Effects of microcorrosion
Gustavo Lopez Badilla, Benjamin Valdez Salas, Michael Schorr Wiener and Carlos Raúl Navarro González	Microscopy and spectroscopy Analysis of MEMS corrosion used in the electronics Industry of the Baja California Region, Mexico	Analysis of mcirocorrosion and their effect in the fabrication of MEMS	Generation of electrical failures in MEMS used in the electronics industry
Gustavo López-Badilla Catalina González-Hernández Antonio Valdez-Ceballos	Análisis de corrosión en MEMS de la industria electrónica en ambientes árido y marino del noroeste de México	Generation of aggressive environments in indoor of industrial plants caused by: - Air pollutants as sulfurs in arid zones and chlorides in marine regions. - Variations of climatic factors as relative humidity and temperature	Low productive yielding, originating economic losses in the economy of this region with deteriorated material, rework, defective products, warranty
Pouponneau P, Savadogo O, Napporn T, Yahia L, Martel S	Corrosion study of single crystal Ni-Mn-Ga alloy and Tb0.27Dy0.73Fe1.95 alloy for the design of new medical microdevices	Analysis of the deterioration of alloys of smart magnetic materials used in the medical activities y the corrosion process	Origination of electrical failures in medical microdevices, caused economic losses by the defective of microcomponents

Table 1. Comparative analyses of microcorrosion

9. Methodology

Water exists in the atmosphere as liquid phase, steam and solid. Moisture levels depend on the steam pressure which exists in a particular environment. The RH and temperature were the most important climatic factors in this study, which promoted the increase very fast the CR. The information of air pollutants, temperature and RH by periods daily, weekly, monthly, seasonally and at the year, was obtained and organized by the EMS (Environmental Monitoring Systems) for the project of the border cities of Mexico and USA installed for the Environmental Protection Agency (EPA) (NAAQS, 2012) in according with the Secretaria de Medio Ambiente y Recursos Naturales (SEMARNAT of Mexico). The proposed methodology used was according to the standards of ASTM (G4) (ASTM, 2001), and the study was made from June of 2009 to August of 2012 with the experimental procedures in indoors of the electronics industry to determine the CL with the ISO 9223 for the CL (ISO/CD 11844-1, 2001, ISO 9223, 1992).

10. Statistical information

The pollutants H2S, SO2, NOX O3 and CO detected by the EMS in the outdoor environments Mexicali, which penetrate at indoors of the electronics industry. This mixture of air pollutants, together with variations in moisture and temperature, rises the CR of MED used in the electronic equipment. For the correlation analysis data were used periods where the RH and temperature were above 70% and 35 ° C, and the concentrations of air pollutants mentioned above that overpass the levels of AQS. The information obtained was organized in tables with the corrosion of metallic surfaces of the MED analyzed to determine the degree of correlation between climatic parameters, contamination and corrosion.

11. Environmental monitoring

The air pollutant data was obtained with specialized equipment containing data acquisition systems specialized and organized on a monthly, controlled by the EPA. The measuring instruments techniques are the chemiluminescence analyzer model 42 to the H2S, the Thermo Environmental Instruments Inc. to detect NOX, 300E model with a filter API (Advanced Pollution Instruments Inc.) for CO, the model photometric analyzer of Thermo Electron Corporation 43C for SO2 and model 400 detectors API for O3. These instruments have special filters to detect gases and particles (Annex A. Mechanisms of measurement).

12. Scanning electron microscopy (SEM) technique

This technique was used to determine the chemical elements from compounds which react with the metallic surfaces of the MED analyzed. The electrons flowing through the low energy generated by the instrument topography showed electrical connections of the MED corroded, indicating the particle analysis, detection of defective MED and the metallurgical defects. With this technique, were determined the chemical composition observed in a spatial distribution of metallic surfaces of MED damaged. This was made in order to know which of the pollutants reacts in the metals to know the chemical agent with major influence in the generation of the corrosion process. A distinction was made between the different possible compounds for the oxidation state of the metal ions involved in the corrosion products. This allowed observe the morphology of corrosion products on the surface depending on the elemental composition. This technique is known in detail, quickly and accurately, to determine the structural form and location of corrosive agents, with which we determined the type of corrosion (uniform and pitting, which are the more common presented in the electronics industry of Mexicali), to suggest the best possible methods and techniques protection for metallic materials of the MED.

13. Results

The origin of corrosion in indoors of the electronics industry was due to variations of the climatic factors that were mentioned above and combined with the presence of sulfurs principally in the Mexicali city. In the winter season, appeared uniform corrosion by the wetting film formed on the metallic surfaces of the MED, and in the summer period, was generated the pitting corrosion in these materials by the occurrence of some droplets that adhere to certain areas of the metal surface. This is partly was due to the wetting time (also called time of wetness (TOW), which is an important factor of the origination of the corrosion phenomena). The TOW was obtained with the evaluation of RH and temperatures above 75% and 35 °C that is common in the city analyzed in the majorly of periods of the year.

14. Evaluation of climatic factors

In this study, the corrosion process initiated in indoor atmospheres of the electronics industry of Mexicali city, influenced by external climatic conditions (Gustavo L. Badilla et al, 2013). The wetting time obtained with the RH and temperature of the city of Mexicali, shows high rates in some winter and summer periods, being an important factor for the type of corrosion that occurs in metals used in the MED of industrial plants, and the levels at which the CR increases. In this process are formed some voltage electrodes, for example in the binding of copper to tin the voltage was 0.2 volts, and caused an electrical malfunction. The minimum wetting time in the winter period was presented in December with a value of 58 cycles and the maximum peak was in February with 163 cycles. This occurred with a temperature range in outdoor environments from 4 ∘C to 24 ∘C, resulting in a negative effect in the indoors environments of the electronics industry. In the summer period from July to August of 2012 the RH is greater than 75%, and temperatures above 35 ∘C, and some droplets were formed in some areas of the metal surface, generating selective corrosion of metal parts, being mainly pitting corrosion. In change, in the winter seasons from December of 2011 to January of 2012 with indices of RH and temperature higher than 90% and lower than 20 °C, occurred the formation of the wetting film in the entire metallic surface presenting the uniform corrosion. The wetting time in the summer period of 2011 was 47 cycles as the minimum level presented in June and the maximum was 135 cycles in August. This factor was very essential to know the kinetics of corrosion in MED, being with major effect in summer than in winter by the presence of pitting corrosion.

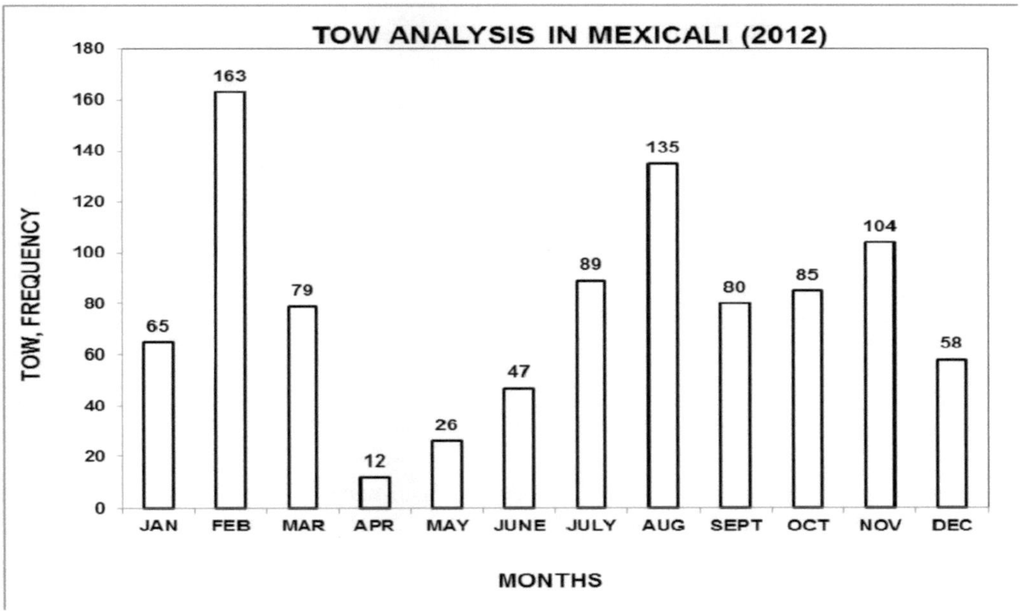

Figure 3. Analysis of TOW in Mexicali by monthly periods (2012)

The TOW influenced in the velocity of deterioration of MED presented a CR higher in the summer periods than in winter seasons due to the effect of temperature (greater than 35 ○C) that is showed in figure 3. The design of electronic components and equipment, such as size, structure and characteristics of the metal surface is essential to prevent the corrosion phenomena. At the beginning of the analysis period was less the TOW, in the summer month with 0.22, but increases from 3AM and 9AM to its maximum value of 0.57 and started down to get close to 0 from 24:00 to 6PM. In February, there were initial values of 1:00 a.m. to 3:00 a.m. 0.25 and maximum of 0.38 at 5AM where it remained and began a descent from 8AM and stay between 0.1 and 0.2 and increases again from 8PM to a value close to 0.3. These cycles of wetting time indicated the duration of the corrosion process in metals generated electronic equipment and the formation of corrosive agents in connections and electrical connectors. The influence of cold air causes the temperature in a metal surface, decreased below the dew point, with the RH levels higher than 80% and was formed on the metal film. This occurred in Mexicali in the period of the study, causing corrosion, which depended on the size, orientation and surface characteristics where the phenomenon caused. With the ISO 9223 standard, levels were determined the wetting time, with the lowest effect T1 and T5 that a major factor in the deterioration of metals used in the electronics industry. In this city, even if the wetting time has levels from T1 to T3 (in annual period), and T1 and T2 in February, August and November, corrosion is generated by regional climatic variations. The annual analysis shows the hours where there were peaks of TH 5AM being at 7AM and in this period were additionally obtained the hourly averages, standard deviation and minimum, to know in more detail the process of film formation on surfaces metallic materials evaluated (Table 2).

Parameters	Peak Hours at Monthly periods	Peak Monthly Periods			Annual
	5, 6, 7AM (Fcia.)	February	August	November	
Average	8 , 9, 7 (T1)	7 (T10	6 (T1)	4 (T1)	63 (T2)
Standard Deviation	4, 5 , 5 (T1)	3.1 (T1)	6.2 (T1)	3.3 (T1)	38.7(T2)
Minima	1, 1, 0 (T1)	3 (T1)	0 (T1)	1 (T1)	27 (T2)
Maxima	15, 17, 17 (T2)	12 (T2)	17(T2)	10 (T2)	134 (T2)
Total	399 (T3)	163 (T2)	135 (T2)	104 (T2)	1504 (T3)

Table 2. Evaluation of TOW by monthly and annual periods in according to ISO 9223

This evaluation was performed for the periods in which need have better control of indoor climate controlled the industrial plants of this city.

15. Analysis of environmental pollution

Data of air pollutants obtained by the EMS showed interesting information to determine the principal chemical agent in the deterioration of the MED. In this section of the study, were evaluated only the pollutants with the different ranges of the NO2, O3 and SO2, and not analyzing the CO for their ranges greater than the by monthly periods. The analysis was made at the times of this chemical agents overpassed the AQS in daily periods representing in percentages days of the 2012 as showed in figure 4. Higher levels present the SO2, NO2 and follows after O3 in 2012 year where the values increased to almost 56 % of times overpassed the AQS for SO2, as the maximum level in January and December and the minimum indices was 8 in July. For NO2, the maximum index with 34 was in January and the minimum level was 4 in July. And to the O3, were 28 in January as the maximum level and 2 in July as the minimum index. The air pollution periods varies in different hourly, daily, weekly, monthly and yearly periods, after by the traffic vehicle and industrial plants mainly, which originates the formation of greenhouse effect principally in the winter season and the dispersion of pollutants in the Mexicali city depending the wind direction and speed. The indoor atmospheric corrosion of the electronics industry has been controlled but not eliminated in cleanrooms, and sometimes metals like copper have spots that indicate the onset of corrosion. The atmospheric corrosion in indoor environments is reduced when they controlled microclimates, which increases the performance of electronic equipment. The environments without climate control, the corrosion rate increases rapidly, damaging metallic materials. The deterioration of metals in February was lower than in August as temperatures above 35 ∘C. In winter SO2 concentrations have a large effect on the poor productive yielding of the MED and thus industrial systems and machinery. It was easily dispersed into the atmosphere and easily adheres to metals, but the corrosion process is slow. The correlation of SO2 at low levels of RH and temperature deteriorates the MED too, mainly in the months of February and August. An analysis of the CR according to the formation of corrosive agents in joints and connections was made, and were higher in August than in February, as a result of higher temperature ranges. In winter we evaluated the wet film formation and corrosive metals and concentration indices of air pollutants.

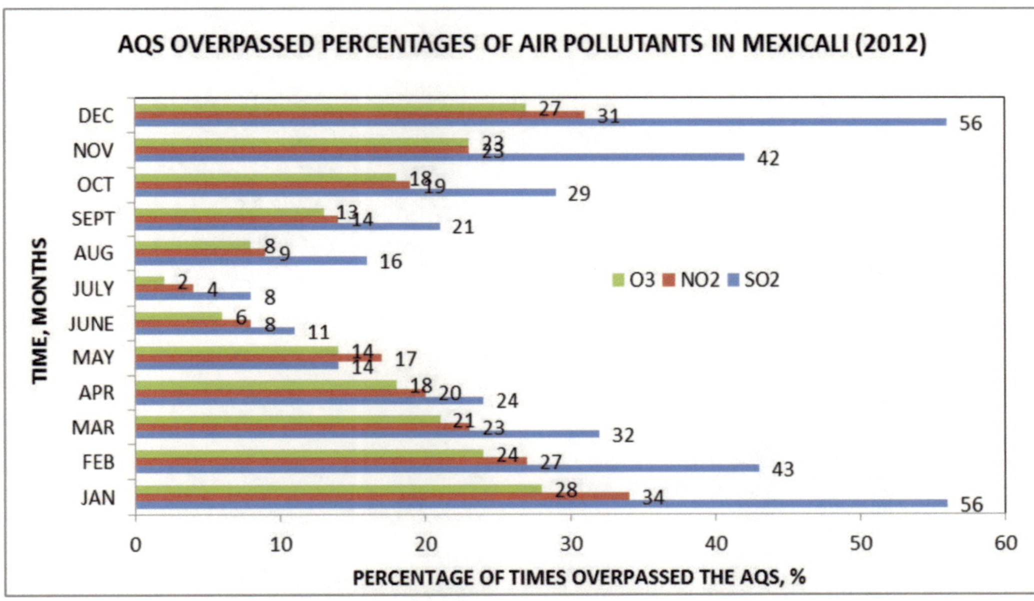

Figure 4. Analysis of TOW in Mexicali by monthly periods (2012)

16. SEM evaluation

The SEM analysis shows the grade of deterioration in the MED electrical connections of the MED. The microphotographs were obtained of the microelectronic devices damaged. An evaluation indicates at least 10 electrical failures at week of the MED as was observed in figures 5 and 6. At macro levels are not observed the deterioration of the materials, but in the SEM analysis were showed the great effect of the corrosion process. To detect very fast the corrosion phenomena, was necessary considering always the use of microanalysis with the SEM technique. Figure 5 shows some MED damaged by corrosion at 500X and after was made an evaluation at 5X to detect with better precision the deterioration of the MED corroded in the spring and summer periods of 2012. At 5X, was observed the formation of the uniform corrosion in spring in the majorly if the metallic surfaces of electrical connectors and connections of the MED. This was promoted by variations of the RH principally. In change in the summer season indicates the presence of in a small section of the metallic surfaces the presence of uniform corrosion and tending to appear the pitting corrosion, being an important factor for it the high levels of temperature.

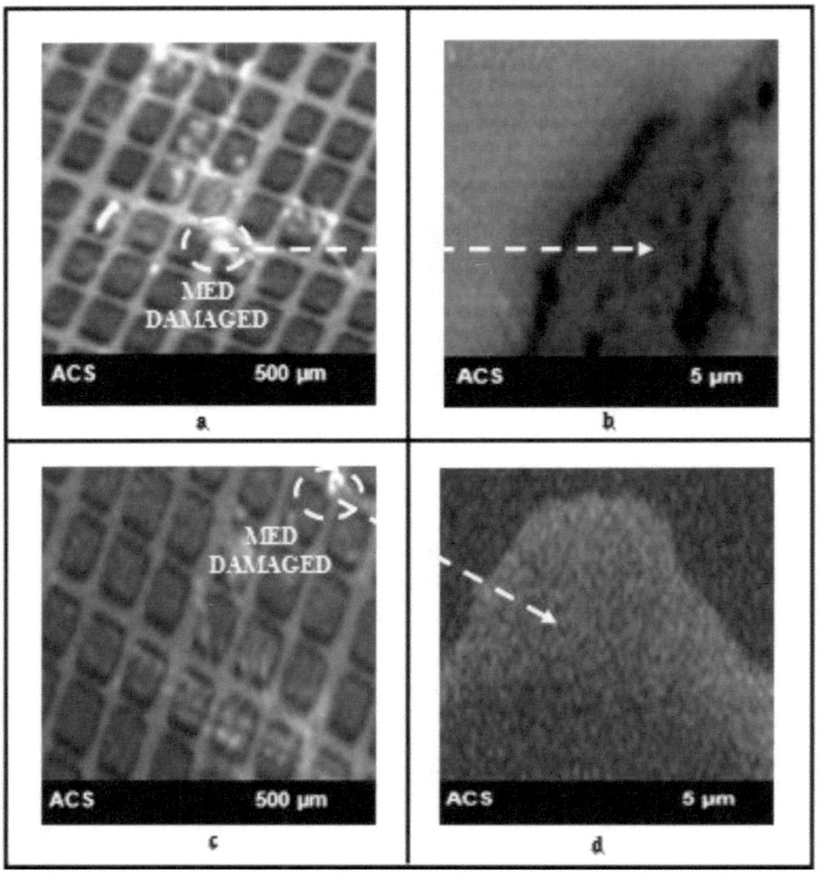

Figure 5. SEM analysis the deterioration of MED in spring at (a) 500 X and (b) 5 X and in summer at (c) 500 X and (d) 5 X in the electronics of Mexicali (2012)

In Figure 6, are showed the analysis of the autumn and winter periods, indicating in the autumn season, a great deterioration represented by the black sections, promoted by the levels higher than 80% of the RH. In the winter period, the appearance of uniform corrosion in the majorly of the metallic surfaces, indicated a similar deterioration as in the spring season by the variations of RH. This shows the kinetic of the corrosion process in different season of 2012 and the mechanisms that promote this electrochemical phenomenon.

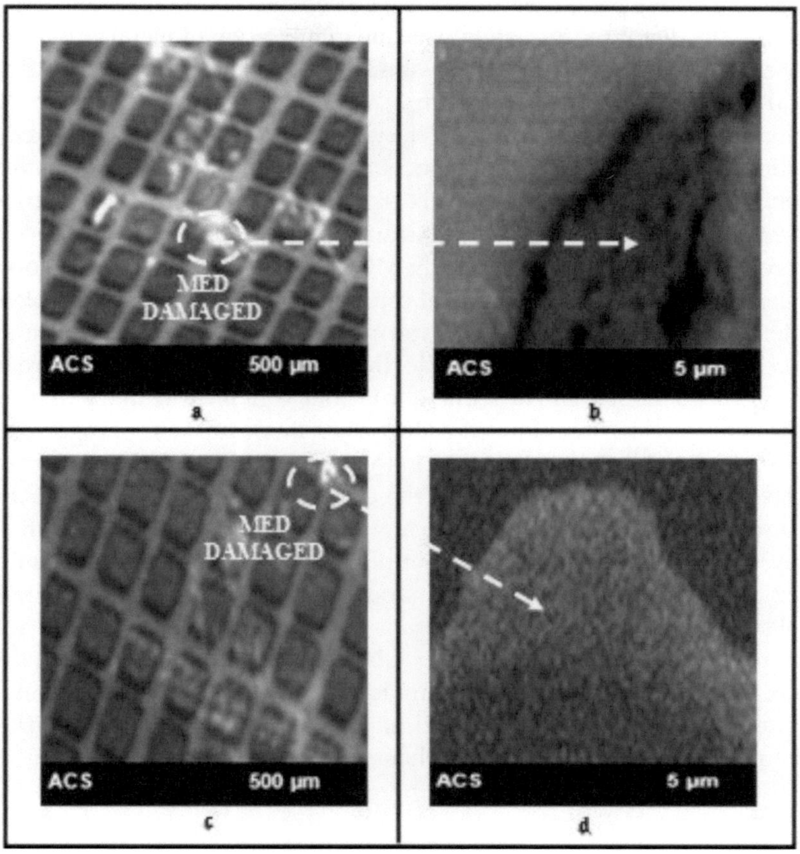

Figure 6. SEM analysis the deterioration of MED in spring at (a) 500 X and (b) 5 X and in summer at (c) 500 X and (d) 5 X in the electronics of Mexicali (2012)

The importance of analyzes the phenomenon of corrosion in metallic materials used in the electronics industry in Mexicali, is to assess the negative effects of corrosivity levels in the indoors of the electronics industry in Mexicali city. The metals used in the MED electrical connections were constituted of copper, gold, silver and tin mainly.

17. Conclusions

The presence of corrosion in the MED caused the low productive yielding in industrial equipments and machinery. The mathematical correlation was made from the information obtained of the pollutants that was adhered to the metallic surfaces. The chemical agents were emitted from outdoor sources and penetrate to the indoor environments, through holes, cracks or air conditioning flow. The uncontrolled climate parameters as humidity and temperature, promoted the formation of a wet film on metal surfaces, generated by the corrosion process. The deterioration of materials generated bad function in the microelectronic devices due to inappropriate conductive paths (dendrites) formed between connections unnecessary electrical connections. The dendrites formation was due to migration of metal ions released by the etching process, which presents a voltage caused by corrosion current as a flow between conductors in the presence of mobile ionic impurities. Thus, electrical currents are generated in the order of micro amperes components alter their function by electrical failures and reducing its operational performance. Most components require specific values of current and voltage for operation, or they stop working. The corrosion phenomena observed in various metallic materials used in the electronics industry was similar in some phases, as was indicated by other studies. However the metals that constitute the MED showed different types of corrosion at the final stage depending on the environmental conditions. Corrosion was a problematic situation in the electronics industry of Mexicali, even in clean rooms. The develop tests to detect the origin and occurrence of corrosion is difficult for the complexity of the microelectronics devices, and lack of controlled indoor climates. Climatic factors, combined with the fine particles and gases such as hydrogen sulfide, sulfur dioxide, carbon monoxide and nitrogen oxides, which sometimes exceed the air quality standards in the region, generating harsh environments in indoor of the electronics industry, causing the corrosion phenomena. One of the principal emission sources which generate aggressive atmospheres in indoor of the electronics industry of this city was the geothermic plant that supplies electricity to Mexicali city and their valley and some cities of United States. Other minimum emission sources are the traffic vehicle, industrial plants, soil erosions and microorganisms. These pollutant particles adhere to the metal surfaces with or without the film formed by climatic variations and start the corrosion process. Also, with levels greater than 80% humidity and 25 ◦C that is common in the Mexicali city was obtained the TOW cycles, which indicates the periods in which a metal surface remains wet enough to give way to a process electrochemical corrosion to occur.

Author details

Gustavo López Badilla[1], María Marcela Acosta Gómez[1], Elizabeth Romero Samaniego[2] and Sandra Luz Toledo Perea[2]

1 Instituto Tecnológico de Mexicali, Mexicali, BC, México

2 Instituto Tecnológico de Ensenada, Ensenada, BC, México

References

[1] Abdulaziz A. And Maher A.; Atmospheric corrosion investigations of iron using quartz crystal microbalance; The Electrochemical Society; ISSN: 1945-7111 online, ISSN: 0013-4651 print; 2003.

[2] ASTM (Annual Book of ASTM Standards); Section Three; Metals Tests Methods and Analytical Procedures; Vol. 03.01; 2001.

[3] ASHRAE; Handbook; Heating, Ventilating, and Air-Conditioning; applications; American Society of Heating, Refrigerating and Air-Conditioning Engineers, Inc.; 1999.

[4] Asociación de Maquiladoras de Mexicali (AMAQ); Departamento de Estadística; Reporte Anual de la Industria en Mexicali; Gobierno Municipal; 2012.

[5] Bella H. Chudnovsky; Degradation of Power Contacts in Industrial Atmhospere: Silver Corrosion and Whiskers; Square D Company Report; 2002.

[6] Cerrud-Sanchez. S.M, Armendáriz J., Ortiz-Prado V.H., Schouwenaars R.; Deterioro por corrosión atmosférica de componentes electrónicos de equipo telefónico; INGENIERIA. Investigación y Tecnología VI, 4. 219-237; ISSN 1405-7743 FI-UNAM; 2005.

[7] Fontana Mars G.; Corrosion Engineering; Editorial McGraw-Hill; 1986.

[8] G. López Badilla, M. M. Acosta Gomez, E. Romero Samaniego, S. L. Toledo Perea, Luz Karina Angulo Balderrama, José Ricardo Silva Talamantes; "Micro and Nanocontaminacion Affects the Operation of Antibacterial Filters in Public Hospitals of Mexicali, B.C., Mexico"; Journal of Chinese Medicine Research and Development; Jan. 2013, Vol. 2 Iss. 1, PP. 15-20, ISSN:2303-9345 (Print), ISSN:2303-9353 (Online), 2013.

[9] Gustavo López B., María Marcela Acosta G., Elizabeth Romero S., Sandra Luz Toledo P., Rodolfo Thomas Guevara G., Carlos Omar Gutiérrez M., Juan Pablo Quezada H.; "Corrosion in Control Systems Decrease the Lifetime of the Electronic Devices of the Industrial Plants of Mexicali, BC, Mexico"; Open Journal of Air Pollution, 2, pp. 29-35; ISSN Print: 2169-2653 ISSN Online: 2169-2661; 2013.

[10] G. L. Badilla, M. M. A. Gomez, E. R. Samaniego, S. L.T. Perea; "Optical MEMS used to detect micro-corrosion in steel cans of food and beverage industries in arid zones"; Research and Precision Instrument and Machinery Journal; ISSN Online: 2327-1132; ISSN Print: 2327-1124; (Accepted, 2013).

[11] Gustavo López-Badilla; María Marcela Acosta-Gómez, Elizabeth Romero-Samaniego, Sandra Luz Toledo-Perea, Mirian Maleny Garcia-Castrellon, Luis Alberto Gameros-Ríos "Evaluation of effect of foucault currents in PLC influenced by atmospheric corrosion in the electronics industry; Revista Ingeniería, Investigación y Tecnología-UNAM; ISSN 1405-7743 FI-UNAM (En revisión, 2013).

[12] G. López-Badilla, M. M. Acosta- Gómez, E. Romero-Samaniego, S. L. Toledo-Perea, L. J. Cruz-Romero, D Hernández-Villasana, C. A. Patiño-Aguirre, Y. Romero-Aguirre; "Analysis in real time and simulation of computer systems damaged by corrosion"; Revista Cientifica; ISSN 1665-0654, ESIME Instituto Politécnico Nacional MÉXICO (En revisión, 2013).

[13] Gustavo L. Badilla, María Marcela A. Gómez, Elizabeth R. Samaniego, Sandra Luz T. Perea, Ricardo C. Soria, Deisy Judith C. Rodriguez, Gisel Guadalupe L. Fernández, Janeth Guadalupe R. Sánchez, Magnolia V. Soto; "Control de humedad para evitar la

generación de la corrosión en interiores de la industria electrónica en la costa de Baja California de México"; Revista Nova Scienta; Universidad LaSalle; León Guanajuato; ISSN 2007 – 0705; (Aceptado, 2013).

[14] G.-López-Badilla, M. M.-Acosta-Gomez, E.-Romero-Samaniego, S.L.-Toledo-Perea; "High-tech industrial development vs. climate change of the Sonora-Baja California region of Mexico"; American Journal of Chemical Engineering; Science Publishing Group; 1(1): 11-16; 2013.

[15] G. López, Ángel Vega M., Diego Millán A., José González K., Guillermo Contreras L.; Revista de Investigación, Ingeniería y Tecnología-UNAM, " Effect of Corrosion in the CS Operation Indoors of the Electronics Industry in the Northwest of Mexico ", Revista de Ingeniería Investigación y Tecnología, volumen XIII (número 4), octubre-diciembre 2012: 461-472; ISSN 1405-7743 FI-UNAM; 2012.

[16] Gustavo Lopez Badilla, Benjamin Valdez Salas, Michael Schorr Wiener and Carlos Raúl Navarro González; "Microscopy and Spectroscopy Analysis of Mems Corrosion Used in the Electronics Industry of the Baja California Region, Mexico"; INTECH Book, Chapter 9, 2012.

[17] Gustavo López-Badilla, Catalina González-Hernández, Antonio Valdez-Ceballos; Análisis de corrosión en MEM de la industria electrónica en ambientes árido y marino del noroeste de México"; Científica, vol.15, núm. 3, pp. 145-150, julio-septiembre 2011.

[18] ISSN 1665-0654, ESIME Instituto Politécnico Nacional MÉXICO

[19] ISO/CD 11844-1, Corrosion of metals and alloys – Classification of low corrositivy in indoor atmospheres – Part 1: Determination and estimation of indoor corrosivity, 2001.

[20] ISO 9223, Corrosion of Metals and Alloys. Corrosivity of Atmospheres. Classification, International Organization for Standardization, Geneve, Switzerland, 1992.

[21] Leidecker H., Brusse J; Tin Whiskers: A History of Document Electrical System Failures; Technical Report to Space Shuttle Program Office; April 2006

[22] NAAQS (National Ambient Air quality Standards); U.S. Environmental Protection Agency (EPA); http://www.epa.gov/air/criteria.html; 2013.

[23] López-Badilla G. Caracterización de la corrosión en materiales metálicos de la industria electrónica en Mexicali, BC, thesis (Ph.D.), Spanish, 2008.

[24] Lopez B.G., Valdez S.B., Zlatev K.R., Flores P.J., Carrillo B.M., Schorr W.M. Corrosion of Metals at Indoor Conditions in the Electronics Manufacturing Industry. Anti-Corrosion Methods and Materials, 2007.

[25] Lopez G., Tiznado H., Soto G., De la Cruz W., Valdez B., Schorr M., Zlatev R. Corrosion de dispositivos electronicos por contaminación atmosferica en intertiores de plantas de ambientes aridos y marinos. Nova Scientia, 2010 (in spanish).

[26] Lopez B.G., Valdez S.B., Schorr W.M., Tiznado V.H., Soto H.G. Influence of Climate Factors on Copper Corrosion in Electronic Equipments and Devices. Anti-Corrosion Methods and Materials, 2010.

[27] López Badilla G, Valdez Salas B, Schorr Wiener M, Zlatev R., Tiznado VH, Soto HG, De la Cruz W. AES in corrosion of electronic devices in arid in marine environments. Anti-Corrosion Methods and Materials. 2011; 6(8): 331-336.

[28] López Badilla G, Valdez Salas B, Schorr Wiener M, Carrillo Beltrán M, Radnev Nedev N, Koytchev Zlatev R, Stoytcheva Stilianova M, Ramos Irigoyen R. Microcorrosión en sensores ópticos usados para detectar microorganismos en industrias de alimentos de Tijuana, México. En Valdez B, & Schorr M (Eds.). Corrosión y preservación de la infraestructura industrial. Barcelona, España: OmniaScience; 2013. pp. 157-173.

[29] Ponchak G.E.; Multichip Module-Dielectric Package; http://parts.jpl.nasa.gov/mmic/9-IV.PDF; 2004.

[30] Pouponneau P, Savadogo O, Napporn T, Yahia L, Martel S.; Corrosion study of single crystal Ni-Mn Ga alloy and Tb0.27Dy0.73Fe1.95 alloy for the design of new medical microdevices; J Mater Sci Mater Med. 2011 Feb;22(2):237-45. doi: 10.1007/s10856-010-4206-2. Epub 2011 Jan 11.

[31] Tahara A. and Shinohara T.; Influence of the alloy element on corrosion morphology of the low alloy stell exposed to the atmospheric environments; Corrosion Science, Vol. 47; pages 2589-2598; ISSN: 0010-938X; 2005.

Chapter 9

Modern Methods for Assessing the Corrosion Resistance of Dental Alloys Used in Dentistry

Ion Patrascu, Vlad Gabriel Vasilescu and
Stefan Milicescu

Additional information is available at the end of the chapter

http://dx.doi.org/10.5772/57418

1. Introduction

The corrosion process involves a sequence of reactions in which a metal or an alloy is attacked by an aggressive agent [4, 13]. Due to their particular electronic structure characterized by the presence of free electrons in the crystalline network, metals and alloys react chemically or electrochemically and thus may suffer corrosion processes [13, p.85]. The stabile chemical state of the metals is the combined one (oxides, hydroxides, other components) as they appear in the ore from which it was derived. In other words, the metals have a spontaneous tendency to return in a thermodynamically stable combined state through corrosion. Chemical Corrosion is the destruction of metallic materials after direct chemical action to occur without an environment changed by the electric charges. The products resulting from the interaction between the metal and environment remain on the metal surface in formats of thin films. The initial oxide layer/film has a thickness of 10 to 20A0, and then corrosive attack depends on its physical properties such as adhesion, compactness, porosity and the volume of this layer in relation to the volume of metal. The nature of the parent metal is determining both the thermodynamic stability and the properties of the oxide film formed on its surface. Some alloys have a corrosion resistance higher than that of the technical metals through their alloying elements (Al, Cr, Ni, Si) which allow the formation of protective oxide coatings. Electrochemical corrosion - represent destructive attack on metals or alloys exercised by the corrosive environment through electrochemical reactions, in which the metal releases electrons and its positive ions enter in solution. The formation of positive ions and electrons create an electrical potential (in volts) called electrode potential [28, 29].

Series electric potentials classify the chemical elements according to their tendency to gain (reduction) or lose (oxidation) electrons in electrolyte solutions, taking as reference the electrical potential of the hydrogen which is conventionally assessed value of 0,000 volts. Elements that have negative values in this series of electric potentials are chemically unstable, highly reactive chemical that is very sensitive to aggressive environment. Instead, elements that have positive values of electrical potential are less reactive metals. Graphical representation of the balance between metal and its various oxidizing species is given in the diagram "potential - pH" in isothermal conditions or thermodynamic stability diagrams (Pourbaix diagram) [13]. They provide thermodynamic data about corrosion phenomena, indicating equilibrium conditions of all the reactions that can take place between the metal and the aggressive environment at a given temperature. Pourbaix Diagrams set only thermodynamic conditions in which chemical reactions can take place between the metal and the corrosive environment. Completing them with kinetic data gives real guidance on the extent of destruction of the metal in the environment.

2. Methods for assessing the corrosion resistance of metals and alloys

For the evaluation of corrosion processes one can use direct methods and indirect methods.

Direct methods consist in weighing or measuring thickness of samples, aggressive agent consumption or gas measurement resulting from corrosion processes. Indirect methods involve electrochemical measurements, electrical, acoustic, optical, etc.

In electrochemical methods the corroded metal is assessed by measuring the amount of current flowing in the process. The corrosion rate is obtained by reporting the amount to the area and time.

3. Corrosion of metals and alloys used as biomaterials

As discussed previously, corrosion is a type of damage in which the metal leaves its metallic state to form cations in solution or is converted into solid compounds (called corrosion products). The corrosion resistance of metallic materials used in oral implants is an important issue, expressing their biocompatibility. The state many metals are in an oxygenated environment or a hydrate one characterized by the lowest energy is that of the oxide. Corrosion occurs when the metal atoms are ionized and go into solution, or they are combining with oxygen or other substances to form a compound that is removed or dissolved. A variety of different chemical reactions take place when a metal is placed in an aqueous environment (Figure 1). The electrolyte containing ions in solution acts to close the electrical circuit. In the oral cavity, corrosion of metals and alloys is a galvanic corrosion which is produced by forming a electrochemical or galvanic cell in which the electrolyte is liquid cavity. Anions are negative ions, which migrate to the anode and the cations are positive ions that migrate to the cathode.

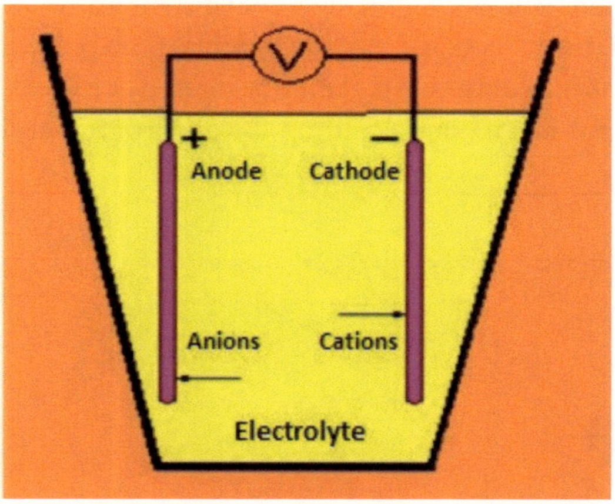

Figure 1. Electrochemic cell [29,p.53]

If two similar metals are present in the same environment, one that is the most negative in galvanic series will become the anode and bimetallic corrosion will occur (or galvanic).

Pourbaix diagram includes regions (zones) of corrosion, passivity and immunity, depending on the electrode potential and pH [13, 14]. Immunity is defined as the equilibrium between the metal and its ions less than 10^{-6} molar [29]. In the immune region the corrosion is energetically impossible. In the passivation domain, a stable solid constituent is an oxide, a hydroxide, a hydrate or a metal salt. Passivity is defined as a balance between a metal and its reaction products (oxide, hydroxide, etc.) at a concentration of less than 10-6 molar [29,p.55] g per liter atom (molar). This applies only if the reaction products (oxides) are adherents. In the field of biomaterials, passivity may be appropriate or not: the destruction of a passive layer can cause an increase in corrosion.

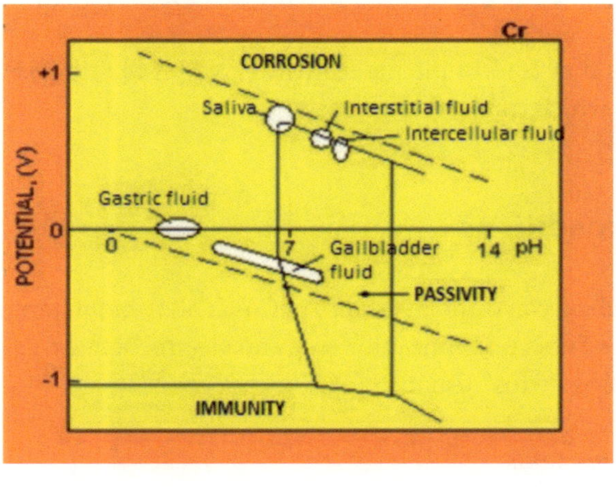

Figure 2. Pourbaix diagram for chromium (Cr) the regions associated with different body fluids (JHDumbleton and J.Black, An Introduction to Orthopaedic Materials, Charles C.Thomas, Springfield, 1975) [29]

As is known oral environment is a highly aggressive chemically, which is characterized by frequent changes in pH due to food and varied microbial flora present. On the other hand, the pH value may change more in the tissue that may be injured or infected. A liquid used in tissue has a pH of about 7.4, but in a wound can drop to values of 3.5, and the infection can reach values of 9.0 [29, p.57]

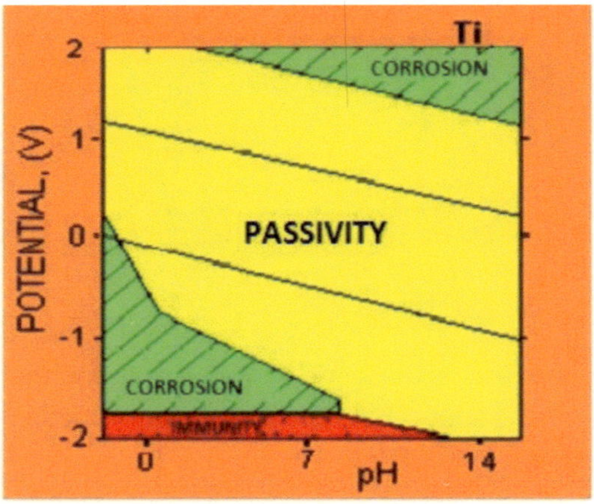

Figure 3. Pourbaix diagram for passive a metal (titanium) (after M.Pourbaix, Atlas of Electrochemical Equilibria in Aqueous Solutions, NACE, Houston / CEBELCOR, Brussels, 1974) [29]

Experience shows that the main ways that may reduce the risk of galvanic corrosion in the oral cavity are:

1. avoid using in the same oral cavity of metals with very different electrical potentials, the more so as they can come in direct contact (eg neighboring teeth);
2. the use of homogeneous alloys, which do not occur with different electrical potentials phase
3. avoid conditions that lead to plaque formation, given that the areas covered plate will present low pH, which can lead to corrosion.

4. Experimental conditions

Experimental Research on corrosion resistance of Co-Cr alloy biomaterial [25] were performed at Competence Center Tribocoroziune interfaces and electrochemical systems (CC-ITES) from the University "Dunarea de Jos" Galati."

4.1. Techniques and methods

The experiments were carried out at room temperature, 25°C, using the following techniques of investigation;

http://dx.doi.org/10.5772/57418

- The variation in time of the open circuit electrode potential (OCP)
- potentiodynamic polarization curves i (E),
- Cyclic voltammetry (CV),
- Records of electrochemical impedance spectra, more complex method for studying interface material / solution (EIS). These techniques have been applied to a potentiostat / galvanostat PGZ Voltalab 100 interface to the PC - soft VoltaMaster 4.

Electrochemical tests were performed on biomaterial, the chemical composition and mechanical properties in accordance with Table 1.

Chemical composition, %									
Specifications	EN 10204 -3.1B	Co	Cr	Mo	Mn	C	Fe	Si	Sonstige
	Max. [%]	63	29, 4	5, 95	0, 6	0, 29	0, 05	0, 7	0, 1

Properties		
Density g/cm3	Hardness HV10	Elasticity modulus, E Mpa
8, 3	420	230.000

Table 1. The chemical composition and mechanical properties of CoCr alloy [25]

We studied the corrosion behavior of Co-Cr alloy for biomedical applications, in aqueous solutions with different pH because it is known that all vital processes in the body is carried to the exact values of pH.

The solutions used in conducting the experiments are shown in Table 2. Were selected three types of media, so-called artificial saliva with different pH values: Hank (pH 7.4), artificial saliva Saliva Fusayama Meyer (pH 5.0) and Ringer's saliva (pH 6.6) seeking behaviors to study corrosion of Co-Cr alloy as metallic biomaterial.

The reason for the use and the choice of other media, such as citric acid is that it is a powerful oxidant and has a pH more acidic (pH 1.8), which can influence (formation, growth) of the oxide layer the surface of the sample.

4.2. Experimental results

The variation in time of the open circuit potential is a method of electrochemical oxidation which indicates the tendency of a material to a corrosive environment. After an immersion time it stabilizes around a stationary value.

To assess the kinetics of corrosion processes for the studied material were recorded under potentiodynamic polarization curves in the potential range from (-1500 mV) versus Ag / AgCl at (+1200 mV) versus Ag / AgCl, at a rate of sweeping the potential of 5 mV / s.

A -B zone is the cathode area, the B - C field the passive one and C – D the transpasivity field.

Potential values in different fields and solutions introduced are in Table 3. Also the table shows the passivation current density of the alloy in the tested solutions for the field of passivation.

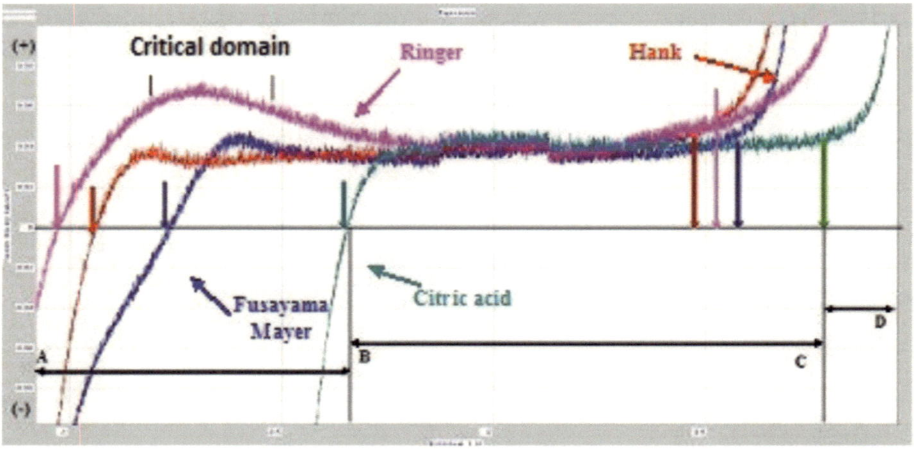

Figure 4. Identify of the passive areas on the potentiodynamic polarization curves for biomaterial Co-Cr immersed in the four environments [25, 26]

Solution	Areas of potential, E [mV vs Ag/AgCl]				
	Cathode	Critical	Passive		Transpassive
			E	$i_{passive}$ μA/cm²	
Fusayama Meyer (SFM)	< -764	-764 la -512	-512 la +569	36	> +569
Ringer	< -1030	-1030 la -390	-390 la +316	42	> +316
Hank	< -940	-940 la -750	-750 la +700	36	> +700
Citric acid	< -344	-344 la -135	-135 la +800	41	> +800

Table 2. The areas of potential and current density for passivation of the Co-Cr alloy [25, 26]

Solution	Polarization resistance R_p [kohms cm²]	The corrosion potential E[mV vs Ag/AgCl]	The corrosion speed i_{cor} [mA/cm²]	
			i_{cor} [mA/cm²]	μm/An
Fusayama Meyer (SFM)	0, 443	-902, 6	14, 16	153, 3
Ringer	3, 08	-1027	36, 80	398, 8
Hank	3, 0	-938	17, 27	187, 0
Citric acid	1, 69	-345	32, 99	357, 2

Table 3. The values of the polarization resistance and the corrosion current density of Co-Cr alloy in the Solutions tested [25, 26]

http://dx.doi.org/10.5772/57418

Areas with aspects of localized pitting corrosion phenomenon highlighted the qualitative analysis by electron microscopy (SEM - Scanning Electron Microscope, fig. 5) were further investigated specific techniques EDS equipment - Energy-dispersive X-ray spectroscopy quantitative analysis.

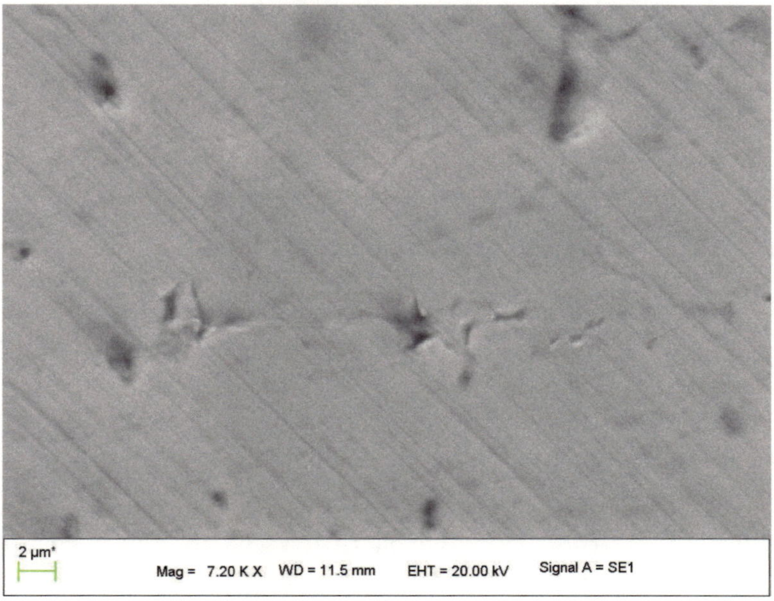

Figure 5. Micrographic aspects - details of the relief surfaces affected by corrosion (Magnification X 5000 using detectors BSE - Back-Scattered Electron Detector and SE-Secondary Electron Detector)[25]

5. Conclusions

- The used electrochemical methods (free potential, potentiodynamic curves, cyclic polarization, electrochemical impedance spectroscopy) are relevant methods for studying the materials/specific operating environment interactions.
- Results regarding the variation in time of the potential in open circuit as well as the results from the polarization curves under potentiodynamic regime show different levels of stabilization of the potential, but also different current densities depending on the pH.
- Polarization curves in potentiodynamic regime indicate the domain in which the passive oxide layer is destroyed through hydrogen release (area of instability), passivity domain of the formed oxide layer and the domain in which the dissolution of the metal occurs through the passive film and a deterioration of it.
- Both in cathode and in anode domain the aspect of the curves is modified, corresponding to the average that simulate solutions/artificial saliva, these being shifted to higher potential values with increasing their pH value.

- Cyclic polarization curves revealed the localized corrosion susceptibility by specific appearance of hysteresis (transpasivity anodic part). The biggest hysteresis was recorded at the tests concerning the Co-Cr alloy in Hank solution, which contains the highest amount of chloride ions.
- By tracing the electrochemical impedance spectroscopy curves it was confirmed the values domain of the polarization resistances obtained for the studied alloy in the tested solutions

Author details

Ion Patrascu[1*], Vlad Gabriel Vasilescu[1] and Stefan Milicescu[1]

*Address all correspondence to: dr.ipatrascu@gmail.com

1 „Carol Davila" University of Medicine and Pharmacy, Bucharest, Romania

References

[1] Anderson JM, et al. Host reactions to biomaterials and their evaluation, Ratner BD, et al., eds. Biomaterials Science: An Introduction to Materials in Medicine London, Elsevier 2004;

[2] A.P.R. Alves*, F.A. Santana, L.A.A. Rosa, S.A. Cursino, E.N. Codaro, A study on corrosion resistance of the Ti–10Mo experimental alloy after different processing methods, Materials Science and Engineering Elsevier, p.693-696;

[3] ASM Handbook Committee, Metals Handbook- Corrosion of Metallic Implants and Prosthetic Devices, vol. 13, ninth ed., American Society for Metals, Metals Park, 1987;

[4] Bratu D., Materiale dentare Bazele fizico-chimice, Ed. Helicon ISBN 975-9159-37-0, Timisoara 1994;

[5] Brunski JB. Metals, Ratner BD, et al., eds. Biomaterial Science: An Introduction to Materials in Medicine, London: Elsevier Academic Press, 2004:137-53;

[6] Christoph Leyens, Manfred Peters, "Titanium and Titanium Alloys-Fundamentals and Applications", ISBN 3-527-30534-3, Ed. WILEY-VCH, 2004 p. 4-16;

[7] Claudia Fleck a, *, Dietmar Eifler, Corrosion, fatigue and corrosion fatigue behaviour of., metal implant materials, especially titanium alloys, International Journal of Fatigue 32 (2010) 929–935;

[8] Gerd Lütjering, James C. Williams, Titanium - Engineering Materials and Processes, ISBN 3-540-42990-5, Springer-Verlag Berlin Heidelberg New York, p. 21-26, 2003;

[9] Craig, R.G. Restorative dental materials, Mosby, Chicago, 1993

[10] Demetrescu, I., Popescu, B., *Comportarea electrochimica a biomaterialelor metalice utilizate in implanturi/cap.5* Bucuresti

[11] E.E. Stanbury, R.A. Buchanan, Fundamentals of Electrochemical Corrosion, Materials Park, OH, ASM International, (2000), p. 219,

[12] Eisenbarth, E., Velten D., M. M.ullera, R.Thull b, J.Breme, *Biocompatibility of b-stabilizing elements of titanium alloys,* Biomaterials 25 (2004) 5705–5713 Elsevier,

[13] Gadea, S., Petrescu, M., Metalurgie fizica si studiul metalelor, vol.3 EDP, Bucuresti 1983

[14] Ghiban, B., Metallic Biomaterials, Ed. Printech Bucuresti 1999

[15] Gotman I. Characteristics of metals used in implants, J. Endourol (1997), 11(6):383-9;

[16] J.V. Bernier, J.-S. Park, A.L. Pilchak, M.G. Glavicic, M.P. MILLER, *Measuring Stress Distributions in Ti-6Al-4V Using Synchrotron X-Ray Diffraction,* Metallurgical and Materials Transactions A, 3120 Volume 39A, 2008,

[17] Karthega, M., Raman, V., Rajendran N., Influence of potential on the electrochemical behaviour of b titanium alloys in Hank's solution, Acta Biomaterialia 3 (2007) Elsevier, p. 1019–1023;

[18] Kenneth J. Anusavice, Phillips' Science of Dental Materials, Tenth Edition ISBN 81-7286-054-4, 1996;

[19] Liliane S. Morais., Glaucio G. Sera., Carlos A. Muller, … Titanium alloy mini-implants for orthodontic anchorage: Immediate loading and metal ion release, Acta Biomaterialia 3 (2007) Elsevier, p. 331– 339;

[20] M.A. Khan., R.L. Williams., D.F. Williams, In vitro corrosion and wear of titanium alloys in the biological environment, Biomaterials 17 (1996), Elsevier, p. 2117-2126;

[21] Nicolas Schift., Brigitte Grosgogeat., Michele Lissac., Francis Dalard, Influence of floride content and pH on the corrosion resistance of titanium and its alloys, Biomaterials Elsevier 23 (2002) p.1995–2002;

[22] Park, J.B., Lakes, R. S., *Biomaterials An Introduction,* Plenum Press, New York 1993

[23] S. Sathish., M. Geetha., N.D. Pandey., C. Richard., R.Asokamani, Studies on the corrosion and wear behavior of the laser nitride biomedical titanium and its alloys, Materials science and engineering C 30 (2010), p. 376 – 382;

[24] S′ergio Luiz de Assis, Stephan Wolynec, Isolda Costa, Corrosion characterization of titanium alloys by electrochemical techniques, Electrochimica Acta 51 (2006) 1815–1819 Elsevier, p. 1815-1819;

[25] Vasilescu, V.G. Studii de biocompatibilitate a materialelor metalice pentru implanturi dentare, lucrare de licenta UMF Carol Davila Bucuresti 2012

[26] Vasilescu, V.G., Benea, L., Vasilescu, E Rezistenta la coroziune a unor aliaje dentare Co-Cr, Conferinta UGALNANO3, Galati 2013;

[27] Vermeşan; H. Coroziune, Ed. Risoprint ISBN:973-656-881-4, Cluj-Napoca 2005;

[28] www.ro.scribd.com/doc/55566517Biomateriale

[29] http://www.scribd.com/doc/80569374/V-Bulancea-Biomateriale

[30] Zamfir, S., Vidu, R., Brînzoi, V., Coroziunea materialelor metalice, EDP Bucureşti1994 ISBN 973-30-2928-9;

[31] Zhuo Caia, *, Ty Shafera, Ikuya Watanabea, Martha E. Nunnb, Toru Okabea, Electrochemical characterization of cast titanium alloys, Biomaterials 24 (2003) Elsevier p. 213–218;

[32] Yoshimitsu Okazaki, Effect of friction on anodic polarization properties of metallic biomaterials, Biomaterials Elsevier 23 (2002) p. 2071–2077

Polyphenols and Herbal-Based Extracts at the Basis of New Antioxidant, Material Protecting Products

Lucia Camelia Pirvu

Additional information is available at the end of the chapter

http://dx.doi.org/10.5772/57183

1. Introduction

Considering that many of the classical corrosion and scale inhibitors are toxic compounds that accumulate in the water and soil, finding natural, eco-friendly inhibitors has represented a constant preoccupation in recent years.

Table 1 shows several examples of vegetal compounds and extracts that have been revealed with anti-corrosive properties.

Given this, it is obvious that the utilization of herbal based extracts as new, ecological, material protecting products is a viable approach mainly because vegetal compounds are able not only to remove oxygen and reactive oxygen species in a biological environment (for example polyphenols, known for their capacity to rapidly oxidize thus being very effective scavenger species), but also to counteract scale formation (for example saponins known for their tensioactive properties) and to inhibit microbial corrosion (the typical example being organosulfur compounds from spices, but also polyphenols). Thus, polyphenols class, especially flavonoids and phenyl-carboxylic acids derivates, likely represent the most feasible alternative of new, natural corrosion inhibitors on basis of their capacity to act as very effective reactive oxygen scavengers, some of them also being antimicrobial species (for example gallic acid, epi(gallo)catechin gallates and tannins). In support, gallic acid has been proved as a double corrosion inhibitor acting as an anodic protector (through capturing oxygen) [36] and also as a microbial development inhibitor by *irreversible changes in membrane properties through hydrophobicity changes, decrease of negative surface charge, and occurrence of local rupture or pore formation in the cell membranes with consequent leakage of essential intracellular constituents* [37].

Moreover, polyphenols compounds have added qualities such as a good thermal stability, high solubility and dispersion into a wide range of solvents, as well as an increasing antioxidant

Investigated sample	Bibliography
Acacia and *Pinus tannins*	1, 2, 3
Quercus (oak) and *Castanea* (chestnut) tannins	4, 5
Mimosa tannins	6
Tamarix articulata tannins	7
Thonningia sanguinea ellagitannins	8
Rizophora (mangrove) tannins and catechins	9, 10, 11
Strychnos nux-vomica alkaloidic fraction	12
Cannabis indica alkaloidic fraction	13
Calotropis procera and *Calotropis gigantea* alkaloidic fraction	14
Solenostemma arghel, Chamommile, Halfabar, Black curmin, Kidney bean, *Lupine* and *Damssisa* extracts	15, 16, 17, 18
Opuntia elatior, Acanthocereus pentagonus, Mimosa tenuiflora, Caesalpinia coriaria, Bumbacopsis quinata and *Acacia mangium* extracts	19
Euphorbia falcata, Rosmarinus officinalis, Zenthoxylum alatum, Hibiscus sabdariffa, Aningeria robusta, Euphorbia hirta and *Dialum guineense* extracts	20, 21, 22, 23, 24, 25, 26
Psidium guajava ***seed extract***	27
Spondias mombin ***fruit extract***	28
Phyllanthus amarus leaves extract	29
Cichorium intybus, Arctium lappa, Centaurea cyanus, Pinus caribea, Eucalyptus citriodora, Piper auritum and *Piper guineense* extracts	30, 31
Cuminum Cyminum/cummin, *Ferula assafoetida*/asea, *Capsicum*/chilli, *Allium sativum*/garlic, *Ocimum basilicum*/basil extracts	32, 33, 34, 35

Table 1. Examples of previous work in the field of herbal-based anti-corrosion products

activity into acid medium (explained through the fact that the resulted, partly hydrolyzed, compounds are often more active scavengers' species than the origin homologues).

Summing, polyphenols class seems to convene most of the demands of a composite anti-corrosion/anti-biodeterioration product, also having the advantage of being less toxic than other natural compounds (for example alkaloids).

As for their role, polyphenols are secondary metabolites of the plants involved in defense against ultraviolet radiation and pathogens aggressions. The richest sources of polyphenols are onion, apple, tea, grapes, red wine and grape juice as well as strawberries, raspberries, blueberries and cranberries fruits. They are usually classified into four different groups by their number of phenol rings and by their different structural elements that bind these rings to one another. The four classes are phenolic acids, flavonoids, stilbenes and lignans, also classified as non-flavonoids and flavonoids. In the following are presented general chemical structures of the

http://dx.doi.org/10.5772/57183

most commonly vegetal polyphenols, respectively flavonoids (flavonols, flavones, flavanones, flavononols, flavanols, anthocyanidin, isoflavones and chalcones derivates) and phenyl-carboxylic acid (protocatechuic and cinnamic acids derivates) (sub)classes.

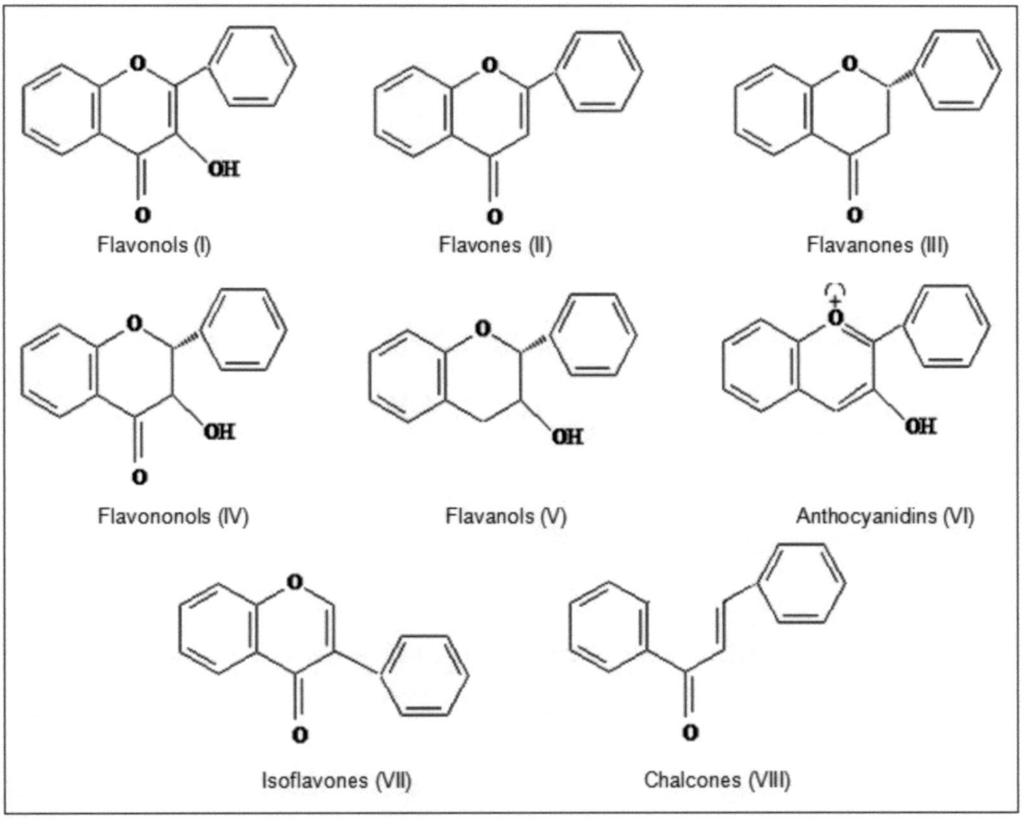

Figure 1. Flavonoids subclasses

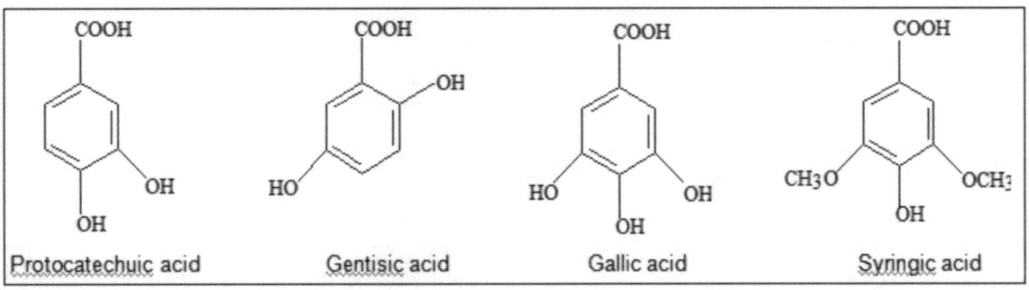

Figure 2. Protocatechuic (phenyl-carboxylic) acid derivates

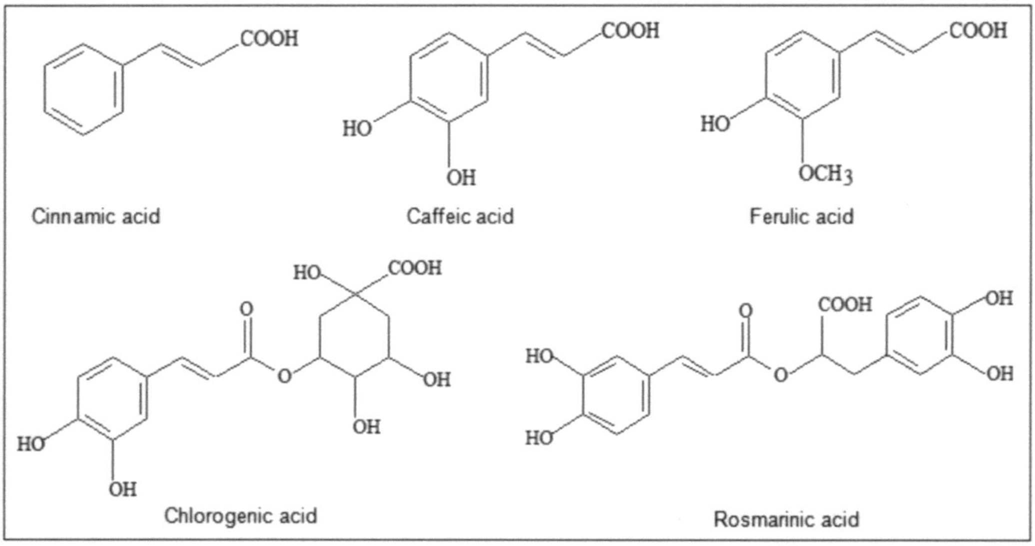

Figure 3. Cinnamic (phenyl-carboxylic) acid derivates

2. Problem statement and application area

Given the complexity of this domain, in fact an inter-, multi- and even trans-disciplinary approach, there is some short-comings need be solved, respectively it should be done much more analytical and microbiological characterization assessments on vegetal extracts tested as new eco-friendly material protecting products for both purposes, practical and scientific, respectively to achieve the quality control of the vegetal extracts and the effectiveness of certain phytocompounds in the ultimate goal of the obtaining of characterized and effective anti-corrosion products.

Data referring to the antioxidant activity of the polyphenols compounds may also be very useful.

For example, some studies [38] carried out on eighteen commonly vegetal polyphenols indicated that the antioxidant activity (AA%) of the flavonoids and phenyl-carboxylic acid compounds depends on both, the number and the position of free hydroxyl (-OH) groups; it should be noticed that studies were made by using chemiluminescence method, luminol/ H_2O_2 system, pH=8.6 [39]. Precisely, studies (see Figure 1) has revealed that if (-)-epicatechin, (+)-catechin (belonging to flavan-3-ol subclass), quercetin and kaempferol (belonging to the flavonols subclass) are the most effective antioxidant species (AA% between 90 and 97%), naringin and naringenin (belonging to the flavanones subclass) are the less active (AA% between 32 and 9%). Gallic acid, chlorogenic acid and rosmarinic acid, belonging to the phenyl-carboxylic acid class, are also good scavenger species their antioxidant activities (AA%) varying between 85 and 89% [39].

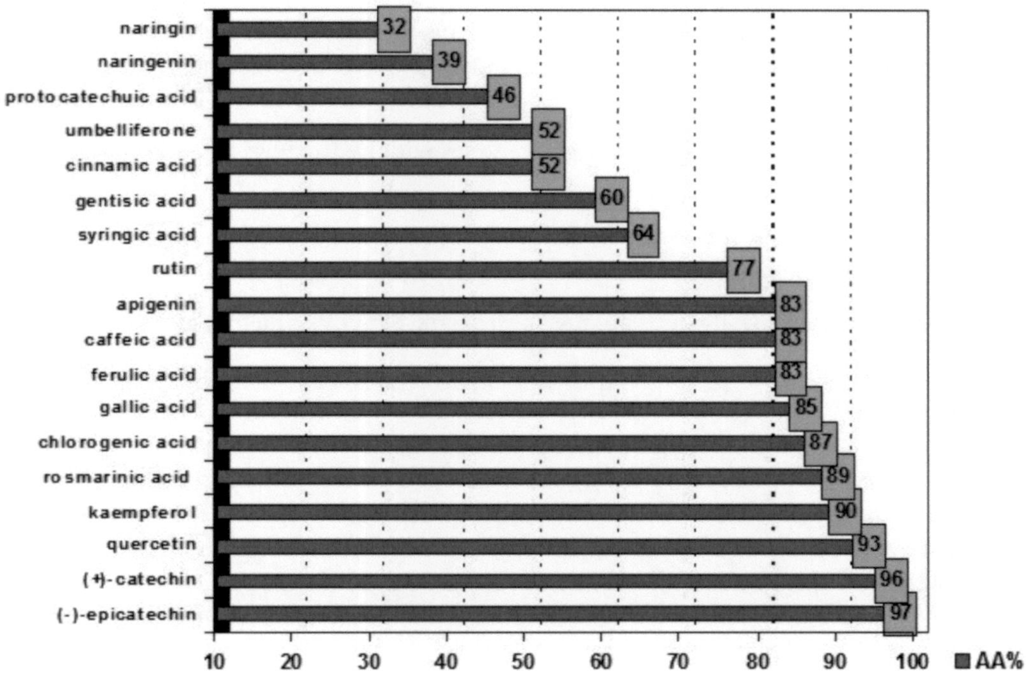

Figure 4. Antioxidant activity (CL method) of eighteen (18) commonly vegetal polyphenols

As for the capacity of the vegetal polyphenols to stop bacterial corrosion, on basis of a comprehensive review [40], structure–activity relationship for antibacterial activity of flavonoids compounds has been summarized as follows:

- 2_,4_- or 2_,6_-dihydroxylation of the B ring and 5,7-dihydroxylation of the A ring in the flavanone structure is important for their antimicrobial/anti-methicillin resistant *Staphylococcus aureus* (MRSA) activity;

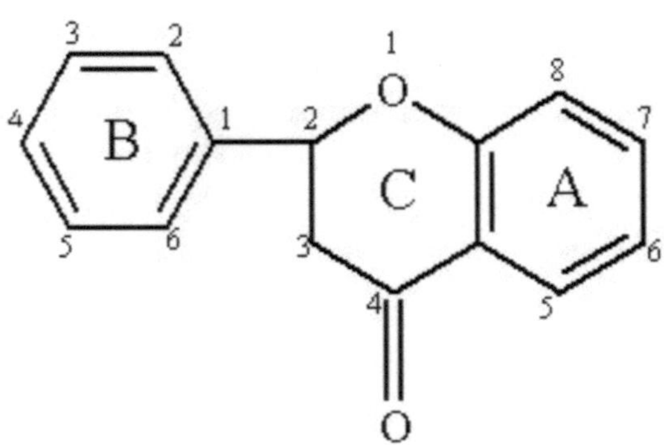

Figure 5. Flavonoid general structure

- substitution at the 6 or 8 position with a long chain aliphatic group such as lavandulyl (5-methyl-2-isopropenyl-hex-4-enyl) or geranyl (trans-3,7-dimethyl-2,6-octadienyl) enhance antimicrobial activity, and substitution with C8 and C10 chains enhanced the activity of flavonoids belonging to the flavan-3-ol class;
- 5-hydroxyflavanones and 5-hydroxyisoflavanones with hydroxyl group at position 2_ are also very active;
- chalcones are more effective against MRSA than flavanones or flavones, hydroxyl groups at the 2_ position being very important for their anti-staphylococcal activity;
- methoxy groups drastically decrease the antibacterial activity of flavonoids.

For example, using *Staphylococcus aureus* to guide the isolation process, some studies [41] on a methanolic extract isolated from dried leaves of *Physena madagascariensis* indicated two monomeric flavanones containing lavandulyl units in the limonene form as being very active against this bacterium. Similarly, studies [42] on some butanol and dichloromethane extracts of root of *Flemingia strobilifera* (a flowering plant in the legume family, *Fabaceae*) indicated flemingiaflavanone (8, 3-diprenyl-5, 7, 4'-trihydroxy flavanone) with significant antimicrobial activity against Gram-positive (*Staphylococcus aureus, Staphylococcus epidermidis*, methicillin resistant *Staphylococcus aureus*/MRSA), Gram-negative bacteria (*Pseudomonas aeruginosa, Escherichia coli*) and fungi (*Candida albicans*); genistin (5, 4 dihydroxy isoflavone7-O-glucoside) showed moderate activity against Gram-positive, Gram-negative bacteria and fungi.

Other studies [43] on some crude methanolic extracts isolated from *Grewia asiatica, Eugenia jambolana* and *Carissa carandas* separated, each one, into four major fractions respectively, 1)phenolic acids, 2)flavanols, 3)flavonols and 4)anthocyanins fractions indicated that, besides being the most active on microbial strains, phenolic acid fractions also inhibited all tested fungal species. Similarly, studies on Tunisian Quince (*Cydonia oblonga* Miller) pulp and peel polyphenolic extracts [44], shown as very reach in caffeoyl derivates demonstrated that chlorogenic acids acts in synergism with other components of the extracts to exhibit their total antimicrobial activities. Other comparative studies [45] on some common phytochemicals respectively, 5 simple phenolics - tyrosol, gallic acid, caffeic acid, ferulic acid, and chlorogenic acid; chalcone - phloridzin; flavan-3-ol - (-)epicatechin; seco-iridoid - oleuropein glucoside; 3 glucosinolate hydrolysis products - allylisothiocyanate, benzylisothiocyanate and 2-phenylethylisothiocyanate, but also on some dual combinations of streptomycin with these phytochemicals against *Escherichia coli, Pseudomonas aeruginosa, Listeria monocytogenes* and *Staphylococcus aureus* indicated that the isothiocyanates had significant antimicrobial activities, while the phenolics were much less efficient; no antimicrobial activity was observed in the case of phloridzin (chalcone derivate).

Differently, studies [46] reported that the marine paint mixed with 2-methoxy-2',4'-dichloro chalcone considerably reduced the formation of biofilm by *Vibrio natriegens*, a marine bacterium, on polycarbonate (PC), polymethylmethacrylate (PMMA) and glass fiber reinforced plastic (GFRP). Precisely, it has been revealed that the surfaces coated with dichloro chalcone containing marine paint had the lowest number of colony forming units (CFU)(1-5×10(6)), proteins (20-30 μg/cm2) and carbohydrates (5-10 μg/cm2) attached to them after 28 days of exposure to the organism when compared to surfaces coated with $CuSO_4$ mixed paint (20-40×10(6) CFU/ml, proteins of 50-60 μg/cm2 and carbohydrates of 40-50 μg/cm2) or plain marine paint (30-40×10(6) CFU/ml, proteins of 120-150 μg/cm2 and carbohydrates of 40-60 μg/cm2). Also, results indicated that the biofilm on PMMA was 7, 10 and 12 μm thick on chalcone, copper and plain paint coated surfaces, respectively. Furthermore, the first two paints increased the surface roughness but decreased the surface hydrophobicity when compared to the plain paint. The obtained results suggested that the low amount of biofilm formed in the presence of dichlorochalcone can be associated to its antibacterial and slimicidal activity and also its ability to reduce the hydrophobicity of the surface.

As for the solvent effectiveness, some studies [30] on different aqueous, acetonic, and ethanol extracts of *Cichorium intybus* L., *Arctium lappa* L., *Centaurea cyanus* L. (Asteraceae), *Allium sativum* L. (Liliaceae), *Pinus caribea* Mor. (Pinaceae), *Eucalyptus citriodora* Hook. (Mirtaceae) and *Piper auritum* Kunth (Piperaceae) against different microorganisms (*Pseudomonas fluorescens, Bacillus cereus, Bacillus polimixa, Enterobacter agglomerans and Streptomyces* sp.,) associated with biodeterioration indicated that the aqueous extracts did not show any antibacterial activity, the antimicrobial activity from bacteria isolated from biofilms being present only in the ethanol extracts.

Referring to the scale issue, with their well known tensioactive properties, triterpenic (acidic) saponins appears as the most viable approach, the neutral ones, respectively sterolic saponins, being very toxic (see bellow the general chemical structure of sterolic and triterpenic saponins, respectively spirostane and oleanolic acid types).

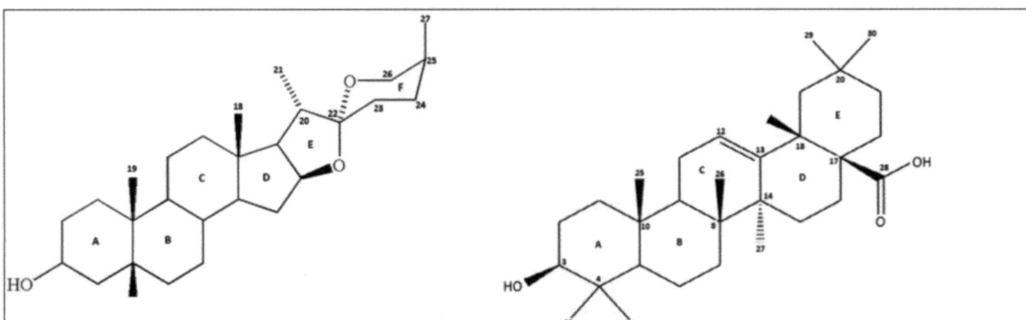

Sterolic saponin (general structure) **Triterpenic saponin, oleanolic acid type (general structure)**

Figure 6.

For example, studies on 39 plant materials indicated that the birch bark (containing betulinic acid), plane bark (also containing betulinic acid), olive leaves, olive pomace, mistletoe sprouts and clove flowers (all containing oleanolic acid), apple pomace (containing ursolic acid) and rosemary leaves (containing an equal mixture of these three triterpene acids) are the richest sources of triterpenic saponins [47]. Other vegetal materials demonstrated as being abundant in triterpenic acids are *Satureja parvifolia* and *Eucalyptus* species. Thus, in the first case, studies on methanolic extracts lead to the isolation of eriodictyol, luteolin and ursolic and oleanolic acids [48]. In the second case, studies on the outer bark of *E. globulus, E. grandis, E. urograndis, E. maidenii* and *E. nitens* indicated triterpenic acids contents varying between 4.5 g/kg in E. urograndis and 21.6 g/kg in E. Nitens, but out of these species, temperate and Mediterranean *E. nitens* and *E. globulus* were also revealed as very rich in triterpenic acids; precisely, E. globulus outer bark was found as the richest source of ursane acids, while E. nitens outer bark was revealed as the richest source of oleanane and lupane acids [49].

Summing, olive oil appears as one of the most complete source of corrosion inhibitors on basis of its content in phenolics (4-hydroxybenzoic acid, protocatechuic acid, syringic acid, 4-hydroxy-phenylacetic acid, homovanillic acid, ferulic acid, sinapic acid), flavonoids (apigenin and luteolin derivates), lignanes, isochromans, tyrosol and hydroxytyrosol derivates [50] as well as in triterpenic acid derivates (oleanolic, ursolic and maslinic acids) [51] theoretically able to manage all, oxidative stress, microbial development, and scale formation.

3. A case study

Below it is presented a case study on four vegetal extracts for the purpose of assessing scavenger/antioxidant activity and corrosion inhibition effectiveness of certain flavonoids and phenyl-carboxylic acid derivates combinations.

Thus, there were compared four series of whole and selective ethanolic extracts, respectively:

- extracts containing [2)]chlorogenic acid derivates aside small quantities of kaempferol, apigenin, quercetin and catechin derivates isolated from leaves of *Fagus sylvatica* L. (see Figure 2, T4-T8 tracks, respectively Table 2);
- extracts containing [1)]quercetin and derivates isolated from scales of *Allium cepae bulbus* (see Figure 2, T11-T15 tracks, respectively Table 2);
- extracts containing a mixture of [3)]quercetin and chlorogenic acid derivates isolated from leaves of *Juglans regia* L. (see Figure 3, T4-T9 tracks, respectively Table 3) and
- extracts containing a mixture of [4)]quercetin, luteolin and apigenin derivates aside small quantities of chlorogenic acid isolated from *Agrimonia eupathoria* L.-herba (see Figure 3, T10-T15 tracks, respectively Table 3).

Chemiluminescence studies (luminol/H_2O_2 system, *p*H=8.6) carried out on these four series of whole and selective vegetal extracts isolated from scales of *Allium cepae bulbus*, leaves of *Fagus sylvatica* [52], leaves of *Juglans regia* and the aerial part (*herba*) of the *Agrimonia eupathoria* [53] indicated maximum antioxidant activities (AA%) of 91% to 97% for *total phenols content* ranging between 3 and 57mg *per* 100mL ethanolic extract (see Table 4).

Subsequently comparative studies [55] on carbon steel corrosion in acidic (0.5M H_2SO_4) solution model indicated that all studied extracts presented anti-corrosion properties, *Fagus sylvatica* L. leaves whole ethanol extract being the most potent anti-corrosion product, also presenting anti-scale properties. Results were patented [56].

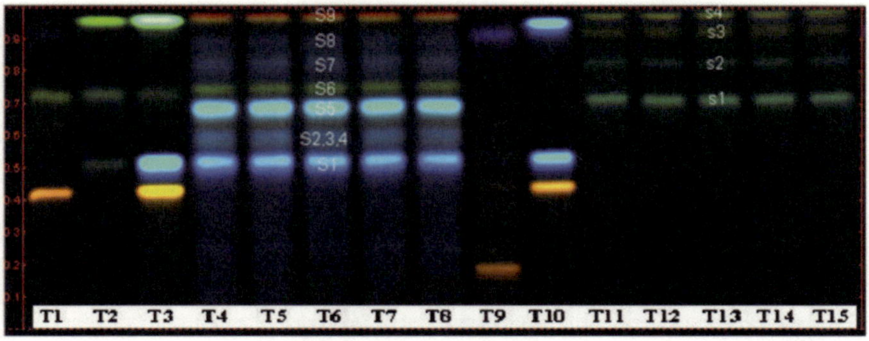

Figure 7. (HP)TLC aspects of *Fagus sylvatica* L. leaves (T4-T8 tracks) and *Alii cepae bulbus* L. scales (T11-T15 tracks) whole ethanol extracts [52]

Tested sample	Spot	Rf~	Colour spot	Attributed compound
Fagus sylvatica L. leaves extracts	S1	0.52	Light blue, fluorescent (fl.)	Chlorogenic acid
	S2	0.56	Brown, non-fl.	Catechin derivate
	S3	0.58	Blue, fl.	Neochlorogenic acid
	S4	0.64	Brown, non-fl.	Catechin derivate
	S5	0.69	Blue-green, fl.	Kaempferol derivate (likely Lespedin)
	S6	0.75	Green, fl.	Apigenin derivate (likely Vitexin)
	S7	0.83	Blue, fl.	Isochlorogenic acids (two isomers)
	S8	0.90		
	S9	0.97	Yellow, fl.	Quercetin
Alii cepae bulbus L. scales extracts	s1	0.72	Green-blue, fl.	Quercetin-4'-O-glycosides (also called spiraeosides)
	s2	0.84	Green-blue, fl.	
	s3	0.94	Green and Yellow fl.	Isorhamnetin and Quercetin
	s4	0.98		

Table 2. Chemical qualitative composition of *Fagus sylvatica* leaves and *Alli cepae bulbus* scales extracts [52]

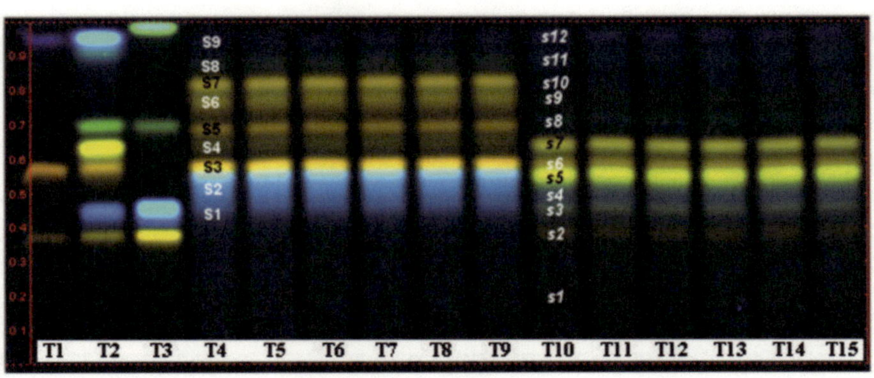

Figure 8. (HP)TLC aspects of *Juglans regia* L. leaves (T4-T9 tracks) and *Agrimonia eupathoria* L. aerial part (T10-T15 tracks) whole ethanol extracts [53]

Tested sample	Spot	Rf~	Colour spot	Attributed compound
Juglans regia L. leaves extracts	S1	0.45	Light-blue, fl.	Chlorogenic acid
	S2	0.50	Light-blue, fl.	Neochlorogenic acid
	S3	0.56	Orange, fl.	Quercetin derivate (likely Hyperoside)
	S4	0.63	Green, fl.	Apigenin derivate (likely Isovitexin)
	S5	0.69	Orange, fl.	Quercetin derivate (likely Isoquercitrin)
	S6	0.76	Orange, fl.	Quercetin derivate (likely Avicularin)
	S7	0.81	Orange, fl.	Quercetin derivate (likely Quercitrin)
	S8	0.87	Green-blue, fl.	Quercetin derivate (likely Juglanin)
	S9	0.96	Dark-blue, fl.	Protocatechuic acid
Agrimonia eupathoria L. aerial part extracts	s1	0.20	Orange, fl.	Quercetin derivate
	s2	0.38	Orange, fl.	Quercetin derivate (likely Rutin)
	s3	0.44	Orange, fl.	Quercetin derivate
	s4	0.50	Light-blue, fl.	Neochlorogenic acid
	s5	0.55	Yellow, fl.	Luteolin derivate (likely Isoorientin)
	s6	0.58	Orange, fl.	Quercetin derivate (likely Hyperoside)
	s7	0.63	Yellow, fl.	Luteolin derivate (likely Orientin)
	s8	0.73	Green, fl.	Apigenin derivate (likely Cosmosiin)
	s9	0.78	Orange, fl.	Quercetin derivate
	s10	0.80	Blue, fl.	Isochlorogenic acid isomer
	s11	0.87	Green, fl.	Apigenin derivate
	s12	0.98	Deep-blue, fl.	Rosmarinic acid

Table 3. Chemical qualitative composition of *Juglans regia* leaves and *Agrimonia eupathoria* aerial part extracts [53]

Parameters	Raw material description							
	Allium cepae bulbus L. scale		*Fagus sylvatica* L. leaves		*Juglans regia* L. leaves		*Agrimonia eupathoria* L. aerial part	
Type of extract: Whole (W) or Selective (S)	W	S	W	S	W	S	W	S
Antioxidant activity: (AA%)	92	95	91	94	95	97	94	97
Total phenols content: mg/100mL (expressed as gallic acid equivalents)	3.1	6.7	20.2	39.2	35.6	64.4	48.4	56.8
Total flavones content: mg/100mL (expressed as quercetin* or rutin equivalents)	2.3*	4.6*	7.2	10	23.2	58.6	21.6	17.6

Note: Analytical measurements were done by using classic methods [54], respectively total phenols content was measured by using *Folin* reagent and the results were expressed as gallic acid equivalents and total flavones content was measured by using $AlCl_3$ in base (sodium acetate) medium and the results were expresses as quercetin*, respectively, rutin equivalents

Table 4. Antioxidant activities of the four series of whole and selective vegetal extracts

4. Results and comments

This case study has revealed the following aspects:

- *whole* vegetal extracts presented similar antioxidant activities (AA%) with selective vegetal extracts but at the lesser amounts of total phenols, thus suggesting the benefit of the vegetal compounds, respectively polysaccharides and proteins, lacking in the selective extracts;

- *Allium* polyphenols (quercetin and derivates) indicated the highest antioxidant effectiveness, meaning that *Allium cepae bulbus* extracts presented the maximum antioxidant activity (respectively, 92%) at the lowest total phenols content (respectively, 3.1 mg/100mL whole ethanol extract).
- *Fagus sylvatica* leaves extracts (containing mainly chlorogenic acid derivates added to only small quantities of flavonoid derivates) also were very effective antioxidant products emphasizing an augmented antioxidant activity (91%) at moderate total phenols content (respectively, 20.2 mg/100mL whole ethanol extract);
- *Juglandis* leaves extracts (containing dominantly quercetin derivates aside moderate levels of chlorogenic acid derivates), similar to *Agrimony herba* extracts (containing a mixture of quercetin, luteolin and apigenin flavonoids aside small quantities of chlorogenic acid derivates) both required high total phenols contents (35.6 mg/100mL and, respectively, 48.4 mg/100mL extract) in order to present the same magnitude of the antioxidant activity (95% and, respectively, 94%);
- *Fagus sylvatica* leaves whole ethanol extract was the most potent anti-corrosion product presenting anti-scale properties, as well.

5. Conclusion

Literature data driven indicated that phytocompounds and herbal-based extracts are of increasing interest in this field of new, eco-friendly material protecting products. Accordingly, it has been revealed that on basis of their capacity to consume oxygen and reactive oxygen species added to the capacity to inhibit microbial development, polyphenols based extracts seems to convene most of the demands of a composite anti-corrosion/anti-biodeterioration product, also having the advantage of being less toxic than other vegetal extracts (for example alkaloids extracts). With effective tensioactive properties, triterpenic saponins also appear as very useful anti-corrosion ingredients by decreasing scale formation and increasing phytocompounds solubility.

In support, a case study aiming the comparison of the antioxidant activity and corrosion inhibition effectiveness of four whole and, respectively, selective vegetal extracts isolated from four vegetal species selected in a manner to contain different combinations of flavonoids and phenyl-carboxylic acid derivates revealed that quercetin compounds had the highest antioxidant/scavenger activity at the lowest concentration in respective environment, thus suggesting high anti-corrosion potential of quercetin based extracts. Subsequently, comparative anti-scale/anti-corrosion studies indicated that, besides containing polyphenols species with high antioxidant/scavenger activity, the co-presence of other protecting, synergetic or boosting compounds seems to be more important for the final anti-corrosion effect. As proof, *Fagus sylvatica* leaves whole ethanol extract abundant in chlorogenic acid derivates aside only small quantities of flavonoid (quercetin, apigenin, kaempferol and catechin) derivates, but containing some saponins derivates [57] offered the most proper protecting conditions on carbon steel corrosion in acidic (H_2SO_4) solution model.

Given these, it has been concluded that it should be done much more analytical and microbiological characterization assessments on vegetal extracts tested as new eco-friendly material protecting products for both purposes, practical and scientific, respectively to achieve the quality control of the vegetal extracts and the effectiveness of certain phytocompounds in the ultimate goal of the obtaining of characterized and effective anti-scale/anti-corrosion products.

Therefore, further studies in this area may be done on those vegetal extracts or combinations of vegetal extracts offering the whole range of material protecting compounds, respectively polyphenols with augmented antioxidant activity and/or polyphenols with strong antimicrobial activity, saponins with tensioactive properties and other vegetal macromolecules, such as polysaccharides and proteins, with protective, synergistic or boosting effects all contributing to the achieving of a highly effective corrosion, biodeterioration and scale inhibitor product.

Acknowledgements

The author is very grateful to Dr Mahmood Aliofkhazraei for analyzing preliminary drafts of the manuscript and subsequently advices.

Author details

Lucia Camelia Pirvu

National Institute of Chemical-Pharmaceutical R&D, Department of Pharmaceutical Biotechnologies, Plant Extraction Laboratories, Bucharest, Romania

References

[1] Gust J. Application of infrared spectroscopy for investigation of rust phase component conversion by agents containing oak tannin and phosphoric acid. Corros NACE 1991;47(6) 453-457.

[2] Matamala G, Smeltzer W, Droguett G. Comparison of steel anticorrosive protection formulated with natural tannins extracted from acacia and from pine bark. Corros Sci 2000;42 1351-1362.

[3] Palma G, Freer J, Baeza J. Removal of metal ions by modified Pinus radiata bark and tannins from water solutions. Water Res 2003;37 4974-4980.

[4] Martinez S, Stalgar I. Correlation between the molecular structure and the corrosion efficiency of chestnut tannin in acidic solutions. J Mol Struct (THEOCHEM) 2003;640 167-174.

[5] Pardini OR, Amalvy JI, Di Sarli AR, Romagnoli R, Vetere VF. Formulation and testing of a waterborne primer containing chestnut tannin. J Coat Technol 2001;3(913) 99-111.

[6] Martinez S, Stern I. Thermodynamic characterization of metal dissolution and inhibitor adsorption processes in the low carbon steel/mimosa tannin/sulfuric acid system. Appl Surf Sci 2002;9149 1 -7.

[7] Mabrour J, Akssira M,Azzi M, Zertoubi M, Saib N, Messaoudi A, Albizane A, Tahiri S. Effect of vegetal tannin on anodic copper dissolution in chloride solutions. Corrosion Science 2004;46(8) 1833-1847.

[8] Ohtani II, Gotoh N, Tanaka J, Higa T, Gyamfi MA, Aniya Y. Thonningianins A and B, New Antioxidants from the African Medicinal Herb Thonningia sanguinea. J Nat Prod 2000;63 676-679.

[9] Afidah AR, Rocca E, Steinmetz J, Kassim MJ, Adnan R, Ibrahim MS. Study of mangrove tannin and flavanoid monomers as alternative steel corrosion inhibitors in acidic medium. Corros Sci 2007;49 402-417.

[10] Afidah AR, Rocca E, Steinmetz J, Kassim MJ. Inhibitive action of mangrove tannins and phosphoric acid on pre-rusted steel via electrochemical methods. Corros Sci 2008;50 1546-1550.

[11] Kang Wei Tan, Mohd. Jain Kassim, Chuan Wei Oo. Possible improvement of catechin as corrosion inhibitor in acidic medium. Corrosion Science 2012;65 152-162.

[12] Bothi Raja P, Sethuraman MG. Strychnos nux-vomica an eco-friendly corrosion inhibitor for mild steel in 1 M sulfuric acid medium. Materials and Corrosion 2008;60(1) 22 – 28.

[13] Abd-El-Nabey BA, Abdel-Gaber AM, Said Ali MEL, Khamis E, El-Housseiny S. Inhibitive action of cannabis plant extract on the corrosion of copper in 0.5 M H2SO4. International Journal of Electrochemical Science 2013;8(4) 5851-5865.

[14] Sudesh Kumar, Surendra Arora, Manish Sharma, Paresh Arora, Suraj Prakash Mathus. Synergistic effect of Caloptris plant in controlling corrosion and mild steel basis solution. J. Chil. Chem. Soc 2009;54(1) 83-88.

[15] Abd-El-Gaber AM, Khamis E, Hefnawy A. Utilizing Arghel Extract as Corrosion Inhibitor for Reinforced Steel in Concrete. Materials and Corrosion 2011;62 1159.

[16] Abd-El-Gaber AM, Abd-El-Naby AM, Sidahmed IM, El-Zayady IM, Saadawy M. Effect of Some Natural Extracts on the Corrosion of Zinc in 0.5 M NaCl. Corrosion science 2006;48(9) 2765-2779.

[17] Abd-El-Naby AM, Abdullatef OA, Abd-El-Gaber AM, Shaker MA, Ismail G. Effect of some natural extracts on corrosion of Zinc in 0.5M NaCl. Int. J. Electrochem. Sci. 2012;7 5864-5879.

[18] Lampart-Szczapa E, Korczak J, Nogala-Kalucka M, Zawirska-Wojtasiak R. Antioxidant properties of lupin seed products. Food Chem 2003;83 279-285.

[19] Jaen JA, Garcia de Saldana E, Hernandez C. Characterisation of reaction products of iron and aqueous plant extracts. Hyperfine Interact 1999;122 139-145.

[20] El Bribri A, Tabyaoui M, Tabyaoui B, El Attari H, Bentiss F. The use of Euphorbia falcata extract as eco-friendly corrosion inhibitor of carbon steel in hydrochloric acid solution. Materials Chemistry and Physics, in press (2013).

[21] Kliskic M, Radosevic J, Gudic S. Aqueous extract of Rosmarinus officinalis L. as inhibitor of Al-Mg alloy corrosion in chloride solution. Journal of applied electrochemistry 2000;30(7) 823-830.

[22] Gunasekaran G, Chauhan LR. Eco friendly inhibitor for corrosion inhibition of mild steel in phosphoric acid medium. Electrochimica acta 2004;49(25) 4387-4395;

[23] El Bribri A, Tabyaoui M, Tabyaoui B, El Attari H, Bentiss F. The use of Euphorbia falcata extract as eco-friendly corrosion inhibitor of carbon steel in hydrochloric acid solution. Materials Chemistry and Physics, in press (2013).

[24] Emeka E. Oguzie. Corrosion Inhibitive Effect and Adsorption Behaviour of Hibiscus Sabdariffa Extract on Mild Steel in Acidic Media, Portugaliae Electrochimica Acta 2008;26 303-314.

[25] Obot IB, Umoren SA, Obi-Egbedi NO. Corrosion inhibition and adsorption behaviour for aluminuim by extract of Aningeria robusta in HCl solution: Synergistic effect of iodide ions. J. Mater. Environ. Sci. 2011;2(1) 60-71.

[26] Nnanna LA, Onwuagba BN, Mejeha IM, Okeoma KB. Inhibition effects of some plant extracts on the acid corrosion of aluminium alloy. African Journal of Pure and Applied Chemistry 2010;4(1) 011-016;

[27] Vinod Kumar KP, Sankara Narayana Pillai M, Rexin Thusnavis G. Seed Extract of Psidium guajava as Ecofriendly Corrosion Inhibitor for Carbon Steel in Hydrochloric Acid Medium. Journal of Materials Science & Technology 2011;27(12) 1143-1149.

[28] Obi-Egbedi NO, Obot IB, Umoren SA. Spondias mombin L. as a green corrosion inhibitor for aluminium in sulphuric acid: Correlation between inhibitive effect and electronic properties of extracts major constituents using density functional theory. Arabian Journal of Chemistry 2012;5(3) 361-373.

[29] Olusegun K. Abiola, Otaigbe JOE. The effects of Phyllanthus amarus extract on corrosion and kinetics of corrosion process of aluminum in alkaline solution. Corrosion Science 2009;51(11) 2790-2793;

[30] Guiamet PS, Gomez de Saravia SG. Natural Products Isolated From Plants Used In Biodeterioration Control. Pharmacologyonline 2006;3 537-544.

[31] Oguzie EE, Ogukwe CE, Ogbulie JN, Nwanebu FC, Adindu CB, Udeze IO, Oguzie KL, Eze FC. Broad spectrum corrosion inhibition: Corrosion and microbial (SRB) growth inhibiting effects. Electrochem. Sci. 2012;7 8543–8559.

[32] Lakshmi Priya, S.; Chitra, A.; Rajendran, S.; Anuradha, K. Corrosion behaviour of aluminium in rain water containing garlic extract. Surface Engineering 2005;21(3) 229-231.

[33] Sangeetha M, Rajendran S, Sathiyabama J, Prabhakar P. Asafoetida Extract (ASF) as green Corrosion Inhibitor for Mild Steel in Sea Water. International Research Journal of Environment Sciences 2012;1(5) 14-21.

[34] Gopal Ji, Sudhish Kumar Shukla, Priyanka Dwivedi, Shanthi Sundaram, Eno E. Ebenso, Rajiv Prakash. Green Capsicum annuum Fruit Extract for Inhibition of Mild Steel Corrosion in Hydrochloric acid solution Int. J. Electrochem. Sci. 2012; 7: 12146 – 12158.

[35] Ambrish Singh, Eno. E. Ebenso, M. A. Quraishi. Theoretical and Electrochemical Studies of Cuminum Cyminum (Jeera) extract as Green Corrosion Inhibitor for Mild Steel in Hydrochloric Acid Solution. Int. J. Electrochem. Sci. 2012;7 8543–8559.

[36] Ostovari A, Hoseinnieh SM, Piekari M, Shadizadeh SR, Hashemi SJ. Corrosion inhibition of mild steel in 1M HCl solution by henna extract: A comparative study of the inhibition by henna and its constituents (Lawsone, Gallic acid, α-D-Glucose and Tannic acid). Corrosion Science 2009;51(9) 1935–1949.

[37] Borges A, Ferreira C, Saavedra MJ, Simoes M. Antimicrobial activity of ferulic and gallic acids against pathogenic bacteria. Microb Drug Resist 2013;19(4):256-65. doi: 10.1089/mdr.2012.0244. Epub 2013 Mar 12.

[38] Pirvu L, Nichita C, Giurginca M, Meghea A. Structure - antioxidant activity relationship between some of the most commonly vegetal polyphenols compounds.Chemistry Magazine 2007;58(9) 914-917.

[39] Meghea A, Iftimie N, Giurginca M, Papadopoulos K. Oxidative stress quantifying in biosystems. I. Chemiluminescence testing of the antioxidant activity of some molecules of biological interest. Chemistry magazine 2003;54(11) 885-887.

[40] Tim Cushnie TP, Andrew J. Lamb. Antimicrobial activity of flavonoids (Review). International Journal of Antimicrobial Agents 2005;26 343–356.

[41] Rahmanullah Siddiqi, Shahina Naz, Samia Ahmad, Syed Asad Sayeed. Antimicrobial activity of the polyphenolic fraction derived from Grewia asiatica, Eugenia jambolana and Carissa carandas. International Journal of Food Science & Technology 2011;46(2) 250-256.

[42] Fattouch S, Caboni P, Coroneo V, Tuberoso CI, Angioni A, Dessi S, Marzouki N, Cabras P. Antimicrobial activity of Tunisian quince (Cydonia oblonga Miller) pulp and peel polyphenolic extracts. J Agric Food Chem. 2007;55(3): 963-9.

[43] Rahmanullah Siddiqi, Shahina Naz, Samia Ahmad, Syed Asad Sayeed. Antimicrobial activity of the polyphenolic fractions derived from Grewia asiatica, Eugenia jambolana and Carissa carandas. International Journal of Food Science & Technology 2011;46(2) 250-256.

[44] Fattouch S. Caboni P, Coroneo V, Tuberoso CI, Angioni A, Dessi S, Marzouki N, Cabras P. Antimicrobial activity of Tunisian quince (Cydonia oblonga Miller) pulp and peel polyphenolic extracts. J Agric Food Chem. 2007;55(3): 963-9.

[45] Saavedra MJ, Borges A, Dias C, Aires A, Bennet RT, Rosa ES, Simoes M. Antimicrobial activity of phenolics and glucosinolate hydrolysis products and their synergy with streptomycin against pathogenic bacteria. Med Chem. 2010;6(3): 174-83.

[46] Sivakumar PM, Prabhawathi V, Mukesh Doble. 2-Methoxy-2',4'-dichloro chalcone as an antimicrofoulant against marine bacterial biofilm. Colloids Surf B Biointerfaces 2010;81(2): 439-46.

[47] Sebastian Jäger, Holger Trojan, Thomas Kopp, Melanie N. Laszczyk, Armin Scheffler. Pentacyclic Triterpene Distribution in Various Plants – Rich Sources for a New Group of Multi-Potent Plant Extracts. Molecules 2009;14, 2016-2031.

[48] van Baren C, Anao I, Leo Di Lira P, Debenedetti S, Houghton P, Croft S, Martino V. Triterpenic acids and flavonoids from Satureja parvifolia. Evaluation of their antiprotozoal activity. Z. Naturforsch C. 2006;61(3-4): 189-92.

[49] Domingues R. M. A., Patinha D. J. S., Sousa G. D. A., Villaverde J. J., Silva C. M., Freire C. S. R., Silvestre A. J. D., Pascoal Neto C. Eucalyptus biomass residue from agro-forest and pulping industries as sources of high-value triterpenic compounds. Cellulose chemistry and Technology 2011;45(7-8) 475-48.

[50] Dimitrios Boskou. Health Properties of olive oil minor constituents. http://www.soci.org/News/~/media/Files/Conference%20Downloads/OFI%20Abu%20Dhabi%20Apr%2008/Boskou.ashx (accessed 21 June 2013).

[51] Allouche Y, Jimenez A, Uceda M, Aguilera MP, Gaforio JJ, Beltran G. Triterpenic content and chemometric analysis of virgin olive oils from forty olive cultivars. J Agric Food Chem 2009;57(9): 3604-10.

[52] Pirvu L, Armatu A, Bubueanu C, Pintilie G, Nita S. Obtaining and chemical characterization of some vegetal extracts with corrosion-scaling inhibition properties. Part I. Fagus sylvatica and Alii cepae bulbus extracts. Romanian Biotechnological Letters 2010;15(6) 5683-5689.

[53] Pirvu L, Barbulescu D, Nichita C, Nita S, Colceru-Mihul S. Obtaining and chemical characterization of some vegetal extracts with corrosion-scaling inhibition properties. Part II. Juglandis folium and Agrimoniae herba extracts. Romanian Biotechnological Letters 2011;6(1) 5937-5943.

[54] Romanian Pharmacopoeia, Xth edn., Bucharest: Medicala; 1993.

[55] Cojocaru A, Maior I, Vaireanu DI, Lingvay C, Lingvay I, Caprarescu S, Badea GE. Ethanol extract of Fagus sylvatica leaves as an eco-friendly inhibitor for carbon steel corrosion in acidic solution. Journal of Sustainable Energy 2010;1(3) 64.

[56] Lingvay I, Pirvu L, Vaireanu DI, Lingvay C, Nita S, Cojocaru A, Colceru-Mihul S, Maior I. Simultaneously corrosion and scale formation controlling ecological inhibitor and the obtaining technology. Patent application no A/00900/13.09(2011).

[57] Romussi G, Bignardi G. Fagus sylvatica L. Terpenoids. Arch. Pharm 1987;320(20) 153-8.

Section 2

Anti-Corrosion Coatings

Chapter 11

Production of Anti-Corrosion Coatings on Light Alloys (Al, Mg, Ti) by Plasma-Electrolytic Oxidation (PEO)

Riyad O. Hussein and Derek O. Northwood

Additional information is available at the end of the chapter

http://dx.doi.org/10.5772/57171

1. Introduction

As the world becomes increasingly more environmentally conscious, many countries are looking for ways to make products that are both safe for the environment and reduce or eliminate any health concerns for their workers. Considerable collaborative work has been done in the academic, industrial and governmental sectors to find environmentally compliant substitutes for chromium, particularly hexavalent chromium [1]. Sol-gel coatings and conducting polymers are being developed as either barrier coatings or reactive inhibitor systems to replace chromates for corrosion protection [2]. Conversion layers provide the ability to modify the substrate surface to give better adhesion, a surface free of contaminants or a coating layer that contains active corrosion inhibitors.

Among the newer surface modification techniques, plasma electrolytic oxidation (PEO) has recently been successfully applied to light-metals. The term 'light-metals' has traditionally been given to magnesium, aluminum and titanium because they are frequently used to reduce the weight of products due to their relative low densities. The plasma-assisted electrochemical surface treatment PEO process is capable of producing very thick coatings on the substrate materials to enhance their hardness, thermal resistance, dielectric strength, friction coefficient, and wear and corrosion resistance during relatively short time, in addition to its excellent throwing power (the ability of the process to deposit metal uniformly on an irregularly shaped cathode [3].

For PEO coating growth there are two simultaneous processes occurring, namely the electrochemical and the plasma chemical reactions. The plasma chemistry of the surface discharges is quite complex in nature, involving, on one hand, charge transfer at the substrate/electrolyte interface, and on the other hand, strong ionization and charge transfer effects between the substrate surface and the electrolyte through the oxide layer with the aid of the plasma [4].

The microstructure and phase content of the coatings are related to the process parameters and to their ultimate corrosion resistance. In this chapter a detailed description of the PEO-coating mechanism will be given supported by detailed diagrams and figures. The effects of the process parameters on the microstructure and associated corrosion resistance of the PEO coatings on Al, Ti and Mg alloys will also be discussed.

2. Surface modification and treatment for lightweight metals

So far, a number of theoretical and experimental studies had been performed addressing different aspects of the corrosion protection of lightweight metals and their alloys. Surface treatment/modification and coating occurs only on/or close to the surface layer of lightweight alloys and the bulk alloy is left unchanged hence the original mechanical or other properties are not changed.

An overview of the different techniques used for treatment or modification of Mg, Al and Ti alloys is shown in Fig.1 [2,5-6] and a comparison of the common surface treatments and coatings is given in Table 1. Generally, coatings can be divided into three classes: conversion coatings, organic and inorganic deposited coatings. Conversion coatings are produced by chemical or electrochemical reactions between the substrate and the aqueous solutions to form an oxide layer that simultaneously grows inwards and outwards. Such conversion coatings represent an effective way to increase the corrosion resistance of aluminum, magnesium and titanium alloys or, as a pre-treatment, to enhance the adhesion of a final deposited coating [2].

Treatment/coating	Drawback (general)	Drawback (environment)
Conversion (chromate, phosphate)	Easily damaged	Toxic, particularly Cr(VI)
Anodizing	Sensitive to impurities in the base metal to be coated	Sulphuric acid baths
Organic/polymer	Weak mechanical and corrosion resistant properties	Poor recyclability
Gas phase deposition (PVD, CVD)	Thin, porous	Chlorine emission
Thermal spray, cold spray	Not suitable for components with complex geometry. Use of more noble materials that can cause corrosion at interface	None
Plasma electrolytic oxidation	Lack of data on coating performance in practice	None

Table 1. A comparison of the common surface treatments and coatings for light-weight alloys. Adapted from [10]

Cost effective organic coating techniques, including sol-gel, painting and powder coating, are typically used after a primary surface-treatment of the substrate. The method most applied to obtain organic coatings is simple dipping in an organic based solution including: gelatin/PLGA particle solution, prospolis and polylactic-co-glycolic acid (PLGA) solution [7,8]. Metallic protective coatings can be produced from inorganic deposition coatings such as from the gas phase or other physical methods including plasma spraying, chemical vapor deposition (CVD), diamond like coatings (DLC) and lasers.

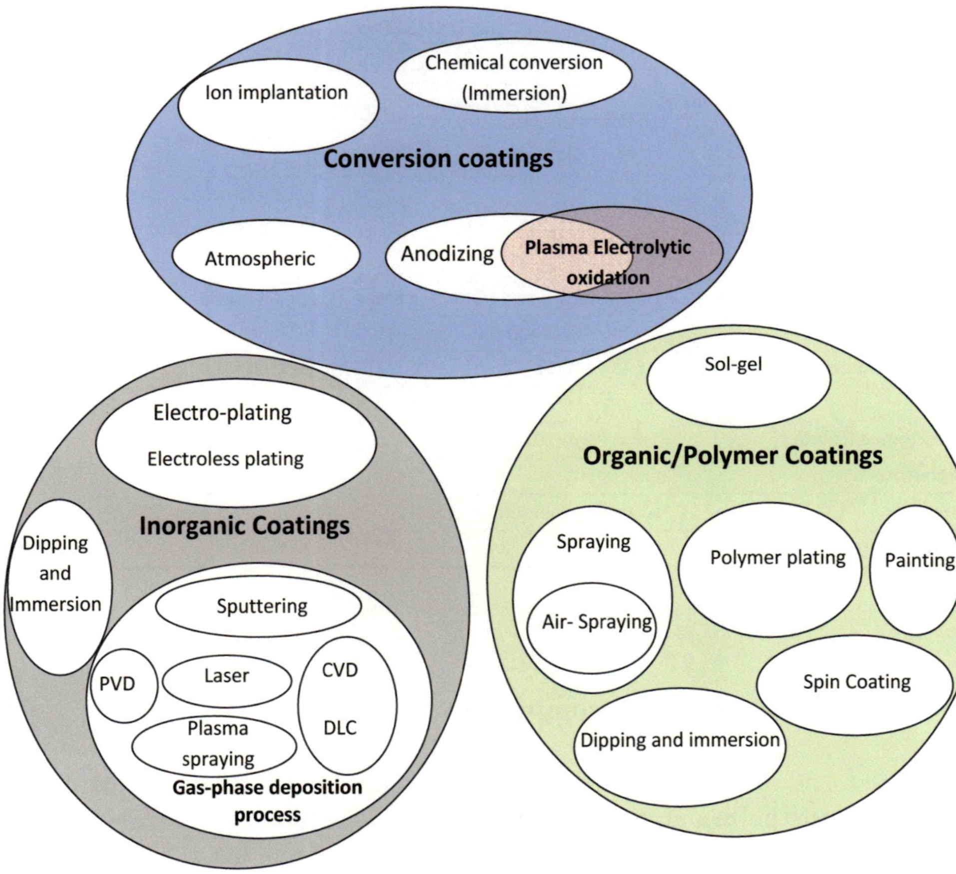

Figure 1. Schematic Diagram showing coating technologies [adapted from [5]].

Gas phase coatings are inappropriate for applications involving geometrically complex components. Moreover, physical deposition methods involve high energy consumption, high costs and complex facilities. The most common conversion layers are from electrochemical anodization. Recently, much research has been focused on the selection of electrolytes, including alkaline solutions with additives of phosphate, silicate, and borate substances. However, the anodizing process sometimes suffers from relatively poor performance. For example, the European Space Agency was having problems with the black anodizing process

[9]. A comparison between PEO and the hard anodization processes is given in Table 2. The PEO process can be considered as a combination of anodizing (electrolytic oxidation) and plasma discharging processes. Where, the main similarities between the PEO and anodizing processes, is that both of them involve oxidation of substrate using electrolytic bath and first stage of the PEO process is an anodization process. It is also common to combine two or more techniques for surface coating and/or modification, to obtain properties that are unattainable using an individual technique.

Process	**PEO**	**HA**
Voltage and current density	High	Low
Deposition rate	Fast (1~2μm/min)	Slow (~0.3μm/min)
Oxidation mechanism	Chemical/ electrochemical and plasma chemical reactions	Chemical/electrochemical reactions
Coating on selected alloys	Practical for various kinds of Al, Mg and Ti alloys	Limited (not used for 2000-series alloys, high zinc or silicon Al alloys and Al casting alloys)
Microstructure	Amorphous and crystalline phase / Inner dense layer and outer porous layer	Amorphous / Columnar porous layer and very thin barrier
Corrosion resistance (Relative)	Excellent (5)	Good (1)
Hardness	High (~Hv1600)	Low (Hv600 max)
Wear resistance (relative)	Excellent (30)	Fair (2)
Thermal protection	Excellent	Good
Electrolyte	Alkaline solution	Acid solution
Dielectric strength	Excellent	Fair

Table 2. Comparison of PEO with HA processes.

3. The PEO process: General comments

The formation of oxide films on metals were first investigated at the beginning of the twentieth century by Günterschulze and then Günterschulze and Betz [11]. They published their first studies on the electrolytic spark discharge produced on aluminum foil in the early 1930s. In the former Soviet Union in the 1960s and 70s, scientists investigated the anodizing process at potentials of over 200 volts for developing parts for the submarine sector and for military purposes. This led to plasma electrolytic oxidation (PEO) technology, which is also known as micro-arc oxidation (MAO) and micro-plasma oxidation (MPO). Hussein et al [12] have provided a detailed historical background of the process. It wasn't until the 1990s that the PEO process gained worldwide recognition as an eco-friendly technology for depositing the tribologically superior and excellent corrosion protection ceramic coatings on aluminum, titanium and magnesium alloys [12]. To date, thousands of a scientific publications dealing with a plasma electrolytic oxidation technology can be found in a wide variety of journals and conference proceedings.

http://dx.doi.org/10.5772/57171

Plasma Electrolyte Oxidation (PEO) is a novel surface engineering technology, considered as one of the most cost-effective and environmentally friendly ways to improve the corrosion and wear resistance of lightweight metals and their alloys [3]. A driving force for PEO developments is the avoidance of expensive equipment required for competing vacuum-based plasma technologies. The substrate is immersed in an aqueous electrolyte (containing neither the concentrated sulphuric acid nor the chromate ions used for hard anodizing) and a high anodic potential is applied to it (typically several hundreds of volts) that trigger numerous micro-discharge events at the metal-electrolyte interface, generating high instantaneous temperatures and pressures ($T > 40000$ K [4]) and $p \sim 100$ GPa, [3]). Different current modes have been utilized in the PEO treatment including, DC, AC, unipolar and bipolar current modes [13] which play important roles in the consequent voltage breakdown, local melting and oxidation of the substrate, quenching and re-crystallization processes. Hence, the formation mechanisms for the coatings are complex due to the involvement of electro-, thermal-, and plasma- chemical reactions in the electrolyte [12,14].

A schematic diagram of a commercial setup for PEO coating is shown in Figure 2 [12]. It consists of a stainless steel bath which contains the electrolyte, and which is often water-cooled, a high power electrical source and a pulse generator. The sample under treatment is connected to the output of the electrical source as one of the electrodes (anode), while being immersed in the bath of electrolyte solution. The stainless steel bath acts as a counter-electrode. The technology can be applied to coat a variety of materials, such as aluminum, magnesium, titanium, and zirconium. It can also be applied to alloys that are difficult to anodize with conventional anodizing processes, such as high copper content and high silicon content aluminum alloys (2000 series and A380), and magnesium alloys, with the deposition of layers such as phosphates, silicates and aluminates oxides. This enables the surfaces of metals such as titanium to be converted into very hard materials like TiN_2 or TiB_2. A key to the development of such coatings lies in the understanding and characterisation of the plasma processes involved in the individual discharges.

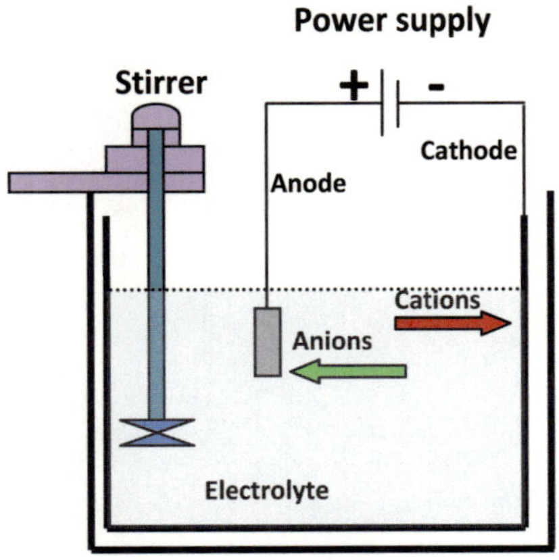

Figure 2. Schematic diagram of the PEO apparatus.

The PEO process can be distinguished from low voltage anodizing in aqueous solutions by its operation at electrode potentials greater than the typical breakdown voltages of the original oxide films (350-600 V) in AC, DC or pulsed AC/DC modes with asymmetric anodic and cathodic potential peak waveforms, depending on the alloy and electrolyte composition [4,12]. PEO also operates differently from high energy plasma coating under dry conditions in a controlled gas pressure chamber. It has been shown that PEO room temperature technology has a number of unique advantages: including technological simplicity and the possibility to coat any size and/or complex structure including welded and riveted joints, and heterogeneous alloys [15]. Pretreatments and post treatments are not strictly necessary for the PEO, except for water rinses; this reduces environmental concerns with respect to pretreatment and post treatment solutions. The PEO process is capable of producing uniform and very thick coatings on the inner and outer surfaces of the substrate materials [16]. One of the main advantages of the PEO coating is that the oxide coating is integral with the metal substrate because the coating is a result of substrate oxidation.

Fig. 3 show a typical Voltage vs. treatment time plots during the PEO processing of (a) AM50 Mg-alloy using a DC current mode and (b) an output anodic (V_A) and cathodic ($-V_C$) voltage vs time curves of an AJ62 Mg-alloy (MgAl6Mn0.3Sr2) using a bipolar current mode. Four consecutive discharge stages can be distinguished, namely: **Stage I:** In the early stage of the process which mainly involves the rapid electrochemical formation of an initial insulating oxide film, a sharp increase in the voltage was seen. In this stage the breakdown voltage is not yet reached. **Stage II:** The rate of the voltage change decreases in this stage, which is characterized by numerous sparks moving rapidly over the whole sample surface area. This indicates a start of the breakdown of the oxide layer, an increase in temperature and, therefore, melting of the substrate metal. **Stage III:** In this stage the rate of voltage increase becomes slow; this stage is characterized by larger but slower moving discharges. As the oxide layer grows, its electrical resistance increases, therefore the nature of the plasma changes. **Stage IV:** In this stage the rate of voltage variation is even slower than that in stage III and concentrated

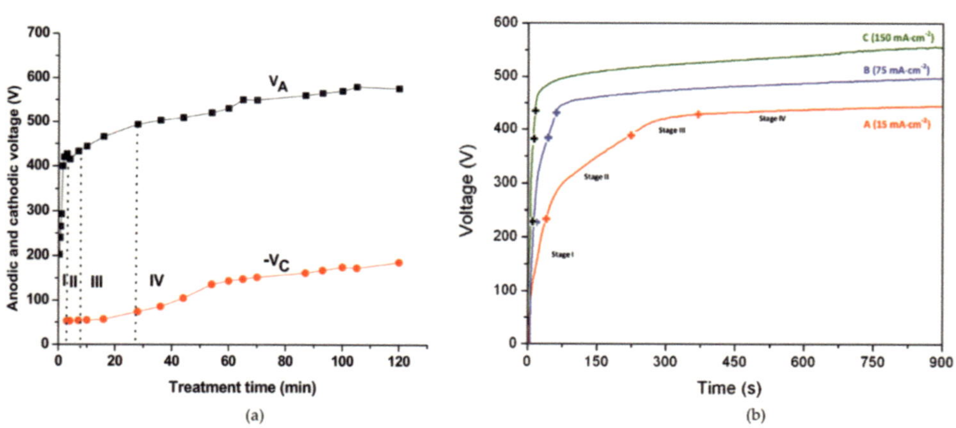

Figure 3. Voltage vs. treatment time during PEO processing of (a) AM50 Mg-alloy using DC current mode [18], (b) AJ62 Mg-alloy using a bipolar current mode showing anodic (V_A) and cathodic ($-V_C$) voltage parts [19].

discharges appear as relatively large and long lasting sparks. For some cases, such strong discharges may cause irreversible damage to the coatings in stage IV. Figure 4 shows a fast video imaging of the PEO coating showing the size and color of the micro discharges changing with time of an Al-alloy [17]. At the early stage of the process, intense gas evolution along with some luminescence at the surface is easily observed; this is followed by sparks flashing randomly all over the aluminum alloy surface, Figure 4(a), while sparks progressively change to micro-arcs (Figure 4(c)). Finally arcs regime, Figure 4(d,e) occurring at later stages of the process causing irreversible damages to the oxide layer.

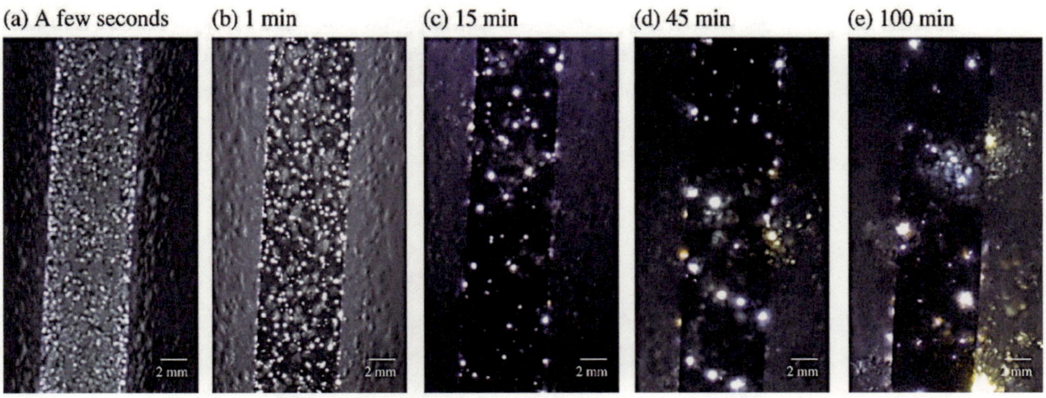

Figure 4. Fast video imaging of the PEO coating showing the size of the micro discharges changing with time [17].

4. PEO applications

Oxide coatings have been deposited on aluminum or magnesium alloys for the purposes of protection against wear and/or corrosion. Coatings on titanium have been investigated for potential applications relying on the biocompatibility or photo-activity of titania (TiO_2). Among the light alloys, Mg alloys are the most reactive and thus susceptible to corrosion, followed by Al alloys and Ti alloys. In general, such alloys need a specific coating thus allowing for a large range of possible applications across all industrial sectors

Depending on where they are used, the coatings should have resistance to chemicals, heat and electricity. PEO-coated products are used in a wide range of applications and industries see Figure 5, including:

i. **Decorative purposes or optical applications:** Depending on electrolyte composition and concentration, PEO processing parameters such as color additives and electrical parameters and chemical composition of substrate, the color of PEO coatings can be black, blue, gray, and white. By designing in porosity, a PEO pre-treatment can enhance the adhesion of paints, sol-gel and powder coatings and produce duplex coatings for enhanced properties particularly in aggressive media [9].

ii. **Thermal Applications:** The thermal conductivity of the PEO oxide layer is low and thermal cycling tests, have shown excellent performance at temperatures between -40 to +100°C. Thus, PEO coatings can be used to increase the thermal shock resistance [20].

iii. **Mechanical Applications (hardness, wear and friction):** Due to the high hardness and superior tribological performance of the oxide layer formed by PEO process, which for example reduces the wear rate of 6061 Al alloy by a factor of about 30 [21] compared to 2 times for the hard anodized coating, allows a replacement of many parts by PEO coated Mg and Al alloys leading to a reduction in fuel consumption in the automotive and aerospace industries. Improved coating performance that can be obtained has yielded applications for PEO technology in the aerospace industry (fasteners, landing gear, blades, discs and shafts of aircraft engines), the automotive industry (seat frames, doors, pistons and cylinder liners), and the gas and oil industries (gears and rotary pumps) [22].

Figure 5. Use of PEO coatings on: (a) engine piston, (b) racing yachts, (c) engine block and (d) dental implants.

i. **Electric and Electronic Applications:** Electrical insulation coatings can be obtained with high dielectric strength for electrical and electronic components. Also PEO is suitable for the hard coating of the inner areas including conic, cavital or cylindrical areas.

ii. **Chemical Applications:** PEO coatings can be used in chemical industries due to the coatings resistance to aqueous environments and resistance to strong acids and strong bases.

iii. **Biomedical applications:** Surface engineered titanium and magnesium alloys for biomedical devices including degradable biomaterials and dental implants [23].

5. Coating surface morphology and microstructure

The surface morphology and coating microstructure will have a significant impact on the corrosion behaviour which is governed by the level of porosity and defects as well as the thickness of the dense layer. Most of the studies discussed in this section used scanning electron microscopy (SEM) and transmission electron microscopy (TEM) to observe the surface and cross sections of the coatings.

The surface of PEO coatings typically show a 'pancake' structure wherein the center of each pancake was a discharge channel through which the molten elements surged out of the channel and quickly solidified leaving distinct boundaries that define each pancake [24-25]. Figure 6 shows a discharge channel from which the molten material is ejected, as result of greater energy input. The relatively large holes in the center of the pancake suggest that there are strong discharges and such holes may penetrate deep into the coating thickness. The discharge channel is shown to be surrounded by solidified material and some localized micro-cracks which may form as result of a rapid cooling by the electrolyte. Where stronger discharges have greater energy inputs which causes a larger amount of substrate and its oxide to melt down, and eventually the molten material is ejected onto the surface, forming larger ceramic particles when it is immediately cooled by the electrolyte. Some microcracks appear on the coating surface: which could be produced by the thermal stresses during the rapid solidification of the molten oxide product in the strong discharge channels [26]. The molten oxides around the pores indicate that the instantaneous temperature in the micro discharge zone might reach several thousand degrees. The surfaces of other coatings were dominated by many randomly arranged curly projections.

Pores of different size are evident on the surface and their sizes reflect the strength of the discharges. Since the creation of porosity cannot be prevented during the PEO process, it is important to find ways to keep the pores as small as possible as well as producing as uniform as possible layers. However, the density of these holes has been reported to decrease substantially with increasing process time [14,27]. The pore size in the coatings varies from 0,01 to a few μm [14]. Furthermore, there are also many very small holes (micro-pores) in the surface, which are due to gas bubbles ejected from surface discharges as shown in Fig. 7(a,b). If necessary, the porosity can be decreased by proper adjustment of the current mode and density, or/and impregnation by various materials. For example, the bipolar and unipolar current modes for the PEO coating of a AJ62 magnesium alloy have been compared and it has been shown that the use of a bipolar mode can reduce the pore density and size, as well as other coating defects [28]. Although a porous microstructure may be beneficial as lubricant reservoir to decrease wear, and for biocompatible applications where it is beneficial for the rapid adhesion and growth of cells, resulting in a significantly stronger bond to the parent tissue [29], it will also result in a decrease in corrosion resistance of the PEO coatings. The surface roughness also increases as the coating thickens. The actual roughness depends on the particular alloy and the treatment parameters.

As the PEO coating thickens, the evolving micro discharges continuously affect the sintering, crystallization and phase transformations in the coating. The PEO coatings are generally described as a multilayer system. Fig. 7(c-d) illustrates the structure of a typical PEO coating (example is for a Mg-alloy). The coating is composed of three layers namely: a porous outer layer, intermediate dense layer and thin inner dense layer. Coating thickness can range from few to more than 200 microns. However, it should be noted that the PEO layer is partly growing into the original substrate surface and is partly built-up on top of it. Coating growth rate, relative sizes, structure and composition of their different regions are greatly influenced by substrate composition, electrolyte composition and current operating modes and magnitude [3,30,31].

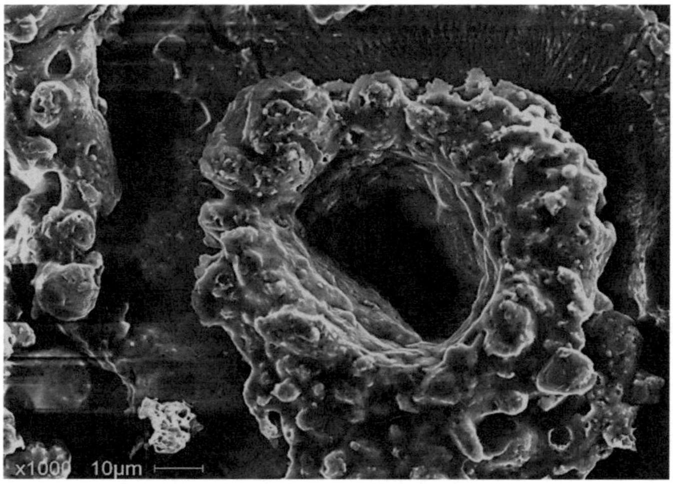

Figure 6. SEM image of a discharge channel; a typical feature of PEO coating processed for 60 minutes at current density of 0.15 A/cm2. Such a pore may penetrate the entire coating thickness to the substrate.

Nie et al [32] studied the microstructure of the inner layer, near the coating/substrate interface, using cross-sectional TEM (Fig. 8). This layer exhibits a number of (predominately amorphous) sub-layers, whilst the lower portion of the intermediate layer has a nano-scaled polycrystalline microstructure. The characteristics of the coating near the interface with the Al 6082 alloy substrate can be divided into hree sub-layers (1, 2 and 3 in Fig. 8(a)).

The PEO coatings on Mg alloys present a porous outer layer, intermediate (or compact) inner layer and dense coating/substrate interface with minimum porosity. The coatings are comprised of amorphous and crystalline phases such as MgO, Mg_2SiO_4, $Mg_3(PO_4)_2$ or Mg_2AlO_4, depending on the different electrolyte [15,33-34]. A typical XRD profile showing the results of phase composition analysis performed on PEO coatings on AZ91 magnesium alloy by Lee et al [35] in three different electrolytes (Bath A contains 0.09 mol/l KOH, 0.5 mol/l KF, 0.1 mol/l $K_4P_2O_7$, Bath B contains an extra 9 g/l ZrO2 while Bath C contains an extra 9 g/l TiO2), is presented in Fig. 9.

Previous work [32,36] had also shown some porous areas near the interface of MgO and Al_2O_3 coatings on Mg and Al substrates using transmission electron microscopy (TEM) [32]. The Al_2O_3 coating was however demonstrated to have superior mechanical and wear properties [32]. On the other hand, such a porous region was not observed in a TiO_2 coating on Ti6Al4V alloys after the PEO process [37].

For the PEO coatings on Al alloys, the porous outer layer consists predominantly of the low temperature modification of alumina, γ-Al_2O_3, with poor mechanical properties. The dense inner layer, on the other hand, is comprised of the mixture of high temperature α- Al_2O_3 modifications of Al_2O_3 and complex Al-X-O phases (X is an element from the electrolyte eg. Si, P), where complex phases of the substrate alloying elements are observed in a thin, interfacial region below the dense layer. The content of the α- Al_2O_3 phase is increased with increasing coating thickness, which may be attributed to the following two reasons. First, any

http://dx.doi.org/10.5772/57171

phase transformation takes place when the discharges are strong enough to heat the oxide coating to the required temperatures, for a time adequate to allow phase transformation. Second, internal residual stresses will be formed as result of the rapid heating and cooling cycles experienced during coating formation. The formation of stable α- Al_2O_3 is favored by relief of these stresses [30]. The aluminum oxide coating morphology and microstructure were also significantly different under different current operating modes [33, 38]. The bipolar current mode could improve the coating quality compared with the unipolar current mode, in terms of surface morphology and cross-sectional microstructure. A dense coating morphology could be achieved by adjusting positive to negative current ratio and their timing to eliminate or reduce the strongest plasma discharges [39]. PEO coatings have excellent adhesion with substrate, which is due to a transition layer on the interface. The transition layer is growing both inside the substrate, and outside, as a result, the coating is integrated with the substrate.

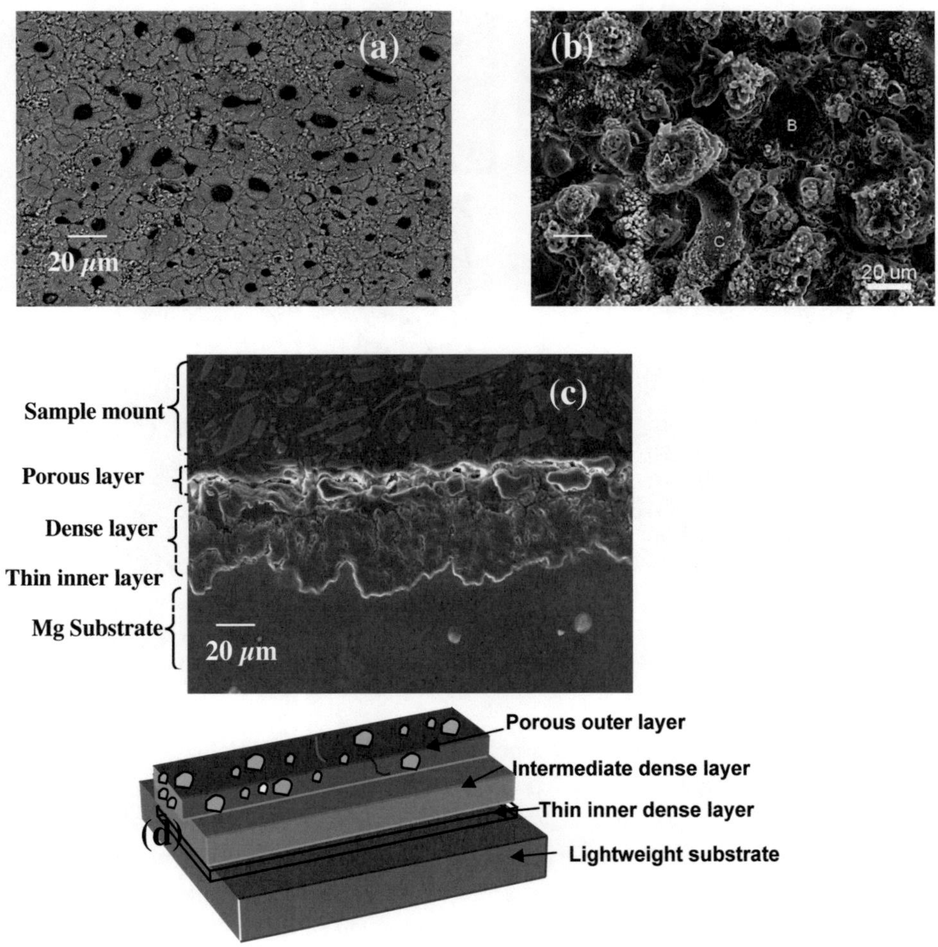

Figure 7. SEM micrographs showing (a,b) the surface morphology, (c) cross-section of PEO coating on AM60B alloy and (d) illustration of the coating layers of PEO coatings.

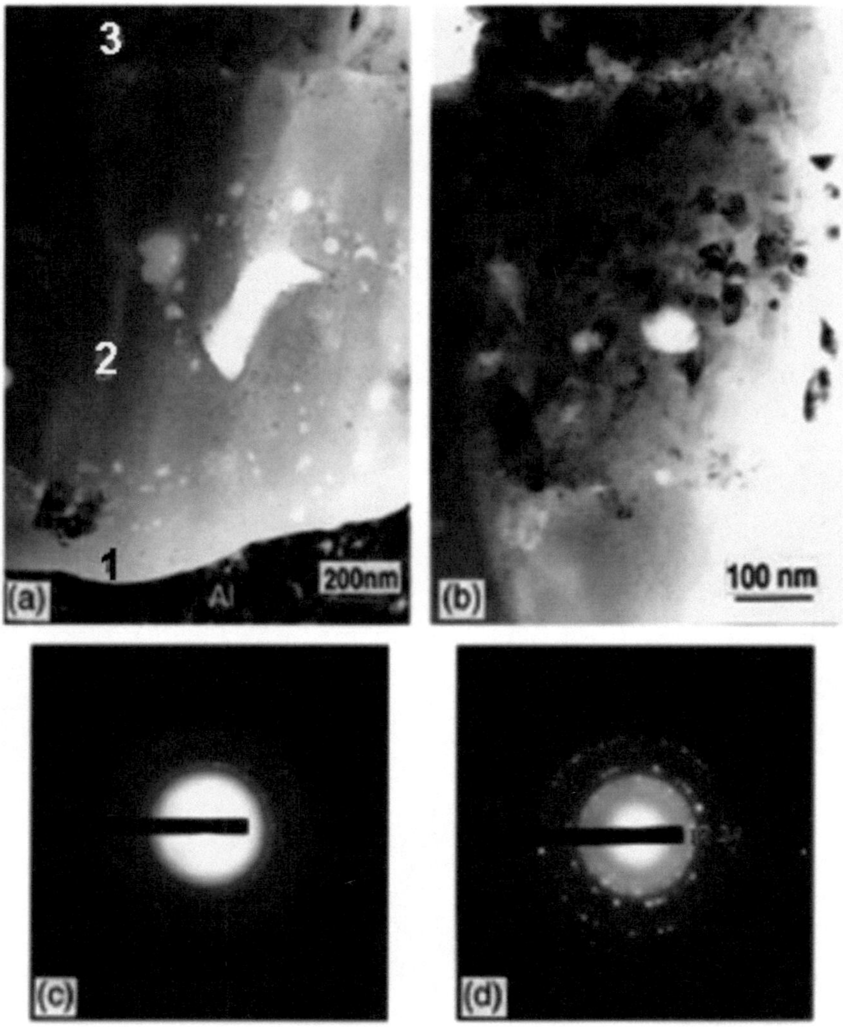

Figure 8. Cross-section TEM images of Al 6082 substrate of (a) the inner layer near coating/substrate interface and (b) the intermediate layer, and SAED patterns taken from intermediate (c) sublayer 2 and (d) sublayer 3, respectively [32].

The effect of the substrate chemical composition on phase evolutions in the PEO coatings formed has been investigated [4, 26, 39]. It has been shown that the amount α-Al_2O_3 phase can be substantially increased, reaching 60 % or more, for coatings produced on aluminum alloys containing 4 – 5 % Cu [4]. Such high fractions of the α- Al_2O_3 phase is generally regarded as responsible for the high hardness, up to 16–25 GPa [26], and low bulk stiffness, of PEO coatings which makes them attractive for wear protection. However, the electrolyte composition is a key factor in promoting any specific phase formation. For instance, it has been shown that, sodium aluminate ($NaAlO_2$) in the electrolyte can be, at least partly, responsible for the formation of more α-alumina [38].

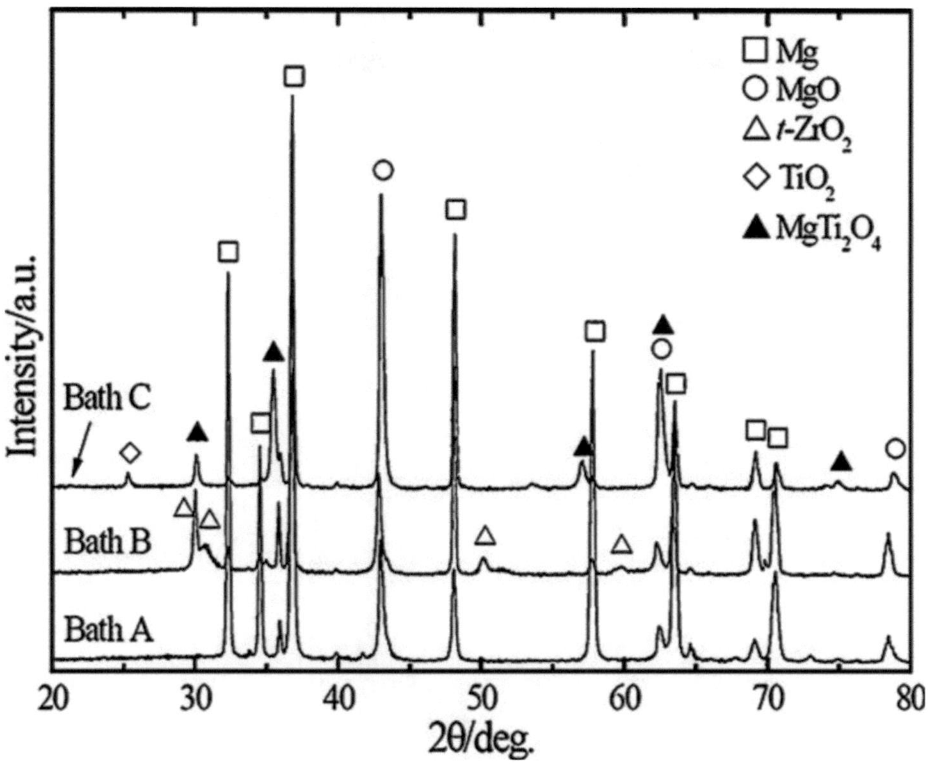

Figure 9. Qualitative XRD patterns of the oxide layers of AZ91 Mg alloy coated in different electrolytes [35].

6. Coating formation mechanisms

6.1. Electrochemistry of plasma electrolytic oxidation (PEO)

The growth of the oxide coating mainly occurs due to three different, yet simultaneous, processes, namely: the electrochemical reactions, the plasma chemical reactions [40-42] and thermal diffusion. The electrochemical formation of surface oxide layers can occur through different mechanisms depending on the electrolyte, eg. silicates, aluminates, phosphates. The PEO coatings are usually produced by AC or bipolar current mode, containing both anodic and cathodic components. An early investigation of the basic electrochemical processes of AC PEO coatings on Ti, has been carried out by Yerokhin et al [43] using AC current mode and a complex aluminate-base electrolyte. According to their study an oxide layer formation is induced both by the ionic component of the current which is transmitted via surface discharges and by the anodizing current passing across the surface which is free of discharges (Figure 10). The other components of the current cause secondary electrochemical processes which lead to liberation of electrode gases (H_2 and O_2) and anodic dissolution of the titanium metal as shown in Figure 10, where the following general reactions normally occur:

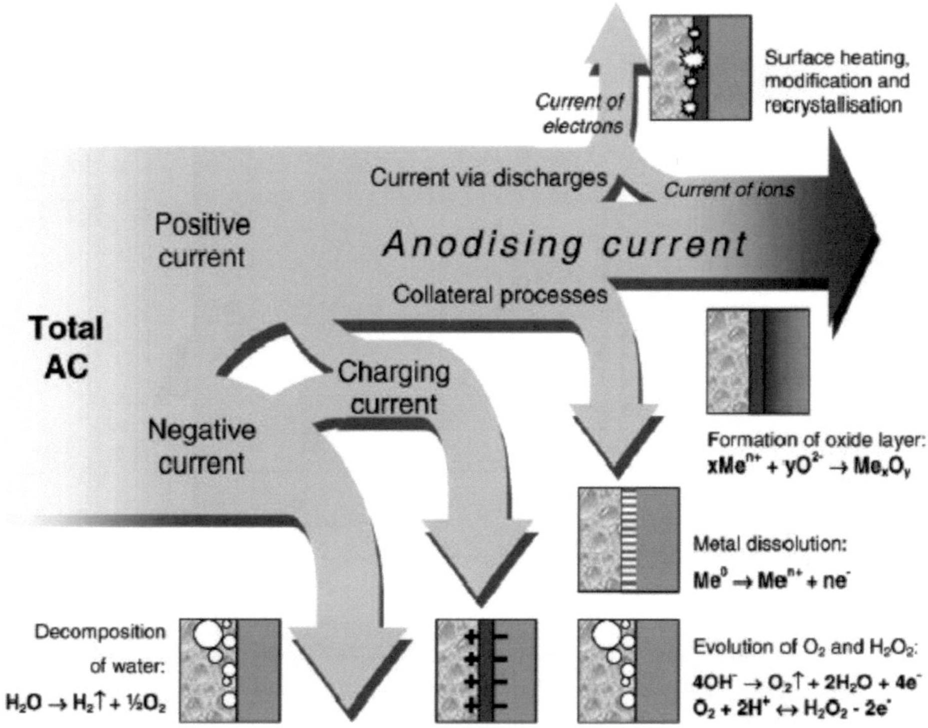

Figure 10. Schematic diagram of current distribution during the PEO treatment of metals in AC mode [43].

- Metal-oxide interface:

$$Ti \longrightarrow Ti4+ +4e- \quad (1)$$

For alkaline aluminate solutions the anions (OH- and AlO2-) can take part in the following processes on the oxide/electrolyte interface [43,44]:

$$Ti4+ +4OH- \longrightarrow TiO2 +2H2O \quad (2)$$

$$Ti4+ +4AlO2- \longrightarrow TiO2 + Al2O3 \quad (3)$$

$$Ti4+ +Al\left(OH\right)4 - \longrightarrow TiO2 + Al2O3 + 2Al\left(OH\right)3 + 5H2O \quad (4)$$

Dissolution and oxygen evolution is quite common in aluminum PEO in alkaline solutions, where the following general reactions normally occur [30]:

http://dx.doi.org/10.5772/57171

- Metal-oxide interface:

i. anodic processes:

$$2Al + 3O^{2-}_{solid} \rightarrow Al_2O_3 + 6e^- \tag{5}$$

$$Al \rightarrow Al^{3+}_{solid} + 3e^- \tag{6}$$

- Oxide-electrolyte interface

i. anodic process

$$4OH^- \rightarrow O_2 \uparrow + 2H_2O + 4e^- \tag{7}$$

ii. alumina chemical dissolution and oxidation of ejected Al:

$$Al_2O_3 + 2(x-3)OH^- + 3H_2O \rightarrow 2[Al(OH)_x]^{n-}_{gel} \tag{8}$$

$$Al^{3+}_{ejected} + xOH^- \rightarrow [Al(OH)_x]^{n-}_{gel} \tag{9}$$

$$[Al(OH)_x]^{n-}_{gel} \rightarrow Al(OH)_3 \downarrow + (x-3)OH^- \tag{10}$$

It has also been found that the film growth rate decreased significantly with increasing electrolyte concentration, since the rate of anodic dissolution increased.

For magnesium alloys the main electrochemical reactions occurring at the metal/oxide and oxide/electrolyte interfaces are as follows [10,15]:

- Metal-oxide interface:

$$Mg \rightarrow Mg^{2+} + 2e^- \tag{11}$$

$$4OH^- \rightarrow 2H_2O + O_2 \uparrow + 4e^- \tag{12}$$

- Oxide-electrolyte interface:

$$Mg^{2+} + 2OH^- \rightarrow Mg(OH)_2 \tag{13}$$

$$Mg(OH)_2 \rightarrow MgO \downarrow + H_2O \tag{14}$$

$$Mg^{2+} + 2AlO_2^- \rightarrow MgAl_2O_4 \downarrow \tag{15}$$

$$2Mg^{2+} + SiO_3^{2-} + 2OH^- \rightarrow Mg_2SiO_4 \downarrow + 2H_2O \tag{16}$$

$$3Mg^{2+} + 2PO_4^{3-} \rightarrow Mg_3(PO_4)_2 \downarrow \tag{17}$$

Reaction (11) is the anodic dissolution and (12) is the oxygen evolution reaction. The cation released from the metal (reaction (11)) combines with the anion in the electrolyte to form compounds $Mg(OH)_2$, Mg_2AlO_4, Mg_2SiO_4 or $Mg_3(PO_4)_2$, depending on the different electrolyte by reactions (13), (15), (16) and (17). The unstable hydroxide $Mg(OH)_2$ dehydrates to MgO by the high temperature (reaction (14)), resulting from the plasma discharge.

6.2. Plasma discharge models and plasma chemistry

Several micro-discharge formation models have been proposed [45-49]. In the first model [46], the micro-discharges appear as a result of the oxide film dielectric breakdown in a strong electric field. The second group of models considers each microdischarges as a gas/glow discharge occurring in a micropore of the oxide film [47]. The formation of a gas phase in the pore (and discharge ignition in it) is believed to be induced by an initial dielectric breakdown of a barrier layer in the bottom of the micropore [45]. The third model [49] assumed the possibility of free electron generation and glow discharge ignition in the gaseous media at the oxide-electrolyte interface, which leads to heating, melting and quenching the underlying oxide layer. Any other model considers the formation of the micro-arc discharge as an electronic 'avalanche', or due to an electronic tunneling effect [48]. Yerokhin et al [45] found that the above models [46-49] do not fit the spatial, temporal and electrical characteristics of microdischarge phenomena which were observed in their investigation. They suggested a new model based on the analogy with contact glow discharge electrolysis. The model assumes the possibility of free electron generation and glow discharge ignition in the gaseous media at the oxide-electrolyte interface, which leads to heating, melting and quenching of the underlying oxide layer.

Recently, optical emission spectroscopy (OES) has been used to investigate the plasma discharge behavior during the PEO process [4,33]. Species from the substrate (Mg, Al and Ti) and the electrolyte (H, OH, Na and K) were found to be involved in the plasma discharge during the PEO process. The evolution of the spectra is considered to reflect the change in mechanism that initiated the plasma discharge, from bound-bound transitions of electrons between atomic level to collision–radiative recombination of electrons (bound-free transitions) and Bremsstrahlung radiation (free-free transitions). According to Dunleavy et al [33], the plasma emission spectra indicated that there were two distinct regions of the plasma, a central core of high temperature (~16, 000 ± 3500 K), with a high electron density (N_e ~5× 10^{17} cm^{-3}) and a peripheral region, probably extending into the surrounding electrolyte, which was much cooler (~3000-4000 K) and less dense (N_e ~5× 10^{15} cm^{-3}). Hussein et al studied the evolution of the emission spectra of plasma discharge during the PEO process [39]. The fluctuations in plasma intensities and temperatures during the plasma discharging as well as the coating morphology were found to be due to the different types of discharge, which originated at the metal/coating interface (type B), within the coating upper layer (type C), or at the coating surface/electrolyte interface (type A) [4]. Type B discharges are responsible for the high temperature spikes (up to 10 000K) present in the electron temperature profiles. On the other hand, type A and C discharges produce the base temperature profile and any small fluctuations around this base line (~ 4500K).

Hussein et al. [10] have described the coating development during PEO processing based on a general theory of the breakdown of a metal/dielectric system in an electric field, optical emission spectroscopy observations and SEM/EDX analysis of the coatings. Their general coatings mechanisms can be applied to the PEO processing of any of the light-weight metals (Mg, Al, Ti or Zr). However process parameters including electrolyte composition or electrical parameters (DC, AC, unipolar, bipolar, constant current or voltage) have a significant effect on such mechanisms. Figure 11 is a schematic of a PEO coating process on a magnesium substrate. For PEO an appropriate electrical potential is applied to substrate to increase the thickness of the thin oxide layer on the surface of the alloy which could provide a very limited protective effect (Fig. 11(a)). As the surface has been passivated by a non-conductive oxide coating, the voltage between the substrate and the electrolyte rapidly rises as the native oxide thickens (Fig 11(b)) and within a few minutes the voltage reaches several hundred volts. The voltage increases until it has become too high for the dielectric coating and a microscopic plasma discharge breaks the coating and generates a large number of very short-lived, very small plasma discharges (Fig 11(c)). These discharges result in localized plasma reactions, with conditions of high temperature and pressure which modify the growing oxide. This break-down results in the formation of a slightly thicker coating, which will be broken again in the course of the next cycle, under a slightly higher potential difference. During the process, the number of discharges decreases but their intensity increases.

An important consequence of the occurrence of those discharges is the development of metallurgical processes in the growing oxide layer, which are induced by the heat liberated in discharge channels from the electron avalanches. Because of the local high temperature ~10^4 K [4,50] and the strong electric field, of the order of ~ 10^6 V/m [3], molten oxide is ejected from the discharge channels in the coating/substrate interface into the coating surface where it is rapidly solidified and re-crystallized by the electrolyte, Fig 11(d). As a result, decomposition of metal hydroxide to oxide and formation of complex compounds can occur, Fig 11(e). The intensity of these processes depend on the density and power of the discharges which are known to be defined by thickness of the oxide layer. Therefore, the thicker the layer, the less frequent, yet more powerful and extended, the discharges become [10].

As shown in the schematic diagram Fig. 11(b), the metal cations that transfer away from the metal substrate react with anions to form a ceramic coating. On the other hand, oxygen anions

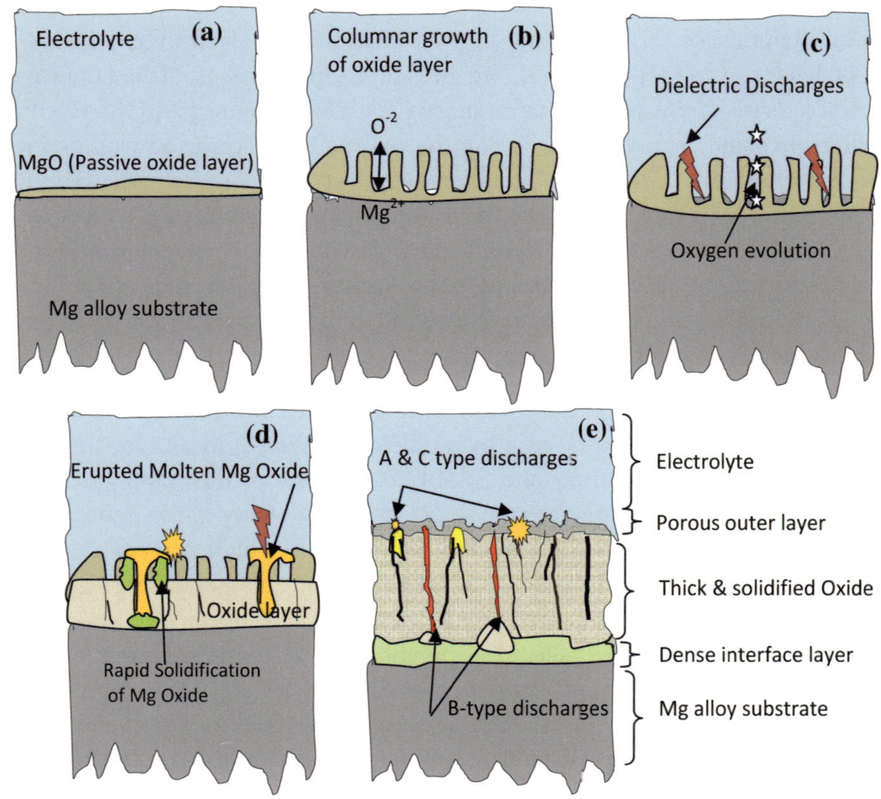

Figure 11. Schematic of coating development during PEO processing.

transfer into magnesium substrate due to the high electric field in the discharge channels and react with Mg^{2+} cations to form a ceramic coating. The instantaneous high temperature and high pressure in the discharge zone greatly enhances inter-diffusion between the oxygen anions and magnesium cations in the coating near the discharge zones. At higher plasma temperatures, the thermal energy supplied to the diffusing ions (O^{2-}) allows the ions to overcome the activation energy barrier and more easily move and interact with the magnesium cations Mg^{2+} produced during the discharges to form MgO.

7. Growth mechanisms

The plasma chemistry of the surface discharges is quite complex in nature, involving, on one hand, charge transfer at the substrate/electrolyte interface, and on the other hand, strong ionization and charge transfer effects between the substrate surface and the electrolyte through the oxide layer with the aid of the plasma [10,51]. Generally, the discharge event tends to occur in the coating–substrate interface or regions near the interface, which are responsible for the thermal and chemical conditions at the metal surface, thus playing an important role in formation, composition, and structure and stress state of phases formed. However, it is worth mentioning that the discharges induce no changes of substrate microstructure or texture. Processes such as melting, melt-flow, re-solidification; sintering and densification of the growing oxide take place during the PEO coating process.

1. **Linearity of the growth rate:** Most PEO studies indicate that the coating thickness increases linearly with coating time [31,52]. However, some research [41,53] shows that such linearity could break down at longer treatment times. According to Sundararajan et al.'s proposed growth mechanism [31,52], the growth of oxide layers results only from molten substrate elements which are oxidized when flowing out through the discharge channels that are created due to the oxide layer breakdown. In this way, an oxide is formed which contributes to the layer when being ejected from the channels and rapidly cooled at the surface–electrolyte interface. Discharge channels are continuously formed and move on the coating surface and since they have a finite life, they are formed and closed continuously through the coating process and contribute to the coating thickness.

2. **Inward and outwards coating growth:** Xue et al [41] describe the growth mechanism of the ceramic coating layer as a combination of inward inwards to the alloy substrate (inner layer) and outwards to the coating surface (outer layer) simultaneously. Fig. 12 is a schematic diagram of the coating development during PEO. The dashed line between Lo and Li in Fig. 12 represents the position of the original surface of the substrate before PEO treatment. During the early stages, the coating grows mainly outwards. After the coating reaches a certain thickness the inner layer grows faster than the outer layer. However, at this time, the coating thickness continues to increase in both directions. The inner growth is attributed to the growth of the compact layer at the film/substrate interface by diffusion or transport of oxygen, while the growth of outer layer onto the surface are due the electrochemical and the plasma chemical reactions.

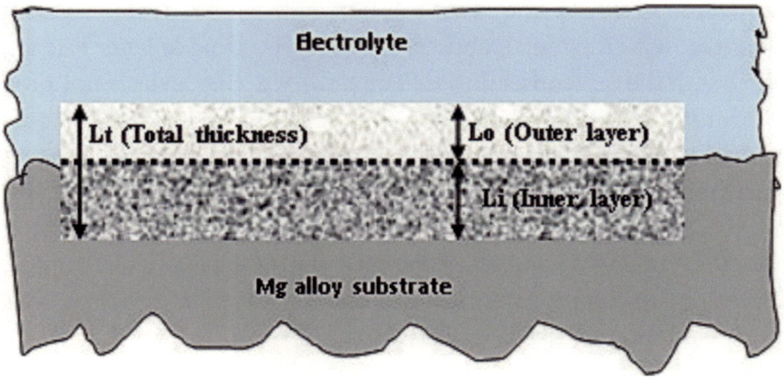

Figure 12. Schematic diagram of dimension changes of magnesium alloy before and after PEO treatment.

3. **Growth of compact layer at film/substrate interface by diffusion or transport of oxygen:** At longer processing times, the inward coating growth rate increases. This change may be connected to the increase of thermally-activated diffusion. For PEO process, both

thermally-activated diffusion and ions transformation have an important contribution leading the coating growth [52]. Inward oxygen diffusion plays a key role in coating growth, and the growth rate of the PEO process coating is controlled by the rate of transferring oxygen towards the magnesium substrate as shown in the schematic diagram of Fig. 11(b-e).

4. **Growth rate dependent on the process parameters:** The rates of growth of the outer and inner oxide layers are process parameter dependent. They result from a combination of three processes namely, (i) discharge processes causing the substrate to melt and oxidize when flowing out through the discharge channels and being rapidly cooled at the surface–electrolyte interface, (ii) partial destruction of the outer layer due to strong discharges and (iii) diffusion of oxygen process from the electrolyte towards the substrate through the coating.

8. Corrosion protection afforded by PEO coatings

8.1. Mg-alloys

Magnesium and its alloys have gained increasing attention for applications requiring high strength to weight ratio, high thermal conductivity, excellent castability, and ease of recycling. However, magnesium alloys exhibit very poor corrosion resistance caused by their chemically active nature, especially galvanic corrosion [2], which can further cause severe pitting corrosion on the metal surface resulting in decreased mechanical stability and an unattractive appearance.

Song and Atrens have concluded that internal galvanic attack, due to potential difference between matrix and precipitates, and the instability of the magnesium hydroxide film formed on the surface of Mg alloys are the two main causes for corrosion of magnesium alloys [54]. When magnesium is exposed to an aqueous solution, both $Mg(OH)_2$ and MgO can be formed: $Mg(OH)_2$ is in contact with the metal, and on top of the hydroxide layer is a MgO layer that has direct contact with the aqueous solution. For pure Mg, this layer is not protective at pH values below 10.5, unless additional alloying elements are added to pure Mg. The corrosion mechanism for the Mg alloys is more complex than that for pure magnesium, due to a multi-phase microstructure. Shi et al [55] pointed out that corrosion rate of Mg alloys is strongly dependent on the composition of the α-Mg matrix and the distribution of the other phases. Such phases have the tendency to accelerate the corrosion of the α-phase. In order for a coating to provide adequate corrosion protection for Mg and Mg alloys, the coating must be uniform with minimum defects and pores and well adhered.

Coating surface morphology, porosity, structure and corrosion behavior of ceramic coatings on Mg-alloys, are affected by many parameters including, electrolyte composition and concentration, the substrate composition, and the process parameters (including current density, current mode, applied voltage, temperature, and treatment time [10]. Test methods most commonly used to determine a corrosion rate of the PEO coatings are immersion tests, potentiodynamic polarization studies and impedance spectroscopy (EIS) [56].

Immersion testing is the most frequently conducted test for evaluating the corrosion of metals in aqueous solutions and can be used for evaluate the resistance of the metal to pitting, crevice corrosion and galvanic corrosion. It is basically depends on total immersion of a test specimen in a corrosive solution for a period of time and then immediately examining it. The electrochemical technique of Tafel extrapolation of polarization curves is also widely used for the evaluation of the corrosion of Mg alloys, in particular coated Mg alloys, because it is simple and fast. Atrens and co-workers [57] have raised a number of issues concerning the reliability of the Tafel extrapolation technique for corrosion measurement of Mg and Mg alloys, and therefore, suggest that Tafel extrapolation should be used with appropriate caution. Electrochemical impedance spectroscopy (EIS) can provide valuable information about surface treatment layers on PEO-coated magnesium, and the interfaces between electrolyte/coating/substrate [10]. It also allows the kinetics of heterogeneous electron-transfer reactions, coupled chemical reactions, or adsorption processes to be studied, and can provide information about pitting and crevice corrosion [58].

Corrosion studies have been mainly performed in aqueous NaCl solutions which can simulate natural seawater of any specified salinity, or in Hanks solution or simulated body fluid (SBF) for biocompatibility studies [59]. Zhang et al. [60] were the first to use Hanks solution to study the corrosion behaviour of PEO-treated Mg alloy samples, followed by Xu et al [61]. All PEO coatings reduced the corrosion rate to a certain extent. However, longer exposure of the PEO coated samples to an aggressive medium will minimize the protection process due to the defects and open pore structures of the PEO layers. Defects always exist as a result of many factors including: discharging process, solidification process, mechanical stresses, and gas evolution during the process. Figure 13 shows the effect of different PEO treatment time on the corrosion properties of AZ91D Mg alloy in 3.5%NaCl solution [62]. The potentiodynamic results show that the corrosion current density (icorr) decreased with PEO processing time. However, the corrosion potential (Ecorr) was relatively constant. The EIS results revealed that ~2 orders of magnitude higher resistance of the inner layer, mainly contributed to the overall resistance of PEO coating.

Using potentiodynamic polarization curves and electrochemical impedance spectroscopy, Barik et al studied the corrosion performance of PEO coatings, and demonstrated that unsealed PEO coating allows permeation of the solution through the pores in the coating [63]. The corrosion rate of the various magnesium alloys coated using either the PEO method and composite polymer-containing coatings has recently, been investigated by Gnedenkov et al [64] using a scanning vibrating probe method (SVP). The SVP method allows the study of the changes in the electrochemical activity of the sample on a micro scale and is particularly useful in investigating the effect of the alloy composition and microstructure on corrosion behavior [62,64-65]. According to their analysis of the SVP data, they conclude that the secondary phases in a VMD10 Mg alloy have a greater effect than α-phase on the corrosion behaviour and accelerate the dissolution of the α-phase.

Comparison of the corrosion protection results offered by PEO is very difficult, due to the variation of the experimental test parameters including solution composition, concentration, test time and other parameters. Ma et al. reported that oxide films grown on a AM50 magne-

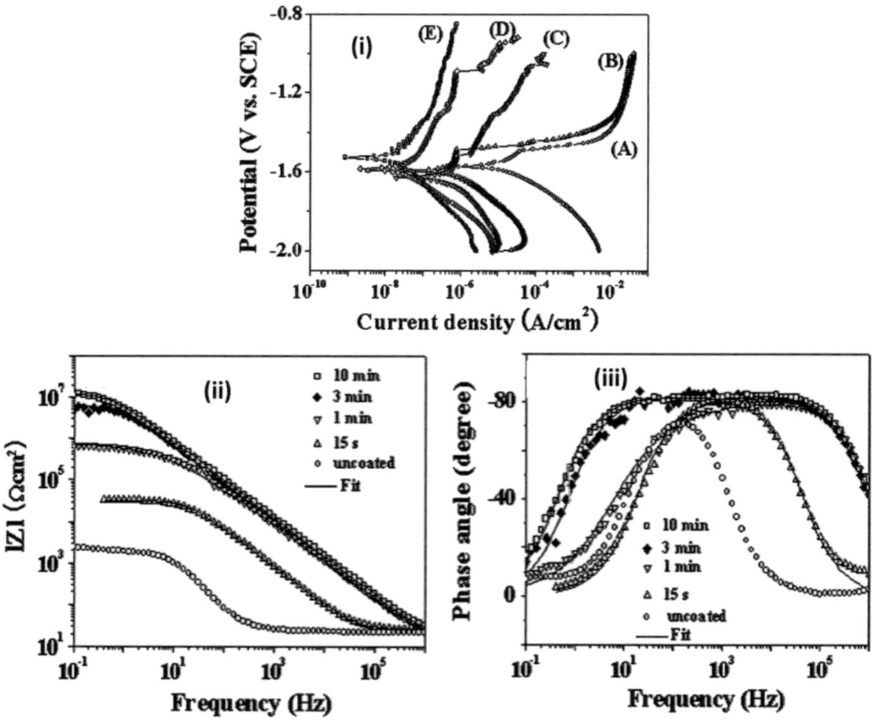

Figure 13. (i) Potentiodynamic polarization curves and (ii and iii) Bode plots of (A) uncoated AZ91D Mg alloy and PEO coatings treated for (B) 15 s, (C) 1 min, (D) 3 min, and (E) 10 min in 3.5 wt.% NaCl solution [62].

sium alloy from a phosphate solution have better corrosion resistance [66] compared with coatings prepared in silicate solutions. However, Liang et al. [67] when comparing the PEO coatings on AM60B magnesium alloy formed in silicate and phosphate electrolytes concluded that PEO coatings produced in a silicate electrolyte are compact and uniform hence exhibit better corrosion resistance thus coatings formed in a phosphate electrolyte, which were relatively porous. Liang et al [68] investigated the electrochemical degradation of a silicate- and phosphate-based PEO coatings on a AM50 magnesium alloy using a pulsed DC power supply in NaCl solutions of different chloride ion concentrations (0.01, 0.1, 0.5 and 1 mol/L). They conclude that the corrosion resistance of the Si-PEO coating was superior and the corrosion deterioration was slower than that of the P-PEO coating in mild corrosive electrolytes (0.01M and 0.1M NaCl). However, in the more concentrated electrolytes (0.5M and 1M NaCl), the Si-PEO and P-PEO coatings cannot provide a long-term protection to the magnesium alloy substrate due to the initiation of localized corrosion, and undergo further deterioration. Ding et al [69] investigated the influences of the addition of Na_2WO_4 to the silicate based electrolyte and breakdown voltage on the PEO coatings microstructure, hardness and wear resistance. In the presence of Na_2WO_4, the coatings have excellent compactability and wear resistance. However, with increasing sodium tungstate concentration, the size of micropores in the PEO coatings prepared in the NaOH electrolyte increased and the corrosion resistance of the PEO coatings decreased [70]. Figures 14 (a,b) show the effect of electrolyte additives (shown in Table

3) on the corrosion performance of PEO coatings formed on magnesium alloy AZ91D [15]. The potentiodynamic polarization results show that the coatings made using electrolyte P had a relatively lower Tafel slope due to a thinner coating with more micro-defects than those made using B- or M solutions. The coating with the M-electrolyte had the highest Tafel slope, which was attributed to a thicker, uniform structure and compact inner barrier layer formed in the bath with additions of fluoride and borate. The different EIS behavior (corrosion resistance) of the PEO films was attributed to their different structures and chemical composition.

Composition (g/l)	Solution P	Solution B	Solution M
$Na_2SiO_3 \cdot 9H_2O$	10–20	10–20	10–20
KOH	3–8	3–8	3–8
$NaH_2PO_4 \cdot 3H_2O$	4–8	N/A	N/A
$Na_2B_4O_7 \cdot 10H_2O$	N/A	5–10	5–10
KF·2H2O	N/A	N/A	5–10

Table 3. Electrolyte composition for PEO processing of magnesium alloy AZ91D [15]

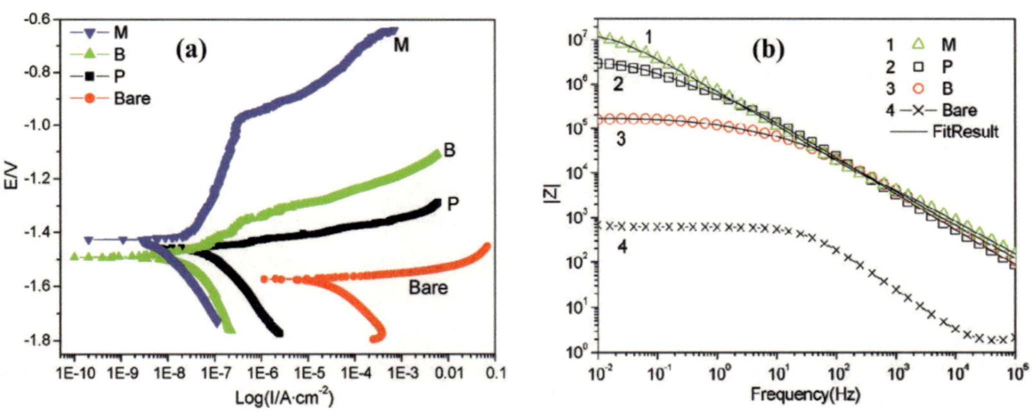

Figure 14. a) Potentiodynamic polarization curves (b) EIS plots of PEO films on AZ91D Mg-alloy treated in solutions containing different additive [15].

Hussein et al [28] studied the PEO-coating corrosion resistance for AJ62 Mg-alloy using both electrochemical impedance spectroscopy (EIS) and potentiodynamic polarization measurements in 3.5%wt NaCl solution. In this study the use of two different current modes (unipolar and bipolar current modes) in sodium aluminates electrolytes for 45 minutes treatment time, were successfully improve the corrosion resistance of AJ62 compared to the uncoated alloy. However, the coating made using the bipolar current mode is more beneficial in improving the corrosion resistance of the PEO coating than the unipolar current mode, see Table 4.

Sample	Current mode	C_R	R_p ($\Omega \cdot cm^2$)	EIS R_{total} (Ω•cm2)
S1	Unipolar	N/A	4.10E+04	3.94E+04
S2	bipolar	0.74	1.24E+06	2.80E+06
S3	bipolar	0.63	1.19E+06	1.80E+06
S4	bipolar	1.0	4.72E+05	2.35E+05

Table 4. PEO Process parameters for coating depositions on AJ62-Mg alloy and the corrosion resistance results from both potentiodynamic polarization (R_p) and EIS test (R_{total}) [28].

The use of a bipolar current mode results in the formation of a dense (minimum defects) coatings with higher corrosion resistance compared to coatings obtained using a unipolar or DC mode. Arrabal et al [71] conclude that the AC PEO coating on Mg alloys reduces the corrosion rate by 2-4 orders of magnitude compared with uncoated alloys. Ma et. al. [66] showed that an enrichment of the $MgAl_2O_4$ spinel phase in the coating, together with the minimum amount of cubic MgO, improves the corrosion resistance of the coating. The study of Barchiche et al [72] shows that increasing the treatment time rather than current density will improve corrosion resistance of the PEO oxide layer on AZ91D as well as a more passive behaviour of anodized layer was obtained with the higher concentrated electrolyte (3M KOH) due to the formation of more compact layer with minimum defects. In a review paper by Gu et al [56], a comparison of the results obtained using immersion tests of Mg and Mg alloys showed that PEO coatings were more effective in terms of corrosion protection than other coatings including alkaline and fluoride treated surfaces and organic based coatings. Hussein et al [12] in their review paper compared the corrosion protection affected to a AZ91D Mg alloy by PEO coatings, see Table 5. A close examination of the data shows a considerable variation in the corrosion current density i_{corr}, which is not directly related to the coating thickness. Other studies have demonstrated the practicality of using PEO coatings as a pre-treatment, followed by a second different coating process to seal any pores and cracks formed during the PEO coating process [56,62]. A number of sealing methods have been used for this purpose including: sealing in an organic polymer and sealing by the use of sols. The use of organic based coating, Sol-gel TiO_2 coating, immersion in SBF and electrophoresis deposition (EPD) of calcium phosphate–chitosan coatings on the top of PEO coatings on pure Mg and Mg alloy shows further improvement of the corrosion resistance compared to the PEO coating on its own [73]. A porous structure of the PEO coating is necessary for such treatment to be effective. Fig. 15 shows that the percentage reduction in corrosion rates of magnesium alloys with different coatings reported in the literature compared with uncoated samples [56]. The reported reduction in corrosion rate using a polymer coatings (chitosan coatings [78], polycaprolactone and dichloromethane coatings [77]), electrodeposition coatings (dicyclopentadiene (DCPD), hydroxyapatite (HA) and fluoridated hydroxyapatite (FHA) coatings [76]), fluoride treatment [75] and alkaline heat treatment [74], around 50–80% compared to more than 90% that can achieved using PEO coatings method [56].

Refs.	PEO operational condition		Coating Thickness μm	Corrosion Data		
	Electrolyte	Applied Power		Testing Method	i_{corr} μA/cm²	E_{corr} V
[102]	NaOH 7 g/L, sodium hexametaphosphate 4 g /L and calcium acetate 0.4 g/L	CV, CC,CP 2 Adm^{-2}	3-15	EIS + P.P. In Hanks solution 37,5 °C	2.4	-1.660
[62]	KF 9 g/L + $NaAlO_2$ 10 g/L	DC pulse 5 A dm^2	15	P.P. in 3.5% NaCl solution	0.01	-1.53
[103]	phosphate		10	phosphate	NA	-1.64
[104]	K_2ZrF_6 =10, $Na_2SiO_3{\cdot}9H_2O$= 10, KOH= 4	CV 300v	10	P.P. in 3.5% NaCl solution	0.05	−1.42
[105]	Na2SiO3·9H2O 10–15 g/L, KOH 2–4g/L, KF·2H2O 3–5 g/L, Na2B4O7·10H2O 2 g/l	CV 400v	20-50	EIS + P.P. 3.5% NaCl solution for 220 h,	7.5	−1.41
[106]	NaOH and Na_2SiO_3	Unipolar 400-460v	3-10	P.P. in 0.9% NaCl solution	0.64	_1.49
[15]	$Na_2SiO_3{\cdot}9H_2O$, KOH, $NaH_2PO_4{\cdot}3H_2O$	350–400V	20	P.P. in 3.5% NaCl solution	0.003	−1.43
[107]	NaAlO2 20 g/l and NaOH 8 g/l	30 mA/cm^2	10	galvanic corrosion in 5%NaCl	1.9x10^4	NA
[60]	2oog/L CrO_3+10g/L $AgNO_3$	NA	NA	P.P. In Hanks solution 37,5 °C	0.204	-0.430

NA= not available; CV= constant voltage; CC= constant current; CP= constant power; P.P.= potentiodynamicl polarization; icorr= corrosion current density; Ecorr= corrosion potential.

Table 5. Examples of plasma electrolytic oxidation of AZ91D Mg alloy. Adapted from [12].

The studies by Arrabal et al [79] shows the effect of the immersion time and the effect of duplex coating combining PEO and polymer layer on AZ31 magnesium alloy as shown in Figs. 16 and 17. PEO treatment significantly increased the corrosion resistance of the AZ31alloy, but only for short immersion times due to the porous nature of the outer region of the PEO coating and deterioration of the inner barrier layer (Fig. 16). As for the PEO + polymer coated specimens, high impedance values indicated almost negligible degradation of the polymer top-layer after 77 days in 5 wt.% NaCl solution and superior corrosion resistance properties of the PEO pretreatment (Fig 17).

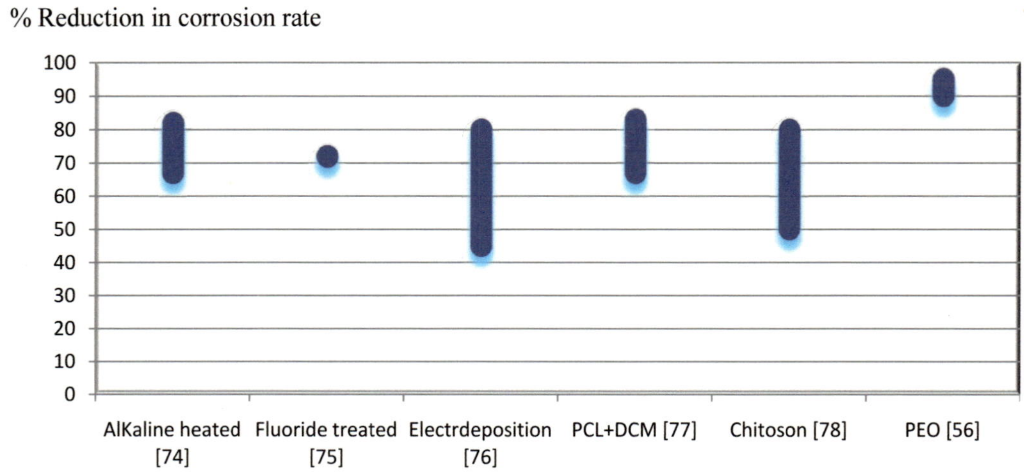

Figure 15. The reduction in corrosion rate for magnesium alloys with different coatings. (adapted from [56]).

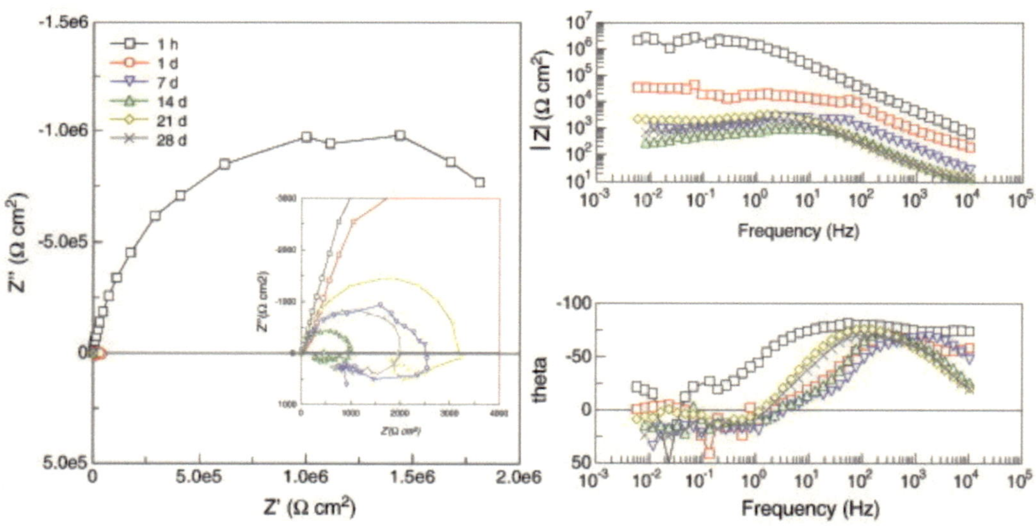

Figure 16. Nyquist and Bode plots of PEO coated AZ31 alloy [79].

http://dx.doi.org/10.5772/57171

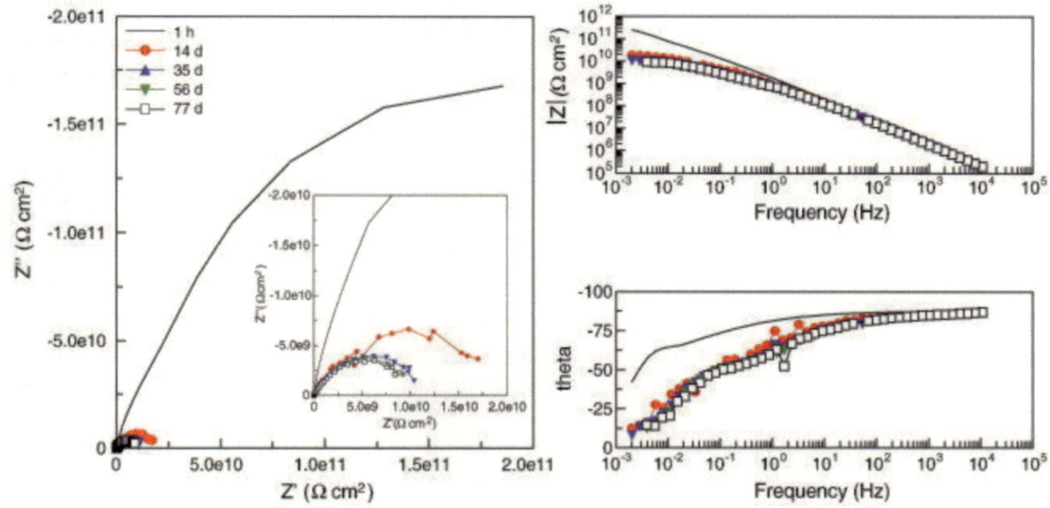

Figure 17. Nyquist and Bode plots of PEO + polymer AZ91 alloy specimen [79].

8.2. Al-alloys

The poor corrosion protection properties of uncoated Al substrates result from the fact that the corrosion resistance considerably decreases after the thin protective oxide film on the uncoated aluminium substrate surface is broken down by the corrosion processes. The corrosion resistance of PEO coatings on aluminum alloys was studied by Nie et al [32], where the effect of thickness on the mechanism and properties of ceramic coatings were measured. They conclude that the PEO-coated 6082 Al alloys exhibited excellent corrosion resistance in 0.5 M NaCl solution, considerably better than even stainless steel. The corrosion protection properties of PEO coated aluminum alloys have been investigated using potentiodynamic polarization curves and electrochemical impedance spectroscopy (EIS) [63,80-81]. The corrosion performance of PEO-coated Al-alloys for long time immersion times can be divided into three stages: (i) the penetration of corrosive medium into aluminum alloy is inhibited effectively by the PEO coating in the initial stage of immersion (less than few hours); (ii) after immersion for more than 10-24 hours, the corrosive medium penetrated into the interface between the alloy and the coating, through the pores and cracks in the coating, and caused corrosion under the coating. Although a similar decreasing trend is observed in the resistance of the compact layer values, they are very high compared with the outer layer values measured for the porous layer at all immersion time.; and (iii) after immersion for more than 24 hours, corrosion processes were controlled by the diffusion of the corrosion products.

In the case of aluminum alloys, the PEO coatings are mainly composed of α-Al2O3 and γ-Al2O3 with some complex Al–Si–O phases also present. The relative phase amount, pore size, corrosion resistance and other properties of the coatings are significantly influenced by the PEO electrolyte concentration and compositions, and on the magnitude of the process parameters. Where the constituents of electrolyte participated in film formation reaction during the PEO process, the structure, composition and other properties of PEO films would vary with the electrolyte compositions. The amount of the harder α-Al2O3 phase could be increased by raising the current density [60]. For PEO coatings, the type of electrolyte is selected according to the substrate involved and the desired properties of the coating [82]. Higher sodium silicate concentration in the electrolyte results in higher growth rates, which may attributed to: more silicate is co-deposited with substrate oxidation and that an increase of the concentrations depresses oxide dissolution and the layer growth rate is increased [30]. It has been reported that phosphate was useful in enhancing the corrosion resistance of inner barrier layer of PEO coatings on aluminum alloys [83]. When the sodium tungstate concentration increases, the coating becomes thicker, contains more α-Al2O3, and has higher micro hardness and corrosion resistance. Li et al [84] obtained hard corrosion resistant alumina coating on Al-Si alloy in borate electrolyte. A similar improvement was also achieved for other aluminum alloys, including Al-Cu, Al-Mg and Al-Zn-Mg [85].

Khan et al [86], and Hussein et al [87], studied the PEO coatings on aluminum formed by unipolar and bipolar pulsed current methods. A bipolar current mode provides better corrosion resistance than unipolar mode due to the reduction of the pores and coating defects. Xue et al [88] through elemental analysis across the cross section of the coating performed after corrosion tests detected the presence of chloride ions only in the porous layer and not in the compact inner layer. Hence the corrosion resistance of the PEO coating is mainly derived from the inner compact layer. The evolution of impedance spectra with time, in general shows good correlation between the morphology of the oxide coating and the corrosion performance over a period of time.

8.3. Ti-alloys

Titanium and Ti alloys are remarkably resistant to corrosion owing to the formation of stable, self-healing oxide films on their surfaces. The major corrosion problem with Ti alloys may be in acidic environments, particularly when the acid concentration and temperature increase. In this case, the oxide film on Ti deteriorates and dissolves and the unprotected metal is oxidized to the soluble trivalent ion [89]. The PEO process has been recently used to prepare a porous and adherent titania coating which enhances corrosion resistance [90–93], bioactivity [94,95], and greater biocompatibility. Because the performance of PEO coatings strongly depends on the experimental conditions, such as the properties of the electrolyte and the PEO operation power, there are many reports on the relationship between the coating performance and the operation conditions. Zhang et al [96] conclude from EIS studies that the addition of calcium hypophosphite has significant influence on the growth process during the PEO coating stages prepared on a Ti6Al4V alloy under DC power. The equivalent circuit shown in Fig. 18 was used to analyze the EIS results in their study, where Rs is the electrolyte resistance; Q1 and Q2 model the electrical properties of the outer and inner layers, respectively; Rpo is the resistance of the electrolyte saturating the outer layer; and R2 represents the resistance of the inner layer. Here, Q, a constant phase element (CPE), is used to replace the pure capacitance. The equivalent circuits employed for curve fitting of the uncoated sample is illustrated in Fig. 18(b) where R_{CT}

represent the charge-transfer resistance of the naturally formed oxide film on Ti. XRD analysis by Li et al [97] shows that the 110 μm ceramic coating prepared on TiAl alloy by PEO method consisted of three layers: dense layer composed of Al2TiO5 and TiO2 rutile phases, intermediate layer and loose layer contained a large amount of amorphous SiO2 besides Al2TiO5 and TiO2 rutile phases. The electrochemical corrosion experiments showed that the coatings produced using both unipolar and bipolar current modes successfully improved the corrosion resistance in comparison with the uncoated material [44]. The corrosion resistance of the PEO-coated specimens was mainly determined by the thickness, composition and quality of the dense inner layer. Hussein et al [44] studies for Ti6Al4V indicated that the PEO coating led not only to decrease of cathodic process, but also appearance of a large passivation plateau on the anodic polarization curves. As the inner layer of PEO coating is a barrier layer to decrease the corrosion rate of the substrate, it is expected that the protection efficiency strongly depended on the thickness and density of the inner layer coatings.

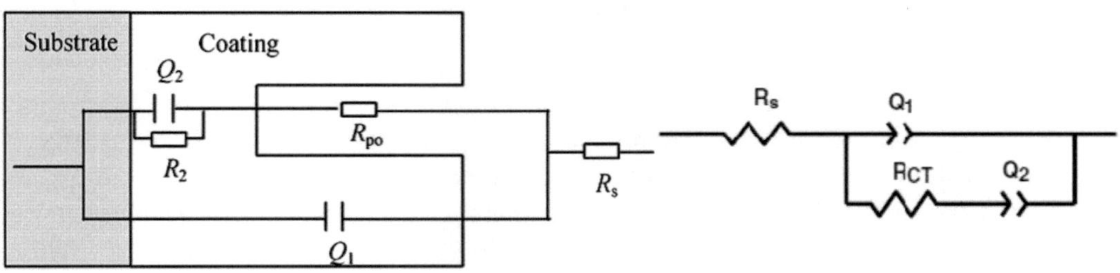

Figure 18. (a) Schematic of coatings and equivalent circuit used to model EIS results [96], (b) the equivalent circuit model for uncoated samples.

High applied voltage during PEO process can accelerate the migration process of the ions in electrolyte, which attracts the Ca2+ and PO43− ions to incorporate into the oxide layer by the electric field on the electrode and promotes the formation of the bioactive apatite [98]. On the other hand, the strong discharge phenomenon generated by high applied voltage increases the pore size which and decreases the density of the coating [99]. The longer immersion in simulated body fluid brings corrosive ions to penetrate through the porous layer resulting in the serious corrosion of the coated titanium alloy [100]. Moreover, cracks and poor adhesion easily arise from high applied voltage. Therefore, it is a key issue how to improve the corrosion resistance of the porous PEO coating with high bioactivity so as to alleviate toxicity and allergic reactions inside a living creature due to the release of metal ions to the adjacent tissue. By adjusting PEO parameters, including: current mode and density, treatment time and electrolyte composition and concentration, the distribution thickness between dense inner layer and porous outer layer and the structure of the coating can be controlled [92]. The compositional and morphological flexibility of coatings characteristic in PEO enables adjusting of calcium and phosphorus contents and structure of coatings favorable for bio adhesion between bone tissue and the implant [101].

9. Conclusions

This chapter has demonstrated that several coating systems can be used to provide improved corrosion protection of (Al, Mg, Ti) alloys. The current coating schemes are complex, multilayer systems that incorporate many different technologies and must be conducted very carefully in order to adequately protect light weight metals from corrosion in harsh service conditions. In order to achieve optimum adhesion, one of the most widely used coating methods for corrosion protection is through the PEO technique. A PEO process can be used to produce an oxide layer on light weight elements using different electrolyte and/or current modes which significantly affect coating morphology and microstructure and hence, corrosion resistance. A dense coating morphology could be achieved by adjusting positive to negative current ratio and their timing to eliminate or reduce the strongest plasma discharges. The rates of growth of the oxide outer and inner layers are processes parameter dependent. They result from a combination of three processes namely, (i) discharge processes causing the substrate to melt and oxidize when flowing out through the discharge channels and rapidly cooled at the surface–electrolyte interface, (ii) partial destruction of the outer layer due to strong discharges and (iii) diffusion processes.

Potentiodynamic polarization corrosion tests and EIS results both showed that PEO coatings significantly increased the corrosion resistance of light weight alloys and the degree of protection is affected by the amount of porosity and other coating defects which control the penetration rate of the electrolyte into the PEO coating. The scope for industrial use of this technology is quite promising due to the process flexibility, low capital cost and the environmentally-friendly precursor materials that are utilized. However, major uncertainties remain concerning process optimization, control and consistency (repeatability). Further improvement and developments are taking place in PEO technology, particularly to produce composite coatings by co-deposition of materials, including polymers and organic materials, within and/or at the top of the PEO layers.

10. Future trends

Research continues to be directed to identifying and developing new coating materials for automotive, aerospace, defense, textile, motor sports, electronics and oil and gas sectors. This is driven by the need for improved performance, reduced weight, increased multi-functionality and greater durability and also for reduced environmental hazards. Corrosion and wear resistance are traditional areas receiving the greatest attention. Both magnesium and aluminum alloys require an appropriate surface treatment to protect than against corrosion. PEO is certainly one of the most promising surface treatments for light weight alloys and is already in use for a number of applications, and as a replacement for hard anodizing especially for aluminum alloys. In addition to the excellent adhesion to the substrate, PEO coatings have higher thermal shock resistance, high dielectric strength and good optical properties. Using PEO coatings as a pre-treatment, followed by a second different coatings process (hybrid coatings) to seal the unavoidable open pores and cracks formed during PEO coating, the corrosion resistance of PEO coatings can be further improved for long immersion times or corrosion in very aggressive environments. However, the porous structure of the PEO coating is necessary because they offer good adhesion if used as a pre-treatment for biocompatibility or decorative applications. While PEO has become increasingly popular in order to compete with other processes, further improvements are essential for optimizing the process and to reduce costs. The capital costs, electricity consumption and process control and reproducibility are critical aspects to be worked on.

Author details

Riyad O. Hussein and Derek O. Northwood

Department of Mechanical, Automotive and Materials Engineering, University of Windsor, Windsor, ON, Canada

References

[1] Park RM, Bena JF, Stayner LT, Smith RJ, Gibb HJ, Lees PSJ (2004). Hexavalent chromium and lung cancer in the chromate industry, a quantitative risk assessment. Risk Anal, 24(5),1099-1108.

[2] Walsh FC, Low CTJ, Wood RJK, Stevens KT, Archer J, Poeton AR, et al. Review. (2009) Plasma electrolytic oxidation (PEO) for production of anodised coatings on lightweight metal (Al, Mg, Ti) alloys. Transactions of the Institute of Metal Finishing 87(3), 122–35.

[3] Yerokhin AL. Nie X, Leyland A, Matthews A and Dowey S J (1999), Plasma electrolysis for surface engineering. Surf. Coat. Technol., 122(2-3), 73-93.

[4] Hussein RO, Nie X, Northwood DO, Yerokhin A, Matthews A (2010), Spectroscopic study of electrolytic plasma and discharging behaviour during the plasma electrolytic oxidation (PEO) process. Journal of Physics D: Applied Physics, 43, 105203.

[5] Hornberger H, Virtanen S, Boccaccini AR (2012), Biomedical coatings on magnesium alloys – A review, Acta Biomaterialia 8, 2442–2455.

[6] Pardo A, Casajús P, Mohedano M, Mohedano M, Coy A E, Viejo F, Torres B, Matykina E, (2009) Corrosion protection of Mg/Al alloys by thermal sprayed aluminium coatings. Applied Surface Science, 255, 6968-6977.

[7] Li JN, Cao P, Zhang XN, Zhang SX, He YH. (2010) In vitro degradation and cell attachment of a PLGA coated biodegradable Mg–6Zn based alloy. J Mater Sci, 45, 6038–45.

[8] Xu X, Lu P, Guo M, Fang M. (2010) Cross-linked gelatin/nanoparticles composite coating on micro-arc oxidation film for corrosion and drug release. Applied Surface Science, 256, 2367–71.

[9] Goueffon Y, Arurault L, Mabruc C, Tonond C, Guiguea P, (2009) Black anodic coatings for space applications, Study of the process parameters, characteristics and mechanical properties, Journal of Materials Processing Technology 209, 5145–5151.

[10] Shrestha S (2010) 'Magnesium and Surface Engineering-Technology Vision', Editorial in Surface Engineering, 26, (313-316).

[11] Gunterschulze A, Betz H, ElektrolytKondensatoren (Herbert Cram, Berlin, 1937; Oborongiz, Moscow, 1938) [in German and in Russian].

[12] Hussein RO, Nie X, Northwood DO (2013), The Application of Plasma Electrolytic Oxidation (PEO) to the Production of Corrosion Resistant Coatings on Magnesium Alloys, a Review. Corrosion and Materials, 38[1], 55-65.

[13] Walsh FC, Low CTJ, Wood RJK, Stevens KT, Archer J, Poeton AR and Ryder A, (2009) Plasma electrolytic oxidation (PEO) for production of anodised coatings on lightweight metal (Al, Mg, Ti) alloys, Transactions of the Institute of Metal Finishing, 87, 122-135.

[14] Sundararajan G and Rama L, (2003) Mechanisms underlying the formation of thick alumina coatings through the MAO coating technology. Surf. Coat. Technol., 167(2-3), 269-277.

[15] Duan H, Yan C, Wang F (2007) Effect of electrolyte additives on performance of plasma electrolytic oxidation films formed on magnesium alloy AZ91D Electrochimica Acta, 52, 3785-3793.

[16] Suminov IV, Apelfeld AV, Ludin VB, Krit BL and Borisov AM, (2005) Microarc oxidation theory, technology and equipment, Moscow Ecomet, ISBN 5-89594-110-9.

[17] Jaspard-Mécuson F, Czerwiec T, Henrion G, Belmonte T, Dujardin L, Viola A and Beauvir J, (2007) Tailored aluminium oxide layers by bipolar current adjustment in the Plasma Electrolytic Oxidation (PEO) process. Surf. Coat. Technol., 201(21), 8677-8682.

[18] Bala Srinivasan P, Liang J, Blawert C, Stormer M, Dietzel W, (2009) Effect of current density on the microstructure and corrosion behaviour of plasma electrolytic oxidation treated AM50 magnesium alloy, Applied Surface Science 255, 4212–4218.

[19] Hussein RO, Nie X, Northwood DO (2013), An investigation of ceramic coating growth mechanisms in Plasma Electrolytic Oxidation (PEO) processing, submitted to Electrochimica Acta, 112, 111-119

[20] Curran J A, Kalkanc H, Magurova Y and Clyne T W (2007), 'Mullite-rich plasma electrolytic oxide coatings for thermal barrier applications', Surf Coat Technol, 201, 8683–8687.

[21] Krishna L R, Purnima A S and Sundararajan G (2006), 'A comparative study of tribological behavior of microarc oxidation and hard-anodized coatings', Wear, 261, 1095–1101.

[22] Zhang P, Nie X, and Hu H, (2009) Wear and Galvanic Corrosion Protection of Mg alloy via Plasma Electrolytic Oxidation Process for Mg Engine Application, SAE Technical Paper 2009-01-0790, doi:10.4271/2009-01-0790.

[23] Frauchiger V M, Schlottig F, Gasser B and Textor M (2004), 'Anodic plasma–chemical treatment of CP titanium surfaces for biomedical applications', Biomaterials, 25, 593–606.

[24] Wang K, Kim Y, Hayashi Y, Lee C, Koo B, (2009) Ceramic coatings on 6061 Al alloys by plasma electrolytic oxidation under different AC voltages, J. Ceramic Processing Research. 10, 562-566

[25] Arrabal R, Matykina E, Hashimoto T, Skeldon P, Thompson GE (2009), Characterization of AC PEO coatings on magnesium alloys, Surf. Coat. Technol., 203, 2207-2220.

[26] Xu W, Jin Q, Zhua Q, Hu M and Ma Y, (2009) Anti-corrosion microarc oxidation coatings on SiCP/AZ31 magnesium matrix composite, Journal of Alloys and Compounds 482, 208-212.

[27] Curran JA, Clyne TW, (2005) Thermo-physical properties of plasma electrolytic oxide coatings on aluminium, Surf. Coat. Technol., 199, 168–176.

[28] Hussein RO, Zhang P, Xia Y, Nie X and Northwood DO (2011) The effect of current mode and discharge type on the corrosion resistance of plasma electrolytic oxidation (PEO) coated magnesium alloy AJ62, Surf. Coat. Technol., 206(7), 1990-1997

[29] Paul W, Sharma CP. (2006) Nanoceramic matrices, biomedical applications. Am J Biochem Biotechnol 2, 41–8.

[30] Yerokhin AL, Shatrov A, Samsonov V, Shashkov P, Pilkington A, Leyland A, Matthews A. (2005) "Oxide ceramic coatings on aluminium alloys produced by a pulsed bipolar plasma electrolytic oxidation process," Surf. Coat. Technol., 199, 150-157.

[31] Sundararajan G and Rama L, (2003) Mechanisms underlying the formation of thick alumina coatings through the MAO coating technology. Surf. Coat. Technol., 167(2-3), 269-277.

[32] Nie X, Meletis E I, Jiang J, Leyland A, Matthews A (2002) Abrasive wear/corrosion properties and TEM analysis of Al_2O_3 coatings fabricated using plasma electrolysis, Surf. Coat. Technol. 149, 245-251.

[33] Dunleavy CS, Golosnoy I O, Curran J A and Clyne T W, (2009) Characterisation of discharge events during plasma electrolytic oxidation. Surf. Coat. Technol., 203(22), 3410-3419.

[34] Ma Y, Nie X, Northwood DO and Hu H, (2004) Corrosion and erosion properties of silicate and phosphate coatings on magnesium. Thin Solid Films, 469-470, 472-477.

[35] Lee K, Lee B, Yoon S, Lee E, Yoo B, Shin D, (2012) Evaluation of plasma temperature during plasma oxidation processing of AZ91Mg alloy through analysis of the melting behavior of incorporated particles, Electrochimica Acta 67, 6-11

[36] Fournier V, Marcus P, Olefjord I, (2002), Oxidation of magnesium, Surface and Interface Analysis 34(1), 494-497.

[37] Nie X, Meletis E I, Leyland A, Matthews A, (2001), Effects of Solution pH and Electrical Parameters on Hydroxyapatite Coatings Deposited by a Plasma-assisted Electrophoresis Technique, J. of Biomedical Material Research 57, 612-618.

[38] Shi-Gang X, Li-Xin S, Rong-Gen Z, and Xing-Fang H, (2005) "Properties of aluminium oxide coating on aluminium alloy produced by micro-arc oxidation," Surf. Coat. Technol., 199, 184-188.

[39] Hussein RO, Nie X and Northwood DO (2010), Influence of process parameters on electrolytic plasma discharging behavior and aluminum oxide coating microstructure, Surf. Coat. Technol., 205, 1659-1667.

[40] Snizhko LO, Yerokhin AL, Pilkington A, Gurevina NL, Misnyankin DO, Leyland A, Matthews A. (2004) Anodic processes in plasma electrolytic oxidation of aluminium in alkaline solutions. Electrochimica Acta., 49[13], 2085-2095.

[41] Xue W, Deng Z, Chen R, Zhang T. (2000) Growth Regularity of Ceramic Coatings Formed by Microarc Oxidation on Al-Cu-Mg alloy. Thin Solid Films,372, 114-117.

[42] Chang L (2009) Growth regularity of ceramic coating on magnesium alloy by plasma electrolytic oxidation. Journal of Alloys and Compounds, 468, 462–465.

[43] Yerokhin AL, Leyland A, Matthews A (2002), Kinetic aspects of aluminium titanate layer formation on titanium alloys by plasma electrolytic oxidation, Applied. Surface. Science, 200, 172-184.

[44] Hussein RO, Nie X, Northwood DO, (2012) A spectroscopic and microstructural study of oxide coatings produced on a Ti-6Al-4V alloy by Plasma Electrolytic Oxidation, Materials Chemistry and Physics 134, 484– 492.

[45] Yerokhin AL, Snizhko LO, Gurevina NL, Leyland A, Pilkington A, Matthews A, (2003) Discharge characterization in plasma electrolytic oxidation of aluminium. J. Phys. D: Appl. Phys. 36, 2110.

[46] Albella J M, Montero I and Martinez-Duart J M, Parkhutik V, (1991) Dielectric breakdown processes in anodic Ta2O5 and related oxides, J. Materials Science, 26[13] 3422-3432

[47] Epelfeld A V, Lyudin V B, Dunkin O N and Nevskaya O S (2000) Bull. Russ. Acad. Sci. Phys. 64, 610

[48] Ikonopisov S, Girginov A and Machkova A (1977b), 'Theory of electrical breakdown during formation of barrier anodic films', Electrochem Acta, 22(10), 1077–1082.

[49] Hickling A and Ingram M D (1964), Contact glow-discharge electrolysis, Trans. Faraday Soc. 60 (496, part 4) 783-793.

[50] Hussein RO, Northwood DO, Nie X, (2010) Coating growth behavior during the plasma electrolytic oxidation process, J. Vacuum Science & Technology A Vacuum Surfaces and Films, 28(4), 766-773.

[51] Rakoch AG, Khokhlov VV, Bautin VA, Lebedeva NA, Magurova YV, Bardin IV, (2006) Model concepts on the mechanism of microarc oxidation of metal materials and the control over this process, Protection of Metals and Physical Chemistry of Surfaces, 42, 158-169.

[52] Krishna LR, Somaraju KRC, Sundararajan G. (2003) The tribological performance of ultra-hard ceramic composite coatings obtained through microarc oxidation. Surf. Coat. Technol. 163 –164, 484–490.

[53] Mécuson F, Czerwiec T, Belmonte T, Dujardin L, Viola A, Henrion G (2005) Diagnostics of an electrolytic microarc process for aluminum alloy oxidation. Surf. Coat. Technol., 200, 804-808.

[54] Song G L, Atrens A, (1999) Corrosion mechanisms of magnesium alloys, Advanced Engineering Materials, 1, 11-33.

[55] Shi Z, Liu M, Atrens A, (2010) Measurement of the corrosion rate of magnesium alloys using Tafel extrapolation, Corrosion Science 52, 579–588.

[56] Gu XN, Li N, Zhou WR, Zheng YF, Zhao X, Cai QZ, (2011) Corrosion resistance and surface biocompatibility of a microarc oxidation coating on a Mg–Ca alloy, Acta Biomaterialia 7, 1880–1889.

[57] Song G, Atrens A, (2003) Understanding magnesium corrosion mechanism: a framework for improved alloy performance, Advanced Engineering Materials 5, 837-858.

[58] Huang VM, Wua S-L, Orazema ME, Pébèreb N, Tribollet B, Vivier V, (2011) Local electrochemical impedance spectroscopy: A review and some recent developments Electrochimica Acta 56, 8048– 8057

[59] Gao JH, Shi XY, Yang B, Hou SS, Meng EC, Guan FX, (2011) Fabrication and characterization of bioactive composite coatings on Mg–Zn–Ca alloy by MAO/ sol-gel. Journal of Material Science: Journal of Materials Science: Materials in Medicine 22, 1681–1687.

[60] Zhang XP, Zhao ZP, Wu FM, Wang YL, Wu J. (2007) Corrosion and wear resistance of AZ91D magnesium alloy with and without microarc oxidation coating in Hank's solution. Journal of Material Science, 42, 8523–8528.

[61] Xu X, Lu P, Guo M, Fang M. (2010) Cross-linked gelatin/nanoparticles composite coating on micro-arc oxidation film for corrosion and drug release. Applied Surface Science 256, 2367–71.

[62] Hyun Sam Ryu, Seong-Jae Mun, Tae Seop Lim, Hong-Chan Kim, Kwang-Seon Shin, and Seong-Hyeon Hong, (2011) "Microstructure Evolution During Plasma Electrolytic Oxidation and Its Effects on the Electrochemical Properties of AZ91D Mg Alloy", Journal of The Electrochemical Society, 158 (9) C266-C273.

[63] Barik RC, Wharton JA, Wood RJK, Stokes KR, Jones RL, 167 (2005) Corrosion, erosion and erosion–corrosion performance of plasma electrolytic oxidation (PEO) deposited Al_2O_3 coatings, Surf. Coat. Technol., 199, 158-167.

[64] Gnedenkov AS, Sinebryukhov SL, Mashtalyar DV, Gnedenkov SV, (2013) Features of the corrosion processes development at the magnesium alloys surface, Surf. Coat. Technol., 225, 112–118.

[65] Williams G, Neil McMurray H, (2008), Localized corrosion of magnesium in chloride-containing electrolyte studied by a scanning vibrating electrode technique, J. Electrochemical Society, 155 (7), C340-C349.

[66] Ma Y, Nie X, Northwood DO and Hu H, (2006) Systematic study of the electrolytic plasma oxidation process on a Mg alloy for corrosion protection, Thin Solid Films 494, 296-301.

[67] Liang J, Hu L, Hao J, (2007) Characterization of microarc oxidation coatings formed on AM60B magnesium alloy in silicate and phosphate Electrolytes, Applied Surface Science, 253, 4490-4496.

[68] Liang J, Srinivasan PB, Blawert C, Dietzel W. (2010) Influence of chloride ion concentration on the electrochemical corrosion behavior of plasma electrolytic oxidation coated AM50 magnesium alloy, Electrochimica Acta, 55, 6802–6811.

[69] Ding J, Liang J, Li TH, Hao J, Xue Q (2007), Effects of sodium tungstate on characteristics of microarc oxidation coatings formed on magnesium alloy in silicate–KOH electrolyte,Transactions of Metals Society of China, 17, 244–249.

[70] Zhao F, Liao AD, Zhang RF, Zhang SF, Wang HX, Shi XM, Li MJ, He XM, (2010) Effects of sodium tungstate on properties of micro-arc coatings on magnesium alloys, Trans. Nonferrous Met. Soc. China, 20, s683- s687.

[71] Arabal R, Matykina E, Hashimoto T, Skeldon P and Thompson G E, (2009) Characterization of AC coatings in magnesium alloys, Surf. Coat. Technol., 203, 2207-2220.

[72] Barchiche C E, Rocca E, Juers C, Hazan J, Steinmetz J, (2007) Corrosion resistance of plasma anodized AZ91D magnesium alloy by electrochemical methods Electrochimica Acta 53, 417–425

[73] Duan H, Du K, Yan C, Wang F. (2006) Electrochemical corrosion behavior of composite coatings of sealed MAO film on magnesium alloy AZ91D, Electrochimica Acta 51, 2898–2908.

[74] Gu XN, Zheng W, Cheng Y, Zheng YF. (2009) A study on alkaline heat treated Mg–Ca alloy for the control of the biocorrosion rate. Acta Biomater, 5, 2790–2709.

[75] Chiu KY, Wong MH, Cheng FT, Man HC. (2007) Characterization and corrosion studies of fluoride conversion coating on degradable Mg implants. Surf. Coat. Technol., 202, 590–598.

[76] Song Y, Zhang S, Li J, Zhao C, Zhang X. (2010) Electrodeposition of Ca–P coatings on biodegradable Mg alloy: in vitro biomineralization behavior. Acta Biomater, 6, 1736–1742.

[77] Wong HM, Yeung KMK, Lam KO, Tam V, Chu PK, Luk KDK, (2010) A biodegradable polymer-based coating to control the performance of magnesium alloy orthopedic implants. Biomaterials,31, 2084–2096.

[78] Gu XN, Zheng YF, Lan QX, Cheng Y, Zhang ZX, Xi TF, Zhang DY, (2009) Surface modification of an Mg-1Ca alloy to slow down its biocorrosion by chitosan, Biomed Mater, 4, 044109.

[79] Arrabal R, Mota JM, Criado A, Pardo A, Mohedano M, Matykina E, (2012) Assessment of duplex coating combining plasma electrolytic oxidation and polymer layer on AZ31 magnesium alloy, Surf. Coat. Technol., 206,4692–4703.

[80] Wen L, Wang Y, Zhou Y, Ouyang J, Guo L, Jia D, (2010) Corrosion evaluation of microarc oxidation coatings formed on 2024 aluminium alloy, Corrosion Science 52, 2687–2696.

[81] Raj V, Ali M, (2009) Formation of ceramic alumina nanocomposite coatings on aluminium for enhanced corrosion resistance, J.Mats. Proc. Techn. 209, 5341–5352

[82] Lv G, Gu W, Chen H, Feng W, Khosa M, Li L, Nie L, Zhang F, Yang G, Si Z, (2006) Characteristics of ceramic coatings on aluminium by plasma electrolytic oxidation in silicate phosphates electrolytes. Appl. Surf. Sci. 253, 2947–2952.

[83] Kurze P, Schrekenbach J, Schwarz T, Krysmann W, (1986) Besehichten durch anodische oxidation unter fuhentludung, Metalloberflaeche, 40, 539-544

[84] LI Hx, Rudnev VS, Zheng XH, Yarovaya TP, Song RG. (2008) Characterization of Al_2O_3 ceramic coatings on 6063 aluminum alloy prepared in borate electrolytes by micro arc oxidation. Journal of Alloys and Compounds, 462, 99-102.

[85] Tillous K, Toll-Douchanoy T, Bauer-Grosse E, Hericher L, Geandier G. (2009) Microstructure and phase composition of micro arc oxidation surface layers formed on aluminium and its alloys 2214-T6 and 7050-T74. Surf. & Coat. Technol,, 203, 2969-2973.

[86] Khan HU, Yerokhin AL, Pilkington T, Leyland A, Mathews A, (2005) Residual stresses in plasma electrolytic oxidation coatings on aluminium alloys produced by pulsed unipolar current. Surf. Coat. Technol. 200, 1580–1586.

[87] Hussein RO, Nie X, Northwood DO, (2010) Influence of process parameters on electrolytic plasma discharging behaviour and aluminum oxide coating microstructure Surf. Coat. Technol., 205, 1659-1667

[88] Xue Wen-bin, Wang Chao, Tian Hua, Lai Yong-chun. (2007) Corrosion behaviours and galvanic studies of micro arc oxidation films on Al-Zn-Mg-Cu alloys, Surf & Coat Technol, 201, 8695-8701.

[89] Donachie MJ, (2000) Titanium: A Technical Guide, 2nd edition, ASM International, Materials Park, OH, p. 225.

[90] Wheeler JM, Collier CA, Paillard JM, Curran JA, (2010) Evaluation of micromechanical behaviour of plasma electrolytic oxidation (PEO) coatings on Ti-6Al-4V, Surf. Coat. Technol. 204, 3399-3409.

[91] Walsh FC, Low CTJ, Wood RJK, Stevens KT, Archer J, Poeton AR, Ryder A, (2009) Plasma electrolytic oxidation (PEO) for production of anodised coatings on lightweight metal (Al, Mg, Ti) alloys, Trans. Inst. Met. Finish. 87 (3), 122–135.

[92] Matykina E, Skeldon P, Thompson GE, (2009) Fundamental and practical evaluations of PEO coatings of titanium, Int. Heat Treat. Surf. Eng. 3 (1,2), 45-51.

[93] Yao Z, Liu Y, Xu Y, Jiang Z, Wang F, (2011) Effects of cathode pulse at high frequency on structure and composition of Al2TiO5 ceramic coatings on Ti alloy by plasma electrolytic oxidation, Mater. Chem. Phys. 126, 227–231.

[94] Frauchigera VM, Schlottigb F, Gasserc B, Textora M, (2004) Anodic plasma-chemical treatment of CP titanium surfaces for biomedical applications, Biomaterials 25, 593–606.

[95] Cui WF, Jin L, Zhou L, (2013) Surface characteristics and electrochemical corrosion behavior of a pre-anodized microarc oxidation coating on titanium alloy, Mater. Sci. Eng., C, 33, 3775-3779

[96] Zhang X L, Jiang Z H, Yao Z P, Wu Z D, (2010) Electrochemical study of growth behaviour of plasma electrolytic oxidation coating on Ti6Al4V: Effects of the additive, Corrosion Science, 52 3465-3473.

[97] Li X, Chenga G, Xuea W, Zheng R, Chengc Y, (2008) Wear and corrosion resistant coatings formed by microarc oxidation on TiAl alloy, Materials Chemistry and Physics 107, 148–152.

[98] Montazeri M, Dehghanian C, Shokouhfar M, Baradaran A, (2011) Investigation of the voltage and time effects on the formation of hydroxyapatite-containing titania prepared by plasma electrolytic oxidation on Ti-6Al-4V alloy and its corrosion behavior, Applied Surface Science, 257 7268-7275.

[99] Li L H, Kong Y M, Kim H W, (2004) Improved biological performance of Ti implants due to surface modification by micro-arc oxidation, Biomaterials 25, 2867-2875.

[100] Bai, YJ, Wang YB, Cheng Y, Deng F, Zheng YF, Wei SC, (2011) Comparative study on the corrosion behavior of Ti-Nb and TMA alloys for dental application in various artificial solutions. Mater. Sci. Eng. C Mater. Biol. Appl., 31, 702–711.

[101] Sul YT, Johansson CB, Petronis S, Krozer A, Jeong Y, Wennerberg A, Albrektsson T, (2002) Characteristics of the surface oxides on turned and electrochemically oxidized pure titanium implants up to dielectric breakdown: the oxide thickness, micropore configurations, surface roughness, crystal structure and chemical composition, Biomaterials, 23, 491– 501.

[102] Yao Z P, Wang D L, Xia Q X, Zhang Y J, Jiang Z H and Wang F P, (2012) Effect of PEO power modes on structure and corrosion resistance of ceramic coatings on AZ91D Mg alloy, Surface Engineering 28, 96-101.

[103] Zhou W, Shan D, Han E, Wei K, (2008) Structure and formation mechanism of phosphate conversion coating on die-cast AZ91D magnesium alloy, Corrosion Science 50, 329–337.

[104] Luo H, Cai Q, Wei B, Yu B, He J, Li D, (2009) Effect of (NaPO3)6 concentrations on corrosion resistance of plasma electrolytic oxidation coatings formed on AZ91D magnesium alloy, Journal of Alloys and Compounds, 474, 551-559.

[105] Luo H, Cai Q, Wei B, Yu B, Li D, He J, Liu Z, (2008) Effect of (NaPO3)6 concentrations on corrosion resistance of plasma electrolytic oxidation coatings formed on AZ91D magnesium alloy, Journal of Alloys and Compounds:464, 537–543.

[106] Yao Z, Li L, Jiang Z, (2009) Adjustment of the ratio of Ca/P in the ceramic coating on Mg alloy by plasma electrolytic oxidation, Applied Surface Science 255, 6724–8.

[107] Song YL, Liu YH, Yu SR, Zhu XY, Wang Q, (2008) Plasma electrolytic oxidation coating on AZ91 magnesium alloy modified by neodymium and its corrosion resistance, Applied Surface Science 254, 3014–3020.

Chapter 12

Corrosion Resistance Through the Application of Anti-Corrosion Coatings

Api Popoola, OE Olorunniwo and OO Ige

Additional information is available at the end of the chapter

http://dx.doi.org/10.5772/57420

1. Introduction

Corrosion is the surface disintegration of metals/alloys within specific environment. Some metals basically exhibit high corrosion resistance than others and this can be attributed to several factors like their chemical constituents, the nature of electrochemical reactions itself and others. The corrosion resistance of metals can be defined in terms of its ability to withstand aggressive conditions. This determines to a large extent the operational lifetime of components in service. However, there are several definitions of corrosion and according to International Union of Pure and Applied Chemistry (IUPAC) "Corrosion is an irreversible interfacial reaction of a material (metal, ceramic, and polymer) with its environment which results in consumption of the material or in dissolution into the material of a component of the environment. Often, but not necessarily, corrosion results in effects detrimental to the usage of the material considered. Exclusively physical or mechanical processes such as melting or evaporation, abrasion or mechanical fracture are not included in the term corrosion" (Heusler *et al.,* 1989). It is realized that this definition virtually include all engineering materials and it is considered as a wide definition. Hence, another definition is given by ISO 8044-1986 which states inter-alia: "Physicochemical interaction between a metal and its environment which results in changes in the properties of the metal and which may often lead to impairment of the function of the metal, the environment, or the technical system of which these form a part". (EFC Working Party 7: Corrosion Education, 2012).

It is realized that the most widely used definition of corrosion is the degradation of material by its reaction with its environment (Trethewey and Chamberlain, 1995). The following are the main concepts in the definition: degradation, material, reaction, and environment. Generally, corrosion is conceived as harmful that is, the degradation/destruction that occurs with corrosion, however, it must be noted that there are numerous advantageous uses in respect of corrosion. Degradation is regarded as the deleterious effects of corrosion and it greatly outweigh their usefulness. Materials are used by human being to achieve their technological prowess but of all the available ones, it is known that metals are "inter pares" among them. It was based on the fact that materials have good physical, mechanical, and chemical properties (Ige, 2007). It has also been observed that the term reaction is also being used in classifying corrosion based on their mechanism; hence it can be physical, chemical or electrochemical (Fontana, 1987; West, 1986). The most significant is the electromechanical reaction, which involves the transfer of electrons between the participants. Environment is described as all species adjacent to the corroding metal at the time of the reaction. Environments that cause corrosion are called corrosive and materials/metals that suffer corrosion are called corrodible (Trethewey and Chamberlain, 1995; West, 1986).

The factors that affect the environment include nature, thermodynamics and the kinetic among others (Ige, 2007) but generally, corrosion resistance or chemical resistance depends on many factors. Its complete and comprehensive study requires knowledge of several fields of scientific knowledge (Fontana, 1987). Some of these factors are as listed: effects of oxygen and oxidizers, temperature, velocity, corrosive concentration, galvanic coupling, metallurgical factors, (Wang, 2009; West, 1986; Tendayi, 2010).

Materials selection is a powerful tool for dealing with severe corrosion. Failures occurring as a result of corrosion attack can be very expensive therefore preventing or reducing these attacks becomes very significant to the industry and households (Fontana, 1987). The materials engineer's solution to the fatal failure induced by corrosion is simply the fabrication of anti-corrosion coatings with superior chemical and mechanical properties than the parent material. The improvement of the lifespan and performance of metallic alloys and components through application of numerous anti-corrosion coatings is highly advantageous (Popoola *et al.*, 2012). These coatings enable more efficient metals/components, efficient industrial operations, cost reduction, saving of scarce material resources and reduction in pollutant emissions. Anti-corrosion coatings often involve development of new surface materials which can impart numerous functional properties unto the surfaces of metals/components. New materials include composites, nano-composites, nano-particles. Numerous industries make use of anti-corrosion coatings since almost all engineering materials (composite, alloys, metals, polymers and ceramics) can be used as reinforcement coating on materials surfaces.

2. Corrosion costs and economics

Corrosion costs society in three ways; viz, it is extremely expensive, it is extremely wasteful of natural resources at a time of increased concern over damage to the environment and it causes considerable inconvenience to human being and sometimes loss of life (Trethewey & Chamberlain, 1995; Ige, 2007). Corrosion is the greatest consumer of metals known to man. Corrosion of industrial metal is one of the oldest problems that have ever challenged the industrial world. Its cost is quite enormous and translates into billions of dollars in many developed nations. It is observed that calculations on a national scale are extremely difficult

http://dx.doi.org/10.5772/57420

and any figure produced for the annual cost of corrosion to a country cannot be considered precise. It can however be taken merely as an indication of the order of protection and prevention that have been added to the cost of deterioration due to corrosion (Ige, 2013). The sum of these costs is a measure of the total demands on the national annual cost of corrosion which varies from 1 - 3.5% of the GNP (Orman, 1976; Gasem, 2013; ASM, 2000). The figures obtained are only the direct economic costs of corrosion.

The indirect costs resulting from actual or possible corrosion are more difficult to evaluate but are probably even greater. It is estimated realistically to be around 300 billion dollars in United States of America (USA), while in developing countries like Nigeria it is assumed to be around 10 billion dollars and it is estimated to be around Rs 2.0 lakh crores per annum in India (Umoru, 2001; Natarajan, 2013). It must be noted that the high degrees of severity of corrosion leads to tremendous economic losses. Aside from its direct costs in monetary term, corrosion is a serious problem because it definitely contributes to the depletion of our natural resources. Also, the rapid industrialization of many countries indicates that there will be competition for metal resources and as such the price will increase (Fontana, 1987). It is observed that the effects of corrosion can be beneficial or deleterious. There are numerous effects of desirable application of corrosion which are available some of them are chemical machining, anodizing, etching, among others. However, the deleterious effects include; aesthetic deterioration, high maintenance and operating cost, plant shutdowns, product contamination, loss of valuable parts, effects on safety and reliability and the recent trend of product liability whereby the onus is on the manufacturers rather than the users. Although corrosion is inevitable, its cost can be considerably reduced (Jones, 1992; Orman, 1976; Fontana, 1987).

3. Mechanism of corrosion

Thermodynamic and electrochemistry are of great importance in understanding and controlling corrosion. Metallurgical factors frequently have a pronounced influence on corrosion resistance. Physical chemistry and its various disciplines are most useful for studying the mechanisms of corrosion reactions, the surface conditions of metals, and other basic properties (Fontana, 1987). As stated earlier, corrosion can be classified into three categories based on the mechanism of their reactions and these are; chemical, physical and electrochemical. Chemical corrosion is purely subjected to the basic laws of chemical kinetics of heterogeneous reactions and refers to cases of corrosion that are not accompanied by generation of electric current for instance, corrosion of metals in non-electrolytes or in dry gases (Koser, 1980). The attack on metal surfaces during etching is also an example of a corrosion process by chemical attack (Shreir, 1994).

The physical mechanism of corrosion is typified in the metallic corrosion of solid metals in contact with liquid metal. In most cases, the solid metal dissolves to form an alloy with the molten metal, while sometimes; attack on solid metal is due to penetration of the liquid metal into the grain boundaries of the solid metal (Umoru, 2001). Electrochemical reactions can be divided into anodic and cathodic reactions. While anodic reaction involves anodic dissolution, at the cathode there is consumption of all the electrons released at the anode. It also depends on the environment and the cathodic reactions include: metal reduction, metal deposition, oxygen consumption, or hydrogen evolution. In some corrosion reactions the oxidation reaction occurs uniformly on the surface, while in other cases it is localized and occurs at specific areas.

3.1. Anti-corrosion coatings: Mechanism

There are so many mechanism of anti – corrosion coatings but generally, mechanism of coating can be differentiated into three, namely; barrier creation between substrate materials and environments, inhibition of the corrosion processes, and coating acting as sacrificial materials. However, recently one of the newest approaches is what is called *"active-passive"*. This involved the coating acting as barrier layers which will not aloe permeation of corrosive agents to the metal surface (passive). While the active approach allows the formation of effective passive layer and this will impedes the corrosion half reactions leading to Schottky barrier at the interface resulting in depletion of electrons (Dennis *et al.*2013).

4. Corrosion thermodynamics and kinetics

Corrosion science involves a study of electrodics, which are electrochemical processes that take place at electrodes. An electrode is essentially the boundary between a solid phase (metal) and a liquid phase (aqueous environment) and these processes take place across the phase boundary. The basic wet corrosion cell consists of four essential components, viz; anode, cathode, electrolyte and connections. The first law of corrosion control is regarded as the removal of any one of the four components of the simple wet corrosion cell and this will stop the corrosion reaction (Ige, 2007).

Corrosion in aqueous solutions has been found to involve electron or charge transfer. A change in electrochemical potential, or the electron activity, or availability on metal surface has profound effect on the rates of corrosion reactions. Thermodynamics give an understanding of the energy changes involved in the electrochemical reactions of corrosion. These energy changes provide the driving force and control the spontaneous direction for a chemical reaction. Thus, thermodynamics show how conditions may be adjusted to make corrosion impossible. When corrosion is possible thermodynamics cannot predict the rate, that is, corrosion may be fast or slow. It is postulated that in an electrochemical reaction, the most negative or active half-cell tends to be oxidized, and the most positive or noble half-cell tends to be reduced. Corrosion thermodynamics can also be used to state the criterion for corrosion. Corrosion will not occur unless the spontaneous direction of the reaction indicates metal oxidation (Umoru, 2001).

From an engineering point of view the major interest in corrosion is the kinetics or rate of corrosion. The principal aim for studying corrosion reaction kinetics are to develop empirical relationship that permit the prediction of corrosion rate under conditions that are different from those originally employed and to determine the mechanism of the overall process.

http://dx.doi.org/10.5772/57420

It is observed that corrosion is thermodynamically possible for most environmental conditions. Thus, it is of primary importance to know how fast corrosion occurs. Fortunately, most alloys corrode only slowly in many environments. Chemical kinetics is a study of the rates of such reactions. Corrosion in aqueous systems is governed primarily by electrochemical reactions and the understanding of the fundamental laws of electrochemical reaction kinetics is thus essential to develop more corrosion – resistant alloys and to improve methods of protection against corrosion (West, 1986). Electrochemical reactions either produce or consume electrons. Thus, the rate of electron flow to or from a reacting interface is a measure of reaction rate and electron flow is conveniently measured as current, I, in amperes.

4.1. Polarization

The rate of an electrochemical reaction is limited by various physical and chemical factors and the reaction are polarized or retarded by these factors. Polarization is defined as the displacement of electrode potential resulting from a net current. Its magnitude is frequently measured in terms of overvoltage. For cathodic polarization, electrons are supplied to the surface, and a buildup in the metal due to the slow reaction rate causes the surface potential, E, to become negative to e. Hence cathodic polarization is negative by definition. For anodic polarization, electrons are removed from the metal, a deficiency results in a positive potential change due to the slow liberation of electrons by the surface reaction, and anodic polarization must be positive. Most often, over potential is also used for polarization. Polarization is commonly classified into the following, viz; activation, concentration, resistance, and combined.

4.2. Anti-corrosion coating: Polarization

Chromium-based and zinc are used mostly as coating materials but due to stringent rules and regulations in terms of Health, Safety and Environment by many agencies, the former usage had declined progressively in the last decade. While the application of zinc as coating is discouraged significantly due to price fluctuation (Dennis *et al,* 2013). Hence alternative materials are sourced aggressively and one of such is the hybrid materials comprising of Unfuctionalised Graphene (UFG) and polyetherimides. This coating provide enhanced corrosion resistance to low alloy steel as depicted by data obtained from Tafel plot shown in Figure 1. It was also reported that the coating works on the active-passive mechanism of corrosion inhibition (Dennis *et al,* 2013).

The study of Wu *et al* (2010) showed that not all the materials evaluated possess corrosion resistance in a given environment. They deposited diamond-like carbon coating (DLC) on AZ31 magnesium alloy by ion beam deposition technique with Cr CrN acting as interlayers. Also evaluated was the effect of interlayers on the corrosion resistance of DLC on magnesium alloy. The results showed that the addition of Cr and CrN as interlayers improved the adhesion between coating and substrate effectively. The polarization result is presented in Figure 2 below. However it did not enhance the corrosion resistance of the DLC/magnesium alloy systems in 3.5 wt.% NaCl solution due to the formation of galvanic cell between substrate and interlayer in the region of through-thickness defects.

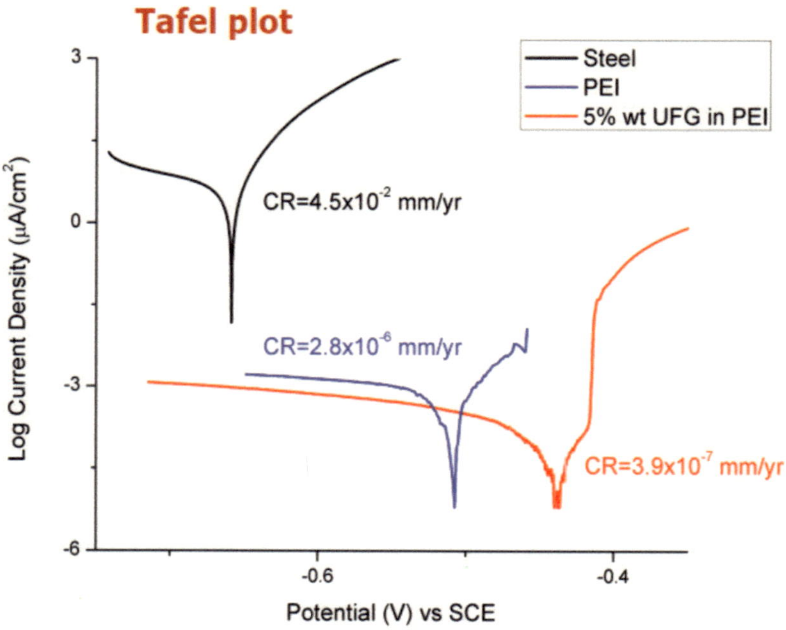

Figure 1. Tafel plot for uncoated and coated substrates (Dennis *et al.,* 2013)

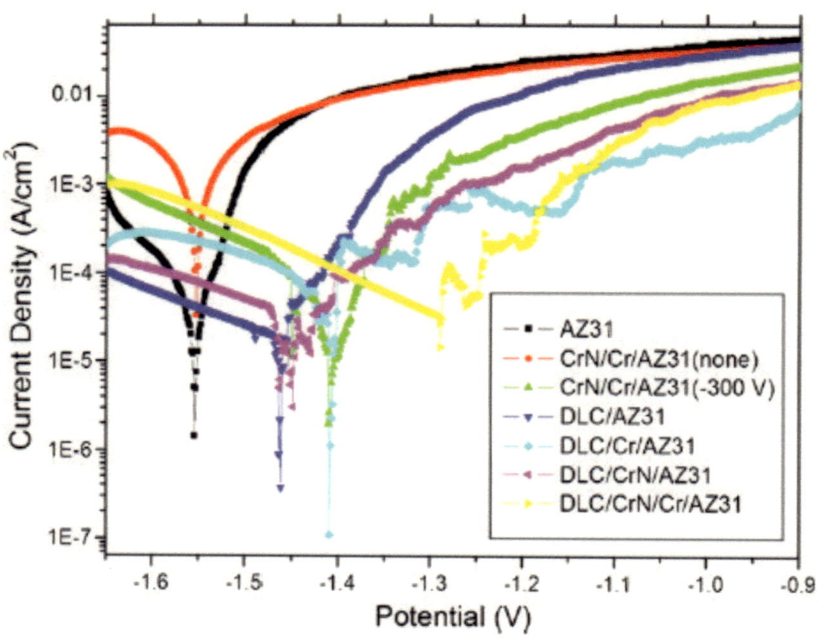

Figure 2. Polarization curves of all samples prepared in the study (Wu *et al.,* 2010)

4.3. Passivation

This is the loss of chemical reactivity experienced by certain metals and alloys under particular environmental conditions. The metals or alloys become inert and act as if they are noble metals such as platinum and gold. The following metals or alloys containing them show passivity; iron, nickel, silicon, chromium, and titanium, while zinc, cadmium, tin, uranium and thallium only show limited passivity under the conditions. Though passivity is difficult to define it is often described quantitatively by characterizing the behavior of metals, which show this unusual effect.

Figure 3 provides a method of defining passivity. A passive metal is one that demonstrates the typical S – shaped dissolution curve with exception of titanium, which does not possess a transpassive region.

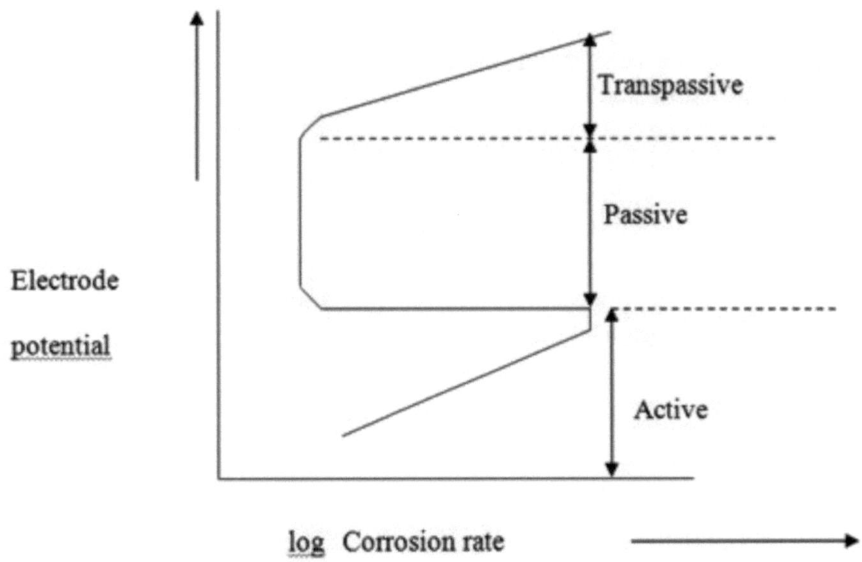

Figure 3. Corrosion characteristics of metal displaying passivity (Ige, 2007)

Passivity may be classified as anodic and chemical. *Anodic passivity* is when a metal is made an anode in an aqueous solution containing only anions with which it forms very sparingly soluble compounds. While *chemical passivity* is slightly more complicated wherein cathodic processes are involved on the same metals in corrosion – inhibitive solutions.

It may be considered as the action of corrosion products on the metal thereby influencing the reaction rate. The corrosion product rate is dependent on the solubility, structure, thickness, adhesion and so on. The film may be thin or thick; also it may be soluble or insoluble. In passivity, two conditions must be satisfied and they are; the redox potential of the solution must be more negative than passivity potential and the rate of the cathodic reaction must be greater than critical current density. Summarily, metals or alloys that possess an active-passive transition become passive or very corrosion resistant in moderate to strongly oxidizing conditions; these materials lose their corrosion-resistance properties. These characteristics have been successfully used to develop new method of preventing corrosion and to predict corrosion resistance.

4.4. Passivation: Anti-corrosion coating

All the polarization resistance techniques (linear, potentiodynamic, cyclic and so on) can be employed to evaluate corrosion kinetics of coating systems; coupled with the coating thickness achievable in a given period. Current density is main parameter used to measure the efficiency of coating kinetics, the higher the current density the poorer the electrochemical behavior. High corrosion rates are characterized with a small passive range and lower pitting potential, suggesting the presence of defects and pores that allow the electrolyte diffusion, thus promoting the failure of the protective barrier (Castro *et al.*, 2005).

The aim of a research work carried out by Hashem (2004) was to prepare new monomer multi-silicon compounds which can acts as more durable adhesive and anticorrosion coating materials. The study also evaluated the structural mechanisms of their fragmentation in order to evaluate the reaction rate. The materials were identified by their retention times, number of cleavages and number of bonds. The results identified the isomers by the differences in the retention times. It was concluded that the isomers were organo-multi-silicon monomer compounds which performed creditably well as durable adhesive and anticorrosion coating materials. One of the rate determining factors in coating systems is the fall of polarization resistance Rp by its delamination with the formation of uniform double layer at the interface. This leads to filliform corrosion among others (Deen *et al.*, 2009).

5. Forms of corrosion

According to ASM (2000) there are basically three factors by which corrosion can be classified, viz; nature of the corrodent, mechanism of corrosion, and appearance of the corroded metal. The latter mode of classification is employed and this is due to the fact that it provides adequate information on the mode of failure associated with the corroded materials. The classification is based on the surface morphology (Figure 4 and Table 1) and it must be noted that the forms of corrosion are distinct in theory but practically, there are cases wherein the corrosion fits in more than one category. The corroded metals can be grouped into eight forms of wet (or aqueous) corrosion and these are (ASM, 2000): uniform or general corrosion, pitting corrosion, crevice corrosion, including corrosion under tubercles or deposits, filiform corrosion, and poultice corrosion, galvanic corrosion, erosion-corrosion, including cavitation erosion, and fretting corrosion, intergranular corrosion, including sensitization and exfoliation, dealloying, including dezincification, and environmentally assisted cracking, including stress cracking corrosion, corrosion fatigue, and hydrogen damage (Effird, 1993; Malka *et* al., 2006; Barker *et* al., 2012; Jiang, et *al, 2005;* Ige, et *al.*, 2012; Colangelo and Heiser, 1974; Fontana, 1987; Chamberlain & Trethewey, 1995; Baker & Castle, 1993; Shreir, 1976).

http://dx.doi.org/10.5772/57420

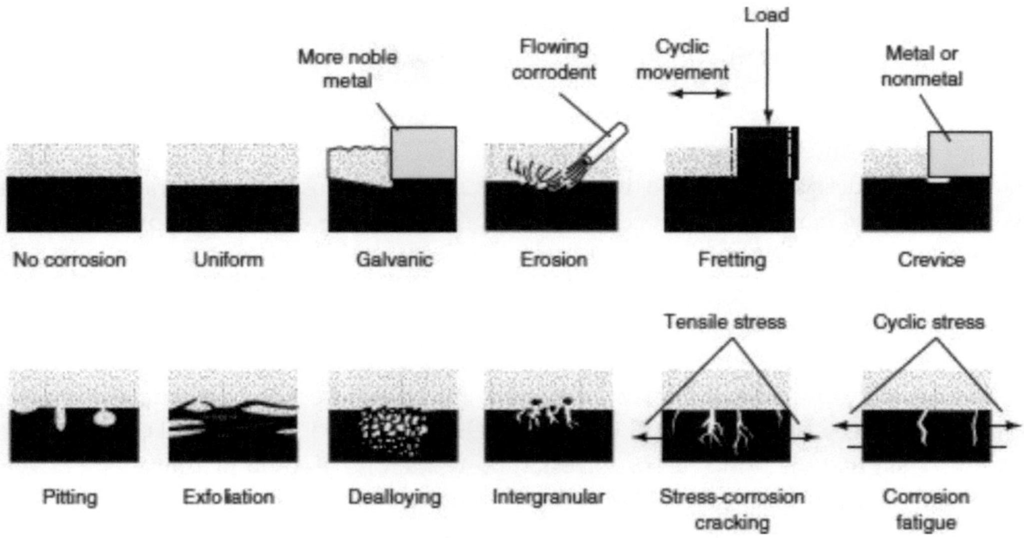

Figure 4. Schematic summary of the various forms of corrosion (ASM, 2000)

General Corrosion:	Localized Corrosion:	Metallurgical Influenced Corrosion:	Mechanically Assisted Degradation:	Environmentally Induced Cracking:
Corrosive attack dominated by uniform thinning	High rates of metal penetration at specific sites	Affected by alloy chemistry and heat treatment	Corrosion with a mechanical component	Cracking produced by corrosion, in the presence of stress
Atmospheric corrosion	Crevice corrosion	Intergranular corrosion	Erosion corrosion	Stress – Corrosion Cracking
Galvanic corrosion	Filiform corrosion	Dealloying corrosion	Fretting corrosion	Hydrogen Damage
Stray-current corrosion	Pitting corrosion		Cavitation and water drop impingement	Liquid metal embrittlement
General biological corrosion	Localized biological corrosion		Corrosion fatigue	Solid metal induced embrittlement
Molten salt corrosion				
Corrosion in liquid metals				
High – temperature corrosion				

Table 1. ASM classifications of corrosion types (Ref: Scully J R, Taylor D. W., Electrochemical Methods of Corrosion Testing, Metals Hand Book Vol 13, 1987)

6. Corrosion prevention and control

The fact that corrosion needs to be controlled cannot be overemphasized in view of the colossal amount of money that is expended on it annually. Materials are protected from corrosion through a wide variety of methods. In some measures, corrosion is avoided by careful choice of material and design of system so that the opportunity for corrosion of a certain specific sort does not arise. In most cases, the occurrence of corrosion is prevented or suppressed by the modification of the environment, modification of the materials properties, application of protective coatings and, cathodic and anodic protection. It must be appreciated that the choice of whether corrosion is to be avoided, suppressed or merely controlled depends on the economics of the processes involved (Ige, 2007). Hung *et al* (2011) highlighted the measures in preventing and control of corrosion as follows: materials selection, cathodic and anodic protection, and coating (Stack, 2002; Birks *et al.,* 1993; Reyes and Neville, 2001). However, it must be noted that the principles of corrosion prevention is based on modification of environments, electrochemical approaches, and corrosion resistant materials. The importance of coating cannot be emphasized because it encompasses all the corrosion control and prevention techniques such as cathodic protection (sacrificial), modification of environment (barrier effects and inhibition), and materials selection (development of corrosion resistant materials such as conducting polymers). The concept of coatings as its affect corrosion resistance is presented in the next section.

7. Coatings

According to Hegedus (2004), there are multi-layered systems in coatings which composed of primer and topcoat. The functions of each layer are specific but each layer is interactive. Coating materials can be metallic, inorganic or organic. Relatively thin coatings of *metallic and inorganic* materials can provide a satisfactory barrier between metal and its environment.

The chief function of such coating is to provide an effective barrier except for sacrificial coatings such as zinc. Metal coatings are applied by electrodeposition, flame spraying, cladding, and hot dipping and vapour deposition but inorganics are applied or formed by spraying, diffusion, or chemical conversion. Spraying is usually followed by baking or firing at elevated temperature. Metal coatings usually exhibit some formability whereas the inorganics are brittle and in both cases a complete barrier must be provided. Porosity or other defects can result in accelerated localized attack on the basic metal because of galvanic effects (www.metalsamples.com). Organic *coatings* involve a relatively thin barrier between substrate material and the environment. Paints, varnishes, lacquers, and similar coatings doubtless protect more metal on a tonnage basis than any other method for combating corrosion. Aside from proper application, the three main factors to consider for organic coatings are listed in order of importance as surface preparation, type of primer or priming coat and the selection of topcoat or coats.

Surface preparation is considered as one of the topmost factors in coating industry. It is known that if the metal surface is not properly prepared, the paint may peel off because of poor bonding, If the primer does not have good adherence or is not compatible with the top coat, early failure occurs. If the first two factors are wrong, the system will fail regardless of the topcoat used. Poor paint performance is in most cases due to poor application and surface preparations. A tremendous variety of paints are available. Asphalt bituminous paints are often used on pipelines. Sometimes a cloth wrapping is used with the coating for reinforcement. Alkyds, glyptols, concrete, lead, iron oxide, phenolics, lithopones, titanium oxide paints and chlorinated rubber are just a few examples. Vinyl and epoxy paints have been widely adopted for corrosion application (www.corrosion-club.com).

There are many traditional techniques of coating and some of them are listed in Table 2 and shown in Figure 5. The techniques are thermal spraying, ion beam, immersion, welding, screen printing, dip-coating, electrogalvanising process, coating under vacuum, such as PVD (Physical Vapour Deposition), CVD (ChemicalVapour Deposition) among others (Wu *et al.*, 2010; metallic coated steel, 2013; ; Popoola *et al.*, 2012; Obadele *et a.*, 2011; Ochonogora *ET AL.*, 2012) and they are still commonly applied in coating metal substrates (Zhang & Tang, 2009). A typical example involved the use of cold tape coating system on new or reconditioned pipelines carrying oil, gas and water (Metallic coated steel, 2013). Their main function is to provide long term corrosion protection from underground elements as well as mechanical protection. The total tape system components consist of a primer applied directly to the pipe surface, an inner-wrap tape layer that provides a corrosion barrier and an outer-wrap tape layer that provides mechanical protection. The pipe coating process is done according to industry standards and some of them include DIN 30672, EN 12068, AWWA C-209, C-214 and C-225 (Figure 6). While Huang and Chen (2012) developed anti-corrosion tape which was effective on mild steel in marine splash zone. The mechanism of the developed tape coating was based on corrosion inhibition.

7.1. Anti – corrosion coating: Metallic materials

Metals and alloys are most commonly chosen for corrosion resistant concern, especially for metallic or ceramic substrates (Zhang and Tang, 2009). Due to serious environmental issues and health concerns, the use of chromating and phosphating phases (Bibber, 2007; Corell, 1998) are being replaced with molybdate, rare earth, silicate and titanium oxides or zirconium oxides (Oki, 2007; Phani *et al.*, 2005; Lunder, 2004; Song and Mansfeld, 2006; Montemora *et al.*, 2009; Wen *et al.*, 2008). For example, Huan and Buchheit (2004) studied a vanadate conversion coating and it was demonstrated that the coating formation offers increase corrosion resistance to pitting and suppress oxygen reduction reactions. However the vanadate coating has potentials of producing adverse health effects. While Guosheng *et al* (2013) concluded that Zn-Ni coating can serve as low potential cathodic coating for steel substrate and has a long life period and they meet the requirements for cathodic protection.

Metallic coated steel can be defined as a steel substrate coated with a layer of zinc, a zinc/aluminium alloy, a zinc/silicon alloy or pure aluminium. Typical coated steel is as illustrated in Figure 7 below. Fayomi and Popoola (2012) investigated the electrochemical behavior and the corrosion properties of Zn coating on steel substrates by means of Vickers microhardness

Applications	Techniques
Gas turbine components	Spraying
Electrolytic cathode for copper refinement	Thermal spraying
Water-cooled stator bars chips used in electrical generators	Chemical vapour deposition, physical vapour deposition
Semiconductor device and the liquid crystal device	Screen printing dip coating thermal spraying
Semiconductor processing equipment	Thermal spraying Sputtering Immersion Chemical vapour deposition, physical vapour deposition
Alkaline-containing environment	Spraying immersion roll coating
High temperature erosion-corrosion environments	Thermal spraying spray and fuse welding

Table 2. The Conventional Application Techniques for Making the Coatings (metallic coated steel, 2013)

and polarization measurements. While, the work of Durodola *et al.,* (2011) demonstrated that Zn coated steel plated performed creditably well in seawater or chloride containing marine environments compared with uncoated steel substrate. The sample concluded that the corrosion resistance of mild steel was improved after zinc deposition and that pitting corrosion observed can be attributed to ionic migrations which are influenced by the applied voltage.

7.1.1. Mechanism of corrosion protection offered by metallic coatings

Unarguably it is assumed that a metallic coating must be self-passivation and has a potential lower than that of the substrate. This allows the coating to have long life circle and high protecting efficiency (Short *et al.,* 1989). At the same time, the potential of coating materials should not be too negative; otherwise hydrogen embrittlement will happen to substrate. Generally, there are two basic mechanisms for metallic coating and they are: barrier effect, and cathodic protection. However a third approaches is by corrosion inhibition.

- Barrier effect: this involved the prevention of corrosion by isolation of the substrate from the corrosive environments. The isolation can be done physically by the presence of the coating or by the formation of corrosion products which provides a protective layer. Also surface passivation can occur due to formation of firmly bonded, hard and non-porous layer. The degree of passivation will depend on the acidity of the environment. The passivation of the surface of the coating will also affect its ability to provide sacrificial cathodic protection. For instance, the aluminium will be transformed into aluminium oxide, thereby creating a barrier that is non-porous, except in saline atmospheres (ABB, 2007; Krishnamurthy *et al.,* 2013; Wessling, 1999; Thiery ad Pommier, 2004; Hashem *et al.,* 2004; Bhadra

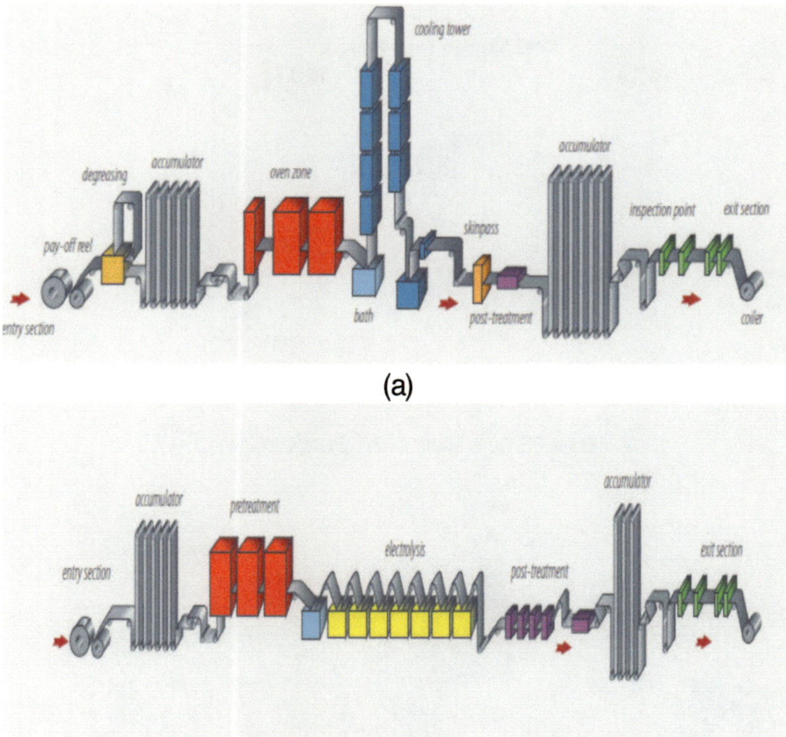

Figure 5. The hot dip coating process (a) and electrogalvanising coating (b) (Wu *et al.*, 2010)

et al., 2010). While Videla and Herrera (2009) stated that typical passivating agents for inhibiting microbial induced corrosion on metallic surfaces include inert metals, conductive polymers, and thiol-based monolayers.

- Cathodic protection: Corrosion is a phenomenon whereby an electrochemical cell is created consisting of an anode which is depleted to the benefit of a cathode. In practice, cells are formed as soon as a lack of surface uniformity appears, creating anode and cathode areas, whereby water acts as the electrolyte. In these cells, the steel will corrode every time it becomes the anode, but when it is the cathode, it will remain intact. The zinc and aluminium, which have a lower electrochemical potential than iron, will oxidise preferentially, thereby protecting the surface of the steel. The coating can therefore be considered to be inert with respect to the metal substrate, and will no longer be sacrificed. Hence, cathodic protection can be attributed to preferential dissolution of the metal substrates and this is due to mechanical damages (scratches, cut edges, holes etc) that may happen to the coating. This process is known as sacrificial protection. The sacrificial potential of a coating will depend on its corrosion products, which vary according to the type of metal in the coating. For example, zinc and other sacrificial coatings can protect steel from corrosive environments by cathodic protection (Mathiazhagan and Joseph (2011).

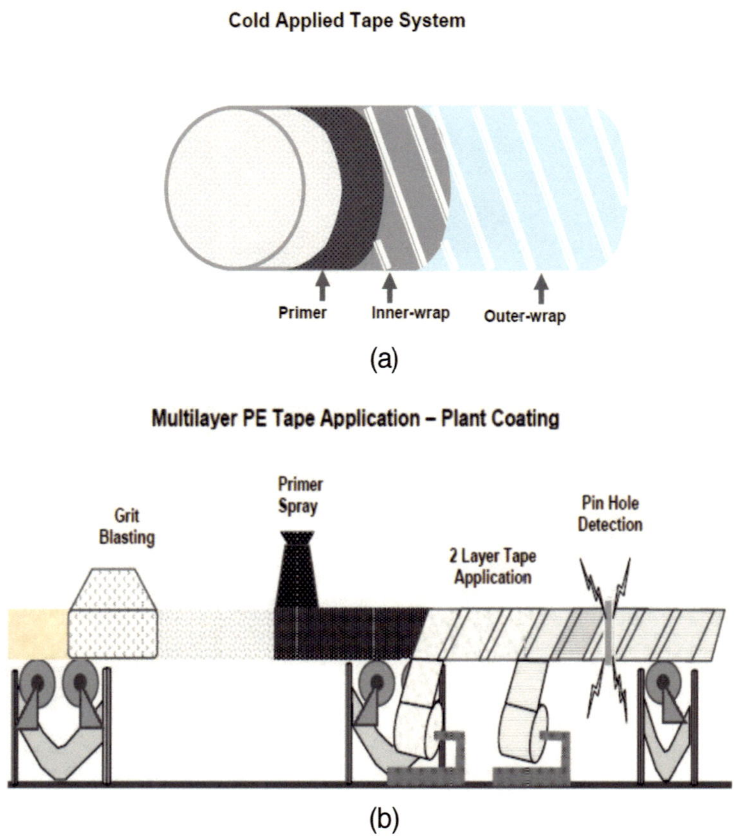

Figure 6. The cold tape coating system (a) and the pipe coating process (b) (Metallic coated steel, 2013)

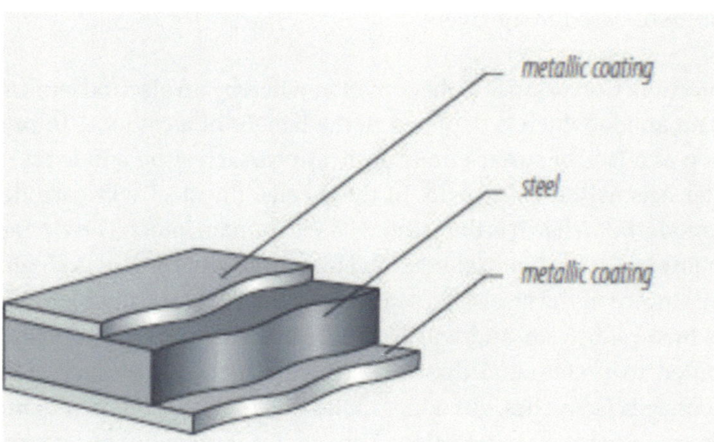

Figure 7. Typical coated steel (Metallic coated steel, 2013)

- Inhibition: The addition of inhibitors into coating systems are regarded as active corrosion protection as against passive corrosion protection as obtained by using barrier film mechanism. The inhibitor reservoir will act as an additional corrosion protection beyond the barrier protection offered by the oxide layer (Buchheit *et al.*, 2002). However the means of deploying the inhibitors are fuzzy and associated with economic and technical difficulties. One of the means of solving this issue is by production of self-healing coatings by integrating the active inhibiting compounds into existing convectional coatings (Shchukin et al., 2006; 2007; Latnikova, 2012). Szabo et al., (2011) showed that there are improved performances in the anti-corrosion property of coating systems inhibited by encapsulation in urea-formaldehyde resin.

7.2. Anti – corrosion coating: Ceramic materials

Several workers have used highly erosion-resistant ceramic coatings such as TiN, CrN in corrosive environments. They however realised that these ceramic materials are brittle and ultimately fail leading to catastrophe. They are very expensive and are used in critical applications; however, the use of novel metallic coatings is still under investigation (Wood and Hutton, 1990; Bousser *et al.*, 2008). Also, numerous patents have been obtained for ceramics materials in the formulation of coatings for anti-corrosion purposes. Some of the ceramics materials are used as corrosion resistant in different application such as: semiconductor industry, fuel cell, and corrosive water containing environments like gas turbine engines, heat exchangers and internal combustion engines among others (Iqhal *et al.*, 2005; Iqhal *et al.*, 2009; Hattori, 2008; Zhang and Tang, 2009).

Krishnamurthy *et al.* (2013) demonstrated the application of graphene as a passivating coating materials retards microbially-induced galvanic corrosion (MIC) of metals. The study was conceived on the basis that microbial fuel cell represents a galvanic cell and that the microbes will accelerates the metallic corrosion in the system. The experimental set up of the galvanic cell is shown in Figure 8. The study observed that graphene coating reduces considerably the MIC by preventing dissolution of soluble Ni and its performance is 10 – fold lower compared with uncoated anode (Figure 9). It was concluded that graphene coating have the advantage of possibility of being grown on large-area substrates by chemical vapour deposition.

Also that graphene coating prevents MIC by forming passivating layer which thereby restricted the movement of the solution to the Ni surface, disallowed the access of microbes to the Ni surface, and protect the Ni surface from microbes by product. With the aid of morphological study, Zaki and Abdul (2009) revealed that both nanostructured TiO_2 and TiO_2 demonstrated a little above average resistance to erosion-corrosion. Expectedly the nanostructured TiO_2 coating performed relatively better than convectional TiO_2 in slurry environment. This was achieved by reducing the gap between the splat boundaries and excluding secondary phase particles. They concluded that by reducing the volume of unmelted particles, number of pores, and provision of optimum surface topography there are possibility for improved performance by the coatings.

The application of talc as catalysts is well documented and it occurs naturally, also it can be synthesized by coprecipitation of divalent and trivalent cation salt solutions. This material was

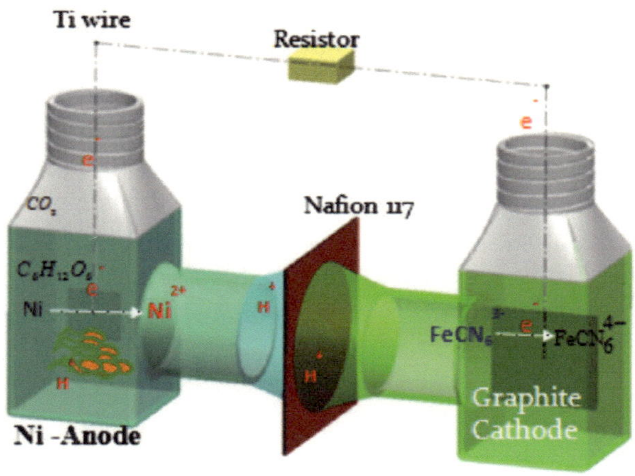

Figure 8. Schematic of a galvanic cell used in the study (Krishnamurthy *et al.*., 2013)

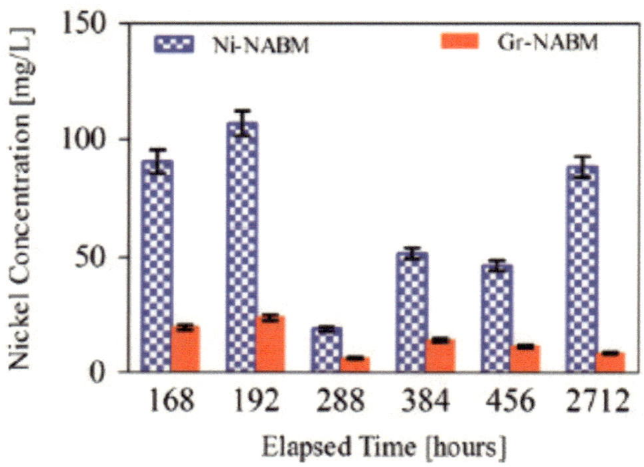

Figure 9. Soluble Ni (mg/L0 in anolyte of Ni – NABM and Gr – NABM (Krishnamurthy *et al.* 2013)

employed as coating in an aggressive environment and the results obtained revealed that it reduced the corrosion rates compared to uncoated surfaces. Post heat treatment increased the corrosion performance. It was also observed that it offers low toxic hazards and the method of production is quite simple. The suspected anticorrosion coating mechanism was due to formation of a barrier film on the substrates (Buchheit *et al.*, 1994).

7.3. Anti – corrosion coating: Polymeric materials

Zhang and Tang (2009) reviewed the patents that have been obtained in the area of polymeric materials as anti-corrosion coatings and they concluded that some polymers act as corrosion resistant materials due to the impervious nature of this material (Lee *et al.*, 2009; Kramer *et al.*, 2007). Wessling (1999) developed new anti-corrosion coating using the chemistry of polyaniline (PAni), an Organic Metal (conductive polymer). The new Organic Metal polyaniline is almost regarded as a noble metal and it passivated metals by shifting their surface potential.

The useage of intrinsically conducting polymers (ICP) as anti-corrosion coating materials on metal surfaces have increased significantly in recent times. (Beck et al., 1994; Ahmed and MacDiarmid, 1996; Sitaram et al., 1997; Camalet et al., 1998; Martins et al., 2002; Tallman et al., 2002; Iroh et al., 2003; Tan and Blackwood, 2003; Zarras et al., 2003). Due to the ease of synthesis, tunable properties, low cost monomer and good thermal stability, polyaniline (PAni) is most applied of the available ICPs. (Genies et al., 1990, Gospodinova and Terlemezyan, 1998; Bhadra et al., 2009; 2011). Another advantage of the ICP is that it can be used as a single coat unlike other traditional materials that requires multi-layer coating before adequate protection of metal surfaces against corrosion can be achieved (Wessling, 1998). While Petersen (2006) evaluated the corrosion protection of anti-corrosive paint systems based on conducting polymers and the conducting polymers are polyaniline and polythophene. The study examined its application in four areas: shopprimers, ballast tank protection, protection in off-shore environment and protection of aluminium. Highlighted were the five possible protection mechanism for the conducting polymers and they are:

- Anodic protection and passivation of the substrate metal
- Formation of a protective metal/polymer complex
- Absorption of OH- and inhibiting of cathodic disbondment
- Inhibition of the cathodic reactions
- Improve the conduction between zinc particles in zinc pigmented paints

Organic coatings are widely used as one of the corrosion protection and control methods for metals. It has vast applications in protection of porous refractory surfaces such as cement mortar or concrete structures. There are several factors that affect selection of suitable coating materials such as environmental factors and the nature of flow medium for pipeline systems. Organic polymeric materials are characterised by different functional groups. Some examples of organic polymeric materials are asphalt mastic and asphalt enamel, coal tar epoxy, extruded polyethylene, fusion bonded epoxy, multilayer polyolefin coating systems including polyethylene or polypropylene and polyurethane. For example, fusion bonded epoxy (FBE) have the following advantages which include: Saving in pipe thickness (or material),(ii) chemical inertness over a broad range of pH, presence of smooth and chemically inert surface, high friction resistance surface, ease in coating of complex shapes, ease in field welding, and ease in transportation of large size pipes and consequent saving in cost ((Vincent, 1999; Kellner *et al.*, 1998; Alexander, 1998) Malik *et al.*, 1999).

7.4. Anti – corrosion coating: Hybrid materials

Several attempts have been deployed to apply composite materials as coatings in the prevention of corrosion in various environments. Zhang and Tang (2009) highlighted the patents in anti-corrosion coatings area which include the work of Hazel *et al.* which provide improve corrosion resistance for gas turbine components, such as the turbine blade as well as the turbine disk, shaft and seal (Suzuki and Watanabe 2007; Nagaraj and Hazel 2007; Buczek *et al.*2009]. Composite materials allow the possibility of having versatile and wide ranging properties to be conferred on the coating materials. Some of the properties include: good adhesion characteristics, strain tolerance, self – healing characteristics, heat conductivity (Lee *et al.*, 2009; Kramer *et al.*, 2007; Dolan and Carson, 2007; Hill, 2009). It can also be due to other factors like technical or social factors and highlighted are two important factors.

Environmental Concerns: although chromate and phosphate phases have excellent corrosion resistance but due to restriction imposed by environment regulations in recent years, their applications as coating materials are progressively declined. Hence some environmental friendly compositions such as fluorometallate, vanadate among many others are investigated (Buchheit et al. 2006; Dolan and Carlson 2007; Mineyski et al., 2006).

Economic consideration: apart from environmental factor, the economy of the coating is another important issue for consideration. As such, studied are currently being organised for cheaper and readily available materials (Hill, 2009; Kiksunov, 2007; Brady, 2006).

A study was carried out to investigate the effectiveness of a novel hybrid ceramic-polymer or creamer coatings on aluminium alloys in industrial atmospheric environments. The aim of the study was to elucidate information about the corrosion intiation mechanism and as well as determine the corrosion performance of the coated substrates in marine environments (Hihara and Kusada, 2011).SRI's anti-corrosion coating technology is based on preceramic polymers that have inorganic skeletons and organic side-groups. This means that at lower temperatures, the material behaves like a polymer: it can take the form of a liquid, can melt, and can form solutions. Then, it can be cured to crosslink at low temperatures to form non-meltable, non-soluble solids. At higher temperatures, it can be pyrolized and converted to a ceramic material. These formulations are inorganicorganic hybrids ("ceramers") in their nature reaching 80 to 90 wt% of inorganic content. Yet, the coatings can be designed to maintain hydrophobic characteristics (SRI, 2010). Hammer *et al.* (2012) determined the optimum concentration of TEOS and MPTS mixtures to be 8. The hybrid coatings were prepared by sol-gel techniques and the coatings were smooth, crack-free, adherent and optically transparent. The coatings present excellent corrosion protection efficiency and they are characterised with high breakdown potential and low current density even for high anodic potentials. The electrochemical data revealed that the coatings can performed exceptionally well in neutral saline and acidic environments.

Based on the work of Shi et al (2009), it was established that the presence of nanoparticles such as Si_2O, Zn, Fe_2O_3 greatly enhanced the anticorrosive performance as well as mechanical properties of epoxy coating. It was noted that SiO_2 nanoparticles is premised on barrier/ passivation mechanism among others. Figure 10 illustrated the effects of corrosion rate on various coating systems on one hand and also as a function of chloride concentration on the other hand. Conclusively, the corrosion rate decreased as the chloride composition in the environment increased and also with the addition of various nanoparticles.

The new vista under study is the investigation of the synergy of various nanoparticles, their mechanism, and dynamics and kinetics in the coating system. Also the potentials of using the nanoparticles as corrosion inhibitors for the anticorrosive properties of the coating system are under consideration (Lamaka *et al.*, 2007).

7.5. Anti – corrosion coating: New and emerging materials

Among new and emerging materials are nanomaterials, biomaterials, biomimetic and biodesign materials, microelectronic mechanical materials and smart materials, just to mention few. It is only smart and nanomaterials that are much interest in the coating industry and a comprehensive and aggressive research are currently being pursued. Some of the relevant studies are reviewed in this section.

Smart coating: It is estimated that 20% of the cost of corrosion in USA is accounted for by coating and allied industrial activities. This leads to the development of smart coating that can either reduces corrosion or predict the onset of corrosion in the system. The latter can be achieved by a simple strategy of change in colour stimuli. A smart coating is defined as any coating that changes material properties in response to an environmental stimulus and the changes can be preferably do so in a reversible manner (Meilunas, 2013). Some school of thoughts believed that terms such as responsive, adaptive, or active are more appropriate to describe these coatings because the coating does not undergo complex-thought process. The assumptions were that in as much as there are various stimulus/response, so also there are potential applications that can be engineered. Some smart coatings have been developed based on their light, temperature, humidity, pressure, electrical current, and many other properties. Smart coating can be differentiated into self-cleaning, self – healing, microcapsule healing, and anti – corrosion coatings.

Smart anti – corrosive coating is believed to be trigger off immediately corrosion commences in the system. A study investigated aluminum 2024 as the substrate for smart coating using sol – gel techniques to form a passive barrier. The coatings contain silicon, zirconium oxides and resin. A blend of polyelectrolytes was used as the sensor based on the pH of the system and in a neutral pH fluid; the polyelectrolyte/inhibitor blend is stable and won't lose its potency. A polyelectrolyte blend that was stable in acidic situations was mostly employed, but which quickly released its inhibitors in a basic environment. Generally, the coating system had strong adhesion to the aluminum substrate and was able to actively reverse the galvanic process at an emerging corrosion site (Meilunas, 2013).

The concept of self – healing anti-corrosion coatings was modelled after wound healing process (Figure 11). Improper wound treatment will leads to inflammation which can be likened to underlying corrosion in the coating systems, whereas wound disinfection can be likened to application of corrosion inhibitors. Introduction of corrosion inhibitors into coating systems for the purpose of self - healing can be achieved by various techniques which range from simple to complex methods. The techniques are limited by inhibitor solubility's, chemical costs, and technical difficulties such as manufacturing process (Latnikova, 2012). *Nanocoating:* Nano-coating is one of the recent trends in the application of nanotechnology in the prevention and control of corrosion protection and it is gaining momentum due to the fact that nanoscale materials have unique physical, chemical and physicochemical properties. Nanocoating involves the incorporation of nanoparticles in the coating formulation which enhance specific features (Mathiaghazan and Joseph, 2011).

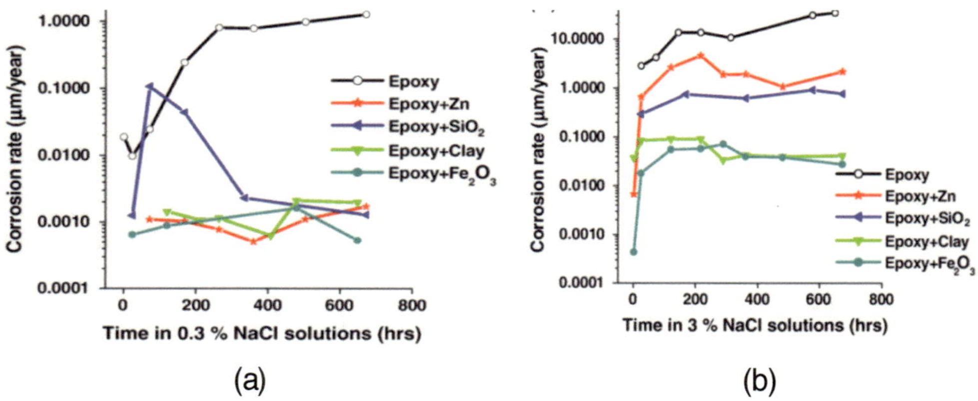

Figure 10. Temporal evolution of corrosion rate of epoxy-coated steel in (a) 0.3 wt.% NaCl solution, and (b) 3 wt.% NaCl solution, as a function of nanoparticles (Shi et al 2009)

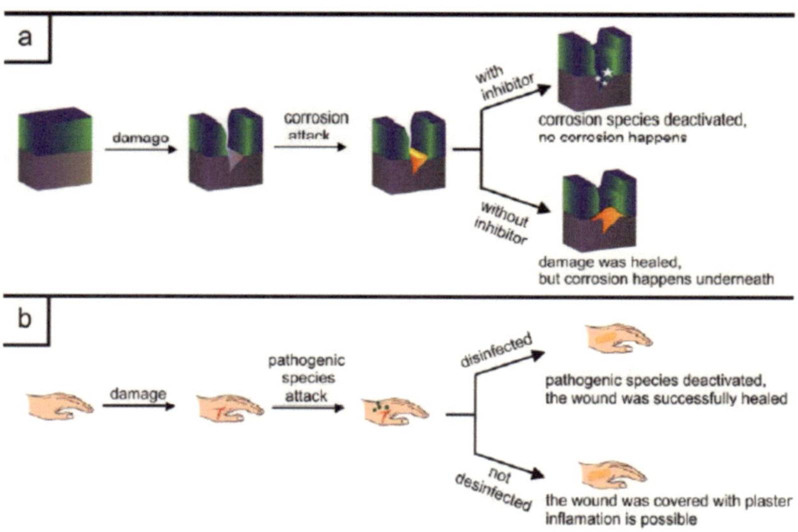

Figure 11. Bio-mimetic concept of self-healing process based on the combination of integrity restoration with application of corrosion inhibitors. (a) healing of a scratch on the surface of a metal substrate covered with anticorrosion coating. (b) healing of a wound (Latnikova, 2012)

Hung *et al* (2011) discussed the anticorrosive properties of non-conjugated polymer nanocomposites in the form of coatings having layered silicates. The results showed that the nanocomposites layered silicates have better corrosion protection compared to bulk polymers. This was obtained from corrosion data in terms of standard electrochemical corrosion measurements, including corrosion potential, polarization resistance, corrosion current, and impedance spectroscopy. While the Tafel plots and Nyquist plots are shown in Figure 12 and it was observed that as the concentration of the clay – nanocomposite increased the corrosion rate decreased. They attributed the performance to the diffusion and permeability characteristics of the O_2 and H_2O in the nanocomposites.

8. Conclusion

- Militating against corrosion is necessary in the industry because the penalties of failures due to corrosion can be very costly including safety hazards.
- Selective modification methods such as dipping, spraying, sol-gel, vapour deposition among others of metals and components is possible through the application of anti-corrosion coatings, therefore only the part which is at risk to degradation needs to be coated, leading to cost reduction as opposed to other traditional prevention methods
- Material loss and component failures can be prevented with proper selection of coatings/ materials in most manufacturing industries. Adequate knowledge of these processes is therefore inevitable.
- Anti-corrosive coatings produced new materials surfaces with numerous functional properties and unique features

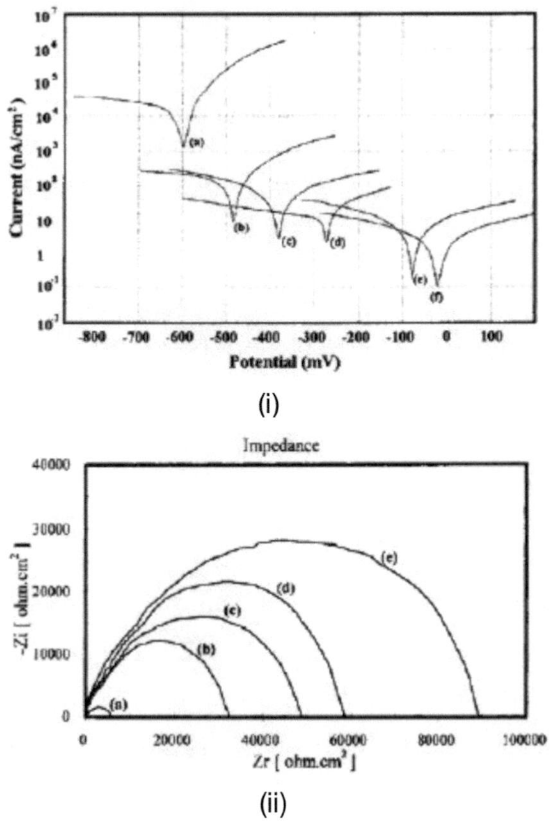

Figure 12. (i) Tafel and (ii) Nyquist plots of five CRS samples in 5 wt% NaCl aqueous solution (a) PMMA-coated, (b) CLMA1-coated, (c) CLMA3-coated, (d) CLMA5-coated, and (e) CLMA10-coated. (Hung et al., 2011)

Author details

Api Popoola[1], OE Olorunniwo[2] and OO Ige[2]

1 Department of Chemical and Metallurgical Engineering, Faculty of Engineering and Build Environment; Tshwane University of Technology, Pretoria, South Africa

2 Department of Materials Science & Engineering, Obafemi Awolowo University, Ile – Ife, Nigeria

References

[1] ABB (2007): Electrodeposited coatings of zinc with passivation in steel, http://www05.abb.com/global/scot/scot235.nsf/veritydisplay/cbd22218b9d065e2c12575460042598b/$file/9adh25_a_en_electrodeposited_coatings_of_zinc_with_passivation_on_steel.pdf, accessed on 11/07/2013

[2] Abboud L. Mamish (2009): Tape coating system for pipeline corrosion protection, http://www.pstc.org/files/public/Mamish09.pdf, accessed on June 28, 2013

[3] Ahmad, Z., and B.J. Abdul Aleem (2009): Erosion-Corrosion Behavior of Plasma-Sprayed Nanostructured Titanium Dioxide Coating in Sodium Chloride-Polystyrene Slurry, CORROSION—Vol. 65, No. 9, 611 – 623

[4] Ahmed, N., and A. G. MacDiarmid (1996): Inhibition of corrosion of steels with the exploitation of conducting polymers, Synth. Met. 78: 103-110

[5] Alexander, M., (1998): 3 Layer Epoxy/Polyethylene. Extruded Coatings for High Temperature Application, Proc. Int. Pipeline Conference Part-2 (2), ASME, Fairfield NJ, USA, 645-652.

[6] ASM International (2000): The effect and economic impact of corrosion, in: Corrosion: Understanding the basics, www.asminternational.org/content/ASM/StoreFiles/06691G_Chapter_1.pdf, accessed on 26/01/2013

[7] Baker, M.A. and Castle, J.E (1993): "The Initiation of Pitting Corrosion at MnS Inclusion", Corrosion Science, Vol 34 (4), 667-682

[8] Barker, R., Hu, X., Neville, A., and Cushnaghan, S. (2012): Inhibition of flow induced corrosion and erosion corrosion for carbon steel pipework from an offshore oil and gas facility, Corrosion, In press

[9] Bernhard Wessling, (1999): Scientific Engineering of Anti-Corrosion Coating Systems based on Organic Metals (Polyaniline), JSCE, Volume 1 Paper 15

[10] Bhadra, S., N. K. Singha, and D. Khastgir (201) Polyaniline based anticorrosive and anti-molding Coating, Journal of Chemical Engineering and Materials Science Vol. 2(1), pp. 1-11

[11] Bhadra, S., Singha, N. K., Lee, J. H., and D. Khastgir (2009): Progress in preparation, processing and applications of polyaniline, Prog. Polym. Sci., 34: 783–810.

[12] Bhadra, S., Singha, N. K., and D. Khastgir (2011): Polyaniline based anticorrosive and anti-molding coating, Journal of Chemical Engineering and Materials Science Vol. 2(1), pp. 1-11

[13] Bibber, J. W. (2009):"Chromium-free conversion coatings for zinc and its alloys, Journal of Applied Surface Finishing, vol. 2, no. 4, pp. 273–275

[14] Birks N., Pettit F.S., Rishel D.M. (1993): Erosion corrosion and wear. Journal de Physique IV, 1993:3. http:11dx.doi.org/10.1051/jp:1993970, accessed on 21/11/2010

[15] Bousser E., Benkahoul M., Martinu L., and Klemberg-Sapieha J.E. (2008): Effect of Microstructure on the erosion resistance of Cr-Si-N coatings, Surface & Coatings Technology, 203, 776 - 780

[16] Brady, B. K. (2006): US20067150918.

[17] Buchheit, R. G., Bode, M. D., and G.E. Stoner (1994): Corrosion-Resistant, Chromate-Free Talc Coatings for Aluminum, Corrosion, Vol. 50, No. 3, 205 – 214

[18] Buchheit, R. G., Mamidipally, S. B., Schmutz, P., and H. Guan (2002): Active Corrosion Protection in Ce-Modified Hydrotalcite Conversion Coatings, Corrosion, Vol. 58, No. 1, 3- 12

[19] Buczek, M. B., Skoog, A. J., and J. A. Murphy (2009): US20097544396

[20] Camalet, J. L., Lacroix, J. C., Aeiyach, S., Ching, K. C., and P. C. Lacaze (1998): Electrosynthesis of adherent polyaniline films on iron and mild steel in aqueous oxalic acid medium, Synth. Met., 93: 133-142.

[21] Castro, Y., Ferrari, B., Moreno, R., and A. Duran (2005): Corrosion behaviour of silica hybrid coatings produced from basic catalysed particulate sols by dipping and EPD, Surface & Coatings Technology, 19, 228– 235

[22] Colangelo, V.J and Heiser, F.A (1974): Analysis of Metallurgical Failures, Wiley- Interscience, New-York.

[23] Correll, D.L. (1998): The role of phosphorus in the eutrophication of receiving waters: A review. J. Environ. Qual. 27:261-266

[24] Creus, J., Brezault, F., Rebere, C., and M. Gadouleau (2006): Synthesis and characterisation of thin cerium oxide coatings elaborated by cathodic electrolytic deposition on steel substrate, Surf. Coat. Technol. 200, 14-15, 4636

[25] Deen, K. M., Ahmad, R., and I.H. Khan (2009): Corrosion Protection Evaluation of Mild Steel Painted Surface by Electrochemical Impedance Spectroscopy, Journal of Quality and Technology Management, Vol V, issue 1, paper 6

[26] Dennis, R. V., Lee, V., Viyannalage, L., Henderson, S. M. and S. Banerjee (2013): Anti-Corrosive Graphene Coatings: An Active—Passive Alternative to Hexavalent Chromium Coatings, http//www.Rit.edu/affiliate/nysp2i/sites/rit.edu.affiliate.nysp2i/files/Banerjee_metal_finishing_final.pdf, accessed on July 1, 2013

[27] Dolan, S. E., and L. R. Carlson (2007): US20077175882

[28] Durodola, B. M., Olugbuyiro, J. A. O., Moshood, S. A., Fayomi, O. S. and A.P.I. Popoola (2011): Study of Influence of Zinc Plated Mild Steel Deterioration in Seawater Environment, Int. J. Electrochem. Sci., 6 5605 – 5616

[29] European Federation of Corrosion Working Party 7: Corrosion Education, 2012: (http://www.efcweb.org/Working+Parties/WP+Corrosion+Education/Definition+of +Corrosion.html)

[30] Effird, K.D., Wright, E.J., Boros, J.A., Hailey, T.G. (1993): Corelation of steel corrosion in pipe flow with jet impingement and rotating cylinder tests, Corrosion, 49, 992 – 1003

[31] Fontana, M.G (1987): Corrosion Engineering, 3rd Ed., Mc Graw-Hill Book Co, Singapore.

[32] Fayomi O. S. I. and A.P.I Popoola (2012): An Investigation of the Properties of Zn Coated Mild Steel, Int. J. Electrochem. Sci., 7 6555 – 6570

[33] Gasem, Z.M. (2013): ME 472: Corrosion engineering 1, ocw.kfupm.edu.sa/ocw.courses/users062/ME4720102/LectureNotes/impact of Corrosion.pdf, accessed on 26/01/2013

[34] Genies, E. M., Boyle, A., Lapkowski, M., and C. Tsintavis (1990): Polyaniline: A historical survey, Synth. Met., 36: 139-182.

[35] Gospodinova, N., and L. Terlemezyan (1998): Conducting polymers prepared by oxidative polymerization, Polyaniline, Prog. Polym. Sci., 23: 1443- 1484.

[36] Guan, H. and R. G. Buchheit (2004): Corrosion protection of aluminium alloy 2024 T3 by vanadate conversion coatings, Corrosion, March 2004, NACE International, Houston, Texas

[37] Guosheng, H. Gu Daming, g., Li Xiangbo, L., and X. Lukuo (2013): Corrosion Behavior of Oxyacetylene Flame Sprayed Zn-Ni Composites Coating with Spray-dried Agglomerating Powders in Natural Seawater, Int. J. Electrochem. Sci., 8 2905 – 2917

[38] Hammer, P., F. C. dos Santos, B. M. Cerrutti, S. H. Pulcinelli and C. V. Santilli (2012).Corrosion Resistant Coatings Based on Organic-Inorganic Hybrids Reinforced

by Carbon Nanotubes, Recent Researches in Corrosion Evaluation and Protection, Prof. Reza Shoja Razavi (Ed.), ISBN: 978-953-307-920-2

[39] Hara, M., Ichino, R., Okido, M., and N. Wadab (2003): AlN formation and enhancement of high-temperature oxidation resistance by plasma-based ion implantation, Surf. Coat. Technol. 169-170, 359-362

[40] Hashem, K. M. E. (2004): Strengthening of Anticorrosion Passivity by Newly Developed Multi-silicon Coatings, Journal of the Chinese Chemical Society, 51, 935-944

[41] Hattori, A. (2008): US20087354652

[42] Hegedus, C. R. (2004): A holistic perspective of coating technology, JCT Research, 1(1),5–19.

[43] Heusler, K. E., Landolt, D., and S. Trasaiti (1989): Electrochemical corrosion nomenclature, Pure & Appl. Chem., Vol. 61, No. 1, pp. 19-22

[44] Hihara, L. H. and K. Kusada (2011): Corrosion of Bare and Coated Al 5052 – H3 and Al 6061 – T6 in Seawater, http://hinmrec.hnei.hawaii.edu/wp-content/uploads/2010/01/Corrosion-of-Aluminum-Alloys-in-Seawater-_-Progress-Report.pdf, accessed on July 7, 2013

[45] Hill, M. (2009): US20097534290

[46] Hunag, Y. and J. Chen (2012): The Development of an Anti-corrosion Wrapping Tape and its Corrosion Protection Effect Evaluation on Mild Steel in Marine Splash Zone, Int. J. Electrochem. Sci., 7, 7121 – 7127

[47] Ige, O.O. (2007): Effects of tin and heat treatment on an alumninum – zinc – magnesium alloy as sacrificial anode in seawater, Unpublished M.Sc Thesis, Obafemi Awolowo University, Nigeria

[48] Ige, O.O., Shittu, M.D., Oluwasegun, K.M., Olorunniwo, O.E. and Umoru, L.E.(2012): Eco-friendly Inhibitors for Erosion-corrosion Mitigation of API-X65 Steel in CO_2 Environment, Ife Journal of Technology, 21 (2) pp 43 – 48

[49] Ige, O. O. (2013): A study of the hydrodynamic effects of erosion corrosion in oil and gas industry, Unpublished PhD Thesis, Obafemi Awolowo University, Nigeria

[50] Iqbal Z., Narasimhan D., Guiheen J. V., and T. Rehg (2005): Corrosion resistant coated fuel cell plate with graphite protective barrier and method of making the same. US Patent 6864007

[51] Iqbal, Z., Narasimhan, D., Guiheen, J. V., and T. Rehg, (2009): US20097482083.

[52] Iroh, J. O., Zhu, Y., Shah, K., Levine, K., Rajagopalan, R., Uyar, T., Donley, M., Mantz, R., Johnson, J., Voevolin, N. N., Balbyshev, V. N., and A. N. Khramov (2003): Electrochemical synthesis: A novel technique for processing multi-functional coatings, Prog. Org. Coat., 47: 365-375

[53] ISO 8044-199: Corrosion of metals and alloys – Basic terms and definitions, ISO/TC 156 Corrosion of metals and alloys, 3rd ed., ISO Publications,

[54] Jiang, X. Zheng, Y.G. and Ke, W. (2005): Effect of flow velocity and entrained sand on inhibitor performances of two inhibitors for CO_2 corrosion of N80 steel in 3% NaCl solution, Corrosion Science 47, 2636-2658

[55] John E. Orth: www.metalsamples.com.

[56] Jones, D.A (1992); Principles and Prevention of Corrosion, Macmillan Publishing Company, New York, p 445-446

[57] Kellner, J.D., Doheny, A.J. and B. B. Patil (1998): A-3 Layer Polyethylene Coating for Plant Applications, Materials Performance, 54, (6), 28-32.

[58] Khaled M. and E. Hashem (2011): Strengthening of Anticorrosion Passivity by Newly Developed Multi-silicon Coatings, Journal of Chemical Engineering and Materials Science Vol. 2(1), pp. 1-11

[59] Kinlen, P. J. (2003): Smart" Anti-Corrosion Coatings Utilizing Conductive Polymers Doped With Inhibitor Molecules, www.electrochem.org/dl/ma/203/pdfs/0322.pdf, accessed on 24/06/2013

[60] Koser, H. J. K. (1980): Corrosion of Metals in Carbon Tetrachloride containing Solvent, Analyst, p. 123

[61] Kramer, T., Gillich, V., and R. Fuchs (2007): US20077182475.

[62] Kriksunov, L. (2007): US20077267869. SRI International (2010): Anti-Corrosion Coatings, http://www.sri.com/sites/default/files/brochures/sri_anticorrosion_coatings.pdf, accessed on July 1, 2013

[63] Krishnamurthy, A., Gadhamshetty, V., Mukherjee, R., Chen, Z., Ren, W., Cheng, H-M, and N. Koratkar (2013): Passivation of microbial corrosion using a graphene coating, Carbon, 56, 45-59

[64] Lee, J. H., Brady, B. K., and R. L.Fuss (2009): US20097494734.

[65] Lamaka, S.V., Zheludkevich, M.L., Yasakau, K.A., Serra, R., Poznyak, S.K., and M.G.S. Ferreira (2007): Nanoporous titania interlayer as reservoir of corrosion inhibitors for coatings with self-healing ability, Prog. Org. Coat., 58, 127 – 135

[66] Latnikova, A. (2012): Polymeric capsules for self-healing anticorrosion coatings, Unpublished PhD thesis, University of Potsdam, Postdam, http://nbn-resolving.de/urn:nbn:de:kobv:517-opus-60432, accessed on June 30, 2013

[67] Lunder, O., Simensen, C., Yu, Y., and K. Nisancioglu (2004): Formation and characterisation of Ti–Zr based conversion layers on AA6060 aluminium, Surf. Coat. Technol. 184, 2-3, 278- 290

[68] Malik, A. U., Ahmad, S., Al-Muaili, I. A. F., Prakash, T. L. and J. O'Hara (1999): Corrosion protection evaluation of some organic coatings in water transmission lines, Technical Report No. TR 3804/APP 95009

[69] Marie Louise Petersen (2006): Anti-corrosive paint systems based on conducting polymers, Unpublished M.Sc Thesis, The Danish Technical University – Marts

[70] Martins, J. I., Bazzaoui, M., Reis, T. C., Bazzaoui, E. A., and L. Martins (2002): Electrosynthesis of homogeneous and adherent polypyrrole coatings on iron and steel electrodes by using a new electrochemical procedure, Synth. Met., 129: 221-228.

[71] Mathazhagan, A. and R. Joseph (2011): Nanotechnology – A new prospective in organic coating – review, International Journal of Chemical Engineering and Applications, Vol. 2 , No. 4, 225 – 237

[72] Meilūnas, E. (2013): Nanotechnology-A New Prospective in Organic Coating – Review, http://www.mesg.nl/wiki/images/2/24/Paper_VGTU_02.pdf, accessed on June 24, 2013

[73] Metallic coated steel (2013): User manual; Flat carbon steel Europe, Luxemborg, http://www.szs.ch/user_content/editor/files/Downloads_Stahlwerkstoffe/metallic %20coated%20steel.pdf, accessed on 25/06/2013

[74] Minevski, Z., Maxey, J., Nelson, C., and C. Eylem (2006): US20067045024.

[75] Montemora, M. F., Simõesa, A. M., and M.G.S. Ferreira, (2002): Composition and corrosion behaviour of galvanised steel treated with rare-earth salts: the effect of the cation, Prog. Org. Coat. 44, 2, 111

[76] Montemora, M. F., Pinto, R., and M.G. S. Ferreira (2009): Chemical composition and corrosion protection of silane films modified with CeO2 nanoparticles Electrochim. Acta 54, 5179.

[77] Natarajan, K. A. (2013): Lecture 1: Corrosion Introduction – Definition and Types; NPTEL course, http://www.nptel.iitm.ac.in/courses/113108051/module1/lecture1.pdf, accessed on June 27, 2013

[78] Nagaraj, B. A., and B. T. Hazel (2007): US20077311940.

[79] Obadele, B. A., Olubambi, P. A., Pityana, S. L.., and A P I Popoola (2011): Evaluation of WC-9Co-4Cr laser surface alloyed coatings on stainless steel, Proceeding of SAIP 2011, the 56th Annual Conference of the South African Institute of Physics, edited by I. Basson and A.E. Botha (University of South Africa, Pretoria, 2011)

[80] Ochonogora, O. F., Meacockb, C., Abdulwahaba, M., Pityanaa, S. and A.P.I. Popoola (2012): Effects of Ti and TiC ceramic powder on laser-cladded Ti–6Al–4V in situ intermetallic composite, Applied Surface Science 263 591–596

[81] Oki, M, J. (2007): Studies on chromium-free conversion coatings on aluminium, Appl. Sci. Environ. Manage. 11, 2, 187

[82] Orman, S. (1976): 'Economic Aspects of Corrosion', ed., by Shreir L. L. in Corrosion, Vol 2, Corrosion Control, Newnes – Butterworths, London, pp 10: 3 – 6.

[83] Phani, A. R., Gammel, F. J., Hack, T., and H. Haefke (2005): Mater. Corros. 56, 2, 77.

[84] Popoola,A. P. I., Fayomi, O. S. I., and O.M Popoola (2012): Comparative Studies of Microstructural, Tribological and Corrosion Properties of Plated Zn and Zn-alloy Coatings, Int. J. Electrochem. Sci., 7 (2012) 4860 – 4870

[85] Reyes M, and Neville A. (2001): Mechanisms of erosion corrosion on a cobalt base alloy and stainless steel UNS S17400 in aggressive slurries, Journal of Materials Engineering and Performance (JMEPEG), 10(6) 723-730

[86] Scully J R, Taylor D. W., Electrochemical Methods of Corrosion Testing, Metals Hand Book Vol 13, 1987

[87] Shchukin, D. G., Zheludkevich, M., Yasakau, K., Lamaka, S., Ferreira, M. G. S., and H. Moehwald (2006): Layer-by-layer assembled nanocontainers for self-healing corrosion protection, Advanced Material, 18, 1672

[88] Shchukin, D. G., and H. Moehwald (2007): Surface-engineered nanocontainers for entrapment of corrosion inhibitors, Advanced Functional Material, 17, 1451-1458

[89] Suzuki, K., and I. Watanabe (2007): US20077291401.

[90] Shi, X., T. A., Nguyen, Suo, Z., Liu, Y., and R. Avci (2009): Effect of nanoparticles on the anticorrosion and mechanical properties of epoxy coating, Surface & Coatings Technology 204 (2009) 237–245

[91] Short, N., Abibsi A., and J. Dennis (1989): Metal finishing, Trans. Inst. Met. Finish, 67, 73.

[92] Shreir, L.L. Jarman, R.A., and Burstein, G.T. (1994): Corrosion, Vol 1 & Vol 2: Metal/ Environmental Reactions, 3rd Edition, Butterworth - Heinemann, Woburn

[93] Sitaram SP, Stoffer JO, OKeefe TJ (1997). Application of conducting polymers in corrosion protection. J. Coat. Technol., 69: 65-69.

[94] Song, Y. K., and F. Mansfeld (2006): Development of a Molybdate–Phosphate–Silane–Silicate (MPSS) coating process for electrogalvanized steel, Corr. Sci. 48, 154.

[95] Stack, M.M. (2002): Mapping tribo-corrosion processes: some new directions for the new millennium, Tribology International, 35, 679-687

[96] Suzuki, K., and I. Watanabe (2007): US20077291401.

[97] Szabo, T., Molnar-Nagy, L., Bognar, J., Nyikos, L., and J. Telegdi (2011): Self-healing microcapsules and slow release microspheres in paints, Progress in Organic Coatings, 72, 52-57

[98] Tallman, D. E., Spinks, G., Dominis, A., and G. G. Wallace (2002): Electroactive conducting polymers for corrosion control, J. Solid State. Electrochem., 6: 73-84.

[99] Tan, C. K., and D. J. Blackwood (2003): Corrosion protection by multi-layered conducting polymer coatings, Corr. Sci., 45: 545-557.

[100] Treacy, G. M., Wilcox, G. D., and M.O.W. Richardson (1999): Behaviour of molybdate-passivated zinc coated steel exposed to corrosive chloride environments, J. Appl. Electrochem. 29, 647 - 654

[101] Tendayi, Z. (2010): "Hydrodynamics on carbon steel erosion-corrosion and inhibitor efficiency in simulated oilfields brines", PhD Thesis, University of Glascow

[102] Thiery, L. and N. Pommier (2004): Hexavalent chromium-free passivation treatments in the automotive industry, http://www.sintef.no/static/mt/norlight/ICEPAM/P10-Pommier_Coventya.pdf, accessed on 11/07/2013

[103] Tomachuk, C. R., Di Sarli, A. R., and C. I. Elsner (2010): Anti-Corrosion Performance of Cr+6-Free Passivating Layers Applied on Electrogalvanized, Materials Sciences and Applications, 1, 202-209

[104] Trethewey, K. R., and Chamberlain, J. (1995): Corrosion for Students of Science and Engineering, 3rd Ed. John Wiley & Sons Inc. New York.

[105] Umoru, L. E. (2001):' Corrosion Studies in a Tar Sand Digester, Unpublished Ph D Thesis, Materials Science and Engineering Dept., O. A. U., Ile – Ife.

[106] Umoru, L. E. (2001): Corrosion Studies in a Tar Sand Digester, Unpublished Ph.D., Thesis, Materials Science and Engineering Dept., O. A. U., Ile – Ife.

[107] Videla H., and L.K. Herrera (2009): Understanding microbial inhibition of corrosion. Beck, F., Michaelis, R., Schloten, F., and B. Zinger (1994): Filmforming electropolymerization of pyrrole on iron in aqueous oxalic-acid, Electrochim. Acta, 39: 229-234.

[108] Vincent, L.D. (1999): Failure Modes of Protective Coating – Who is at Fault, Materials Performance, 38, (4), 38.

[109] Wang, S. (2009): Effect of Oxygen on CO_2 Corrosion of Mild Steel, M.Sc Thesis, Russ College of Engineering and Technology, Ohio University

[110] Wei-I Hung, Kung-Chin Chang, Ya-Han Chang and Jui-Ming Yeh (2011). Advanced Anticorrosive Coatings Prepared from Polymer-Clay Nanocomposite Materials, Advances in Nanocomposites - Synthesis, Characterization and Industrial Applications, Dr. Boreddy Reddy (Ed.), ISBN: 978-953-307-165-7, InTech

[111] Wen, N. T., Lin, C. S., Bai, C.Y., and M.D. Ger, (2008): Structures and characteristics of Cr(III)-based conversion coatings on electrogalvanized steels, Surf. Coat. Technol. 203, 317

[112] Wessling, B. (1999): Scientific Engineering of Anti-Corrosion Coating Systems based on Organic Metals (Polyaniline), The Journal of Corrosion Science and Engineering, Vol 1, Paper 15

[113] West, J.M. (1986): Basic Corrosion and Oxidation, 2nd Ed., Ellis Horwood, p. 312 – 337

[114] Wood, R.J.K., and Hutton, S.P. (1990): The synergistic effect of erosion-corrosion: trends in published results, Wear, 140, 387 – 394

[115] Wu G., Sun L., Dai W., Song L., and A. Wang (2010): Influence of interlayers on corrosion resistance of diamond-like carbon coating on magnesium alloy, Surface & Coatings Technology, 204, 2193–2196

[116] Zarras, P., Anderson, N., Webber, C., Irvin, D. J., Irvin, J. A., Guenthner, A., and J. D. Stenger-Smith (2003): Progress in using conductive polymers as corrosion-inhibiting coatings, Rad. Phys. Chem., 68: 387-394.

[117] Zhang, T and D. (2009): Tang Current Research Status of Corrosion Resistant Coatings, Recent Patents on Corrosion Science, 1, 1-5

[118] Zhu, D. Q., and W.J. van Ooij (2004): Enhanced corrosion resistance of AA 2024-T3 and hot-dip galvanized steel using a mixture of bis-[triethoxysilylpropyl]tetrasulfide and bis-[trimethoxysilylpropyl]amine, Electrochim. Acta 49, 1113 - 1125

Chapter 13

Hybrid Conducting Nanocomposites Coatings for Corrosion Protection

M. Federica De Riccardis and Virginia Martina

Additional information is available at the end of the chapter

http://dx.doi.org/10.5772/57278

1. Introduction

In the last years, conducting polymers (CPs) have arisen wide interest as electrode materials and they have reached a state of development that allow their use in various applications such as sensors, membranes for the separation of gas mixtures, corrosion protection, and so forth.

Recently, several studies in the fields of new materials have introduced the possibility to use CPs as suitable matrices to disperse nanostructured elements, such as nanoparticles or nanotubes. It has been shown that the introduction of nanostructures into a polymer matrix can improve the electrical conductivity and the mechanical properties of the original polymer matrix. The combination of conductive polymers with conductive particles (better if carbon nanostructures), producing a new class of materials known as *hybrid conducting nanocomposites*, has already shown some synergistic properties, with a variety of applications in the energy field.

The most interesting property of conducting polymers is their high (almost metallic) conductivity, which can be changed by simple oxidation or reduction, and also by bringing the material into contact with different compounds. Conducting polymers usually have a good corrosion stability when in contact with solution or/and in the dry state. For instance, polyaniline is stable in its leucoemeraldine and emeraldine states, even in 10 mol/l acid solutions. Furthermore, CPs can be deposited from a liquid phase, even in complex topographies. Redox processes combined with the intercalation of anions or cations can therefore be used to switch the chemical, optical, electrical, magnetic, mechanic and ionic properties of such polymers. These properties can be modified by varying the anion size and preparation techniques or by including other chemical species. As regards the preparation, CPs can be synthesised chemically or through electrochemical polymerisation. The electrochemical method is used more often because it is environmental friendly and presents several advantages. In fact polymerisation media can be used repeatedly, and polymerisation rate can be controlled by varying potential values.

Among the CPs, polyaniline (PANI) has been one of the most studied because of its facile synthesis, electrical conductivity, low cost and environmental stability. It is characterised by different domains of conductivities, which refer to different oxidation states of the polymer. Although the synthesis of a PANI coating is direct and controllable, particular care has to be taken on choosing the proper polymerisation solution and electrochemical parameters, which affect the growth and properties of PANI. Moreover, PANI has good anticorrosion properties, depending on oxidation states, but the mechanism underlying the protective behaviour of PANI is still not completely clear. Most studies agree that PANI films have an active role in keeping the passivity of the metal substrate in acid solution, whereas others suggest the non conductive state of PANI performs better than that of a conductive one.

Recently CPs have been also used as the matrix of hybrid conducting nanocomposites containing carbon nanotubes (CNTs). Many efforts have been made to optimise the preparation of CPs-CNTs composites and different interaction mechanisms have been hypothesised and discussed in literature. Recent studies have shown that PANI could be used to functionalise and solubilise CNTs via formation of donor-acceptor complexes. Moreover, the possible presence of functional groups on the CNTs' surface can favour a chemical interaction of both the monomer and the polymer during its generation. Whatever interaction theory is adopted, the combination of PANI and CNTs has surely favorable effects on properties, also on corrosion protection.

This contribution will be divided in four sections. In the first part, conjugated polymers and their conducting mechanism will be treated. Then, a review of conjugated polymers applied to corrosion protection will be reported. Moreover, the proposed anticorrosion mechanisms occurring in conducting polymers and their composites will be reported by considering the most recent literature. In the third part, two methods used by the authors to obtain PANI-CNTs composite films will be described. The first method consists of an electropolymerisation process (EP) of ANI monomer during which the incorporation of CNTs in PANI matrix occurs. In the second method, PANI chains are treated to be able to encapsulate CNTs and so to make possible the formation of a nanocomposite film by Electrophoretic Deposition (EPD). Finally in the last part, the properties of the hybrid conducting nanocomposite coatings produced by both the methods, will be reported and discussed, also by comparing them with similar materials prepared by usual methods.

2. Conducting polymers: properties and synthesis mechanisms

Over the last twenty five years, the study and applications of conducting and electro-active conjugated polymers have reached a development state that enabled their use in various fields of technology, in form of electronic and optical devices, rechargeable batteries, sensors and biosensors, anticorrosion coatings, solar cells, etc. [1-3].

Conducting (or conjugated) polymers possess strong appeal as modifiers for electrode surfaces due to their electrical conductivity when doped by oxidant or reducing agents. This property is due to an extended π-electron system, over which electrons can be delocalised.

2.1. Conducting mechanism in conjugated polymers

A key requirement for a polymer to become electrically conducting is an overlap of molecular orbitals suitable to allow the formation of delocalised molecular wave functions. Besides this, molecular orbitals should be only partially filled, so that a free movement of electrons through the lattice is possible. In particular, a π-bonding, in which carbon orbitals are in the sp^2p_z configuration and the orbitals of the successive carbon atom along the backbone overlap, leads to electron delocalisation along the backbone of the polymer. The π-electron systems is composed by single and double bonds alternating along the polymer chain, and therefore responsible for electronic properties unusual for polymers, such as electrical conductivity, low ionisation potential, and high electron affinity. Thanks to their electronic configuration, CPs show a slight conductivity even in the neutral state but, when suitably doped, *i.e.* either partially oxidised or reduced, mobile charge carriers are generated and the electronic conductivity results to be strongly enhanced.

The conductivity of these materials can be explained by the "band theory". According to this view, polymerisation causes the highest occupied molecular orbital (HOMO) and the lowest unoccupied molecular orbital (LUMO) of each structural unit to form, altogether, π and π* bands develop (Figure 1). In the terminology of solid-state physics these are the valence and conduction bands, respectively.

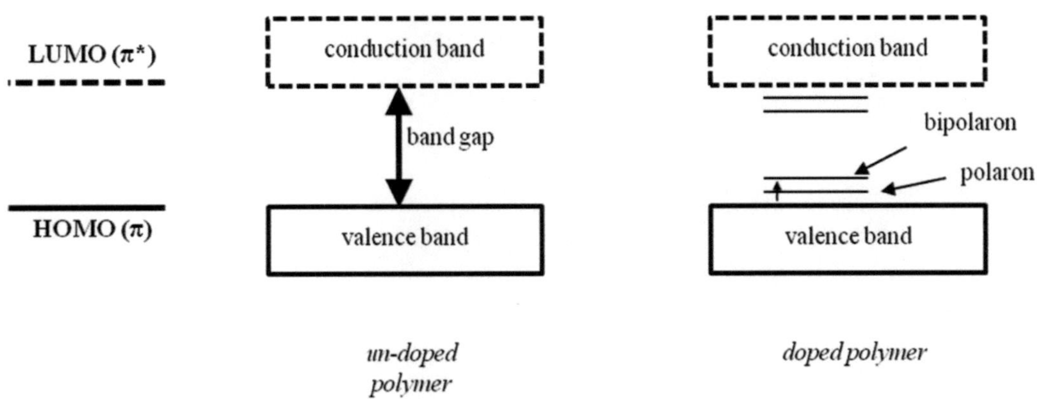

Figure 1. Schematic diagram of the evolution of the band structure of a CP.

In the neutral form, the valence band is filled and the conduction band is empty, the band-gap being typically 2-3 eV, which permits a low intrinsic conductivity. In simple terms, p-doping (by oxidation) can be viewed as the creation of 'mobile' holes in the valence band, and n-doping (by reduction) as the addition of 'mobile' electrons to the conduction band. However, these modification actually change the band structure, creating various midgap electronic states. By p-doping, the removal of one electron from a segment of the chain creates a radical cation, i.e. a mobile polaron, generally delocalised over four to five structural units. The removal of a second electron creates a dication, so the combination of two polarons generates a mobile bipolaron. The formation of these charge carriers causes local distortions in the geometry of the chain and creates electronic states energetically located above the valence band. Similar valence and conduction bands arise when the polymer is n-doped [4-5].

All CPs, e.g. Polypyrrole, Polythiophene, Polyfuran, Polyaniline, Poly(phenylenevinylene), etc. (Figure 2) and most of their derivatives, undergo either p- and/or n-redox doping process during which the number of electrons associated with the polymer backbone changes. The electrical conductivity results from the existence of charge carriers (through doping) and their ability to move along the π-bonded "highway". Reversible doping of CPs, with associated control of the electrical conductivity over the full range from insulator to metal, can be accomplished either by chemical or electrochemical doping.

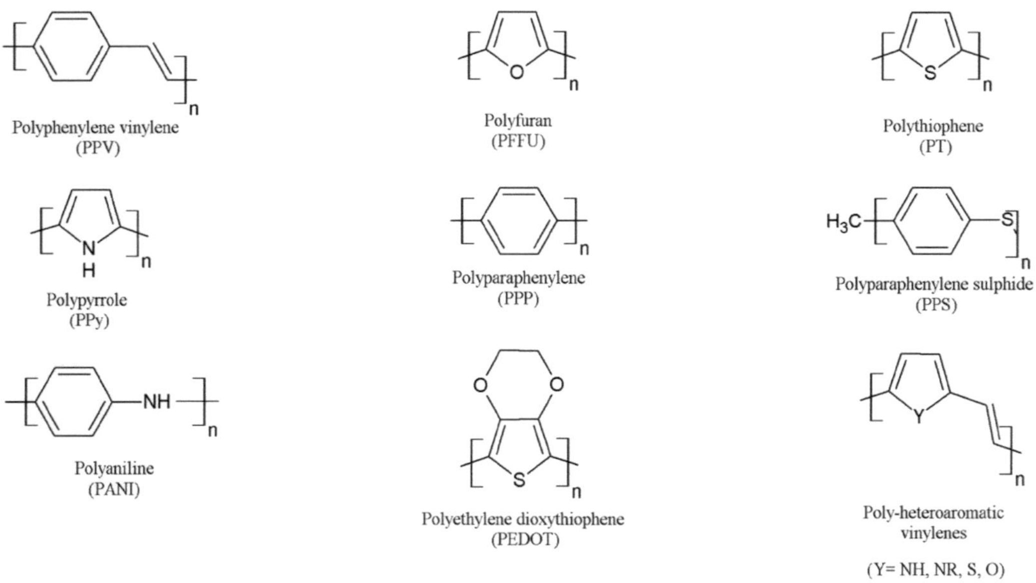

Figure 2. Chemical structures of some conjugated polymers.

2.2. Conducting polymers on electrodes – Electropolymerisation

To modify electrode surfaces, the CPs can be conveniently produced directly onto the electrode through an electrochemical polymerisation process. Various mechanisms have been proposed to explain the growth of CPs at electrodes. The first one was initially proposed by Geniès et al. [6] and finally proved by Andrieux et al. about 10 years later [7], consisting of an oxidative coupling mechanism.

Owing to the chemical diversity of compounds studied, a general scheme cannot be provided. However, it has been shown that the first step is the oxidation of monomers to form radical cations, electrochemical step (*E*). The second step is of chemical nature (C) and involves the coupling of two radicals to produce a di-hydro dimer dication. This leads to a neutral dimer by loss of two protons and concomitant re-aromatisation of the system, which constitutes the driving force of the whole chemical step. The oligomers produced are more easily oxidised than the starting monomer, so that they are converted into the relevant radicalic cationic form (*E*) and couple either with each other (C), or with monomer radical cations (C). Polymerisation proceeds then through successive electrochemical and chemical steps according to an $E(CE)_n$ mechanism, until the oligomer becomes insoluble in the electrolytic medium and precipitates onto the electrode surface (Figure 3) [8].

The chemical polymerisation proceeds similarly but it is commonly induced by a chemical oxidant, such as $FeCl_3$.

$$M \longrightarrow M^{+\cdot} + e^- \qquad (E)$$

$$M^{+\cdot} + M^{+\cdot} \longrightarrow M + 2H^+ \qquad (C)$$

$$M{-}M \longrightarrow M{-}M^{+\cdot} + e^- \qquad (E)$$

$$M{-}M^{+\cdot} + M^{+\cdot} \longrightarrow M{-}M{-}M \qquad (C)$$

etc.

Figure 3. Mechanism of the first steps of the electrooxidation of electronically CPs.

Advantages of the electrochemical polymerisation over the chemical one, consists in reaction proceeding at room temperature and in controlling carefully the reaction rate. The electrode potential at which the monomer and the subsequently generated oligomers are oxidised is high enough to induce polymer p-doping. The coating progressively deposited becomes conductive and additional monomers in the solution are oxidised as well, so that the growth goes on. The thickness of the deposit can be more or less roughly controlled from few nanometers to more than 0.1 mm [9] by varying the experimental conditions, e.g. deposition potential and time, current intensity, number of cycles and potential scan rate, depending on technique chosen.

Cyclic voltammetry is most commonly used as a first approach to the study of the polymerisation process of a monomer, since it allows to 'follow' the different stages of the process. During cyclic voltammetry, the potential is varied linearly from a value at which no electrode reaction takes place to a value at which monomer oxidation and polymer deposition-doping occur. Reversing the scan to negative potentials, the discharge, i.e. the neutralisation of the p-doped polymer on the electrode, is observed in correspondence to a backward cathodic peak. In the following scans, the occurrence of an anodic/cathodic peak system at less anodic potentials than the monomer oxidation, growing in height scan by scan, can be observed. This has to be ascribed to the charge/discharge of the polymer coating progressively formed on the electrode surface. Indeed, the occurrence of the polymerisation and film deposition is indicated by the increasing peak currents relative to oxidation/neutralisation of the polymer on subsequent cycles.

Electrodeposition of polymers by potential step techniques leads to a well-defined chronoamperometric response, showing a characteristic rising current-time transient in the initial stage, in most cases followed by a decay to an approximately constant value of the current. On a first approximation, charge collected at the potential at which monomer oxidation occurs is directly proportional to the amount of deposited polymer. At the same potential the polymer is oxidised and, therefore becomes conductive, allowing the current to flow and the deposit to grow.

Electrogeneration of CPs can be also performed through galvanostatic method. In this case, polymer formation and deposition occur by applying a constant current over a pre-fixed time length. The potential is concurrently modulated by an electrochemical apparatus. The coatings obtained through this technique are more homogeneous, especially when very low current densities are applied. Also in this case, the total charge spent in the electrochemical process is directly related to the amount of polymer chains deposited, i.e. with the thickness of the resulting film, reasonably hypothesising that the yield of polymer formation is constant over time.

2.3. Conducting polymers and their electrochemical doping — The polyanilines

The electrochemical doping of CPs was discovered by the MacDiarmid and co-workers in 1980, opening a new scientific direction [10]. The doping process definitely distinguishes CPs from all other types of polymers. Usually, an organic polymer, either an insulator or semiconductor, has low conductivity (10^{-10} - 10^{-5} S/cm). The doping process reversibly converts a polymer from insulator, or semiconductor, into a 'metallic' conducting regime (~1-10^{4} S/cm), with little or no degradation of the polymer backbone. Both the doping and the un-doping processes involve dopant counterions which stabilise the charge on the polymer backbone.

Among the CPs, Polyaniline (PANI) began to be investigated extensively some decades ago and attracted interest as a conducting material because of its low cost, straightforward synthesis that proceeds with high yield, environmental stability, and electrical conductivity. It is characterised by a relatively wide potential stability, a reproducible synthesis and a well-behaved electrochemistry showing different domains of conductivities, which refer to different oxidation states of the polymer.

The generalised formula of the base form of PANI consists of alternating reduced and oxidised repeat unit (Figure 4). The terms *'leucoemeraldine'*, *'emeraldine'* and *'pernigraniline'* refer to the different oxidation states of the polymer where y = 1, 0.5, and 0, respectively. PANI can be rendered conducting through two independent ways: oxidation of the leucoemeraldine base, or protonation of the emeraldine base. Depending on the oxidation state and the degree of protonation, PANI can be either an insulator or a conductor with different conductivity [11-12].

http://dx.doi.org/10.5772/57278

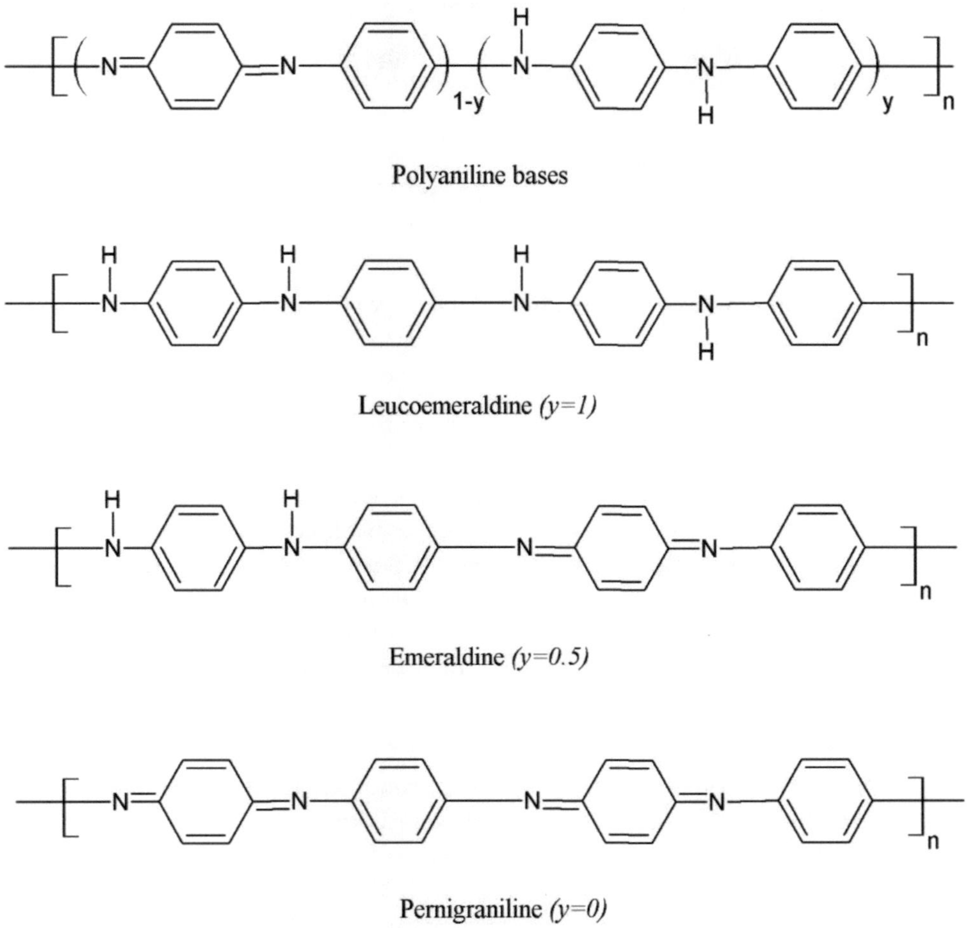

Figure 4. Different forms of PANI.

As an electrochromic polymer, PANI is one of the most promising materials because its colour can be controlled electrically. However, the electrochemical deposited PANI exhibits a significant resistivity, which is attributed to the lack of conducting pathways at the nanoscale associated with random deposition morphology. PANI is also characterised by an appreciable supercapacitive behaviour, although its mechanical stability is not so much good.

As mentioned before, PANI has shown to have good anticorrosion properties, also in controlling pitting corrosion resulting from the permeation and breakdown of the protective coating. In recent years, several methodologies have been proposed for the application of PANI coatings: (i) as a primer alone; (ii) as a primer coating with conventional topcoats or with other CPs films; (iii) blended with conventional polymer coatings, such as epoxy or polyurethane; (iv) as an anticorrosive additive to the paint formulation; (v) as a matrix where nanoparticles or nanostructures materials are incorporated.

3. Applications of conducting polymers as corrosion protective coatings

Since the first reported publications, the discussion on the potential of conducting polymers as promising candidates for a new generation of corrosion control coatings has opened. Although about more than a thousand publications have been published on this topic, their technical application has not yet made its breakthrough. The reason for this is that most coatings based on conducting polymers fail so far under practical corrosion conditions, as their efficacy depends on how they are applied.

In the recent years many papers have reported the use of CPs as materials for anti-corrosion coatings, applied on both ferrous and non ferrous metals, providing a potential cheap alternative to chromium and phosphate treatments and their pollution.

It is well known the corrosion process involves one or more oxidation reactions (affecting the metal under corrosion) and one or more reduction reactions (affecting O_2 or H ions and/or water). Both the reactions occur at the metal surface where a corrosion cell is established. To reduce the corrosion rate it is necessary to control the dynamics of the corrosion process, for example applying one or more coatings on the metal. Such a coating acts as a barrier between the metal and its environment, slowing down the rate at which water, oxygen, or ions from the environment reach the metal surface. Moreover, the coating can act as an active layer when it consists of or contains a material that can interact chemically or electrochemically with the metal, modifying the corrosion process or its rate.

The corrosion protection of metals can be done by a CP coating, such as highly conjugated organic polymer with some degree of electronic conductivity. As mentioned above, the conductivity of these polymers at the neutral state is typically quite low, ranging from insulating to semiconducting. After a partial oxidation (often referred to as p-doping), these polymers become highly conductive, as a metal. The partial reduction (referred as to n-doping) is possible although hard to maintain, especially in air. However, the use of n-doped conjugated polymers in corrosion protection is not diffused.

There are some possible interactions between an active metal and a CP. As the oxidized form of the polymer is electronically conductive, a non-redox interaction occurs when the metal is brought into electrical contact with the CP. In this case the electron moves from the metal, that is more active, into the CP, until the Fermi energy of the two phases are equal.

In the presence of an aerated electrolyte, the metal and the CP becomes a galvanic couple, where oxygen reduction can occur at the CP surface. In addition, an electron transfer from the metal (which becomes oxidized) to the CP (which becomes reduced) might be expected. In this case the oxidation capacity of the bulk polymer is significant. The electronic interactions between metal and CP depend on the quality of the mutual contact, influenced also by the eventual presence of an oxide layer interposed between them. On the other hand, the chemical interactions between metal and polymer or dopant anions occurring at the interface are equally important, because CP can assume the role of inhibitor coating.

Since the interactions between CP and metal are numerous and complex, it is hard to optimise coatings based on CPs that might successfully act for corrosion protection. So, initially in this part of chapter, the proposed corrosion mechanisms occurring when CPs coat a metal will be reviewed. In literature several cases of ferrous and non ferrous metals coated by mono- and bi-layers based on CPs and their composites are reported. Our efforts will be devoted to report the most recent results, with particular attention to polyaniline, because of its potential practical applications. It is worth noting that a wide range of experimental methods were used to investigate the corrosion protection behaviour, it is therefore difficult to estimate the correct mechanism, also because opposite theories were depicted on the basis of experimental results obtained by different procedures. For each kind of structures based on CPs, the main data are reported in summarising tables, placed at the end of each sections (Tables 1-5).

3.1. CPs for protective mono-layer coatings

The first anticorrosion coatings based on CPs were made of pure polymers, in doped or undoped form. Their ability to transport and store charge has been believed to be the principal reason for their reliability to anodically protect metals against high corrosion rates. However the use of conducting polymers on active metals has been not common because of high positive potential required to form the polymers by electrochemical methods. For instance, to polymerize aniline potentials of the order of 1 V are needed. At these potentials, the most part of metals are corroded too rapidly to allow the formation of a conducting polymer on their surface. For this reason, the first CPs coatings were applied on inert metals, such as stainless steels, and it was only in the second half of the 1990s that PPy and PANI started to be grown on aluminum and iron electrodes.

The most ascertained mechanism of corrosion protection by a pure CP layer is based on two aspects: the physical barrier effect and the anodic protection. The first one is similar to that produced by a paint coating which inhibits the corroding substance from penetrating the substrate. So regarding the anodic protection, the conducting polymer works as an oxidant for the metal substrate, shifting the potential towards the passive state. However, these two aspects are not completely distinct, since in the presence of a CP coating the maximum current in the active-passive transition is limited by the barrier effect, and then the potential can be easily shifted to a higher potential in the passive state by the strongly oxidative property of the conducting polymer. Therefore, both the barrier effect and the oxidative capacity induce the anodic protection of the sustaining substrate.

An experimental evidence of this affirmation was given by Fang et al. [13] who described very clearly the mechanism of corrosion protection of PANI coating on stainless steel (SS). Moreover they demonstrated the mechanism of coating failure by means of experimental results. They affirmed that the electropolymerisation of PANI on SS substrate favours the passivation of the metal, that is held by the emeraldine state of PANI. Therefore, the corrosion protection ability of the PANI is mainly due to two factors: the positive potential of the emeraldine (EM) state of PANI, at which the SS substrate can be passivated, and the physical distance existing between metal substrate and environment.

Unfortunately, the electrodeposited PANI coating is intrinsically porous. This fact favours the access of corrosive species to the SS surface, where they react and produce some corrosion products. Then the adherence of the PANI coating to the substrate is worsened. The evidence of the formation of an oxide layer at the interface between metal and CP coating was demonstrated by Hermas [14] by means of XPS analysis on SS after peeling off two different polymer coatings made of PANI and PoPD (Poly(o-phenylenediamine)). The passive films formed on SS during anodic polarization in a sulphuric acid solution is thinner than those formed on coated stainless steel under the same exposure condition. The polymer layer enhances the enrichment of chromium and nickel in the entire passive oxide, forming a more protective film than that formed during anodic polarization. Moreover, the type of the polymer influences the composition of the passive film. Indeed the amount of chromium, the main passivating constituent, is increased in the oxide film in the order: anodically passivated SS << PANI-SS <PoPD-SS, indicating that the conductive polymers induce a good passivation of the stainless steels, and that the best passivation is obtained by PoPD film. The passive films, particularly the outer layer of modified steels, contain a very low amount of water and sulphate species, revealing also a good resistance to pitting corrosion.

A scheme proposed for the corrosion inhibition of PANI on active metals is related to O_2 oxidising PANI to a partially oxidised and conducting form. The oxidised polymer acts as a catalyst between O_2 and the metal surface, removing electrons from the metal substrate and forming a passive oxide layer. Some evidence that this situation was not stable was reported. Kilmartin [15] used PANI and Poly(methoxyaniline) (also known as Poly-anisidine) to protect SS. He retained that PANI cannot remain in a partially oxidised state on pure iron in acid solutions, since the rate at which electrons are passed to PANI due to metal corrosion is greater than the rate at which the conducting polymer can be oxidised by O_2. However, even in a predominantly reduced state, any positively charged Polyaniline units can eventually pass their charge to substrate, forming a passivating iron oxide layer which serves to lower the rate of corrosion. In the presence of chloride, with higher rates of corrosion, the charge generated within the polymer film by the action of O_2 is not always produced rapidly enough to repair the passive film at the rate at which it is being broken down under the influence of chloride ions. In this case, the more catalytic material is available to passivate the metal, and the longer the system remains protected prior to the onset of pitting corrosion. Therefore there is a competition between Polyaniline being oxidised by dissolved O_2 and the underlying metal using up the Polyaniline charge to maintain a passive oxide film (Figure 5).

The majority of studies on corrosion protection by CPs reports on the doped form of the polymer, where an anion (dopant ion) is incorporated on oxidation, and is released upon reduction. The dopant ion has an important role in the control of corrosion mechanisms. When metal oxidation and polymer reduction result in a damage to the coating, a "special" dopant ion with inhibition function can be released by the coating and stops or slows down the corrosion [16] (Figure 6). If the anion is released as a consequence of anion exchange process, it can work as a physical barrier to prevent penetration of aggressive ions or to inhibit oxygen reduction.

Dominis et al. [17] studied how the type of the dopant anion used in the emeraldine salt (ES) form of PANI affects the corrosion rate of carbon steel. The differences in corrosion rate induced by different dopants used with the ES appear to be correlated with the galvanic activity between ES and carbon steel. Their results indicated that the coupling with PANI–ES

http://dx.doi.org/10.5772/57278

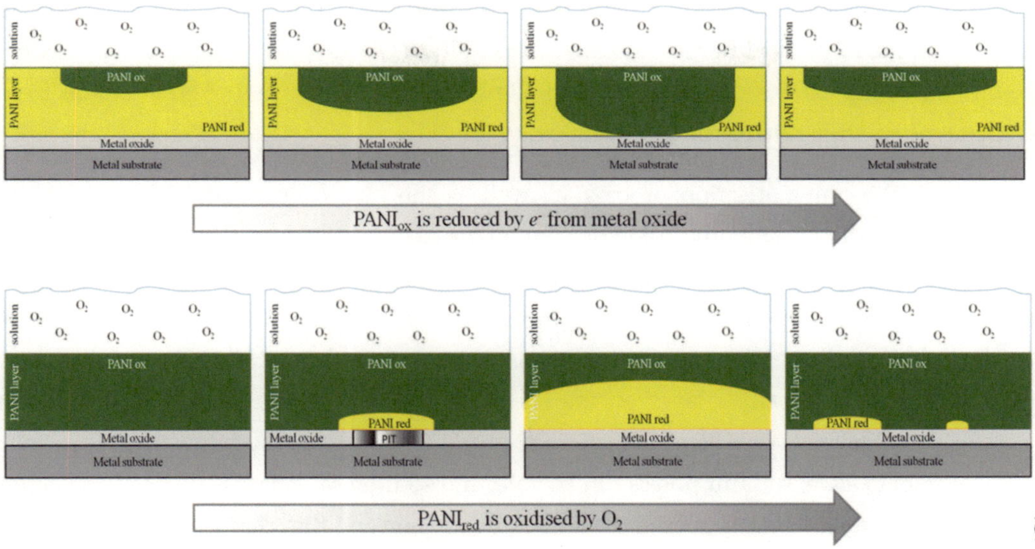

Figure 5. Schemes proposed for the corrosion inhibition of Polyaniline on active metals in chloride solutions.

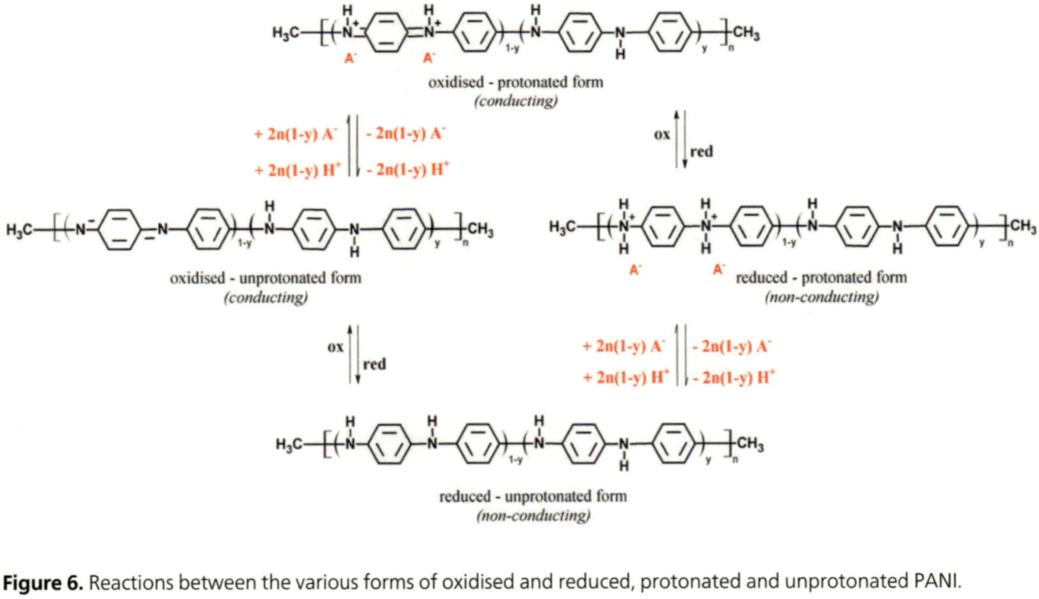

Figure 6. Reactions between the various forms of oxidised and reduced, protonated and unprotonated PANI.

does not passivate the steel under the test conditions used. Instead the steel behaves like a non-passivating metal with faster reduction of the polymer corresponding with a higher oxidation rate of the steel.

Different kinds of dopants, including organic and inorganic ions, have been used, often incorporated in the CP during its electrochemical synthesis. Among organic anions, doped aliphatic and aromatic sulfonates provide good electrical and mechanical properties and good stability for the PPy layer. Inorganic ions, such as ClO^-_4 or Cl^-, have also used as dopant in PPy, being easily released during reduction, and allowing the incorporation of other anions from an aqueous solution during the following oxidation [18].

In general, the polymerisation process, as well as the morphology and the adhesion to substrate, are influenced by the anion present on the electrode surface, the redox potential of the anion, and the ionic charge. The size of the counterions plays a particularly important role in the exchange of doped anions. In fact, the protection mechanism by dopant ions release works if the mobility of the anion is high enough to migrate through the CP film in a short time and to form a passive layer on the corroding metal [19]. In reference [20] different electrolytes in deposition of PPy films on oxidizable metals, such as copper, were used. Oxalic acid, sodium oxalate, sodium/potassium tartrate, and sodium salicylate were used as electrolytes and compared in terms of easiness for PPy deposition and film characteristics. Copper is efficiently passivated before Polypyrrole electrodeposition and a thin passivating layer is formed prior to the pyrrole electropolymerisation and prevents copper corrosion without inhibiting the polymer formation.

Mono-layer coatings based on CPs				
Type of polymer coating (s) and substrate	**Method of coating deposition**	**Method of corrosion tests**	**REF.**	**Results**
PANI on Zn-Co plated carbon steel	Cyclic voltammetry	EIS and Anodic polarisation in 3.5% NaCl	[85]	The corrosion potential for PANI on ZnCo plated steel shifts to more positive values with respect to uncoated steel and ZnCo coated steel. The polarisation resistance is maximum for PANI/ZnCo/steel and minimum for uncoated steel.
PANI on Al	Galvanostatic polarisation	Tafel polarisation and EIS in 1% NaCl	[89]	The pure PANI coating has a poor corrosion resistant behaviour due to galvanic action of PANI. A post – treatment in a Cerium salt solution increases the PANI coating performance.
Poly(aniline-co-m-amino benzoic acid) on steel	Cyclic voltammetry	Tafel polarisation and EIS in 1N HCl	[90]	The protection efficiency of copolymer coating is three times higher than that of pure PANI.
Poly(N-ethylaniline) (PNEA) on copper	Cyclic voltammetry	Potentiodynamic polarisation and EIS in 0.1M H_2SO_4	[91]	The corrosion rate of Cu is reduced by PNEA coatings electrodeposited at the lowest upper potential limit for shorter periods, by PNEA coatings electrodeposited at the highest upper potential limit for longer periods.
PPy on Al alloys	Galvanostatic activation and Galvanostatic deposition	Single-cycle anodic polarisation and subsequent repassivation Monitoring of OCP in 0.6M NaCl.	[92]	The nature of the substrate influences the structure and morphology of PPy films. The performance of PPy films depends on the corrosion test used and by the structural characteristic of films.

Mono-layer coatings based on CPs				
Type of polymer coating (s) and substrate	**Method of coating deposition**	**Method of corrosion tests**	**REF.**	**Results**
Poly (o-ethyl aniline) (POEA) on copper	Cyclic voltammetry	Potentiodynamic polarisation and EIS in 3% NaCl	[93]	The corrosion rate of POEA coated Cu is about 70 times lower than that of uncoated Cu.
PNEA on steel	Cyclic voltammetry	Tafel polarisation in 1M H_2SO_4	[94]	The PNEA monolayer coating is less efficient than the bilayer coatings based on PNEA and PPy (see Table 2 – ref [94]).
PANI on steel	Cyclic voltammetry	Potentiodynamic polarisation in 0.2M H_2SO_4	[13]	The corrosion protection ability of PANI increases with the increase of the thickness.
	Cyclic voltammetry	EIS and monitoring of OCP in 0.5M H_2SO_4 and 0.5M HCl	[15]	The corrosion protective behaviour of PANI film depends on its oxidation state. The OCP oscillates after some hours of exposition to acid environment meaning a change from reduced to oxidised state of PANI.
	Cyclic voltammetry	Tafel polarisation in 1M H_2SO_4	[21]	The corrosion performance of monolayer PANI coating is worse than that of bilayer PPy/PANI and PANI/PPy coatings (see Table 2 – ref [21]).
	Cyclic voltammetry	Potentiodynamic polarisation. EIS and monitoring OCP in 3% NaCl	[32]	The single PANI layer coating provides a worse protection than the bilayer PANI+POA coatings (see Table 2 – ref [32]). The corrosion rate of PANI coated steel is 35 times lower than that of uncoated steel.
	Galvanostatic deposition	Tafel polarisation in 3.5% NaCl	[95]	The corrosion rate of steel increases when the concentration of Aniline in the electrolytic solution increases from 0.1 to 0.4 M.
	Oxidative polymerisation	Potentiodynamic polarisation in 3.5% NaCl	[47]	The corrosion current on steel coated by PANI is better than uncoated steel but it is worse than PANI+TiO_2 coatings (see Table 3 – ref [47, 48]).
	Cyclic voltammetry	Potentiodynamic polarisation and EIS in 0.5M NaCl	[96]	PANI acts as a worse protective layer than PoPD on steel against pitting corrosion.
	Oxidative polymerisation	Tafel polarisation and EIS in 3.5% NaCl	[97]	The corrosion current on PANI coated steel is 20 times lower than that on uncoated steel. The corrosion potential PANI coated steel is more positive of 300 mV than the uncoated steel.
	Cyclic voltammetry	Tafel polarisation in 0.1M HCl EIS in 0.5M NaCl, 0.1M HCl Potentiodynamic polarisation in 0.5M H_2SO_4	[111]	All corrosion tests show the worst performance of PANI single coating with respect the single PPy coating, PANI/PPy and PPy/PANI bilayer coatings (see Table 2 – ref [111]).
PPy on steel	Cyclic voltammetry	Tafel polarisation in 1M H_2SO_4	[21]	The corrosion performance of monolayer PPy coating is worse than that of bilayer PPy/PANI and PANI/PPy coating (see Table 2 – ref [21]).
	Galvanostatic deposition	Potentiodynamic polarisation, monitoring OCP, and EIS in 3.5% NaCl	[36]	PPy coating has worse anticorrosion properties than PPy + ZnO coating with 10 wt% of ZnO relative to PPy (see Table 3 – ref [36]).

Mono-layer coatings based on CPs				
Type of polymer coating (s) and substrate	**Method of coating deposition**	**Method of corrosion tests**	**REF.**	**Results**
	Cyclic voltammetry	Tafel polarisation in 0.1M HCl EIS in 0.5M NaCl, 0.1M HCl Potentiodynamic polarisation in 0.5M H_2SO_4	[111]	All corrosion tests show the best performance of PPy single coating with respect to the single PANI coating, and PANI/PPy and PPy/PANI bilayer coatings.
Poly(o-anisidine) (POA) on steel	Cyclic voltammetry	Potentiodynamic polarisation, EIS and monitoring OCP in 3% NaCl	[32]	The single POA layer coating provides a worse protection than the bilayer PANI + POA coatings (see Table 2 – ref [32]). The corrosion rate of PANI coated steel is 9 times lower than that of uncoated steel.
Poly(o-methoxyaniline) on steel	Cyclic voltammetry	EIS in 0.5M H_2SO_4	[15]	The corrosion protective behaviour of Poly(o-methoxyaniline) film depend on its oxidation state. The OCP oscillates after some hours of exposition to acid environment meaning a change in its oxidation state.
Poly (o-toluidine) (POT) on steel	Cyclic voltammetry	Potentiodynamic polarisation in 3% NaCl	[98]	The incorporation of CdO nanoparticles improves the corrosion protection of POT coating (see Table 3 – ref [98]).
PANI on iron	Oxidative polymerisation	Tafel polarisation and Monitoring OCP in 3.5% NaCl	[99]	The corrosion current on iron coated by PANI coating is almost 10 times lower than the uncoated iron
	Casting	Tafel polarisation in 1M H_2SO_4, 1M HCl and 3.5% NaCl	[100]	The corrosion protection of pure aniline is worse than the composite layer containing PANI and Clinoptilolite (see Table 3 – ref [100]).
Poly(o-phenylenediamine) (PoPD) on steel	Cyclic voltammetry	Potentiodynamic polarisation and EIS in 0.5M NaCl	[96]	PoPD acts as a better protective layer than PANI on steel against pitting corrosion.
	Cyclic voltammetry	Monitoring OCP in 3% NaCl	[101]	OCP is positive and increases gradually in time.

Table 1. Some examples of single layer coatings based on CPs for corrosion protection.

3.2. Layered structures based on CPs as protective coatings

A lot of attention has been devoted to study and improve the performances of coatings based on pure CP (doped or undoped), but their application to corrosion protection is subject to some limitations. There are meanly two reasons: the charge stored in the polymer layer (used to oxidise base metal and to produce passive layer) can be irreversibly consumed during the system redox reactions, and some intrinsic porosity cannot be avoided. Consequently, protective properties of the polymer coating may be lost with time.

In recent years, systems consisting of more than one layer have been adopted, generally formed by an inner highly electro-active layer and an outer more compact layer. The role of the superimposed layer should be not to block the corrosion reaction by isolation, but rather to lengthen the diffusion path of the corroding species to lower the corrosion rate at the interface metal/coating. Often the inner layer and the outer layer interact mutually improving the performances of the single layers.

A two-layer system that has been diffusely studied is that formed by PANI and PPy films deposited in alternating way [21-26]. In reference [27], PPy was electropolymerised in 0.25 M monomer and oxalic acid solution. From experimental results they argued a model pursuant to which $Fe_2(C_2O_4)_3$ is formed from oxidation of Fe^{2+} into Fe^{3+}, that derives in turn from $FeC_2O_4 \cdot 2H_2O$, produced in the cyclic voltammograms of steel in the presence of oxalate ions. Simultaneously, monomer oxidation starts on the surface, resulting a monomer cation radical attached to the steel surface while FeC_2O_4 dissolves. After complete dissolution of FeC_2O_4, the steel surface is re-protected by the formed polymer, and the polymer including doping oxalate ions covers completely the iron substrate. On the basis of these theory, Hasanov [21] sustained that the oxidised form of PANI retaining dopant oxalate ions is a good conductor and helps to protect steel in the corrosive medium (Figure 7). The PANI and PPy layers are connected to each other by polar groups in the polymer matrix, forming a PANI/PPy coating on the steel surface with strong adsorption to steel. At the potential of the electropolymerisation, the iron oxalate is removed from the steel surface and therefore incorporated in the polymer matrix as dopant ions, and not bound directly to iron as iron oxalate. Moreover, Fe^{2+} ions from the dissolution of steel together with oxalate ions in the polymer matrix form iron oxalate on the steel surface in possible cracks or crevices, preventing further steel corrosion. Definitively, iron oxalate has a good diffusion barrier properties and is impermeable to corrosion-causing agents.

This mechanism proposed for the PPy/PANI coatings is arguable for other bilayer or multilayer conductive polymer coatings. The protection properties depend on the chemical structure of polymer coatings, on their adsorption on the metal surface, on the permeability of ions/molecules, on the transformation between redox states of polymers, and on their deposition order.

In reference [23] bilayer coatings, consisting of the inner conducting PPy doped with large dodecylsulfate ions (DS^-) and the outer conducting PANI doped with small SO_4^{2-} ions, were

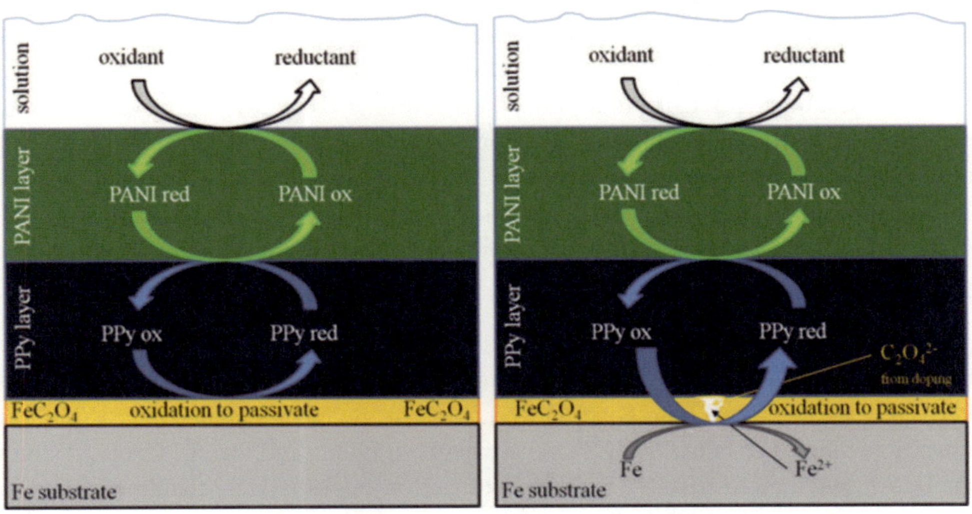

Figure 7. Proposed mechanisms of corrosion in PPy/PANI coated steel.

deposited on 304 SS for the bipolar plates of a PEMFC. The composite coatings inhibit the corrosion of the alloy significantly better than the single PPy coating. In fact, the composite coating shows a positive shift of the corrosion potential and a significant positive shift in the pitting corrosion potential with respect to the bare alloy. However, the free corrosion current density for the composite coatings is close to that of the bare steel, but significantly higher than that of the single PPy coatings, probably related to a contribution from the oxidation/reduction reaction of the polymer coatings, rather than to the corrosion of the substrate alloy. It was found that the composite coatings exhibited a greater chemical stability in HCl than the single PPy coating. This may be due to the bilayer coating acting as better barriers to trap chloride and more efficient oxidisers to maintain the metal in the passivity domain. The different ions permselectivity of the PPy underlayer and the PANI top layer increases the barrier effects of the composite coating. Moreover, the external PANI having a higher redox potential can oxidise the internal PPy layer in situ, thus maintaining the anodic protection effect of the composite coating.

In a proton exchange membrane fuel cell (PEMCF) stack, a good bulk conductivity is required because bipolar plates separate cells and carry current from cell to cell. A good surface conductivity is required and hence a low resistance surface contact must be kept. For this reason, the usual methods for corrosion protection producing insulating surfaces are not applicable to PEMFC. The use of conducting polymers should be useful but these polymers should not be faced with a passive oxide films, as usually occurs in the cases described above.

Recently many efforts have been made to use aluminum instead of steel as a material for bipolar plates. In addition to low cost and weight, the aluminum can provide bulk corrosion resistance, although even slight corrosive attack from free ions that contaminate the Nafion® membrane in the cell can occur. In addition to some evidences of corrosion of aluminum bipolar plates, the surface oxide which gives aluminum corrosion resistance prevents good electrical contact between the fuel cell electrode and the aluminum bipolar plate. For all these reasons it should be favourable to protect aluminum from corrosion by coatings based on conducting polymers. In reference [22] it is reported the electrochemically deposition of PPy and PANI on 6061 Al, after galvanostatic activation of aluminum. Whereas the corrosion resistance of the PPy coated samples shows no significant improvement from those of the uncoated plates, the PANI coated samples demonstrate a very good corrosion resistance and only a slightly higher contact resistance than that of the standard graphite plates. Since the conductivity of Polyaniline increases with its degree of oxidation, it is expected that in the acid environment of a fuel cell the contact resistance will be even lower.

Several PANI derivates have been studied [15, 28-31]. In [32] it was reported on a bi-layered composites of PANI and Poly(o-anisidine) (POA). PANI coatings provide much better protection to stainless steel than Poly(2-anisidine) films. Both combinations of the bi-layered composite coatings, i.e., POA/PANI (POA on top of the PANI) and PANI/POA (PANI on top of the POA) give a more effective protection to carbon steel substrate than a single layered PANI or POA coating. However, the corrosion protection offered to carbon steel depends on the deposition order of polymer layers in the composite. The PANI/POA composite provides a better protection to carbon steel against corrosion than POA/PANI coating, and anyway the bi-layered composites prevent steel corrosion for longer time.

As previously mentioned, for corrosion protection by conductive polymers, two mechanisms have been proposed: one is the physical barrier effect, and the other is the anodic protection. On the barrier effect, the polymer coating works as a barrier against the penetration of oxidants and aggressive anions, protecting the substrate metals. On the anodic protection, the conducting polymer with the strongly oxidative property works as an oxidant to the substrate steel, potential of which is shifted to that in the passive state.

There is a wide risk of the localized corrosion connected with the anodic protection, because when steels covered with the conducting polymer are immersed in a solution containing chloride or bromide ions, the anions doped in the CP can be exchanged with the chloride/bromide anions in the solution. The chloride/bromide anions penetrate the CP towards the substrate steels, and then induce the breakdown of the surface passive oxide film, followed by a pitting corrosion.

When looking at the CP as a charged membrane, the immobile anions with a large mass can be considered to have fixed sites with a negative charge. In the channel between the negatively charged sites, the cations can be mobile whereas the movement of the anions is greatly inhibited. Therefore, the membrane exhibits cationic permselectivity. In such a way, under the situation where the dopant anions are large enough, the anions in the solution are excluded from the CP and the metal substrate is protected against the pitting corrosion. An example of this mechanism was depicted by Kowalski [33-34] who protected a steel substrate by a PPy layer, with the inner part doped with phosphomolybdate ions (PMo). PMo works as a passivator which stabilises the passive state of steels and facilitates the formation of passive oxide. For the outer layer, the PPy doped with dodecylsulfate (DoS) is used. The outer layer can inhibit the anions from penetrating in the PPy layer. The results show that the steel covered with the bilayered PPy, about 5 μm thick consisting of PPy-PMo/PPy-DoS, exhibits a good passivation and no corrosion products are observed. It was assumed that PMo ions doping the inner PPy stabilises the passive oxide and helps the maintenance of the passive state of the substrate steel.

This kind of bilayer coatings also show an interesting self-healing property in which the passive oxide is spontaneously repaired after it develops small defects. After the coating and passive oxide are locally flawed, PMo in the PPy layer is hydrolysed and decomposes to molybdate and phosphate ions, and then both ions arrive at the flawed sites. The molybdate ions react with ferric ions on the flawed site to produce the ferric molybdate film. The salt film gradually changes into the passive oxide on the damaged site, according to following reactions [35]:

$$\begin{aligned} &Fe \rightarrow Fe^{3+} + 3e \\ &2Fe^{3+} + 3MoO_4^{2-} \rightarrow Fe_2\left(MoO_4\right)_3 \end{aligned} \quad (1)$$

In the experiment conducted by Kowalsky, a PPy layer receives the small flaw and then is immersed in a 3.5% sodium chloride solution. Initially, the open circuit potential fall down. When the corrosion is continuing at the defect site, the potential decreases to that of bare steel. The potential, however, rises up and recovers in the passive potential region. After that, the potential maintains the high potential in the passive region. Raman scattering spectroscopy reveals the molybdate salt is formed and a salt layer of ferric molybdate is reformed on the site of flaw.

One can argue that in designing a bilayer system coating for the corrosion protection, two important factors have to be considered: one is the stabilization of the passive film on the steel by action of dopant ions in the inner CP layer, and the other is the control of ionic permselectivity by organic dopant acid ions in the outer CP layer.

3.3. Composite structures based on CPs as protective coatings

Porosity and anion exchange properties of CPs could be disadvantageous, particularly when pitting corrosion starts caused by small aggressive anions. An interesting alternative, that has been established more recently, has been to consider composite systems based on conducting polymers.

Composite materials usually consist of a polymer matrix in which fibres and/or small filler particles are thoroughly dispersed. Silicon dioxide or titanium dioxide particles, for example, comprise some of the common fillers in composite materials such as plastics and films. Recently, conducting polymer/inorganic nanocomposites have also attracted more and more attention. These composite materials have shown better mechanical, physical and chemical properties, due to combining the qualities of CP and inorganic particles.

Fillers play an important role in the improvement of polymers matrix properties and each filler, with its volume fraction, shape and size, may have an effect on one or more particular properties. Size reduction of filler particles to nano-scale causes them to show different properties compared to those exhibited in micro scale sizes. At the same way, the reduction of filler particle size in polymer composites strongly affects the final properties of composites.

Nanocomposite materials based on PPy were studied in [36]. PPy-ZnO nanocomposite films with 10%wt nanorod ZnO loading are prepared by electropolymerisation from a monomer solution on mild steel. The role of the nano particles is to increase the barrier effect of the polymer matrix, improving its protecting properties. The reported excellent anticorrosion protection is attributed to the size and shape of nanorod, which is small in size and has a large aspect ratio. For small particles, free space between particles and polymer is far lesser than that of larger particles. Thus electrolyte is harder to penetrate through the pores in coating film where nano-pigment are added. In addition, due to longer diffusion path in nanocomposite coating, the water and ions need more time to arrive at the substrate. Moreover, from Electromechanical Impedance Spectroscopy (EIS) measurements it was found that although the initial value of coating resistance for the PPy and PPy-ZnO coatings are close to each other, the resistance of the PPy coating decreases sharply and becomes much lower than that of the PPy-ZnO coating. Therefore, it was supposed that the protection life-time of the PPy-ZnO nano-

Layered structures based on CPs				
Type of polymer coating (s) and Substrate	**Method of coating deposition**	**Method of corrosion tests**	**REF.**	**Results**
PANI/POA on low carbon steel	Cyclic voltammetry	Potentiodynamic polarisation, EIS and monitoring OCP in 3% NaCl	[32]	The bilayer coating provides a better protection than the single layered PANI or POA coatings. The corrosion rate of PANI/POA coated steel is 88 times lower than that of uncoated steel.
POA/PANI on low carbon steel	Cyclic voltammetry	Potentiodynamic polarisation, EIS and monitoring OCP in 3% NaCl	[32]	The bilayer coating provides a better protection than the single layered PANI or POA coatings. The corrosion rate of POA/PANI coated steel is 44 times lower than that of uncoated steel.
PANI/PPy on steel	Cyclic voltammetry	Tafel polarisation in 1M H_2SO_4	[21]	The corrosion performance of bilayer coating is better than that of monolayer PPy and PANI coating.
	Cyclic voltammetry	Tafel polarisation in 0.1M HCl EIS in 0.5 M NaCl, 0.1M HCl Potentiodynamic polarisation in 0.5M H_2SO_4	[111]	All corrosion tests show a better performance of PANI/PPy bilayer coating with respect to the single PANI coating.
PANI/PPy on zinc	Galvanostatic deposition	Potentiodynamic polarisation and EIS in 3.5% NaCl	[25]	The corrosion potential and the corrosion current depend on the galvanostatic deposition parameters. The protection efficiency is 94.3% for the bilayer film and 92.9% for single PPy film.
PPy/PANI on steel	Cyclic voltammetry	Tafel polarisation in 1M H_2SO_4	[21]	The corrosion performance of bilayer coating is better than that of monolayer PPy and PANI coatings. The highest inhibition efficiency is that of PPy/PANI coating.
	Galvanostatic deposition for PPy and Cyclic voltammetry for PANI	Potentiodynamic polarisation, EIS and monitoring OCP in 0.3M HCl	[23]	The bilayer coating shows a better corrosion protection than the single PPy coating, with a shift of about 500 mV of pitting corrosion potential and a good stability for up 36 days.
	Potentiostatic deposition	EIS and monitoring OCP in 3.5% NaCl	[24]	The bilayer film has a better anticorrosive behaviour than the monolayer films based on PPy and PANI.
	Cyclic voltammetry	Tafel polarisation in 0.1M HCl EIS in 0.5M NaCl, 0.1M HCl Potentiodynamic polarisation in 0.5M H_2SO_4	[111]	All corrosion tests show a worse performance of PPy/PANI bilayer coating with respect to single PANI coating and PANI/PPy bilayer coating.
PPy-PMo/PPy-DoS on steel	Oxidative polymerisation	Monitoring OCP in 3.5% NaCl	[19]	The steel coated by the PPy based bilayer coating is kept passive for about 200 hours in NaCl environment.
PPy/PNEA on steel	Cyclic voltammetry	Tafel polarisation in 1M H_2SO_4	[94]	The bilayer coating is more efficient than PNEA monolayer coating.
PNEA/PPy on steel	Cyclic voltammetry	Tafel polarisation in 1M H_2SO_4	[94]	The bilayer coating is more efficient than PNEA monolayer coating.

Table 2. Some examples of layered coatings based on CPs for corrosion protection.

composite coating is longer than that of the pure PPy coating for the following reasons: (a) the existence of ZnO nanorods increases the deviousness of diffusion pathway of corrosive species; (b) PPy-ZnO coating is more compact compared with the PPy coating.

A particular use of PPy is as modifier of alumina nanoparticles, before using as coating filler [37]. It is known that additives, such as nanoparticles or nanoflakes, prevents the uniform diffusion through intermolecular pores into an anticorrosion coating. Tallman suggested to take a vantage from the capacity of CPs changing their volume during undoping. Therefore, modifying alumina particles with PPy, and inducing a swelling of the PPy, the voids formed in the coating during its lifetime by ageing, are filled. At the same time, the swollen PPy absorbs additional moisture that penetrates from the outside and blocks transfer of ion permeating inside coating.

The enhanced barrier effect is also the reason of the incorporation of a layered material such as smectite clays (e.g. montmorillonite) in CPs. Nanocomposed based on clays and polymer have been diffusely prepared in past decades due their improved gas barrier, thermal stability, mechanical strength, fire retardant and anticorrosive properties. It has been shown that the montmorillonite (MMT) enhances the anticorrosive effect of Polyaniline and Polypyrrole coatings because of the enhancement of the barrier property of the CP [38-40]

Generally, two kinds of MMT are used: organophilic montmorillonite (O-MMT) and hydrophilic montmorillonite (Na-MMT). The preparation of PANI/MMT nanocomposites with hydrophilic and organophilic nanoclay particles is successfully performed by in situ polymerisation method [41] and therefore different structures of PANI/MMT nanocomposites are obtained. It was found that the incorporation of MMT nanoparticles in PANI matrix, promotes the anticorrosive efficiency of PANI/MMT nanocomposite coatings on iron samples. Comparison of the corrosion rate shows that the anticorrosive properties of PANI/Na-MMT nanocomposite coating is better than PANI/O-MMT and pure Polyaniline coating in 3.5% NaCl but not in 1M H_2SO_4, where the corrosion rate of PANI/O-MMT nanocomposite coated samples is lower. This behaviour is attributed to the hydrolysis of Na-MMT in H_2SO_4 solution which causes changes in the structure of MMT.

However, enhanced corrosion protection of PANI/MMT nanocomposite compared to pure PANI coated samples results from silicate nanolayers of clay dispersed in PANI matrix which increase the deviousness of diffusion pathway of corrosive agents such as oxygen gas, hydrogen and hydroxide ions.

An improvement of anticorrosive properties of PANI is also observed after introducing metal particles such as Zn particles. Zinc, as a transition metal, has conducting and semiconducting properties together with potential ability to cathodic sacrificial protection of metals against corrosion. More recently zinc rich coatings have been applied for the protection of mild steel against corrosion. Also zinc dust and inorganic zinc salts have been used as anticorrosive fillers in conducting PANI coating matrices [42-46]. It has been demonstrated that zinc particles can improve the barrier properties of a PANI coating on copper by the formation of voluminous zinc corrosion products within the pores of PANI coating [43].

Also in cathodic protection of metals by zinc particles, the conducting PANI can play the role of conductance between the zinc particles and the metal surface. Olad [39-40] found out that incorporation of zinc nanoparticles and zinc micro-size particles produces an effective PANI/Zn nanocomposite and PANI/Zn composite coatings on iron, respectively. The electrical conductivity of both nanocomposite and composite systems are correlated with the zinc content, and it is higher when Zn particles are nanosized.

A similar behaviour has been found in the anticorrosive properties of PANI/Zn coatings because the synergetic effect of zinc nanoparticles is more than that of micro sized particles.

Among various inorganic particles, titanium dioxide (TiO_2) nanoparticles have attracted attention due to their excellent properties such as a charge carrier, oxidizing power, non toxicity, chemical and photo stability. Nanostructured TiO_2 can be used to develop multifunctional devices with unique mechanical, chemical, electrical, optical and magnetic properties. Conductive PANI/TiO_2 nanocomposites combine the qualities of PANI and nanocrystalline TiO_2 within a single material, thereby developing multifunctional materials with combined properties which have very strong potential applications.

Rathod [47] and Radhakrishnan [48] are in agreement when suggesting that the corrosion protection of PANI/TiO_2 coating is based on almost three factors. The first is the barrier effect formerly cited, preventing oxygen and moisture from reaching the metal substrate. In the specific case of TiO_2 is increased by the nano scale of the additive. Moreover, since that PANI polymerisation is carried out in the presence of a nano-TiO_2 dispersed in the reaction medium, the PANI forms around these particles giving a core-shell type structure: the core being TiO_2 with the shell formed by PANI. This type of structure give rise to higher surface area for PANI as compared to the particles made wholly from PANI.

The second factor relies on the role of PANI in corrosion protection, that is its redox behaviour. PANI captures the ions liberated during the corrosion reaction of steel in the presence of NaCl, water and oxygen, becomes doped and liberates the dopant ions which form a passivating layer even when the corrosion process at the substrate has started. Thus, it acts as a self healing coating with improved corrosion resistance.

As third factor, there is also a charge trapping process because TiO_2 is n-type with a band gap of 3.13 eV while PANI is p-type with band gap of 2.1 eV, and this gives rise to potential barrier formation at the interface. It is quite likely that both these species may form p–n junctions, which prevent charge transport across the layer. In other words, the synergetic action between TiO_2 and PANI hinders the process of electron or charge to transfer across the film, the barrier property of the coating get enhanced, and therefore the flow of electron or ions in and through the film is hindered. All the factors lead to much better performances than the single component system, that are pure nano-TiO_2 and pure PANI coatings.

A diffused use of CPs, and in particular of PANI, with epoxy has been made in the last five years. PANI has been introduced in a resin in form of pigments [49-50], fibres [51-52], nanoparticles [53], or as additive [54-55]. Moreover, the effectiveness of both the emeraldine base (EB) and the emeraldine salt (ES) form of PANI for corrosion protection has been studied. In refererence [56] it was observed that from corrosion tests with an aggressive saline solution, the epoxy + PANI-EB formulation provides more protection than the epoxy + PANI-ES. The comparison of these results with those obtained for other epoxy coatings containing conducting polymers as anticorrosive additives, indicates that the protection mechanism of PANI-EB is based on the ability of this polymer to store charge. Moreover, as the highest protection is observed by epoxy + PANI-EB paint, it is concluded that the mechanism based on the electroactivity of partially oxidised polymers is more effective than that based on the interception and transport of electrons.

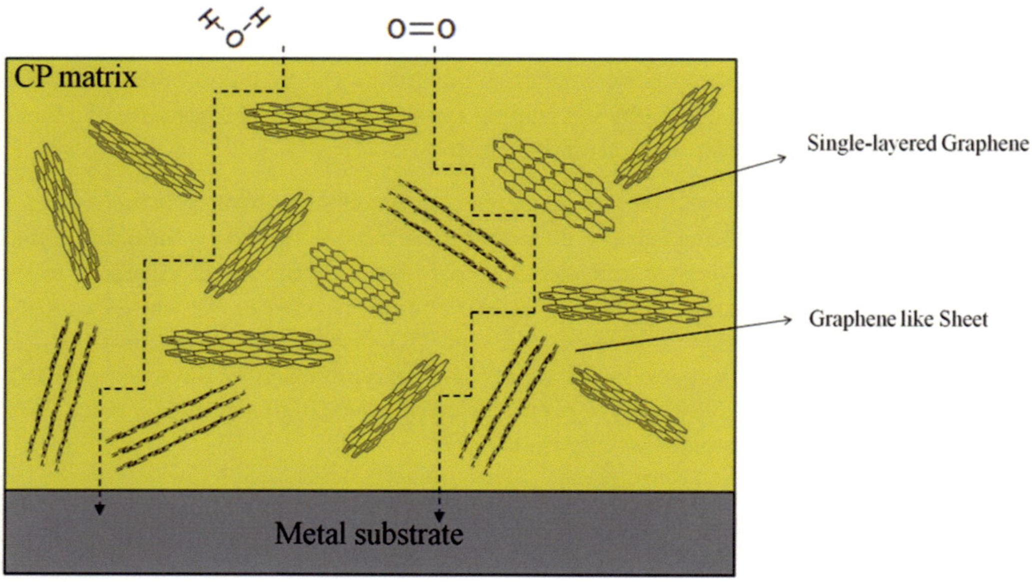

Figure 8. Scheme of the hypothesised retarding effect of graphene on gas diffusion.

More recently carbon nanostructures have been used as fillers in CP coatings, starting *de facto* the age of "hybrid conducting nanocomposites". The first study on PANI/graphene composites as anticorrosion coating material is that of Chang et al. [57]. They obtained graphene-like sheets grafted with 4-aminobenzoyl groups by Friedel–Crafts acylation with 4-aminobenzonic acid (ABA) in polyphosphoric acid (PPA)/phosphorous pentoxide (P_2O_5). The subsequent chemical oxidation polymerisation of the aniline monomers with different amounts of 4-aminobenzoyl group- functionalised graphene-like (ABF-G) sheets is conducted in ammonium persulfate (APS)/1.0 M aqueous hydrochloric acid to yield Polyaniline /graphene composites (PAGCs). By comparing gas barrier properties of PAGCs with those of Polyaniline/clay composites (PACCs), it is argued PAGCs are more suitable anticorrosion coating materials than PACCs. This also suggests that polymer/graphene composites are a potential advanced anticorrosion coating candidate to replace traditional polymer/clay composites, where the diffusing gas molecule would encounter a more tortuous path in traversing the polymer/graphene composite coatings (Figure 8).

Also carbon nanotubes (CNTs) were introduced in PANI matrix, to form composite materials with enhanced electrical properties [58-62] by facilitating charge-transfer processes between the two components. Hermas [63] obtained anticorrosion coatings on a stainless steel surface by in situ electropolymerisation of PPy-oxidised multi-walled carbon nanotubes and PPy- oxidised single-walled carbon nanotubes composites on from 0.1 M oxalic acid by using cyclic voltammetry. The results show that the addition of the oxidised carbon nanotubes greatly enhances the electropolymerisation process, especially in the case of oxidised single walled carbon nanotubes.

Similar results are reported in [64], referring to electrodeposition of a nano-composite coating made of oxidised CNTs and Poly(o-phenylenediamine) (PoPD) on a stainless steel. Also in this case the presence of the CNTs enhances the deposition of the PoPD and this enhancement is more significant in the presence of single walled carbon nanotubes (SWCNTs) by comparison to multi-walled carbon nanotubes (MWCNTs). The nanocomposite coating keeps the stainless steel in a passive state in an acidic solution.

The functionalisation of CNTs plays an important role in the preparation process of a CP-CNT nanocomposite material. For example, if single-walled carbon nanotubes (SWCNTs) are covalently functionalised with a water soluble conducting polymer, they acquire an excellent solubility in water and a high potential for covalent immobilisation of various dopants.

In [65] nanocomposite films based on Polyaniline, functionalised single-walled carbon nanotubes and different dopants are studied. These nanoporous composite films are grown electrochemically from aqueous solutions such that constituents are deposited simultaneously onto a substrate electrode. The composite films consist of nanoporous networks of SWCNTs coated with polymeric film. Cyclic voltammetry and electrochemical impedance spectroscopy demonstrate that these composite films have similar electrochemical response rates to pure polymeric films but a lower resistance and much improved mechanical integrity. The negatively charged functionalised carbon nanotubes serve as anionic dopants during the electro polymerisation to synthesise polymer-CNTs composite films. The specific electrochemical capacitance of the composite films has a significantly greater value than that for pure polymer films prepared similarly. The higher capacitance of the composite films results obviously from the contribution of the embedded functionalised SWCNTs that provide interconnected pathways for electrons through the functionalised SWCNTs and ions through the pore network or the direct interaction between the delocalised electrons on polymer chains and the functionalised SWCNTs. For this reason, PANI-CNTs composites have been particularly studied for using as supercapacitors. Nevertheless, these hybrid nanocomposite materials have shown to have good corrosion protection properties, as reported in the following section.

Composite layers based on CPs				
Type of polymer coating (s) and Substrate	**Method of coating deposition**	**Method of corrosion tests**	**REF.**	**Results**
PPy+ZnO on steel	Galvanostatic deposition	Potentiodynamic polarisation, monitoring OCP, and EIS in 3.5% NaCl	[36]	PPy+ZnO coating with 10wt% of ZnO relative to PPy has better anticorrosion properties than the pure PPy coating.
PPy+MMT on steel	Dip coating	Tafel polarisation in 5% NaCl	[39]	Increasing the MMT content in PPy from 0 to 10wt%, the corrosion current decreases and the corrosion potential shifts to nobler potential values.
PPy +Al_2O_3 on Al alloy 2024T3	Painting	EIS in DHS	[37]	The coating impedance increases when the Alumina nanoparticles content increases.
PANI+Na-MMT on iron	Casting	Tafel polarisation in 1M H_2SO_4 and 3.5% NaCl, Monitoring OCP in 1M HCl	[41]	The anticorrosion behaviour of PANI+Na-MMT is better than PANI and PANI+O-MMT in NaCl, but worse than PANI and PANI+O-MMT in H_2SO_4
PANI+O-MMT on iron	Casting	Tafel polarisation in 1M H_2SO_4 and 3.5% NaCl, Monitoring OCP in 1M HCl	[41]	The anticorrosion behaviour of PANI+O-MMT is better than PANI and PANI+Na-MMT in H_2SO_4, but worse than PANI and PANI+Na-MMT in NaCl .
PANI+ Sulfonated chitosan (S-CTS) on steel	Chemical synthesis	EIS and polarisation curves in 3.5% NaCl	[105]	The best corrosion inhibition is obtained for a S-CTS/ Aniline ratio equal to 2:1. The maximum value of the inhibitive efficiency is 92.3%.
Dodecylbenzenesulfonic acid-doped Polyaniline nanoparticles [n-PANI (DBSA)] on steel	Casting	EIS in 3.5% NaCl	[106]	n-PANI (DBSA) coating shows a coating resistance and a coating capacitance stable up to 77 days.
Benzoate-doped PANI on steel	Galvanostatic deposition	Potentiodynamic polarisation in 3% NaCl, outdoors and in atmosphere	[107]	Benzoate-doped PANI coatings could protect steel in the three different environments for a limited period of time, even when the coatings are partially applied.
PANI+Zn on iron	Casting	Tafel polarisation and Monitoring OCP in 0.1M HCl	[42]	The corrosion rate values of PANI+Zn coatings with a different Zn content are lower than of pure PANI, but their kinetic anticorrosive properties get worse with increasing the zinc content.
	Casting	Tafel polarisation and Monitoring OCP in 0.1M HCl, 0.1M H_2SO_4, and 3.5% NaCl	[43]	In all the corrosive environments, the incorporation of Zn nanoparticles increases the anticorrosive efficiency on PANI coating.
Anodic Alumina with PANI+TiO2 on Al alloy AA2024T3	Potentiostatic deposition	Potentiodynamic polarisation in 5mM NaCl and 0.1M Na_2SO_4	[108]	The coating containing TiO_2 nanoparticles protects the substrate against the corrosion better than pure PANI and coatings based on PANI containing ZrO_2 particles.

Composite layers based on CPs				
Type of polymer coating (s) and Substrate	**Method of coating deposition**	**Method of corrosion tests**	**REF.**	**Results**
Anodic alumina with PANI+ZrO_2 on Al alloy AA2024T3	Potentiostatic deposition	Potentiodynamic polarisation in 5mM NaCl and 0.1M Na_2SO_4	[108]	The coating containing ZrO_2 nanoparticles protects the substrate against the corrosion better than pure PANI but worse than coatings based on PANI containing TiO_2 particles.
PANI+TiO_2 on steel	Dip coating	Tafel polarisation and Monitoring OCP in 3.5% NaCl	[48]	The corrosion protection property of PANI+TiO_2 coatings depends on TiO_2 content. When the TiO_2 content increases, the OCP shifts to nobler potentials but the corrosion current increases.
	Oxidative polymerisation	Potentiodynamic polarisation in 3.5% NaCl	[47]	The corrosion current of PANI+TiO_2 coatings on steel is better than the single component system (pure PANI or nano-TiO_2). The corrosion protection property of PANI +TiO_2 coatings depends on TiO_2 content and it is better when TiO_2/aniline weight ratio is 0.05.
POT + CdO on steel	Cyclic voltammetry	Potentiodynamic polarisation in 3% NaCl	[98]	The incorporation of CdO nanoparticles improves the corrosion protection of POT coating.
PANI+ZnO on iron	Oxidative polymerisation	Tafel polarisation and Monitoring OCP in 3.5% NaCl	[99]	The corrosion current on iron coated by PANI+ZnO coating is almost 30 times lower than the uncoated iron.
PANI+PVC+ZnO on iron	Oxidative polymerisation	Tafel polarisation and Monitoring OCP in 3.5% NaCl	[99]	The corrosion current on iron coated by PANI+PVC+ZnO coating is almost 200 times lower than the uncoated iron.
PANI + Clinoptilolite (Clino) on iron	Casting	Tafel polarisation in 1M H_2SO_4, 1M HCl and 3.5% NaCl	[100]	The encapsulation of PANI in the Clino channels and the dispersion of Clino layers in PANI matrix enhances the corrosion protection of composite layer with respect to pure Aniline.
Methyl Orange (MO)-doped PANI on steel	Painting	Monitoring OCP in 5% HCl, 3.5% NaCl and 5% NaOH	[109]	MO composite coatings act as "corrosion indicators".
MO-PANI/Castor Oil Polyurethane (COPU) on steel	Painting	Monitoring OCP in 5% HCl, 3.5% NaCl and 5% NaOH	[109]	MO composite coatings act as "corrosion indicators". The corrosion protection is improved by the MO-PANI nanoparticles in COPU.
PANI+SiO_2 on Al	Dipping	Potentiodynamic polarisation in 0.5M NaCl	[112]	The corrosion protection ability of PANI+SiO_2 coating decreases when the TEOS (precursor of SiO_2) increases.

Table 3. Some examples of composite coatings based on CPs for corrosion protection.

CPs mixed to Epoxy				
Type of polymer coating (s) and Substrate	**Method of coating deposition**	**Method of corrosion tests**	**REF.**	**Results**
PANI with epoxy on steel	Immersion and solvent evaporation	Immersion in 3.5% NaCl	[54]	The epoxy coating containing a very low PANI concentration (0.3wt%) provides the best corrosion protection with respect to the unmodified epoxy coating.
	Air spray	EIS in 3.5% NaCl	[50]	The PANI pigmented paint coating shows a capacitive behaviour.
	Immersion and solvent evaporation	Immersion in 3.5% NaCl	[56]	The epoxy coating modified with PANI-EB performs better than both the unmodified coating and the epoxy coating modified with inorganic corrosion inhibitors.
	Painting and solvent evaporation	EIS and Tafel polarisation curve in 0.5% NaCl	[52]	PANI fibres having different morphology are prepared in different medium: HCl, HNO_3, H_2SO_4, and H_3PO_4. H_3PO_4-doped PANI shows the best protective effect, followed by HNO_3-doped PANI, H_2SO_4-doped PANI , and HCl-doped PANI.
	Film application	EIS in 3.5% NaCl	[55]	EB-PANI is added to the hardener. The coating resistance and the charge transfer resistance are the highest for 2.5wt% PANI-EB, and the lowest for epoxy coating without EB-PANI.
	Painting	EIS in 3.5% NaCl	[53]	Epoxy modified with PANI nanoparticles shows an increase in coating resistance for longer immersion times.
	Dipping	Monitoring OCP and EIS in 3% NaCl	[102]	Epoxy coating with 2.5wt% PANI-ES offers lower corrosion protection than the coating with 2.5wt% PANI+P-PVA.
	Painting and solvent evaporation	EIS in 3.5% NaCl	[51]	Epoxy containing PANI sulfonated has the best performance in the protection of steel with respect to coatings containing PANI or zinc phosphate.
PANI with epoxy on magnesium alloy	Painting	EIS in 3.5% NaCl	[40]	The corrosion protection of PANI coating on AZ91D magnesium alloy is better with respect to the varnish coating but worsens in presence of O_2.
PANI+Zn with epoxy on iron	Casting	Tafel polarisation in HCl	[103]	Epoxy is used as an additive. The addition of both Zn particles and epoxy resin in PANI coating improves the overall anticorrosion performance of coating. PANI+Zn+Epoxy nanocomposite coating shows the best anticorrosion performance by the addition of 4wt% Zn particles and 3-7wt % epoxy.
PANI with epoxy and Polyurethane on steel	Painting	Accelerated corrosion in 5% NaCl EIS in 3.5% NaCl	[49]	In epoxy coating, PANI pigments increase the capacitance and decrease the OCP. In Polyurethane coating, PANI has an adverse effect.
PANI with chlorine ether resin on steel	Painting	Tafel polarisation curves in 5% H_2SO_4, 5% NaOH, 3.5% NaCl	[104]	The best performance of PANI + resin coating is obtained for PANI concentration between 15% and 25%.
PANI+O-MMT with epoxy on	Painting	EIS in 3.5% NaCl	[40]	The corrosion protection of PANI+O-MMT coating on AZ91D magnesium alloy is better than pure PANI coating.

CPs mixed to Epoxy				
Type of polymer coating (s) and Substrate	**Method of coating deposition**	**Method of corrosion tests**	**REF.**	**Results**
magnesium alloy				
PANI+P-PVA with epoxy on steel	Dipping	Monitoring OCP and EIS in 3% NaCl	[102]	Epoxy coating with 2.5wt% PANI+P-PVA offers higher corrosion protection than the coating with 2.5wt% PANI-ES .
PPy with epoxy on steel	Immersion and solvent evaporation	Immersion in 3.5% NaCl	[54]	The epoxy coating containing PPy (with concentration from 0.3 to 1.5wt%) provides a better corrosion protection with respect to the unmodified epoxy coating, but a worse one with respect to PANI coating with epoxy.

Table 4. Some examples of coatings based on CPs and Epoxy for corrosion protection.

4. Preparation of hybrid conducting nanocomposites based on PANI and CNTs

In this section, two methods successfully used to prepare PANI-CNTs nanocomposite coatings on steel will be described: Electropolymerisation (EP) and Electrophoretic Deposition (EPD) (Figure 9).

Electropolymerisation is the usual method used for the synthesis of PANI coatings. In general, this method consists of applying a potential to a working electrode immersed in an electrolyte solution. As mentioned before, the potential can be applied through CV, potentiostatic, or galvanostatic methods. The EP method used in this work was CV. During the EP process, the monomer is electrochemically oxidised in an acid solution (usually HCl or H_2SO_4) and polymerisation occurs on the electrode surface where the deposition of the polymer film takes place. The main advantage of this method is the accurate control of the polymerisation rate and of the state of material generated by varying potential values. Moreover, EP is environmental friendly and polymerisation media can be used repeatedly. However, this method is not applicable to large areas, due to problems related to solubility, film quality and coverage. Therefore, with a view to industrial applications of CPs coatings, these drawbacks are particularly relevant.

The synthesis of PANI-CNTs nanocomposite films via EP is an electrochemical co-deposition wherein the CNTs are incorporated in the PANI matrix. In other words, the formation of the nanocomposite occurs during the electrogeneration-deposition of the polymer just near the electrode surface, in the solution containing both CNTs and monomer (in this case Aniline). In such a way, CNTs are functionalised by PANI in the solution and then, after being covered by PANI, deposit onto the electrode surface.

Hybrid conducting nanocomposites based on CPs				
Type of polymer coating (s) and Substrate	**Method of coating deposition**	**Method of corrosion tests**	**REF.**	**Results**
PANI+graphene on steel	Casting	Tafel polarisation in 3.5% NaCl	[57]	The corrosion current on steel coated by PANI +graphene coating is better than both on the uncoated and on PANI coated steel.
PEDOT+graphene on ITO	Casting	EIS in 2M HCl and 2M H_2SO_4	[110]	(Study of capacitive/supercapacitive properties).
PANI+CNTs on platinum	Cyclic voltammetry	EIS and Cyclic voltammetry in 0.1M H_2SO_4	[86]	(Study of capacitive/supercapacitive properties).
	Cyclic voltammetry	EIS and Cyclic voltammetry in 0.1M H_2SO_4	[87]	(Study of capacitive/supercapacitive properties).
	Cyclic voltammetry	EIS in 0.5M H_2SO_4	[59]	(Study of capacitive/supercapacitive properties).
PANI+CNTs on steel	Potentiostatic deposition	Potentiodynamic polarisation in 3% NaCl	[70]	The corrosion rate of steel coated by PANI+CNTs coating is better than both on the uncoated and on PANI coated steel.
	Cyclic voltammetry	Potentiodynamic polarisation in 3.5% NaCl	Current work	PANI+CNTs nanocomposite coating deposited on steel shows the same corrosion potential of PANI on steel.
	EPD	Potentiodynamic polarisation in 3.5% NaCl	Current work	PANI+CNTs nanocomposite coating shows a more positive corrosion potential than that PANI coating obtained in the same deposition conditions.
PANI on CNTs on steel	Cyclic voltammetry	Potentiodynamic polarisation in 3.5% NaCl	Current work	PANI coating on CNTs shows a more positive corrosion potential than that PANI coating obtained in the same deposition conditions.
	EPD	Potentiodynamic polarisation in 3.5% NaCl	Current work	PANI coating deposited on CNTs on steel shows the same corrosion potential of PANI on steel.
PoPD+CNTs on steel	Cyclic voltammetry	Monitoring OCP	[64]	The OCP remains stable for longer time when CNTs where added to PoPD.

Table 5. Some examples of hybrid conducting nanocomposites coatings based on CPs for corrosion protection.

Recent studies have shown that the functionalisation of CNTs by means of PANI occurs via formation of donor-acceptor complexes. In fact CNTs act as good electron acceptors, while PANI is a fairly good electron donor [66]. The π-bonded surface of the CNTs interacts strongly with the conjugated structure of PANI, especially through the quinoid ring. Such a selective interaction of the CNTs with the quinoid ring of PANI has been reported in literature [67-68]. Moreover, the presence of functional groups on the CNTs surface can favour a chemical interaction of both the monomer and the polymer during its generation [69].

Analysing the cyclic voltammograms acquired from EP PANI in H_2SO_4, it can be observed that, at the anodic potential sweep on the anodic branch of the voltammograms, three anodic oxidation peaks appear and the correspondent three reduction cathodic peaks appear at the

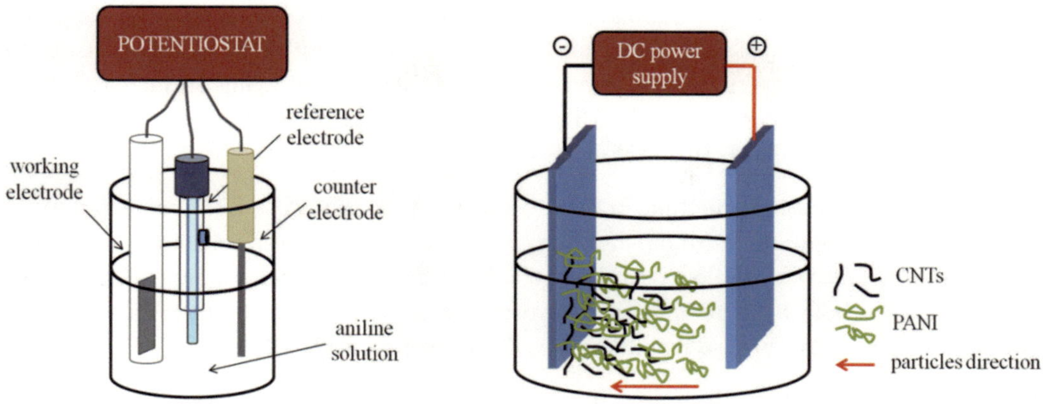

Figure 9. Scheme of an EP (on the left) and an EPD (on the right) cell.

reverse potential sweep, on the cathodic branch of the voltammograms (Figure 10). This fact points to the existence of three redox processes, which take place in the PANI film on different potential ranges. The first redox peak is commonly assumed to correspond to the electron transfer from/to the PANI film. In order to compensate the charge of the PANI film, anion doping/dedoping of the PANI film occurs. The peak in the potential range of 700÷900 mV is probably due to a side reaction in the PANI film, whereas the third redox peak corresponds to deprotonation and protonation process.

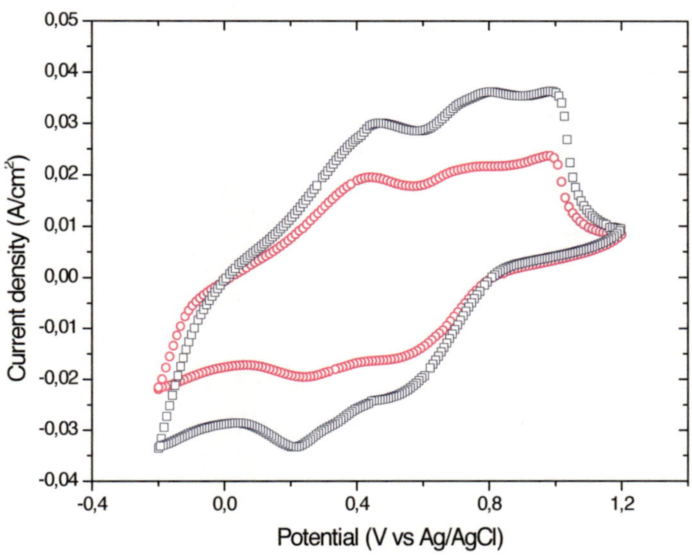

Figure 10. Cyclic voltammetric curve in 0.5 M H_2SO_4 relative to PANI (magenta circles) and PANI+CNTs (blue squares) obtained by EP.

The same peaks are present in voltammograms acquired on PANI-CNTs composite films. The difference is in the recorded current, higher when CNTs are present in aniline solution, due to the effect of CNTs acting as condensation nuclei and therefore allowing an increase of the amount of electropolymerised PANI. This consideration has been also reported by other authors for PANI [70], PPy [71], and PoPD [64].

As an alternative deposition method, EPD has been used to obtain PANI-CNTs nanocomposite coatings. EPD is one of the most outstanding coating techniques based on electrodeposition. It is a traditional method employed to obtain ceramic coatings. Nowadays, increasing interest has been shown both by academics and industrialists, due to its wide potential in coating processing technology, as described, for example, in references [72-74].

The main advantages of this technique are high versatility, since it can be used with different materials and their combinations, and cost effectiveness, because it requires simple and cheap equipment. Moreover, it can be used both on a large scale also when coating objects with a complex geometrical form, and on a smaller scale, to fabricate composite micro- and nano-structures, as well as near net-shape objects having accurate dimensions (micro- and nano-manufacturing).

EPD is a two-step process. In the first step, charged particles, suspended in a liquid medium, move towards the oppositely charged electrode under the effect of an externally applied electric field (electrophoresis). In the second step, the particles deposit on the electrode forming a more or less thick film, depending on process conditions (concentration of particles in solution, applied electric field, time). The substrate acts as an electrode and the deposit of particles is the coating (Figure 9).

Of course, the process parameters, mainly the formulation of the electrophoretic suspension containing the particles, have to be suitable in relation to the coating material and the coating application.

Each of the two steps mentioned above requires special attention. The main requirement to obtain an efficient EPD process is to use suitable suspensions where particles are well suspended and dispersed in a proper liquid medium. The stability of a suspension is essentially due to two possible phenomena occurring at the surface of solids in suspension: the dissociation of functional groups in ionic charges (electrostatic stability), or the adsorbing of ionic surfactants (steric or electrosteric stability). Both the phenomena produce a superficial charge that also allows the movement of solids in the electric field.

With regards to the second step of EPD, the deposition, the deposited mass per area unit in a cell with planar geometry is given by the expression [75]:

$$m = C_S \mu A E t \tag{2}$$

where C_S, solids concentration in the suspension; t, deposition time; μ, electrophoretic mobility; E, electric field strength; A, surface area of the electrode. In this expression, the critical parameter is the electrophoretic mobility, defined as:

$$\mu = \frac{\varepsilon\zeta}{4\pi\eta} \quad (3)$$

where ε, dielectric constant; ζ, zeta potential; η, viscosity of the liquid medium. It is evident that a particle has a good mobility, if the zeta potential is high. As a consequence, the efficiency of the deposition process can be evaluated through zeta potential measurements.

Recently the EPD process has been applied to suspension containing polymers both in the form of small solid particles and as more or less long chains [58,76]. For preparing EPD suspensions based on Polyaniline, PANI is dissolved in a mixture of $CHCl_3$/EtOH containing m-cresol and dodecylbenzenesulfonic acid (DBSA). DBSA is used because it increases significantly the mean molecular area, provided the sub-phase is maintained sufficiently acidic to keep the PANI protonated and the DBSA negatively charged. The addition of m-cresol as a co-solvent is believed to straighten the PANI chains and is therefore responsible for increasing the mean molecular area per PANI unit [77]. Since PANI is in the protonated state, the possible interactions between PANI and CNTs are:

- electrostatic interaction between negative functional (usually $–COO^-$ or $–SO_4^{2-}$) groups of CNTs (residual of the purification treatment) and $–NH^+$ of PANI;
- π-stacking, referring to the attractive no-covalent interaction between aromatic rings of PANI and CNTs;
- hydrogen bonding between –NH of PANI and –OH of CNTs (deriving from functionalising medium).

On average, all these interactions do not neutralise the total positive charge of PANI chains, since the amount of CNTs in the PANI suspension is quite low. This affirmation is confirmed by measuring the zeta potential. In effect, after adding CNTs to the PANI suspension, the zeta potential increases in positive correlation with the CNTs content, meaning that CNTs in suspension contribute to the improvement of the efficiency of the EPD process, through an increased conductivity of the electrophoretic suspension. This consideration is in agreement with some results obtained by Dhand et al. [78-79].

Voltammograms acquired on EPD PANI in the same conditions of EP PANI show similar peaks both in anodic and in cathodic scans, but the recorded current is significantly lower in EPD PANI than in EP PANI (Figure 11). Moreover, a shift of about 200 mV towards more cathodic potentials is revealed with respect EP PANI. The position of EPD PANI peaks is almost the same as chemically synthesised PANI reported in [80]. On the other hand, PANI used for the preparation of EPD suspension is a purchased chemical and presumably it has been chemically synthesised.

It is worth to note that, similarly to PANI-CNTs composite obtained by EP, the recorded current in EPD PANI-CNTs film is higher than in PANI film. This increase of the current could be indicative of fast charge transport in parallel paths to surface of PANI-CNTs film.

In reality, only a few authors have published referring to EPD of PANI [81-84]. In references [82-85] PANI, also containing CNTs, has been electrophoretically deposited, to obtain sensors and biosensors. EPD films based on PANI and CNTs are produced by starting with a suspension containing PANI in ES form and CNTs functionalised by $-COO^-$ groups, dissolved in formic acid and acetonitrile. Investigations by FT-IR and UV-visible spectra acquired on PANI-CNTs films evidence a doped state of PANI induced by the interaction between imine sites of PANI and carboxyl groups in CNTs. Moreover, from EIS investigation it results that the charge transfer resistance for ES-CNTs composite film is much lower than that of the ES film.

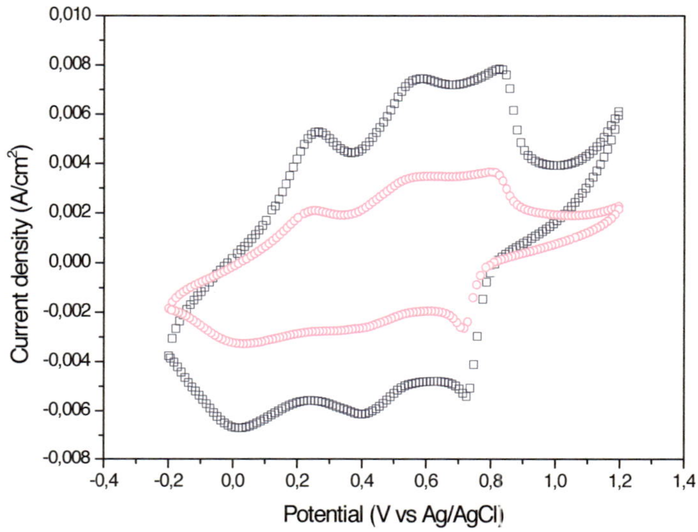

Figure 11. Cyclic voltammetric curve in 0.5 M H_2SO_4 relative to PANI (magenta circles) and PANI+CNTs (blue squares) obtained by EPD.

The enhanced charge transfer is attributed to grain-to-grain wiring of the PANI chains with the uniformly distributed CNTs in the film. Besides this, the presence of CNTs provides "conducting bridges" between ES conducting domains and functions such as charged "jumping centres" facilitating a fast charge transfer via electrons from one polymer chain to another, and finally to the electrode surface. This hypothesis does not disagree with the interaction mechanisms supposed above between PANI chains and CNTs and with the increased electroactivity evaluated through voltammograms.

A third kind of nanocomposite material based on PANI and CNTs has been prepared by using EP and EPD to deposit PANI on a EPD deposit of CNTs. In other words, firstly CNTs have been deposited by EPD on an electrode and then a PANI film has been deposited on CNTs by EP or by EPD. In such a way two other types of samples have been produced. In these cases, the interactions between ANI (in EP process) or PANI (in EPD process) and CNTs are different in respect to those occurring when CNTs are present in an ANI or a PANI solution. Similarly to other PANI-CNTs nanocomposites, voltammograms (Figure 12) show redox peaks relative to the oxidation states of PANI at the same potentials recorded for EP PANI and EPD PANI, respectively. The recorded current is higher for EP PANI than for EPD PANI. In Figure 13 it is reported a scheme of the three different kinds of CP based nanocomposites considered in this discussion.

At the best knowledge of the authors, only one paper reports on PANI deposited onto CNTs [86]. In that case, a CNTs solution was cast on the surface of an electrode that was dried at room temperature. Then PANI was deposited on CNTs by a conventional CV electropolymerisation. Only capacitive characteristics of this kind of layered material were studied.

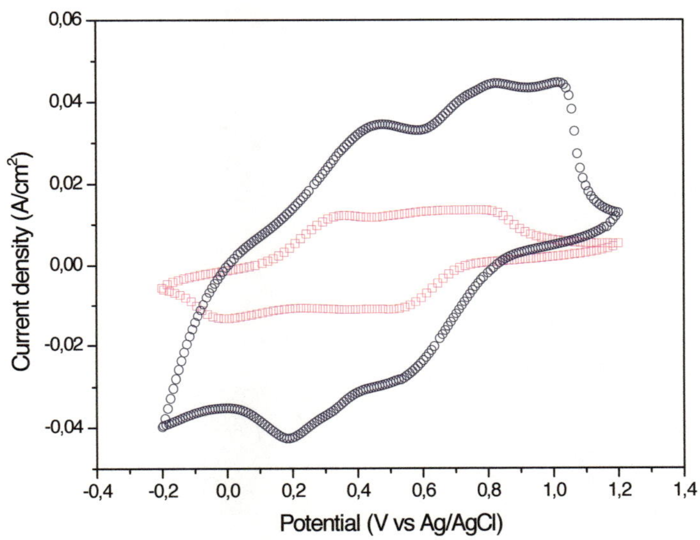

Figure 12. Cyclic voltammetric curve in 0.5 M H_2SO_4 relative to EPD PANI (magenta squares) and EP PANI (blue circles) deposited on CNTs.

5. Anticorrosion behaviour of hybrid conducting nanocomposites based on PANI and CNTs

In order to investigate the anodic performance of the PANI-CNTs nanocomposites, potentiodynamic scans have been acquired in a NaCl solution (Figure 14). The potentiodynamic curves relative to PANI and PANI-CNTs obtained by EP are overlapping, meaning that only PANI is active towards electrolytic species in the test solution. As mentioned before, CNTs are completely coated by PANI during the EP process and therefore their surface is not exposed to electrolytic solution. Moreover, this means PANI is very efficient in blocking the aggressive species, as no evidence of a contribution of CNTs to corrosion protection is observed. From a corrosion point of view, the highest amount of PANI electropolymerised in the presence of CNTs in solution does not influence the protective effect of the coating.

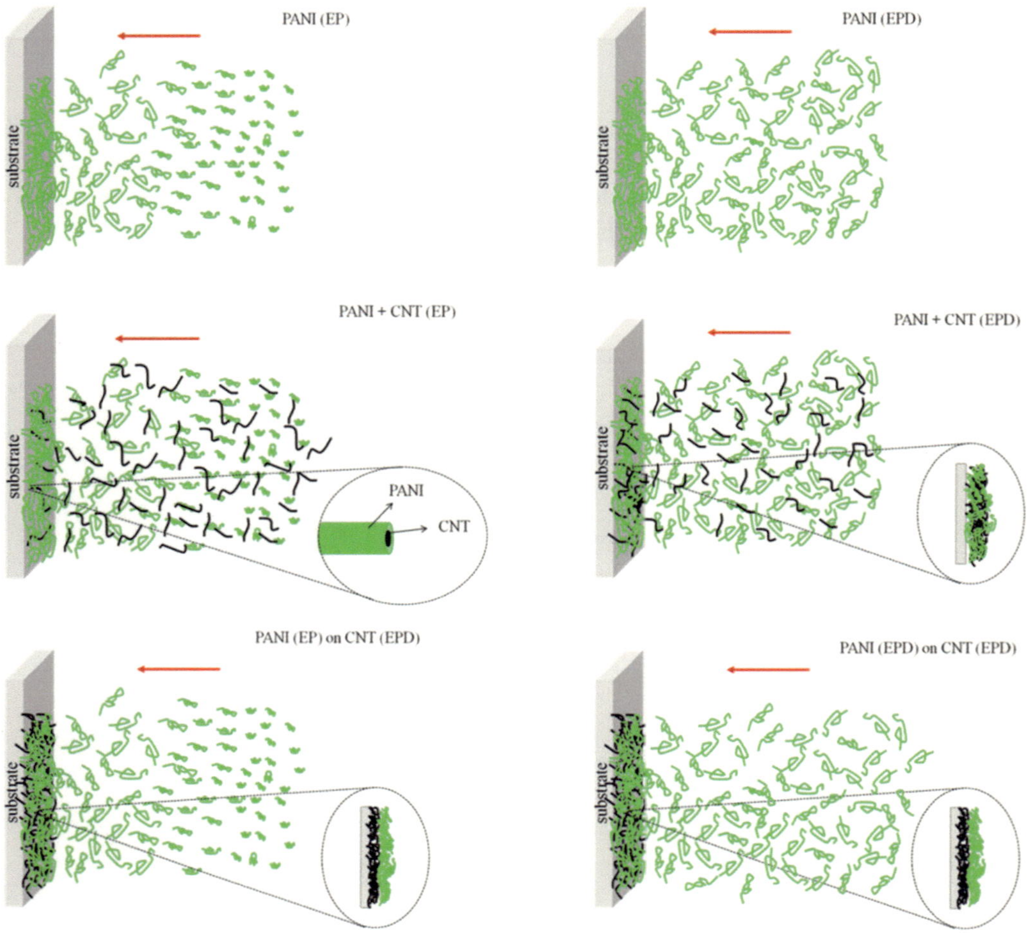

Figure 13. Scheme of the PANI-CNTs based samples. The arrow shows the direction of the polymer chains or the CNTs movement, from the solution towards the electrode by effect of the electric field.

On the contrary, corrosion potential is shifted towards more positive values when PANI is deposited on previously electrophoretically deposited CNTs. In this case, the exposed area is wider than a flat substrate, and therefore the amount of PANI electropolymerised is higher. It is reasonable to suppose that the increased deposited amount of conducting polymer causes the improved corrosion protection of PANI deposited on CNTs and therefore the corrosion potential results nobler.

In [87], EIS measurements on composites containing PANI and functionalised CNTs, deposited by EP, are reported. The aim of that work was to investigate the capacitance of PANI-CNTs composites and therefore the results there described are devoted to the understanding of the behaviour of the composite in consideration to this application. It is reported the real impedance at low frequencies, where the capacitive behaviour dominates, is an indication of the combined resistance of the electrolyte and the film including both electronic and ionic contributions. The values of the real impedance at 0.01 Hz of the PANI-functionalised CNTs films are significantly lower in resistance than PANI films. This result is attributed both to higher overall conductivity offered by PANI-functionalised CNTs than the PANI film, and to the increased porosity of composite material, that improves ionic accessibility. Moreover, it is suggested that the embedded CNTs provide interconnected pathways for electrons through the CNTs and for ions through the pore network or the direct interaction between the delocalised electrons on polymer chains and the CNTs. In principle, these hypotheses are prejudicial to corrosion protection because of the favourite access of aggressive species to coating-electrode interface.

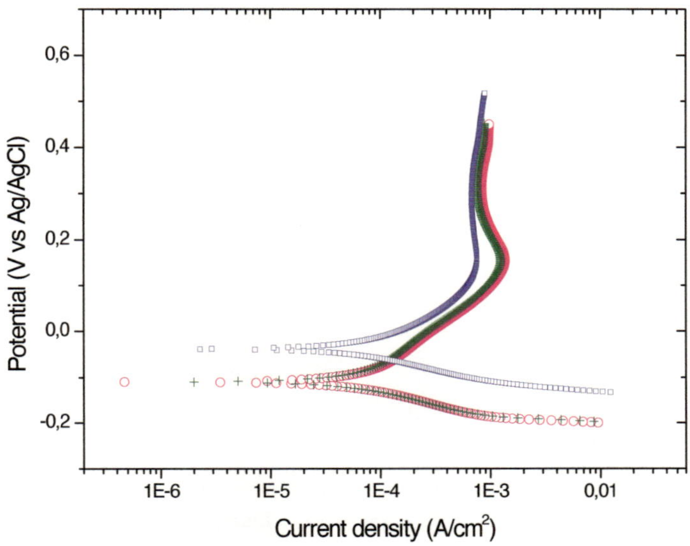

Figure 14. Potentiodynamic curves acquired in 3.5 % NaCl relative to EP PANI (green crosses), EP PANI-CNTs nanocomposite composite (magenta circles), and EP PANI on CNTs (blue squares)

Same authors investigated a similar composite material based on PANI and CNTs functionalised by Poly(m-aminobenzenesulfonic acid) (PABS) [70]. Differently from our case, they found that PANI-PABS-CNTs composite has a better anticorrosion behaviour than PANI. In fact, CNTs functionalisation highly influences the interaction with PANI and, as a consequence, the electrochemical behaviour of PANI-CNTs nanocomposite is modified. As an evidence of this affirmation, the authors observed that in CV performed on PANI electropolymerised with CNTs functionalised by COO_- and SO_4^{2-} group, the recorded current in the case of CNTs functionalised by SO_4^{2-} is almost three times higher than that of CNTs functionalised by COO_- group (Figure 15). In reference [87] it is shown that also the morphology of PANI-CNTs films is modified by the different functionalisation of CNTs.

The nanocomposite coatings based on PANI and CNTs prepared by EPD have an absolutely dissimilar behaviour. Both PANI films deposited on the bare electrode and on the modified electrode by CNTs have a similar E_{corr} and I_{corr}, whereas the nanocomposite film formed by

CNTs co-deposited with PANI has the best anticorrosive behaviour. The difference is attributed to a different interaction between CNTs and PANI (Figure 16).

In the EPD process, when PANI chains, more or less entangled, arrive at the electrode, they can coagulate and deposit onto it as a results of van der Waals attractive forces conveying at close distance [72]. As these forces are accomplished by any strong chemical bond, there is no difference if the electrode is bare or modified by CNTs previously deposited on it. Moreover, the PANI surface exposed to electrolyte solution does not show morphological differences with respect to PANI covering the modified electrode by CNTs (Figure 17).

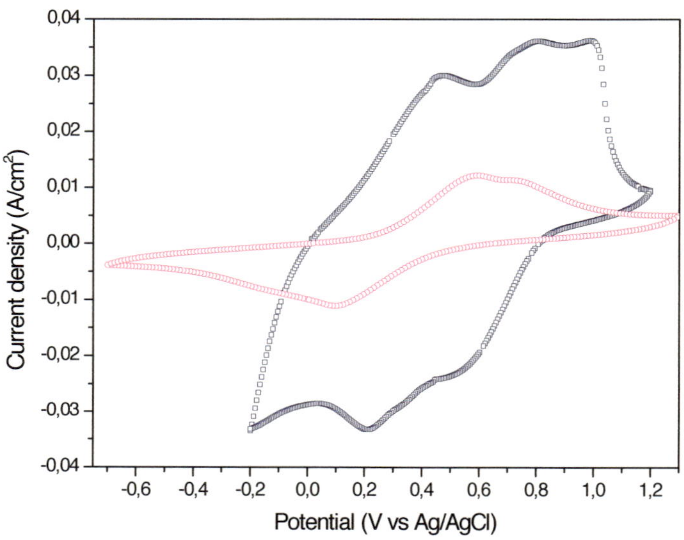

Figure 15. Cyclic voltammetric curve in 0.5 M H_2SO_4 relative to EP PANI-CNTs nanocomposite. CNTs were treated with H_2SO_4 (blue squares) or HNO_3 (magenta circles).

On the contrary, when CNTs and PANI chains are in the eletrophoretic suspension, chemical interactions occur, as depicted before. This fact entails two effects. The first one is an increased mass deposited by EPD, as demonstrated by the increased zeta potential and by CV acquired on PANI in the presence of CNTs. This result can be explained considering that a single CNT manages to catch a certain number of PANI chains, which are dragged towards the electrode during electrophoresis. The second effect consists of an increased electroactivity of PANI chains functionalising the CNTs surface, due to a faster charge transport in the composite film than in the PANI film, as emphasised before. For both these reasons, the corrosion potential relative to the PANI-CNTs coating is shifted towards less negative potentials. Definitively, the addition of CNTs to PANI is advantageous for the anticorrosion properties of the PANI based coating. In fact, CNTs allow to increase the amount of PANI deposited both in electropolymerisation and in electrophoretic co-deposition, and moreover, in the case of EPD process, improve the electroactivity of the nanocomposite coating.

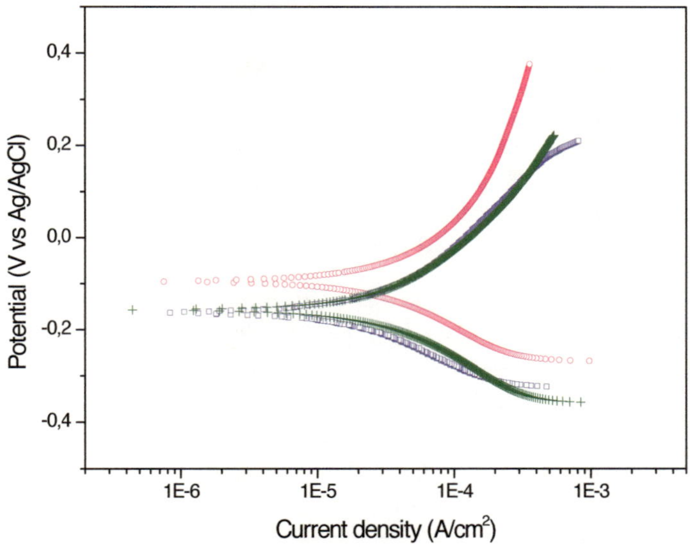

Figure 16. Potentiodynamic curves acquired in 3.5 % NaCl relative to EPD PANI (green crosses), EPD PANI-CNT nano-composite composite (magenta circles), and EPD PANI on CNTs (blue squares).

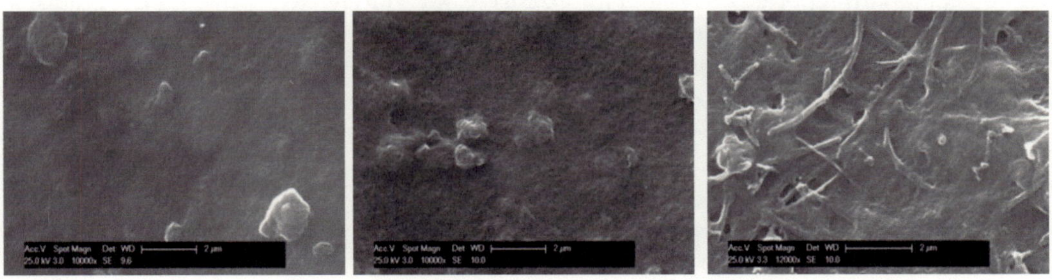

Figure 17. SEM images of the surface of EPD PANI (on the left), EPD PANI on CNTs (in the middle), and EPD PANI+CNTs (on the right).

6. Conclusions

The use of CPs as corrosion protection layers is partly motivated by the desire to replace coatings that are hazardous to the environment and to human health. Since the equilibrium potentials of several CPs are positive relative to those of iron and aluminum, they should provide anodic protection effects comparable to those provided by chromate (VI) or similar inorganic systems.

Either electropolymerisation or chemical oxidation of the respective monomer can be used to form the coating. Cheap and effective polymers, such as PANI, PPy, PEDOT, and PoPD (and their derivatives), have mostly been used to protect steel, aluminum, and copper, but the provided efficiency and the mechanism of the corrosion protection are not yet fully clarified. Several authors have discussed anodic protection on iron, others have proposed that the passivation is achieved because the doped emeraldine salt form of PANI keeps the potential of the underlying stainless steel in the passive region. However, other authors claim that the mechanism by which PANI protects the underlying metal surface from corrosion is independent of the doping level. Due to the redox processes taking place, more or less thick layers of iron oxide are formed and are stabilised against dissolution and reduction. Inhibition is also reasonable due to physical blocking and reduction of the active surface.

Another strategy for corrosion prevention using CPs is to incorporate inhibitor anions into the polymer coatings. CPs have been also used as primary films under an epoxy layer or additive in paint coatings. Very often it has been referred to the healing effect of CPs upon surface passivation along a defect or scratches. Most recent approaches consider the use of composite coatings based on CPs, containing nanoparticles or carbon nanostructures. CNTs have been incorporated in CPs matrices (PPY, PoPD, and PANI) but the principal characteristic studied has been the capacitive behaviour. Nevertheless, PANI-CNTs nanocomposites have demonstrated to have good corrosion protection properties, although the results are not fully concordant. The use of EPD allows us to produce PANI-CNTs nanocomposite coatings, with characteristics comparable with those of analogous nanocomposites obtained by conventional electrochemical methods.

All considerations reported above and the extensive literature related to this subject induce to conclude that the potential of conducting polymer coatings for corrosion protection is wide but also controversial. In general, the efficacy of CPs very much depends on how they are applied and on the conditions of the corrosion experiment. Therefore many efforts still have to be made in investigations and comprehensions of protective mechanisms in CPs and in hybrid conductive nanocomposites coatings.

Author details

M. Federica De Riccardis* and Virginia Martina

*Address all correspondence to: federica.dericcardis@enea.it

ENEA-Italian National Agency for New Technologies, Energy and Sustainable Economic Development, Technical Unit for Materials Technologies- Brindisi Research Centre (UTT-MATB), Brindisi, Italy

References

[1] Angelopoulos M. Conducting Polymers in Microelectronics. IBM Journal of Research and Development 2001;45 57-75.

[2] Cosnier S. Recent Advances in Biological Sensors Based on Electrogenerated Polymers: A Review. Analytical Letters 2007;40 1260-1279.

[3] Gurunathan K, Murugan AV, Marimuthu R, Mulik UP, Amalnerkar DP. Electrochemically Synthesized Conducting Polymeric Materials for Applications Towards Technology in Electronics, Optoelectronics and Energy Storage Devices. Materials Chemistry and Physics 1999;61 173-191.

[4] Brédas JL, Street GB, Polarons, Bipolarons, and Solitons in Conducting Polymers. Accounts of Chemical Research 1985;18 309-315.

[5] Waltman RJ, Bargon J. Electrically Conducting Polymers: A Review of the Electropolymerisation Reaction, of the Effects of Chemical Structure on Polymer Film Properties, and of Applications Towards Technology. Canadian Journal of Chemistry 1986;64 76-95.

[6] Geniès EM, Bidan G, Diaz AF. Spectroelectrochemical Study of Polypyrrole Films. Journal of Electroanalytical Chemistry 1983;149 101-113.

[7] Andrieux CP, Audebert P, Hapiot P, Savéant JM. Identification of the First Steps of the Electrochemical Polymerization of Pyrroles by Means of Fast Potential Step Techniques. Journal of Physical Chemistry 1991;5 10158-10164.

[8] Audebert P, Miomandre F. Electrochemistry of Conducting Polymers. In: Skotheim TA, Reynolds JR (eds.) Conjugated Polymers. Theory, Synthesis, Properties, and Characterization. Boca Raton: Taylor & Francis Group; 2007. p18-1/18-40.

[9] Pickup PG. Electrochemistry of Electronically Conducting Polymer Films. In: Ralph E. White et al. (Eds.) Modern Aspects of Electrochemistry, number 33. New York: Kluwer Academic/Plenum Publishers; 1999.

[10] Nigrey PJ, MacDiarmid AG, Heeger AJ. Electrochemistry of Polyacetylene CHx: Electrochemical Doping of CHx Films to the Metallic State. Journal of the Chemical Society, Chemical Communications 1979;14 594-595.

[11] Stilwell DE, Park SM. Electrochemistry of Conductive Polymers. III. Some Physical and Electrochemical Properties Observed from Electrochemically Grown Polyaniline. Journal of the Electrochemical Society 1988;135 2491-2496.

[12] Huang WS, Humphrey BD, MacDiarmid AG. Polyaniline: A Novel Conducting Polymer. Morphology and Chemistry of Its Oxidation and Reduction in Aqueous Electrolytes. Journal of Chemical Society Faraday Transactions I: Physical Chemistry Condensed Phases 1986;82 2385-2400.

[13] Fang J, Xu K, Zhu L, Zhou Z, Tang H. A Study on Mechanism of Corrosion Protection of Polyaniline Coating and Its Failure. Corrosion Science 2007;49 4232-4242.

[14] Hermas AA. XPS Analysis of the Passive Film Formed on Austenitic Stainless Steel Coated with Conductive Polymer. Corrosion Science 2008;50 2498–2505.

[15] Kilmartin PA, Trier L, Wright GA. Corrosion Inhibition of Polyaniline and Poly(O-Methoxyaniline) on Stainless Steels. Synthetic Metals 2002;131 99–109.

[16] Kendig M, Hon M, Warren L. 'Smart' Corrosion Inhibiting Coatings. Progress in Organic Coatings 2003;47 183–189.

[17] Dominis AJ, Spinks GM, Wallace GG. Comparison of Polyaniline Primers Prepared with Different Dopants for Corrosion Protection of Steel. Progress in Organic Coatings 2003;48 43–49.

[18] Ohtsuka T, Iida M, Ueda M. Polypyrrole Coating Doped by Molybdo-Phosphate Anions for Corrosion Prevention of Carbon Steels. Journal of Solid State Electrochemistry 2006;10 714–720.

[19] Ohtsuka T. Corrosion Protection of Steels by Conducting Polymer Coatings. International Journal of Corrosion 2012; Article ID 915090 doi:10.1155/2012/915090

[20] Martins dos Santos LM, Lacroix JC, Chane-Ching KI, Adenier A, Abrantes LM, Lacaze PC. Electrochemical Synthesis of Polypyrrole Films on Copper Electrodes in Acidic and Neutral Aqueous Media. Journal of Electroanalytical Chemistry 2006;587 67–78.

[21] Hasanov R, Bilgic S. Monolayer and Bilayer Conducting Polymer Coatings for Corrosion Protection of Steel in 1M H2SO4 Solution. Progress in Organic Coatings 2009;64 435–445.

[22] Joseph S, McClure JC, Sebastian PJ, Moreira J, Valenzuela E. Polyaniline and Polypyrrole Coatings on Aluminum for PEM Fuel Cell Bipolar Plates. Journal of Power Sources 2008;177 161–166.

[23] Ren YJ, Zeng CL. Effect of Conducting Composite Polypyrrole/Polyaniline Coatings on the Corrosion Resistance of Type 304 Stainless Steel for Bipolar Plates of Proton-Exchange Membrane Fuel Cells. Journal of Power Sources 2008;182 524–530.

[24] Panah NB, Danaee I. Study of the Anticorrosive Properties of Polypyrrole/Polyaniline Bilayer Via Electrochemical Techniques. Progress in Organic Coatings 2010;68 214–218.

[25] Pruna A, Pilan L. Electrochemical Study on New Polymer Composite for Zinc Corrosion Protection. Composites: Part B 2012;43 3251–3257.

[26] Johnston JH, Moraes J, Borrmann T. Conducting Polymers on Paper Fibres. Synthetic Metals 2005;153 65-68.

[27] Su W, Iroh JO. Electrodeposition Mechanism, Adhesion and Corrosion Performance of Polypyrrole and Poly(N-Methylpyrrole) Coatings on Steel Substrates. Synthetic Metals 2000;114 225-234.

[28] Bhandari H, Sathiyanaranayan S, Choudhary V, Dhawan SK. Synthesis and Characterization of Proccessible Polyaniline Derivatives for Corrosion Inhibition. Journal of Applied Polymer Science 2009;111 2328–2339.

[29] Yin P, Kilmartin PA. Formation of Poly-2,5-Dimethoxyaniline on Steels. Current Applied Physics 2004;4 141–143.

[30] Ma L, Huang C-Q, Gan MY. Synthesis and Anticorrosion Properties of Poly-2(3-dimethylaniline) Doped with Phosphoric Acid. Journal of Applied Polymers Science 2013;127 3699–3704.

[31] Bereket G, Hur E, Sahin Y. Electrochemical Synthesis and Anti-Corrosive Properties of Polyaniline, Poly(2-Anisidine) and Poly(Aniline-Co-2-Anisidine) Films on Stainless Steel. Progress in Organic Coatings 2005;54 63–72.

[32] Chaudhari S, Patil PP. Corrosion Protective Bi-Layered Composites of Polyaniline and Poly(O-Anisidine) on Low Carbon Steel. Journal of Applied Polymer Science 2008;109 2546–2561.

[33] Kowalski D, Ueda M, Ohtsuka T. Corrosion Protection of Steel by Bi-Layered Polypyrrole Doped with Molybdophosphate and Naphthalenedisulfonate Anions. Corrosion Science 2007;49 1635–1644.

[34] Kowalski D, Ueda M, Ohtsuka T. The Effect of Counter Anions on Corrosion Resistance of Steel Covered by Bi-Layered Polypyrrole Film. Corrosion Science 2007;49 3442-3452.

[35] Kowalski D, Ueda M, Ohtsuka T. Self-Healing Ion-Permselective Conducting Polymer Coating. Journal of Materials Chemistry 2010;20 7630–7633.

[36] Hosseini MG, Bagheri R, Najjar R. Electropolymerization of Polypyrrole and Polypyrrole-ZnO Nanocomposites on Mild Steel and Its Corrosion Protection Performance. Journal of Applied Polymer Science 2011;121 3159–3166.

[37] Tallman DE, Levine KL, Siripirom C, Gelling VG, Bierwagen GP, Croll SC. Nanocomposite of Polypyrrole and Alumina Nanoparticles as a Coating Filler for the Corrosion Protection of Aluminum Alloy 2024-T3. Applied Surface Science 2008;254 5452–5459.

[38] Yeh JM, Chin CP. Structure and Properties of Poly(O-Methoxyaniline)–Clay Nanocomposite Materials. Journal of Applied Polymer Science 2003;88 1072–1080.

[39] Yeh JM, Chin CP, Chang S. Enhanced Corrosion Protection Coatings Prepared from Soluble Electronically Conductive Polypyrrole-Clay Nanocomposite Materials. Journal of Applied Polymer Science 2003;88 3624-3672.

[40] Zhang Y, Shao Y, Zhang T, Meng G, Wang F. High Corrosion Protection of a Polyaniline/Organophilic Montmorillonite Coating for Magnesium Alloys. Progress in Organic Coatings 2013;76 804–811.

[41] Olad A, Rashidzadeh A. Preparation and Anticorrosive Properties of PANI/Na-MMT and PANI/O-MMT Nanocomposites. Progress in Organic Coatings. 2008;62 293-298.

[42] Olad A, Barati M, Shirmohammadi H. Conductivity and Anticorrosion Performance of Polyaniline/Zinc Composites: Investigation of Zinc Particle Size and Distribution Effect. Progress in Organic Coatings 2011;72 599– 604.

[43] Olad A, Rasouli H. Enhanced Corrosion Protective Coating Based on Conducting Polyaniline/Zinc Nanocomposite. Journal of Applied Polymer Science 2010;115 2221–2227.

[44] Kumar AS, Meenakshi SK, Sankaranarayanan TSN, Srikanth S. Corrosion Resistant Behaviour of PANI-Metal Bilayer Coatings. Progress in Organic Coatings 2008;62 285–292.

[45] Meroufel A, Deslouis C, Touzain S. Electrochemical and Anticorrosion Performances of Zinc-Rich and Polyaniline Powder Coatings. Electrochimica Acta 2008;53 2331–2338.

[46] Sathiyanarayanan S, Azim SS, Venkatachari G. Corrosion Protection of Galvanized Iron by Polyaniline Containing Wash Primer Coating. Progress in Organic Coatings 2009;65 152–157.

[47] Rathod RC, Umare SS, Didolkar VK, Shambharkar BH, Patil AP. Production and Characterization of PANI/TiO2 Nanocomposites: Anticorrosive Application on 316LN SS. Transactions of the Indian Institute of Metals 2013;66 97–104.

[48] Radhakrishnan A, Siju CR, Mahanta D, Patil S, Madras G. Conducting Polyaniline–Nano-TiO2 Composites for Smart Corrosion Resistant Coatings. Electrochimica Acta 2009;54 1249–1254.

[49] Diniz FB, De Andrade GF, Martins CR, De Azevedo WM. A Comparative Study of Epoxy and Polyurethane Based Coatings Containing Polyaniline-DBSA Pigments for Corrosion Protection on Mild Steel. Progress in Organic Coatings 2013;76 912–916.

[50] Akbarinezhad A, Ebrahimi M, Faridi HR. Corrosion Inhibition of Steel in Sodium Chloride Solution by Undoped Polyaniline Epoxy Blend Coating. Progress in Organic Coatings. 2009;64 361-364.

[51] Baldissera AF, Ferreira CA. Coatings Based on Electronic Conducting Polymers for Corrosion Protection of Metals. Progress in Organic Coatings 2012;75(3) 241-247.

[52] Ge CY, Yang XG, Hou BR. Synthesis of Polyaniline Nanofiber and Anticorrosion Property of Polyaniline-Epoxy Composite Coating for Q235 Steel. Journal of Coating Technology Research 2012;9(1) 29-69.

[53] Panah NB, Payehghadr M, Danaee I, Nourkojouri H, Sharbatdaran M. Investigation of Corrosion Performance of Epoxy Coating Containing Polyaniline Nanoparticles. Iranian Polymer Journal (English Edition) 2012;21 747-754.

[54] Armelin E, Pla R, Liesa F, Ramis X, Iribarren JI, Alemán C. Corrosion Protection with Polyaniline and Polypyrrole as Anticorrosive Additives for Epoxy Paint. Corrosion Science 2008;50 721–728.

[55] Zaarei D, Sarabi AA, Sharif F, Gudarzi MM, Kassiriha SM. A New Approach to Using Submicron Emeraldine-Base Polyaniline in Corrosion-Resistant Epoxy Coatings. Journal of Coatings Technology Research 2012;9(1) 47-57.

[56] Armelin E, Aleman C, Iribarren JI. Anticorrosion Performances of Epoxy Coatings Modified with Polyaniline: A Comparison Between the Emeraldine Base and the Salt Forms. Progress in Organic Coatings 2009;65 88-93.

[57] Chang CH, Huang TC, Peng CW, Yeh TC, Lu HI, Hung WI, et al. Novel Anticorrosion Coatings Prepared from Polyaniline/Graphene Composites. Carbon 2012;50 5044-5051.

[58] Martina V, De Riccardis MF, Carbone D, Rotolo P, Bozzini B, Mele C. Electrodeposition of Polyaniline–Carbon Nanotubes Composite Films and Investigation on Their Role in Corrosion Protection of Austenitic Stainless Steel by SNIFTIR Analysis. Journal of Nanoparticles Research 2011;13 6035–6047.

[59] Guo DJ, Li HL. Well-Dispersed Multi-Walled Carbon Nanotube/Polyaniline Composite Films. Journal of Solid State Electrochemistry 2005;9(6) 445 -449.

[60] Xu J, Yao P, Liu L, Jiang Z, He F, Li M, Zou J. Synthesis and Characterization of an Organic Soluble and Conducting Polyaniline-Grafted Multiwalled Carbon Nanotube Core–Shell Nanocomposites by Emulsion Polymerization. Journal of Applied Polymer Science 2010;118 2582–2591.

[61] Saini P, Choudhary V, Singh BP, Mathur RB, Dhawana SK. Polyaniline–MWCNT Nanocomposites for Microwave Absorption and EMI Shielding. Materials Chemistry and Physics 2009;113 919–926.

[62] Massuyeau F, Zhao Y, El Mel AA, Yaya A, Geschier F, Gautron E, et al. Improved Photoconductive Properties of Composite Nanofibers Based on Aligned Conjugated Polymer and Single-Walled Carbon Nanotubes. Nano Research 2013;6(2) 149–158.

[63] Hermas AA, Al-Juaid SS, Al-Thabaiti SA, Qusti AH, Abdel Salam M. In Situ Electropolymerization of Conducting Polypyrrole/Carbon Nanotubes Composites on Stainless Steel: Role of Carbon Nanotubes Types. Progress in Organic Coatings 2012;75 404–410.

[64] Abdel Salam M, Al-Juaid SS, Qusti AH, Hermas AA. Electrochemical Deposition of a carbon Nanotube-Poly(O-Phenylenediamine) Composite on a Stainless Steel Surface. Synthetic Metals 2011;161 153–157.

[65] Branzoi F, Branzoi V, Musina A. Coatings Based on Conducting Polymers and Functionalized Carbon Nanotubes Obtained by Electropolymerization. Progress in Organic Coatings 2013;76 632–638.

[66] Sun Y, Wilson SR, Schuster DI. High Dissolution and Strong Light Emission of Carbon Nanotubes in Aromatic Amine Solvents. Journal of the American Chemical Society 2001;123 5348-5349.

[67] Zengin H, Zhou W, Jin JY, Czerw R, Smith DW, Echegoyen L, et al. Carbon Nanotube Doped Polyaniline. Advanced Materials 2002;14 1480-1483.

[68] Cochet M, Masser WK, Benito AM, Callejas MA, Martinèz MT, Benoit JM, et al. Synthesis of a New Polyaniline/Nanotube Composite: "In-Situ" Polymerisation and Charge Transfer Through Site-Selective Interaction. Chemical Communications 2001;16 1450-1451.

[69] Peng C, Jin J, Chen GZ. A Comparative Study on Electrochemical Co-Deposition and Capacitance of Composite Films of Conducting Polymers and Carbon Nanotubes. Electrochimica Acta 2007;53 525-537.

[70] Ionita M, Branzoi IV, Pilan L. Multiscale Molecular Modeling and Experimental Validation of Polyaniline-CNTs Composite Coatings for Corrosion Protecting. Surface and Interface Analysis 2010;42 987–990.

[71] Hughes M, Chen GZ, Shaffer MSP, Fray DJ, Windle AH. Electrochemical Capacitance of a Nanoporous Composite of Carbon Nanotubes and Polypyrrole. Chemistry of Materials 2002;14, 1610-1613.

[72] De Riccardis MF. Ceramic Coatings Obtained by Electrophoretic Deposition: Fundamentals Models Post-Deposition Processes and Applications. In: Feng Shi ed.; Ceramic Coatings - Applications in Engineering: InTech;2012. p 43-68. Available from http://www.intechopen.com/books/ceramic-coatings-applications-in-engineering/ceramic-coatings-obtained-by-electrophoretic-deposition accessed 24 February 2012

[73] Boccaccini AR, Zhitomirsky I. Applications of Electrophoretic Deposition Techniques in Ceramic Processing. Current Opinion in Solid State & Materials Science 2002;6(3) 251-260.

[74] Boccaccini AR, Roether JA, Thomas BJC, Shaffer MSP, Chavez E, Stoll E, Minay EJ. The Electrophoretic Deposition of Inorganic Nanoscaled Materials. Journal of The Ceramic Society of Japan 2003;114(1) 1-14.

[75] Hamaker HC. Formation of a Deposit by Electrophoresis. Transactions of The Faraday Society 1940;35 279-287.

[76] De Riccardis MF, Martina V, Carbone D. Study of Polymer Particles Suspensions for Electrophoretic Deposition. Journal of Physical Chemistry B 2013;117(6) 1592-1599.

[77] Riul Jr.A, Mattoso LHC, Telles GD, Herrmann PSP, Colnago LA, Parizotto NA, et al. Characterization of Langmuir-Blodgett Films of Parent Polyaniline. Thin Solid Films 1996;285 177-180.

[78] Dhand C, Arya SK, Singh SP, Singh BP, Datta M, Malhotra BD. Preparation of Polyaniline/Multiwalled Carbon Nanotubes Composite by Novel Electrophoretic Route. Carbon 2008;46 1727-1735.

[79] Dhand C, Pratima R, Solanki R, Datta M, Malhotra BD. Polyaniline/Single-Walled Carbon Nanotubes Composite Base Triglyceride Biosensor. Electroanalysis 2010;22(22) 2683–2693.

[80] Jeon IY, Tan LS, Baek JS. Synthesis and Electrical Properties of Polyaniline/Polyaniline Grafted Multiwalled Carbon Nanotube Mixture Via In Situ Static Interfacial Polymerization. Journal of Polymer Science: Part A: Polymer Chemistry 2010;48 1962–1972.

[81] Li G, Martinez C, Semantick S. Controlled Electrophoretic Patterning of Polyaniline From a Colloidal Suspension. Journal of The American Chemical Society 2005;127 4903-4909.

[82] Dhand C, Singh SP, Arya SK, Datta M, Malhotra BD. Cholesterol Biosensor Based on Electrophoretically Deposited Conducting Polymer Film Derived from Nano-Structured Polyaniline Colloidal Suspension. Analytical Chimica Acta 2007;602 244–251.

[83] Dhand C, Sumana G, Datta M, Malhotra BD. Electrophoretically Deposited Nano-Structured Polyaniline Film for Glucose Sensing. Thin Solid Films 2010;519 1145–1150.

[84] Dhand C, Solanki PR, Sood KN, Datta M, Malhotra BD. Polyaniline Nanotubes for Impedimetric Triglyceride Detection. Electrochemistry Communications 2009;11 1482–1486.

[85] Dhand C, Arya SK, Datta M, Malhotra BD. Polyaniline–Carbon Nanotube Composite Film for Cholesterol Biosensor. Analytical Biochemistry 2008;383 194–199.

[86] Branzoi V, Branzoi F, Pilan L. Electrochemical Fabrication and Capacitance of Composite Films of Carbon Nanotubes and Polyaniline. Surface and Interface Analysis 2010;42 1266–1270.

[87] Branzoi F, Branzoi V, Musina A. Coatings Based on Conducting Polymers and Functionalized Carbon Nanotubes. Progress in Organic Coatings 2013;76 632– 638.

[88] Ozyilmaz AT, Akdag A, Karahan IH, Ozyilmaz G. The Influence of Polyaniline (PANI) Coating on Corrosion Behaviour of Zinc-Cobalt Coated Carbon Steel Electrode. Progress in Organic Coatings 2013;76(6) 993-997.

[89] Kamaraj K, Sathiyanarayanan S, Venkatachari G. Electropolymerised Polyaniline Films on AA 7075 Alloy and Its Corrosion Protection Performance. Progress in Organic Coatings 2009;64 67–73.

[90] Kamaraj K, Karpakam V, Sathiyanarayanan S, Venkatachari G. Electrosysnthesis of Poly(Aniline-Co-M-Amino Benzoic Acid) for Corrosion Protection of Steel. Materials Chemistry and Physics 2010;122 123–128.

[91] Duran B, Turhanb MC, Bereketa G, Sarac AS. Electropolymerization, Characterization and Corrosion Performance of Poly(N-Ethylaniline) on Copper. Electrochimica Acta 2009;55 104–112.

[92] Rizzi M, Trueba M, Trasatti SP. Polypyrrole Films on Al Alloys: The Role of Structural Changes on Protection Performance. Synthetic Metals 2011;161 23–31.

[93] Shinde V, Patil PP. Evaluation of Corrosion Protection Performance of Poly(O-Ethyl Aniline) Coated Copper by Electrochemical Impedance Spectroscopy. Materials Science and Engineering B 2010;58 142–150.

[94] Hasanov R, Bilgic S, Gece G. Experimental and Theoretical Studies on the Corrosion Properties of Some Conducting Polymer Coatings. Journal of Solid State Electrochemistry 2011;15 1063-1070.

[95] El-Shazly AH, Wazan AA. Investigation of the Performance of the Galvanic Coupling of Polyaniline Coated Steel and Zinc in Seawater. International Journal of Electrochemical Science 2011;6 337–347.

[96] Dhanabal T, Amirthaganesan G, Ravichandran J. Pitting Corrosion Protection of Low Nickel Stainless Steel by Electropolymerized Conducting Polymer Coating in 0.5 M NaCl Solution. Bulletin of Materials Science 2011;34 563-569.

[97] Peng CW, Chang KC, Weng CJ, Lai MC, Hsu CH, Hsu SC, et al. Nano-Casting Technique to Prepare Polyaniline Surface with Biomimetic Superhydrophobic Structures for Anticorrosion Application. Electrochimica Acta 2013;95 192–199.

[98] Chaudhari S, Gaikwad AB, Patil PP. Synthesis and Corrosion Protection Aspects of Poly(O-Toluidine)/CdO Nanoparticle Composite Coating on Mild Steel. Journal of Coating Technology and Research 2010;7 119-129.

[99] Olad A, Nosrati R. Preparation and Corrosion Resistance of Nanostructured PVC/ZnO-Polyaniline Hybrid Coating. Progress in Organic Coatings 2013;76 113-118.

[100] Olad A, Naseri B. Preparation, Characterization and Anticorrosive Properties of a Novel Polyaniline/Clinoptilolite Nanocomposite. Progress in Organic Coatings 2010; 67 233-238.

[101] Hermas AA. Protection of Type 430 Stainless Steel Against Pitting Corrosion by Ladder Conductive Polymer. Progress in Organic Coatings 2008;61 95-102.

[102] Chen F, Liu P. Conducting Polyaniline Nanoparticles and Their Dispersion for Waterborne Corrosion Protection Coatings. Applied Materials & Interface 2011;3 2694-2702.

[103] Olad A, Barati M, Behboudi S. Preparation of PANI/Epoxy/Zn Nanocomposite Using Zn Nanoparticles and Epoxy Resin as Additives and Investigation of Its Corrosion Protection Behavior on Iron. Progress in Organic Coatings 2012;74(1) 221-227.

[104] Chen S, Zhu J, Zhou T, He B, Huang W, Wang B. Preparation and Properties Study of Polyaniline Conductive Anti-Fouling Coatings. International Journal of Electrochemical Science 2012;7 8170– 8184.

[105] Yi Y, Liu G, Jin Z, Feng D. The Use of Conducting Polyaniline as Corrosion Inhibitor for Mild Steel in Hydrochloric acid. International Journal of Electrochemical Science 2013;8 3540 – 3550.

[106] Arefinia R, Shojaei A, Shariatpanahi H, Neshati J. Anticorrosion Properties of Smart Coating Based on Polyaniline Nanoparticles/Epoxy-Ester System. Progress in Organic Coatings 2012;75 502–508.

[107] Elkais AR, Gvozdenović MM, Jugović BZ, Grgur BN. The Influence of Thin Benzoate-Doped Polyaniline Coatings on Corrosion Protection of Mild Steel in Different Environments. Progress in Organic Coatings 2013;76(4) 670-676.

[108] Zubillaga O, Cano FJ, Azkarate IS, Molchan IS, Thompson GE, Skeldon P. Anodic Films Containing Polyaniline and Nanoparticles for Corrosion Protection of AA2024T3 Aluminum Alloy. Surface & Coatings Technology 2009;203 1494-1501.

[109] Alam J, Riaz U, Ahmad S. Development of Nanostructured Polyaniline Dispersed Smart Anticorrosive Composite Coatings. Polymers for Advanced Technologies 2008;19 882-888.

[110] Alvi F, Ram MK, Basnayaka PA, Stefanakos E, Goswami Y, Kumar A. Graphene-Polyethylenedioxythiophene Conducting Polymer Nanocomposite Based Supercapacitor. Electrochimica Acta 2011;56 9406-9412.

[111] Yagan A, Pekmez NO, Yildiz A. Inhibition of Corrosion of Mild Steel by Homopolymer and Bilayer Coatings of Polyaniline and Polypyrrole. Progress in Organic Coatings 2007;59 297–303.

[112] Yu Q, Xu J, Liu J, Li B, Liu Y, Han Y. Synthesis and Properties of PANI/SiO2 Organic-Inorganic Hybrid Films. Applied Surface Science 2012;263 532-535.

Chapter 14

An ABS Recycled Coating for Corrosion Protection and Conservation of Copper and Alloys of Cultural or Historic Value

C. Menchaca-Campos, M. Hernández-Escampa,
F. Rodríguez-Acuña, F. Millán-Cruz,
P. Rodríguez-Rojas, M. Hernández-Gallegos,
R. Guardian and J. Uruchurtu

Additional information is available at the end of the chapter

http://dx.doi.org/10.5772/57059

1. Introduction

During the 19th century, interest to establish chronologies to explain human past developments appeared. Such efforts were based on socio-cultural aspects, sustaining the argument of consecutive social phases such as: savagery, barbarian and civilization stages [1]. In 1836, Christian Jurgensen Thomsen [1] proposed an alternative conception based on industrial and technological development aspects [2]. The new created system proposed three stages to understand Prehistory: Stone, Bronze and Iron ages, which although no longer unanimously accepted, still illustrate the close relationship between the development of metallurgy and human culture. Therefore the history of metal technologies is considered a fundamental part of human history [3].

Cultural heritage conservation constitutes a relevant field of research, which frequently relies on material science to solve the challenges posed by the degradation of artifacts, architectural structures and ornaments [4-10]. Historic, artistic and archaeological items are frequently built of metals (Figure 1). Through the ages, copper and its alloys (bronze) constitute a particular set of metals widespread used in such manufactures [11-15]. Even when these materials tend to form protective patinas due to their interactions with the environment [16-20], such natural protective layers may fail, especially in highly polluted atmospheres [21-23].

The first step in the care of collections is to understand and minimize or eliminate conditions that can cause damage. The second step is to follow basic guidelines for care, handling and cleaning. Collections of un-conserved or ill-conserved archaeological artifacts might be unstable if they do not receive appropriate conservation treatment. It is not within the scope of this chapter to address in detail the problems associated with outdoor bronze sculptures [24].

Two or more metals are combined to form an alloy. Brass and bronze are copper alloys obtained by mainly adding zinc and tin respectively. Alloys generally have a different appearance or working properties compared to their component metals. Other alloys of copper include "gunmetal", bell metal, and German silver, also called nickel silver or "paktong" [24]. Brass is an alloy of about 70% copper with about 30% zinc. It may be cast into solid shapes, rolled into thin sheets, spun into vessel shapes or drawn wire. Brass is often found in scientific instruments, scale models and historical apparatus of all kinds. In its natural state it is typically light yellow-gold in appearance [25, 26].

Historic cast bronze has been widely used since antiquity to create weapons, sculptures and decorative objects. Bronzes are traditionally patinated and in such state they appear anywhere from light green to dark brown in color. Patinas are sometimes described as any controlled corrosion that imparts an aesthetically pleasing color and/or texture to the artifact. Patinas may be applied using chemicals or may have accumulated over time naturally [24]. Ancient cast bronze is usually composed of 90% copper, 6% tin and 4% zinc. However, different percentages of tin content have been reported between 11% [27] up to 25% [6].

Another category of artifacts that should be included in any discussion of brass and bronze is "bronzed" and plated finishes. Sometimes called "French Bronze", in the nineteenth century, paint or varnish containing bronze pigments were applied to cast iron or soft white metal [28]. These painted finishes may develop copper oxidation, but should not be polished. Another method of bronzing involved applying very thin copper plating to a metal surface; this kind of finish does not hold up to polishing either. Colored lacquers were sometimes applied to scientific instruments, and should be carefully considered as an original patina on a bronze artifact [25].

1.1. Causes of damage

Corrosion, poor handling and inappropriate storage are the major causes of damage to metal artifacts. Copper alloys are relatively stable, corrosion-resistant materials if they are properly cared for. Active corrosion causes a loss of metal from the object, and is most closely associated, with an environment which is heavily polluted, or subjected to salt water evaporation. Salts, oils and moisture can cause corrosion development staining copper alloys [24]. Highly polluted urban or industrial atmospheres are prone to produce acidic rainfall. Under such low pH conditions crystallized copper sulfates dissolve from the corrosion crust initiating complex corrosion processes [29, 30]. Long term corrosion of bronze archaeological artifacts in soil has been related to acidification and salt pollution [31, 32]. Other alloys as nickel-aluminum bronzes have shown increased corrosion rates in sulfide polluted seawaters [33].

1.2. Corrosion

Uncoated copper and alloys artifacts that have been kept clean and dry will usually develop stable surfaces. Stable surfaces may appear reddish, black to brown, or green to blue. The brownish and black colors may result from natural, non-destructive oxidation of the copper, which is sometimes named "tarnish". Thick crusts may accumulate, but dense corrosion that does not progress may actually protect the object if it is left intact. Purposely applied, "patinas", which may be any variety of colors, usually act as a protection to the metal [24, 25].

The development of small spots of light green powder that grow rapidly, is indicative that the artifact is suffering from what is commonly called "bronze disease" and indicates active corrosion [34, 35]. As mentioned above, green surfaces are often seen but may not be cause for alarm, if the corrosion layer is continuous and does not flake off easily. Bright or 'waxy" whitish-green powder forming in small areas on the surface, or in cavities on the metal surface, indicates advanced active corrosion. If left in its current state, this type of corrosion can cause a significant amount of damage to historic copper alloys, as it causes pitting of the surface and a continuous loss of metal [36]. This condition is sometimes noticed when green powder falls from an undisturbed object. It is caused by the presence of salts in the air, or by deposits left behind from inappropriate cleaning or handling and progresses when the air is humid. High levels of ammonia pollution in the air will cause bluish deposits of corrosion. In rare cases, patinas contain corrosive ingredients that may contribute to corrosion [25].

Corrosion will progress on copper alloys in the presence of high relative humidity (RH), normally above 70%. However, an upper safe limit of 46% is considered to avoid bronze disease [34, 37]. As for most metals, the critical RH may be lower in polluted atmospheres. Dust and grime left to accumulate on metal artifacts will actually hold moisture to the surface, and may induce corrosion even where the humidity is not that high.Varnished or lacquer brass and bronze can usually withstand corrosion as long as the coating is not broken. Where there are losses in the coating, corrosion will progress, usually seen as dark streaking where unprotected areas have oxidized [26, 38].

Polishing exposes fresh, reactive metal to the atmosphere and, therefore, to further oxidation. It might be desirable to coat objects that will be displayed at a museum exhibition. A lacquer coating to valuable silver and brass generally is applied, for their protection [39]. This also mitigates the need for frequent polishing, which inevitably wears away the metal surface if done frequently. A spray applied lacquer or wax coating are used commonly on large areas of metal as the best coatings. Wax provides a relatively flexible coating that is easily applied and that can be renewed. It can be used on top of original patinas and lacquers that are not meant to be disturbed, as long as they are cleaned first [40, 41].

On the other hand, acrylonitrile-butadiene-styrene (ABS) waste and residues represent a well-known discarded by-product from the automotive industry [42, 43]. Since law usually forbids the recycling of this material because all pieces must be brand new, considerable amounts of ABS become unused waste and a nuisance to dispose due to environmental restrictions. Therefore, research has been done trying to find usable recycling applications with added value, for this and other polymer discards. Due to the finite reserve of hydrocarbon in the

(a) (b) (c)

Figure 1. Copper and bronze historical and cultural artifacts.(a) Monument to the Revolution, Mexico City. (b) Morelos Monument, Universidad Autónoma del Estado de Morelos, Cuernavaca, Mexico. (c) 19th Bronze Bell, Cathedral of Cuernavaca, Mexico.

world, a proposal has been to obtain degradation oils from plastics, potentially useful as fuel [44]. However, such processes cannot avoid the presence of heteroatoms such as chlorine from poly(vinylchloride) (PVC) or nitrogen from ABS. These heteroatoms relate to corrosion of metallic parts and participate in the formation of dangerous compounds such as cyanic products, SOx or NOx, in the combustion process [45, 46].

The mechanical and chemical properties of ABS have been studied demonstrating its durability, elongation to break and impact resistance. However, thermo-oxidation processes can degrade this polymer due to hydrogen subtracted by oxygen [46]. Two models explain the ABS degradation: heat aging or physical aging [47-50]. It is well known that this polymer can be dissolved in organic solvents; also, thin layers become practically transparent. The application of coatings to both, clean or patina covered metallic surfaces has proved to be useful to preserve cultural heritage [51]. Two highly desirable conditions of conservation 'oriented' coatings are: transparency in order not to affect the aesthetical traits of the artifacts, as well as reversibility, which refer to an easy removal process [52]. Therefore, an ABS coating potentially fulfills some of the major requirements for coatings used in cultural heritage conservation. Hence, a combination of adequate properties and ecological values due to waste recycling and added value might be implied in the utilization of this coating for the stated purpose.

2. ABS coating

A coating development has been proposed recurring to the dissolution properties of ABS in acetone, applied by immersion of coupons. Characterization and evaluation of the protective properties of recycled ABS thin films on copper and bronze surfaces were performed. This was done simulating the degradation of the coatings exposed to a highly corrosive ambient. A comparison was made with the behavior of aliphatic polyurethane, a commonly used commercial coating for conservation purposes, as reported [51].

The ABS waste solution preparation consisted on cutting the polymer into small pieces and afterwards, dissolving them in acetone under constant stirringat room temperature. In this

http://dx.doi.org/10.5772/57059

way, the obtained solution was ready for the metal samples to be dip coated and then removed, allowing the acetone solvent to evaporate, at various cycles and times of immersion.The metallic coupons were immersed in consecutive: one (ABS-1), two (ABS-2) and three (ABS-3) thirty seconds per immersion cycles in the ABS solution. So the best preparation conditions were attained [51].

2.1. Characterization and evaluation

The recycled ABS coating was characterized in two different states: in raw form and in solution. Thermal analysis and Fourier Transformed Infrared Spectroscopy (FTIR) with Attenuated Total Reflectance (ATR) were used, being extremely reliable and recognized fingerprint characterization methods [53].

2.2. Polymer characterization

Figure 2a shows the TGA/DTA measurement results for recycling ABS, as it is discarded. The TG curve shows one stage weight decrease between 300 and 500°C, due to thermal decomposition with the onset in 400°C, remaining just 10% at 500°C, which corresponds to additives, fillers and other aggregates present in the copolymer because of the application for which it was tailored for. At the same time, the DTA curve shows two main peaks, a small one around 220°C due to melting of the elastic butadiene, and a bigger one that initiates at 420°C corresponding to the ABS copolymer decomposition [54].

The IR spectra of ABS in raw and in solution are shown in Figure 2b. Both present characteristic peaks of the three main components, but in solution they are clearly seen: the nitrile band of acrylonitrile at 2237 cm-[1], the aromatic ring of styrene at 1602, 1494, 761, and 699 cm-[1], and the double bond of butadiene at 967 cm-[1] trans and 911 cm-[1] vinyl. The rest of non-identified peaks in the spectra correspond to acetone major absorptions [55].

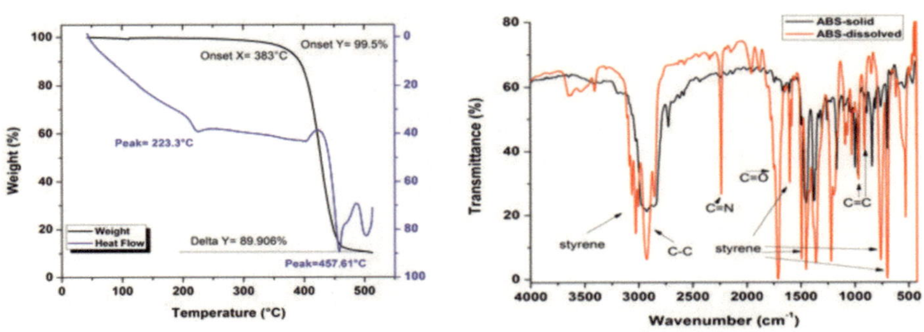

(a) Thermal analysis (b) Infrared spectroscopy

Figure 2. ABS characterization.

ABS shows accelerated thermal degradation due to aging at elevated temperatures above 80 °C. The impact resistance decreases dramatically beyond a critical aging time at 120 °C and this reduction depends on surface property modifications during aging. Visual examination of specimen cross-sections after aging, verifies that discoloration is limited to a surface layer, which is characteristic of degradation where oxygen diffusion into the bulk is limited [55]. Also, degradation causes an increase in Young's modulus at the specimen surface, which in turn promotes brittle failure. Absorbance bands from Fourier transform infrared spectroscopy indicate that surface degradation proceeds by chain scission and cross-linking in the polybutadiene (PB) phase of aged ABS specimens [54]. Dynamic mechanical thermal analysis also supports the occurrence of cross-linking, as shown by an increase in the glass transition temperature of the PB phase after aging. Although degradation in the styrene–acrylonitrile (SAN) phase is less significant to the reduction in overall mechanical properties of ABS compared to the PB phase, an assessment of SAN copolymer indicates that heat aging decreases impact resistance [55]. As temperature and aging time influence degradation, it is proposed that at ambient service temperatures (40 °C or less), the degradation mechanism differs to that at elevated temperatures, and comprises both surface and bulk polymer degradation effects. These characteristics support the idea of the application of ABS as a coating for ambient temperature, as obtained and reported [54-56].

To evaluate the corrosion performance, electrochemical methods were performed in commercial copper and bronze (73% Cu, 25% Sn and, 2% Zn by wt.) coupons. All electrochemical evaluations (Figure 2) performed include: corrosion potential, electrochemical noise measurement, impedance spectroscopy, and potentiodynamic polarization, the details being reported elsewhere [57-61].

A first group of samples were used as blank to evaluate the electrochemical properties of the base materials in two different solutions. As a reference, a second group of coupons were coated with commercial aliphatic polyurethane (varnish). The application consisted of a single layer using a small brush, simulating the manual procedure, frequent in conservation.

A third group of samples were coated by three successive dip coating immersions in the ABS solution. Coupons with different immersion cycles (ABS-1, ABS-2, and ABS-3) were evaluated electrochemically as a function of time for two weeks in the test solutions.The first solution, Na_2SO_4 0.1M, is mildly aggressive while the second one, Na_2SO_4 0.1M + 3% wt. NaCl, presents highly corrosive properties, simulating an aggressive urban-marine atmosphere. Also, ABS coated copper and bronze samples were scribed with a cutter to evaluate the coating adherence, using the *"pull-up"* test, after 2 weeks of immersion.

2.3. Coating thickness measurements

Scanning electron microscope (SEM) was used to determine the thickness of the coatings for each case. Alternatively, five measurements over the coated coupons surface were made using a coating thickness gauge, obtaining an average coating thickness over the metal sample.

Figure 3a presents the coating thickness after one immersion with an interval between 14 to 22 μm, looking not very compact and disbonded. After the first immersion, the procedure was repeated once (ABS-2) (Figure 3b) and twice (ABS-3) for 30 s (Figure 3c). The coating looks compact but maintaining a thickness around 13 to 16 μm. These results suggest that immersion

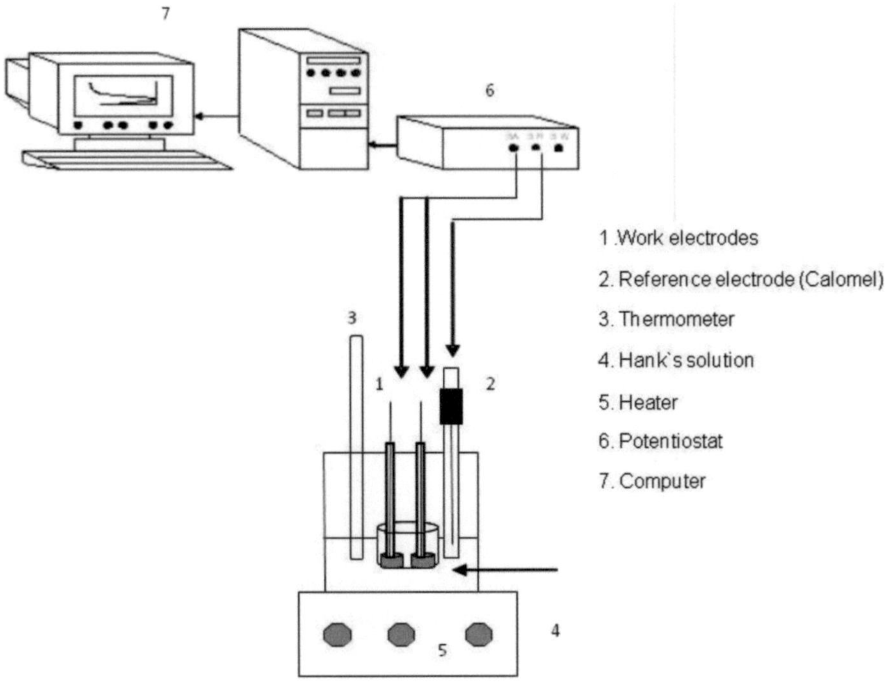

Figure 3. Electrochemical setup.

cycles did not increase the coating thickness, but rather, pores present after the first immersion were sealed and the compactness was improved (see below). Also during the second and third immersions, the ABS solution promotes partial removal of the top of the coating due to the action of acetone.

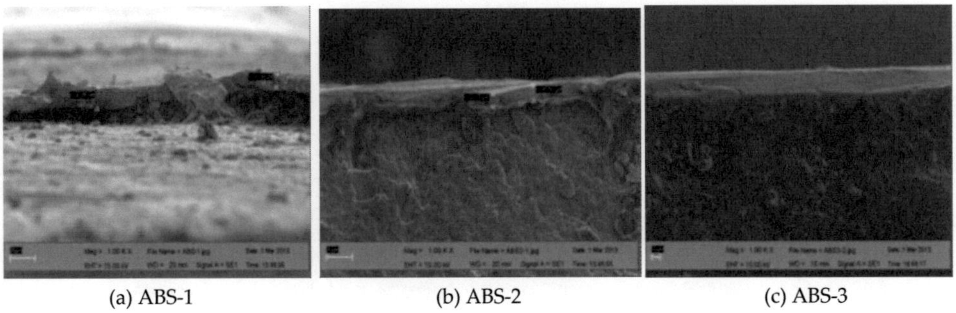

(a) ABS-1 (b) ABS-2 (c) ABS-3

Figure 4. Coating thickness measurement by SEM.

For varnish brush painted metal surfaces, the average thickness was 144 and 297 μm for copper and bronze samples, respectively. For ABS-3 deposited coatings the average thickness after three immersion cycles, was 36.8 and 41.2 μm, for copper and bronze respectively (table 1). The varnish coatings were an order of magnitude thicker as compared to the ABS-3 ones.

Sample	Coating Thickness (μm)					Average (μm)
Cu-ABS-3	40	34	35	34	41	36.8
Cu-varnish	134	133	144	112	142	133
Bronze-ABS-3	44	37	28	56	41	41.2
Bronze-varnish	290	329	134.7	422	311	297

Table 1. Coating thickness

3. Results and discussion

3.1. Electrochemical evaluation

Table 2 presents free corrosion potentials as a function of time of immersion for the metal coating schemes tested. The potential evolution reflects the degradation process of the coating, starting with noble potential values becoming more active as time elapses. ABS-3 coated metal samples presented more negative values than varnish coatings, especially for bronze coated coupons. Values are similar to the ones registered for bare (blanks) copper and bronze samples [6, 60, 61].

Time (h) /Coating System	Copper-Varnish	Copper-ABS-3	Bronze-Varnish	Bronze-ABS-3
	E_{corr} (mV)			
0	-49	-98	-3.3	-108
24	-108	-127	-3.2	-134
48	-111	-133	-3.1	-211
168	-114	-241	-16	-382
360	-324	-246	-336	-557
Rest Potential. Polished surface in Na_2SO_4+NaCl	-254		-299	

Table 2. Rest potentials of copper and bronze coupons in time.

Figure 4 shows the polarization curves of the copper and bronze coupons used as blank obtained in Na_2SO_4 0.1M and Na_2SO_4 0.1M + 3% wt. NaCl solutions. It also includes the curve corresponding to the varnish coated coupons in the aggressive solution. The corrosion rates of copper increase two orders of magnitude in comparison to the mild solution. In contrast, the copper coated with varnish decreases the corrosion rates as much as three orders of magnitude compared to the uncoated copper in the aggressive solution, demonstrating the protectiveness of the coating. The performance of the varnish is considered as reference to compare the protectiveness of the polymer coating under evaluation.

http://dx.doi.org/10.5772/57059

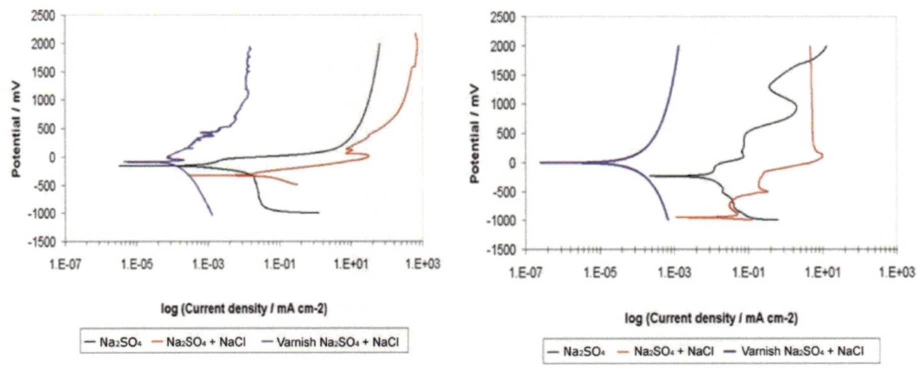

(a) Copper (b) Bronze

Figure 5. Polarization curves of blanks and coated with varnish samples in Na_2SO_4, Na_2SO_4 + 3 wt. % NaCl solution.

In Figure 5 it is shown the polarization curves of the varnish coated copper (Figure 5a) and bronze (Figure 5b) coupons in aggressive solution for different times of immersion up to two weeks. As time elapses, the corrosion rates increases suggesting the degradation of the coating, in the case of copper samples. For bronze is the other way around. It is coherent with the presence of passivation regions after 24 to 48 hours probably due to the oxidation of the base material and corrosion products are formed, as the solution reaches it through the coating pores or cracks. Nevertheless, the current densities observed are in the same order of magnitude as expected [6].

A high tin content in a bronze tends to inhibit corrosion, a fact which is suggested to be connected with the precipitation of insoluble SnO_2xH_2O on anodic sites. On the other hand, an alloy with high zinc and low lead content tends to exhibit the highest corrosion rate in some corrosive environments [6, 62-64].

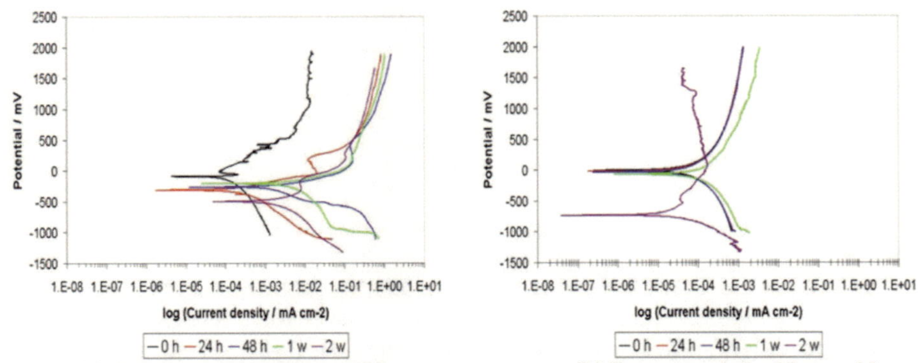

(a) Copper coated with varnish (b) Bronze coated with varnish

Figure 6. Polarization curves for samples in Na_2SO_4 + NaCl solution, as a function of time.

Figure 6 presents the electrochemical noise and impedance measurements for the varnish coated copper and bronze coupons. The Nyquist impedance diagrams show that the total impedance values decrease in time although there is an increment after two weeks, which once again is related to the corrosion products formed as the coating deteriorates. This behavior was registered for both metals. Such oxidation can start as early as 24 to 48 hours according to the polarization curves. The electrochemical noise resistance values show similar behavior as the impedance measurements, although an increase can be seen from week one. For the bronze sample (Figure 6b) the noise impedance at the time of immersion presents the lowest value in contrast with the copper electrode (Figure 6a).Therefore, the results of all the electrochemical techniques are coherent to each other since they show similar trends (see below Figure 10).

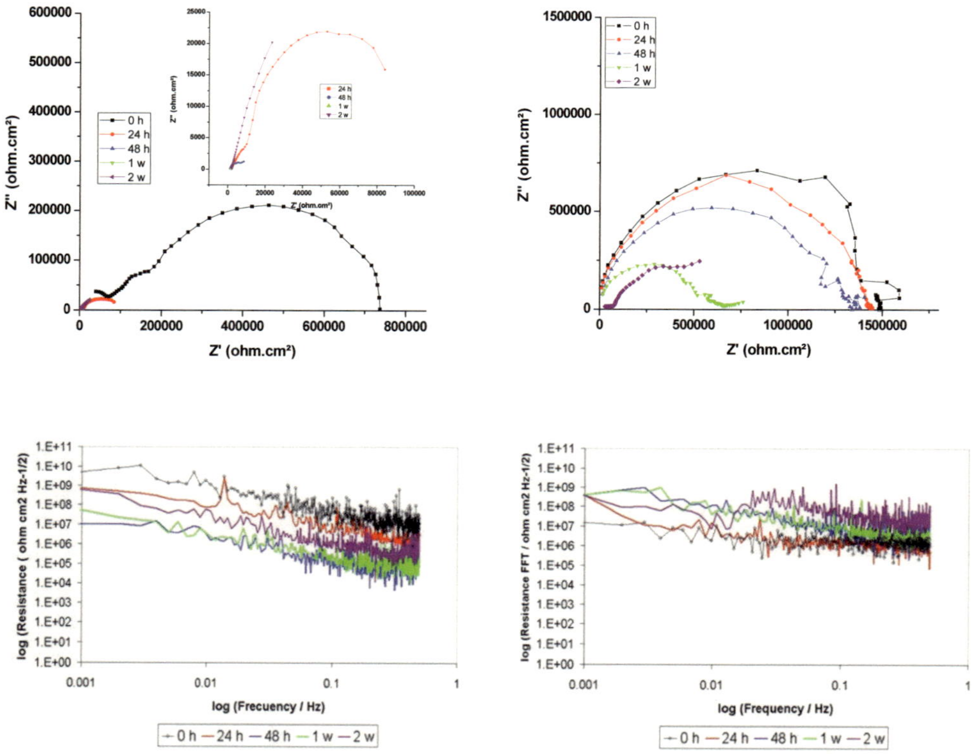

(a) Copper coated with varnish

(b) Bronze coated with varnish

Figure 7. Nyquist and electrochemical noiseimpedance diagrams for samples in Na_2SO_4 + 3 wt. % NaCl solution, as a function of time.

Figure 7 shows the Nyquist diagrams for each of the three coating cycles: ABS-1, ABS-2, and ABS-3. For ABS coated copper electrodes (Figure 7a), even when no significant increase of thickness was achieved with the consecutive immersions, the total impedance increased as a function of the number of immersion. This can be related to the fact that each immersion sealed the pores of the polymer layer. The total impedance values of the ABS-3 coupons reached similar values compared to the varnish coated coupons. In the case of bronze samples (Figure 7b), total impedance values are lower than the copper samples, but with a similar trend. Nevertheless, the impedance behavior observed for the first two immersions suggest a highly capacitive response, therefore a good protective coating film.

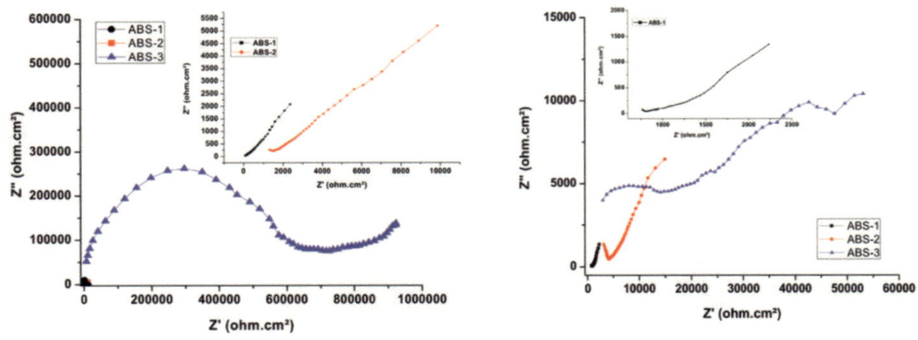

(a) ABS coated copper (b) ABS coated bronze

Figure 8. Nyquist impedance diagrams for samples in Na_2SO_4 + 3 wt. % NaCl solution.

Figure 8 shows the polarization curves of the ABS-3 copper and bronze coupons. In the initial immersion time of the ABS-3 copper sample (Figure 8a), there is a well-defined passivation region suggesting corrosion and film formation process in the surface of the base material. As time elapses, the passivation region tend to diminish and the corrosion rates increase as the coating deteriorates. Corrosion potentials become more negative in time indicating surface activation of the metal. For coated bronze samples (Figure 8b), the corrosion potential becomes active and the passive current increases as a function of time, so the corrosion rate indicating a deterioration of the coating film.

Figure 9 presents the impedance and electrochemical noise measurements as a function of time for the ABS-3 coated copper and bronze coupons. The Bode and noise impedance diagrams (Figure 9a) show that the total impedance values decrease in time without any increment as observed in the case of the varnish sample.The electrochemical noise impedance values also show a general decrease, compared with the initial time of immersion. However, there is an increase in impedance values after one week. The same trend was observed for the impedance values of the ABS-3 coated bronze electrodes (Figure 9b). This behavior is similar to the one observed for the varnish coated samples and suggest that after coatings deteriorate, a protective film over the metal samples is developed blocking the pores formed over the coating.

Figure 10 compares the total impedance (Z_t) and noise resistance (R_n) values in time for both, the varnish and ABS-3 copper coupon. In both cases, the behaviors decrease as time goes by and the coatings deteriorate. The resulting values are similar in magnitude and trend comparing each other and hence, it is possible to suggest that the protective performance of the coatings is comparable. In the case of Z_t, the ABS coating stays with higher protective values

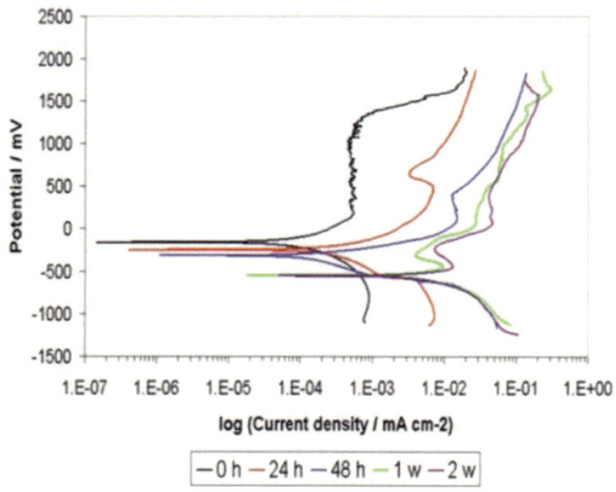

(a) ABS-3 coated copper

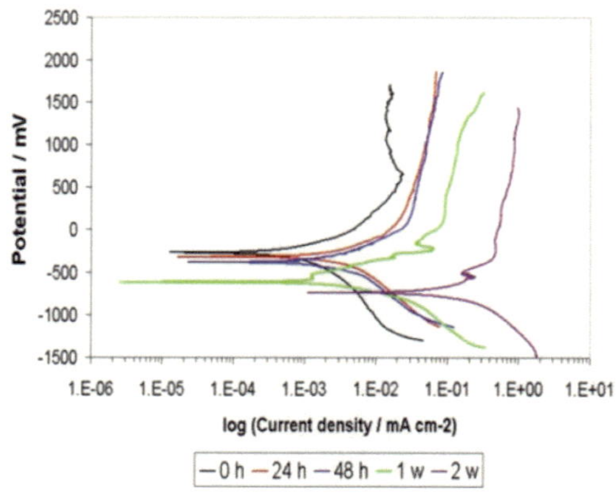

(b) ABS-3 coated bronze

Figure 9. Polarization curves of ABS-3 samples in Na_2SO_4 + 3 wt.% NaCl solution, as a function of time.

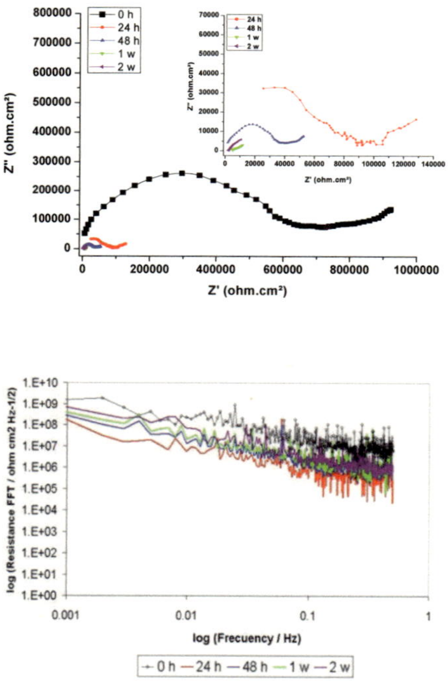

(a) ABS-3 coated copper

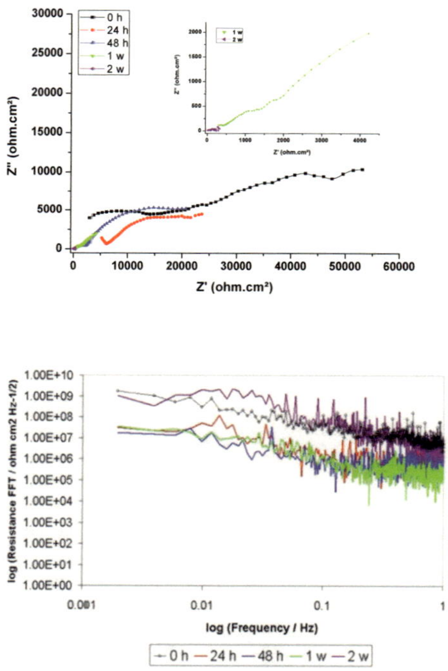

(b) ABS-3 coated bronze

Figure 10. Nyquist and noise impedance spectra for ABS-3 samples in Na_2SO_4 + 3 wt. % NaCl solution, as a function of time.

during a considerable time. R_n values indicate that even when initially the protection of the ABS coating is slightly below in relation to the varnish, at some point in time, the parameter values are inverted and both coatings reach very similar values. Being the noise resistance independent of the frequency, the values obtained and presented in the noise impedance spectra are related totally to the coating performance [47, 48].

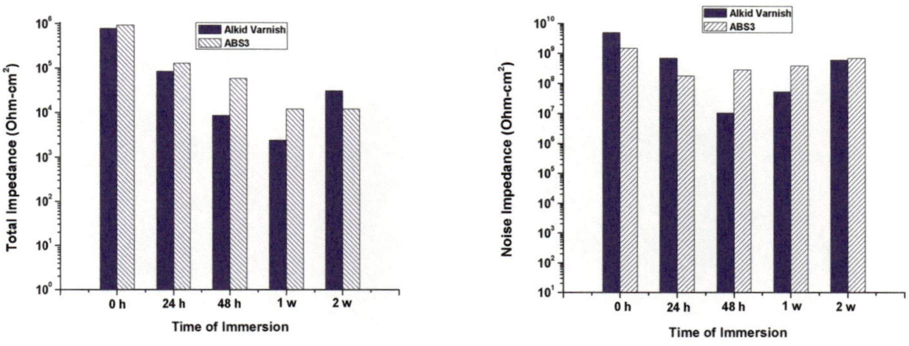

(a) Z_t (b) R_n

Figure 11. Comparative values of total and noise impedance for varnish and ABS-3 coated copper as a function of time.

Figure 11 compares the macroscopic aspect of coupons coated with varnish and ABS-3 after two weeks in the aggressive solution. The coupon protected with varnish shows blue-green corrosion products while the one coated with ABS-3 stayed in the copper color range. This suggests that the pores and/or failures in the varnish allow diffusion of bigger ions in solution such as sulfates reaching the metallic surface, which combined with chloride and oxygen may form $CuSO_4$, as compared to those in the ABS coating. In the latter, probably smaller pores allow diffusion only of oxygen, generating a highly adherent and protective copper oxide (Cu_2O and CuO) patina whose properties combine with those of the ABS-3 coating. Also, the difference observed in both samples corroborates the suggestion made as to the immersion cycles and pore sealing presented and discussed in Figure 2 above.

(a) Varnish (b) ABS-3

Figure 12. Macroscopic aspect of coupons coated after two weeks in the aggressive solution [51].

Figure 12(a) presents the general aspect of the ABS-3 coated metal immersed in the electrolyte for two weeks. A good surface condition can be observed through the transparent ABS-3 coating, suggesting good corrosion protection. In Figure 12(b) the scribed section shows good adhesion of the coating after the *pull up* test, presenting corrosion products only within it.

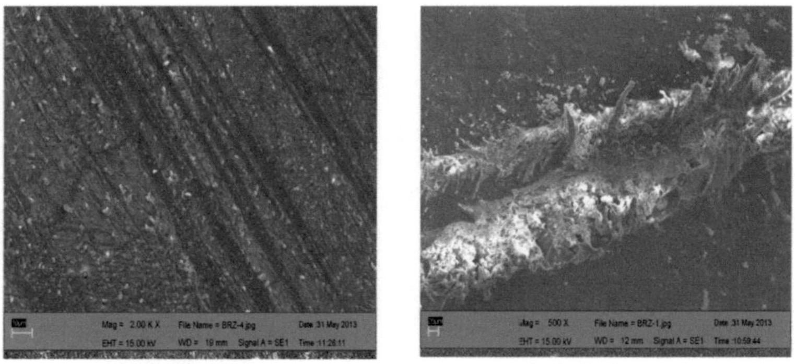

(a) General view (b) Scribed section

Figure 13. SEM micrographs of ABS-3 coating.

Atmospheric corrosion of copper reaction is [49, 50]:

$4Cu + O_2 + 2H_2O => 2Cu_2O + 2OH + H_2$

Under polluted conditions:

$Cu_2O+0.5SO_2+1.5H_2O+0.75O_2=>2Cu(OH)_{1.5}(SO_4)_{0.25}$

$Cu_2O+Cl+1.5H_2O+0.25O_2=>2Cu(OH)_{1.5}Cl_{0.5}$

Copper and alloys such as bronze, show moderate stability under atmospheric conditions, but due to increasing pollution and acid rain, in urban environments exposed bronze undergoes into accelerated corrosion, expressed by some impairment of appearance. Corrosion of copper and its alloys form complex patinas over the surface, being more or less protective depending upon the environmental conditions and the nature of the pollutants present. In the atmosphere, the range in corrosion rates is in the order of less than 0.1μm/year for rural atmospheres, to approximately 5.6μm/year for industrial-marine environment [49]. Electrochemical measurements immersed in neutral solution on samples with patinas formed after long exposures under different atmospheres, rendered polarization resistance values of 5.4 kohms-cm^2 for rural atmospheres up to 35.2 kohms-cm^2 for industrial-marine polluted atmospheres. For comparison bare copper, presents polarization resistance values in the range from 1.5 to 6 kohms-cm^2 [50]. These results show low atmospheric corrosion rates for rural atmosphere being higher in polluted atmospheres. Nevertheless the patinas formed under contaminated conditions are more protective according to the electrochemical results.

In comparison under similar experimental conditions, for copper samples coated with a commercial varnish used for restoration and conservation of pieces of works of art of cultural and historic heritage, polarization resistance values obtained were in the range from 10 to 100

http://dx.doi.org/10.5772/57059

kohms-cm^2 [24]. Results obtained and presented in this work, compares favorably with the results previously reported for this type of coatings. When using a copper corrosion inhibitor such as benzotriazole a polymeric oxide is formed [53, 54], and the polarization resistance values obtained in chloride solution, were around 50 kohms-cm^2. According to the results obtained in this work, the ABS-3 coating presents good performance when compared to other corrosion protection systems [48, 49].

Work Title	Types of studied coatings	Reference
An Analysis of Polyethylene Coating Corrosion in Oil and Gas Pipelines	Three Polyethylene layers (Epoxy primer plus Adhesive (polyglue) plus PE Top Coating)	65
Electrochemical evaluation of patinas formed on nineteenth century bronze bells	Patinas on bronze	6
Electrochemical assessment of the restoration and conservation of a heavily corroded archaeological iron artifact.	Rust converter	7
Corrosion resistance properties of electroless nickel composite coatings	Electroless nickel (EN) composite coatings incorporated with PTFE and/or SiC particles	66
Corrosion protection by multilayered conducting polymer coatings	Multilayered coatings consisting of combinations of conducting polymers polyaniline (Pani) and polypyrrole (Ppy),	67
Effect of Mg and Si on the microstructure and corrosion behavior of Zn-Al hot dip coatings on low carbon steel.	Zn and Zn-6Al Coatings with or without 0.5 %wt Mg and 0.1%wt Si	68
Electroactive conducting polymers for corrosion control	Electroactive conducting polymer coatings.	69
Studies of a new accelerated evaluation method for coating corrosion resistance—thermal cycling testing	Aerospace electrodeposition-coat primer plus plasma polymer pretreatment $SrCrO_4$-pigmented epoxy-polyamide primer (PEPP)	70
Characterization of the corrosion behavior of Zn and Zn alloy electrodeposits: Atmospheric and accelerated tests.	Zn, Zn-Co (0.6%), Zn-Fe (0.4%) and Zn-Ni (12%) coatings	71
Effect of the amine functional group on corrosion rate of iron coated with films of organofunctionalsilanes	Coatings made with two silanes, one with a functional group (γ-aminopropyltriethoxysilane [γ-APS]) and one without a functional group (bis-triethoxysilyl ethane, BTSE),	72
Corrosion protection using polyaniline coating formulations	Polyaniline (PANI) coatings	73
Polyaniline/epoxy coatings with good anti-corrosion properties	Polyaniline/epoxy blend coatings	74

Work Title	Types of studied coatings	Reference
Use of electrochemical impedance spectroscopy for the study of corrosion protection by polymer coatings	Alkid coatings	75
Passivation of metals by coating with polyaniline: corrosion potential shift and morphological changes	Coatings with polyaniline dispersions	76
Evaluation of corrosion behavior of coated metals with AC impedance measurements	Polybutadiene coating of 8 ± 2 μm thickness applied by spin coating	77

Table 3. Examples of surface treatments in order to improve protective characteristics.

4. Summary

Historically, copper and its alloys constitute a relevant group of metals used in the manufacture of cultural heritage. Even when these materials tend to be self-protective under atmospheric conditions, pollution, acid rain and other factors can disrupt their natural protective patinas. Therefore, several procedures have been designed from the materials science perspective in order to protect these and other metallic materials. Such treatments vary in efficiency and not all may fulfill the requirements for conservation purposes.

The ABS films fulfill two major requirements for cultural heritage conservation purposes that are transparency and reversibility. Transparency is usually required due to the need of not altering the visual properties of the artifacts. Reversibility has been considered an ideal goal because in reality, any intervention will leave some degree of impact or residues in the materials. However, a tendency towards reversibility is highly desirable. Both properties are achieved by dissolving ABS in organic solvents such as acetone, combined with a highly acceptable protective capacity of the material. Small, delicate artifacts might be more suitable to undergo the proposed process. Tests were done simulating aggressive ambient, hardly found within the security of museums or similarly protected surroundings.

The proposed application for conservation adds value to an industrial waste and contributes to find recycling alternatives for this discarded polymer, hence collaborating not only with the conservation science, but also with the ecological concerns and needs. In the present, all kind of residues contribute to the ambient degradation and therefore any recycling proposal has an intrinsic value. Indicated high performance of the coatings needs further studies to characterize and optimize the system, still the application becomes reasonable valid and its use in further conservation cases represents a viable option.

Acknowledgements

To CONACYT and PROMEP for grants and project funding, and to Dr. M. García-Sánchez, Departamento de Química, Universidad AutónomaMetropolitana-Iztapalapa, for FTIR analysis.

Author details

C. Menchaca-Campos[1], M. Hernández-Escampa[2], F. Rodríguez-Acuña[1], F. Millán-Cruz[1], P. Rodríguez-Rojas[1], M. Hernández-Gallegos[2], R. Guardian[1] and J. Uruchurtu[1*]

*Address all correspondence to: juch25@uaem.mx

1 CIICAp, Universidad Autónoma del Estado de Morelos, Cuernavaca, México

2 Facultad de Ingeniería, Universidad Nacional Autónoma de México, México, D.F., México

References

[1] Morgan L.H., 1877. *Ancient Society or Researches in the Lines of Human Progress from Savagery through Barbarism to Civilization*. London: MacMillan & Company.

[2] Heizerr F., 1962. "The Background of Thomsen's Three-Age System", *Technology and Culture*, vol. 3, no. 3, pp. 259-266

[3] Diamond J.M., Ordunio D., 1997. *Guns, germs, and steel*. New York: Norton.

[4] Torraca G., 1982. *Porous building materials-materials science for architectural conservation*. Iccrom, Rome, Italy.

[5] Scherer G.W., Flatt R., Wheeler, G., 2001. "Materials science research for the conservation of sculpture and monuments". *MRS Bulletin*, vol. 26, no. 1, pp. 44-50.

[6] Rodríguez-Acuña F., Genescá J., Uruchurtu J., 2010. "Electrochemical evaluation of patinas formed on nineteenth century bronze bells". *Journal of Applied Electrochemistry*, vol. 40, no. 2, pp. 311-320.

[7] Hernandez-Escampa M., Gonzalez J., Uruchurtu-Chavarin J., 2010. "Electrochemical assessment of the restoration and conservation of a heavily corroded archaeological iron artefact". *Journal of Applied Electrochemistry*, vol. 40, no. 2, pp. 345-356.

[8] Janssens K., Van Grieken R., (ed.). 2004. *Non-destructive micro analysis of cultural heritage materials*. Elsevier.

[9] Giorgi R., Dei L., Ceccato M., Schettino C., Baglioni P., 2002. "Nanotechnologies for conservation of cultural heritage: Paper and canvas deacidification". *Langmuir*, vol. 18, no. 21, pp. 8198-8203.

[10] Baglioni P., Giorgi R., 2006. "Soft and hard nanomaterials for restoration and conservation of cultural heritage". *Soft Matter*, vol. 2, no. 4, pp. 293-303.

[11] Coghlan H.H., Voce E., Penniman T.K., 1951. "Notes on the Prehistoric Metallurgy of Copper and Bonze in the Old World". *Occasional papers on Technology*, no. 4, pp. 1-131.

[12] Hosler D., Macfarlane A., 1996. "Copper sources, metal production, and metals trade in Late Postclassic Mesoamerica". *Science*, vol. 273, pp. 1819-1824.

[13] Griffin J.B., (ed.). 1961. *Lake Superior copper and the Indians: miscellaneous studies of Great Lakes prehistory*. Ann Arbor: University of Michigan.

[14] Chernykh E.N., Wright S., 1992. *Ancient metallurgy in the USSR: The early metal age*. Cambridge University Press.

[15] Barnard N., 1993. "Thoughts on the emergence of metallurgy in pre-Shang and early Shang China and a technical appraisal of relevant bronze artifacts of the time". *Kinzoku Hakubutsukankiyo= Bulletin of the Metals Museum*, vol. 19, pp. 3-48.

[16] Soto L., Franey J.P., Graedel T.E., Kammlott G.W., 1983. "On the corrosion resistance of certain ancient Chinese bronze artifacts". *Corrosion Science*, vol. 23, no. 3, pp. 241-250.

[17] Cicileo G.P., Crespo M.A., Rosales B.M., 2004. "Comparative study of patinas formed on statuary alloys by means of electrochemical and surface analysis techniques", *Corrosion Science*, vol. 46, Issue 4, pp. 929-953.

[18] Fitzgerald K.P., Nairn J., Atrens A., 1998. "The Chemistry of Copper Patination", *Corrosion Science*, vol. 40, Issue 12, pp. 2029-2050.

[19] Franey J.P., Davies, M.P., 1987. "Metallographic studies of the copper patina formed in the atmosphere", *Corrosion Science*, vol. 27, Issue 7 pp. 659-668.

[20] Chiavari C., 2007. "Composition and electrochemical properties of natural patinas of outdoor bronze monuments". *Electrochimica Acta*, vol. 52, no. 27, pp. 7760-7769.

[21] Marbán-Salgado J.A., Uruchurtu-Chavarín J., Mayorga-Cruz D., 2012. "*Observation of Copper Corrosion Oxide Products Reduction in Metallic Samples by Means of Digital Image Correlation*". *International Journal of Electrochemical Science*, vol. 7, pp. 1107-1117.

[22] Mareci D., Chelariu R., Sutiman D., Rusu I., 2011. "Electrochemical characteristics of electrochemically patinated bronze". *European Journal of Science and Theology*, vol. 7, no. 4, pp. 121-129.

[23] Marušić K., Ćurković H.O., Takenouti H., 2013. "Corrosion Inhibition of Bronze and Its Patina Exposed to Acid Rain". *Journal of the Electrochemical Society*, vol. 160, no. 8, pp. C356-C363.

[24] Brown B.F., Burnett H.C., Chase W. (eds). 1991. *Corrosion and Metal Artifacts: A Dialogue Between Conservators and Archaeologists and Corrosion Scientists*. NACE International: The Corrosion Society.

[25] Loyen F., 1978. *The Thames and Hudson Manual of Metal Working*. W.W. Norton & Company.

[26] Macleish A.B., 1986. *The Care of Antiquities and Historical Collections.* Second Edition. Altamira Press.

[27] Begemann F., Schmitt-Strecker S., Pernicka E., Lo Schiavo F., 2001. "Chemical composition and lead isotopy of copper and bronze from Nuragic Sardinia". *European Journal of Archaeology*, vol. 4, no. 1, pp. 43-85.

[28] Watt A., 1889. *Electro-Metallurgy Practically Treated*. vol. 135. C Lockwood and Son.

[29] Lins A., Power, T., 1994. "The corrosion of bronze monuments in polluted urban sites: a report on the stability of copper mineral species at different pH levels". En *Ancient and historic metals: conservation and scientific research: proceedings of a symposium organized by the J. Paul Getty Museum and the Getty Conservation Institute, November 1991*. Getty Conservation Institute. pp. 119-151.

[30] Kucera V., Fitz S., 1995. "Direct and indirect air pollution effects on materials including cultural monuments". *Water, Air, and Soil Pollution*, vol. 85, no. 1, pp. 153-165.

[31] Nord A.G., Mattsson, E., Tronner, K., 2005. "Factors influencing the long-term corrosion of bronze artefacts in soil". *Protection of metals*, vol. 41, no. 4, pp. 309-316.

[32] Tronner K, Nord A.G., Borg G.Ch., 1995. "Corrosion of archaeological bronze artefacts in acidic soil". *Water, Air, and Soil Pollution*, vol. 85, no. 4, pp. 2725-2730.

[33] Schüssler A., Exner H.E., 1993. "The corrosion of nickel-aluminium bronzes in seawater-II. The corrosion mechanism in the presence of sulphide pollution". *Corrosion Science*, vol. 34, no. 11, pp. 1803-1815.

[34] Scott D.A., 1990. "Bronze disease: A review of some chemical problems and the role of relative humidity". *Journal of the American Institute for Conservation*, vol. 29, no. 2, pp. 193-206.

[35] Gilberg M., 1988. "History of bronze disease and its treatment". In *Early Advances in Conservation*. British Museum. pp. 59-70.

[36] Hongfan Z., Hao Z., 1999. "A Study of the Corrosion on Bronze". *Electrochemistry.* vol. 3, pp. 014.

[37] Erhardt D., Mecklenburg M., 1994. "Relative humidity re-examined". *Studies in Conservation*, vol. 39, no. Supplement 2, pp. 32-38.

[38] Plenderleith H.J., Werner A.E.A., 1972. *The Conservation of Antiquities and Works of Art: Treatment, Repair and Restoration*. London: Oxford University Press. Second Edition.

[39] Wharton G., 1989. "The cleaning and lacquering of museum silver". *WAAC newsletter*, vol. 11, no. 1, pp. 4-5.

[40] Sandwith H., Stainton S., 1986. *The National Trust Manual of Housekeeping*, USA: Penguin Books.

[41] United States Government Printing Office., 1999. Metals in America's Historic Buildings: Uses and Preservation Treatments. January 8.

[42] Gonzales J.E., Krejchi M.T., Mafoti R.M., Odstrcil K.W., Hoelscher F.E., Kendall E.W., Oriseh A.S. 1999. *U.S. Patent No. 6,007,005*. Washington, DC: U.S. Patent and Trademark Office.

[43] Karvelas D.E., Jody B.J., Poykala J.A. Jr., Daniels E.J., Arman B., 1996. *Recovery and separation of high-value plastics from discarded household appliances* (No. ANL/ES/CP--87943; CONF-960202--6). Argonne National Lab., IL (United States). Energy Systems Div.

[44] Sakata Y. 1996. "Thermal and catalytic degradation of municipal waste plastics into fuel oil". *Polymer Recycling*, vol. 2, no. 4, pp. 309-315.

[45] Özdogan S., Uygur S., Egrican N., 1997. "Formation and dispersion of toxic combustion by-products from small-scale combustion systems". *Energy*, vol. 22, no. 7, pp. 681-692.

[46] Radtke F.,Köppel R.A., Baiker A., 1995. "Formation of undesired by-products in the NO_x catalysis by hydrocarbons". *Catalysis today*, vol. 26, no. 2, pp. 159-167.

[47] Wolkowicz M.D., Gaggar S.K., 2004. "Effect of thermal aging on impact strength acrylonitrile-butadiene-styrene (ABS) terpolymer". *Polymer Engineering & Science*, vol. 21, no. 9, pp. 571-575.

[48] Gesner B.D. 1965. "Environmental surface effects on ABS resins". *Journal of Applied Polymer Science*, vol. 9, pp. 3701-3706.

[49] Shimada J., Kabuki K., 1968. "The mechanism of oxidative degradation of ABS resin. Part II. The mechanism of photo-oxidative degradation". *Journal of Applied Polymer Science*, vol. 12, pp. 671-682.

[50] Wyzgoski M.G., 1976. "Effects of oven aging on ABS, poly(acrylonitrile-butadiene-styrene)". *Polymer Engineering & Science*, vol. 16, no. 4, pp. 265-269.

[51] Hernández-Escampa M., Rodríguez-Acuña F., Millán-Cruz F., Rodríguez-Rojas P., Hernández-Gallegos M., Covelo A., Menchaca-Campos C., Uruchurtu, J. 2013. "Electrochemical Evaluation of a Recycled Copolymer Coating for Cultural Heritage Conservation Purposes". *International Journal of Polymer Science*, vol. 2013, 9 pages.

[52] Appelbaum B., 1987. "Criteria for treatment: reversibility". *Journal of the American Institute for Conservation*, pp. 65-73.

[53] Salman R., 1991. "On the Optical Properties of ABS Plastics". *Polymer-Plastics Technology and Engineering*, vol. 30, no. 4, pp. 343-349.

[54] Suzuki M., Wilkie C.A., 1995. "The thermal degradation of acrylonitrile-butadiene-styreneterpolymei as studied by TGA/FTIR". *Polymer Degradation and Stability*, vol. 47, no. 2, pp.217-221.

[55] Shuying Y., Castilleja J.R., Barrera E.V., Lozano K., 2004. "Thermal analysis of ana-crylonitrile–butadiene–styrene/SWNT composite", *Polymer Degradation and Stability*, vol. 83, pp. 383-388.

[56] Burn L.S., Davis P., Hill A.J., 2002. "Thermal degradation of acrylonitrile–butadiene styrene (ABS) blends", *Polymer Degradation and Stability*, vol. 76, no. 3, pp. 425–434.

[57] Uruchurtu-Chavarin J., Malo J.M. 1997. "Electrochemical noise as a powerful electrochemical technique for corrosion studies". *Trends in Corrosion Research*, vol. 2, pp. 49-58.

[58] ASTM-G-106., 2010. "Standard Practice for Verification of Algorithm and Equipment for Electrochemical Impedance Measurements".

[59] Hernández M., Genescá J., Uruchurtu J., Barba, A. 2009. "Correlation between electrochemical impedance and noise measurements of waterborne coatings", *Corrosion Science*, vol. 51, no. 3, pp. 499-510.

[60] Morcillo M., Almeida M.E.M., Rosales B.M., Marrocos M., Uruchurtu J., 1999. *Corrosión y Protección de Metales en las Atmósferas de Iberoamérica*, Madrid: CYTED.

[61] Mariaca-Rodríguez L., Genescá-Llongueras J., Uruchurtu-Chavarín J., Hernández L.S. (Eds.) 1999. *Corrosividad Atmosférica* (MICAT-México), México: Plaza y Valdes, México.

[62] Strandberg H., Johansson L.G., Lindqvist O. 1997. "The atmospheric corrosion of statue bronzes exposed to SO_2 and NO_2". *Materials and Corrosion.* vol. 48, no. 11, pp. 721–730.

[63] Kosec T., Merl D.K., Milosev I., 2008. "Impedance and XPS study of benzotriazole films formed on copper, copper-zinc alloys and zinc in chloride solution," *Corrosion Science*, vol. 50, no. 7, pp. 1987–199.

[64] Finzgar M., Milosev I., 2010. "Inhibition of copper corrosion by 1,2,3-benzotriazole: a review," *Corrosion Science*, vol. 52, no. 9, pp. 2737–2749.

[65] Samimi A., Zarinabadi S., 2011. "An Analysis of Polyethylene Coating Corrosion in Oil and Gas Pipelines". *Journal of American Science, USA*. vol. 7 no. 1, pp. 1032-1036.

[66] Huang Y.S., Zeng X.T., Hu X.F., Liu F.M. 2004. "Corrosion resistance properties of electroless nickel composite coatings". *Electrochimica Acta*, vol. 49, no. 25, pp. 4313-4319.

[67] Tan C.K., Blackwood D.J., 2003. "Corrosion protection by multilayered conducting polymer coatings". *Corrosion Science*, vol. 45, no. 3, pp. 545-557.

[68] Tanaka J., Ono K., Hayashi S., Ohsasa K., Narita T., 2002. "Effect of Mg and Si on the microstructure and corrosion behavior of Zn-Al hot dip coatings on low carbon steel". *ISIJ international*, vol. 42, no. 1, pp. 80-85.

[69] Tallman D.E., Spinks G., Dominis A., Wallace, G.G., 2002. "Electroactive conducting polymers for corrosion control". *Journal of Solid State Electrochemistry*, vol. 6, no. 2, pp. 73-84.

[70] Bierwagen G.P., He L., Li J., Ellingson L., Tallman D. E., 2000. "Studies of a new accelerated evaluation method for coating corrosion resistance-thermal cycling testing". *Progress in Organic Coatings*, vol. 39, no. 1, pp. 67-78.

[71] Ramanauskas R., Muleshkova L., Maldonado L., Dobrovolskis P. 1998. "Characterization of the corrosion behaviour of Zn and Zn alloy electrodeposits: Atmospheric and accelerated tests". *Corrosion Science*, vol. 40, no. 2, pp. 401-410.

[72] Subramanian V., Van Ooij W.J., 1998. "Effect of the amine functional group on corrosion rate of iron coated with films of organo-functional silanes". *Corrosion*, vol. 54, no. 3, pp. 204-215.

[73] Kinlen P.J., Silverman D.C., Jeffreys, C.R. 1997. "Corrosion protection using polyaniline coating formulations". *Synthetic Metals*, vol. 85, no. 1, pp.1327-1332.

[74] Talo A., Passiniemi P., Forsen O., Yläsaari S. 1997. "Polyaniline/epoxy coatings with good anti-corrosion properties". *Synthetic Metals*, vol. 85, no. 1, pp. 1333-1334.

[75] Mansfeld F., 1995. "Use of electrochemical impedance spectroscopy for the study of corrosion protection by polymer coatings". *Journal of Applied Electrochemistry*, vol. 25, no. 3, pp. 187-202.

[76] Wessling B. 1994. "Passivation of metals by coating with polyaniline: corrosion potential shift and morphological changes". *Advanced Materials*, vol. 6, no. 3, pp. 226-228.

[77] Mansfeld F., Kendig M.W., Tsai S. 1982. "Evaluation of corrosion behavior of coated metals with AC impedance measurements". *Corrosion*, vol. 38, no. 9, pp. 478-485.

Chapter 15

Study on the Anticorrosive Behavior of New Hygiene Structured Pigment Based on Waste Core and Nano Shell in Alkyd Paints

Nivin M. Ahmed and
Hesham Tawfik M. Abdel-Fatah

Additional information is available at the end of the chapter

http://dx.doi.org/10.5772/57243

1. Introduction

Core-shell structured particles are gaining lots of importance recently due to their exciting applications in different fields; these particles are constructed of cores and shells with different chemical compositions which show ultimately distinctive properties of varied materials different from their counterparts, to manipulate the surface functions that can meet diverse application requirements. The purpose of coatings on core-particles is carried out for different reasons like surface modification, increasing functionality, stability, dispersibility, and control release of the core material. The presence of shell can alter some properties of core like surface charge, functionality and reactivity. These new core-shell particles are widely applied in different applications [1-6].

Silica fume is a byproduct of the smelting process in the ferrosilicon industry. The reduction of high-purity quartz to silicon at temperatures up to 2000°C produces SiO_2 vapors, this leads to the oxidation and condensation in low temperature zones to tiny particles consisting of non-crystalline silica.

Silica fume products have very high purity resulting in highly consistent physical properties and particle size distributions. Their complete inertness and neutral pH make these products not capable of altering or initiating reactions when incorporated in catalyzed or multi-component chemical systems. Silica fume is also considered as eco-friendly material as using it converts a waste product into a useful marketable product which leads to a reducing cost [7].

The extremely fine and non-porous nature of pyrogenic silica or fumed silica as it is also known, make it an ideal thickening and bulking agent with outstanding thixotropic qualities. Thixotropic refers to a substances characteristic of reducing viscosity or thickness with extended agitation or shaking. This makes fumed silica ideal filler for paints, thus causing them to thin out during application and regain their viscosity when left to stand preventing drips and runs [8-10].

Titanium dioxide (TiO_2) is the most important white pigment used in the coatings industry. It is widely used because it efficiently scatters visible light, thereby imparting whiteness, brightness and opacity due to its high refractive index; it improves the light and dirt resistance of a formulation. Titanium oxide also possesses low oil absorption, high tinting strength, and inert chemical properties [9-11]. Recently titanium dioxide was found attractive as photo-catalyst, although high photo-catalytic activity is required for titanium dioxide when used as photo-catalysts, low photo-catalytic activity is sometimes preferred for titanium dioxide additives to avoid the degradation of matrix. The catalytic activity of titanium dioxide particles can be enhanced or suppressed by precipitating them on a core of other appropriate pigment to control this phenomenon. These cores layers like silica (SiO_2), alumina (Al_2O_3), or a polymer are often expected to suppress the catalytic activity of titanium dioxide particles [12-15].

Titanium dioxide is commercially available in two crystal structures, anatase and rutile. Rutile TiO_2 pigments is preferred because they scatter light more efficiently. They are more stable and durable than anatase pigments [16]

Titanium dioxide pigments are insoluble in coating vehicles; their performance properties, e.g., chemical, photochemical, and physical characteristics, are determined principally by the particle size of the pigment and the chemical composition of its surface [17].

Although silica fume is a fine lightweight fluffy amorphous powder that possesses suitable values of specific gravity, bulking value and oil absorption which makes it easy to find a new market in different industries, this nature causes the appearance of some problems when applied in paint formulations as they form agglomerations and poor wetting with the binder [18-20].

In this work there is a trial to modify these defects in using silica fume as filler in paints by precipitating very thin layers of a nano-titanium dioxide on its surface in order to get over the defaults caused by silica fume in paint formulations and also to promote its surface properties. Following this trend, silica fume waste was covered with three different layers of nano-titanium dioxide. The effect of these new pigments on different mechanical and corrosion properties of paint films containing them was studied; also the effect of shell-layer thickness on the anticorrosive properties was estimated.

Nano-titanium dioxide properties are different from that of titanium dioxide itself, because materials reduced to the nano-scale can suddenly show very different properties compared to what they exhibit on a micro-scale because of two effects [20-21];

Surface effects [20,21]

1. More percent of atoms on surface compared to inner atoms can be found.
2. More surface free energy is available: this increased surface area and surface atoms results in the increase of surface energy associated with the particles.
3. Increasing the surface area of a substance generally increases the rate of a chemical reaction.

Volume effects

1. Lower wavelength (higher frequency and higher energy).
2. The average energy spacing increases as the number of atoms are reduced; this enhances the catalytic properties of nanoparticles.

Although these two oxides (SiO_2 and nano-TiO_2) have no significant role as anticorrosive materials [21], the new prepared pigments exhibited good inhibition effect in protecting metal surfaces from corrosion. The use of these pigments can bring about a significant savings besides being eco-friendly by converting a waste product into a useful marketable one which leads to the preparation of a reducing cost anticorrosive pigment. Also, the new formed pigments will combine the properties of both materials and overcome their deficiencies. This study proved that, as the nano-titanium dioxide shell increases, better protection was offered by the prepared pigments.

2. Experimental

2.1. Materials

All the employed pigments, extenders, resins, solvents, additives and chemicals were products of different local and international companies.

Silica fume waste was obtained from ferrosilicon factories in Aswan, Egypt. Its chemical composition is represented in Table (1), the waste was collected from the chimneys of the factory and this may be the cause of its gray color as it contains a very high percentage of carbon resulting from the smelting of this alloy preparation.

Concentration of main constituents	Wt., %
SiO_2	97.98
Al_2O_3	0.39
Fe_2O_3 ^{Total}	1.04
MnO	0.08
CuO	0.02
MgO	0.83
CaO	0.32
SrO	0.003
Na_2O	0.53
K_2O	0.75
P_2O_5	0.07
SO_3	0.005
L.O.I	4.93

Table 1. Chemical composition of silica fume

2.2. Preparation of TiO_2/SiO_2 fume core-shell pigments

Titanium tetrachloride was added in three concentrations 5-10ml to 100ml hydrochloric acid. Silica was impregnated in these three solutions and left for sometime to assure complete covering while stirring. Ammonia solution was then added drop-wisely to these impregnated silicas to adjust their pH; till complete precipitation of the pigments. The formed paste is then filtered through a buchner system and washed very well. This paste was then calcined at 650-750°C.

The preparation of the pigments is represented in the following scheme;

$$TiCl_4/SiO_2 \xrightarrow[NH_4OH]{HCl} TiOH_4/SiO_2 + NH_4Cl \xrightarrow[65C-750^{\circ}C]{\Delta} TiO_2/SiO_2 \quad (1)$$

Three concentrations of titanium dioxide and consequently of titanium were deposited on silica surface, these three concentrations were determined using XRF technique. Table (2) includes these results which show that concentrations are ascending from TiO_2/SiO_2 (1) to TiO_2/SiO_2 (3).

Concentration of main constituents (wt. %)	Silica fume	TiO_2/SiO_2 (1)	TiO_2/SiO_2 (2)	TiO_2/SiO_2 (3)
SiO_2	97.98	92.48	89.61	87.88
TiO_2	----	7.52	10.19	12.42
Al_2O_3	0.39	0.36	0.35	0.33
Fe_2O_3	1.04	1.43	1.27	1.23
MgO	0.08	0.60	0.60	0.57
CaO	0.02	0.31	0.33	0.32
Na_2O	0.83	0.18	0.17	0.18
K_2O	0.32	0.95	0.98	0.86
P_2O_5	0.003	0.06	0.06	0.06
SO_3	0.53	0.005	0.005	0.007
L.O.I	0.75	0.62	0.53	0.76

Table 2. XRF analysis results of the prepared TiO_2/SiO_2 pigments

2.3. Methods of instrumental analysis

SEM/EDAX analysis

Energy-dispersive X-ray analysis technique, and scanning electron microscopy, (JEOL JX 840), micro-analyzer electron probe, was used in this work to estimate the particle shapes and to determine the elements deposited on silica surface to estimate the formation of the new pigments.

Transmission electron microscopy

Various pigments were examined using (JEOL JX 1230) technique in order to estimate the particle shapes and to determine the particle sizes of the prepared pigments.

X-ray fluorescence

The different concentrations of each element in the prepared pigments were determined using Axios, sequential WD-XRF spectrometer, PANalytical 2005.

2.4. Formulation of paints based on alkyd resin

2.4.1. Preparation of anticorrosive paint formulations

The prepared pigments were tested in 20 paint formulations. These paint formulations contain silica fume, commercial titanium dioxide, and silica fume covered with different layer thicknesses of nano-titanium dioxide denoted as $SiO_2/TiO_2(1)$, $SiO_2/TiO_2(2)$, and $SiO_2/TiO_2(3)$ expressing concentration of nano-titanium dioxide layers in ascending order from the lower to the higher concentration. All paint formulations were based on medium oil-modified soya-bean dehydrated castor oil alkyd resin. The formulations were divided into four groups; each is based on different concentration of the pigments under investigation ranging from 15-30, and all the groups have the same P/B which is 2.175, this ratio was selected from previous work concerning silica fume in alkyd-based paints [6]. Paint formulations are given in Tables (4-7), while the physical-mechanical and corrosion properties are represented in Table (8). The corrosion features of the paint films are shown in Figs. (4-7).

2.4.2. The effects of the prepared pigments on the mechanical properties of paints

Because the aim of this study was to formulate a pigment whose properties would contribute to the improvement of the mechanical and corrosion qualities of the paints, selected physical-chemical and corrosion tests were carried out.

- **Determining the resistance of paints against impact (ASTM D 5638-00, 2007)**

The result of this test reveals at what height (in cm) of the free fall of a weight onto the paint, the paint film has not yet been disturbed.

- **Determination of paint resistance against cupping in Erichsen apparatus (ASTM D 5638-00, 2007)**

The result of this test reveals the cupping of the test panel with a coating in mm at which the first impairment of the paint occurred.

- **Determination of paint Hardness (ASTM D 6577-00, 2007)**

The result of this test indicates the elasticity of paint film by using the simple pendulum test with a needle, when this needle reaches a groove which is made in the paint film, the time it takes in seconds is the measure of the hardness of paint films.

- **Determining the degree of coating adhesion by means of a cross-cut test (ASTM D 3359-97, 2007)**

The test was performed with a special cutting knife whose edges are 2mm apart. The cut of the created grate was evaluated according to a Gt0–Gt4 scale.

2.4.3. Overall evaluation of the selected physical-mechanical properties of the paints

The physical-mechanical tests nature including the adhesion of the paint film identified by means of a cross-cut test, impact resistance, and resistance against cupping indicating the elasticity of the paint film were measured. Also, good surface resistance is important for the resistance of the paints against mechanical impacts, scratches, and erosion caused by abrasion of dust particles in the outside atmospheric environment. The determination of paint film surface hardness on glass was carried out by means of a pendulum apparatus, the test consists of measuring the number of oscillations of the pendulum that bears onto the paint film with two steel balls. The unit of measuring of hardness is a percentile value related to the hardness of a glass standard that equals to 100%.

The detected results of the mechanical tests were assigned with the corresponding numerical values from the scale for the determination of physical-mechanical properties. The high value of the different mechanical properties means that the pigment contained in the paint has a positive influence on its mechanical properties. This can be clearly seen from data represented in Table (8).

2.4.4. The effect of the prepared pigments on the anticorrosion performance and the chemical resistance of the paints

The primary goal of this study was to prepare a pigment that would increase the anticorrosion performance of the paints. The determination of the prepared pigments properties by means of laboratory tests and electrochemical measurements in corrosive environments was a priority assignment.

2.4.5. Corrosion test evaluation methods

The evaluation of coatings after exposure to the corrosion tests followed methods based on the (ASTM D 714-87) for degree of blistering, (ASTM D 6294-98) for degree of rusting, and (ASTM D 2803-93) for photographic inspection.

2.4.6. Electrochemical evaluation method

Electrochemical Impedance Spectroscopy (EIS) is one of the most modern techniques available to characterize the electrical properties of organic coatings and their adhesion to metal surfaces [22-25]. It is an extremely useful technique in generating quantitative data that relates to the quality of the coat on a metal substrate. EIS is a very sensitive detector of a coated metal condition, its response can indicate changes in the coating long before any visible damage occurs, and it is not an absolute measurement.

In the present study, EIS was employed in order to investigate the effect of different concentration of SiO_2/TiO_2 core-shell pigments in anticorrosive paint formulations based on medium oil alkyd resin using the same P/B which is 2.175. The choice of this P/B was based on a previous study which indicated that this P/B was the best for the performance for similar pigments. The test was carried out in 3.5 wt% NaCl at different immersion times [6].

EIS experiments were carried out using a conventional three-electrode cell. The working electrode was the coated specimen, using a saturated calomel electrode as reference electrode, and a platinum foil (1cm^2) as counter electrode. The used electrolyte was 3.5wt.% NaCl solution. EIS measurements were carried out using AC signals of amplitude 5mV peak to peak at the open circuit potential in the frequency range between 15kHz and 0.3Hz. EIS data was collected using Gamry PCI300/4 Potentiostat/Galvanostat/Zra analyzer, EIS300 Electrochemical Impedance Spectroscopy software, and Echem Analyst 5.21 for results plotting, graphing, data fitting & calculating.

3. Results and discussion

3.1. The physical–chemical properties of the prepared pigments

Figures (1 and 2) show the morphology of the prepared pigments using SEM and TEM. From the featured photos it can be seen that, silica fume possesses a spherical particle shapes, while titanium dioxide possesses platelet structure. The new pigments possess the two shapes altered on each other, i.e. micrographs showed that the spheres of silica are carrying on its surface and around the spheres the tiny plates of nano-titanium dioxide that appear as a roughness on the sphere surfaces.

From these two figures, it can be detected that the prepared pigment particles are formed with two shapes and two particle sizes. Both SEM and TEM showed the size of titanium dioxide in the shell to be in the nano-scale indicating the formation of nano-titanium dioxide on micronized silica fume.

These altered shapes of nano-TiO_2/SiO_2 pigments can provide better corrosion protection properties than its individual constituents if applied in paint formulations as these tiny plates are expected to block the voids between the spheres of silica, thus protecting the substrate by providing a tortuous path of the corroding materials diffusion through the pigment particles till they reach the metal surfaces in a very weak condition, and thus exhibiting better protection properties [26].

Fig. (3) shows the EDAX analysis of prepared pigments. EDAX or energy dispersive X-ray analysis can detect the elements on the surface up to one micron depth. As featured from the chart, titanium was detected; this revealed its presence on silica surfaces, and its concentration can be seen in the Table below.

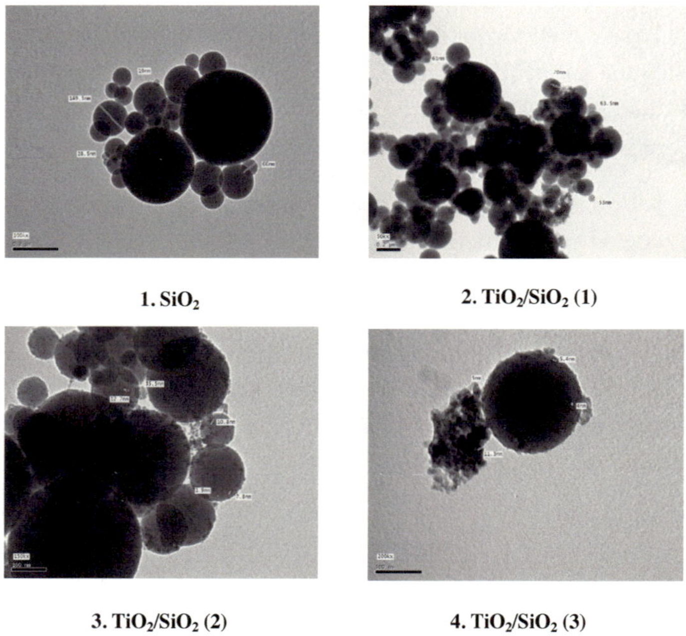

Figure 1. TEM micrographs of silica fume and covered silica with titanium dioxide at magnification of $15x10^3X$

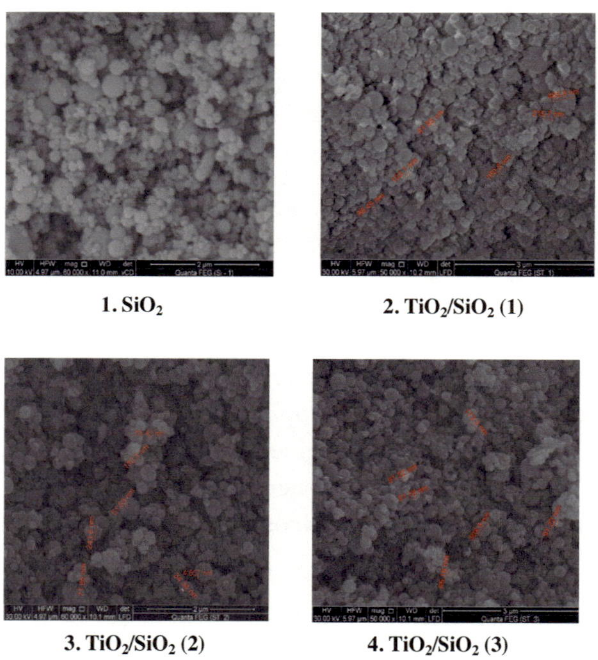

Figure 2. SEM micrographs of silica fume and covered silica with titanium dioxide at magnification of $50x10^3X$

http://dx.doi.org/10.5772/57243

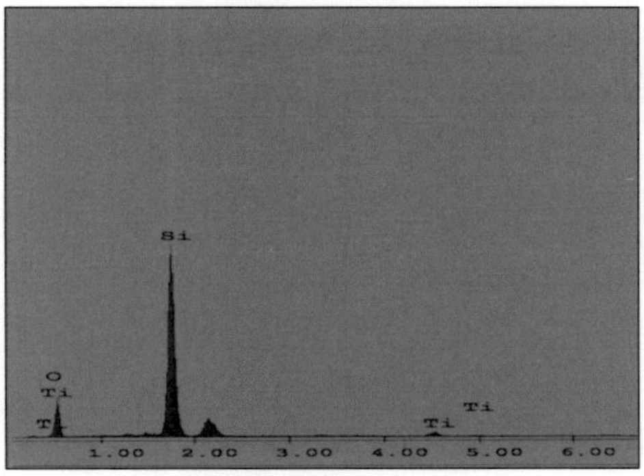

Element	Wt %	At %	K-Ratio	Z	A	F
O K	43.47	57.83	0.1068	1.0219	0.2401	1.0007
Si K	50.42	37.23	0.4036	0.9861	0.7519	1.0002
Ti K	6.11	3.94	0.0169	0.8790	0.9091	1.0000
Total	100.00	100.00				

Figure 3. EDAX analyses data for TiO_2/SiO_2 core-shell pigments

Constituent (%wt.)	6	7	8	9	10
Fe_2O_3 (Hematite)	18.5	18.5	18.5	18.5	18.5
Kaolin	20	20	20	20	20
$BaSO_4$	10	10	10	10	10
SiO_2	20	---	---	---	---
TiO_2	---	20	---	---	---
TiO_2/SiO_2 (1)	---	---	20	---	---
TiO_2/SiO_2 (2)	---	---	---	20	---
TiO_2/SiO_2 (3)	---	---	---	---	20
Total pigment	68.5	68.5	68.5	68.5	68.5
Total binder	31.5	31.5	31.5	31.5	31.5
P/B	2.175	2.175	2.175	2.175	2.175
Wetting & dispersing agent	1	1	1	1	1
Drier	0.5	0.5	0.5	0.5	0.5
Total	100	100	100	100	100

Table 3. Paint formulations of TiO_2/SiO_2 pigments with medium oil alkyd resin (Group I)

Constituent (%wt.)	1	2	3	4	5
Fe_2O_3 (Hematite)	23.5	23.5	23.5	23.5	23.5
Kaolin	20	20	20	20	20
$BaSO_4$	10	10	10	10	10
SiO_2	15	---	---	---	---
TiO_2	---	15	---	---	---
TiO_2/SiO_2 (1)	---	---	15	---	---
TiO_2/SiO_2 (2)	---	---	---	15	---
TiO_2/SiO_2 (3)	---	---	---	---	15
Total pigment	68.5	68.5	68.5	68.5	68.5
Total binder	31.5	31.5	31.5	31.5	31.5
P/B	2.175	2.175	2.175	2.175	2.175
Wetting & dispersing agent	1	1	1	1	1
Drier	0.5	0.5	0.5	0.5	0.5
Total	100	100	100	100	100

Table 4. Paint formulations containing TiO_2/SiO_2 pigments with medium oil alkyd resin (Group II)

Constituent (%wt.)	11	12	13	14	15
Fe_2O_3 (Hematite)	13.5	13.5	13.5	13.5	13.5
Kaolin	20	20	20	20	20
$BaSO_4$	10	10	10	10	10
SiO_2	25	---	---	---	---
TiO_2	---	25	---	---	---
TiO_2/SiO_2 (1)	---	---	25	---	---
TiO_2/SiO_2 (2)	---	---	---	25	---
TiO_2/SiO_2 (3)	---	---	---	---	25
Total pigment	68.5	68.5	68.5	68.5	68.5
Total binder	31.5	31.5	31.5	31.5	31.5
P/B	2.175	2.175	2.175	2.175	2.175
Wetting & dispersing agent	1	1	1	1	1
Drier	0.5	0.5	0.5	0.5	0.5
Total	100	100	100	100	100

Table 5. Paint formulations containing TiO_2/ SiO_2 Pigments with medium oil alkyd resin (Group III)

Constituent (%wt.)	16	17	18	19	20
Fe_2O_3 (Hematite)	8.5	8.5	8.5	8.5	8.5
Kaolin	20	20	20	20	20
$BaSO_4$	10	10	10	10	10
SiO_2	30	---	---	---	---
TiO_2	---	30	---	---	---
TiO_2/SiO_2 (1)	---	---	30	---	---
TiO_2/SiO_2 (2)	---	---	---	30	---
TiO_2/SiO_2 (3)	---	---	---	---	30
Total pigment	68.5	68.5	68.5	68.5	68.5
Total binder	31.5	31.5	31.5	31.5	31.5
P/B	2.175	2.175	2.175	2.175	2.175
Wetting & dispersing agent	1	1	1	1	1
Drier	0.5	0.5	0.5	0.5	0.5
Total	100	100	100	100	100

Table 6. Paint formulations containing TiO_2/ SiO_2 Pigments with medium oil alkyd resin (Group IV)

3.2. The effect of prepared pigments on mechanical properties of paints

Table (8) shows the results of determining hardness of organic coatings by means of pendulum apparatus, impact resistance and ductility. Paint films containing silica showed the best mechanical properties among the group, while those containing titanium dioxide showed the lowest values, this is due to that fine silica fume is used as additive that enhance the surface smoothness and dispersability of the media into which they are added, these fine silica particles have a lower coefficient of friction than titanium dioxide and other popular fillers. On a molecular level, these particles are perfectly spherical in shape and move more freely to provide a superior tactile feel [27], and thus better mechanical properties were obtained.

In general, TiO_2/SiO_2 pigments showed almost similar mechanical properties as silica but with lower values, and better than paint films containing titanium dioxide. This can be attributed to that the mixed pigments have a wide range of particle sizes, where the smaller nano-titanium dioxide particles may actually fit into the interstices between the larger particles of silica fume, thus taking up volume originally filled by the binder with pigment volume. In this case, the additional pigment does not compromise the mechanical properties of the film, also nano-titanium dioxide pigments did not provide the necessary range in particle size required to achieve a high degree of pigment packing leading to less mechanical properties than paint

Parameter	Group I					Group II					Group III					Group IV				
Drying time, hr.	**1**	**2**	**3**	**4**	**5**	**6**	**7**	**8**	**9**	**10**	**11**	**12**	**13**	**14**	**15**	**16**	**17**	**18**	**19**	**20**
Surface dry	1 - 2																			
Thorough dry	3 – 4																			
Adhesion	Gt0	Gt0	Gt0	**Gt0**	**Gt0**	Gt1	Gt0	Gt1	**Gt1**	**Gt1**	Gt1	Gt0	Gt1	**Gt1**	**Gt1**	Gt1	Gt0	Gt1	**Gt1**	**Gt1**
Hardness, Sec (90μ)	57	67	75	73	72	60	65	67	65	62	78	82	88	86	82	78	82	88	82	80
Ductility, Mm	6.3	3	6.7	6.5	6.4	6.7	2.7	6.4	6.1	5.8	6.4	4.3	6.6	6.2	5.6	6.4	4.3	5.4	5.2	5.1
Impact resistance, g.m.	0.74	0.33	0.74	0.75	0.64	0.75	0.28	0.93	0.88	0.75	0.74	0.26	0.96	0.88	0.74	0.74	0.26	0.74	0.64	0.54
Corrosion resistance																				
Degree of Blistering	6D	8M	4M	6MD	6M	4MD	8F	8F	8MD	8F	4D	6D	8MD	8M	8M	8MD	8M	8M	8F	8F
Degree of rusting	3-P	2-G	3-G	3-G	3-S	5-G	6-G	4-S	3-S	3-G	5-P	4-P	7-G	7-G	8-G	5-S	4-S	9-S	9-G	10
Filiform corrosion	Fig.1-FA																			

Table 7. Physical, mechanical and corrosion results of paint formulations containing TiO_2/ SiO_2 Pigments with medium oil alkyd resin

films containing silica fume. Also, the presence of the upper layers of nano-titanium dioxide precipitated on silica surface disturb the texture of silica by altering their platelet particles between the silica spheres leading to less homogenous texture and thus less mechanical properties. Paint films containing titanium dioxide exhibit poor mechanical properties among the entire group.

In general the mechanical properties of paint films in all the groups can be arranged as follow;

Paint films containing SiO_2> $TiO_2/SiO_2(1)$> $TiO_2/SiO_2(2)$> $TiO_2/SiO_2(3)$> TiO_2

3.3. The effect of prepared pigments on anticorrosive properties of paints

Table (8) features the results of determining the anticorrosion performance of paint films through blistering on paint surface and rust under film. The method classifies the osmotic blisters to the groups according to their sizes designated by numbers 2, 4, 6, and 8 (2 denotes the largest size, 8 the smallest one).

To the blister size information on the frequency of occurrence is given. The highest occurrence density of blisters is designated as D (dense), the lower ones as MD (medium dense), M (medium) and F (few). In such away a series from the surface area attacked at least by the osmotic blisters up to the heaviest occurrence can be formed as follows: 8F-6F-4F-2F-8M-6M-4M-2M-8MD-6MD-4MD-2MD-8D-6D-4D-2D.

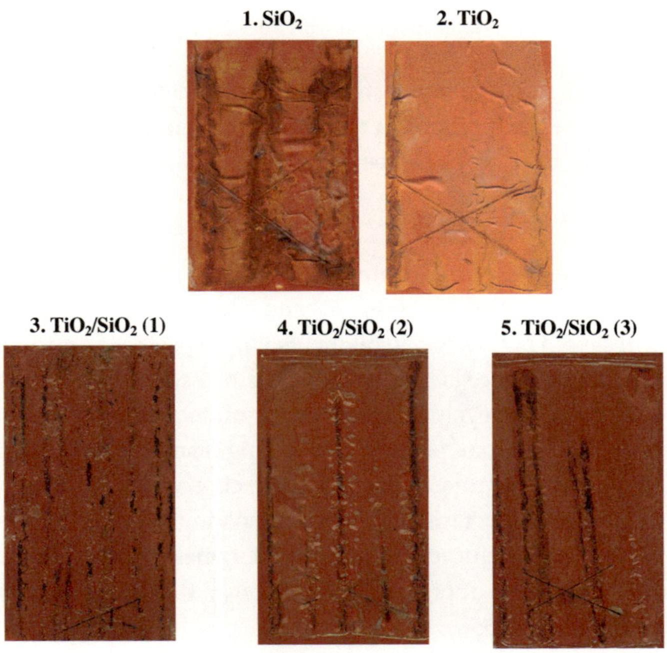

Figure 4. Photos of paint films of Group I after immersion in 3.5% NaCl after 28 days

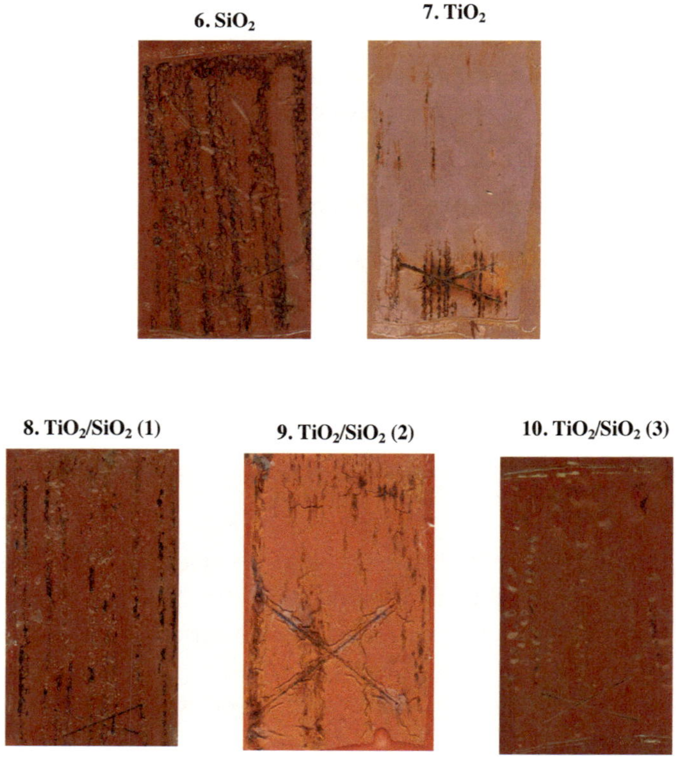

Figure 5. Photos of paint films of Group II after immersion in 3.5% NaCl after 28 days

As can be detected from Table (8) and Figures (4-7) that paint films of group IV were better in their performance than the other groups in rust under paint films. In general the performance in corrosion protection of the groups can be arranged as follows;

Group IV > Group III > Group II > Group I

Also it can be detected that, paint films containing SiO_2/TiO_2 (3) were better in its anticorrosive performance than those containing SiO_2/TiO_2 (2) and both were better than SiO_2/TiO_2 (1), this may be due to that in case of anticorrosive pigments whose protective effect is located on their surfaces where nano-titanium dioxide is located, the direct contact with the matrix and hence their interaction with it will be the main reason of their protection properties. It was found that polymer/nano-particle have unique properties resulting from the large fraction of atoms that reside at the surface leading to strong interfacial interactions with the polymer matrix, which can improve the protective characteristics of organic coatings [28-32, 33]. Also, it was considered that the pigments, which induced a very strong polymer-pigment interaction, resulted in coatings with a better dispersion of pigments and stability, and were resistant to corrosion by inducing barrier properties [34, 35].

Also, the previous results can be dedicated to that nano-titanium dioxide particles forming the upper layers of the pigment are arranged in alignment on silica surfaces and sometimes in

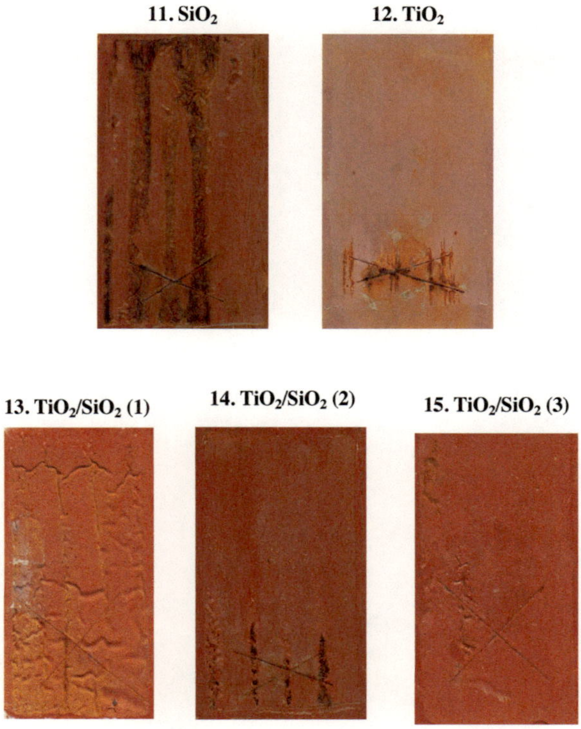

Figure 6. Photos of paint films of Group III after immersion in 3.5% NaCl after 28 days

between its particles. In case of low concentration of nano-titanium dioxide, its small platelet particles order themselves in the interstices positions between the spherical silica particles, these spaces formed between the binder and the pigment particles, even under the best circumstances give the chance for areas arise on the surface of the pigment particle where the binder and the particle may be in extremely close physical proximity but are not chemically bonded. This area between binder and pigment can be a potential route for water molecules to slip through the cured film [2, 36]. As the concentration of nano-titanium dioxide increases, a more close-pack texture locking almost all these positions will be formed leaving no routes for the corroding materials to reach the metal surface and thus better corrosion inhibition can be detected [37, 38].

3.4. Electrochemical Impedance studies

Electrochemical impedance spectroscopy (EIS) was used for the evaluation of the electrochemical properties of painted mild steel with formulations containing different concentrations of TiO_2/SiO_2 core-shell pigments. The painted specimens were tested in 3.5 wt% NaCl at different immersion times.

Equivalent circuit modeling is a standard tool for the interpretation of impedance data. Equivalent circuit consists of R_s which is the electrolyte/solution resistance, followed by a

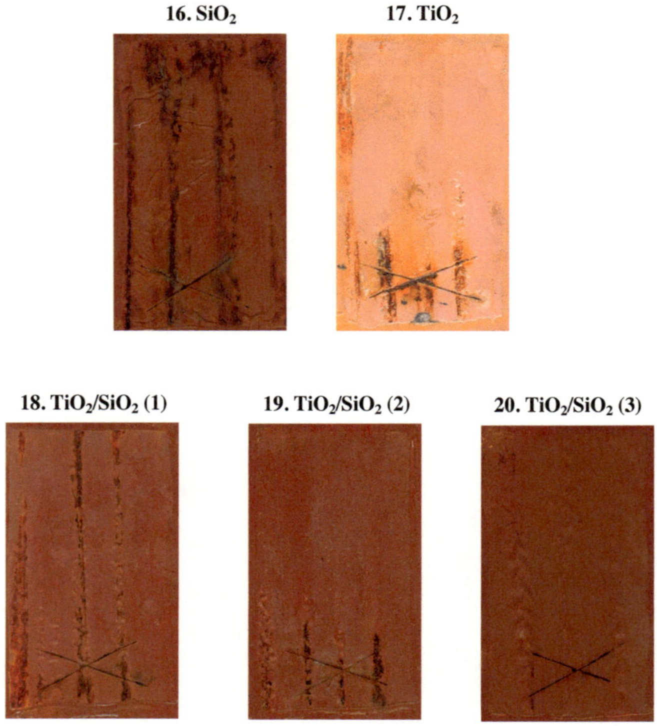

Figure 7. Photos of paint films of Group IV after immersion in 3.5% NaCl after 28 days

coating capacitance C_c which is in parallel with R_{po} which is the pore resistance; and Warburg impedance Z_w (diffusional element) is also included to represent the electrochemical processes taking place at the paint-metal interface. This element Z_w is often represented as C_{dl} (the double layer capacitance) which is in parallel with R_{ct} (charge transfer resistance) in corroding coated metals.

The values of pore resistance (R_{po}) obtained from impedance data, at different immersion times 1, 7, 14, 21 and 28 days, are listed in Table (8) and shown in Fig. 8.

Group	Paint no.	Pigments in paint formulations	R_{po} (Kohm.cm^2)				
			1 day	7 day	14 day	21day	28 day
I	3	[TiO_2/SiO_2(1)]	0.291	0.182	0.164	0.122	0.113
	4	[TiO_2/SiO_2(2)]	0.865	0.409	0.243	0.220	0.191
	5	[TiO_2/SiO_2(3)]	0.758	0.566	0.350	0.290	0.274
II	8	[TiO_2/SiO_2(1)]	0.307	0.191	0.175	0.140	0.109
	9	[TiO_2/SiO_2(2)]	2.680	0.887	0.884	0.412	0.409
	10	[TiO_2/SiO_2(3)]	9.520	3.010	2.770	0.906	0.618

Group	Paint no.	Pigments in paint formulations	R_{po} (Kohm.cm^2)				
			1 day	7 day	14 day	21day	28 day
III	13	[TiO_2/SiO_2(1)]	0.867	0.522	0.490	0.306	0.272
	14	[TiO_2/SiO_2(2)]	2.700	1.000	0.770	0.456	0.408
	15	[TiO_2/SiO_2(3)]	1.686	3.550	1.568	1.080	0.764
IV	18	[TiO_2/SiO_2(1)]	1.211	1.140	0.566	0.406	0.367
	19	[TiO_2/SiO_2(2)]	3.680	1.255	1.170	0.612	0.491
	20	[TiO_2/SiO_2(3)]	126.0	37.80	10.93	4.840	1.800

Table 8. EIS results for different paint formulations in 3.5 wt% NaCl at different immersion times

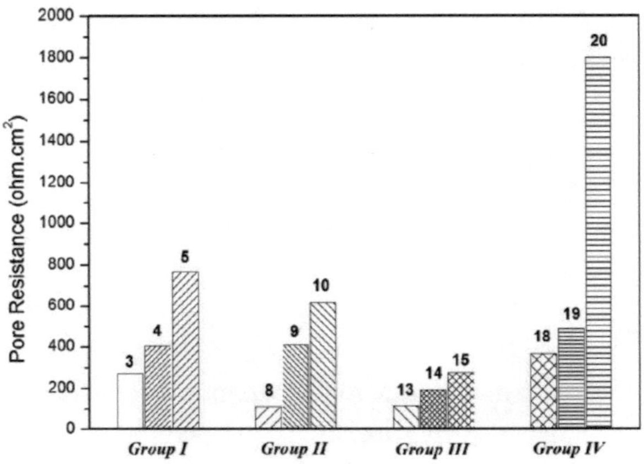

Figure 8. Pore resistance of TiO_2/SiO_2 paint formulations in 3.5 wt% NaCl after immersion for 28 days

Results represented in this table indicated the following;

a. The pore resistance (R_{po}) is clearly higher in case of group IV paint films which contains 30% of TiO_2/SiO_2 pigments, than that of other groups at all immersion times. It is obvious that the rate of corrosion follows the order;

Group IV > Group III > Group II > Group I

b. Concerning the upper layer thickness it was found that the pore resistance (R_{po}) increases in the following order;

Paint films containing TiO_2/SiO_2(3) > TiO_2/SiO_2(2) > TiO_2/SiO_2(1)

c. As shown in Fig. 9, the paint film containing TiO_2/SiO_2 (3) in group IV (paint no. 20) offered the best corrosion resistance even after 28 days immersion in 3.5wt% NaCl solution (1.8K ohm), indicating that paint formulations loaded with 30% of TiO_2/SiO_2(3) pigment is the best among the four groups and they can provide effective protection to carbon steel.

d. The pore resistance (R_{po}) is inversely proportional to immersion times. As paint no. 20 was the best among the groups it was taken as an example to show in Fig. 9 the typical Nyquist plots after different immersion time in 3.5 wt% NaCl.

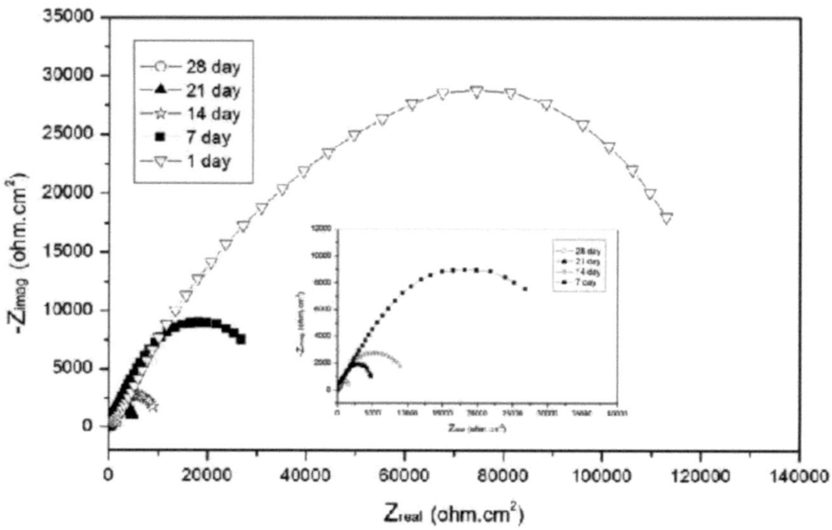

Figure 9. Nyquist plots of paint formulations containing TiO_2/SiO_2 (Sample 20) in 3.5 wt% NaCl at different immersion times 1, 7, 14, 21 and 28 days,

The decrease in pore resistance may be due to the penetration of water and movement of ionic species among the coating layer, increasing the coating conductivity. This can be explained according to the following steps;

- Initially, the electrolyte penetrates through the coating layer, and sets up conducting paths at different depths within the coating [39]. With increase immersion time, the electrochemical reactions at the interface between the coating and the metal surface make progress where the electrolyte phase meets the metal/oxide interface and a corrosion cell is then activated [40].
- This step is followed by that the barrier properties of the coating are decreased, suggesting a decrease in the coating resistance, i.e. decrease the radius of the semi-circle [41-43].

Generally, electrochemical studies were in high accordance with accelerated laboratory test results which revealed that group IV was the best and paint films containing TiO_2/SiO_2(3) showed the best corrosion inhibition behaviour among all paint films of the groups.

4. Conclusions

Hybrid pigments were prepared based on silica fume waste in the bulk covered with different layers of nano-titanium dioxide to overcome the deficiency of forming silica fume agglomerations due to its fine particles and promote its surface properties. The pigments were tested as anticorrosive pigments in paints based on medium oil alkyd resin. The study showed that as the thickness of the nano-titanium dioxide layers increases better corrosion protection performance was exhibited by paint films, and the study also revealed that the pigment loading in paint formulations is directly proportional to corrosion inhibition.

Electrochemical studies showed that pore resistance was in inverse relationship with the concentration of the prepared pigments, showing that as pigment concentration increases less pore size was detected and hence better corrosion protection was performed and these results are in high agreement with the accelerated laboratory test.

Acknowledgements

This work was supported by Science and Technology Development Fund (STDF) under project number 1242.

Author details

Nivin M. Ahmed[1] and Hesham Tawfik M. Abdel-Fatah[2]

1 Polymers and Pigments Department, National Research Centre, Dokki, Cairo, Egypt

2 Central Chemical Laboratories, Egyptian Electricity Holding Company, Sabtia, Cairo, Egypt

References

[1] Ahmed, N.M., Selim, M.M. Anticorrosive performance of titanium dioxide-talc hybrid pigments in alkyd paint formulations for protection of steel structures. Anti-Corrosion Methods and Materials 2010; 57 (3) 133-141.

[2] Ahmed, N.M., Selim, M.M. Innovative titanium dioxide-kaolin mixed pigments performance in anticorrosive paints. Pigment & Resin Technology 2011; 40 (1) 4-16.

[3] Ahmed, N.M., Mohamed, H.A. Performance of phosphate-alumina pigments in waterborne paints for protection of cold-rolled steel. J. Coating Technology and Research 2011; 8 (2) 201-210.

[4] Caruso, R.A., Susha, A., Caruso, F. Structural Control of TiO_2 Networks Using Polymer Gel Templates. Chem. Mater. 2001; 13 (1) 400-409.

[5] Daniel, M.C., Astrue, D. Gold Nanoparticles: Assembly, Supramolecular Chemistry, Quantum-Size-Related Properties, and Applications toward Biology, Catalysis, and Nanotechnology. Chem. Rev. 2004; 104 (2) 293-346.

[6] Ahmed, N.M., Abd El-Fattah, H.T.M. The Role of Silica Fume Pigments in Corrosion Protection of Steel Surfaces. In: Reza Shoja Razavi (Ed.), Recent Researches in Corrosion Evaluation and Protection, InTech; 2012. p. 67-90.

[7] Abu-Ayana, Y.M., Youssef, E.A.M., El-Sawy, S.M. Silica fume-formed during the manufacture of ferrosilicon alloys-as an extender pigment in anticorrosive paints". Anti-Corrosion Methods and Materials 2005; 52 (3) 345-352.

[8] Youssef, E.A.M., Ahmed, N.M., Abd El-Ghaffar, M.A. Characterization and evaluation of silica fume as an extender pigment for surface coating applications. Pigment & Resin Technology 1998; 27 (2) 88-98.

[9] Siddique, R. Utilization of silica fume in concrete: Review of hardened properties. Resources, Conservation and Recycling 2011; 55 (11) 923-932.

[10] Boddy, A.M., Hooton, R.D., Thomas, M.D.A. The effect of the silica content of silica fume on its ability to control alkali-silica reaction. Cement and Concrete Research 2003; 33 (8) 1263-1268.

[11] Panjehpour, M., Ali, A.A.A., Demirboga, R. A Review for characterization of silica fume and its effects on concrete properties. International Journal of Sustainable Construction Engineering & Technology 2011, 2 (2) 1-7.

[12] Ahmed, N.M., Abdel-Fatah, H.T.M., Youssef, E.A. Corrosion studies on tailored Zn Co aluminate/kaolin core–shell pigments in alkyd based paints. Progress In Organic Coatings 2012, 73 (1) 76-87.

[13] Pagliaro, M., Ciriminna, R., Palmisano, G. Silica-Based Hybrid Coatings. Journal of Materials Chemistry 2009; 19 (1) 3116-3126.

[14] Giner, V.T., Ivorra, S., Baeza, F.J., Zornoza, E., Ferrer, B. Silica fume admixture effect on the dynamic properties of concrete. Construction and Building Materials 2011; 25 (8) 3272-3277.

[15] Vila-Gonzalez, C.A., Cruz-Silva, R., Menchaca, C., Sepulveda-Guzman, S., Uruchurtu, J. Use of Silica Tubes as Nanocontainers for Corrosion Inhibitor Storage. Journal of Nanotechnology 2011; 11 (1) 1-9.

[16] Tarrío-Saavedra, J., López-Beceiro, J., Naya, S., Gracia, C., Artiaga, R. Controversial effects of fumed silica on the curing and thermomechanical properties of epoxy composites. eXPRESS Polymer Letters 2010; 4 (6) 382-395.

[17] Xianming, S., Tuan, A.N., Zhiyong, S., Yajun, L., Avci, R. Effect of nanoparticles on the anticorrosion and mechanical properties of epoxy coating. Surface & Coatings Technology 2009; 204 (2) 237-245.

[18] Peter, H., Robert, V. High grade kaolin fillers–production review: Industrial Minerals. Applied Clay Science 1999, 54 (1) 25-37.

[19] Salamone, J.C. Concise polymeric materials encyclopedia, CRC press 1999. p. 477.

[20] Shao, Y., Li, Y., Du, Y., Wang, F. Enhancement of the Protectiveness of Epoxy Coatings with Surface-Modified Nano-Titanium Particles. Corrosion 2006; 62 (6) 483-490.

[21] Lewis, O.D., Critchlow, G.W., Wilcox, G.D., De Zeeuw, A., Sander, J. A study of the corrosion resistance of a waterborne acrylic coating modified with nano-sized titanium dioxide. Progress in Organic Coatings 2012; 73 (1) 88-94.

[22] Scully, J.R., Hensley, S.T. Lifetime Prediction of Organic Coatings on Steel and a Magnesium Alloy using Electrochemical Impedance. Corrosion 1994; 50 (9) 705-716.

[23] Wei, D.F., Chatterjee, I., Jones, D.A. Evaluation of Corrosive Degradation in Coated Steel Using Alternating Current Impedance Spectroscopy. Corrosion 1995; 51 (2) 97-104.

[24] Tsai, C.H., and Mansfield, F. Determination of Coating Deterioration with EIS: Part II. Development of a Method for Field Testing of Protective Coatings. Corrosion Sciences 1993; 49 (9) 726-737.

[25] Mansfield, F., Kendig, M.W., and Tsai, S. Evaluation of corrosion behavior of coated metals with AC impedance measurements. Corrosion 1982; 38 (3) 478-484.

[26] Walter, G. Laboratory Simulation of Atmospheric Corrosion by SO_2-II. Electrochemical Mass Loss Comparisons. Corrosion Sciences 1991; 32 (15) 1041-1048.

[27] Lambourne, R. Paint and surface coatings, Ellis Horwood Ltd. Pub. 1987, p. 131.

[28] Guin, A.K., Nayak, S.K., Rout, T.K., Bandyopadhyay, N., Sengupta, D.K. Corrosion behavior of nanohybrid titania–silica composite coating on phosphated steel sheet. J. Coat. Technol. Res. 2012; 9 (1) 97-106.

[29] Chen, C., Khobaib, M., Curliss, D. Epoxy Layered-Silicate Nanocomposites. Prog. In Org. Coat. 2003; 47 (3) 376-383.

[30] Voevodin, N.N., Balbyshev, V.N., Khobaib, M., Donley, M.S. Nanostructured coatings approach for corrosion protection. Prog. In Org. Coat. 2003; 47 (4) 416-423.

[31] Kirubaharan, A.M.K., Selvaraj, M., Maruthan, K., Jeyakumar, D. Synthesis and characterisation of nanosized titanium dioxide and silicon dioxide for corrosion resistance applications. J. Coat. Technol. Res. 2012; 9 (2) 163-170.

[32] Mathiazhagan, A., Joseph, R. Nanotechnology-A New Prospective in Organic Coating – Review. International Journal of Chemical Engineering and Applications 2011; 2 (4) 225-237.

[33] Vesely, D., Kalendova, A. Anticorrosion efficiency of $Zn_xMg_yAl_2O_4$ core-shell spinels in organic coatings, *Progress in Organic Coatings* 2008; 62 (1) 5-20.

[34] Hegedus, C.R., Kamel, I.L. Polymer-Filler Interaction. Effects on Coating Properties. J. Coat. Technol. 1993; 65 (822) 37-43.

[35] Rong, M.Z., Zhang, M.Q., Zheng, Y.X., Zeng, H.M., Walter, R., Friedrich, K. Improvement of tensile properties of nano-SiO_2 /PP composites in relation to percolation mechanism. Polymer 2001; 42 (1) 167-183.

[36] Quanxiang, Y., Yun, L., Jizu, Y. Preparation of titanium dioxide compound pigments based on kaolin substrates. J. Coat. Technol. Res. 2010; 7 (2) 229-237.

[37] Sorensen, P.A., Kiil, S., Dam-Johansen, K., Weinell, C.E. Anticorrosive coatings: a review. J. Coat. Technol. Res. 2009; 6 (1) 135-176.

[38] Barranco, V., Feliu Jr., S., Feliu, S. EIS study of the corrosion behavior of zinc-based coatings on steel in quiescent 3% NaCl solution. Part 2: coatings covered with an inhibitor-containing lacquer. Corrosion Science 2004; 46 (9) 2221-2240.

[39] Del Amo, B., Romagnoli, R., and Vetere, V.F. Steel Corrosion Protection by Means of Alkyd Paints Pigmented with Calcium Acid Phosphate. Ind. Eng. Chem. Res. 1999; 38 (3) 2310-2314.

[40] Shao, Y., Jia, C., Meng, G., Zhang, T., Wang, F. The role of a zinc phosphate pigment in the corrosion of scratched epoxy-coated steel. Corrosion Science 2009; 51 (1) 371-379.

[41] Panjehpour, M., Ali, A.A., Demirboga, R. A Review for characterization of silica fume and its effects on concrete properties. International Journal of Sustainable Construction Engineering & Technology 2011; 2 (2) 1-7.

[42] Torii, K., Kawamura, M. Pore structure and chloride ion permeability of mortars containing silica fume. Cement and Concrete Composites 1994, 16 (4) 279-286.

[43] Kaiser, J.P., Zuin, S., Wick, p. Is nanotechnology revolutionizing the paint and lacquer industry? A critical opinion. Science of the Total Environment 2013; 442 (1)282-289.

Section 3

Corrosion Inhibitors

Chapter 16

Corrosion Inhibitors – Principles, Mechanisms and Applications

Camila G. Dariva and Alexandre F. Galio

Additional information is available at the end of the chapter

http://dx.doi.org/10.5772/57255

1. Introduction

Corrosion processes are responsible for numerous losses mainly in the industrial scope. It is clear that the best way to combat it is prevention.

Among the various methods to avoid or prevent destruction or degradation of metal surface, the corrosion inhibitor is one of the best know methods of corrosion protection and one of the most useful on the industry. This method is following stand up due to low cost and practice method. [1] [2] [3] [4]

Important researches have being conducted with government investment mainly in large areas such as development construction of new pipelines for shale gas and growth in construction. The focus of these researches has being the inhibitors applications in water and concrete for the protection of metals. [5]

Historically, inhibitors had great acceptance in the industries due to excellent anti-corrosive proprieties. However, many showed up as a secondary effect, damage the environment. Thus the scientific community began searching for friendly environmentally inhibitors, like the organic inhibitors. [6] [7] [8] [9] [10] [11] [12] [13]

This chapter presents a revision of the corrosion inhibitors applications mainly the novel compositions environmentally friendly. Is describes the mechanisms of action of inhibitors, main characteristics, environmental impact, technical analysis and calculation of efficiency.

1.1. Mechanisms of actions of inhibitors

Inhibitors are substances or mixtures that in low concentration and in aggressive environment inhibit, prevent or minimize the corrosion. [2]

Generally the mechanism of the inhibitor is one or more of three that are cited below:

- the inhibitor is chemically adsorbed (chemisorption) on the surface of the metal and forms a protective thin film with inhibitor effect or by combination between inhibitor ions and metallic surface;
- the inhibitor leads a formation of a film by oxide protection of the base metal;
- the inhibitor reacts with a potential corrosive component present in aqueous media and the product is a complex. [4] [14] [15]

1.2. Historic review

There are many industrial systems and commercial applications that inhibitors are applicable, such as cooling systems, refinery units, pipelines, chemicals, oil and gas production units, boilers and water processing, paints, pigments, lubricants, etc. [16]

There are evidences of the use of inhibitor since the early XIX century. On that time they were already used to protect metals in processes such as acid picking, protection against aggressive water, acidified oil wells and cooling systems. Since years 1950's and 1960's, there was significant advances in the development of technology for corrosion inhibitor as the application of electrochemistry to evaluate corrosion inhibitors. [17]

Recent studies estimate that the U.S. demand for corrosion inhibitors will rise 4.1% per year to USD$ 2.5 billion in 2017. In 2012 they estimated that the market demand of inhibitors was divided on 26.6% to refining petroleum, 16.9% utilities, 16.7% gas and oil production, 15.3% chemical, 9.5% metals, 7.1% pulp and paper and 8.0% other. [5]

Now a days, due to changes occurred on the market of corrosion inhibitors, some industrial corrosion inhibitors are being unused. Due to high toxicity of chromate, phosphate and arsenic compounds, related to various environmental and health problems, strict international laws were imposed. Reducing the use of these and therefore increasing the need for the development of other inhibitor to supply the lack in this area. Should, however, present a similar anti corrosive properties similar than a chromate inhibitor. [18]

An important number of papers have been published with the intention of develop an environmentally friendly corrosion inhibitors and a lot of research has been doing to development of the called "green" corrosion inhibitors.[19] Also, has been increasing research in natural products, such as plant extracts, essential oils and purified compounds to obtain environmentally friendly corrosion inhibitors. [20]

The first evidence of natural product use as corrosion inhibitors is 1930's. When extracts of *Chelidonium majus* (Celadine) and other plants were used on the first time in H_2SO_4 pickling baths. Successful developments of researches to obtain natural corrosion inhibitors are growing as quickly as the environmental consciousness is gaining ground.[21]

Chromates as active inhibitors are being replaced by other components such as molybdate compounds and rare earth metal salt, like cerium chloride. Also, drugs have been studied as corrosion inhibitors. [22] [23] [24]

2. Inhibitors classifications

The corrosion inhibitors can be chemicals either synthetic or natural and could be classified by:

- the chemical nature as organic or inorganic;
- the mechanism of action as anodic, cathodic or a anodic-cathodic mix and by adsorption action, or;
- as oxidants or not oxidants. [4]

In general, the inorganic inhibitors have cathodic actions or anodic. The organics inhibitors have both actions, cathodic and anodic and the protective by a film adsorption.

This chapter is subdivided in according to the classification of the inhibitors shown on the Figure 1

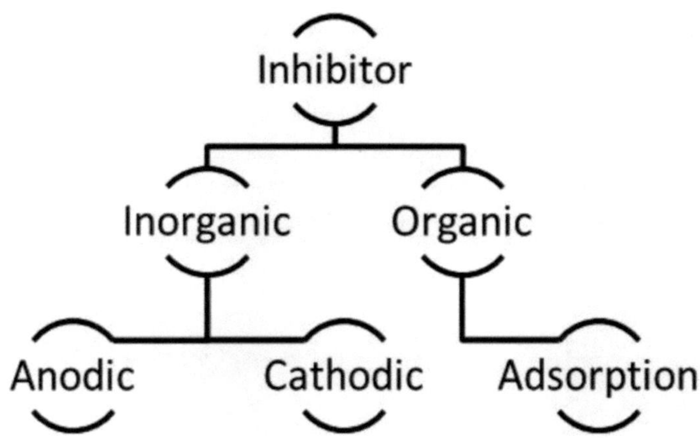

Figure 1. Classification of inhibitors

2.1. Inorganic inhibitors

2.1.1. Anodic inhibitors

Anodic inhibitors (also called passivation inhibitors) act by a reducing anodic reaction, that is, blocks the anode reaction and supports the natural reaction of passivation metal surface, also, due to the forming a film adsorbed on the metal. In general, the inhibitors react with the corrosion product, initially formed, resulting in a cohesive and insoluble film on the metal surface. [4] [24]

Figure 2 shows a potentiostatic polarization diagram of a solution with behavior inhibitor anodic. The anodic reaction is affected by the corrosion inhibitors and the corrosion potential of the metal is shifted to more positive values. As well, the value of the current in the curve decreases with the presence of the corrosion inhibitor.

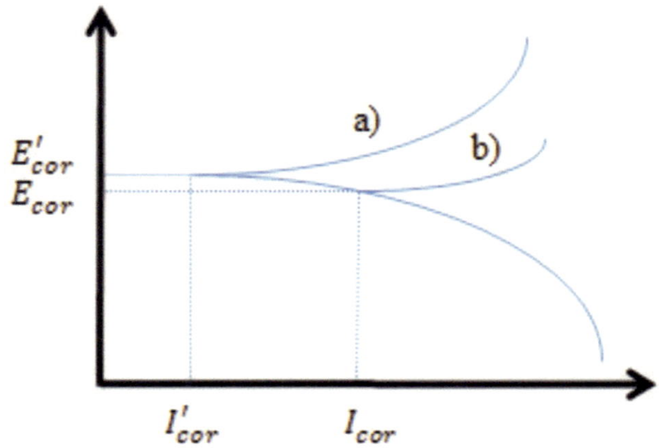

Figure 2. Potentiostatic polarization diagram: electrochemical behavior of a metal in a solution with anodic inhibitor (a) *versus* without inhibitor (b).

The anodic inhibitors reacts with metallic ions Me^{n+} produced on the anode, forming generally, insoluble hydroxides which are deposited on the metal surface as insoluble film and impermeable to metallic ion. From the hydrolysis of inhibitors results in OH^- ions. [4] Figure 3 shows how is the mechanism of the anodic inhibitory effect.

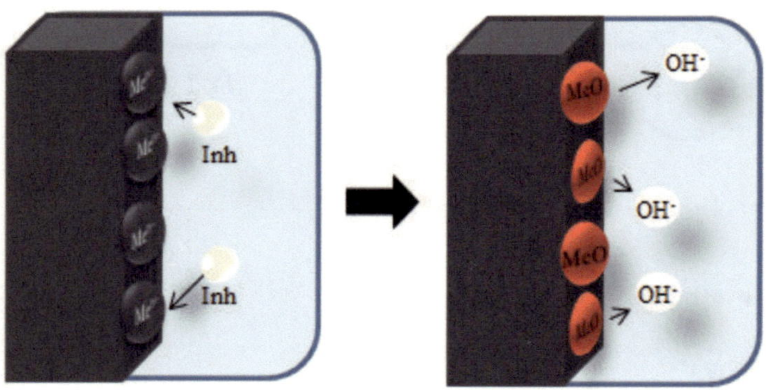

Figure 3. Illustration of anodic inorganic inhibitors effect and their mechanism of action.

When the concentrations of inhibitor becomes high enough, the cathodic current density at the primary passivation potential becomes higher than the critical anodic current density, that is, shift the potential for a noble sense, and, consequently, the metal is passivated. [25] [26]

For the anodic inhibitors effect, it is very important that the inhibitor concentrations should be high enough in the solution. The inappropriate amount of the inhibitors affects the formation of film protection, because it will not cover the metal completely, leaving sites of the metal exposed, thus causing a localized corrosion. [4] [25] [26]

Concentrations below to the critical value are worse than without inhibitors at all. In general can cause pitting, due reduction at the anodic area relative to cathodic, or can accelerate corrosion, like generalized corrosion, due to full breakdown the passivity. [25]

Some examples of anodic inorganic inhibitors are nitrates, molybdates, sodium chromates, phosphates, hydroxides and silicates.

2.1.2. Cathodic inhibitors

During the corrosion process, the cathodic corrosion inhibitors prevent the occurrence of the cathodic reaction of the metal. These inhibitors have metal ions able to produce a cathodic reaction due to alkalinity, thus producing insoluble compounds that precipitate selectively on cathodic sites. Deposit over the metal a compact and adherent film, restricting the diffusion of reducible species in these areas. Thus, increasing the impedance of the surface and the diffusion restriction of the reducible species, that is, the oxygen diffusion and electrons conductive in these areas. These inhibitors cause high cathodic inhibition. [4] [24] [27]

The Figure 4 shows an example of a polarization curve of the metal on the solution with a cathodic inhibitor. When the cathodic reaction is affected the corrosion potential is shifted to more negative values.

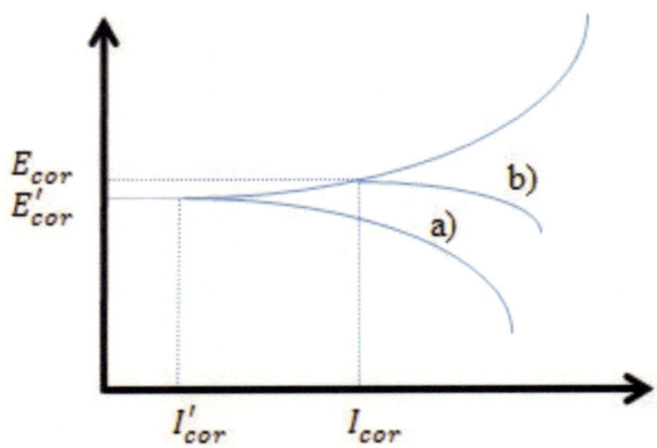

Figure 4. Potentiostatic polarization diagram: electrochemical behavior of the metal in a cathodic inhibitors solution (a), as compared to the same solution, without inhibitor (b).

The cathodic inhibitors form a barrier of insoluble precipitates over the metal, covering it. Thus, restricts the metal contact with the environment, even if it is completely immersed, preventing the occurrence of the corrosion reaction. Due to this, the cathodic inhibitor is independent of concentration, thus, they are considerably more secure than anodic inhibitor. The Figure 5 shows the illustration of mechanical effect of cathodic inhibitors to restrain the corrosion process. [4]

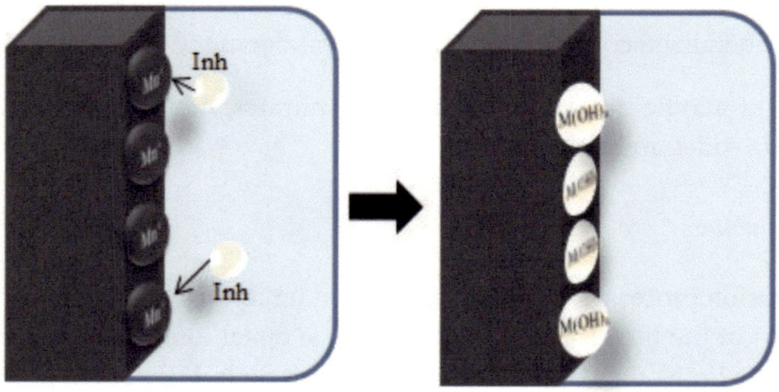

Figure 5. Illustration has shown the mechanism of actuation of the cathodic inhibitors.

Some examples of inorganic cathodic inhibitors are the ions of the magnesium, zinc, and nickel that react with the hydroxyl (OH^-) of the water forming the insoluble hydroxides as ($Mg(OH)_2$, $Zn(OH)_2$, $Ni(OH)_2$) which are deposited on the cathodic site of the metal surface, protecting it. [4] Also can be cited polyphosphates, phosphonates, tannins, lignins [15] and calcium salts as examples that presents the same reaction mechanism.

It seen in hard waters a kind of this mechanism of inhibiting, due to the effect of the magnesium or calcium bicarbonate on it. When temporary hard water flows over the metal it can assist on the nucleation of carbonates, allowing the reactions near to the equilibrium and forming precipitations on the metal surface. These precipitations, like a $CaCO_3$, cover the cathodic area, protecting the metal. So these cathodic inhibitor depends only on the chemistry of the water, is not due the metal composition, because of this they are applicable to all metals. [4] [25]

As example, may be mentioned the oxides and salts of antimony, arsenic and bismuth, which are deposited on the cathode region in acid solutions. These cathodic inhibitors minimize the release of hydrogen ions due to a phenomena that can difficult the discharge of the hydrogen, called overvoltage.

2.2. Organic inhibitor

Organic compounds used as inhibitors, occasionally, they act as cathodic, anodic or together, as cathodic and anodic inhibitors, nevertheless, as a general rule, act through a process of surface adsorption, designated as a film- forming. Naturally the occurrence of molecules exhibiting a strong affinity for metal surfaces compounds showing good inhibition efficiency and low environmental risk. [28] These inhibitors build up a protective hydrophobic film adsorbed molecules on the metal surface, which provides a barrier to the dissolution of the metal in the electrolyte. They must be soluble or dispersible in the medium surrounding the metal. [4]

In the Figure 6, that shows a theoric potentiostatic polarization curve, it can be seen that the effect of the solution containing organic inhibitor on the metal presents an anodic and cathodic behavior. After the addition of the inhibitor, the corrosion potential remains the same, but the current decreases from I_{cor} to I'_{cor}.

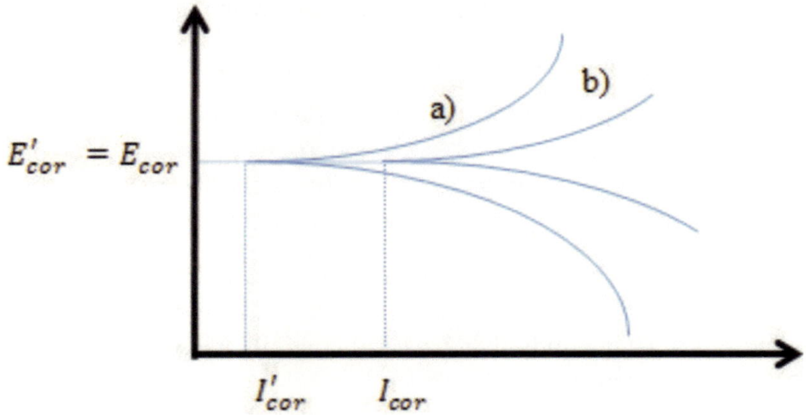

Figure 6. Theoretical potentiostatic polarization diagram: electrochemical behavior a metal on a solution containing a cathodic and anodic inhibitor (a) compared to the same solution without the inhibitor (b).

Is showed in Figure 7 the mechanism of actuation of organic inhibitors, when it is adsorbed to the metal surface and forms a protector film on it.

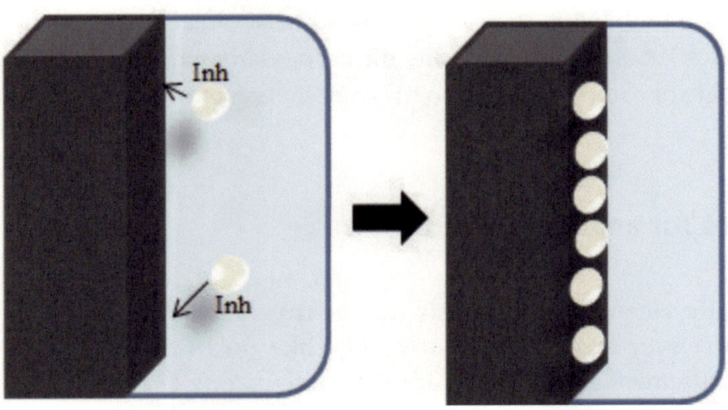

Figure 7. Illustration of the mechanism of actuation of the organic inhibitor: acting through adsorption of the inhibitor on the metal surface. Where the Inh represents the inhibitor molecules.

The efficiency of an organic inhibitor depends of the:

- chemical structure, like the size of the organic molecule;
- aromaticity and/or conjugated bonding, as the carbon chain length;

- type and number of bonding atoms or groups in the molecule (either π or σ);
- nature and the charges of the metal surface of adsorption mode like bonding strength to metal substrate;
- ability for a layer to become compact or cross-linked,
- capability to form a complex with the atom as a solid within the metal lattice;
- type of the electrolyte solution like adequate solubility in the environment. [16]

The efficiency of these organic corrosion inhibitors is related to the presence of polar functional groups with S, O or N atoms in the molecule, heterocyclic compounds and pi electrons, generally have hydrophilic or hydrophobic parts ionizable. The polar function is usually regarded as the reaction center for the establishment of the adsorption process.[4] [28]

The organic acid inhibitor that contains oxygen, nitrogen and/or sulfur is adsorbed on the metallic surface blocking the active corrosion sites. Although the most effective and efficient organic inhibitors are compounds that have π-bonds, it present biological toxicity and environmental harmful characteristics. [29]

Due to the metal surface covered is proportional to the inhibitor concentrates, the concentrations of the inhibitor in the medium is critical. [24] [30]

Some examples are amines, urea, Mercaptobenzothiazole (MBT), benzotriazole e toliotriazol, aldehydes, heterocyclic nitrogen compounds, sulfur-containing compounds and acetylenic compounds and also ascorbic acid, succinic acid, tryptamine, caffeine and extracts of natural substances. [4] [25] [28]

There are still some inhibitors that act in vapor phase (volatile corrosion inhibitor). Some examples are: dicicloexilamônio benzoate, diisopropylammonium nitrite or benzoate, ethanolamine benzoate or carbonate and also the combination of urea and sodium nitrite. [4] [24]

3. Techniques for analysis of inhibitors

The most usefully technique to analysis the effectiveness of an inhibitor are weight loss experiment and electrochemical measurements, like polarization curve method and the impedance measurement analyzing. In addition, microscopy techniques are used to characterize the corrosion process.

4. Considerations to employ inhibitors

For all types of inhibitors, we should consider some environmental actions factors because some elements such as metals, pH, composition, impurities, agitation, temperature, geometry of the system, the concentration of inhibitor and the mixture of one or more inhibitors may change the anti-corrosive mechanism. [4] [23] [31] [33]

To employment of the inhibitors is quite satisfying that certain factors should be seen as the real cause of the corrosion, the cost X benefit and possible interactions of the inhibitor with the environment, such as the influence of a catalyst, deposition or contamination. Four fundamental aspects must be analyzed to obtain a satisfactory result from the use of the inhibitor.

5. Inhibitor efficiency

The inhibitor efficiency could be measured by the follow equation:

$$E_f = \frac{R_i - R_o}{R_o} \times 100 \qquad (1)$$

where, E_f is inhibitor efficiency (percentage), R_i is corrosion rate of metal with inhibitor and R_o is corrosion rate of metal without inhibitor.[38]

6. Industrial application

Acid pickling: Prevent the attack in the metal due to the acid solution in which metal gets cleaned of mill scale (bark lamination), and also prevented the subsequent hydrogen evolution inhibitors are added, typically organic, must be soluble or dispersed in the solution. Examples: thiourea and amino and its derivatives, propargyl alcohol. [4]

Oil industry: sodium carbonates or organic amines complex are employed to reduce the corrosive effect of CO_2, H_2S and organic acids, enabling the use of more cheaper materials and less resistant to corrosion in wells extracting crude oil. Pipes for gasoline and kerosene are employed sulphonated oils, sodium nitrite. Oil well uses up fatty amines, fatty acids, imidazolines and quaternary ammonium salts. Internal pipe corrosion occurs in wet gas transportation due to condensation of water containing dissolved corrosive gases. Corrosion is caused by the dissolution of the corrosive gases, such as carbon dioxide and hydrogen sulfide as well as condensation of acid vapors [42].

Water transmission and distribution systems: is used corrosion inhibitor in combination with pH adjusters and alkalinity control towards an efficient protection [32] The most common inhibitors are phosphates, amines volatiles (cyclohexylamine, morphine) [4]

Concrete: To improve the durability of reinforced concrete structures, which are impaired due the high alkalinity, are used corrosion inhibitors, mixed with cement or concrete paste. An example is phosphate ion. [33]

Boiler: Thermoelectric use, in general, Ammonia, Cyclohexylamine, alkanol and Morpholine as inhibitors in boilers in various processes. The inhibitors, also, are added by the hydrochloric acid used for the solubilization of limescale to prevent the attack on pipes.

7. Conclusion

Inhibitors are a great method of preventing corrosion and are easy to apply. Has application in a wide range of sectors

The knowledge of the method of the action, facilitates the choice of the inhibitors, improves efficiency, avoids the process is impaired and side effects.

It is important in the choice of inhibitor, ascertain the subsequent effects of this towards the environment.

The search for environmental friendly inhibitors has shown excellent results, outperforming conventional inhibitors.

Reference organic inhibitors newly developed

alloy	environmental	pH	inhibitor	concentration	efficiency	reference
AA2024-T3	0,1 M NaCl	7	$K_2Cr_2O_7$	10^{-4} M	97.31	[23]
		7	Cerium dibutylphosphate ($Ce(dbp)_3$)	10^{-4} M	94.10	
		7	Cerium chloride ($CeCl_3$)	10^{-4} M	73.90	
AA5754	3% NaCl	6.3	Laurus nobilis L. oil	50ppm	84.3	[43]
Al	3% NaCl	6.3	Laurus nobilis L. oil	50ppm	89.9	[43]
AA5083	3.5% NaCl	5.5	$CeCl_3$	500ppm	297	[44]
			$LaCl_3$	500ppm	113	[44]
Steel	1M HCl	-	N. cadamba Bark extract	5mgL-1	91.0	[45]
Steel	1 M HCl	-	VSBH ($C_{31}H_{47}O_4N_2Br$)	400 ppm	95.0	[46]
Steel	88% phosphoric acid	-	Dried Zenthoxylum-alatum plant fruits	2400ppm	98.0	[47]
Steel	0.5 M H_2SO_4	-	Polyacrylamide grafted Okra mucilage(O-*g*-PAM)	100ppm	94.4	[48]
Copper	0.5M HCl	-	Chitosan	8×10^{-6}	92.0	[29]
Cu-10Al-5Ni alloy	3.5% NaCl	-	Cysteine	6mM/L	96.0	[49]
			N-acetylcysteine	6mM/L	87.6	[49]

alloy	environmental	pH	inhibitor	concentration	efficiency	reference
			Methionine	6mM/L	76.7	[49]
Cu	0.6M NaCl	-	Cysteine	16mM	76.5	[50]
Cu	1M HCl	-	Cysteine	18mM	84.13	[50]
AZ91D	0.05 wt.% NaCl	9	Paeonol	50ppm by wt	90	[51]
AZ91D	ASTM D1384-87	8.2	8-hydroxyquinoline	Saturated with 8HQ	82	[52]
AZ91D alloy	0.05 wt.% NaCl	6.8	5,10,15,20-tetraphenylporphyrin (TPP)	5ppm	84	[53]

Table 1. Inhibitors organics for aluminum, steel, copper, magnesium and its alloys at room temperature.

Author details

Camila G. Dariva and Alexandre F. Galio*

*Address all correspondence to: afgalio@ufrgs.br

PPEng - CAPES and Universidade Federal do Pampa Bagé/RS, Brazil

References

[1] M. S. Al-Otaibi, A. M. Al-Mayouf,. M. Khan, A. A. Mousa, S. A. Al-Mazroa e H. Z. Alkhathlan, "Corrosion inhibitory action of some plant extracts on the corrosion of mild steel in acidic media," Arabian Journal of Chemistry, pp. 1-7, 2012.

[2] I.B. Obot, N.O. Obi-Egbedi, S.A. Umoren, "Antifungal drugs as corrosion inhibitors for aluminium in 0.1 M HCl," Corrosion Science, vol. 51, issue 8, pp. 1868-1875, 2009.

[3] A. Yıldırım, M. Çetin, "Synthesis and evaluation of new long alkyl side chain acetamide, isoxazolidine and isoxazoline derivatives as corrosion inhibitors," Corrosion Science, vol. 50, issue 1, pp.155-165, 2008.

[4] V. Gentil, Corrosão, 4ª ed., Rio de Janeiro: LTC, 2003.

[5] P. Finishing, "pfonline," Finishing Industry, 03 06 2013. [Online]. Available: http://www.pfonline.com/news/us-demand-for-corrosion-inhibitors-to-reach-25-billion-in-2017. [Access on 10 07 2013].

[6] Negm, Nabel A.; Kandile, Nadia G.; Badr, Emad A.; Mohammed, Mohammed A. Gravimetric and electrochemical evaluation of environmentally friendly nonionic

corrosion inhibitors for carbon steel in 1 M HCl. Corrosion Science, vol. 65, pp. 94-103, 2012.

[7] A. M. Abdel-Gaber, B.A. Abd-El-Nabey, E. Khamis, D.E. Abd-El-Khalek. A natural extract as scale and corrosion inhibitor for steel surface in brine solution. Desalination. vol. 278, Issue 1-2, pp. 337-342, 2011.

[8] Salasi, M.; Sharabi, T.; Roayaei, E.; Aliofkhazraei, M. "The electrochemical behaviour of environment-friendly inhibitors of silicate and phosphonate in corrosion control of carbon steel in soft water media," Materials Chemistry and physics, vol. 104, Issue 1, pp. 183-190, 2007.

[9] G. Blustein, R. Romagnoli, J.A. J. "Zinc basic benzoate as eco-friendly steel corrosion inhibitor pigment for anticorrosive epoxy-coatings," Colloids and Surfaces A: Physicochemical and Engineering Aspects, vol. 290, Issue 1-3, pp. 7-18, 2006.

[10] P. Bommersbach, C. Alemany-Dumont, J.P. Millet, B. Normand. "Formation and behaviour study of an environment-friendly corrosion inhibitor by electrochemical methods," Electrochimica Acta, vol. 51, Issue 6, pp. 1076- 1084, 2005.

[11] A. Lecante, F. Robert, P.A. Blandinières, C. Roos. "Anti-corrosive properties of S. tinctoria and G. ouregou alkaloid extracts on low carbon steel," Current Applied Physics, vol. 11, Issue 3, pp. 714-724, 2011.

[12] Radojčić, K. Berković, S. Kovač, J. Vorkapić-Furač. "Natural honey and black radish juice as tin corrosion inhibitors," Corrosion Science, vol. 50, Issue 5, pp. 1498-1504, 2008.

[13] P.C. Okafor, M.E. Ikpi, I.E. Uwah, E.E. Ebenso, U.J. Ekpe, S.A. Umoren. "Inhibitory action of Phyllanthus amarus extracts on the corrosion of mild steel in acidic media," Corrosion Science, vol. 50, Issue 8, pp. 2310-2317, 2008.

[14] Hong Ju, Zhen-Peng Kai, Yan Li, "Aminic nitrogen-bearing polydentate Schiff base compounds as corrosion inhibitors for iron in acidic media: A quantum chemical calculation," Corrosion Science, vol. 50, Issue 3, pp. 865-871, 2008.

[15] L. V. Ramanathan, Corrosão e seu controle, São Paulo: Hemus, 1988.

[16] B. SANYAL, "Organic compounds as corrosion inhibitors in different environments - A review," Progress in Organic Coatings, vol. 9, pp. 165-236, 1981.

[17] M. Pourbaix. "Applications of electrochemistry in corrosion science and in practice," Corrosion Science, vol. 14, pp. 25-82, 1974.

[18] V. S. Saji, "A review on recent patent in corrosion inhibitor," Recent Patents on Corrosion Science, vol. 2, pp. 6-12, 2010.

[19] A. El Bribri, M. Tabyaoui, B. Tabyaoui, H. El Attari, F. Bentiss,"The use of Euphorbia falcata extract as eco-friendly corrosion inhibitor of carbon steel in hydrochloric acid solution," Materials Chemistry and Physics, vol. 141, issue 1, pp. 240-247, 2013.

[20] M. Znini, L. Majidi, A. Bouyanzer, J. Paolini,. J. -M. Desjobert, J. Costa e B. Hammouti, "Essential oil of salvia aucheri mesatlantica as a green inhibitor for the corrosion of steel in 0.5M H2SO4," Arabian Journal of Chemistry, vol. 5, n. 4, pp. 467-474, 2012.

[21] Pandian Bothi Raja, Mathur Gopalakrishnan Sethuraman,"Natural products as corrosion inhibitor for metals in corrosive media — A review," Materials Letters, vol. 62, issue 1, pp. 113-116, 2008.

[22] R.L. Twite, G.P. Bierwagen,"Review of alternatives to chromate for corrosion protection of aluminum aerospace alloys," Progress in Organic Coatings, vol. 33, issue 2, pp. 91-100, 1998.

[23] S.J. García, T.H. Muster, Ö. Özkanat, N. Sherman, A.E. Hughes, H. Terryn, J.H.W. de Wit, J.M.C. Mol, "The influence of pH on corrosion inhibitor selection for 2024-T3 aluminium alloy assessed by high-throughput multielectrode and potentiodynamic testing," Electrochimica Acta, vol. 55, pp. 2457-2465, 2010.

[24] P. R. Roberge, Handbook of corrosion engineering, New York: Mc Graw Hill Handbook, 1999.

[25] A. C. Dutra e L. D. P. Nunes, Proteção catódica técnicas de combate a corrosão, 5 ed.,Rio de Janeiro: interciências, 2011.

[26] E. Bardal, Corrosion and protection, London: Springer, 2004.

[27] D. Talbot e J. Talbot, Corrosion science and technology, Florida: CRC Press, 2000.

[28] Aprael S. Yaro, Anees A. Khadom, Rafal K. Wael, "Apricot juice as green corrosion inhibitor of mild steel in phosphoric acid," Alexandria Engineering Journal, vol 52, issue 1, pp. 129-135, 2013.

[29] Mahmoud N. El-Haddad, "Chitosan as a green inhibitor for copper corrosion in acidic medium," International Journal of Biological Macromolecules, vol. 55, pp 142-149, 2013.

[30] El-Sayed M. Sherif, "Effects of 2-amino-5-(ethylthio)-1,3,4-thiadiazole on copper corrosion as a corrosion inhibitor in 3% NaCl solutions," Applied Surface Science, vol 252, issue 24, pp. 8615-8623, 2006.

[31] Daobing Huang, Junying Hu, Guang-Ling Song, Xingpeng Guo, "Inhibition effect of inorganic and organic inhibitors on the corrosion of Mg–10Gd–3Y–0.5Zr alloy in an ethylene glycol solution at ambient and elevated temperatures," Electrochimica Acta, vol. 56, pp. 10166-10178, 2011.

[32] G. H. Koch, M. P. H. Bronger, N. G. Thompson, Y P. Virmani, J.H. Payer, "Nace international," Corrosion cost and preventive strategies in the United States. Avaliable: http://nace.org/uploadedFiles/Publications/ccsupp.pdf. [Access on 07 10 2013].

[33] L. Yohai, M. Vázquez, M.B. Valcarce, "Phosphate ions as corrosion inhibitors for reinforcement steel in chloride-rich environments," Electrochimica Acta, vol 102, pp. 88-96, 2013.

[34] E. Samiento-Bustos, J.G. González Rodriguez, J. Uruchurtu, G. Dominguez-Patiño, V.M. Salinas-Bravo, "Effect of inorganic inhibitors on the corrosion behavior of 1018 carbon steel in the LiBr + ethylene glycol + H_2O mixture," Corrosion Science, vol. 50, issue 8, pp.2296-2303, 2008.

[35] Mohammad M. Fares, A.K. Maayta, Mohammad M. Al-Qudah, "Pectin as promising green corrosion inhibitor of aluminum in hydrochloric acid solution," Corrosion Science, vol. 60, pp.112-117, 2012.

[36] Serpil Şafak, Berrin Duran, Aysel Yurt, Gülşen Türkoğlu, "Schiff bases as corrosion inhibitor for aluminium in HCl solution," Corrosion Science, vol. 54, pp. 251-259, 2012.

[37] M. Elayyachy, A. El Idrissi, B. Hammouti, "New thio-compounds as corrosion inhibitor for steel in 1 M HCl," Corrosion Science, vol 48, issue 9, pp. 2470-2479, 2005.

[38] M.S. Al-Otaibi, A.M. Al-Mayouf, M. Khan, A.A. Mousa, S.A. Al-Mazroa, H.Z. Alkhathlan, "Corrosion inhibitory action of some plant extracts on the corrosion of mild steel in acidic media," Arabian Journal of Chemistry, In Press, Available online 8 February 2012.

[39] M Lagrenée, B Mernari, M Bouanis, M Traisnel, F Bentiss, "Study of the mechanism and inhibiting efficiency of 3,5-bis(4-methylthiophenyl)-4H-1,2,4-triazole on mild steel corrosion in acidic media," Corrosion Science, vol. 44, issue 3, pp. 573-588, 2002.

[40] Sk.A. Ali, M.T. Saeed, S.U. Rahman, "The isoxazolidines: a new class of corrosion inhibitors of mild steel in acidic medium," Corrosion Science, vol. 45, issue 2, pp. 253-266, 2003.

[41] D. Seifzadeh, H. Basharnavaz, A. Bezaatpour, "A Schiff base compound as effective corrosion inhibitor for magnesium in acidic media," Materials Chemistry and Physics, vol 138, issue 2-3, pp. 794-802, 2013.

[42] Corrosion 2013 Program Preview, Materials Performance, Florida: Nace, 2013. Available: http://mp.epubxp.com/i/105569. [Access on 07/10/2013].

[43] Jasna Halambek, Katarina Berković, Jasna Vorkapić-Furač, "Laurus nobilis L. oil as green corrosion inhibitor for aluminium and AA5754 aluminium alloy in 3% NaCl solution," Materials Chemistry and Physics, vol. 137, Issue 3, pp. 788-795, 2013.

[44] A Aballe, M Bethencourt, F.J Botana, M Marcos, "CeCl 3 and LaCl 3 binary solutions as environment- friendly corrosion inhibitors of AA5083 Al– Mg alloy in NaCl solutions," Journal of Alloys and Compounds, vol. 323–324, pp. 855- 858, 2001.

[45] Pandian Bothi Raja, Ahmad Kaleem Qureshi, Afidah Abdul Rahim, Hasnah Osman, Khalijah Awang, "Neolamarckia cadamba alkaloids as eco -friendly corrosion inhibitors for mild steel in 1 M HCl media," Corrosion Science, vol. 69, pp. 292-301, 2013.

[46] N.A. Negm, N.G. Kandile, I.A. Aiad, M.A. Mohammad, "New eco -friendly cationic surfactants: Synthesis, characterization and applicability as corrosion inhibitors for carbon steel in 1 N HCl," Colloids and Surfaces A: Physicochemical and Engineering Aspects, vol. 391, Issues 1–3, pp. 224-233, 2011.

[47] G Gunasekaran, L.R Chauhan, "Eco friendly inhibitor for corrosion inhibition of mild steel in phosphoric acid medium," Electrochimica Acta, vol. 49, Issue 25, pp. 4387-4395, 2004.

[48] Sitashree Banerjee, Varsha Srivastava, M.M. Singh, "Chemically modified natural polysaccharide as green corrosion inhibitor for mild steel in acidic medium," Corrosion Science, vol. 59, pp. 35-41, 2012.

[49] Ghada.M. Abd El-Hafez, Waheed A. Badawy, "The use of cysteine, N-acetyl cysteine and methionine as environmentally friendly corrosion inhibitors for Cu-10Al-5Ni alloy in neutral chloride solutions," Electrochimica Acta, In Press, Accepted Manuscript, Available online 1 July 2013.

[50] Khaled M. Ismail, "Evaluation of cysteine as environmentally friendly corrosion inhibitor for copper in neutral and acidic chloride solutions," Electrochimica Acta, vol. 52, Issue 28, pp. 7811-7819, 2007.

[51] Junying Hu, Dezhi Zeng, Zhi Zhang, Taihe Shi, Guang-Ling Song, Xingpeng Guo, "2-Hydroxy-4-methoxy-acetophenone as an environment-friendly corrosion inhibitor for AZ91D magnesium alloy," Corrosion Science, vol. 74, pp. 35-43, 2013.

[52] H. Gao, Q. Li, Y. Dai, F. Luo, H.X. Zhang, "High efficiency corrosion inhibitor 8-hydroxyquinoline and its synergistic effect with sodium dodecylbenzenesulphonate on AZ91D magnesium alloy," Corrosion Science, vol. 52, Issue 5, pp. 1603-1609, 2010.

[53] Junying Hu, Daobing Huang, Guoan Zhang, Guang-Ling Song, Xingpeng Guo, "Research on the inhibition mechanism of tetraphenylporphyrin on AZ91D magnesium alloy," Corrosion Science, vol. 63, pp. 367-378, 2012.

Chapter 17

Palm oil as Corrosion Inhibitor for Aluminium Car Radiator

Junaidah Jai

Additional information is available at the end of the chapter

http://dx.doi.org/10.5772/57273

1. Introduction

Organic compounds are found to be effective corrosion inhibitors due to the adsorption of molecules and ions on the metal surface. As reviewed by Maayta and Al-Rawashdeh [1], the extent of adsorption of an inhibitor depends on many factors such as the nature of the surface charge of the metal, the mode of the adsorption of the inhibitor, the inhibitor's chemical structure and the type of the corrosive solution. The presence of large molecules with functional groups containing heteroatoms (such as oxygen, nitrogen, sulphur, phosphorus), triple bonds or aromatic rings in the inhibitor's chemical structure enhances the adsorption process [2]. There has been a growing trend on the use of natural resources as corrosion inhibitors, which are environmentally friendly, cheap and readily available. Table 1 lists out some works done to evaluate various natural resources as corrosion inhibitors.

Generally, not many works have been done on the application of natural oils as corrosion inhibitors. One of the common and abundant natural oils in Asian countries is palm oil. Crude palm oil (CPO) contains equal amounts of saturated and unsaturated fatty acids which are palm stearin and palm olein, respectively. Palm olein contains monounsaturated and polyunsaturated acids consisting of oleic and linoleic acids. Both oleic and linoleic acids contain carbonyl groups which gives palm olein the potential to act as a corrosion inhibitor.

Currently, due to its lightweight property and corrosion resistance, Al alloy is used to replace copper as the material for car radiators. Numerous methods have been applied to protect Al against corrosion. One of the attempts is to use palm olein as an environmentally friendly corrosion inhibitor. However little or no work on this matter has been reported. Therefore, this work focuses on the effect of palm olein as a corrosion inhibitor for Al which would be suitable for application in Al car radiators.

Natural resources	References
natural honey	El-Etre & Abdallah, 2000 [3]
Vanillin	El-Etre, 2001 [4]
opuntia ficus mill (family of cactaceae)	El-Etre, 2003 [5]
nypa fruticans wurmb	Orubite & Oforka, 2004 [6]
lawsonia (henna) extract	El-Etre, Abdallah, & El-Tantawy, 2005 [7]
olive leaves	El-Etre, 2007 [2]
fenugreek leaves	Noor, 2007 [8]
musa sapientum peels	Eddy & Ebenso, 2008 [9]
pennyroyaloil from *mentha pulegium*	Bouyanzer et al., 2006 [10]
artemisia oil	Benabdelah, Benkaddour, & Hammouti, 2006 [11]
	Ouachikh et al., 2009 [12]
	Kalaiselvi et al., 2010 [13]
	Bammou et al., 2011 [14]
	Garai et al., 2012 [15]
	Huang et al., 2013 [16]
fennel seed	Fouda et al., 2013 [17]

Table 1. Natural resources as corrosion inhibitor

2. Materials

Aluminium alloy (Al 6061) sheet with the composition listed in Table 2 was used. Its composition was determined by X-ray Fluorescence (XRF). Al 6061 was utilized in this work as it is a commonly used alloy in manufacturing automobiles, particularly used as the main material for car radiators.

Element	Si	Fe	Cu	Mn	Mg	Cr	Zn	Ti	Al
% (w/w)	0.549	0.269	0.205	0.004	6.240	0.118	0.008	0.008	92.599

Table 2. Chemical composition of aluminium alloy (Al 6061)

Palm oil is abundant in Malaysia and its new application has to be explored. Crude palm oil (CPO) contains equal amounts of saturated and unsaturated fatty acids which are palm stearin and palm olein, respectively. Palm olein (PO) can be separated from palm stearin by centrifuging. Composition of the PO is stated in Table 3, as determined through Gas Chromatograph/ Mass Spectroscopy (GCMS) (Agilent Technologies 6890N). The main components of palm olein are oleic acids ($C_{18}H_{34}O_2$), hexadecanoic acid ($C_{16}H_{32}O_2$) and stigmasterol ($C_{29}H_{48}O$). The minor components are nonadecene, octadecanoic acid, tetracosahexaene, campesterol, cyclopentane, tricosene and cyclooctacosanetetrone.

Components	Composition (%)
Oleic acid	18.10
Hexadecanoic acid	15.72
Stigmasterol	11.01
Nonadecene	9.74
Octadecanoic acid, methyl ester	6.34
Tetracosahexaene	2.15
Campesterol	1.46
Cyclopentane	0.73
Tricosene	0.35
Cyclooctacosanetetrone	0.33
Free fatty acid	34.00

Table 3. Composition of palm olein from crude palm oil

All fatty acids listed in Table 3 have a carboxylic group in their molecular structure. However, they are differentiated by the presence of other functional groups, such as single double bond in oleic acid and aromatic hydroxyl group in stigmasterol (Table 4).

Different types of additives can be used to enhance inhibition efficiency of any corrosion inhibitor. In this work, poly(oxyethylene)x-sobitane-monolaurate commercially known as Tween 20 (T20), hexane and diethyene triamine (DETA) were used for this purpose.

3. Formulation of corrosion inhibitor

Corrosion inhibitor should be formulated to suit the condition where it would be placed. For car radiators, the formulated corrosion inhibitor should be soluble in the coolant, thermally stable and effective within the range of the radiator's working pressure and temperature. In this work, the presence of an emulsifier, such as Tween 20, and a stabilizing agent, such as hexane, in the formulation is important, as palm olein is inherently unsoluble in water. Furthermore, the solubility and stability of the formulated corrosion inhibitor should be investigated by testing the corrosion inhibitor at different pH and temperatures since both factors give significant influence to the formulation. Diethylene triamine (DETA) was used to enhance the inhibition efficiency (IE) of the inhibitor at elevated temperatures.

Blends of palm olein and water in the presence of emulsifier and stabilizing agents at certain pH and temperature can produce stable emulsions with two separate layers; thick and dilute emulsions. Only the soluble and stable emulsion can be taken as the palm olein (PO) inhibitor. The solubility of an emulsion is considered good if there is no oily layer formed, whereas the stability of an emulsion is considered good if there is no oily layer formed after a prolonged period of time.

Analyses on the formulated PO inhibitor's concentration, pH, as well as micelle size and shape are very important since the inhibition and adsorption mechanism of the inhibitor on the aluminium surface can be predicted and understood from them.

Molar concentration of the PO corrosion inhibitor can be determined through acid-base titration method. The shape of micelles in the formulated PO corrosion inhibitor can be observed under optical microscopy (Axioskop 40, Zeiss) at 100 x magnification. As for the micelles size, it can be measured using particle size analyzer (Malvern Instrument, Mastersizer 2000).

Components	Molecular structures
Oleic acid	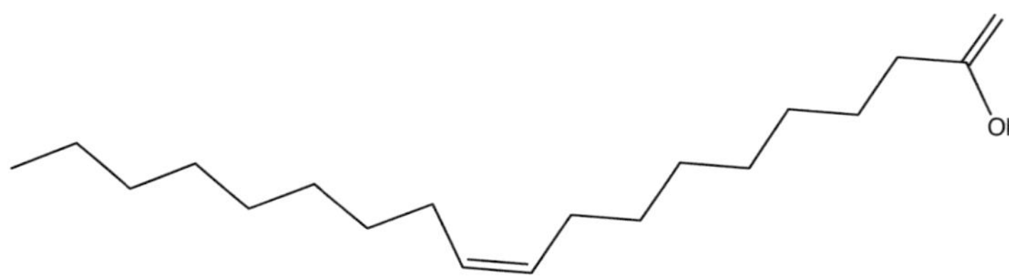
Hexadecanoic acid	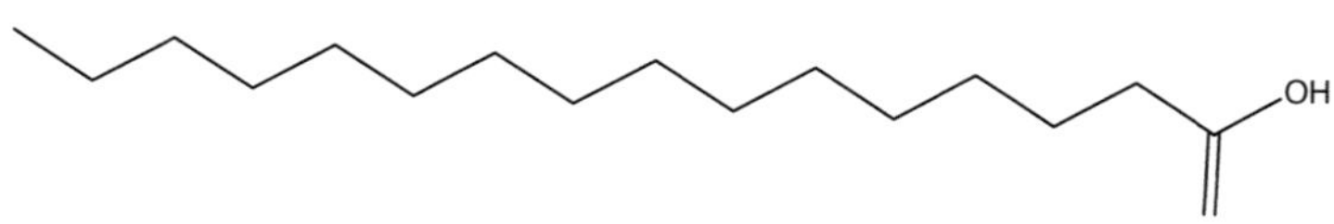
Stigmasterol	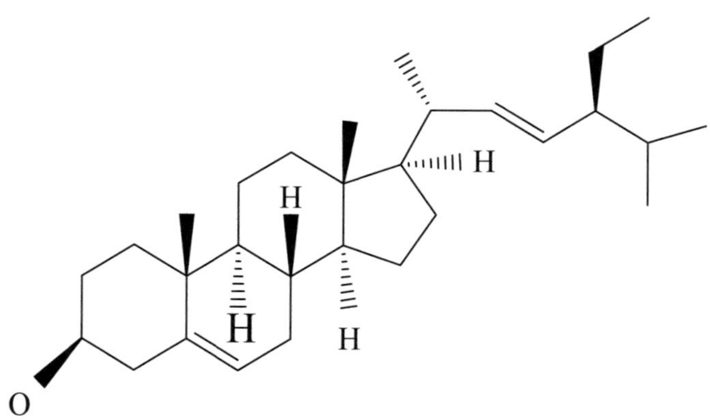

http://dx.doi.org/10.5772/57273

Components	Molecular structures
Cyclooctacosanetetrone	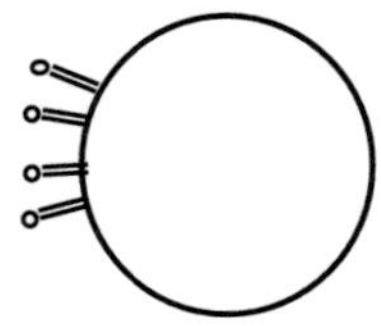
Cyclopentane	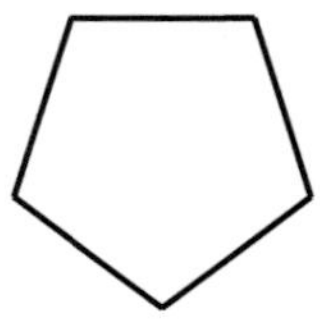
Nonadecene	
Octadecanoic acid	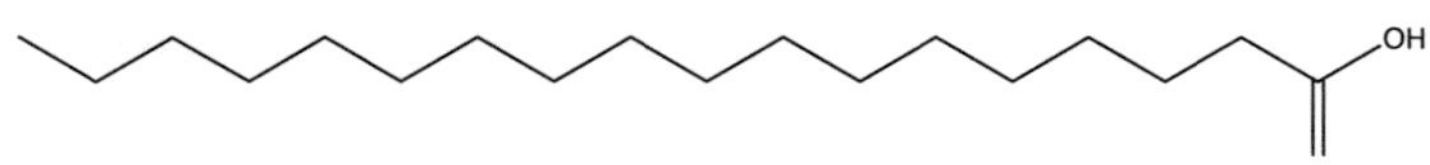
Tetracosahexaene	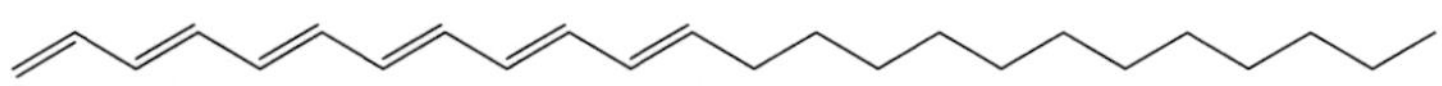
Tricosene	

Table 4. Molecular structures of the components in palm olein [18]

4. Corrosion study

Performance of the formulated inhibitor should be evaluated by several corrosion tests such as weight loss, potentiodynamic polarization and electrochemical impedance spectroscopy. For corrosion tests, 1 M HCl solution is commonly used as the corrosive media especially for aluminium alloy since it is easily attacked by Cl^- ions. In order to study the inhibition efficiency and inhibiting behaviour of the formulated corrosion inhibitor, metal has to be exposed to the corrosive media in the absence and presence of the inhibitor at different temperatures and concentrations of inhibitor. The chosen temperature range of the testing has to be based on the application of the inhibitor.

4.1. Weight loss (WL) study

Cleaned metal sheet with desired dimensions of 2 cm x 3 cm x 0.3 cm are used as test plates. Dimension and weight of the plate should be accurately measured prior to exposing the plate in a corrosive solution. The cleaned plate has to be suspended in the corrosive solution for several hours according to the desired exposure time. The plates were collected and retrieved at intervals of 1, 3, 6, 12, 24 and 48 hours for data collection. The collected plates should be cleaned before the plate was weighed. The metal sample preparation and cleaning techniques are according to ASTM G1-90 [19]. Each experiment should be triplicated for significant result. Corrosion rate can be determined from the weight loss data using the following formula [20];

$$Corrosion\ rate\ \left(millimeter\ per\ year\right) = 87.6\ \frac{W}{DAT} \tag{1}$$

where W is the weight loss (mg), D is the density of the Al sample (g/cm^3), A is the area of sample (cm^2) and T is the exposure time (h). The percentage of inhibition efficiency IE% for the weight loss method was calculated as follows;

$$IE\% = (\frac{W^o - W}{W^o})\ x\ 100 \tag{2}$$

where W and W^o are the weight loss of the Al 6061 with and without the inhibitor, respectively.

4.2. Potentiodynamic polarization (PP) study

Sample preparation is very important in corrosion studies. For PP test, the sample should be prepared as in Figure 1. Normally the metal sheet has to be cut into sample pieces having dimensions such as 1 cm x 1 cm x 0.03 cm. A copper wire has to be attached to one side of the flat surface of the sample for electrical connection before the samples are cold mounted in a blend of resin and hardener. Normally only 1 cm^2 surface area is exposed to the corrosive media. This surface should be mechanically polished with sandpaper grade of 180 followed by 600. The polished surface has to be cleaned with distilled water followed by acetone and finally dried.

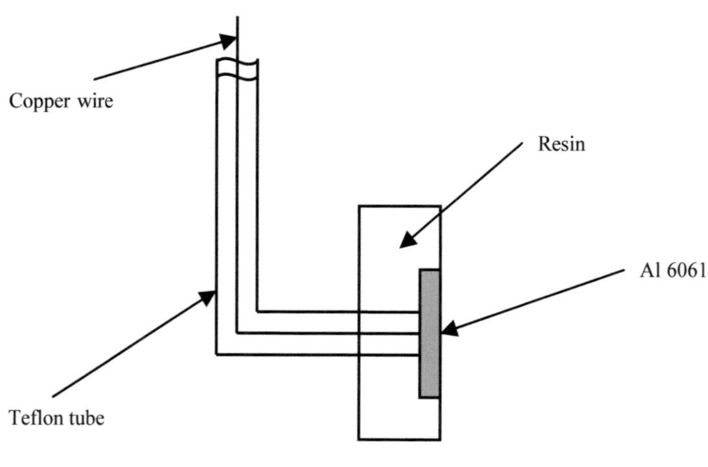

Figure 1. Sample assembly for potentiodynamic polarization measurement

Polarization measurement is carried out in a three-electrode electrochemical cell with consists of counter, working and reference electrodes. Normally, platinum mesh of 2 cm^2 and saturated calomel electrode (SCE) are used as the counter electrode and reference electrode, respectively, while the metal sample acts as the working electrode. An electrochemical station such as Voltalab (PGP201) apparatus can be used as a potential source. Figure 2 shows a schematic diagram of the electrochemical cell for the polarization test. Prior to measurement, the electrode is immersed in the test solution for 60 minutes at an open circuit condition until a steady state condition is achieved. Subsequently, PP measurements are taken at a scanning rate of 1 mV/s. For aluminium samples, the common potential range starts from -1200 to +200 mV versus the SCE.

For the PP study, corrosion behaviour of the sample has to be analyzed using corrosion potential (E_{corr}), corrosion current density (i_{corr}), polarization resistance (R_p), anodic Tafel slope (β_a), cathodic Tafel slope (β_c) and corrosion rate (CR). The IE for the PP study method can be calculated as follows;

$$IE\% = \left(\frac{i^o_{corr} - i_{corr}}{i^o_{corr}} \right) x\ 100 \tag{3}$$

where i_{corr} and i^o_{corr} are the corrosion current density of the Al 6061 in the presence and absence inhibitor, respectively. Every experiment was repeated several times to make sure its reproducibility and the best are reported here.

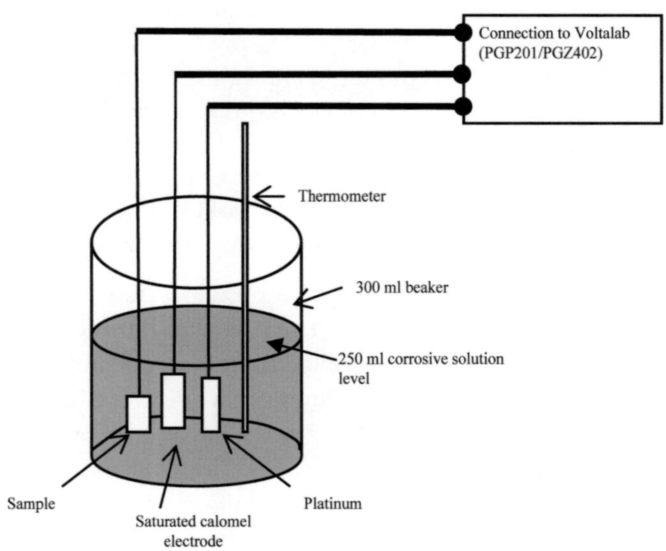

Figure 2. Schematic diagram of the electrochemical cell for polarization and electrochemical impedance spectroscopy tests

4.3. Electrochemical impedance spectroscopic (EIS) study

Sample preparation and EIS measurement are similar to that of the PP study. The EIS measurements were performed using AC signals of amplitude 10 mV peak to peak at the open circuit potential in the frequency range of 100 kHz to 100 mHz. Nyquist diagram can be determined from the experiment and the inhibition mechanism of the inhibitor can be explained from the calculated inductance or capacitance signals. The experiment should be repeated several times to make sure its reproducibility.

In confirming the corrosion behaviour of the inhibitor from corrosion tests, surface corrosion analysis using scanning electron microscope (SEM) should be done on the corroded surfaces. Performance test of the formulated inhibitor as anticorrosion for car radiator should also be done. As coolant and anticorrosion for car radiator, 95 wt% of coolant (glycerin) and the balance 5 wt% being the formulated inhibitor can be used.

5. Palm olein as anticorrosion for aluminium car radiator

5.1. Formulation of palm olein corrosion inhibitor

In formulating the PO inhibitor, emulsifier was added to enhance the solubility of palm olein in water. Figure 3 shows the corrosion rate of Al 6061 immersed in 1 M HCl solution containing different weight ratios of PO to T20 which were 5:0.5, 5:1.0 and 5:1.5. Within 1 to 6 hours of immersion time, the corrosion rates of all solution ratios were almost constant and similar to each other. Nevertheless, the 5:1.0 had shown the lowest corrosion rate followed by the 5:1.5 and 5:0.5 ratio solutions. A further increase in the immersion time from 6 to 24 hours had shown sharp increase in the corrosion rates for all ratios, with the 5:1.0 being the lowest followed by the 5:1.5 and 5:0.5. However, an increase in the immersion time from 24 to 48 hours had shown gradual reduction in the corrosion rate for all ratios. The results showed that despite the different weight ratios of PO to T20, all solutions produced almost similar corrosion rates. However, the solution with 5:1 ratio produced slightly lower corrosion rates than the other two solutions. The 5:1 ratio solution might have had reached its critical micelle concentration (CMC), upon which a further increase in the amount of emulsifier would not change or increase the corrosion rate, as explained by Al-Rawashdeh, and Mayata [21]. Thus, the POT20 stock solution with the weight ratio of PO to T20 as 5:1 was selected and subsequently used in this work.

The initial pH of 25% (v/v) POT20 in water was 4.5. Table 5 shows the effect of pH on the solubility of 25% (v/v) POT20 in distilled water. After 1 hour, the solution settles into two separate layers; an oily layer at the top and another dilute layer of emulsion at the bottom. The same finding was recorded for pH 5 and 11 solutions. Two separate layers indicated that PO was not fully soluble in water. On the other hand, three separate layers were observed for pH 7 solution; an oily layer at the top, a thick emulsion in the middle and a dilute emulsion at the bottom. After 24 hours, it was observed that pH 7 solution had the thinnest oily later. This finding suggested that at pH 7, the solubility of the PO in water had improved due to the better stability and smaller micelle size of the produced emulsion [22]. Therefore the pH 7 solution was used in the preparation of the formulation. However, the stability of the solution was still low. Table 6 shows the effect of temperature on the solubility and stability of the 25% (v/v)

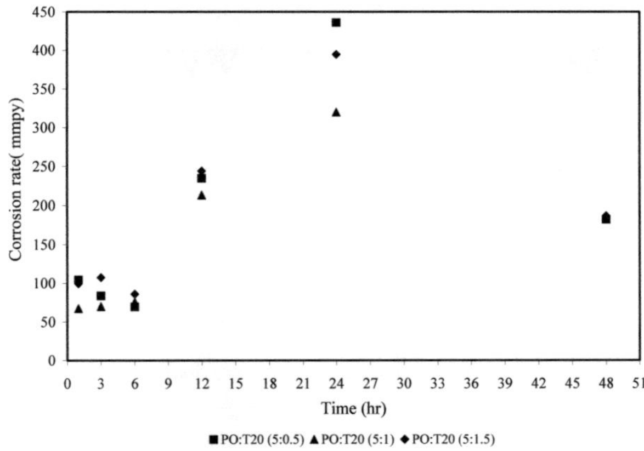

Figure 3. Corrosion rate of Al 6061 immersed in 1 M HCl with different weight ratio of PO to T20 at room temperature from weight loss test

POT20 in water. After one hour of settling down at 299 K, the three layers were observed. Similarly, an oily layer appeared at the top, a thick emulsion in the middle and a dilute emulsion at the bottom. However, at 323 K, two separate layers were observed with thick emulsion at the top and dilute emulsion at the bottom and no oily layer. The absence of an oily layer suggested that at 323 K, two forms of emulsions at formed; thick and dilute emulsions as agreed by Goyal, and Aswal [23]. In other words, the solubility of PO in water at 323K is higher than it is at 299 K. However, the stability of the solution at this temperature was still not satisfactory since after 24 hours of settling down, three separate layers were observed at both temperatures. Nevertheless, the 323 K was taken as the working temperature in the preparation of the formulation with some additives added to stabilize the formulation.

POT20, % (v/v)	25	25	25
Distilled water, % (v/v)	75	75	75
pH of the solution	pH 11	pH 7	pH 5
After 1 hour			
	Two separated layers observed; oily layer at the top and dilute emulsion at the bottom.	Three separated layers observed; oily layer at the top, thick emulsion in the middle and dilute emulsion at the bottom.	Two separated layers observed; oily layer at the top and dilute emulsion at the bottom.
After 24 hours			
Oil, % (v/v)	24	10	23
Thick emulsion, % (v/v)	0	14	0
Dilute emulsion, % (v/v)	76	76	77

Table 5. The effect of pH on the solubility and stability of POT20 in water

POT20, % (v/v)	25	25
Distilled water, % (v/v)	75	75
pH of the solution	pH 7	pH 7
Temperature	323 K	299 K
After 1 hour		
	Two separated layers were observed, thick emulsion at the top and dilute emulsion at the bottom	Three separated layers were observed, oily layer at the top, thick emulsion at the middle and dilute emulsion at the bottom.
After 24 hours		
Oil, % (v/v)	4	6
Thick emulsion, % (v/v)	22	22
Dilute emulsion, % (v/v)	74	72

Table 6. The effect of temperature on the solubility of POT20 in water

Table 7 shows the effect of different concentrations of hexane varying at 5, 3, 1, 0.5 to 0%, (v/v) on the stability of 25% (v/v) POT20 in distilled water. The stability of the PO in water with 5, 3 and 1% (v/v) hexane remained unchanged until 9 days whilst with 0.5% (v/v), lasted for 14 days. The solution without hexane showed 12 days stability which was better than those in 5, 3 and 1% (v/v) hexane. This finding showed that the saturation concentration had been reached at above 1% (v/v) hexane; whereby further increase in hexane concentration did not improve the stability of the emulsion. Therefore, the optimum suitable volume ratio of hexane to POT20 was 0.5 to 25. Thus, this volume ratio was used in this work.

POT20, % (v/v)	25	25	25	25	25
Hexane, % (v/v)	5	3	1	0.5	0
Distilled water, % (v/v)	70	70	70	70	70
Stability, days	9	9	9	14	12

Table 7. Different amount of hexane on the stability of PO in water

The solubility and stability study revealed that the formulated PO inhibitor consists of two types of emulsion; thick emulsion and dilute emulsion. Dilute emulsion was soluble and stable in water, whereas the thick emulsion was insoluble in water. As such, the dilute emulsion was used as the corrosion inhibitor in this work and was labeled as POT20H. The thick emulsion was kept for future work.

http://dx.doi.org/10.5772/57273

5.2. Inhibition Efficiency (IE) of the palm olein corrosion inhibitor

Table 8 shows the IE of different concentrations of POT20H in 1 M HCl solution at 299 K as determined from the WL test. It was evident from the data that IE was directly proportional to POT20H concentration, but inversely proportional to immersion time. It increased with POT20H concentration from 10% (v/v) to 50% (v/v) but decreased with the increase of immersion time. Furthermore, IE of 50% (v/v) POT20H had remained 100% even after 48 hours of immersion time. In summary, at 299 K, 50% (v/v) POT20H consistently recorded better IE than any other concentration despite increasing immersion time from 1 to 48 hours.

Table 9 shows the IE of 50% (v/v) POT20H in 1 M HCl solution at different temperatures, 299, 323 and 343 K. Evidently, an increase in temperature reduced the IE of the inhibitor. POT20H seemed to perform the best at 299 K. In order to enhance POT20H's IE elevated temperatures, DETA was added in the formulation.

POT20H, % (v/v)	50	40	30	20	10
1 M HCl, % (v/v)	50	60	70	80	90
Immersion time (h)	Inhibition efficiency (%)				
1	100	97	95	94	83
3	100	99	99	99	96
6	100	99	95	94	91
12	100	94	86	71	68
24	100	84	60	34	22
48	100	56	36	13	6

Table 8. Inhibition efficiency of different concentrations of POT20H at 299 K from weight loss test

POT20H	50% (v/v)		
1 M HCl	50% (v/v)		
Immersion time (h)	Inhibition efficiency (%)		
	299 K	323 K	343 K
1	100	97	64
3	100	91	36
6	100	82	10
12	100	68	6
24	100	60	4
48	100	54	1

Table 9. Inhibition efficiency of POT20H solution at different temperatures from weight loss test

Figure 4 shows the corrosion rate determined from the WL test of Al 6061 in 1 M HCl and 50% (v/v) POT20H at 323 K with different concentrations of DETA. An addition of 3% (v/v) DETA into the 50% (v/v) POT20H solution remarkably reduced the corrosion rate as compared to that of the control 50% (v/v) POT20H. The corrosion rate of the control 3% (v/v) DETA was slightly higher than that of the 50% (v/v) POT20H with 3% (v/v) DETA solution which confirmed that inhibition was not solely due to DETA but the combination of POT20H and DETA. Reduction in the concentration of DETA from 3 to 2% (v/v) slightly reduced the corrosion rate. However, further reduction in the concentration from 2% (v/v) to 1% (v/v) followed by 0.5% (v/v) had markedly increased the corrosion rate. In other words, 50% (v/v) POT20H containing 2% (v/v) DETA showed the best IE. Consequently, for this study the best volume ratio of POT20H to DETA was 50 to 2 and this formulation was labeled as POT20HA.

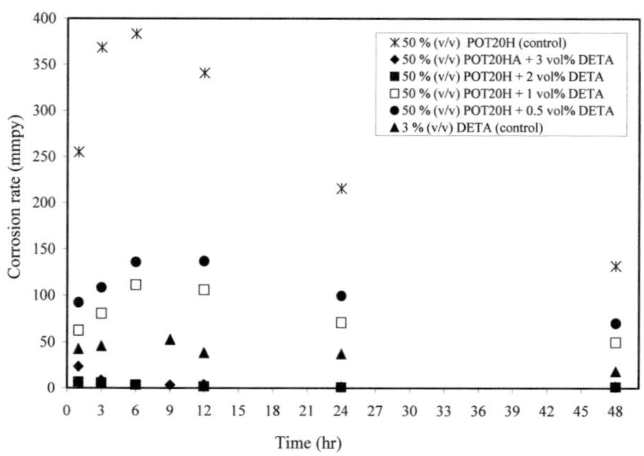

Figure 4. Corrosion rate from weight loss test of Al 6061 immersed in 1 M HCl and 50% (v/v) POT20H with the presence and absence of DETA at 323 K

5.3. Analysis of the palm olein corrosion inhibitor

The acid-base titration method was used to determine the molar concentration of the POT20HA. The initial pH of 30% (v/v) POT20HA in distilled water was 12.5 pH and 0.1 M HCl solution was used to neutralize the POT20HA. To calculate the molar concentration of the POT20HA the following reaction was considered; whereby 1 mol of POT20HA reacted with 1 mol of HCl [18],

$$\text{POT20HA:} + \text{H-Cl} \longrightarrow \text{Cl}^- + \text{POT20HA-H} \tag{4}$$

Hence, in 30% (v/v) POT20HA, 0.1 M POT20HA reacted with 0.1 M of HCl. Table 10 shows the concentration of POT20HA which was extrapolated from the titration result.

Density of POT20HA was 0.98 g/cm3, as calculated from its weight and volume. From its concentration and density, the molar mass of POT20HA was calculated as 2969.70 g/mol. A thin film of POT20HA was scanned under XRD at the phase angle of 10 to 70 degree. However, only one significant peak was observed, specifically at the phase angle of 2 to 5 degree. The XRD spectrum predicted that POT20HA contains n-phenyl-n-dodecanamide ($C_{18}H_{29}NO$) (illustrated in Figures 5). The compound reveals the presence of CON functional group, thus, it is classified as the amide family. The amide was synthesized from the reaction of carboxylic acid with an amine involving condensation process. Figure 6 shows the general reaction between a carboxylic acid and an amine to form an amide. Solubility of amide in water was almost similar to the solubility of ester in water; this explains POT20HA's slight solubility in water.

% (v/v)	Molarity (M)
10	0.03
20	0.07
30	0.10
40	0.13
50	0.17
100	0.33

Table 10. Concentrations of POT20HA

Optical microscopy reveals the solubility of POT20HA in water as shown in Figure 7. Particles with almost spherical in shape were observed in the emulsion, confirming the presence of micelles. The result shows that POT20HA is soluble in water in the form of micelles. Particles size of the micelles were distributed in 4 ranges, 0.04 to 0.6, 0.7 to 8, 9 to 50 and 60 to 300 μm. Less than 50% of the particles were 2.12 μm in size as presented in Figure 8.

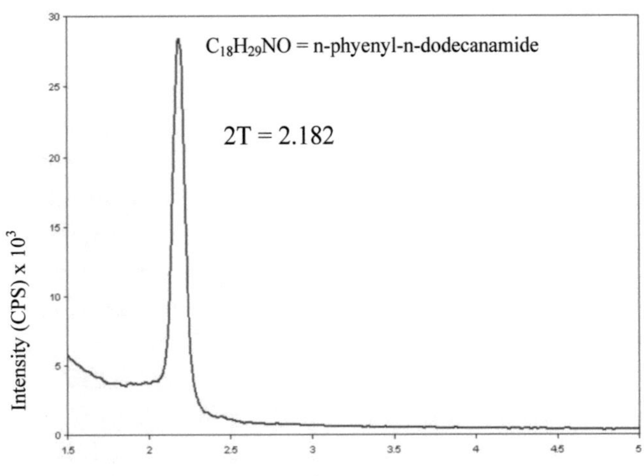

Figure 5. The XRD spectrum of POT20HA

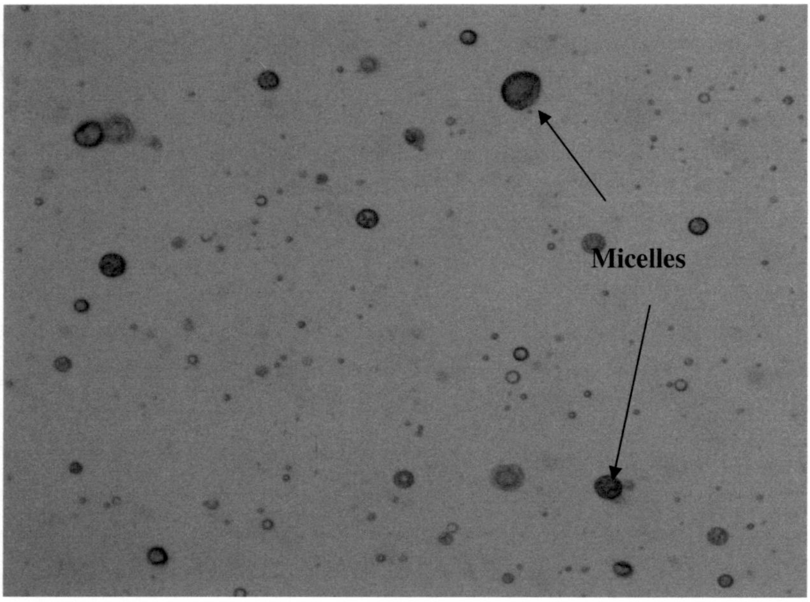

Figure 7. Optical microscopy observation on the POT20HA (100 x magnification)

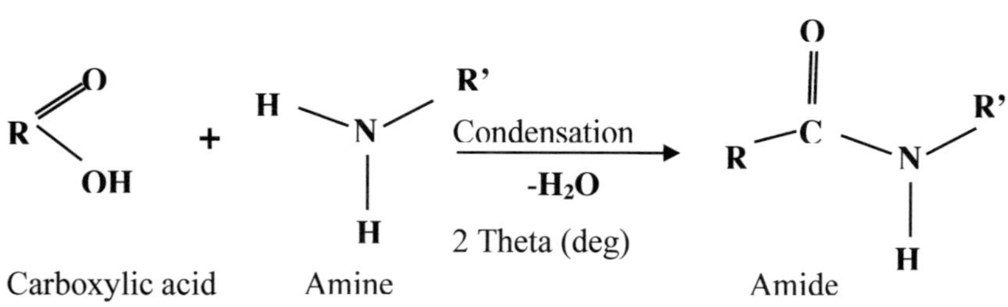

Figure 6. General reaction of carboxylic acid and amine to produce amide

Results had shown that T20 was a suitable emulsifier for PO in water with the weight ratio of PO to T20 set at 5:1. Hence the formulation is labeled as POT20. Besides, hexane had shown to be an excellent stabilizing agent with the volume ratio of hexane to POT20 set at 0.5:25. Consequently, solution with pH 7 and temperature of 323 K were found to be the most suitable condition in the preparation of the formulation. The mixture stirring speed was 125 rpm for the duration of 1 hour. As mentioned earlier, the formulation produced two emulsions i.e. dilute and thick emulsions. The dilute emulsion known as POT20H was used as the PO inhibitor due to its solubility in water. In enhancing the IE of the POT20H at elevated temperature, DETA was added in the formulation and the volume ratio of POT20H to DETA was 50:2 and was labelled as POT20HA. The POT20HA was chosen as the PO corrosion inhibitor in this study. As predicted, POT20HA consists of n-phenyl-n-dodecanamide compound which is the

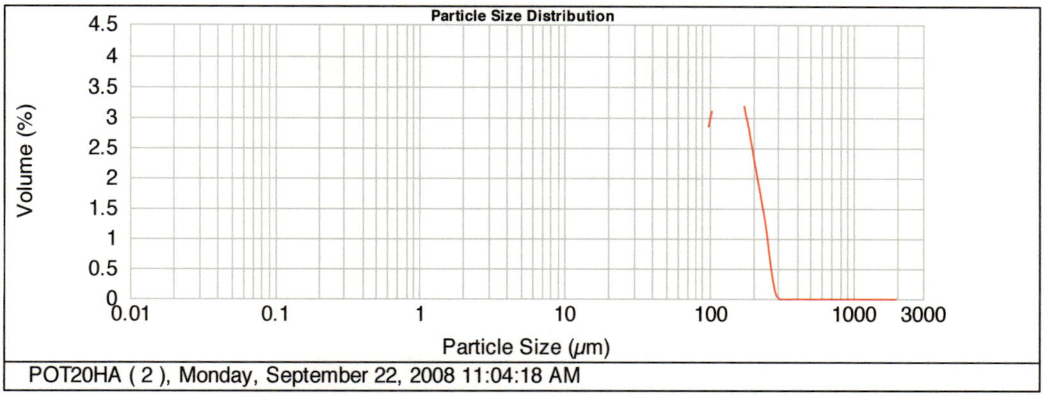

Figure 8. Distributions of POT20HA micelles size

family of amide and exists in the form of spherical micelle. The Inhibition behaviour of POT20HA was subsequently investigated through corrosion tests.

5.4. Corrosion evaluation

They are several corrosion test techniques which can be used to study the inhibition behaviour of an inhibitor. However, the most common methods used are weight loss, potentiodynamic polarization as well as electrochemical impedance spectroscopy.

5.4.1. Weight loss measurement

Figure 9 shows the corrosion rates of Al 6061 in 1 M HCl solution in the presence of POT20HA at 299K. The corrosion rate of Al 6061 was very much higher in the absence of POT20HA as compared to the corrosion rates measured in the presence of varying levels of POT20HA. Corrosion occurred due to the presence of water, air, Cl^- and H^+, which accelerated the corrosion process of the Al [20]. Increasing immersion time from initial to 3 hours significantly increased the corrosion rate. On the hand, further increase in the immersion time from 6 to 12 and 24 hours resulted in decrease in the corrosion rate. After 24 hours, the corrosion rate had reached its steady state. The same corrosion rate behaviour was obtained by Radzi [24] when Zn-Al coated low carbon steel wires were fully immersed in 3.5% NaCl solution. Nevertheless, increasing the immersion time from 3 to 48 hours reduced the corrosion rate by 78%. The reduction was due to the presence of aluminium hydroxide, which covered the Al 6061 surface [25].

The observed reduction in corrosion rates of Al 6061 in response to the increase of POT20HA concentrations indicated the positive effect of the inhibitor, as shown in Figure 9(b). Test solutions containing 0.03 and 0.07 M POT20HA showed similar corrosion behaviour, whereby a reduction in the corrosion rate was observed for the first three hours of immersion. This shows the ability of POT20HA to form a protective layer on the Al 6061 surface. However, an increase in the immersion time from 3 to 6 and 12 hours had shown slight increase and reduction in the corrosion rate, respectively.

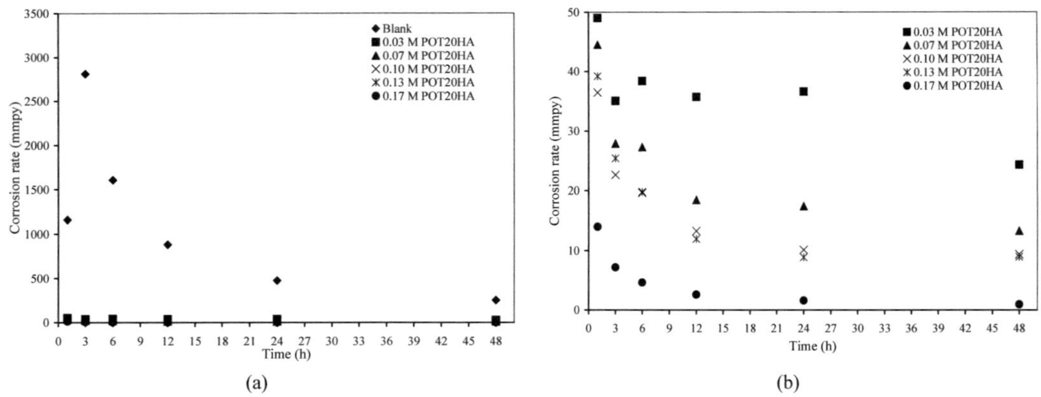

Figure 9. Corrosion rate of Al 6061 in 1 M HCl solution (a) with and without the presence of POT20HA and (b) with different concentrations of POT20HA at 299 K

Subsequently, after 24 hours, there was a small decrease in the corrosion rate of Al 6061 in the 0.07 M POT20HA but a slight increase in the 0.03 M. The increase and decrease of the corrosion rate suggested that there was insufficient surface coverage by the POT20HA on the Al 6061 surface. Diffusion of chloride ions (Cl^-) and hydrogen ions (H^+) through the pores of the uncovered surface led to the dissolution of Al [26, 27]. Consequently, the Al dissolution eventually led to the formation of corrosion product which repassivated the Al surface, which in turn reduced the corrosion rate. Tests on 0.10, 0.13 and 0.17 M POT20HA showed a direct reduction in corrosion rates from initial to 24 hours of immersion time. Eventually after 24 hours, the corrosion rate leveled as it reached its steady state. Therefore, it was evident that 0.10, 0.13 and 0.17 M POT20HA had produced excellent surface coverage which was not easily penetrated by Cl^- ions. In other words, 0.10, 0.13 and 0.17 M POT20HA produced better surface coverage than 0.03 and 0.07M POT20HA.

The corrosion rates of Al 6061 immersed in different concentrations of POT20HA at 323 and 343 K are shown in Figures 10 (a) and (b), respectively. Corrosion behaviour of Al 6061 was almost similar at both temperatures. For every concentration at both temperatures, an increase in the immersion time from initial to 24 hours had shown continuous gradual decrease in the corrosion rate. In each case, corrosion rate reached its steady state after 24 hours. In summary, 0.17 M POT20HA showed the lowest corrosion rate at both temperatures under study.

Table 11 reveals the IE of different concentrations of POT20HA at different temperatures. The IE increased with increasing concentration of POT20HA at 299, 323 and 343 K. However, the IE decreased with increasing temperature. At all temperatures and concentrations under study, an increase in the immersion time from initial to 3 hours had shown gradual increase in the IE. On the other hand, further increase in the immersion time from 3 to 24 hours had shown reduction in the IE. Nevertheless, IE of POT20HA at 299 K was the highest followed by those at 323 and 343 K. The IE was considered excellent when the value was equal or higher than 95%. The 0.07, 0.10, 0.13 and 0.17 M POT20HA at 299 K had shown excellent IE. On the other hand, at higher temperatures of 323 and 343 K, only 0.17 M POT20HA showed excellent IE. In general, the 0.17 M POT20HA had exhibited excellent IE at all temperatures under study.

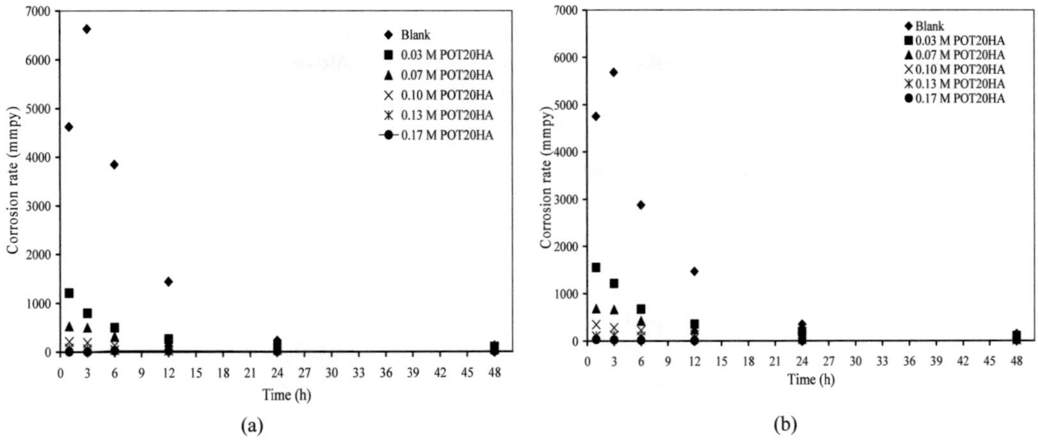

Figure 10. Corrosion rate of Al 6061 in 1 M HCl solution with the absence and presence of different concentrations of POT20HA at (a) 323 K and (b) 343 K

	Inhibition Efficiency (%)				
Time (hour)	0.03 M POT20HA	0.07 M POT20HA	0.10 M POT20HA	0.13 M POT20HA	0.17 M POT20HA
			299 K		
1	96	96	97	97	99
3	99	99	99	99	100
6	98	98	99	99	100
12	96	98	98	99	100
24	92	96	98	98	100
48	90	95	96	96	100
			323 K		
1	74	89	95	98	100
3	88	92	97	99	100
6	87	92	97	99	99
12	82	86	92	98	98
24	32	45	70	85	95
48	21	38	50	69	92
			343 K		
1	67	86	93	98	99
3	79	88	95	98	100
6	77	85	92	97	100
12	76	84	91	96	99
24	44	56	72	91	98
48	30	36	73	90	93

Table 11. Inhibition efficiency (%) of different concentrations of POT20HA at 299, 323 and 343 K from weight loss test

5.4.2. Potentiodynamic polarization study

Table 12 shows the electrochemical parameter of Al 6061 immersed in 1 M HCl in the presence and absence of different concentrations of POT20HA and temperatures. At all temperatures under study, an increase in concentration of POT20HA has reduced the corrosion potential (E_{corr}), corrosion current (i_{corr}) and the corrosion rate (CR); thus, increasing the polarization resistance (R_p). With 0.17 M POT20HA, the R_p increased remarkably as compared to those of the lower concentrations. The R_p is associated with the resistance action of POT20HA towards corrosion reaction. Accordingly, the 0.17 M POT20HA had shown the highest IE as compared to those of the lower concentrations at 299, 323 and 343 K. Furthermore, an addition of inhibitor to the corrosive solution had changed the value of the anodic and cathodic Tafel slopes, β_a and β_c, respectively, as compared to those values in the absence of inhibitor. Thus, it shows that the inhibitor controls both the cathodic and anodic corrosion reactions.

It is observed that the E_{corr} of 0.03 M POT20HA shifted slightly to the left while those of the higher concentrations were shifted to the right in all of the temperatures under study. As determined from the data, E_{corr} difference between the POT20HA and the blank was less than 85 mV. Hence, the POT20HA is considered to be of the mixed type of inhibitor with predominantly anodic action, except for the 0.03 M POT20HA, which was more towards cathodic [28, 29]. This confirmed the results of β_a and β_c, which generally revealed the ability of POT20HA in protecting both the anodic and cathodic reactions of the corrosion process.

Parameter	E_{corr} (mV)	icorr (mA/cm²)	Rp (ohm.cm²)	β_a (mV)	β_c (mV)	Corrosion rate (mm/y)	IE (%)
Concentration of POT20HA				299 K			
Blank	-759	33.504	5.65	922.9	-1209	375.20	-
0.03 M	-779	0.639	14.56	131.4	-197	7.16	98
0.07 M	-784	0.417	15.79	84.8	-170	4.67	99
0.10 M	-750	0.394	31.20	159.8	-303	4.41	99
0.13 M	-756	0.158	38.56	90.4	-272	1.77	100
0.17 M	-722	0.020	792.44	37.0	-554	0.219	100
				323 K			
Blank	-806	37.823	2.54	571.7	-715	423.600	-
0.03 M	-809	5.000	5.33	110	-287	61.280	86
0.07 M	-775	4.464	40.03	336.4	-405	14.760	97
0.10 M	-761	0.673	43.69	229.9	-287	6.781	98
0.13 M	-754	0.775	129.89	308.1	-369	7.812	98
0.17 M	-736	0.006	1530	74.8	-959	0.006	100

				343 K			
Blank	-831	30.957	1.98	347.2	-453	346.700	-
0.03 M	-849	3.945	14.64	193.2	-342	44.180	87
0.07 M	-809	2.733	29.21	553.8	-475	30.600	91
0.10 M	-803	1.245	20.53	101.3	-262	13.940	96
0.13 M	-800	1.270	64.68	4.3	-452	14.220	96
0.17 M	-740	0.493	774.20	321.7	-31	4.966	99

Table 12. Polarization parameters and IE of different concentrations of POT20HA at 299, 323 and 343 K from polarization test

5.4.3. Electrochemical Impedance Spectroscopy (EIS) study

The general shape of the curve was almost similar for all concentrations of POT20HA, indicating that almost no change in the corrosion mechanism occurred when concentrations were varied [30]. Adding and increasing of POT20HA increased the capacitive semicircle loop diameter which indicated the increase in the IE (Figure 11 and 12). Furthermore, the 0.03 and 0.07 M POT20HA had shown almost similar capacitive values as shown in Figure 12 (a).

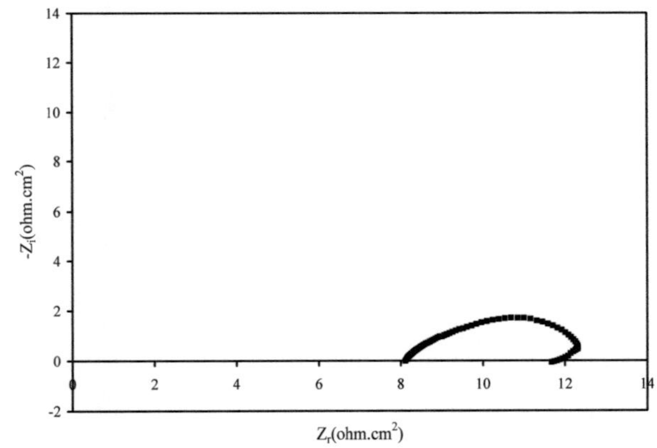

Figure 11. Electrochemical impedance plot of Al 6061 at 299 K in blank 1 M HCl solution

The impedance of Al 6061 in 1 M HCl solution with different concentrations of POT20HA at 323 K (Figure 13) was almost similar to that of the 299 K especially in the blank, 0.03, 0.07, 0.10 and 0.13 M POT20HA solution. The 0.03 and 0.07 M POT20HA had shown almost similar capacitive values; a similar behaviour was found at 299 K. Nonetheless, the impedance behaviour of 0.17 M POT20HA was rather different than those of the lower concentrations and the 0.17 M at 299 K. There was no inductance loop observed at the low frequency, which indicated the absence of passive film dissolution.

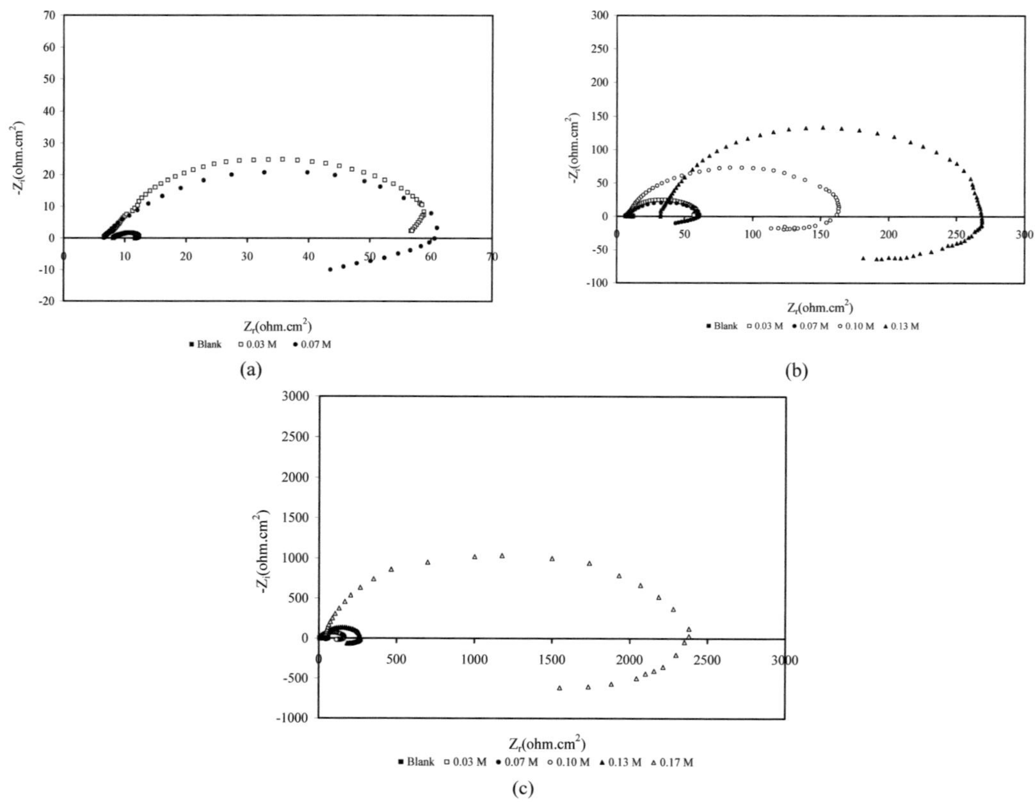

Figure 12. Electrochemical impedance plot of Al 6061 in 1 M HCl solution at 299 K zooming to (a) 0.03 and 0.07 M POT20HA and (b) 0.10 and 0.13 M POT20HA Electrochemical impedance plot of Al 6061 at 299 K in 1 M HCl solution with 0.17 M POT20HA

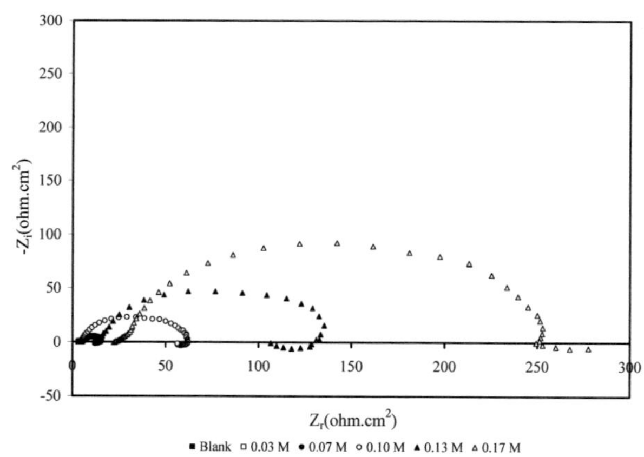

Figure 13. Electrochemical impedance plot of Al 6061 in 1 M HCl solution at 323 K at different concentration of POT20HA

The impedance of Al 6061 in 1 M HCl solution with different concentrations of POT20HA at 343 K was almost similar to those of the lower temperatures under study. The presence of inhibitor had increased the capacitive value. However, the 0.17 M POT20HA had shown different impedance behaviour than those of the lower temperatures. There was no inductance loop obtained at the low frequency and the formation of capacitive loop was not well defined as shown in Figure 14. Similar type of curve was observed by Sherif & Park and Mabrour et al. [31, 32], the significant increase in the loop size and the absence of inductance loop as compared to those of the lower concentrations of POT20HA solution indicate that the 0.17 M POT20HA had better inhibitive behaviour with no dissolution of passive film.

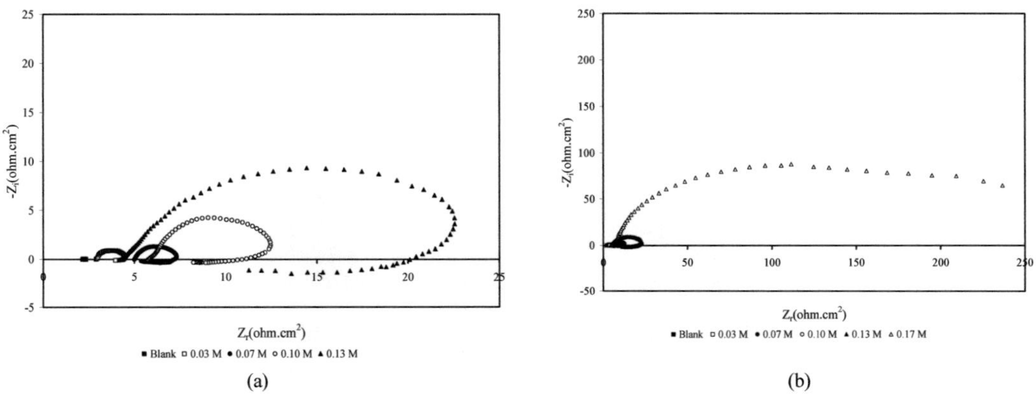

Figure 14. Electrochemical impedance plot of Al 6061 in 1 M HCl solution at 343 K at different concentration of POT20HA

Finally, results from the corrosion tests revealed the inhibition behaviour of the POT20HA towards Al 6061 at different temperatures and concentrations. The weight loss (WL) test showed that IE increased with increasing concentration but decreased with increasing of immersion time and temperature. However, the IE of the 0.17 M POT20HA remained high even after 48 hours of immersion time. The potentiodynamic polarization (PP) test revealed that the POT20HA acted as a mixed type of inhibitor which was capable in inhibiting the anodic and cathodic sides of corrosion process. Consequently, EIS results showed the ability of POT20HA in forming a protective passive film on Al 6061 surface. The thickness of the passive film increased with increasing concentration but decreased with increasing temperature. On the other hand, dissolution of the passive film occurred at elevated temperature and insufficient levels of POT20HA, causing the Al 6061 to eventually dissolve. EIS confirmed that a stable passive film was formed by 0.17 M POT20HA.

5.5. Surface corrosion analysis

SEM micrograph images, as shown in Figure 15, 16 and 17, revealed the effect of different concentrations of POT20HA on the Al 6061 after 3 hours of immersion time in 1 M HCl solution at 299 K. Morphology of the Al 6061 in 0.03 M POT20HA was almost similar to that of in the blank solution as shown in Figure 15 and 16(a). Homogeneous corroded area throughout the surface sample showed that the corrosive solution had attacked the entire grain boundaries of the Al 6061 surface. Further increase in the concentration from 0.03 M to 0.07, 0.10 and 0.13 M showed a reduction in the surface attack as illustrated in Figure 16(b), 17(a) and 17(b), respectively. However, some of the surface was unattacked (grey area), while some intergranular corrosion was observed propagating along the grain boundaries of the Al 6061 surface. Length of the propagating intergranular corrosion reduced with the increase of concentration as clearly observed in 0.07 M and 0.13 M. Further increase in the concentration to 0.17 M showed only the presence of minor pitting corrosion as observed in Figures 18(a) and 18(b) which, revealed one of the pitting corrosion on the Al 6061 surface (at point A). The increase in concentration of POT20HA from 0.03 to 0.17 M had changed the corrosion pattern from general to intergranular and finally to pitting corrosion. Similar corrosion pattern were observed at 323 and 343 K.

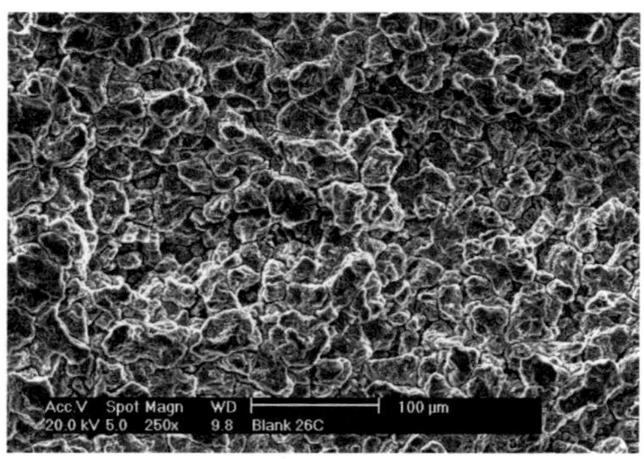

Figure 15. SEM micrograph image of the Al 6061 surface after 3 hours of immersion in blank 1 M HCl solution at 299 K

(a) (b)

Figure 16. SEM micrograph images of the Al 6061 surface after 3 hours of immersion in 1 M HCl solution with different concentrations of POT20HA at 299 K (a) 0.03 M and (b) 0.07 M

http://dx.doi.org/10.5772/57273

Figure 17. SEM micrograph images of the Al 6061 surface after 3 hours of immersion in 1 M HCl solution with different concentrations of POT20HA at 299 K (a) 0.10 M and (b) 0.13 M

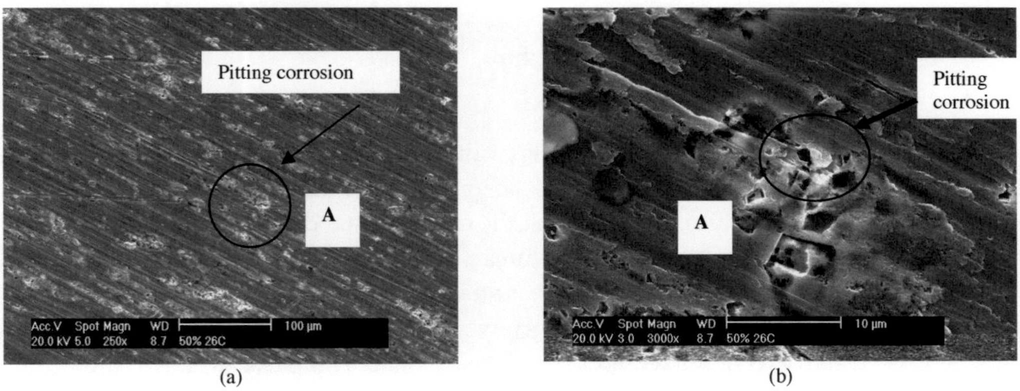

Figure 18. SEM micrograph images of (a) the Al 6061 surface after 3 hours of immersion in 1 M HCl solution with 0.17 M POT20HA at 299 K and (b) enlargement of point A

The SEM observation confirmed the results of WL, PP and EIS tests. The 0.17 M POT20HA solution showed the formation of minor pitting corrosion in all temperatures under study, thus, revealing the ability of this concentration in protecting the Al 6061 surface at these temperatures. At lower concentrations, Cl^- ions could penetrate the Al 6061 surface due to the incomplete surface coverage by the inhibitor. The penetration leads to the formation of pitting corrosion. The lower the concentration of the inhibitor, the more surface area were exposed to the Cl^- ions, hence more pitting corrosion formed which lead to the formation of intergranular and finally general corrosion. The increase of temperature would increase the activation energy of the Cl^- ions, hence increased the aggressiveness of corrosion. On the other hand, an increase in the immersion time will increase the contact between the Cl^- ions and the Al 6061 surface, thus promoting the corrosion process. However, with the 0.17 M POT20HA solution, the aggressiveness of the corrosion process due to the temperature and immersion time can be eliminated.

5.6. Adsorption isotherm relationship

The correlation between the concentration of POT20HA (C) and surface coverage (θ) was determined by fitting the experimental data on a suitable adsorption isotherm, such as Frumkin, Temkin and Langmuir relationships. The adsorption isotherm relationships were expressed by the following equations;

$$\text{Frumkin relationship} \quad \ln\left(\frac{\theta}{C(1-\theta)}\right) = \ln K + 2a\theta \tag{5}$$

$$\text{Temkin relatioship} \quad \theta = (\tfrac{1}{f})\ln(KC) \tag{6}$$

$$\text{Langmuir relationship} \quad \frac{C}{\theta} = \frac{1}{K} + C \tag{7}$$

Table 13 shows the correlation coefficient (R^2) values of the experimental data (weight loss measurement) according to the respective adsorption isotherm relationship at different immersion time. The R^2 values indicated that the experimental data were in agreement with the Langmuir relationship at all temperatures under study. This condition shows that POT20HA has formed monolayer film that was attached to the Al 6061 surface without lateral interaction between the adsorbed inhibitor. However, at the 323 and 343 K, the data for the immersion time above 24 hours would fit the Temkin relationship. According to the Temkin relationship, the f values at this condition are positive. The attraction between Al 6061 and the adsorbed inhibitor is as explained by Noor [8]. The experimental data did not fit the Frumkin relationship. This finding reveals the transition from Langmuir to Temkin isotherm relationship, hence, shows that the adsorption behaviour of an inhibitor is strongly influenced by the temperature.

The inhibition adsorption of POT20HA on Al 6061 surface was determined by the standard free energy of adsorption (ΔG^o_{ads}). The ΔG^o_{ads} value was calculated using the determined K value from the Langmuir relationship and expressed in the following equation;

$$\ln K = \ln\frac{1}{55.5} - \frac{\Delta G^o_{ads}}{RT} \tag{8}$$

Figure 19 shows the ΔG^o_{ads} values from WL (1h), PP and EIS at different temperatures according to the Langmuir relationship. The three tests showed almost similar values of ΔG^o_{ads}, which

Relationship	Temperature(K)	1h	6 h	24 h	48h
Langmuir	299	0.9996	0.9999	0.9999	0.9993
	323	0.9997	0.9994	0.7747	0.4303
	343	0.9995	0.9980	0.9074	0.2831
Temkin	299	0.6359	0.8980	0.9848	0.9606
	323	0.9796	0.9616	0.9220	0.8827
	343	0.9843	0.9841	0.9113	0.8447
Frumkin	299	0.0029	0.1376	0.4069	0.2582
	323	0.5352	0.6209	0.7965	0.9252
	343	0.6430	0.7965	0.9097	0.8032

Table 13. The correlation coefficient (R^2) of the experimental data from different adsorption isotherm relationships

ranged from -22 to -26 kJ/mol. These values indicated physical adsorption on the transfer of unit mole of the inhibitor from solution onto the metal surface [33]. The negative sign of the free energy of adsorption indicated the adsorption of the inhibitor at the metal surface was a spontaneous process [34]. For WL and PP, an increase of temperature from 299 K to 323 K has increased the ΔG^o_{ads} values (less negative). However further increase in the temperature to 343 K has slightly decreased (more negative) the value of ΔG^o_{ads}. On the other hand, the opposite trend of the ΔG^o_{ads} values on the EIS was observed as compared to those of the WL and PP. This finding could be due to the measurement technique that was used in the EIS. Therefore, the ΔG^o_{ads} values from the EIS are not involved in the following discussion.

Similar behaviour of ΔG^o_{ads} had been observed by Noor [8] where 2 M HCl solution in the presence of inhibitor at various temperatures were studied. The increase in ΔG^o_{ads} with increasing temperature indicates the occurrence of exothermic process at which adsorption is unfavourable which cause desorption of inhibitor from the Al 6061 surface. On the other hand, the decrease in ΔG^o_{ads} value with increasing temperature indicates the occurrence of endothermic process which promotes adsorption of the inhibitor on the Al 6061 surface. When both conditions are observed within the temperature range under study it shows the occurrence of both the exothermic and endothermic adsorption processes.

5.7. Inhibition mechanism

Since the POT20HA was in the form of dilute emulsion, the adsorption mechanism of the inhibitor on the Al 6061 surface in 1 M HCl was determined through an emulsion analysis. As previously mentioned, POT20HA existed as spherical micelles with particle size ranging from 0.04 to 300 μm. The pH of emulsion was 12.5 which indicated that the micelles were negatively charge [35], thus are capable in forming electrostatic bonding or being physically adsorbed onto the positively charged Al 6061 surface. The pH of the emulsion without the presence of

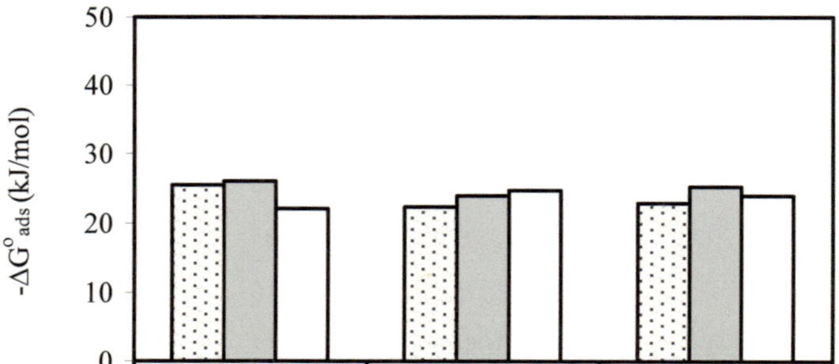

	299 K	323 K	343 K
WL(1h)	26	22	23
PP	26	24	25
EIS	22	25	24

Temperature (K)

Figure 19. The ΔG^{o}_{ads} values of POT20HA from different corrosion tests and temperatures

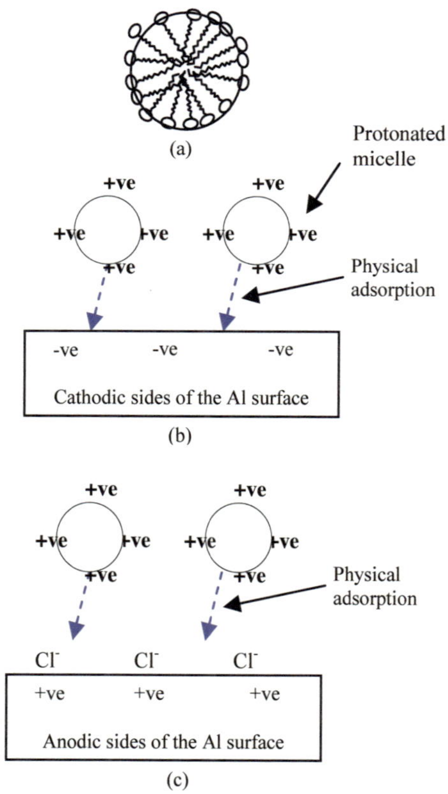

Figure 20. Adsorption of POT20HA in 1 M HCl solution; (a) spherical micelle, (b) the cathodic sides of the Al surface, and (c) the anodic sides of

DETA was about 7; therefore, it was shown that the increase in the basicity of the emulsion was due to the presence of DETA. Results showed that the inhibition was due to the presence of an amide compound (Figure 20).

The PP test had revealed that POT20HA acts as a mixed type of inhibitor which controls both the anodic and cathodic sides of corrosion reaction. Consequently, adsorption isotherm had shown that POT20HA was adsorbed on the Al surface through physical adsorption via weak van der Waal forces. In 1 M HCl solution, POT20HA which was in the form of micelles were expected to be fully protonated [36]. The carbonyl functional group of the amide family of POT20HA was expected to be protonated by the HCl solution and acted as cationic species. Thus, the protonated micelles adsorbed on the cathodic sides of the corroding surface and blocked the H_2 evolution. Figure 20(a) shows the schematic illustration of the spherical micelles, while Figure 20(b) illustrates the attachment of protonated micelles on the cathodic sides of the Al surface. Subsequently, the protonated micelles were expected to be adsorbed onto the anodic sides of the Al surface by means of electrostatic interaction. The electrostatic interaction was formed due to the presence of Cl^- ions in the 1 M HCl solution which were likely to be adsorbed onto the anodic sides of the Al surface and form negatively charged surface which then promoted the adsorption of the protonated micelles. This phenomenon is illustrated schematically by Figure 20(c). Hence, the presence of POT20HA in the corrosion system of Al in 1 M HCl had controlled the dissolution of Al as well as the evolution of H_2.

5.8. Performance evaluation

From the above result and discussion on the formulated corrosion inhibitor, the POT20HA would be able to protect the Al 6061 surface through physical adsorption. Hence, assumption can be made that the POT20HA is capable to work as an anticorrosion in car radiator, in which the temperature is at 88±2°C. In responding to this assumption, POT20HA was used as anticorrosion in a simulated condition of car radiator according to JIS K 2234.

The JIS K 2234 specification was taken as the reference in evaluating the performance of the PO anticorrosion-coolant in the circulation test. Table 14 shows the JIS K 2234 specification and changes of pH, mass of Al, Cu and Fe after the circulation test. In the test, the Al, Cu and Fe were immersed in a solution containing 30% (v/v) of PO anticorrosion-coolant and the balance "adjusting" water at 88±2°C for 1000±2 hours at a circulation rate of 1.5 liter/min. The result showed that the POT20HA had successfully protected the Cu, Fe and Al from corrosion within the JIS K 2234 specification for the circulation test.

		JIS K 2234 specification	After the circulation test
Change of mass (mg/cm^2)	Al	±0.60	0.26
	Cu	±0.30	0.12
	Fe	±0.30	0.16
Initial pH		7 to 11	10.40
Final pH		6.5 to 11.0	9.43

Table 14. Mass and pH changes after circulation test

5.9. Physical properties of PO anticorrosion-coolant

Table 15 shows the physical properties of both the PO and the commercial anticorrosion-coolants. The density and boiling points of these anticorrosion-coolants were within the acceptable range. The boiling point of PO anticorrosion-coolant was significantly much higher than those of the commercial anticorrosion-coolant and the specified values in JIS K 2234. However, the freezing point of the PO anticorrosion-coolant was slightly higher than the specified value.

Performance of the formulated PO antirust-coolant followed the specification of the JIS K 2234 and is almost similar to the well known commercial product. Furthermore, PO showed better boiling point than the commercial product, even though the freezing point was slightly lower. Therefore, the formulated PO antirust-coolant was an excellent anti-boil, anticorrosion and significant enough as anti-freeze which can be used throughout the year, especially in tropical and temperate countries.

Physical Properties	JIS K 2234	PO	Commercial
Density	1.112 (min)	1.239	1.12
Boiling Point	152°C(min)	230°C	192°C
Freezing point (50%(v/v) anticorrosion-coolant)	-34°C(max)	-26°C	-36.5°C

Table 15. The physical properties of the anticorrosion-coolant

6. Conclusions

The formulated palm olein inhibitor (POT20HA) was an amide compound which is partially soluble in water. The temperature of (50°C) 323 K and pH 7 were found to be the suitable condition for formulation preparation. The suitable emulsifier, stabilizing agent and enhancing agent were Tween 20, hexane and DETA, respectively.

The weight loss study showed that the IE was found to be concentration dependent. However the IE is inversely proportional to the immersion time and temperature. The potentiodynamic polarization study showed that the POT20HA was a mixed type of inhibitor. The electrochemical impedance spectroscopy results indicated the ability of POT20HA in forming protective passive film on Al 6061 surface. The thickness of passive film increased together with increasing concentration of POT20HA but decreased with increasing temperature. SEM observation showed a smooth and homogeneous inhibited Al 6061 surface with 0.17M POT20HA.

The Langmuir isotherm relationship fitted the data of all temperatures and immersion times under study. The ΔG^o_{ads} and E_a values for all concentrations and temperatures under study confirmed that the POT20HA was adsorbed on the Al surface through physical adsorption. Adsorption mechanism study had shown that the protonated POT20HA micelles in 1 M HCl was adsorbed onto the cathodic sides of the corroding surface and blocked the H_2 evolution. Subsequently, the protonated micelles of POT20HA were adsorbed onto the anodic sides of the Al surface by means of electrostatic interaction of the Cl^- ions.

The performance tests had shown that POT20HA had the ability to function as an anticorrosion material when mixed with glycerin in Al car radiator. Performance of the formulated PO anticorrosion-coolant fulfilled the specification of the JIS K 2234 and matched the standard of a well known commercial product.

Acknowledgements

The Ministry of Education of Malaysia is acknowledged for providing Fundamental Research Grant Scheme (FRGS) for this work. A special gratitude also goes to the Research Management Institute (RMI) of Universiti Teknologi MARA, Malaysia for the continuous supports. Special thanks to the Dean of Faculty of Chemical Engineering, Universiti Teknologi MARA, Malaysia for the support and encouragement. Finally, appreciations also go to those who are directly and indirectly involved in this work.

Author details

Junaidah Jai*

Address all correspondence to: junejai@salam.uitm.edu.my

Faculty of Chemical Engineering, University of Technology MARA, Shah Alam, Malaysia

References

[1] Maayta A, Al-Rawashdeh N. Inhibition of acidic corrosion of pure aluminum by some organic compounds. Corrosion Science. 2004;46(5):1129-40.

[2] El-Etre A. Inhibition of acid corrosion of carbon steel using aqueous extract of olive leaves. Journal of Colloid and Interface Science. 2007;314(2):578-83.

[3] El-Etre A, Abdallah M. Natural honey as corrosion inhibitor for metals and alloys. II. C-steel in high saline water. Corrosion Science. 2000;42(4):731-8.

[4] El-Etre A. Inhibition of acid corrosion of aluminum using vanillin. Corrosion Science. 2001;43(6):1031-9.

[5] El-Etre A. Inhibition of aluminum corrosion using Opuntia extract. Corrosion Science. 2003;45(11):2485-95.

[6] Orubite K, Oforka N. Inhibition of the corrosion of mild steel in hydrochloric acid solutions by the extracts of leaves of Nypa fruticans Wurmb. Materials Letters. 2004;58(11):1768-72.

[7] El-Etre A, Abdallah M, El-Tantawy Z. Corrosion inhibition of some metals using lawsonia extract. Corrosion Science. 2005;47(2):385-95.

[8] Noor EA. Temperature effects on the corrosion inhibition of mild steel in acidic solutions by aqueous extract of fenugreek leaves. International Journal of Electrochemical Science. 2007;2(12).

[9] Ebenso E, Eddy N, Odiongenyi A. Corrosion inhibitive properties and adsorption behaviour of ethanol extract of Piper guinensis as a green corrosion inhibitor for mild steel in H_2SO_4. African Journal of Pure and Applied Chemistry. 2008;2(11):107-15.

[10] Bouyanzer A, Hammouti B, Majidi L. Pennyroyal oil from Mentha pulegium as corrosion inhibitor for steel in 1M HCl. Materials Letters. 2006;60(23):2840-3.

[11] Benabdellah M, Benkaddour M, Hammouti B, Bendahhou M, Aouniti A. Inhibition of steel corrosion in 2M H_3PO_4 by artemisia oil. Applied surface science. 2006;252(18): 6212-7.

[12] Ouachikh O, Bouyanzer A, Bouklah M, Desjobert J-M, Costa J, Hammouti B, et al. Application of essential oil of Artemisia herba alba as green corrosion inhibitor for steel in 0.5 MH 2 SO 4. Surface Review and Letters. 2009;16(01):49-54.

[13] Kalaiselvi P, Chellammal S, Palanichamy S, Subramanian G. Artemisia pallens as corrosion inhibitor for mild steel in HCl medium. Materials Chemistry and Physics. 2010;120(2):643-8.

[14] Bammou L, Mihit M, Salghi R, Bouyanzer A, Al-Deyab S, Bazzi L, et al. Inhibition Effect of Natural Artemisia Oils Towards Tinplate Corrosion in HCl solution: Chemical Characterization and Electrochemical Study. Int J Electrochem Sci. 2011;6:1454-67.

[15] Garai S, Garai S, Jaisankar P, Singh J, Elango A. A comprehensive study on crude methanolic extract of Artemisia pallens (Asteraceae) and its active component as effective corrosion inhibitors of mild steel in acid solution. Corrosion Science. 2012;60:193-204.

[16] Huang J, Cang H, Liu Q, Shao J. Environment Friendly Inhibitor for Mild Steel by Artemisia Halodendron. Int J Electrochem Sci. 2013;8:8592-602.

[17] Fouda A, Rashwan S, Abo-Mosallam H. Fennel seed extract as green corrosion inhibitor for 304 stainless steel in hydrochloric acid solutions. Desalination and Water Treatment. 2013 (ahead-of-print):1-12.

[18] McMurry J. Organic Chemistry, United States of America: Thomson Learning. Inc; 2004.

[19] Testing ASf, Staff M. Vol 03.02 2003: Astm Intl; 2003.

[20] Fontana MG. Corrosion engineering: McGraw-Hill; 1986.

[21] Al-Rawashdeh N, Maayta A. Cationic surfactant as corrosion inhibitor for aluminum in acidic and basic solutions. Anti-Corrosion Methods and Materials. 2005;52(3): 160-6.

[22] Guzey D, Kim H, McClements DJ. Factors influencing the production of o/w emulsions stabilized by β-lactoglobulin–pectin membranes. Food Hydrocolloids. 2004;18(6):967-75.

[23] Goyal P, Aswal V. Micellar structure and inter-micelle interactions in micellar solutions: Results of small angle neutron scattering studies. CURRENT SCIENCE-BANGALORE-. 2001;80(8):972-9.

[24] Toff MRM. A comparison of the corrosion behaviour of zinc and zinc-5% aluminium coatings. United Kingdom: University of Sheffield; 1999.

[25] Sharma S. A study on stress corrosion behavior of Al6061/albite composite in higher temperature acidic medium using autoclave. Corrosion Science. 2001;43(10):1877-89.

[26] Branzoi V, Golgovici F, Branzoi F. Aluminium corrosion in hydrochloric acid solutions and the effect of some organic inhibitors. Materials Chemistry and Physics. 2003;78(1):122-31.

[27] Khaled K, Al-Qahtani M. The inhibitive effect of some tetrazole derivatives towards Al corrosion in acid solution: Chemical, electrochemical and theoretical studies. Materials Chemistry and Physics. 2009;113(1):150-8.

[28] Ferreira E, Giacomelli C, Giacomelli F, Spinelli A. Evaluation of the inhibitor effect of L-ascorbic acid on the corrosion of mild steel. Materials Chemistry and Physics. 2004;83(1):129-34.

[29] Yan Y, Li W, Cai L, Hou B. Electrochemical and quantum chemical study of purines as corrosion inhibitors for mild steel in 1M HCl solution. Electrochimica acta. 2008;53(20):5953-60.

[30] Rosliza R, Wan Nik W, Senin H. The effect of inhibitor on the corrosion of aluminum alloys in acidic solutions. Materials Chemistry and Physics. 2008;107(2):281-8.

[31] Sherif E, Park S-M. Effects of 1, 4-naphthoquinone on aluminum corrosion in 0.50 M sodium chloride solutions. Electrochimica acta. 2006;51(7):1313-21.

[32] Mabrour J, Akssira M, Azzi M, Zertoubi M, Saib N, Messaoudi A, et al. Effect of vegetal tannin on anodic copper dissolution in chloride solutions. Corrosion Science. 2004;46(8):1833-47.

[33] Cheng S, Chen S, Liu T, Chang X, Yin Y. Carboxymenthylchitosan as an ecofriendly inhibitor for mild steel in 1 M HCl. Materials Letters. 2007;61(14):3276-80.

[34] El-Etre A. Khillah extract as inhibitor for acid corrosion of SX 316 steel. Applied surface science. 2006;252(24):8521-5.

[35] Liu Y, Guo R. pH-dependent structures and properties of casein micelles. Biophysical chemistry. 2008;136(2):67-73.

[36] Singh P, Bhrara K, Singh G. Adsorption and kinetic studies of L-leucine as an inhibitor on mild steel in acidic media. Applied surface science. 2008;254(18):5927-35.

Chapter 18

Identification and Application of Corrosion Inhibiting Long-Chain Primary Alkyl Amines in Water Treatment in the Power Industry

Peter Kusch, Gerd Knupp, Marian Kozupa,
Jolanta Iłowska and Maria Majchrzak

Additional information is available at the end of the chapter

http://dx.doi.org/10.5772/57355

1. Introduction

Heated boilers are an essential element of industrial processes. They need to be reliable and kept in good working order. With skyrocketing fuel and energy costs, maintaining the reliability and consistent performance of a boiler while minimizing energy costs, is challenging for any industrial works. Since boiler systems are constructed primarily of carbon steel and the medium for heat transfer is water, the potential threat for corrosion, scale forming and bio-fouling is great. The build up of corrosion can result in a forced shutdown of the boiler and the whole industrial process [1].

Certain chemical and thermal conditions favor the occurrence of corrosion processes on the surfaces of heated boilers. This phenomenon is a result of chemical reactions taking place in non-homogenous steam-water environment, which have often a local nature and, additionally, are difficult to determine in qualitative analysis of working medium [2]. In high pressure boiler drums, where feed water is even one hundred times more concentrated (even more in regions where steam bubbles are generated) it is difficult to avoid scaling and corrosion of boiler's construction walls. Even a relatively slight decrease in boilers efficiency caused by scale leads to increase in fuel consumption up to hundreds of tons annually [2]. Finding ways to operate a large cooling water system economically while maximizing heat transfer is a complex and challenging task.

One of the effective ways to improve installation economics and reliability is protective water treatment with chemical formulations containing corrosion inhibitors [2]. In the past, the efficient protective treatments of carbon steels were based on inorganic inhibitors like ammonia, nitrites, phosphates or chromates. Today, their use is restricted due to their high toxicity and environmental impact. The most common protective water treatment methods base on the use of neutralizing amines. These chemicals, such as morpholine, cyclohexylamine and *N,N*-diethylhydroxylamine (DEHA) neutralize the carbonic acid (formed in the reaction of carbon dioxide with water) and increase the pH of the condensate [3].

In the first half of the 20th century the aliphatic monoamines [$CH_3(CH_2)_nNH_2$, $n = 10 \div 20$] and alkyl polyamines [$R\text{-}(NH\text{-}CH_2\text{-}CH_2\text{-}CH_2\text{-})_nNH_2$, $n = 1 \div 5$] drew researchers' attention due to their exceptional ability to form thin adhesive films, usually monomolecular, which strongly bond to the metal surface. In this layer, referred to as an amine film, amine groups ($-NH_2$) are bonded to a metal surface, while long hydrophobic hydrocarbon chains are oriented in opposite direction, which works as a barrier protecting the metal surface against gaseous molecules (O_2, CO_2) as well as ionic pollutants (*e.g.* Cl^-, SO_4^{2-}) [2].

Despite the fact that use of organic chemical agents in anti-corrosion protection is not recommended by several international regulations, many film-forming amines (FFA) have been used in the power industry and in heat engineering [2-4]. Just like in classical treatment concepts, treatment with FFA must be carefully monitored in order to ensure successful treatment and a high degree of operational safety. Sampling points are the boiler feed-water, the boiler water and the condensate. The condensate should have a slightly alkaline pH and it should be possible to detect an excess of free FFA [4].

At the end of 1980 the Instytut Ciężkiej Syntezy Organicznej "Blachownia" (Institute of Heavy Organic Synthesis "Blachownia") (Kędzierzyn-Koźle, Poland) along with ZPBE ENERGOPOMIAR Ltd. (Gliwice, Poland) and P.U.B. "Ekochem" Ltd. (Gliwice, Poland) worked out a group of multifunctional amine formulations named *Kotamina* [5-7]. These formulations contain an appropriately composed mixture of alkyl amines (main component) with different partition coefficients, which enables control of corrosion kinetics within a whole system through dosing of these formulations into condensate, feed water or make-up water. The presence of alkalizing amines in the formulation helps in keeping constant pH in a desired range, whereas alkyl amines, both render the water alkaline and limit the corrosive impact of ionic and gaseous impurities by formation of monomolecular protective films. The alkyl amines were chosen on the basis of results obtained during investigation of their influence on the adsorption on metal surfaces. The utility properties of these formulations as well as experience related to their application in Polish power industry for over twenty years were presented in several papers [5-7].

2. Thermostable alkyl amines as a base for anti-corrosive and anti-scaling formulations

Alkyl amines used at present in correction formulations exhibit limited stability toward higher temperatures and their decomposition starts at temperatures slightly above 300 °C. One disadvantage of alkyl amine formulations application in the field was the high concentration of ammonia in the condensate and in the steam caused by decomposition of amines of higher molecular weight. Utilization of thermostable alkyl amines in water treatment technology, whose decomposition is above 500 °C, causes decrease in ammonia levels and improves anti-corrosive protection of the brass elements [2].

Achieving further improvement in correction technology of steam-water systems, additional investigations with usage of thermostable, branched long-chain alkyl amines (as a base for correction formulations) were developed. The research on the application of thermostable alkyl amines in water treatment was performed within the framework of the international project EUREKA E!2426 BOILTREAT "New technology of boiler water treatment" with partners from research institutes, industry and academia from Poland, Lithuania, Romania, France and Germany [8]. Thus, a new anticorrosion agent called *Kotamina Plus* was formulated, which contains branched long-chain primary alkyl amines of the Primene JM-T™ type [9-11].

Others multifunctional anti-corrosive and anti-scaling formulations based on alkyl amines were described in several publications or patents [12-18].

3. Identification and determination of corrosion inhibiting long-chain primary alkyl amines of the Primene JM-T™ type by gas chromatography (GC) and gas chromatography - mass spectrometry (GC/MS)

3.1. Chemicals and samples

Alkyl amines Primene JM-T™ obtained from Rohm and Haas France S.A.S. (Valbonne, France), dicyclohexylamine for synthesis from Merck-Schuchardt (Hohenbrunn, Germany), *n*-hexane for LC (Biosolve B.V., Valkenswaard, the Netherlands), methanol LiChrosolv gradient grade for HPLC (Merck, Darmstadt, Germany), tetrahydrofurane (THF) LiChrosolv for LC (Merck), trifluoroacetic anhydride (TFAA) from Macherey-Nagel (Düren, Germany), 5% dimethydi-chlorosilane (DMDCS) in toluene (Sylon CT) for glass silylation from Supelco (Bellefonte, PA, U.S.A.) and deionized water, purity Type 1 according to USP 25-NF 20 from the water purification system Milli-Q (Millipore Corp., Bedford, MA, U.S.A.) were used.

To prevent adsorption of alkyl amines on the glass surface, spiked water samples of 10 mg L^{-1} to 100 mg L^{-1} of Primene JM-T™ were prepared and stored at ambient temperature in a 1 L polypropylene (PP) volumetric flasks (Kartell, Noviglio, Italy).

For the evaluation of the quantitative determination, various standard solutions of 12.5 mg L^{-1}, 25 mg L^{-1}, 50 mg L^{-1}, 75 mg L^{-1} and 100 mg L^{-1} Primene JM-T™-TFA with each 10 mg L^{-1} of dicyclohexylamine (internal standard, I.S.) in *n*-hexane/THF 95/5 (v/v) were prepared in 10 cm^3 silylated volumetric flask.

For solid phase extraction (SPE) of organic compounds from water samples LiChrolut RP-18 E (octadecyl, endcapped, 500 mg/3 cm^3) cartridges supplied by Merck were used.

3.2. Derivatization procedure

Amines are generally known to be very difficult to analyze by gas chromatography (GC) due to their basic character [19, 20]. In addition to the basic character, the amino group introduces a large dipole in the molecule. This dipole is responsible for strong interaction with silanol groups and siloxane bridges in the structure of the stationary phase of the GC capillary column. This often results in nonlinear adsorption effects and can be seen as strong tailing peaks in the chromatogram. The best way to prevent interaction of the strong dipole is to derivatize the amine. Derivatization reduces the polarity of the molecule making it more retentive in chromatographic analysis. The conversion of compounds enhances GC performance as the analyte volatility is increased and peak shape improved because of reduced surface adsorption. Derivatized analyte offer a greater response to the chromatographic detection system than the parent compounds. The choice of a derivatizing reagent is based on the functional group requiring derivatization, the presence of other functional groups in the molecule, and the reason for performing the derivatization. We have selected the acylation with trifluoroacetic acid anhydride (TFAA) as derivatization method for alkyl amines. A benefit of acylation is the formation of fragmentation-directing derivatives for GC/MS analysis [19].

A compact ultrasonic bath Sonorex Super RK 31H from Bandelin electronic (Berlin, Germany) and reaction vessels of 5 cm^3 with solid cap and PTFE liner (Supelco) deactivated with 5% DMDCS in toluene (Sylon CT) were used for derivatization. Approximately 10 mg of the investigated alkyl amines were dissolved in 0.5 cm^3 THF in a 5 cm^3 glass micro-reaction vessel and 100 μL of TFAA were added. The sealed vessel was placed in the ultrasonic bath and agitated by heating at 60 °C for 15 min. Excess of reagent, released trifluoroacetic acid and THF were evaporated with a gentle steam of nitrogen at room temperature or by using a vacuum pump. The resulting product was dissolved in 1 cm^3 THF and analysed by GC or GC/MS.

3.3. Solid-phase extraction (SPE) of water samples from the power plant

Solid-phase extraction (SPE) is a form of digital (step-wise) chromatography designed to extract, partition, and/or adsorb one or more components from a liquid phase (sample) onto stationary phase (sorbent or resin). Over the last thirty years, SPE has become the most powerful technique available for rapid and selective sample preparation prior to analytical chromatography.

Water samples of boiler water, superheated steam and condensate from the power plant were stored at ambient temperature in 5 L PP bags (Bürkle, Lörrach, Germany) and extracted by solid-phase extraction from a 1 L PP volumetric flask (Kartell). For conditioning the SPE LiChrolut® RP-18 E cartridge packing, the tube was rinsed with 6 cm^3 methanol followed with 6 cm^3 deionised water. After the conditioning step 1–2 L of the investigated water sample were percolated at a flow rate of 5 cm^3 min^{-1} through the SPE tube. After washing the tube with 5 cm^3 of deionised water, the adsorbed organic compounds were eluted in a mixture of 5 cm^3 *n*-hexane/THF 95/5 (v/v) and collected in a deactivated micro-reaction vessel. After elution, the solvent was evaporated with a gentle steam of nitrogen at room temperature or by using a vacuum pump. Then, the extract was derivatized by acylation according to the procedure described in 3.2. The resulting product was dissolved in 1 cm^3 of the internal standard solution, containing 10 - 12 mg L^{-1} of dicyclohexylamine in *n*-hexane and analyzed by GC-FID.

3.4. Instrumentation

Gas chromatographic (GC) analyses were performed with an *Autosystem* or an *Clarus 500* gas chromatograph from PerkinElmer Instruments (Norwalk, CT, U.S.A.), both equipped with a split/splitless injector at 290 °C and a flame-ionization detector (FID) operated at 320 °C. Helium 5.0 grade (Westfalen AG, Münster, Germany) was used as a carrier gas. The helium inlet pressure was 140 kPa and the split flow was 20 cm^3 min^{-1}. The fused silica capillary column used in this investigation was 60 m x 0.25 mm I.D., film thickness 0.25 μm *DB-5ms* from J&W Scientific (Folsom, CA, U.S.A.). The oven temperature was programmed from 60 °C (1 min hold) at 6 °C min^{-1} to 280 °C (hold 50 min). Chromatographic data were processed with *TotalChrom Workstation* software, version 6.3. (PerkinElmer Instruments).

GC/MS measurements were made using also two apparatus. The first was an *ThermoQuest Trace 2000* gas chromatograph (ThermoQuest CE Instruments, Milan, Italy) interfaced to a ThermoQuest/Finnigan Voyager quadrupole mass spectrometer (ThermoQuest/Finnigan MassLab Group, Manchester UK) operated in electron impact ionization (EI) mode, negative chemical ionization (NCI) mode and positive chemical ionization (PCI) mode with an ThermoQuest *Xcalibur* data system, the *NIST 05* mass spectra library, and a *CombiPAL* autosampler (CTC Analytics AG, Zwingen, Switzerland). The fused silica capillary columns, 60 m long, 0.25 mm I. D. with *DB-5ms* (J&W) or *Elite-5ms* (PerkinElmer Instruments) stationary phases, film thickness 0.25 μm were used. The temperature of the columns was programmed from 60 °C (1 min hold) at 3 °C min^{-1} to 280 °C (30 min hold) or from 150 °C (1 min hold) at 3 °C min^{-1} to 280 °C (50 min hold). Helium 5.0 grade (Westfalen AG) was used as a carrier gas. Constant flow of helium of 1 cm^3 min^{-1} was used during the whole analysis. The temperature of the split/splitless injector was 250 °C and the split flow was 20 or 100 cm^3 min^{-1}. The transfer line temperature was 280 °C. The EI ion source temperature was kept at 250 °C. The ionization occurred with a kinetic energy of the impacting electrons of 70 eV. The emission current was 150 μA. The detector voltage was 350 V. Methane grade 4.5 (Westfalen AG) was used as reagent gas for PCI and NCI. The positive chemical ionization occurred with a kinetic energy of 70 eV. The emission current was also150 μA. The PCI ion source temperature was 250 °C. Mass spectra and reconstructed chromatograms (total ion current [TIC]) were obtained after eluting of solvent (10 min) by automatic scanning in the mass range m/z 35 – 750 u.

The second GC/MS was a *7890A* gas chromatograph with a series *7683B* autosampler and a series *5975C* quadrupole mass spectrometer (Agilent Technologies Inc., Santa Clara, CA, U.S.A.) operated in electron impact ionization (EI) mode. The fused silica capillary column, 30 m long, 0.25 mm I. D. with *HP-5MS* stationary phase, film thickness 0.25 μm was used. GC/MS data were processed with the *ChemStation* software (Agilent Technologies) and the *NIST 05* mass spectra library (Agilent Technologies). The temperature of the column was programmed from 150 °C (1 min hold) at 3 °C min^{-1} to 280 °C (50 min hold). Helium 5.0 grade (Westfalen AG) was used as a carrier gas. Constant flow of helium of 1.0 cm^3 min^{-1} was used during the whole analysis. The temperature of the split/splitless injector was 250 °C and the split flow was 10 cm^3 min^{-1}. The transfer line temperature was 280 °C. The EI ion source temperature was kept at 230 °C. The ionization occurred with a kinetic energy of the impacting electrons of 70 eV. The quadrupole temperature was 150 °C. Mass spectra and reconstructed chromatograms (total ion current [TIC]) were obtained after eluting of solvent (10 min solvent delay) by automatic scanning in the mass range m/z 35-750 u.

3.5. Results and discussion

3.5.1. Structure elucidation of long-chain primary alkyl amines of the Primene JM-T™ type

The general reaction for acylation of the investigated long-chain primary alkyl amines with trifluoroacetic anhydride (TFAA) is shown in equation (1):

$$(R_1)(R_2)(R_3)C\text{-}NH_2 + (CF_3CO)_2O \longrightarrow (R_1)(R_2)(R_3)C\text{-}NH\text{-}C(O)CF_3 + CF_3COOH \qquad (1)$$

Figure 1 shows a typical total ion current GC/MS chromatogram of trifluoroacetylated (TFA) derivative of Primene JM-T™ dissolved in *n*-hexane. The obtained electron impact (EI) mass spectra of the separated compounds from Primene JM-T™-TFA as well as the mass spectra recorded in the negative chemical ionization mode (NCI) and in the positive ionization (PCI) mode, respectively are presented in our previous publications [20-23].

Table 1 summarized the significant *m/z* fragments in the recorded mass spectra of the investigated TFA derivative of Primene JM-T™.

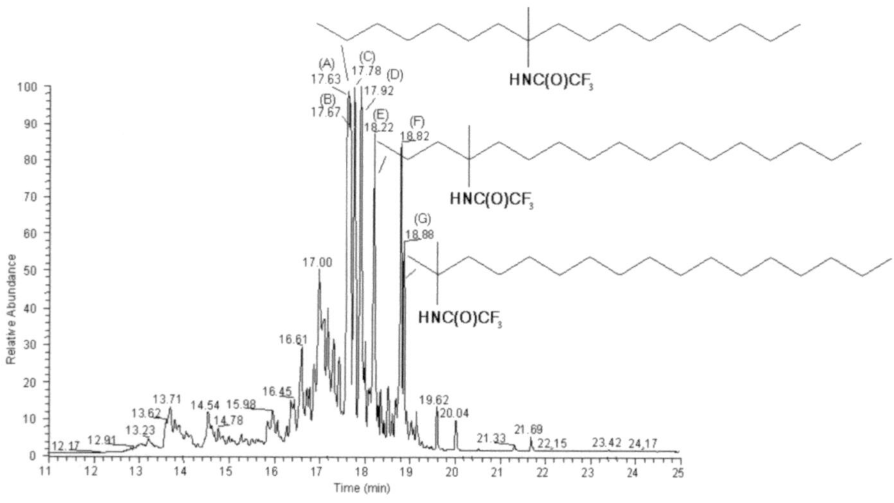

Figure 1. Total ion current GC/MS chromatogram of trifluoroacetylated (TFA) derivative of Primene JM-T™ dissolved in *n*-hexane. Fused silica capillary column: *DB-5ms*, 60 m x 0.25 mm I.D., film thickness 0.25 μm. Column temperature programmed from 150 °C (1 min hold) at 6 °C min^{-1} to 280 °C (hold 30 min). Split/splitless injector: 250 °C. Helium constant flow 1 cm^3 min^{-1}, split flow 100 cm^3 min^{-1}. Peak identification: t_R = 13.0 min – 15.5 min: isomers of octadecene, (A) t_R = 17.63 min: 8-methyl-8-heptadecaneamine-TFA; (B) t_R = 17.67 min: 7-methyl-7-heptadecaneamine-TFA; (C) t_R = 17.78 min: 6-methyl-6-heptadecaneamine-TFA; (D) t_R = 17.92 min: 5-methyl-5-heptadecaneamine-TFA; (E) t_R = 18.22 min: 4-methyl-4-heptadecaneamine-TFA; (F) t_R = 18.82 min: 3-methyl-3-heptadecaneamine-TFA; (G) t_R = 18.88 min: 2-methyl-2-heptadecaneamine-TFA [23]. Publishing with licence number 3185340683087 from Elsevier.

http://dx.doi.org/10.5772/57355

Mass spectrum	Significant fragments (m/z)			Chemical structure
	EI	NCI	PCI	
A	$[CF_3]^+$ (69); $[CF_3C(O)]^+$ (97); $[CF_3C(O)NHCHCH_3]^+$ (140); $[M - C_9H_{19}]^{+\cdot}$ (238); $[M - C_8H_{17}]^{+\cdot}$ (252); $[M - C_7H_{15}]^{+\cdot}$ (266);	$[M - 1 - C_9H_{19}]^{-\cdot}$ (237); $[M - 1 - C_8H_{17}]^{-\cdot}$ (251); $[M - 1 - C_7H_{15}]^{-}$ (265); $[M - 1]^{-\cdot}$ (364);	$[M + H - C_9H_{20}]^+$ (238); $[M + H - CF_3C(O)NH_2]^+$ (253); $[M + H - C_7H_{16}]^+$ (266); $[M + H - CH_4]^+$ (350); $[M - H]^+$ (364); $[M + H]^+$ (366);	8-methyl-8-heptadecaneamine-TFA
B	$[CF_3]^+$ (69); $[CF_3C(O)]^+$ (97); $[CF_3C(O)NHCHCH_3]^+$ (140); $[M - C_{10}H_{21}]^{+\cdot}$ (224); $[M - C_6H_{13}]^{+\cdot}$ (280);	$[M - 1 - C_{10}H_{21}]^{-\cdot}$ (223); $[M - 1 - C_6H_{13}]^{-\cdot}$ (279); $[M - 1]^{-\cdot}$ (364);	$[M + H - C_{10}H_{22}]^+$(224); $[M + H - CF_3C(O)NH_2]^+$ (253); $[M + H - C_6H_{14}]^+$ (280); $[M + H - CH_4]^+$ (350); $[M - H]^+$ (364); $[M + H]^+$ (366);	7-methyl-7-heptadecaneamine-TFA
C	$[CF_3]^+$(69); $[CF_3C(O)]^+$ (97); $[CF_3C(O)NHCHCH_3]^+$ (140); $[M - C_{11}H_{23}]^{+\cdot}$ (210); $[M - C_5H_{11}]^{+\cdot}$ (294);	$[M - 1 - C_{11}H_{23}]^{-\cdot}$ (209); $[M - 1 - C_5H_{11}]^{-\cdot}$ (293); $[M - 1]^{-\cdot}$ (364);	$[M + H - C_{11}H_{24}]^+$ (210); $[M + H - CF_3C(O)NH_2]^+$ (253); $[M + H - C_5H_{12}]^+$ (294); $[M + H - CH_4]^+$ (350); $[M - H]^+$ (364); $[M + H]^+$ (366);	6-methyl-6-heptadecaneamine-TFA
D	$[CF_3]^+$ (69); $[CF_3C(O)]^+$ (97); $[CF_3C(O)NHCHCH_3]^+$ (140); $[M - C_{12}H_{25}]^{+\cdot}$ (196); $[M - C_4H_9]^{+\cdot}$ (308);	$[M - 1 - C_{12}H_{25}]^{-\cdot}$ (195); $[M - 1 - C_4H_9]^{-\cdot}$ (307); $[M - 1]^{-\cdot}$ (364);	$[M + H - C_{12}H_{26}]^+$ (196); $[M + H - CF_3C(O)NH_2]^+$ (253); $[M + H - C_4H_{10}]^+$ (308); $[M + H - CH_4]^+$ (350); $[M - H]^+$ (364); $[M + H]^+$ (366);	5-methyl-5-heptadecaneamine-TFA
E	$[CF_3]^+$ (69); $[CF_3C(O)]^+$ (97); $[CF_3C(O)NHCHCH_3]^+$ (140); $[M - C_{13}H_{27}]^{+\cdot}$ (182); $[M - C_3H_7]^{+\cdot}$ (322);	$[M - 1 - C_{13}H_{27}]^{-\cdot}$ (181); $[M - 1 - C_3H_7]^{-\cdot}$ (321); $[M - 1]^{-\cdot}$ (364);	$[M + H - C_{13}H_{28}]^+$ (182); $[M + H - CF_3C(O)NH_2]^+$ (253); $[M + 1 - C_3H_8]^+$ (322); $[M + H - CH_4]^+$ (350); $[M - H]^+$ (364); $[M + H]^+$ (366);	4-methyl-4-heptadecaneamine-TFA
F	$[CF_3]^+$ (69); $[CF_3C(O)]^+$ (97); $[CF_3C(O)NHCHCH_3]^+$ (140); $[M - C_{14}H_{29}]^{+\cdot}$ (168);	$[M - 1 - C_{14}H_{29}]^{-\cdot}$ (167); $[M - 1 - C_2H_5]^{-\cdot}$ (335); $M - 1]^{-\cdot}$ (364);	$[M + H - C_{14}H_{30}]^+$ (168); $[M + H - CF_3C(O)NH_2]^+$ (253); $[M + H - C_2H_6]^+$ (336); $[M + H - CH_4]^+$ (350); $[M - H]^+$ (364);	3-methyl-3-heptadecaneamine-TFA

Mass spectrum	Significant fragments (m/z)			Chemical structure
	EI	NCI	PCI	
	$[M - C_2H_5]^{+\cdot}$ (336);		$[M + H]^+$ (366);	
G	$[CF_3]^+$ (69); $[CF_3C(O)]^+$ (97); $[CF_3C(O)NHCHCH_3]^+$ (140); $[M - C_{15}H_{31}]^{+\cdot}$ (154); $[M - CH_3]^{+\cdot}$ (350);	$[M - 1 - C_{15}H_{31}]^{-\cdot}$ (153); $[M - 1 - CH_3]^{-\cdot}$ (349); $[M - 1]^{-\cdot}$ (364);	$[M + H - C_{15}H_{32}]^+$ (154); $[M + H - CF_3C(O)NH_2]^+$ (253); $[M + H - CH_4]^+$ (350); $[M - H]^+$ (364); $[M + H]^+$ (366);	2-methyl-2-heptadecaneamine-TFA

Table 1. Significant fragments and identification of trifluoroacetylated (TFA) derivatives of *tert*-octadecylamines in Primene JM-T™ [21, 23]. Publishing with licence number 3185351058212 from Elsevier.

The proposed PCI fragmentation pattern for *N*-(*tert*-octadecyl)-trifluoroacetamides is shown in Figure 2 [22, 23]. The proposed chemical structures of the trifluoroacetylated *tert*-octadecylamines are summarized in Table 1.

3.5.2. *Quantitative determination of Primene JM-T™ in water samples from the power plant*

Gas chromatography with flame ionization detection (GC-FID) was used for the quantitative determination of Primene JM-T™ in boiler water, condensate and superheated steam samples from the power plant. The linearity of FID for the quantitative determination of Primene JM-T™ was evaluated by consecutive injecting of standard solutions (see 3.1). Each standard solution was injected in triplicate and the mean value of the total peak area ratio Primene JM-T™- TFA/I.S. was taken for construction of the calibration line. The total peak area of Primene JM-T™-TFA means the sum of all peaks area of trifluoroacetylated *tert*-octadecylamines in the sample. The calibration graph obtained in Figure 3 shows the relationship between the obtained total peak area ratio Primene JM-T™-TFA/I.S. and the concentration of Primene JM-T™-TFA in the standard solutions for FID. The quantity of Primene JM-T™ in boiler water, condensate and superheated steam samples from the power plant was calculated from results of chromatographic analyses and results of detectors calibration using the equation (2):

$$C_{\text{i sample}} = \left[\left(A'_{\text{sample}} - b\right) / \, a \cdot f \cdot R\right] \cdot 100\% \qquad (2)$$

where $C_{\text{i sample}}$ is the concentration (mg L^{-1}) of Primene JM-T™ in the sample, A'_{sample} is the total peak area ratio Primene JM-T™-TFA/I.S. in the sample, *a* is the slope of the calibration line, *b* is the *y*-intercept of the calibration line, *f* is the pre-concentration factor (1000 – 2000), and *R* is the average SPE yield of Primene JM-T™ from the water sample.

The total peak area ratio Primene JM-T™- TFA/I.S. in the sample (A'_{sample}) was calculated from the equation (3):

$$A'_{sample} = \left(A_{i\ sample} \cdot m_{I.S.\ sample}\right) / \left(A_{I.S.\ sample} \cdot m_{I.S.\ cal}\right) \quad (3)$$

where $A_{i\ sample}$ is the total peak area of all TFA derivatives of *tert*-octadecylamines in the sample, $A_{I.S.\ sample}$ is the peak area of the internal standard (I.S.) in the sample, and $m_{I.S.\ sample}$ and $m_{I.S.\ cal}$ are the masses of the internal standard in the sample and in the standard solution used for detector calibration, respectively.

Figure 4 shows the SPE-GC/FID chromatograms of the investigated water samples from the Power Plant Białystok (Poland). The concentrations of Primene JM-T™ determined as the sum of *tert*-octadecylamines in boiler water, condensate and superheated steam samples from the same power plant were 89 μg L^{-1} ($n = 5$, RSD = 7.8 %), 37 μg L^{-1} ($n = 5$, RSD = 2.3 %) and 45 μg L^{-1} ($n = 5$, RSD = 5.4 %), respectively.

4. Corrosion studies

Corrosion studies were performed at the Lithuanian Energy Institute in Kaunas [24]. The results obtained in experiments performed within a temperature range between 20 and 500 ^{0}C, show that the anticorrosion protection increases as temperature becomes higher [2]. In the presence of the new formulation *Kotamina Plus*, the corrosion rate was seven times lower than for non-inhibited water at 90 ^{0}C, and two and half times lower than that in water inhibited with sodium hexametaphosphate (Na $PO_3)_6$. It is important to note that *Kotamina Plus* appears to be over 30% more efficient than *Kotamina*. The corrosion rate at 500 ^{0}C for a boilers' steel equals 0.34 μm/year (see Figure 5) [24].

5. Industrial applications

Alkyl amines with branched structure are more thermally stable than linear alkyl amines, and exhibit better adsorption and effectiveness in surface shielding protection [2]. One of the basic features of high molecular mass alkyl amines as components of correction formulations is their influence on improvement of heat exchange in boiler and in turbine condensers. In order to determine this influence for *Kotamina Plus* appropriate investigations in the Engineering Department of Polish Academy of Sciences (PAN) were performed. Results of these experiments show that application of *Kotamina Plus* leads to increase of the convective heat transfer coefficient during boiling by 55% and to increase in heat flux by over 120%. Finally, heat flux during condensation is increased by over 120% in comparison to demineralised water [2]. This improves the efficiency of the entire power unit and lowers the temperature of boiler pipes, helping to avoid overheating.

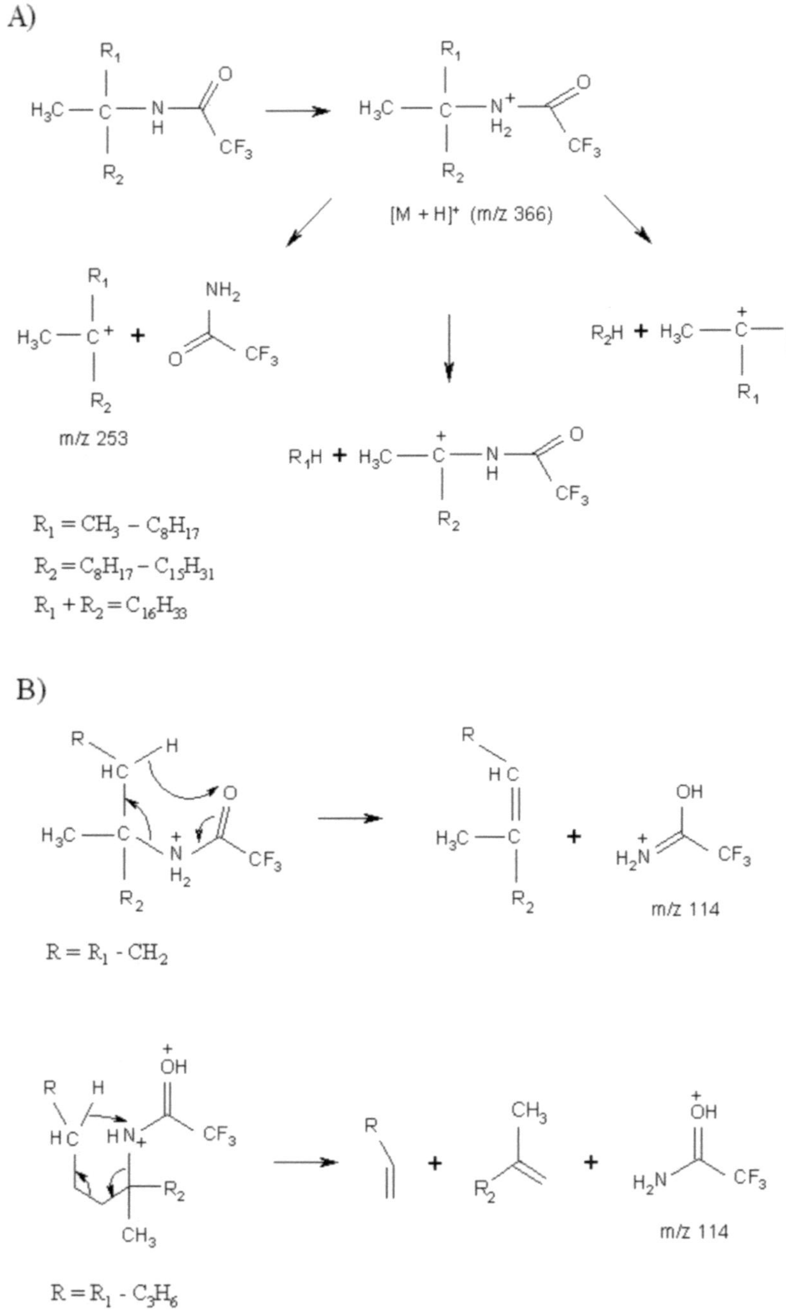

Figure 2. Proposed PCI-MS fragmentation pattern for investigated *N*-(*tert*-octadecyl)-trifluoroacetamides. A) Cleavage mechanism, B) McLafferty rearrangement [22, 23]. Publishing with licence number 3185330243225 from Elsevier.

In the industrial investigations performed in Polish power plants as well as in thermal-electric power plants, *Kotamina Plus* exhibited excellent efficiency in anti-corrosive protection assuring high purity of circulating medium, and therefore meeting requirements related to boilers exploitation. Experiments performed on a WT 230 boiler's circulating system showed that substitution of *Kotamina* by *Kotamina Plus* results in decrease of iron content by 60% in boiler water, by 30% in saturated and superheated steam, by 25% in condensate and by 40% in make-up water (Figure 6). As a result of *Kotamina Plus* application in WT 650 boiler, the total iron content in particular streams was decreased by over 10% and modification of formulation (*Kotamina Plus/P*) caused further decrease by ca. 30% in boiler water and from 15 to 25% in other streams (Figure 7). The ammonia content was significantly decreased in all streams to 1/3 of non-normalized value but the widely accepted upper limit is 500 μg/dm^3. This is important for protection against corrosion of brass elements, especially in turbine condensers. Low ammonia concentration positively influences stabilization of pH value in the system and considerable influences mating of boiler and sub-turbine heat exchangers.

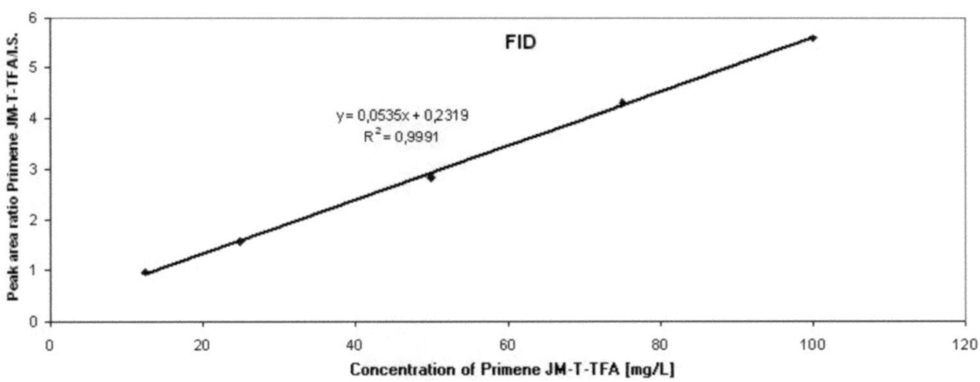

Figure 3. Total peak area ratio Primene JM-T™-TFA/I.S. *vs.* concentration of Primene JM-T™- TFA in the standard solutions for FID [20]. Publishing with licence number 3185340683087 from Elsevier.

Application of alkyl amine formulations instead of phosphate and hydrazine leads to considerable conductivity drop of the boiler water. *Kotamina Plus* introduction, as a result of decreased ammonia concentration, caused further conductivity drop in boiler water as well as in feed water. It allows to lower the desalination of the boiler and to save make-up water as well as energy needed for its heating. Lower ammonia concentration additionally stabilizes the pH value for particular streams. In WT 650 boilers for both blocks (B1 and B2) the pH was raised and for live steam and condensate pH was lowered. This is advantageous because it counteracts the wear of constructional material of a turbine condenser, which possesses brazen piping (Figure 8).

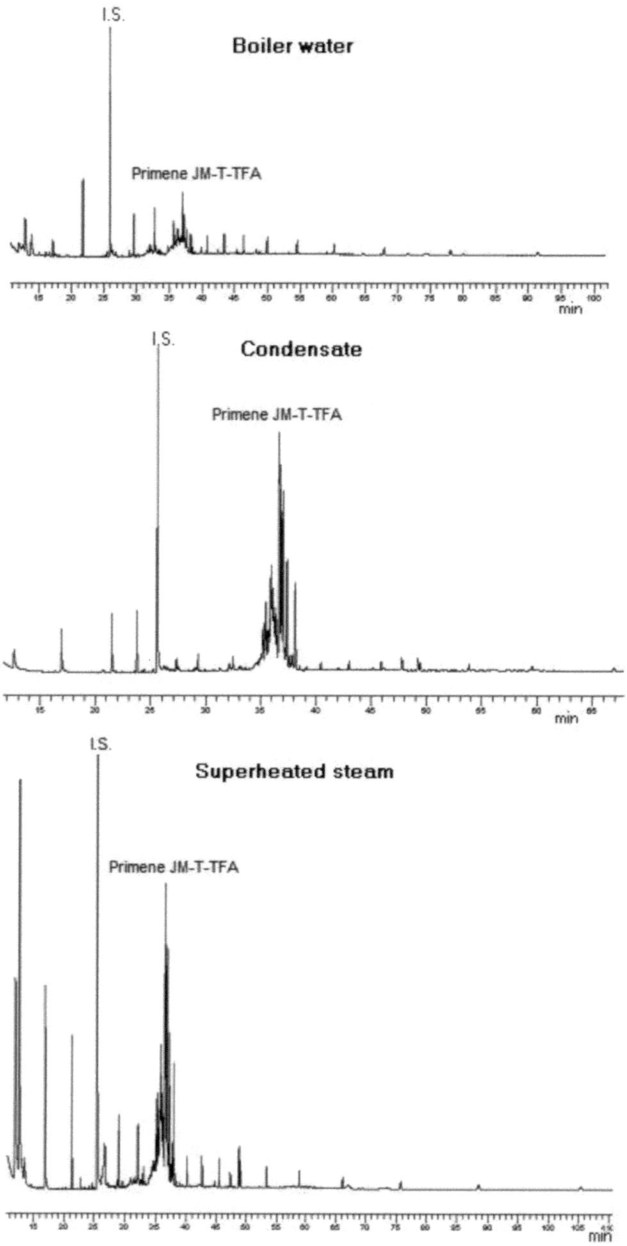

Figure 4. SPE-GC/FID chromatograms of boiler water (top), condensate (middle) and superheated steam (bottom) samples from Power Plant Białystok (Poland) [23]. Fused silica capillary column: *DB-5ms*, 60 m x 0.25 mm I.D., film thickness 0.25 µm. Column temperature programmed from 60 °C (1 min hold) at 6 °C min^{-1} to 280 °C (hold 50 min). Split/splitless injector: 290 °C. Helium constant pressure 140 kPa, split flow 20 cm^3 min^{-1}. FID: 320 °C. Peak identification: t_R = 25.46 min: dicyclohexylamine (internal standard, I.S.); t_R = 36.52 min: 8-methyl-8-heptadecaneamine-TFA; t_R = 36.58 min: 7-methyl-7-heptadecaneamine-TFA; t_R = 36.70 min: 6-methyl-6-hetadecaneamine-TFA; t_R = 36.87 min: 5-methyl-5-heptadecaneamine-TFA; t_R = 37.22 min: 4-methyl-4-heptadecaneamine-TFA; t_R = 37.95 min: 3-methyl-3-heptadecaneamine-TFA; t_R = 38.05 min: 2-methyl-2-heptadecaneamine-TFA.

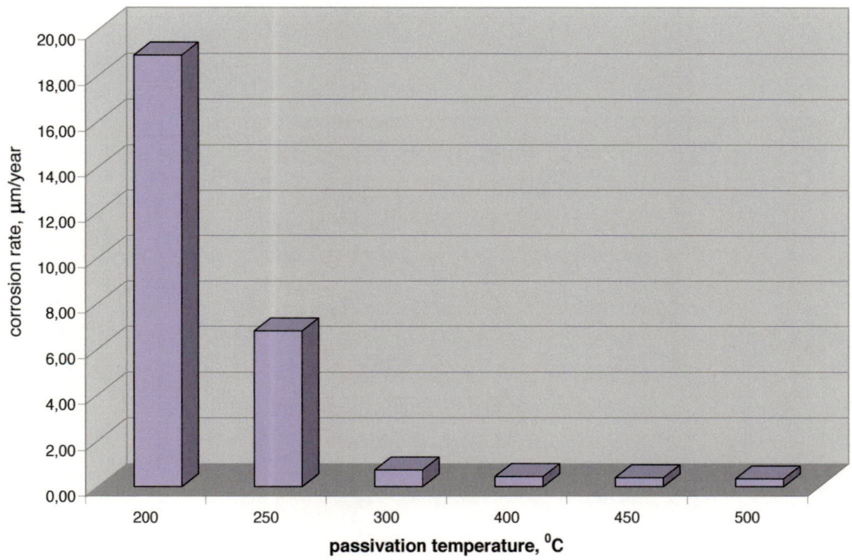

Figure 5. Dependence of corrosion rate on temperature for boilers' steel [2, 24].

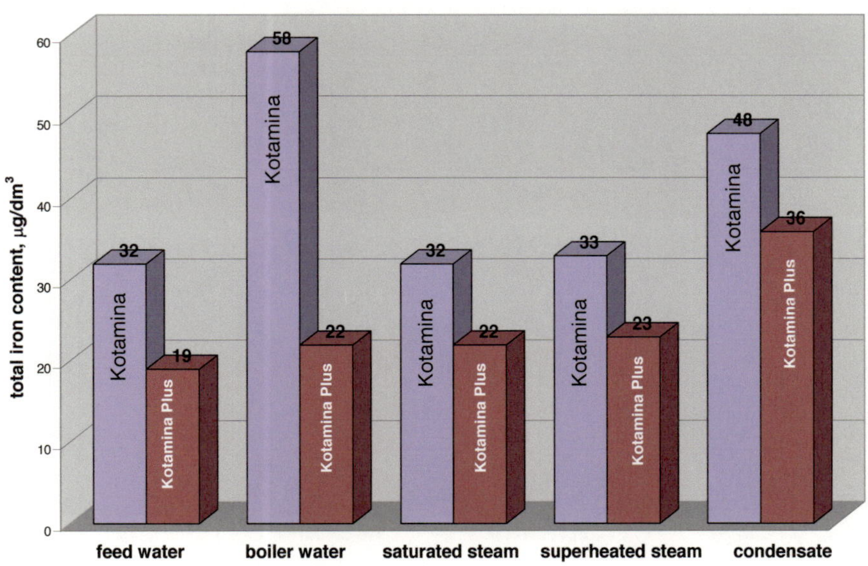

Figure 6. Total iron content in particular streams – WT 230 boiler [2, 24].

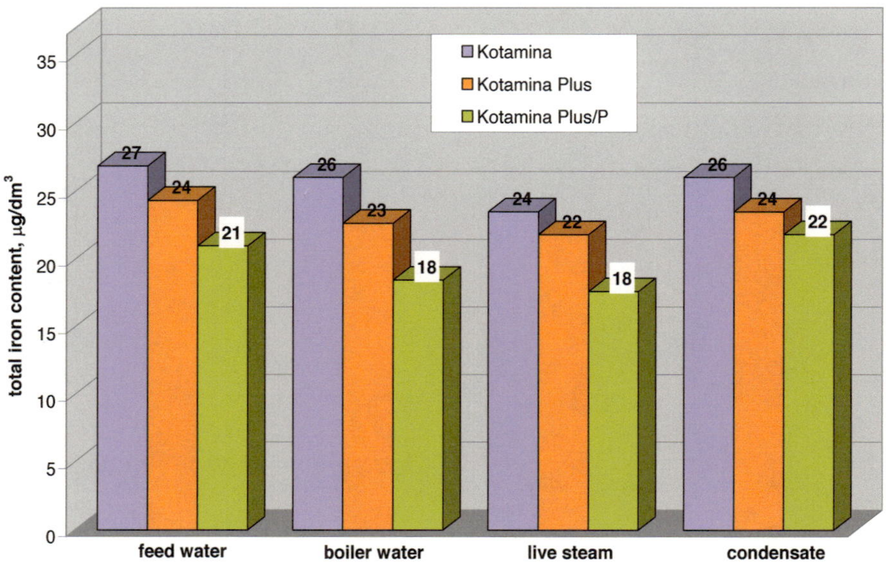

Figure 7. Total iron content in individual streams – average value [2, 24].

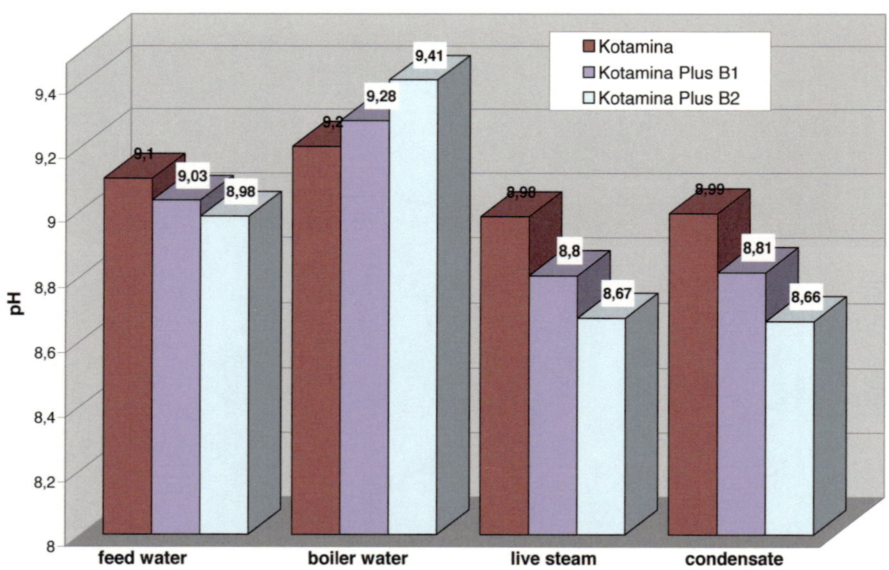

Figure 8. pH values in individual streams – average value [2].

6. Conclusions

Gas chromatography with flame-ionization detection (FID) and gas chromatography-mass spectrometry (GC/MS) with electron impact ionization (EI) and chemical ionization (PCI and NCI) were successfully used for separation and identification of commercially available long-chain primary alkyl amines. The investigated compounds were used as corrosion inhibiting and antifouling agents in a water-steam circuit of energy systems in the power industry. Solid-phase extraction (SPE) with octadecyl bonded silica (C_{18}) sorbents followed by gas chromatography were used for quantification of the investigated Primene JM-T™ alkyl amines in boiler water, condensate and superheated steam samples from the power plant.

Amine formulations from *Kotamina* group favor formation of protective layers on internal surfaces and keep them free from corrosion and scale. Alkyl amines contained in those formulations both render the environment alkaline and limit the corrosion impact of ionic and gaseous impurities by formation of protective layers. Moreover, alkyl amines limit scaling on heating surfaces of boilers and in turbine, ensuring failure-free operation. Application of alkyl amine formulation enhances heat exchange during boiling and condensation processes. Alkyl amines with branched structure are more thermally stable than linear alkyl amines, exhibit better adsorption and effectiveness of surface shielding. As a result, application of thermostable long-chain branched alkyl amines increases the efficiency of anti-corrosive protection. Moreover, the concentration of ammonia content in water and in steam was also considerably decreased.

Author details

Peter Kusch[1], Gerd Knupp[1], Marian Kozupa[2], Jolanta Iłowska[2] and Maria Majchrzak[2]

1 Bonn- Rhein-Sieg University of Applied Sciences, Department of Applied Natural Sciences, Rheinbach, Germany

2 Institute of Heavy Organic Synthesis "Blachownia", Kędzierzyn-Koźle, Poland

References

[1] EUREKA Success Story E!2426 BOILTREAT: Fighting Industrial Boiler Corrosion: http://www.eurekanetwork.org/showsuccessstory?p_r_p_564233524_articleId=105563&p_r_p_564233524_groupId=10137 (accessed 28 June 2013).

[2] Kozupa M., Sajewicz J. Application of Alkyl Amines as a Base for Anti-corrosive and Anti-Scaling Formulation. In: Demadis K. (ed.) Water Treatment Processes. New York: Nova Science Publishers; 2012. p. 105-115.

[3] Luong J., Shellie R.A., Cortes H., Gras R., Hayward T. Ultra-Trace Level Analysis of Morpholine, Cyclohexylamine, and Diethylaminoethanol in Steam Condensate by Gas Chromatography with Multi-Mode Inlet, and Flame Ionization Detection. Journal of Chromatography A 2012; 1229 223-229.

[4] Stiller K., Wittig T., Urschey M. The Analysis of Film-Forming Amines – Methods, Possibilities, Limits and Recommendations. PowerPlant Chemistry 2011; 13(10) 602-611.

[5] Kozupa M., Wpływ Alkiloamin na Adsorpcję i Wymianę Ciepła w Procesach Wrzenia i Kondensacji (Influence of Alkyl Amines on Adsorption and Heat Exchange in Boiling and Condensation Processes) (in Polish). Doktor-Thesis. Poznań University of Technology, Poznań, Poland; 2002.

[6] Twardowski S., Maciejko M., Kozupa M. Modern Amine Agents for Correction of the Working Medium in Power Unit Loops (in Polish). Energetyka 2000; (9) 75-79.

[7] Twardowski S., Kozupa M. Experience from 10-Year Use of Polish Amine Preparation for Correction of Water-Steam Cycles (in Polish). Energetyka 2002; (3) 155-159.

[8] http://www.eurekanetwork.org/project/-/id/2426 (accessed 2 July 2013).

[9] Jerzykiewicz W., Kozupa M., Twardowski S., Maciejko M. Corrosion and Scale Preventing Agent (in Polish). Polish Pat. 2007, PL 193491 B1 20070228.

[10] Kozupa M., Jerzykiewicz W., Majchrzak M., Haliniak B., Zdunek A., Twardowski S., Maciejko M., Szymonik M. Anti-corrosive and Anti-Scaling Agent (in Polish). Polish Pat. 2009, PL 202585 B1 20090731.

[11] Kozupa M., Jerzykiewicz W., Sajewicz J., Gniady J., Fiszer R., Twardowski S., Maciejko M., Szymonik M. Anticorrosion and Anti-Scaling Agent for Water System Equipment (in Polish). Polish Pat. 2010, PL 207236 B1 20101130.

[12] Coleman G. N., Endicott D. L., Jakush E. A., Nikolov A. Shear Mixing for Preparation of Stable Fuel Emulsions of Internal-Combustic Multiple In-line Blending Stations for Adding Additives and Water. US Pat. 2003, No. B1 6607566.

[13] Montgomerie H., Obeyesekere N. U. Corrosion Inhibitors with Low Environmental Toxicity. Eur. Pat. 2000, EP 1 043 423 A2.

[14] Darling D., Rakshpal R. Green Chemistry Applied to Corrosion and Scale Inhibitors. Materials Performance 1998; 37(12) 42-45.

[15] Migahed M.A., Abd-El-Raouf M., Al-Sabagh A. M., Abd-El-Bary H. M. Corrosion Inhibition of Carbon Steel in Acid Chloride Solution Using Ethoxylated Fatty Alkyl Amine Surfactants. Journal of Applied Electrochemistry 2006; 36 395-402.

[16] Abasq P. Usefulness of a Single Water-Conditioning Product for the Modern Operation of Boilers. Eau, l´Industrie, les Nuisances 1989; 132 63-64.

[17] Pirogov N. V., Levashova V. I., Oparina F. R., Nafikova E. V. Bactericidal and Corrosion-Inhibiting Activity of Quaternary Ammonium Salts Synthesized from Aliphatic Amines and Alkenyl Halides. Petroleum Chemistry 2013; 53(1) 65-69.

[18] Shcherban M. G., Chekanova L. G., Radushev A. V., Plotnikova M. D., Koltashev D. V., Nasrtdinova T. Yu. Colloidal-Chemical and Inhibiting Properties of N-(2-Hydroxyethyl)alkylamines. Russian Journal of Applied Chemistry 2012; 85(3) 385-390.

[19] Zaikin V., Halket J. A Handbook of Derivatives for Mass Spectrometry. Chichester: IM Publications LLP; 2009. p. 33-56.

[20] Kusch P., Knupp G., Hergarten M., Kozupa M., Majchrzak M. Solid-Phase Extraction– Gas Chromatography and Solid-Phase Extraction – Gas Chromatography - Mass Spectrometry Determination of Corrosion Inhibiting Long-Chain Primary Alkyl Amines in Chemical Treatment of Boiler Water in Water-Steam Systems of Power Plants. Journal of Chromatography A 2006; 1113(1-2) 198-205.

[21] Kusch P., Knupp G., Hergarten M., Kozupa M., Majchrzak M. Erratum to "Solid-Phase Extraction-Gas Chromatography and Solid-Phase Extraction - Gas Chromatography-Mass Spectrometry Determination of Corrosion Inhibiting Long-Chain Primary Alkyl Amines in Chemical Treatment of Boiler Water in Water-Steam Systems of Power Plants". Journal of Chromatography A 2009; 1216(38) 6671.

[22] Kusch P., Knupp G., Hergarten M., Kozupa M., Majchrzak M. Identification of Corrosion Inhibiting Long-Chain Primary Alkyl Amines by Gas Chromatography and Gas Chromatography-Mass Spectrometry. International Journal of Mass Spectrometry 2007; 263(1) 45-53.

[23] Kusch P., Knupp G., Kozupa M., Majchrzak M. Gas Chromatographic Separation and Mass Spectrometric Identification of Corrosion Inhibiting Long-Chain Primary Alkyl Amines and Alkyl Diamines Applied in Water Treatment in the Power Industry. In: Demadis K. (ed.) Water Treatment Processes. New York: Nova Science Publishers; 2012. p. 117-143.

[24] Minkevičius A., Levinskas R., Lukošiūte I., Kviklys A. Stainless 08X18H10T and Low-Alloyed 15X1M1Φ Steels Protection with Emulsions of Thermostable Amines. Medžiagotyra (Materials) 2003; 9(4) 342-346.

Chapter 19

Environmentally Friendly Corrosion Inhibitors

Rafael Martinez Palou, Octavio Olivares-Xomelt and
Natalya V. Likhanova

Additional information is available at the end of the chapter

http://dx.doi.org/10.5772/57252

1. Introduction

In most industries whose facilities are constituted by metallic structures, the phenomenon of corrosion is invariably present. This problem originates very important material and economic losses due to partial or total replacement of equipment and structures, and plant-repairing shutdowns.

Material losses and corrosion consequences are priced so high that in some countries like the U.S. and England these factors have been estimated from 3 to 4% of the GDP.

Corrosion not only has economic implications, but also social and these engage the safety and health of people either working in industries or living in nearby towns. The oil industry in Mexico is one of the most affected by corrosion because this phenomenon exerts its effects from the very moment of oil extraction on, causing a constant struggle against it.

The use of corrosion inhibitors (CIs) constitutes one of the most economical ways to mitigate the corrosion rate, protect metal surfaces against corrosion and preserve industrial facilities [1, 2].

Inorganic CIs are those in which the active substance is an inorganic compound. This is one of the simplest ways to improve the passivity of a metal by adding electropositive metal salts to the medium. These metal ions must have a more positive redox potential more positive than the metal constituting the surface to be protected and also a more positive potential than that required for discharging a proton so that the electropositive metal to be reduced is deposited on the surface.

The deposited metal promotes the cathodic depolarization by overvoltage reduction and formation of an adherent deposit. Among the metals used for this purpose are: mercury (Hg), palladium (Pd), iridium (Ir), platinum (Pt), rhodium (Rh) and rhenium (Re).

Moreover, there are inorganic anions providing passivation protection to metal surfaces through their incorporation into the oxide layer; the most widely used of these are: chromate (CrO_4^{2-}), nitrate (NO_2^-), molybdate (MoO_3^-), phosphate ($H_2PO_3^-$) and silicates [3].

Organic inhibitors have been the most widely used in petroleum refining processes because of their ability to form a protective layer on the metal surface in media with high hydrocarbons content. At present there are a number of organic inhibitors belonging to different chemical families i.e. fatty amides [4, 5], pyridines [6-8], imidazolines [9-12] and other 1,3-azoles [13-15] and polymers [16] have showed excellent performance as CIs (Table 1) [17].

Chemical family	Structure	Main application
Primary amines and diamines	Alkylamines (n = 2-12) $CH_3\text{-}(CH_2)_n\text{-}NH_2$ Diamines (n = 2-8) $H_2N\text{-}(CH_2)_n\text{-}NH_2$ Cycloalkylic 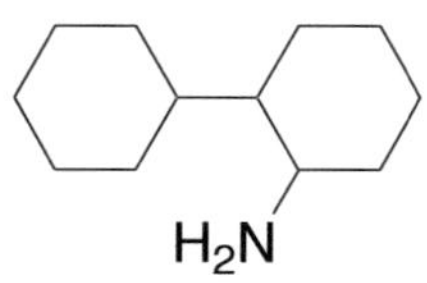 Aromatic (X = H, NO_2, CH_3, Cl, COOH) 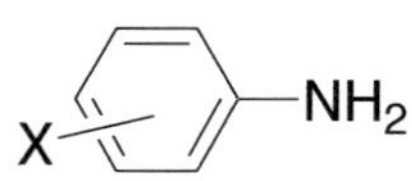	CIs for acid media
Secundary amines	Benzilamines 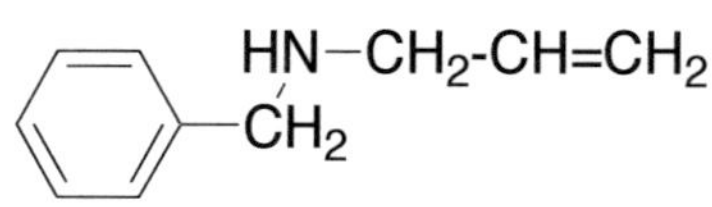 Etoxilated amines $CH_3\text{-}(CH_2)_n\text{-}NH\text{-}(OCH_2CH_2)_n$	CIs for carbon steel in acid media
Oximes	Alkyloximes 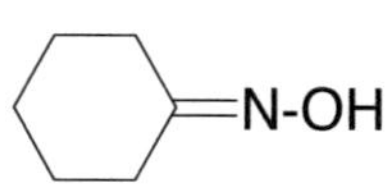 Aromatics 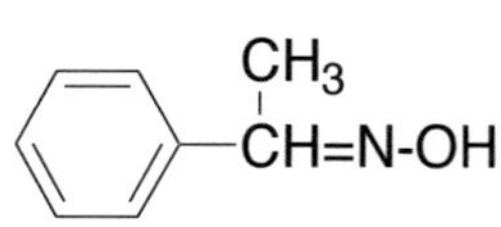	CIs for carbon steel in acid media

http://dx.doi.org/10.5772/57252

Chemical family	Structure	Main application
Nitriles	Alkylnitriles $C_{17}H_{35}$-CN Aromatics 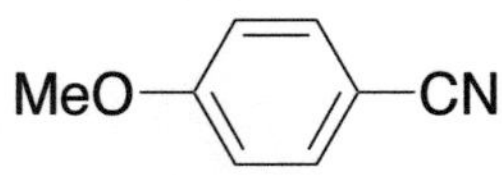	CIs for carbon steel in acid media
Ureas y thioureas	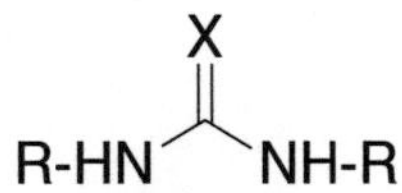 X= O, S, R = alkyl, aryl	CIs for copper alloys and carbon steel in acid media
Amides y thioamides	Amides 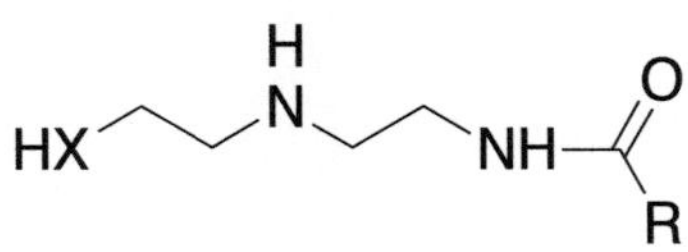 Thioamides 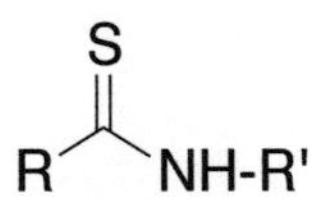 R, R' = alkyl	CIs for carbon steel in acid media
Imidazoles	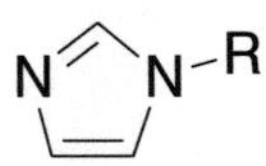 R = alkyl, aryl	CIs for copper alloys and carbon steel in basic media
Benzoazoles	 X = N-R, S, O	CIs for copper alloys and carbon steel in basic media
Imidazolines	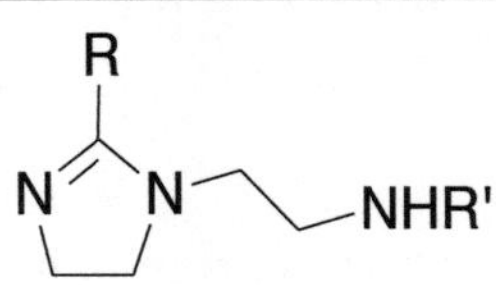 R = alkyl, aryl; X = NH_2, NHR, OH	CIs for carbon steel in acid media

Chemical family	Structure	Main application
Pyridines	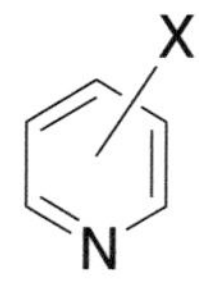 $X = CH_3$, Br, OR	CIs for carbon steel in acid media
Triazoles	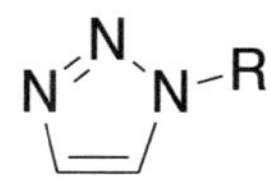 R = alkyl, aryl	CIs for copper alloys in basic media
Benzotriazoles	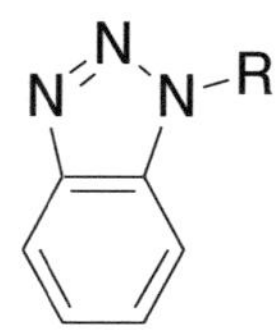 R = alkyl, aryl	CIs for copper alloys in basic media
Tetrazoles	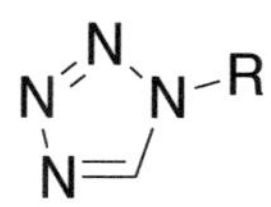 R = alkyl, aryl	CIs for copper alloys in basic media
Polyvinyls	$R\text{-}(\underset{R'}{C}H{=}CH)_n$ R, R′ = alkyl, aryl, heterocyclics	CIs for carbon steel in acid media
Polyesters	$R\text{-}(OCH_2CH_2)_n$ R = alkyl, aryl	CIs for carbon steel in acid media

Table 1. Organic corrosion inhibitors widely used in petroleum refining processes

The aim of adding inhibitors in low concentrations to corrosive media is to delay the reaction between the metal and the corrosive species in the medium. CIs act by adsorbing either ions or molecules onto the metal surface, generally reducing the corrosion rate by blocking the anodic and/or cathodic reactions.

In spite of much inorganic, organic and polymeric compounds have been showed good performances as CIs for different metals and alloys, many of these compounds are toxic and do not fulfill completely the requirements imposed by the environmental protection standards. The new generation of environmental regulations requires the replacement of toxic chemicals with the so-called "Green chemicals". The final choice of the inhibitor for a particular application is restricted by several factors, including increased environmental awareness and the need to promote environmentally friendly processes, coupled with the specificity of action of most acid inhibitors, which often requires the combined action of compounds to achieve effective corrosion inhibition. This is the reason why in the last years big efforts have been made by researchers in this area to develop new environmentally friendly CIs (EFCIs).

In this chapter, generalities about the corrosion phenomena and CIs are presented and a review of research papers describing the development of novel EFCIs, both natural and synthetic, for several corrosive environments and different metals and alloys are discussed in detail, especially for the applications in the Oil Industry.

2. Generalities about corrosion [2]

The term corrosion can be defined as the interaction (electrochemical reaction) of a metal with the surrounding environment, causing a slow, steady, and irreversible deterioration in the metal, in both physical and chemical properties.

The corrosion causes very important material and economical losses due to partial or total replacement of equipment and structures, and plant-repairing shutdowns.

Corrosion not only has economic implications, but also social and these engage the safety and health of people either working in industries or living in nearby towns. The petroleum industry is one of the most affected by corrosion due to the presence of many corrosive substances in the crude oil, which affect equipments and pipelines from the extraction of crude oil to the transportation of final products.

The factors that can cause corrosion can be identified as:

- Physical
- Chemical
- Electrochemical
- Microbiological

Physical corrosion is caused by impact, stress or exhaustion of the material. Chemical corrosion is caused by oxygen, sulfur, fluorine, chlorine or other gases, which act directly on the metal under environmental conditions that facilitate this phenomenon. Electrochemical corrosion is a spontaneous process that denotes the existence of anodic and cathodic zones, and an electrolyte; electrical contact between the anodic and cathodic zones is also required (Figure 1).

Microbiological corrosion is the deterioration of a metal that occurs directly or indirectly as a result of the activity of microorganisms such as bacteria and algae. These microorganisms are deposited on the metal, creating a "live" area, using nitrogen, oxygen, hydrogen, and/or carbon from the environment for their metabolic activities, producing metabolites, which can be deposited on the metal promoting corrosion. Biological activity may cause corrosion in a variety of media such as natural water, sea water, petroleum products and oil emulsions.

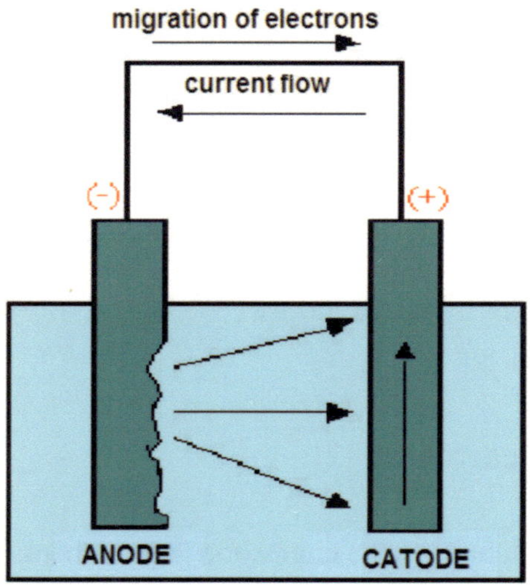

Figure 1. Representation of electrochemical corrosion.

According to the environment to which materials are exposed, there are various forms of corrosion: uniform or general, bite, erosion, stress, cavitation, galvanic and hydrogen embrittlement-blistering. Knowing how corrosion works helps to understand the phenomenon and provide possible solutions to counter the corrosive process.

a. Uniform or general corrosion is the most common, which is characterized by the fact that corrosion occurs uniformly over the metal surface and has a high corrosion rate; the loss of the metal surface occurs through an anodic site, and the appearance of the corroded surface is relatively uniform, but manifests roughness (Figure 2) [18].

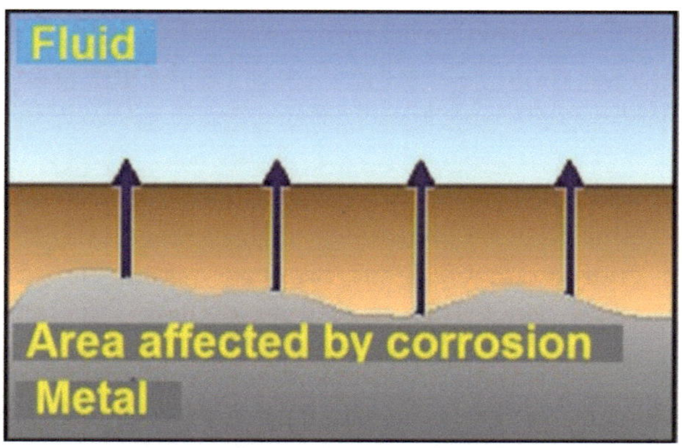

Figure 2. Uniform Corrosion

b. Pitting corrosion: Is a localized attack, where some parts of the metal surface are free of corrosion, but small localized areas are corrode quickly; this occurs when any solid

http://dx.doi.org/10.5772/57252

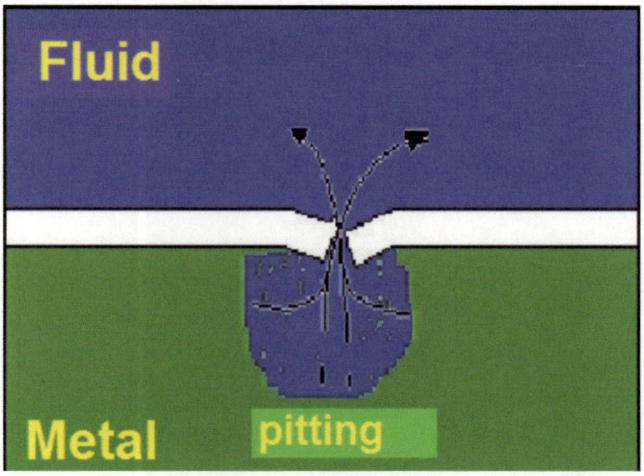

Figure 3. Pitting corrosion.

corrosion product or neutralization salts are located on the metal surface, causing deep holes, which is known as pitting (Figure 3); these areas are the most susceptible to the corrosion process [19].

c. Corrosion by erosion: This type of corrosion provokes uniform thinning of the metal surface, which is associated with the exposure to a high velocity fluid, which causes the corrosion product to be stripped from the metal surface, resulting in the exposure of the bare metal, which can be corroded again, causing an accelerated attack, (Figure 4). This type of corrosion is further exacerbated when fluids contain solid particles that are harder than the metal surface, which hit constantly the metal [20].

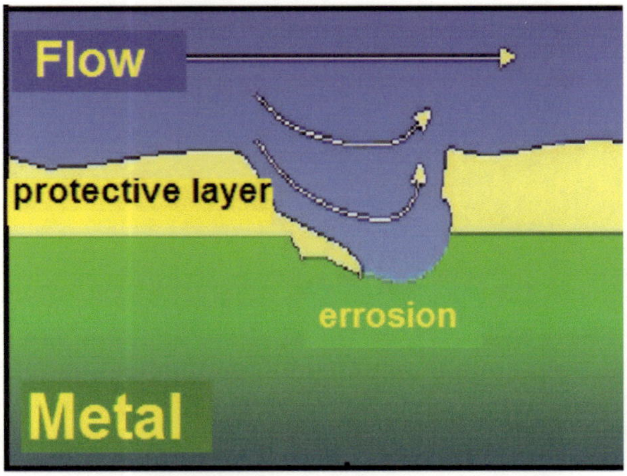

Figure 4. Corrosion by erosion

d. Stress corrosion cracking: This type of corrosion promotes the formation of a fracture in the metal structure due to mechanical stress and a chemically aggressive medium (Figure 5) [21].

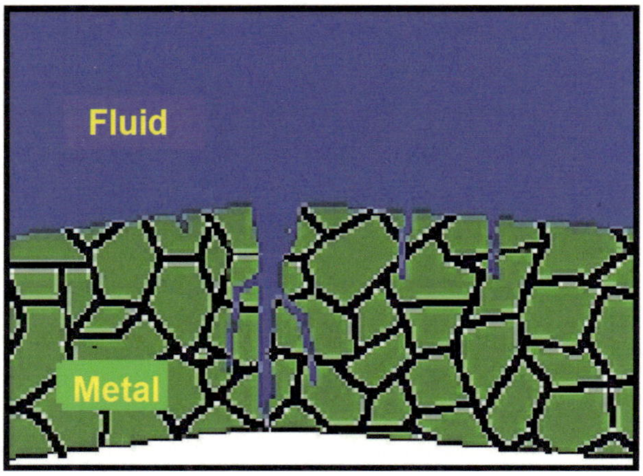

Figure 5. Stress corrosion cracking

e. Galvanic or bimetallic corrosion occurs when there is a potential difference between dissimilar metals immersed in a corrosive solution; the potential difference produces a flow of electrons between the metals, where the less resistant metal is the anode (metal active), and the most resistant is the cathode (noble metal). This attack can be extremely destructive, dramatically accelerating the corrosion rate of the most reactive metal, but the severity degree of galvanic corrosion depends not only on the potential difference between the two metals, but also on the involved surface area ratios, (Figure 6) [22].

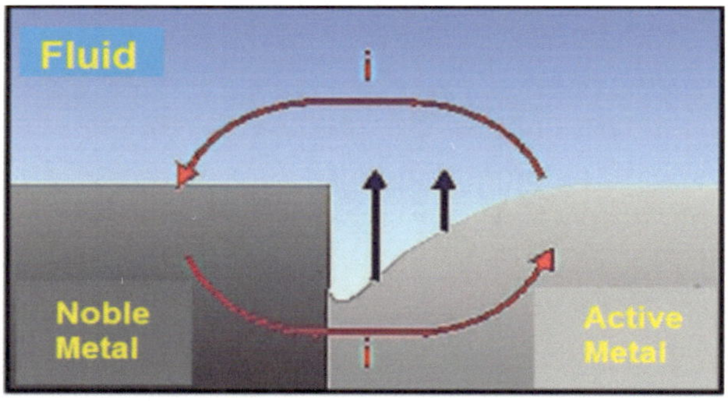

Figure 6. Galvanic corrosion

f. Corrosion by cavitation is a form of erosion caused by the formation and rupture of vapor bubbles in the fluid near the metal surface, causing a sequence of pits in the form of small, but deep cracks (Figure 7) [23].

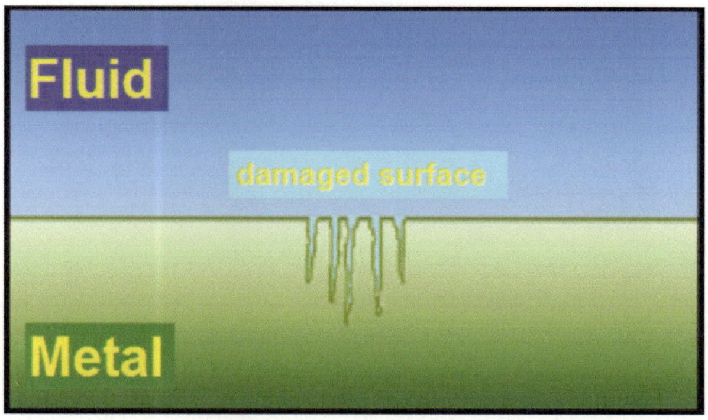

Figure 7. Corrosion by cavitation

g. Corrosion via hydrogen embrittlement and blistering is associated with the hydrogen atoms that are produced on the metal surface in an aqueous medium; a reduction reaction when atomic hydrogen penetrates the metal takes place; the presence of defects allow the interaction between the hydrogen atoms and the metal, forming molecular hydrogen, which being trapped by the metal, provides enough pressure to form blisters, resulting in microcracks, (Figure 8). This type of failure occurs mainly in basic media, where there are compounds such as sulfides and/or cyanides; this corrosion process is also present in plants with catalytic refining processes.

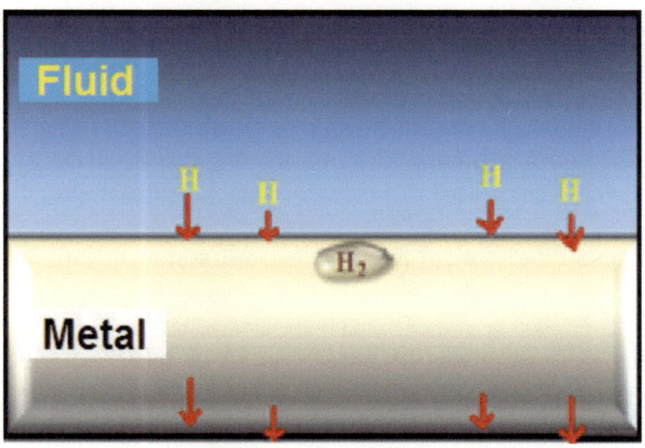

Figure 8. Corrosion by hydrogen embrittlement and blistering.

In this kind of corrosion process, some hydrogen atoms diffuse through steel and become retained, where they recombine with each other, forming a very strong internal pressure that exceeds the strength of steel, forming blisters.

In most oil refining plants, the reactive metal is iron, which is the major component of the steel present in pipelines and equipment; the electrolyte is water and the corrosive or oxidizing agent is formed by acids, salts, bases, oxygen, etc. One of the most common methods used to reduce corrosion in petroleum refining processes is the application of corrosion inhibitors, which are specific for each process phase, medium and corrosion type [24].

3. Corrosion control [25]

In order to control some of the corrosion problems, several preventive measures are taken:

a. Cathodic protection. This is an effective method to control corrosion on structures either buried or immersed in an electrolyte; according to the operation mode, anodes are classified as impressed current and sacrificial.

b. Protection with anticorrosive coating. This is mainly used to form a physical barrier between the corrosive environments to protect the structure. It is used mainly with metallic elements exposed to the atmosphere.

c. Corrosion Inhibitors. These are substances that added in small concentrations (parts per million, ppm) to a corrosive environment decrease the corrosion rate effectively. This method has its main application in the interiors of pipelines, vessels and equipments.

A corrosion inhibition program should be monitored continuously to ensure that it is achieving the desired protection.

The corrosion measurement is the quantitative method by which we know the effectiveness of the control that is being carried out, and provides feedback that makes possible to optimize the control and corrosion prevention methods.

Particularly in the Petroleum Industry, the monitoring can be done by using the following methods:

- Monitoring feedstocks by chemical analysis to find some of their features and corrosive contents.
- Monitoring corrosives by analysis of bitter waters of batteries (pH, chlorides, sulfides, ammonium thiocyanate and cyanide).
- Corrosion Monitoring: Be made in the following ways:

a. Using gravimetric coupons located at places where corrosion is to be measured (Figure 9).

b. With corrosimetric specimens. These probes are installed at the places to be monitored. A corrosometer connected to a probe detects a current amount and depending on it, it is known if there is corrosion and the communication speed.

c. Analyzing the iron and copper contents in the bitter waters of accumulators.

d. By placing hydrogen probes at the absorber tower.

http://dx.doi.org/10.5772/57252

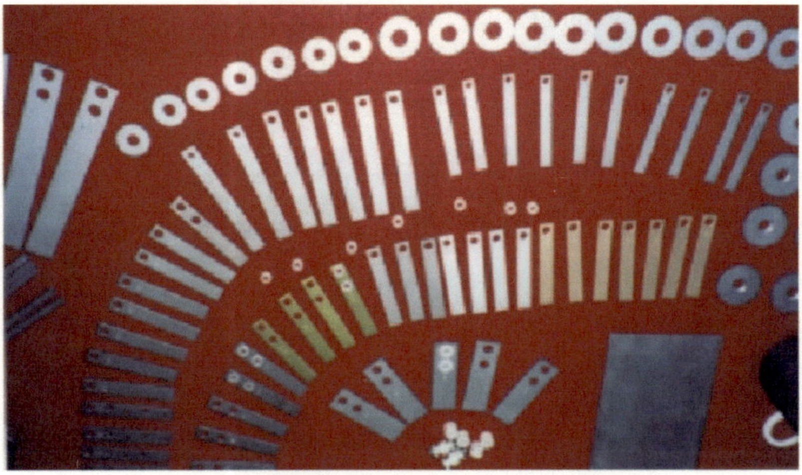

Figure 9. Coupons to measure corrosion.

4. Corrosion Inhibitors (CIs)

CIs are either organic or inorganic chemicals, or more commonly, formulations thereof that are added in small amounts (parts per million, ppm) to a corrosive environment in order to delay or decrease the corrosion process of the surface to be protected.

Due to the fact that equipment constructed with materials resistant to corrosion is very expensive, it is common to use corrosion inhibitors as a practical, economical and simple alternative.

A recent study in the United States indicated that their industries spent about $276 billion/year (on what?) and around 900 million/year on about 200 million tons of CIs. This market is shared by about 40% of inorganic inhibitors such as sulfonates and phosphonates (for cooling towers) and 60% of organic inhibitors, for example amines, cyclic amines, quaternary amidoamines, dietilamines, imidazolines and fatty acids, which are primarily used as CIs in the Petroleum Industry, in the production of gas, refineries, oil pipelines and products [26].

The CI formulations generally are made up of one or more active ingredients and suitable vehicles (other additives and solvents) that encourage compatibility with the environment and make viable the active transport to the area to be protected (metal surface).

The properties that must be met by a CI are [27, 28]:

a. Capability of reducing corrosion rates.

b. The active principle of the CI must be in contact with the metal to be protected.

c. Must not have side effects.

Sometimes, two components or active ingredients in a formulation may have a higher efficiency when they are mixed than that obtained from the sum of the efficiencies that are obtained when they are used individually at the same concentration. This effect is known as synergy or synergistic effect and is widely used in the formulation of CIs.

The CI can be classified in different ways [29, 30].

According to the specific application within the oil refining processes:

a. Embedding inhibitors.

b. Blistering inhibitors.

c. High temperature inhibitors.

d. Inhibitors for acidic media.

e. Inhibitors for basic media.

f. Inhibitors for cooling water.

The CI can also be classified according to the type of material to be protected. In the oil refining processes, CIs are of special interest for carbon steel, in which the major component is iron; and inhibitors for copper-zinc alloys (Admiralty), which are the most common materials used in the design of refineries.

CIs can be classified as anodic, which are those that inhibit oxidation of the metal; cathodic, which inhibit the reduction of oxygen; and mixed inhibitors, which inhibit both processes.

CIs can also be classified according to the type of compound that forms the active ingredient in the formulation as inorganic, organic and biocides.

5. Inhibitor mechanism

The action mechanisms of CIs are [31]:

- By adsorption, forming a film that is adsorbed onto the metal surface.
- By inducing the formation of corrosion products such as iron sulfide, which is a passivizing species.
- By changing media characteristics, producing precipitates that can be protective and eliminating or inactivating an aggressive constituent.

It is well known that organic molecules inhibit corrosion by adsorption, forming a barrier between the metal and the environment. Thus, the polar group of the molecule is directly attached to metal and the nonpolar end is oriented in a vertical direction to the metal surface, which repels corrosive species, thus establishing a barrier against chemical and electrochemical attack by fluids on the metallic surface (Figure 10).

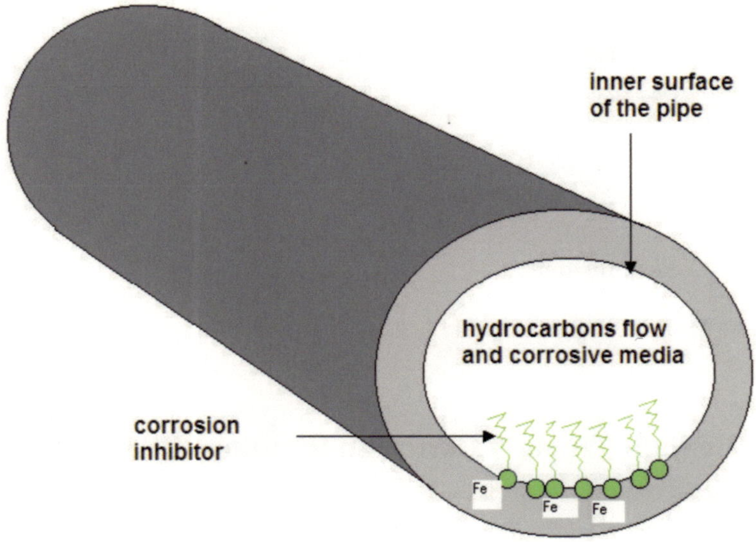

Figure 10. Representation of a CI adsorbed into a metal surface.

An inhibitor may be effective in one system, while in another it is not, (Table 1); therefore, it is convenient to consider the following factors:

- Chemical structure of the inhibitor component.
- Chemical composition of the corrosive medium.
- Nature of the metal surface.
- Operating conditions (temperature, pressure, pH, etc.).
- Thermal stability of the inhibitor. - Corrosion inhibitors have temperature limits above which lose their effectiveness because they suffer degradation of the containing components.
- Solubility of the inhibitor in the system. - The solubility of the inhibitor in the system is required to achieve optimum results in the metal surface protection; this depends on the length of the hydrocarbon chain.
- The addition of surfactants to enhance the dispersibility or solubility of inhibitors.
- Modification of the molecular structure of the inhibitor by ethoxylation to increase the polarity, and thus reach its solubility in the aqueous medium.

The main features of an inhibitor are:

- Ability to protect the metal surface.
- High activity to be used in small quantities (ppm).

- Low cost compound(s).
- Inert characteristics to avoid altering a process.
- Easy handling and storage.
- Preferably with low toxicity.
- Non-contaminant.
- It should act as an emulsifier.
- It should act as a foaming agent.

6. Environmentally Friendly Corrosion Inhibitors (EFCIs)

In recent years, owing to the growing interest and attention of the world towards the protection of the environment and the hazardous effects of using chemicals on the ecological balance, the traditional approach on CIs has gradually changed. As mentioned before, for an inhibitor to be an effective protector against metal corrosion, it should be readily adsorbed on the metal surface through either physisorption or chemisorption processes. Either of these adsorption processes depends primarily on the physicochemical properties of the inhibitor group such as functional groups, electronic density at the donor atom, molecular structure, etc. For instance, organic molecules, which have had a wide applicability and that have been extensively studied and used as CIs, often contain nitrogen, oxygen, and sulfur atoms, as well as multiple bonds in their molecules.

6.1. Evaluating the toxicity of CIs

Aspects to be taken into account in the development of CIs are their toxicity and impact on environmental pollution of both the active and other components of the formulation.

The European Economic Community assigned the Paris Commission (PARCOM) the task of providing guidance for environmental pollution control, protection of the ecosystems and the evaluation of the toxicity of raw materials and industrial waste products.

The PARCOM Environmental has developed a standardized test that covers three aspects:

1. Toxicity: This must be determined for the formulation as a whole.

Toxicity should be measured by using either the 50 Lethal Concentration (LC50), which is the concentration at which 50% of the test organisms are killed, or the EC50, which is the concentration that can cause an adverse organism affection, e.g. the concentration decreases the emission intensity of luminescent bacteria by 50% or the concentration decreases the growth or average weight of certain microorganisms by 50%.

The toxicity degree may be classified according to the LC50 value, where these categories are described in Table 2 [32].

CATEGORY	LC_{50}
Supertoxic	5 mg/kg of weight or less
Extremely toxic	5-50 mg/kg
Highly toxic	50-500 mg/kg
Moderately toxic	0.5-5 g/kg
Slightly toxic	5-15 g/kg
Practically non-toxic	More than 15 g/kg

Table 2. Classification of the toxicity of chemical compounds according to the LC50.

Toxicity testing for corrosion inhibitors to be measured in at least three different species and for the optimum established time, (Table 3).

Group	Preferred especies	Test
Algae	*Skeletonema costatum*	72-hour EC_{50}
Fish and crustaceans	*Acartia tonsa*	48-hour LC_{50}
Parasites	*Coropium voluntaros*	10-day LC_{50}

Table 3. Parameters for developing standardized toxicity tests.

2. Biodegradation: It must be determined for all the formulation components.

This test measures the persistence in the environment of the formulation components. The standard test that should be applied is the marine OECD. The allowable limit is more than 60% after 28 days.

3. Bioaccumulation: This test measures the level of product buildup in the body. Bioaccumulation is measured by the partition coefficient (eq. 1), as this parameter can be correlated with the cell interface/water ratio.

$$\mathrm{Po/w} = \frac{\text{concentration in octanol}}{\text{concentration in water}} \tag{1}$$

This means that the greater the partition coefficient, the more likely it is that the compound passes through the cell membrane, being bioaccumulated.

EFCIs can be arbitrarily divided into two categories: natural products and low toxicity synthetic products. In this last category, special attention has been paid to a new class of low toxicity organic compounds known as ionic liquids.

There are a few studies where CIs are evaluated according to the methodology described in this section and designed as low toxicity CIs or EFCIs. Most of the inhibitors that receive this rating are based on products that are derived from natural sources that are considered as compatible, biodegradable or environmentally friendly, although strictly, their toxicity has not been assessed by following the testing protocol presented above.

In the next sections, a brief overview of recent research works on the study of EFCIs with particular interest in those with potential applications in the Petroleum Industry is given.

6.2. Natural products as EFCIs

Natural products have been studied extensively as corrosion inhibitors both in product mixtures extracted from natural sources such as plants or essentially pure products derived from animals or plants (i.e. vitamins and aminoacids).

From the economic and environmental view points, plant extracts are an excellent alternative as inhibitors because of their availability and biodegradability. These extracts can be obtained in a simple way and purification methods are not required. The extracts are generally obtained from cheap solvents that are widely available, at a low cost and with low toxicity; the aqueous extract is more relieved, but due to the low solubility of many natural products in water, common ethanol extracts are also obtained. These extracts contain a variety of natural products such as essential oils, tannins, pigments, steroids, terpenes, flavones and flavonoids, among other well-known active substances used as CIs. In general, these compounds present conjugated aromatic structures, long aliphatic chains such as nitrogen, sulfur, and oxygen heteroatoms with free electron pairs that are available to form bonds with the metal surface; in most cases, they act synergistically to exhibit good efficiency regarding the corrosion protection. This can be demonstrated in the case of *Ginkgo biloba* in which the main components (flavonoids and terpenoids) have been identified (Figure 11). This extract has demonstrated excellent efficiency as CI with potential applications in the Oil Industry concerning the corrosion inhibition of Q235A steel. The antibacterial activity of the extracts against oil field microorganisms (SRB, IB and TGB) has also been proved [33].

The main disadvantage of using plant materials as CIs is their frequently low stability, they are readily biodegradable; however, this disadvantage can be minimized or avoided by adding biocides such as *N*-cetyl-*N,N,N*-trimethyl ammonium bromide.

In the last years, Umoren and Obot's research group has published several papers about the evaluations of plant extracts as CIs, for example, *Phyllanthus amarus* [34], *Pachylobus edulis* [35], *Raphia hookeri* [36], *Ipomoea involcrata* [37] and *Spondias mombin L.* [38].

Recently, this group described the inhibitive action of ethanolic extracts from leaves of *Chlomolaena Odorata L.* (LECO) as eco-friendly CI of acid corrosion of aluminum in 2 M HCl, using hydrogen evolution and thermometric techniques [39]; and more recently, for corrosion of mild steel in H_2SO_4 solutions [40]. In this last paper, the obtained results showed that LECO functioned as a CI and its efficacy increased with the extract concentration, but decreased with temperature. At a concentration as low as 5 %v/v of the extract, the inhibitory efficiency reached about 95% at 303 K, and 89% at 333 K.

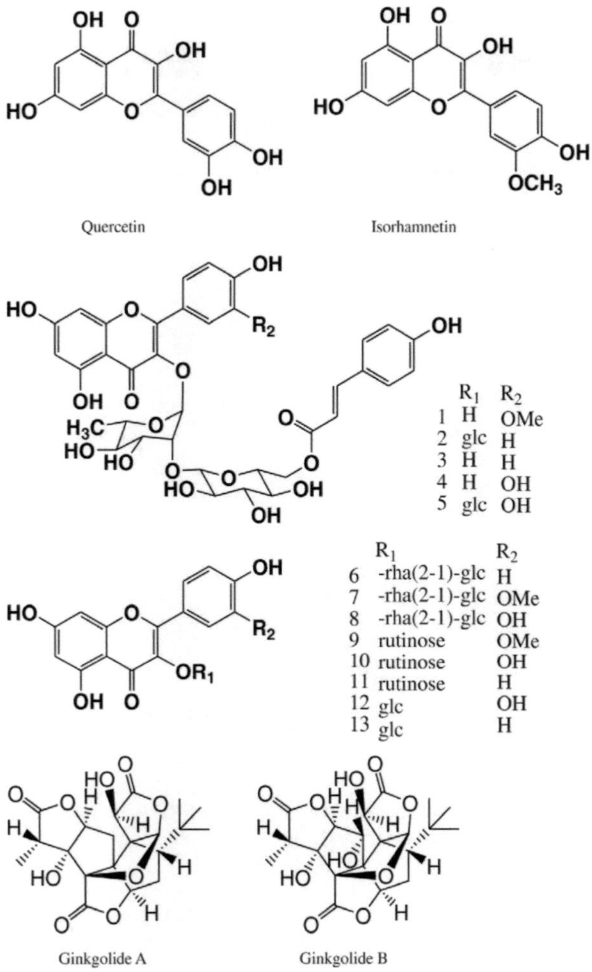

Figure 11. Structures of flavonoids and terpenoids found in *Ginkgo biloba*.

In another interesting work, this group showed the excellent inhibitory properties of Coconut coir dust extract (CCDE) as corrosion inhibitor of aluminum in 1 M HCl, using weight loss and hydrogen evolution techniques at 30 and 60°C by monitoring the volume of evolved hydrogen gas at fixed time intervals (Figure 12). The representative plots of the volume of the evolved hydrogen gas as a function of the reaction time at 30 and 60°C for Al in 1 M HCl, in the absence and presence of different concentrations of the CCDE, showed a remarkable increase in the volume of evolved H_2 gas in the blank acid solution at both studied temperatures. As for the introduction of CCDE into the corrosive medium, it is seen that there is a considerable reduction in the volume of evolved hydrogen gas, suggesting that the CCDE components were adsorbed onto the metal surface, and blocked the electrochemical reaction efficiently by decreasing the available surface area [41].

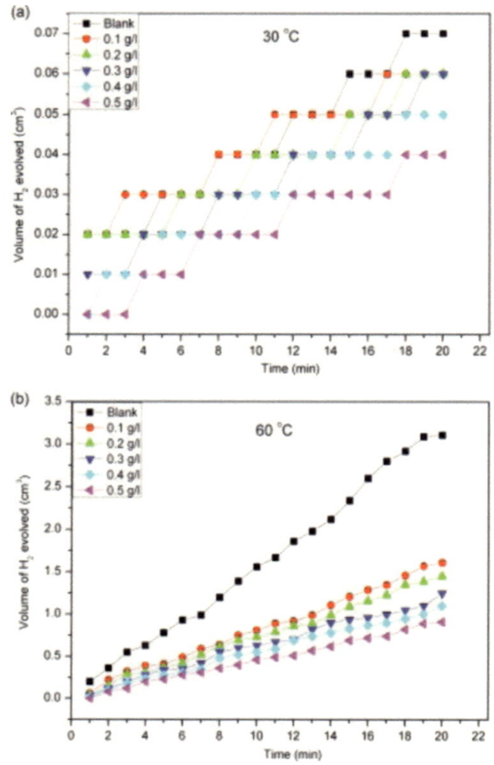

Figure 12. Plot of evolved hydrogen volume against time for Al in 1 M HCl with and without different CCDE concentrations at (a) 30°C and (b) 60°C. Reprinted from ref. [41].

Table 4 summarizes examples of natural extracts that have been evaluated in recent years as CIs [42, 43].

In the category of natural isolated products, aminoacids and their derivatives are some of the most studied pure compounds as EFCI. These natural compounds and derivatives have been used as good CIs for mild steel [75-80], carbon steel [81-83], stainless steel [84], iron [85], nickel [86], copper [87-91], aluminum [92], and alloys [93] and [94] in different aggressive solutions.

Very recently, the inhibitory properties of l-histidine on the corrosion of carbon steel in weak acid media containing acetic acid/sodium acetat have been tested. The inhibition efficiencies obtained by weight loss measurements are in good agreement with values given by the Tafel method and electrochemical impedance spectroscopy. The adsorption of l-histidine obeys the Langmuir isotherm; the negative values of the Gibbs energy indicate the nature of the interactions between the inhibitor molecules and metal surface. Further, the inhibition effect was studied by using scanning electron microscopy and energy dispersive X-ray analysis [95].

Other low-toxicity natural products including natural polymers described as CIs are summarized in Table 5.

Reference	Natural product source	Metal or Alloy protected
[44]	Polysaccharide (galactomannan) extracted from the endosperms of some leguminosae plants as coffee beans and chickpeas.	Concrete Armor Protection
[45]	*Chenopodium ambrorsioides* extract	Carbon steel
[46]	Ethanolic Extracts from Seeds of *Garcinia kola*	Mild steel
[47]	*Musa acuminata* flower extract	Mild steel
[48]	*Cannabis* plant extract	Copper
[49]	Mexican Argemona plant extract	Mild steel
[50]	*Lavandula dentata* aqueous extract	Mild steel
[51]	*Aframomum melegueta* extracts	Mild steel
[52]	*Osmanthus fragran* leaves extract	Carbon steel
[53]	*Neolamarckia cadamba* extract (bark, leaves) and pure alkaloid (3 beta-isodihydrocadambine)	Mild steel
[54]	*Murraya koenigii* leaves extract	Mild steel
[55]	Berberine extracted from *Coptis chinensis*	Mild steel
[56]	*Hibiscus sabdariffa* extract	Mild steel
[57]	Artemisia oil	Steel
[58]	Ethanolic extract of *Musa* species	Mild steel
[59]	Aqueous extract of *Hibiscus rosa-sinensis Linn*	Carbon steel
[60, 61]	Opuntia-Ficus-Indica (Nopal)	Aluminum and Steel
[62]	Tobacco extract	Steel and Aluminum
[63]	Aqueous extract of rhizome powder (*Curcuma longa L*)	Carbon steel
[64]	Onion juice	Zinc
[65]	Vernonia amygdalina	Aluminum
[66]	Mangrove tannin	Copper
[67]	*Punica granatum* extract	Brass
[68]	*Phoenix dactylifera L.* fruit juice	Aluminum
[69]	Rain water containing garlic extract	Aluminum
[70]	Polyphenols extracted from olive mill wastewater	Carbon steel
[71]	*Morinda tinctoria* leaves extract	Mild steel
[72]	Aqueous extract of Creosote Bush (*Larrea tridentata*) leaves	Carbon steel
[73]	Prosopis Laevigata	Aluminum
[74]	*Lanvandula stoekas* leaves extract	Stainless steel

Table 4. Extracts of natural products described as CIs.

Reference	Natural product	Metal or Alloy protected
[96]	caffeine	Carbon steel
[97]	caffeine	Copper
[98]	Purine and adenine	Copper
[99]	Vitamin B_1	AISI 4130 steel alloy
[100]	Vitamin B_1	Copper
[101, 102]	Vitamin C	Steel
[103]	Pteroyl-L-glutamic acid (Vitamin M)	Scale inhibitor for oil wells of carbonate reservoir
[104]	Citric acid	Aluminum
[105]	Benzoic acid	Iron and aluminum
[106]	Vitamin B_1, B_6 and C	Nickel
[107]	Peptin	Aluminum
[108]	lignin terpolymer	Aluminum
[109]	cassava starches	Aluminum
[110]	Carrageenan (polysacharide polymer)	Aluminum
[111]	Chitosan	Steel

Table 5. Natural products, including natural polymers described as CIs.

6.3. Synthetic compounds as EFCIs

Numerous inorganic and organic compounds have been reported as CIs for metals in different media, but the toxic nature of many of them limits their application. In the last decades, much attention has been focused on the need to design and develop synthetic non-toxic corrosion organic inhibitor to replace toxic ones for a sustainable development, for example, 12-aminododecanoic acid has been described as "green" CI for carbon steel [112], while imidazole derivatives [113] and guanidine [114] have been reported as non-toxic CIs for copper in acid media.

Several human drugs, including diuretics and barbiturates have shown good performance as CIs of metals in acid media [115-128]. In this sense, two well-known diuretics, Furosemide and Torsemide were recently evaluated as CIs of mild steel in hydrochloric acid medium. From the two inhibitors, Torsemide performance is superior to that of Furosemide due to its high electron density, which favors its adsorption on the metal surface. Polarization studies revealed that the inhibiting action of the compounds is under mixed control. The free adsorption energy and the temperature influence on the adsorption of inhibitors onto a mild steel surface have been reported. The adsorption of the compounds was found to obey the Langmuir adsorption isotherm. The inhibition and formation mechanisms of the Fe–inhibitor complex were confirmed by FT-IR and UV–visible absorption spectral analysis. The scanning electron microscopy (SEM) and atomic force microscopy (AFM) results established the formation of a protective layer on the mild steel surface. Quantum chemical calculations were applied to correlate the inhibition performance of inhibitors with their electronic structural parameters [129].

In 2011 the application of drugs as promising novel EFCIs was reviewed [130].

Several synthetic polymers have also been designed as efficient EFCIs for carbon steel in alkaline solution [131] and for calcium carbonate scale inhibitor for cooling water [132, 133].

Ionic liquids (ILs) deserve particular attention due to the rapid grown in the number of applications in the Oil Industry [134], and particularly within the topic of synthetic EFCIs, these have shown an effective performance as inhibitors of various metals and alloys [135-151].

ILs are ionic compounds, showing anisotropic molecular shape; their structure contains both organic and inorganic type components with various functional groups. They offer novel physical and chemical properties like low toxicity, high chemical stability, low vapour pressure and high electrical conductivity [152-157].

Quite a number of ILs do behave not only as green solvents suitable for the electrochemical devices and methods but also as unique and robust electrolytes with high stability Also, these compounds present structure properties suitable to be absorbed on metal surfaces and some derivatives of these families have proved that they can act as EFCIs in acid and basic corrosive environments.

Likhanova et al. have published a paper about the inhibitory action of 1,3-dioctadecylimidazolium bromide ($ImDC_{18}Br$), *N*-octadecylpyridinium bromide ($PyC_{18}Br$) in 1 M H_2SO_4 on mild steel at room temperature. The effect of the concentration of inhibitor compounds was investigated by electrochemical tests, whereas the surface analysis was performed at 100 ppm for both compounds. In the case of $ImDC_{18}Br$, corrosion products were additionally studied by X-ray diffraction and Mössbauer spectroscopy. The results revealed that ILs act as corrosion inhibitors with 82-88% at 100 ppm to protect the mild steel corrosion in the aqueous solution of sulfuric acid; their efficiencies are increased with the inhibitor concentration in the range 10–100 ppm range. SEM-EDX, XRD and Mössbauer analysis indicated the presence of carbon species and iron sulfates in the presence of ILs; whereas corrosion products such as iron oxyhydroxides were present in the absence of the ILs; this behavior was described by the proposed corrosion inhibition mechanism [158].

The same research group published a related paper in 2011 where five imidazolium-type ILs containing N1-vynil and N3-long alkyl saturated chains as cation and bromide as anion were synthesized under microwave irradiation and evaluated as CIs for acid environment. Weight loss and electrochemical polarization techniques were used to test the inhibitory properties of these compounds in AISI 1018 carbon steel immersed in 1.0 M H_2SO_4. All the studied ILs showed inhibitory properties dependent on the chain length linked to N3. The highest efficiency of IL4 was confirmed by Scanning Electron Microscopy (SEM)/Energy-Dispersive X-ray spectroscopy (EDX) and Atomic Force Microscopy (AFM) images of film formation. SEM revealed revealed that the surface morphology was strongly damaged in the CI absence, but in the presence of 100 ppm of CI, damage was considerably diminished, which confirmed the high efficiency of 1-Vinyl-3-octadecylimidazolium bromide at this concentration (Figure 13) [159].

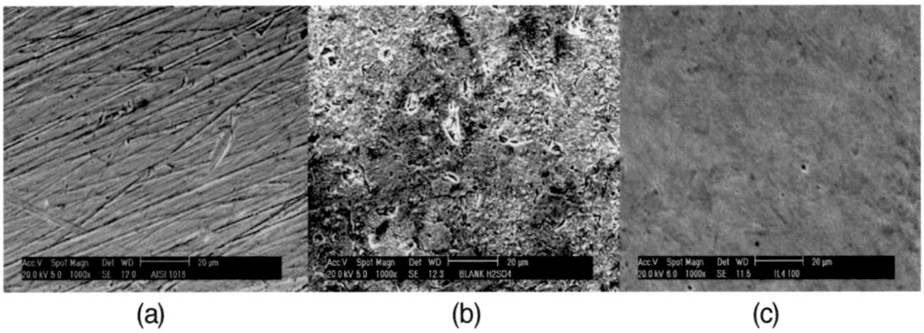

Figure 13. SEM images (1000X) of metallic surfaces: (a) after polishing, (b) after 6 hours of immersion in the corrosive medium without inhibitor, (c) after 6 hours of immersion in the corrosive medium with 100 ppm of 1-Vinyl-3-octadecylimidazolium bromide. Reprinted (adapted) with permission from ref. [159]. Copyright (2011) American Chemical Society.

ILs have also been employed to prepare a thin protective aluminum layer on carbon steel surface by electroreduction and electrodeposition of 1-butyl-3methyl-imidazolium chloroaluminate ($AlCl_3$/[BMIM]Cl) [160, 161].

Recently, the inhibition effect of 1-ethyl-3-methylimidazolium dicyanamide (EMID) against steel corrosion was investigated. In this study, EMID was evaluated as corrosion inhibitor for steel and then it was fixed in the polymer film. EMID is able to assemble a protective film on steel surface, under acidic conditions. The results of SEM analysis and potentiodynamic studies also showed that this inhibitor film is stable around corrosion potential. The steel surface becomes positively charged during inhibitor adsorption and the anionic part of EMID plays the major adsorption role. The inhibitor was fixed within polypyrrole coating on steel, and it was shown that this addition could improve the protection efficiency of the coating [162].

Junguroya et al. reported another recent interesting result. They found that water containing traces of hydrophobic ILs ([BMIM]Cl and [BMIM]NTf_2) exhibit unusual corrosion inhibiting behavior by protecting metal copper and nickel from electrochemical corrosion under aerobic conditions. The anodic dissolution of a copper electrode results in the formation of Cu (I) species. The simultaneous re-electrodeposition of nanocrystalline copper on the cathode occurred without additives to the resulting electrolyte [163]. Also, the high-temperature corrosion behavior of several metals (Ni, Cu, and alloys) in [BMIM]NTf_2 under aerobic conditions has been investigated by electrochemical methods [164].

The development of EFCIs based on organic rare earth compounds such as salycilates, phosphates, chromates and cinnamates was reviewed in 2011 [165].

7. Conclusions

As it has been seen through this chapter, corrosion inhibitors are economically feasible to mitigate the problems caused by corrosion. Environmental regulations in industrialized countries are increasing the pressure to eliminate, in the short term, a number of compounds widely used in industrial to prevent corrosion. A number of alternatives of EFCIs are currently emerging, oriented towards minimizing environmental impact providing effective corrosion inhibition. EFCIs include natural products, extracts from plants, and synthetic low-toxicity compounds. We hope that these products will be able to replace, in the near future, the toxic commercial products that are still being used by many industries worldwide.

Author details

Rafael Martinez Palou[1], Octavio Olivares-Xomelt[2] and Natalya V. Likhanova[1]

1 Instituto Mexicano del Petróleo, Dirección de Investigación y Posgrado, México, México

2 Benemérita Universidad Autónoma de Puebla, Facultad de Ingeniería Química, Puebla, México

References

[1] Sastri VS. Green Corrosion Inhibitors. Theory and Practice. John Wiley & Sons: Hoboken, NJ; 1998.

[2] Sastri VS. Corrosion Inhibitors Principles and Applications. John Wiley & Sons: New York; 1998.

[3] Bethencourt M. Lanthanide compounds as environmental friendly corrosion inhibitors of aluminium alloys: a review. Corrosion Science 1998;40(11) 1803-1819.

[4] Olivares-Xometl O, Likhanova NV, Gómez B, Navarrete J, Llanos-Serrano ME, Arce E, Hallen JM. Electrochemical and XPS studies of decylamides of alpha-amino acids adsorption on carbon steel in acidic environment. Applied Surface Science 2006;252(6) 2894-2909.

[5] Olivares-Xometl O, Likhanova NV, Domínguez-Aguilar MA, Arce E, Dorante H, Arellanes-Lozada P. Synthesis and corrosion inhibition of alpha-amino acids alkylamides for mild steel in acidic environment. Material Chemistry Physics 2008;110(2-3) 344-351.

[6] Abd El-Maksoud SA, Fouda AS. Some pyridine derivatives as corrosion inhibitors for carbon steel in acidic medium. Material Chemistry Physics 2005;93: 84-90.

[7] Ergun Ü, Yüzer D, Emregül KC. The inhibitory effect of bis-2,6-(3,5-dimethylpyrazolyl)pyridine on the corrosion behaviour of mild steel in HCl solution. Material Chemistry Physics. 2008;109: 492-499.

[8] Noor EA. Evaluation of inhibitive action of some quaternary N-heterocyclic compounds on the corrosion of Al–Cu alloy in hydrochloric acid. Material Chemistry Physics 2009;114: 533-541.

[9] Cruz J, Martínez-Palou R, Genesca J, García-Ochoa, E. Experimental and theoretical study of 1-(2-ethylamino)-2-methylimidazoline as an inhibitor of carbon steel corrosion in acid media. Journal of Electroanalytical Chemistry 2004;566(1) 111-121.

[10] Martinez-Palou R, Rivera J, Zepeda LG, Rodríguez AN, Hernandez MA, Marín-Cruz J, Estrada A. Evaluation of corrosion inhibitors synthesized from fatty acids and fatty alcohols isolated from sugar cane wax. Corrosion 2004;60(5) 465-470.

[11] Olivares-Xometl O, Likhanova NV, Martínez-Palou R, Dominguez-Aguilar MA. Electrochemistry and XPS study of an imidazoline as corrosion inhibitor of mild steel in an acidic environment. Materials and Corrosion 2009;60(1) 14-21.

[12] Liu FG, Du M, Zhang J, Qiu M. Electrochemical behavior of Q235 steel in saltwater saturated with carbon dioxide based on new imidazoline derivative inhibitor. Corrosion Science 2009;51(1) 102-109.

[13] Likhanova NV, Martínez-Palou R, Veloz MA, Matías DJ, Reyes-Cruz VE, Olivares-Xometl O. Microwave-assisted synthesis of 2-(2-pyridyl)azoles. Study of their corrosion inhibiting properties. Journal of Heterocyclic Chemistry 2007;44(1) 145-153.

[14] Popova A, Christov M, Zwetanova A. Effect of the molecular structure on the inhibitor properties of azoles on mild steel corrosion in 1 M hydrochloric acid. Corrosion Science 2007;49(5) 2131-2143.

[15] Antonijevic MM, Milic SM, Petrovic MB. Films formed on copper surface in chloride media in the presence of azoles. Corrosion Science 2009;51(6) 1228-1237.

[16] Tallman DE, Spinks G, Dominis A, Wallace GG. Electroactive conducting polymers for corrosion control Part 1. General introduction and a review of non-ferrous metals. Journal of Solid States Electrochemistry 2002;6(2) 73-84. b) Spinks G, Dominis A, Wallace GG, Tallman DE. Electroactive conducting polymers for corrosion control - Part 2. Ferrous metals. Journal of Solid States Electrochemistry 2002;6(2) 85-100.

[17] Chemical Inhibitors for Corrosion Control. The Royal Society of Chemistry; 1990.

[18] The Multimedia Corrosion Guide [CD-ROM]. INSA, Lyon; 2013.

[19] Marcus P, Maurice V, Strehblow HH. Localized corrosion (pitting): A model of passivity breakdown including the role of the oxide layer nanostructure. Corrosion Science 2008;50(9) 2698-2704.

[20] Levy AV. The erosion-corrosion behavior of protective coatings. Surface and Coatings Technology. 2002;36(1-2) 387-406.

[21] Sieradzki K, Newman RC. Stress-corrosion cracking. Journal of Physics and Chemistry of Solids. 1987;48(11) 1101-1113.

[22] Song G, Johannesson B, Hapugoda S, StJohn D. Galvanic corrosion of magnesium alloy AZ91D in contact with an aluminium alloy, steel and zinc. Corrosion Science 2004;46(4) 955-977.

[23] Al-Hashem A, Riad W. The role of microstructure of nickel–aluminium–bronze alloy on its cavitation corrosion behavior in natural seawater. Material Characterization 2002;48(1) 37-41. b) Neville A, McDougall BAB. Erosion-and cavitation-corrosion of titanium and its alloys. Wear 2001;250(1-12) 726-735.

[24] González JL, Ramirez R, Hallen JM, Guzman RA. Hydrogen-Induced Crack Growth Rate in Steel Plates Exposed to Sour Environments. Corrosion 1997;53(12) 935-943.

[25] Revie W, Uhlig HH. Corrosion and corrosion control: An introduction to corrosion science and engineering. Wiley-Interscience: New York; 2008. https://www.corrosioncost.com (accessed 1 august 2013), Ref. 2, p537-566.

[26] Distasio JI. Chemical for Oil Field Operations. Ed. Noyes Data Corp., New Jersey, E.U; 1981.

[27] Salensky G. Organic Corrosion Inhibitors. In: Salensky, G. (ed.) Handbook of Coatings Additives; 1987. p340-356. Ref. 2, p681-688.

[28] Corrosion Mechanisms in Theory and Practice, 3rd Ed. Marcus P (Ed.), Taylor and Francis Group, CRC, Boca Raton; 2012. Ref. 2, p885-894.

[29] Chen G, Zhang M, Zhao J, Zhou R, Meng Z, Zhang Z. Investigation of Ginkgo biloba leave extracts as corrosion and oil field microorganism inhibitors. Chemistry Central Journal 2013;7(Article Number: 83).

[30] Okafor PO, Ikpi ME, Uwah IE, Ebenso EE, Ekpe UJ, Umoren SA. Inhibitory action of Phyllanthus amarus extracts on the corrosion of mild steel in acidic media. Corrosion Science 2008;50(8) 2310-2317.

[31] Umoren SA, Ekanem UF. Inhibition of Mild Steel Corrosion in H2SO4 using Exudate Gum from Pachylobus Edulis and synergistic Potassium Halides Additives. Chemical Engineering Communications 2010;197(10) 1339-1356.

[32] Umoren SA,Obot IB, Ebenso EE,Obi-Egbedi NO. The Inhibition of aluminium corrosion in hydrochloric acid solution by exudate gum from Raphia hookeri. Desalination 2009; 247(1-3) 561-572.

[33] Obot IB, Obi-Egbedi NO, Umoren SA, Ebenso EE. Synergistic and Antagonistic Effects of Anions and Ipomoea invulcrata as Green Corrosion Inhibitor for Aluminium Dissolution in Acidic Medium. Internation Journal of Electrochemical Science 2010;5(7) 994-1007.

[34] Obi-Egbedi NO, Obot IB, Obot IB, Umoren SA. Spondias mombin L. as a green corrosion inhibitor for aluminium in sulphuric acid: Correlation between inhibitive effect

and electronic properties of extracts major constituents using density functional theory. Arabian Journal of Chemistry 2012;5(3) 361-373.

[35] Obot IB,Obi-Egbedi NO. An interesting and efficient green corrosion inhibitor for aluminium from extracts of Chlomolaena odorata L. in acidic solution. Journal of Applied Electrochemistry 2010;40(11) 1977-1984.

[36] Obot IB, Ebenso EE, Gasem ZM. Eco-friendly Corrosion Inhibitors: Adsorption and Inhibitive Action of Ethanol Extracts of Chlomolaena Odorata L. for the Corrosion of Mild Steel in H2SO4 Solutions. Journal of Electrochemical Science 2012;7(3) 1997-2008.

[37] Umoren SA, Eduok UM, Israel AU, Obot IB, Solomon MM. Coconut coir dust extract: a novel eco-friendly corrosion inhibitor for Al in HCl solutions. Green Chemistry Letters and Reviews 2012;5(3) 303-313.

[38] Rani BEA, Basu BBJ. Green inhibitors for corrosion protection of metals and alloys: An overview. International Journal of Corrosion 2012; articule ID: 380217.

[39] Rajendran S, Sri VG, Arockiaselvi J, Amalraj AJ. Corrosion inhibition by plant extracts - an overview. Bulletin of Electrochemistry 2005;21(8) 367-377.

[40] Lame A, Kokalari E, Jano A. Use of Green Inhibitors for Concrete Armor Protection Against H2SO4 Corrosion. Asian Journal of Chemistry 2013;25(7) 4017-4021.

[41] Belkhaouda, M.; Bammou, L.; Zarrouk, A, Salghi R, Ebenso EE, Zarrok H, Hammouti B, Bazzi L, Warad I. Inhibition of C-steel Corrosion in Hydrochloric Solution with Chenopodium Ambrorsioides Extract. International Journal of Electrochemical Science 2013;8(5) 7425-7436.

[42] Ikeuba AI, Okafor PC, Ekpe UJ, Ebenso EE. Alkaloid and Non-Alkaloid Ethanolic Extracts from Seeds of Garcinia Kola as Green Corrosion Inhibitors of Mild Steel in H2SO4 Solution. International Journal of Electrochemical Science 2013;8(5) 7455-7467.

[43] Gunavathy N, Murugavel SC. Studies on Corrosion Inhibition of Musa acuminata Flower Extract on Mild Steel in Acid Medium. Asian Journal of Chemistry 2013;25(5) 2483-2490.

[44] Abd-El-Nabey BA, Abdel-Gaber AM, Ali M. El. Said, Khamis E, El-Housseiny S. Inhibitive Action of Cannabis Plant Extract on the Corrosion of Copper in 0.5 M H2SO4. International Journal of Electrochemical Science 2013; 8(5) 7124-7137.

[45] Ji G, Shukla S, Dwivedi P, Sundaram S, Prakash R. Inhibitive Effect of Argemone mexicana Plant Extract on Acid Corrosion of Mild Steel. Industrial & Engineering Chemistry Research 2011;50(21) 11954-11959.

[46] Zarrok H, Zarrouk A, Salghi R, Ebn Touhami M, Oudda H, Hammouti B, Touir R, Bentiss F, Al-Deyab SS. The Anti-Corrosion Behavior of Lavandula dentata Aqueous

Extract on Mild Steel in 1M HCl. International Journal of Electrochemical Science 2013;8(4) 6005-6013.

[47] Oguzie EE, Iheabunike ZO, Oguzie KL, Ogukwe CE, Chidiebere MA, Enenebeaku CK, Akalezi CO. Corrosion Inhibiting Effect of Aframomum melegueta Extracts and Adsorption Characteristics of the Active Constituents on Mild Steel in Acidic Media. Journal of Dispersion Science and Technology 2013;34(4) 516-527.

[48] Li LJ, Zhang XP, Lei JL, He JX, Zhang ST, Pan FS. Adsorption and corrosion inhibition of Osmanthus fragran leaves extract on carbon steel. Corrosion Science 2012;63: 82-90.

[49] Raja PB, Qureshi AK, Rahim AA, Osman H, Awang K. Neolamarckia cadamba alkaloids as eco-friendly corrosion inhibitors for mild steel in 1 M HCl media. Corrosion Science 2013;69: 292-301.

[50] Quraishi MA, Singh A, Singh VK, Yadav DY, Singh AK. Green approach to corrosion inhibition of mild steel in hydrochloric acid and sulphuric acid solutions by the extract of Murraya koenigii leaves. Material Chemitry Physics 2010;122(1) 114-122.

[51] Li Y, Zhao P, Liang Q, Hou BR. Berberine as a natural source inhibitor for mild steel in 1 M H_2SO_4. Applied Surface Science 2005;252(5) 1245-1253.

[52] Oguzie EE. Corrosion inhibitive effect and adsorption behaviour of Hibiscus sabdariffa extract on mild steel in acidic media. Portugaliae Electrochimica Acta 2008;26(3) 303-314.

[53] Bouyanzer A, Hammouti B. A study of anti-corrosive effects of Artemisia oil on steel. Pigment and Resin Technology. 2004;33(5) 287-292.

[54] Eddy NO, Odoemelam SA, Odiongenyi AO. Ethanol extract of musa species peels as a green corrosion inhibitor for mild steel: Kinetics, adsorption and thermody-namic considerations. Electronic Journal of Environmental, Agricultural and Food Chemistry 2008;8(4) 243-255.

[55] Anuradha K, Vimala R, Narayanasamy B, Selvi J, Arockia RS. Corrosion inhibition of carbon steel in low chloride media by an aqueous extract of hibiscus rosa-sinensis Linn. Chemistry Engineering Communication 2008;195(3) 352-366.

[56] El-Etre AY. Inhibition of aluminum corrosion using Opuntia extract. Corrosion Science 2003;45(11) 2485-2495.

[57] Torres-Acosta AA. Opuntia-Ficus-Indica (Nopal) mucilage as a steel corrosion inhibitor in alkaline media. Journal of Applied Eclectrochemistry 2007;37(7) 835-841.

[58] Davis GD. The Use of Extracts of Tobacco Plants as Corrosion Inhibitors. http://www.electrochem.Org/dl/ma/ 202/pdfs/0340.PDF.

[59] Rajendran S, Shanmugapriya S, Rajalakshmi T, Raj AJA. Corrosion inhibition by an aqueous extract of rhizome powder. Corrosion 2005;61(7) 685-692.

[60] El-Etre AY. Natural onion juice as inhibitors for zinc corrosion. Bulletin of Electrochemistry 2006;22(2) 75-80.

[61] Popoola API, Fayomi OSI. Electrochemical Study of Zinc Plate in Acid Medium: Inhibitory Effect of Bitter Leaf (vernonia Amygdalina). International Journal of Electrochemistry Science 2011;6(10) 4581-4592.

[62] Shah AM, Rahim AA, Hamid SA, Yahya S. Green Inhibitors for Copper Corrosion by Mangrove Tannin. International Journal of Electrochemistry Science 2013;8(2) 2140-2153.

[63] Deepa PR, Selvaraj S. Inhibitive and adsorption properties of punica granatum extract on brass in acid media. Journal of Phytology. 2010;2(11) 58-64.

[64] Gerengi H. Anticorrosive Properties of Date Palm (Phoenix dactylifera L.) Fruit Juice on 7075 Type Aluminum Alloy in 3.5% NaCl Solution. Industrial & Engineering Chemistry Research 2012;51(39) 12835-12843.

[65] Priya SL, Chitra A, Rajendra S, Anuradha, K. Corrosion behaviour of aluminium in rain water containing garlic extract. Surface Engineering 2005;21(3) 229-231.

[66] Larif M, Elmidaoui A, Zarrouk A, Zarrok H, Salghi R, Hammouti B, Oudda H, Bentiss F. An investigation of carbon steel corrosion inhibition in hydrochloric acid medium by an environmentally friendly green inhibitor. Research on Chemical Intermediates 2013;39(6) 2663-2677.

[67] Krishnaveni K, Ravichandran J, Selvaraj A. Effect of Morinda Tinctoria Leaves Extract on the Corrosion Inhibition of Mild Steel in Acid Medium. Acta Metallurgica Sinica-English Letters 2013;26(3) 321-327.

[68] García-Inzunza R, Valdez-Salas B, Schorr Wiener M, Beltran M, Koytchev RZ, Stilianova M, Ramos-Irigoyen R, Vargas-Osuna L, Terrazas-Gaynor J. Aqueous Extract of Creosote Bush (Larrea tridentata) Leaves as Green Inhibitor for Carbon Steel in Hydrochloric Acid Solution. International Journal of Electrochemistry Science 2013;8(5) 6864-6877.

[69] Ramirez-Arteaga, M.; Valladares, M. G.; Gonzalez Rodriguez, J. G. Use of Prosopis Laevigata as a Corrosion Inhibitor for Al in H2SO4. International Journal of Electrochemistry Science 2013; 8(5) 6864-6877.

[70] Boudalia M, Guenbour A, Bellaouchou A, Laqhaili A, Mousaddak M, Hakiki A, Hammouti B, Ebenso EE. Corrosion Inhibition of Organic Oil Extract of Leaves Of Lanvandula Stoekas on Stainless Steel in Concentrated Phosphoric Acid International Journal of Electrochemistry Science 2013;8(5) 7414-7424.

[71] Fu J, Li S, Wang Y, Cao L, Lu L. Computational and electrochemical studies of some amino acid compounds as corrosion inhibitors for mild steel in hydrochloric acid solution. Journal of Material Science 2010;45(22) 6255-6265.

[72] Ashassi-Sorkhabi H, Majidi MR, Seyyedi K. Investigation of inhibition effect of some amino acids against steel corrosion in HCl solution. Applied Surface Science 2004;225(1-4) 176-185.

[73] Olivares-Xometl O, Likhanova NV, Dominguez-Aguilar MA, Arce E. Synthesis and corrosion inhibition of α-amino acids alkylamides for mild steel in acidic environment. Material Chemistry Physics 2008;110(2-3) 344-351.

[74] Morad MS. Corrosion inhibition of mild steel in sulfamic acid solution by S-containing amino acids. Journal of Applied Electrochemistry 2008;38(11) 1509-1518.

[75] Oezcan M. AC impedance measurements of cysteine adsorption at mild steel/sulphuric acid interface as corrosion inhibitor. J. Solid State Electrochemistry 2008;12(12) 1653-1661.

[76] Morad MS. Effect of amino acids containing sulfur on the corrosion of mild steel in phosphoric acid solutions containing Cl^-, F^- and Fe^{3+} ions: behaviour under polarization condition. Journal of Applied Electrochemistry 2005;35(9) 889-895.

[77] Olivares O, Likhanova NV, Gómez B, Navarrete J, Llanos-Serrano ME, Arce E, Hallen JM. Electrochemical and XPS studies of decylamides of α-amino acids adsorption on carbon steel in acidic environment. Applied Surface Science 2006;252(6) 2894-2909.

[78] Fu J, Li S, Cao L, Wang Y, Yan L, Lu L. L-Tryptophan as green corrosion inhibitor for low carbon steel in hydrochloric acid solution. Journal of Material Science 2010;45(4) 979-986.

[79] Cui R, Gu N, Li C. Polyaspartic acid as a green corrosion inhibitor for carbon steel. Materials and Corrosion-Werkstoffe und Korrosion 2011; 62(4) 362-369.

[80] Silva AB, Agostinho SML, Barcia OE, Cordeiro GGO, D'Elia E. The effect of cysteine on the corrosion of 304L stainless steel in sulphuric acid. Corrosion Science 2006;48(11) 3668-3674

[81] Amin MA, Khaled KF, Mohsen Q, Arida HA. A study of the inhibition of iron corrosion in HCl solutions by some amino acids. Corrosion Science 2010;52(5) 1684-1695.

[82] Gece G, Bilgic S. A theoretical study on the inhibition efficiencies of some amino acids as corrosion inhibitors of nickel. Corrosion Science 2010;52(10) 3435-3443.

[83] Moretti G, Guidi F. Tryptophan as copper corrosion inhibitor in 0.5 M aerated sulfuric acid. Corrosion Science 2002;44(9) 1995-2011.

[84] Zhang D, Gao L-X, Zhou G-D. Inhibition of copper corrosion in aerated hydrochloric acid solution by amino-acid compounds. Journal of Applied Electrochemistry 2005;35(11) 1081-1085.

[85] Radovanovic MB, Petrovic MB, Simonovic AT, Milic SM, Antonijevic MM. Cysteine as a green corrosion inhibitor for Cu37Zn brass in neutral and weakly alkaline sulphate solutions. Environmental Science and Pollution Research 2013;20(7) 4370-4381.

[86] Zhang DQ, Cai QR, He XM, Ga LX, Zhou GD. Inhibition effect of some amino acidsof copper corrosion in aerated hydrochloric acid solution by amino-acid compounds. Material Chemistry Physics 2008;112(2) 353-358.

[87] Barouni K, Bazzi L, Salghi R, Milhit M, Hammouti B, Albourine A, El Issami S. Some amino acids as corrosion inhibitors for copper in nitric acid solution. Material Letters 2008;62(19) 3325-3327.

[88] Ashassi-Sorkhabi H, Ghasemi Z, Seifzadeh D. The inhibition effect of some amino acids towards the corrosion of aluminum in 1 M HCl + 1 M H_2SO_4 solution. Applied Surface Science 2005;249: 408-418.

[89] Ghasemi Z, Tizpar A. The inhibition effect of some amino acids towards Pb–Sb–Se–As alloy corrosion in sulfuric acid solution. Applied Surface Science 2006;252(10) 3667-3672.

[90] Kiani MA, Mousavi MF, Ghasemi S, Shamsipur M, Kazemi SH. Inhibitory effect of some amino acids on corrosion of Pb–Ca–Sn alloy in sulfuric acid solution. Corrosion Science 2008;50, 1035-1045.

[91] Bobina M, Kellenberger A Millet J.-P., Cornelia M, Nicolae V. Corrosion resistance of carbon steel in weak acid solutions in the presence of L-histidine as corrosion inhibitor. Corrosion Science 2013;69: 389-395.

[92] Rajendran S, Amalraj AJ, Joice MJ, Anthony N, Trivedi DC, Sundaravadivelu M. Corrosion inhibition by the caffeine - Zn2+ system. Corrosion Reviews 2004;22: 233-248.

[93] Fallavena T, Antonow M, Goncalves RS. Caffeine as non-toxic corrosion inhibitor for copper in aqueous solutions of potassium nitrate. Applied Surface Science 2006;253(2) 566-571.

[94] Scendo M. Inhibition of copper corrosion in sodium nitrate solutions with nontoxic inhibitors. Corrosion Science 2008;50(6) 1584-1592.

[95] Hoseinzadeh AR, Danaee I, Maddahy MH. Thermodynamic and Adsorption Behaviour of Vitamin B1 as a Corrosion Inhibitor for AISI 4130 Steel Alloy in HCl Solution. International Journal of Research in Physical Chemistry & Chemical Physics 2013;227(4) 403-417.

[96] Abiola OK, John MO, Asekunowo PO, Okafor PC, James OO. 3-[(4-amino-2-methyl-5-pyrimidinyl) methyl]-5-(2-hydroxyethyl)-4-methyl thiazolium chloride hydrochloride as green corrosion inhibitor of copper in HNO3 solution and its adsorption characteristics. Green Chemistry Letters and Reviews 2011;4(3) 273-279.

[97] Fuchs-Godec R, Pavlovic MG, Tomic MV.The Inhibitive Effect of Vitamin-C on the Corrosive Performance of Steel in HCl Solutions. International Journal of Electrochemical Science 2013;8(1) 1511-1519.

[98] Valek L, Martinez S, Mikulic D Brnardic I. The inhibition activity of ascorbic acid towards corrosion of steel in alkaline media containing chloride ions. Corrosion Science 2008;50(9) 2705-2709.

[99] Kumar T, Vishwanatham S, Kundu SS. A laboratory study on pteroyl-L-glutamic acid as a scale prevention inhibitor of calcium carbonate in aqueous solution of synthetic produced water. Journal of Petroleum Science and Engineering 2010;71(1-2) 1-7.

[100] Solmaz R, Kardas G, Yazici B, Erbil M. Citric acid as natural corrosion inhibitor for aluminium protection. Corrosion Engineering Science and Technology. 2008;43(2) 186-191.

[101] Sibel Z. The effects of benzoic acid in chloride solutions on the corrosion of iron and aluminum. Turkish Journal of Chemistry 2002;26(3) 403-408.

[102] Malhotra S, Singh G. Vitamins: potential inhibitors for nickel in acidic media. Surface Engineering 2005;21(3) 187-192.

[103] Fare MM, Maayta AK, Al-Qudah MM. Pectin as promising green corrosion inhibitor of aluminum in hydrochloric acid solution. Corrosion Science 2012;60: 112-117.

[104] Ren Y, Luo Y, Zhang K, Zhu G, Tan X. Lignin terpolymer for corrosion inhibition of mild steel in 10% hydrochloric acid medium. Corrosion Science 2008;50: 3147-3153.

[105] Bello M, Ochoa N, Balsamo V, Lopez-Carrasquero F, Coll S, Monsalve A. González G. Modified cassava starches as corrosion inhibitors of carbon steel: An electrochemical and morphological approach Carbohydrate Polymers 2010;82(3) 561-568.

[106] Fares MM, Maayta AK, Al-Mustafa JA. Corrosion inhibition of iota-carragenan natural polymer on aluminum in presence of zwitterion mediator in HCl media. Corrosion Science 2012;65: 223-230.

[107] Zheludkevich ML, Tedim J, Freire CSR, Fernandes SCM, Kallip S, Lisenkov A, Gandini A, Ferreira MGS. Self-healing protective coatings with "green" chitosan based pre-layer reservoir of corrosion inhibitor. Journal of Material Chemistry 2011;21(13) 4805-4812.

[108] Ghareba S, Omanovic S. Interaction of 12-aminododecanoic acid with a carbon steel surface. Towards the development of 'green' corrosion inhibitors. Corrosion Science 2010;52(6) 2104-2113.

[109] Stupnisek-Lisac E, Gazivoda A, Madzarac M. Evaluation of non-toxic corrosion inhibitors for copper in sulphuric acid. Electrochimica Acta 2002;47(26) 4189-4194.

[110] Khaled KF. Adsorption and inhibitive properties of a new synthesized guanidine derivative on corrosion of copper in 0.5 M H2SO4. Applied Surface Science 2008;255(5) 1811-1888.

[111] Obot IB, Ebenso EE, Obi-Egbedi NO, Afolabi AS, Gasem, ZM. Experimental and theoretical investigations of adsorption characteristics of itraconazole as green corrosion

inhibitor at a mild steel/hydrochloric acid interface. Research on Chemical Intermediates 2012;38(8) 1761-1779.

[112] Ahamad I, Quraishi MA. Mebendazole: New and efficient corrosion inhibitor for mild steel in acid medium Corrosion Science 2010;52(2) 651-656.

[113] Shukla, S. K.; Singh, A. K.; Ahamad, I.; Quraishi, M. A.Streptomycin: A commercially available drug as corrosion inhibitor for mild steel in hydrochloric acid solution Mater. Lett. 2009;63(9-10) 819-822.

[114] Verma C, Quraishi MA, Ebenso EE. Electrochemical Studies of 2-amino-1, 9-dihydro-9-((2-hydroxyethoxy) methyl)-6H-purin-6-one as Green Corrosion Inhibitor for Mild Steel in 1.0 M Hydrochloric Acid Solution. International Journal of Electrochemical Science 2013;8(5) 7401-7413.

[115] Oezcan M, Solmaz R, Kardas G, Dehri, I. Adsorption properties of barbiturates as green corrosion inhibitors on mild steel in phosphoric acid. Colloids and Sufaces A. 2008;24(3-4) 57-63.

[116] Kardas G, Solmaz R. Electrochemical investigation of barbiturates as green corrosion inhibitors for mild steel protection. Corrosion Reviews 2006;24(3-4) 151-171.

[117] Prabhu RA, Shanbhag AV, Venkatesha TV. Influence of tramadol [2-[(dimethylamino)methyl]-1-(3-methoxyphenyl)cyclohexanolhydrate] on corrosion inhibition of mild steel in acidic media. Journal of Applied Electrochemistry 2007;37(4) 491-497.

[118] El-Naggar M. Corrosion inhibition of mild steel in acidic medium by some sulfa drugs compounds. Corrosion Science 2007;49(5) 2226-2236.

[119] Morad MS. Inhibition of iron, corrosion in acid solutions by Cefatrexyl: Behaviour near and at the corrosion potential Corrosion Science 2008;50(2) 436-448.

[120] Singh AK, Quraishi MA. Effect of Cefazolin on the corrosion of mild steel in HCl solution. Corrosion Science 2010;52(1) 152-160.

[121] Shukla, SK, Quraishi MA. Cefotaxime sodium: A new and efficient corrosion inhibitor for mild steel in hydrochloric acid solution Corrosion Science 2009;51(5) 1007-1011.

[122] Obot IB, Obi-Egbedi NO, Umoren SA. Antifungal drugs as corrosion inhibitors for aluminium in 0.1 M HCl. Corrosion Science 2009;51(8) 1868-1875.

[123] Ahamad I, Prasad R, Quraishi MA. Inhibition of mild steel corrosion in acid solution by Pheniramine drug. Experimental and theoretical study Corrosion Science 2010;52(9) 3033-3041.

[124] Singh AK, Quraishi MA. Inhibitive effect of diethylcarbamazine on the corrosion of mild steel in hydrochloric acid. Corrosion Science 2010; 52(4) 1529-1535.

[125] Kumar SH, Karthikeyan S. Torsemide and Furosemide as Green Inhibitors for the Corrosion of Mild Steel in Hydrochloric Acid Medium. Industrial & Engineering Chemistry Research 2013;52(22) 7457-7469.

[126] Gece G. Drugs: A review of promising novel corrosion inhibitors. Corrosion Science 2011;53(12) 3873-3898.

[127] Al Juhaiman LA, Abu Mustafa A, Mekhamer WK. Polyvinyl pyrrolidone as a green corrosion inhibitor for carbon steel in alkaline solutions containing NaCl. Anti-Corrosion Methods and Materials 2013;60(1) 28-36.

[128] Liu GQ, Huang JY, Zhou YM, Yao QZ, Yang Y, Ling L, Wang HC, Wu WD, Sun W, Hu ZH. Fluorescent-tagged acrylic acid-allylpolyethoxy carboxylate copolymer as a green inhibitor for calcium phosphate in industrial cooling systems. Designed Monomer Polymers 2012;16(1) 89-98.

[129] Liu G, Huang J, Zhou Y, Yao Q, Wang H, Ling L, Zhang P, Cao K, Liu Y, Wu W, Sun,W. Carboxylate-Terminated Double-Hydrophilic Block Copolymer Containing Fluorescent Groups: An Effective and Environmentally Friendly Inhibitor for Calcium Carbonate Scales. International Journal of Polymeric Materials and Biopolymeric Materials 2013; 62(13) 678-685.

[130] Martínez-Palou R., Flores P. Perspectives of Ionic Liquids for Clean Oilfield Technologies. In Kokorin A (ed.) Perspectives of Ionic Liquids for Clean Oilfield Technologies. Rijeka: InTech; 2011. p567-530. Available from http://www.intechopen.com/books/ionic-liquids-theory-properties-new-approaches.

[131] Khaled KF. The inhibition of benzimidazole derivatives on corrosion of iron in 1 M HCl solutions. Electrochimica Acta 2003;48: 2493-2503.

[132] Zhou X, Yang HY, Wang FH. [BMIM]BF4 ionic liquids as effective inhibitor for carbon steel in alkaline chloride solution. Electrochimica Acta 2011;56(11) 4268-4275.

[133] Cho E, Mun J, Chae OB, Kwon OM, Kim HT, Ryu JH, Kim YG, Oh SM. Corrosion/passivation of aluminum current collector in bis(fluorosulfonyl) imide-based ionic liquid for lithium-ion batteries. Electrochemistry Communication 2012; 22: 1-3.

[134] Palomar ME, Olivares-Xometl O, Likhanova NV, Pérez-Navarrete JB. Imidazolium, Pyridinium and Dimethyl-Ethylbenzyl Ammonium Derived Compounds as Mixed Corrosion Inhibitors in Acidic Medium. Journal of Surfactants and Detergents. 2011; 14(2) 211-220.

[135] Gabler C, Tomastik C, Brenner J, Pisarova L, Derr N, Allmaier G. Corrosion properties of ammonium based ionic liquids evaluated by SEM-EDX, XPS and ICP-OES. Green Chemistry 2011;13(10) 2869-2877.

[136] Zhang QB, Hua YX. Corrosion inhibition of mild steel by alkylimidazolium ionic liquids in hydrochloric acid. Electrochimica Acta 2009;54(6) 1881-1887.

[137] Zhang QB, Hua YX. Corrosion inhibition of aluminum in hydrochloric acid solution by alkylimidazolium ionic liquids. Material Chemistry and Physics 2009; 119(1-2) 57-64.

[138] Ashassi-Sorkhabi H, Es'haghi M. Corrosion inhibition of mild steel in acidic media by [BMIm]Br Ionic liquid. Material Chemistry and Physics 2009;114(1) 267-271.

[139] Perez-Navarrete JB, Olivares-Xometl CO, Likhanova NV. Adsorption and corrosion inhibition of amphiphilic compounds on steel pipeline grade API 5L X52 in sulphuric acid 1 M. Journal of Applied Electrochemistry 2010;40(9) 1605-1617

[140] Morad MS, Hermas AA, Obaid AY, Qusti, AH. Evaluation of some bipyridinium dihalides as inhibitors for low carbon steel corrosion in sulfuric acid solution. Journal of Applied Electrochemistry 2008;38(9) 1301-1311.

[141] Saleh MM, Atia AA. Effects of structure of the ionic head of cationic surfactant on its inhibition of acid corrosion of mild steel. Journal of Applied Electrochemistry 2006;36(8) 899-905.

[142] Wang H, Liu SQ, Huang KL, Liu Y, Li Z. Ethylbenzotriazolium Bromide Ionic Liquid: A New Water Soluble Inhibitor for Corrosion of Mild Steel in Acid Media. Asian Journal of Chemistry 2013;25(2) 954-956.

[143] Zarrouk A, Messali M, Zarrok, H, Salghi, R, Ali, A. Al-Sheikh, Hammouti, B, Al-Deyab SS, Bentiss, F. Synthesis, Characterization and Comparative Study of New Functionalized Imidazolium-Based Ionic Liquids Derivatives Towards Corrosion of C38 Steel in Molar Hydrochloric Acid. International Journal of Electrochemical Science 2012;51(10) 13282-13299.

[144] Likhanova NV, Olivares-Xometl O, Guzmán-Lucero D, Domínguez-Aguilar MA, Nava N, Corrales-Luna M, Mendoza, M. C. Corrosion inhibition of carbon steel in acidic environment by imidazolium ionic liquids containing vinylhexafluorophosphate as anion International Journal of Electrochemistry Science 2011;6(109) 4514-4536.

[145] Qi-Bo Z, Yi-Xin H. Effect of alkylimidazolium ionic liquids on the corrosion inhibition of copper in sulfuric acid solution Acta Physico Chimica Sinica 2011;27(3) 655-663.

[146] Murulana LC, Singh, AK, Shukla, SK, Kabanda MM, Ebenso EE. Experimental and Quantum Chemical Studies of Some Bis(trifluoromethyl-sulfonyl) Imide Imidazolium-Based Ionic Liquids as Corrosion Inhibitors for Mild Steel in Hydrochloric Acid Solution. Industrial & Engineering Chemical Research 2012;51(40) 13282-13299.

[147] Rogers RD, Seddon KR., editors. Ionic Liquids: Industrial Applications for Green Chemistry. ACS: Boston; 2002.

[148] Rogers RD, Seddon KR., editors. Ionic Liquids as Green Solvent: Progress and Prospects. ACS: Boston; 2003.

[149] Wasserscheid P., Keim W., editors. Ionic Liquids in Synthesis, Wiley-VCH: Wenheim; 2004.

[150] Martínez-Palou R. (2006). Química en Microondas. (E-book). CEM Publishing: Matthew, NC; 2006. p131-154.

[151] Martínez-Palou R. (2007). Ionic liquids and Microwave-assisted Organic Synthesis. A "Green" and Synergic Couple. Journal of Mexican Chemical Society 2007;51(4) 252-264.

[152] Martínez-Palou R. Microwave-assisted synthesis using ionic liquids. Molecular Diversity 2010;14(4) 3-25.

[153] Freemantle M. An Introduction to Ionic Liquids. RSC Press. Cambridge, UK; 2009.

[154] Likhanova NV, Domínguez-Aguilar MA, Olivares-Xometl O, Nava-Entzana N, Arce E, Dorante H. The effect of ionic liquids with imidazolium and pyridinium cations on the corrosion inhibition of mild steel in acidic environment. Corrosion Science 2010;52(6) 2088-2097.

[155] Guzmán-Lucero D, Olivares-Xometl O, Martínez-Palou R, Likhanova NV, Domínguez-Aguilar MA, Garibay-Febles V. Synthesis of Selected Vinylimidazolium Ionic Liquids and Their Effectiveness as Corrosion Inhibitors for Carbon Steel in Aqueous Sulfuric Acid. Industrial & Engineering Chemical Research 2011;50(12) 7129-7140.

[156] Caporali S, Fossati A, Lavacchi A, Perissi I, Tolstogouzovm A, Bardi U. Aluminium electroplated from ionic liquids as protective coating against steel corrosion. Corrosion Science 2008;50(2) 534-539.

[157] Yue GK, Lu XM, Zhu Y, Hang XP, Zhang SJ. Surface morphology, crystal structure and orientation of aluminium coatings electrodeposited on mild steel in ionic liquid. Chemical Engineering Journal 2009;147(1) 79-86.

[158] Tuken T, Demir F, Kicir N, Sigircik G, Erbil M. Inhibition effect of 1-ethyl-3-methylimidazolium dicyanamide against steel corrosion. Corrosion Science 2012;59: 110-118.

[159] Lebedeva O, Junguroya G, Zakharov A, Kultin, D, Chernikova E, Kustov, L. Water as an Inhibitor of Metal Corosion in Hydrophobic Ionic Liquids. Journal of Physical Chemistry C 2012;116(42) 22526-22531

[160] Perissi I, Bardi U, Caporali S, Lavacchi A. High temperature corrosion properties of ionic liquids. Corrosion Science 2006;48(9) 2349-2362.

[161] Forsyth M, Seter M, Hinton B, Deacon G, Junk P. New "Green Corrosion Inhibitors" Based on Rare Earth Compounds. Austrlian Journal of Chemistry 2011;64(6) 812-819.

Chapter 20

The Corrosion Inhibition of Aluminium by Some of 3-alkyloxyaniline Monomeric Surfactants and Their Analogues Polymers in 0.5 M HCl Solution

S.M. Sayyah, S.S.Abd El-Rehim, M.M. El-Deeb and S.M. Mohamed

Additional information is available at the end of the chapter

http://dx.doi.org/10.5772/57350

1. Introduction

Corrosion is the destructive result of reaction between a metal or metal alloy and its environment [1-3]. Metal atoms in nature are present in chemical compounds (i.e, minerals). The same amounts of energy needed to extract metals from their minerals are emitted during the chemical reactions that produce corrosion. The corrosion process (anodic reaction) of the metal dissolving as ions generates some electrons, are consumed by a secondary process (cathodic reaction). These two processes have to balance their charges. The sites hosting these two processes can be located close to each other on the metal's surface, or far apart depending on the circumstances.

The thermodynamic or chemical energy stored in a metal or that is freed by its corrosion varies from metal to metal. It is relatively high for metals such as magnesium, aluminium, and iron, and relatively low for metals such as copper, silver and gold [4].

There are several factors effects on the corrosion process, which classified to factors associated mainly with the metal such as:

1. Effective electrode potential of a metal in a solution [5].
2. Overpotential of hydrogen on the metal [6-8].
3. Chemical and physical homogeneity of the metal surface [9].
4. Inherent ability to form an insoluble protective film [10].
5. Hydrogen-ion concentration (pH) in the solution [11-16].

and factors associated mainly with the environment such as [16]:

1. Influence of oxygen in solution adjacent to the metal [18].
2. Specific nature and concentration of other ions in solution [19].
3. Rate of flow of the solution in contact with the metal [20-22].
4. Temperature [23].
5. Contact between dissimilar metals or other materials [24].

Some chemicals react with a metallic surface, or metal environment giving the surface a certain level of protection named corrosion inhibitor. Corrosion inhibitors are commonly added in small amounts to corrosion medium, either continuously or intermittently to prevent serious corrosion [25].

Some inhibitors retard corrosion by adsorption to form a thin invisible film, others form visible bulky precipitates which coat the metal and protect it from attack. Other inhibitors, when added to an environment, retard corrosion but do not interact directly with metal surface.

The four components of a corrosion cell (anode, cathode, electrolyte and electronic conductor), three may be affected by a corrosion inhibitor to retard corrosion. The inhibitor may cause:

1. Increase Polarization of the anode (anodic inhibitor).
2. Increase Polarization of the cathode (cathodic inhibitor).
3. zncrease the electrical resistance of the circuit.

Four classes of inhibitors will be discussed as follow:

1. *Anodic Inhibitors [26-28]:*

Anodic inhibitors which cause a large shift in the corrosion potential which called Passivating inhibitors. They are also called dangerous inhibitors, because if used in insufficient concentrations, they cause pitting or sometimes an increase in corrosion rate. There are two types of Passivating inhibitors: (i) oxidizing anions such as chromate, nitrite and nitrate which can passivate the working electrode in the absence of oxygen; and (ii) non-oxidizing ions such as phosphate, tungstate, and molybdate which require the presence of oxygen to passivate the working electrode.

2. *Cathodic Inhibitors [29]:*

Cathodic inhibitors either slow the cathodic reaction itself, or they selectively precipitate on cathodic areas to increase circuit resistance and restrict diffusion of reducible species to the cathodes.

Some cathodic inhibitors make the discharge of hydrogen gas more difficult, and they are increase the hydrogen over voltage. Another possible cathodic reaction is the reduction of oxygen.

3. *Organic and polymer inhibitors:*

Organic compounds constitute a broad class of corrosion inhibitors which can not be designed specifically as either anodic or cathodic. Anodic and cathodic effects alone are sometimes observed in the presence of organic inhibitors, but, as a general rule, organic inhibitors affect the entire surface of a corroding metal when present in sufficient concentration. Both anodic and cathodic areas probably are inhibited, by diffrent depending on the potential of the metal, chemical structure of the inhibitor molecule, and size of the molecule. The inhibition is the result of adsorption of inhibitor on the metal surface. The film formed by adsorption of soluble organic inhibitors is invisible.

Polymers are used as corrosion inhibitors for different metals. Polyaniline as an electronically conductive polymer has attracted considerable attention. Because of its excellent environmental stability in the electro-conducting form, it has many potential applications; unique electrical and optical properties [31-41]. The applications of polyaniline have been limited due to its poor processability [42], which is true for most conducting polymers. Several studies have been done in order to improve the solubility of polyaniline, among them, using functionalized protonic acids as dopant, like p-toluene-sulphonic acid, octyl-benzene-sulphonic acid, dodecyl benzene-sulphonic acid [43–45], poly(styrene) sulphonic acid [46,47], and phosphoric acid esters [48]. An alternative method to obtain soluble conductive polymers is the polymerization of aniline derivatives. The most studied aniline derivatives are alkyl [49,50], alkyloxy [51–54], hydroxy [55,56], chloroaniline [57–59]. Also, substitution at the nitrogen atom was reported by Sayyah et al. [60] to improve the solubility of polyaniline.

This study has been done in order to improve the solubility of polyaniline by using terminal side chain containing 6-12 carbon atom and the presence of the hydrophilic SO_3Na anionic group at the end of side the alkyl chain group. By this way, the solubility of poly aniline conducting polymer has been improved and have the properties of surface active agent (surfactant).

Surfactants (surface active agents) are amphiphilic molecules that consist of a non-polar hydrophobic portion, usually a straight or branched hydrocarbon chain, which is attached to a polar or ionic portion (hydrophilic). The hydrophilic molecule has a tendency to interact with or be dissolved by water and other polar substances. The hydrocarbon chain interacts weakly with the water molecules, due to the cooperative action of dispersion and hydrogen bonding between the water molecules tends to squeeze the hydrocarbon chain out of the water and hence these chains are referred to hydrophobic, while the polar or ionic head group interacts strongly with water molecules via dipole or ion–dipole interactions. Therefore, a surfactant contains both a water insoluble component and a water soluble component [61]. The balance between hydrophobic and hydrophilic parts of the molecule gives these systems their special properties, e.g. accumulation at various interfaces and association in solution (to form micelles). Adsorption of surfactant molecules at the interface between metal and solution lowers the surface tension and higher the surfactant adsorption. The degree of surfactant adsorption at the interface depends on surfactant structure and the nature of the two phases that meet the interface [62]. The surfactants can be classified based on the nature of the hydrophilic group as follow:

1. *Anionic Surfactants*

These are the most widely used class of surfactants in industrial applications due to their relatively low cost of manufacture and they are used in practice as detergent [63]. For optimum detergency the hydrophobic chain is a linear alkyl group with a chain length in the region of 12–16 carbon atoms. The most commonly used hydrophilic groups are carboxylates, sulphates, sulphonates and phosphates.

2. *Cationic Surfactants:*

The most common cationic surfactants are the quaternary ammonium compounds with the general formula R0R00R000R0000NX, where X is usually chloride ion and R represents alkyl groups [64,65]. A common class of cationic is the alkyl trimethyl ammonium chloride, where R contains 8 –18 carbon atoms, e.g. dodecyl trimethyl ammonium chloride, $C_{12}H_{25}(CH_3)_3NCl$. Another widely used cationic surfactant class is dialkyl dimethyl ammonium chloride, with the alkyl groups having a chain length of 8–18 carbon atoms.

3. *Amphoteric Surfactants [66]:*

These are surfactants containing both cationic and anionic groups. The most common amphoterics are the derivatives of trimethyl glycine $(CH_3)_3NCH_2COOH$ (described as betaine). An example of betaine surfactant is lauryl amido propyl dimethyl betaine $C_{12}H_{25}CON(CH_3)_2CH_2COOH$. The main characteristic of amphoteric surfactants is solubility in solution.

4. *Nonionic Surfactants [67]:*

The most common nonionic surfactants are those based on ethylene oxide, such as ethoxylated surfactants. Several classes can be distinguished: alcohol ethoxylates, alkyl phenol ethoxylates, fatty acid ethoxylates, monoalkaolamide ethoxylates, sorbitan ester ethoxylates and ethylene oxide–propylene oxide copolymers (sometimes referred to as polymeric surfactants).

5. *Polymeric Surfactants:*

There has been considerable recent interest in polymeric surfactants due to their wide applications in the oil industry, preparation of emulsions, suspensions, stabilization, wetting spreading and adhesion.

Aluminum and its alloys have been widely used as a material in the fields of transport, storage of liquefied industrial gases, building, electrical engineering, household appliances, containers, aircraft, chemical equipments including heat exchangers, pressure vessels, condensers, rotatory dryers, tanks, portable containers, valves, and piping. Aluminium is particularly suitable for neutral or oxidizing substances such as paraffin's, alcohols, nitric acid, hydrogen peroxide,.. etc. This is because they are characterized by low density, low cost, excellent electrical and thermal conductivities, good corrosion resistance, good appearance, and high ductility [68–74].

The corrosion resistance of aluminum and its alloys comes from the continuous film that is developed on their surfaces upon exposure to the atmosphere or aqueous solutions [75]. This film is not guaranteed to protect the aluminum surface in corrosive media. The most widely used HCl acid solution, so this medium has induced a great deal of research on aluminium [76]. Most of the effective inhibitor have hetero atom such as O, N, S containing multiple bonds in their molecules through which they can adsorb on the metal surface. The sites of these elements have higher electron density, making them the reaction centers. Other articles on the corrosion inhibition of aluminium by different type of inhibitors are summarized in Table (1).

In the present study, 3-alkyloxy aniline sodium sulfonate monomeric surfactants and their analogues polymers as mixed–type inhibitors for the corrosion of aluminium in 0.5M HCl. This performance will be investigated via weight loss and potentiodynamic polarization techniques. In addition this work will extend to compare the experimental data obtained from weight loss and potentiodynamic polarization technique with several adsorption isotherms at different temperatures in order to determine the thermodynamic functions for the adsorption process and to get more information about the mode of adsorption of the inhibitors on the surface of aluminium.

Author	**Metal type**	**Inhibitor**	**Medium**	**Method of Investigation**	**General Remarks**
B. Muller et al. [77] (1994)	Aluminum	Low and high molecular weight Polyacrylic acids (PAAC)	Water and butyl glycol V/V 9:2 at pH=10	Atomic absorption spectroscopy (AAS) by measuring soluble Al acrylate in solution	• In the concentration range from 0.3-0.5 wt% of PAAC, low molecular weight PAAC has markedly inhibition effect than high molecular weight. • In case of the concentration range 0.05-0.1 wt%, low molecular weight PAAC has no effect where the high molecular weight of PAAC has an inhibition effect.
B. Muller et al. [78] (1995)	Aluminum	High molecular weight of styrenemaleic acid copolymers	Mixture of water and butyl glycol V/V 9:2 at pH=10	AAS by measuring soluble Al(ɯ) acrylate in solution And volumetric measurement of hydrogen	• The inhibition efficiency is measured by volumetric measurement of the hydrogen produced by the corrosion. • The lower acid number of the copolymers, the lower is the volume of hydrogen.
Ogurtsov, N.A., et al. [79] (2004)	Aluminum alloy	Undoped and doped polyaniline (PANI)	0.1N HCl and 3.5% NaCl solutions	Electrochemical impedance spectroscopy (EIS)	• Polyaniline can provide corrosion protection of Al alloy in both NaCl and in dilute HCl solutions. • The degree of the corrosion protection of the Al alloy by the undoped PANI is higher than that of the doped PANI.

Author	Metal type	Inhibitor	Medium	Method of Investigation	General Remarks
					• Based on the obtained corrosion currents, it is possible to believe that, the thickness of the surface protective aluminum oxide layer plays a key role in the inhibiting effect.
et.al. [80] (2004)	Aluminum	Polyamide	Oxalic acid	Potentiostatic and potentiodynamic anodic polarization techniques	• The inhibition efficiency increases with increasing polyamide concentration until a critical value and then starts to decrease. • The inhibitive behavior of the polymer is due to adsorption of the polyamide compounds on the metal surface and formation of insoluble complexes. • The adsorption process is found to obey Temkin adsorption isotherm.
et al. [81] (2008)	Aluminum	Polyvinyl pyrrolidone (PVP) and Polyacrylamide (PA)	HCl	Weight loss, hydrogen evolution and thermometric techniques	• The inhibition efficiency increased with increasing inhibitor concentrations but decrease with increasing temperature. • The inhibition efficiency of PVP higher than PA which may be due to the differences in their molecular structures play a significant role in the adsorption process. • The inhibitors (PVP and PA) obey Freundlich, Temkin and Flory-Huggins adsorption isotherms. • Results obtained from the kinetic and thermodynamic parameters indicate spontaneous and physical adsorption of the inhibitors.
Amin, M. A., et al. [82] (2009)	Aluminum	Polyacrylic acids with different molecular weights (PAA_1 = 1800, PAA_2 = 11,000 and PAA_3 = 14,000 g mol^{-1})	Alkaline solutions (pH 8 and 10)	Weight loss, potentiodynamic polarization and impedance techniques.	• The polymers inhibit the alkaline corrosion of aluminium. • The inhibition efficiencies of inhibitors increase with increasing concentration, molecular weight, immersion time and act as mixed-type inhibitors.

Author	Metal type	Inhibitor	Medium	Method of Investigation	General Remarks
					• The results obtained from the chemical and electrochemical measurements are in good agreements.
Umoren, S.A., et al. [83] (2009).	Aluminum alloy 3SR	Polyvinylpyrroli-done and poly acrylamide	HCl	Weight loss, hydrogen evolution and thermometric methods	• The polymers inhibite acid induced corrosion of aluminum. • PVP was found to be a better corrosion inhibitor than PA. • Adsorption of these inhibitors follows Temkin and El-Awady adsorption isotherm models. • Kinetic/thermodynamic parameters (E_a, K_{ads}, ΔG°_{ads}} of adsorption of the studied inhibitors reveal that the adsorption was physical in nature and spontaneous.
, B.. et al. [84] (2011).	Aluminum alloy	Polypyrrole (PPy) and poly(pyrrole-co-o-anisidine)	3.5% NaCl	Oopen circuit potential(E_{ocp}). electrochemical impedance and anodic polarization techniques.	• Thermogravimetric results were indicated that copolymer film has higher thermal stability than PPy. • The synthesized copolymer film decreased the corrosion of Al by acting as physical barrier on the surface and exhibiting anodic protective behavior. • The E_{ocp}–time curves showed that the barrier effect of copolymer coating increased during immersion period.
Oliveira, M.A.S., et. al. [85] (2011).	Aluminum alloy	Polyaniline doped with poly(methylmetha crylate-co-acrylic acid) (PAni-PMMA-co-A)	NaCl, pH 5.8	Impedance studies	• The films, PAni-PMMA-co-AA and epoxy resin, formed on the Al alloy surfaces were porous and non-uniform allowing electrolyte permeation and oxidation of the metal inside the pores. • The PAni-PMMA-co-AA films offer better corrosion protection to aluminum alloy than the commercial epoxy resin films.
Zaafarany.I, [86] (2012)	Aluminum	Alginate and pectate water-soluble natural polymer	NaOH	Gasometric and weight-loss techniques	• The alginate and pectate are effective inhibitors for corrosion of aluminum in alkaline medium.

Author	Metal type	Inhibitor	Medium	Method of Investigation	General Remarks
					• Pectates are more active for corrosion inhibition than alginates, this may be due to the geometrical configuration of the functional groups, which play an important role in the magnitude of inhibition efficiency. • The natural or synthetic polymers containing the same functional groups will have the same corrosion behavior and, hence, may proceed through similar corrosion mechanism.
Ghoreishi, S.M., et al. [87] (2012).	Aluminum alloy	Poly(o-anisidine)	3.5% NaCl	Potentiodynamic polarization technique and electrochemical impedance spectroscopy (EIS).	• The homogeneous and adherent poly(o-anisidine) were successfully synthesized on Al alloy. • The EIS results are in good agreement with the potentiodynamic polarization measurements. • This study reveals that the poly(o-anisidine) coating has excellent corrosion protection properties and can be considered a potential coating material to protect aluminum alloy against corrosion.
Awad , M.K. et al. [88] (2013)	Aluminum	Poly ethylene glycol	HCl	Weight loss and polarization techniques.	• Poly ethylene glycols with different number of repeating unit can work as chemical inhibitor for corrosion aluminum surface. • The adsorption of these polymers on metal surface is physical adsorption and obeys Langmuir adsorption isotherm. • The inhibition efficiency increases with both number of repeating unit and concentration but decreases with increasing temperature.
et al. [89] (2013)	Aluminum	Water-soluble natural poly) pectates ((PEC)	HCl	Gasometric and weight loss techniques	• The inhibition efficiency was found to increase with increasing inhibitor concentration and decrease with increasing temperature.

Author	Metal type	Inhibitor	Medium	Method of Investigation	General Remarks
					• The inhibition action of PEC on Al metal surface was found to obey the Freundlich isotherm. • Factors such as the concentration and geometrical structure of the inhibitor, concentration of the corrosive medium, and temperature affecting the corrosion rates were examined. • The kinetic parameters were evaluated and a suitable corrosion mechanism consistent with the kinetic results is discussed

Table 1. Some physical data of using water soluble polymers as corrosion inhibitors

2. Expermental

2.1. Weight loss measurements

Aluminium metal provided by the Egyptian Aluminium Company, Naghammady of the following chemical composition: 99.57 % Al, 0.31 % Fe, 0.07 % Si, 0.015 % Ti %, 0.0016 % Zn, 0.0003 % Cr, 0.0019 % Mg, 0.0021 % Mn and 0.0007 Cu. The dimensions of the tested samples are 2 x 2 (cm^2), 0.1cm (thickness) is used in this study. The samples were polished successively with fine grade emery papers, cleaned with acetone, washed with doubly distilled water and finally dried, weighed and then introduced into test solution.

2.2. Potentiodynamic polarization measurements

The working electrode was made from aluminium rod has the same composition as mentioned in point 2.1. The rod is axially embedded in araldite holder to offer an active flat disc shaped surface of an area 0.785 cm^{-2} as shown in scheme (1). Prior to each experiment, the working electrode was polished successively with fine emery paper. The polished metal surface was rinsed with acetone and distilled water before dipping it into the electrolytic cell. A platinum wire was used as the counter electrode and a saturated calomel electrode as a reference electrode to which all potentials are referred.

The electrochemical experiments are performed using the Potentiostat /Galvanostat Wenking PGS95, connected to computer. The I-E curves are recorded using computer software (model ECT). The experiments were carried out by changing the electrode potential automatically from the starting potential towards more positive values at the required scan rate till the end of the experiments. The complete polarization cell is shown in scheme (2). It is a multi-necks

250 ml flask. The multiple necks were used to introduce working (W), counter (C) electrode and Luggin probe entered the cell through the clamped ball and socket joint.

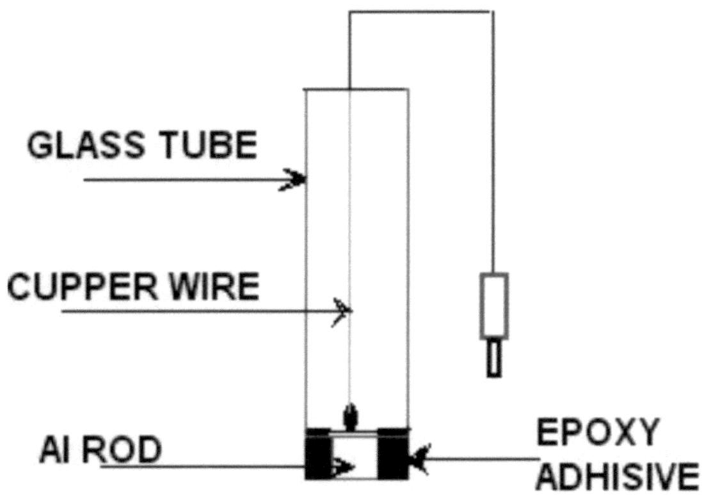

Scheme 1. The working electrode.

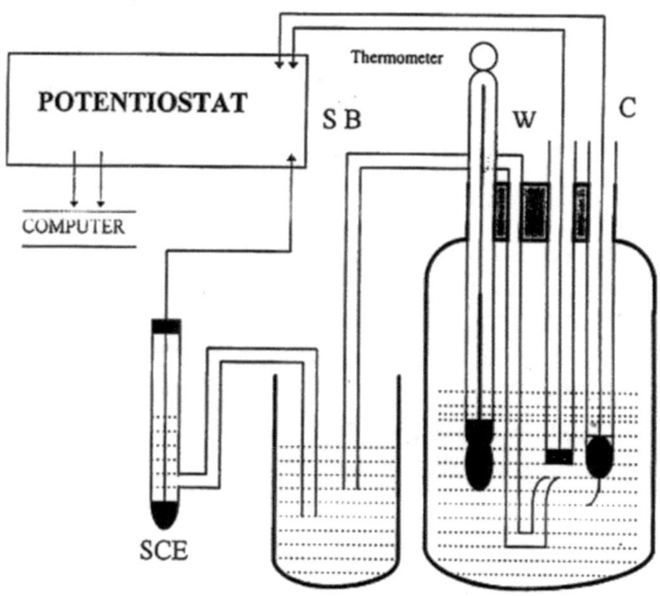

Scheme 2. Electrode cell and the circuit diagram for Potentiodynamic polarization measurements.

http://dx.doi.org/10.5772/57350

2.3. Materials

3-Aminophenol provided by Aldrich chemical Co. Sodium sulfite used in the synthetic process was obtained from Merck chemical Co., (Germany). Concentrated hydrochloric acid was of chemically pure grade products provided by Prolabo-Chemical Co., (U.K.). Doubly distilled water was used to prepare all solutions. Potassium persulfate and 1,6- dibromohexane, 1,10-dibromodecane and 1, 12- dibromododecane were obtained from Aldrich Chemical Co., (England).

2.4. Synthesis of monomeric (3-alkyloxy anilines sodium sulfonate)

3-(6-Bromohexayloxy) aniline, 3-(10-bromodecyloxy) aniline and 3-(12-bromododecyloxy) aniline were prepared by the reaction of (1:1) 3-aminophenol with 1, 6- dibromohexane, 1, 10-dibromodecane and 1, 12- dibromododecane respectively, in the presence of sodium ethoxide.

3-(6-bromohexayloxy) aniline, 3-(10-bromodecane) aniline and 3-(12-bromododecane) aniline were reacted with sodium sulfite to produce 3(6- sodium sulfonate hexacyloxy) aniline (MC_6), 3(10- sodium sulfonate decyloxy) aniline (MC_{10}) and 3(12- sodium sulfonate dodecyloxy) aniline (MC_{12}).

2.5. Synthesis of polymeric surfactants

The amount of monomers of 3(6- sodium sulfonate hexacyloxy) aniline, 3(10- sodium sulfonate decyloxy) aniline and 3(12- sodium sulfonate dodecyloxy) aniline were dissolved in 25 ml of 2.0 M HCl solutions in a well stoppered conical flasks of 250 ml capacity followed by the addition of the required amounts of potassium persulfate (0.15 M) dissolved in 25 ml of (2M) HCl solutions to the reaction mixture under nitrogen atmosphere. The order of addition of substances was kept constant in all the performed experiments. The stoppered conical flasks were then placed in automatically controlled thermostat at 25^0C. The flasks were shaken (50 shakings/10 s/15 min) for one hour by using an automatic shaker. The flasks were left for 72 hour at room temperature to continue the reaction and then filtered using a Buchner funnels. The precipitate were washed with the distilled water, and finally dried under vacuum at room temperature until constant weight.

2.6. Elemental and spectroscopic analysis

The elemental analysis of the prepared monomeric and polymeric surfactants were carried out in the micro analytical laboratory at Cairo University by using oxygen flask combusition and a dosimat E415 titirator (Switzerland).

The ultraviolet-visible absorption spectra of the prepared monomeric and polymeric surfactants were measured using Shimadzu UV spectrophotometer (M 160 PC) at room temperature in the range 200-400 nm using dimethylformamide as a solvent and reference.

Scanning electron microscopy and X-ray diffractometer (philip1976. model1390) was operated for the polymer samples under the following conditions which were kept constant for all of the analysis processes Cu X-ray tube, scan speed =8/min, current=30mA, voltage =40kv and preset time=10

2.7. Critical micelle concentration and surface tension

Critical micelle concentration and surface tension were measured by using K100 Tensiometer (Kruss Type, Germany) for the hydrochloric form of synthesized monomeric and polymeric surfactants using different concentrations.

3. Results and discussion

3.1. The elemental analysis and spectroscopic analysis of the prepared monomeric surfactants and their polymers

The elemental analysis of the prepared monomeric and polymeric surfactants are summarized in Table (2).The data show that, there is a good agreement with the calculated one for the suggested structures present in scheme (3).

The UV-visible spectra of the prepared monomeric surfactants and their polymers are represented in figure (1); the spectra show the following absorption bands:

1. The two absorption bands appear at λ_{max} = 214 and 227 nm In case of MC_6, which may be attributed to π-π^* transition (E_2-band) of the benzene ring and the β-band for π-π^* transition ($A_{1g} - B_{2u}$), appears at λ_{max} = 206 and 210 nm in case of MC_{10} and at λ_{max} = 207 and 211 nm in cace of MC_{12}.

2. In case of PC_6, two absorption bands appear at λ_{max} = 211 and 225 nm which may be attributed to π-π^* transition showing a bathochromic shift, appears at λ_{max} = 211 and 253 nm in case of PC_{10} and appears at λ_{max} = 210 and 240 nm in case of PC_{12}. Beside these bands, an absorption band appears in the visible region at λ_{max} = 352, λ_{max} = 348 and λ_{max} = 344 nm in case of PC_6,PC_{10} and PC_{12} respectively which may be due to the high conjugation of the aromatic polymeric chain.

The X-ray diffraction patterns of the three prepared polymers (PC_6, PC_{10} and PC_{12}) are represented in figure (2). The figure shows that one of the prepared polymers (PC_{12}) is amorphous while the polymer PC_6 and PC_{10} give peak at 2-Theta equal to 19.835 degree with d-spacing (4.4726) and the peak intensity in case of PC_6 is higher than that of polymer PC_{10} has a small portion of crystallinity which are also confirmed by the electron microscopic picture represented in figure (3). In case of (PC_{12}) there is no characteristic peak in the X-ray diffraction patterns in the region of crystalline organic compounds indicating that the polymer is amorphous and from the electron microscopic picture it is clear that the amorphous grain particles of the polymers (PC_{12}) are ranged from spherical to elongated particles.

http://dx.doi.org/10.5772/57350

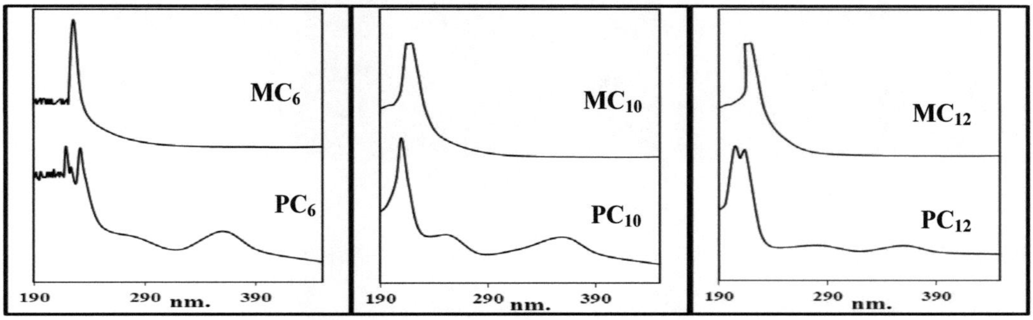

Figure 1. UV spectrum of the prepared monomeric and polymeric surfactants.

Name	C %		H %		Cl %		S %		N %	
	Calc.	Found	Calc.	Found	Calc.	Found	Calc.	Found	Calc.	Found
MC_6	48.81	48.12	6.10	5.99	–	–	10.85	10.65	4.75	4.58
PC_6	46.27	46.92	6.72	6.54	5.70	5.75	10.28	10.36	4.50	4.53
MC_{10}	65.08	64.12	8.81	8.73	–	–	10.85	10.73	4.75	4.61
PC_{10}	52.28	52.14	7.49	7.54	4.83	4.68	8.71	8.75	3.81	3.72
MC_{12}	73.22	72.93	10.17	10.02	–	–	10.85	10.63	4.75	7.66
PC_{12}	54.65	55.37	7.97	7.69	4.44	4.51	8.10	8.04	3.54	3.52

Table 2. Elemental analysis of the prepared monomeric and polymeric surfactants.

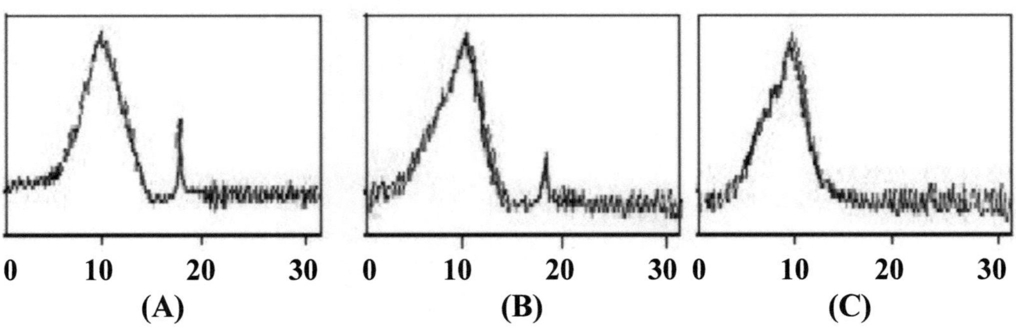

Figure 2. X-Ray of the prepared polymer samples PC_6 (A), PC_{10} (B) and PC_{12} (C).

(A) (B) (C)

Figure 3. Electron microscope picture of the prepared polymer samples PC_6 (A), PC_{10} (B) and PC_{12} (C).

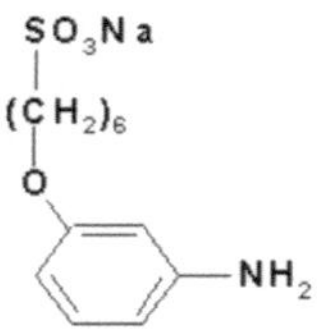

3(6- Sodiumsulfonate hexayloxy) aniline (MC_6).

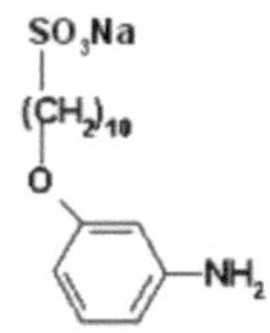

3(10- Sodiumsulfonate decyloxy) aniline (MC_{10}).

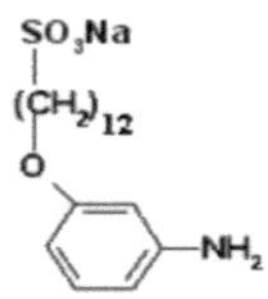

3(12- Sodiumsulfonate dodecyloxy) aniline (MC_{12}).

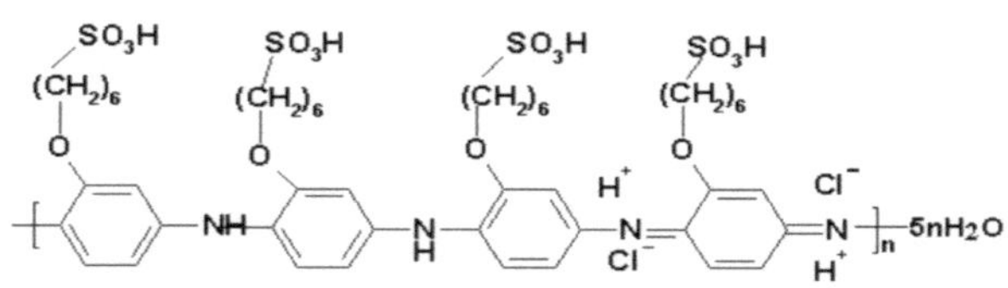

Poly 3-(hexayloxy sulfonic acid) aniline (PC_6)

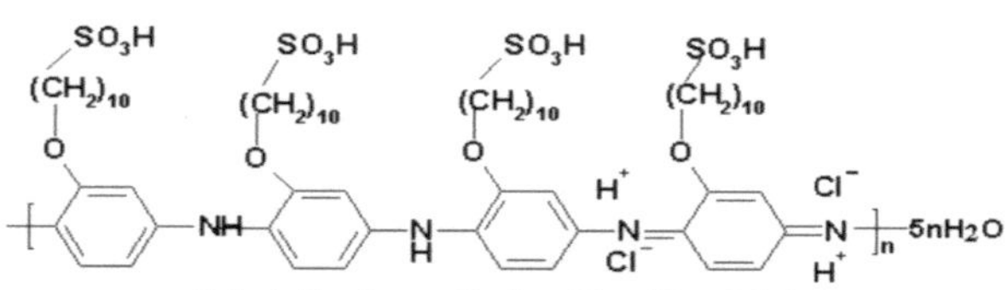

Poly 3-(decyloxy sulfonic acid) aniline (PC_{10}).

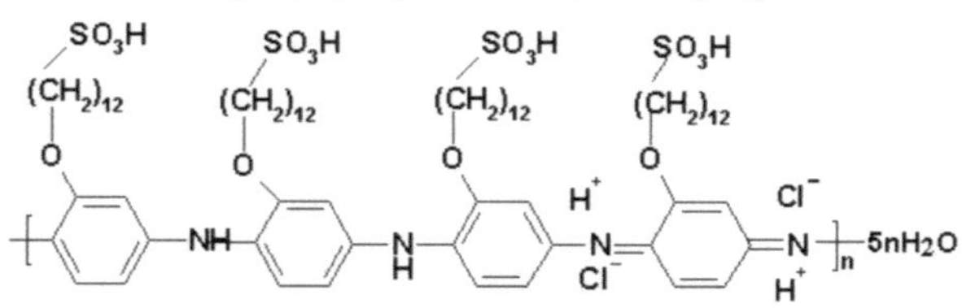

Poly 3-(dodecyloxy sulfonic acid) aniline (PC_{12})

Scheme 3. Structure of monomeric surfactants and their polymers.

3.2. Critical micelle concentration and surface tension

3.2.1. Critical micelle concentration (CMC) of the prepared surfactants

The relations between the surface tension (γ) and the different concentrations of the prepared monomeric and their analogues polymeric surfactants at 25 ⁰C are represented in figure (4). From figure (4), it is clear that the surface tension (γ) decreases with the increase of the prepared monomer and polymer concentrations which means also the increase of adsorption at air / water interface. The critical micelle concentration (CMC) of the prepared monomeric and their analogs polymeric surfactants was determined from the abrupt change in the slope of the corresponding plot of the surface tension (γ) versus concentration. The CMC results in Table (3) show that the CMC values decrease as the alkyl chain moiety increase from C_6-C_{12} for all samples under investigation. The lower values of CMC for the prepared surfactants indicate the ability of these surfactants to dissolve in water solution which is due to the presence of the hydrophilic SO_3Na anionic group of the alkyl chain moiety, which increases the solvation of H_2O molecules surround it and increase the solubility of the surfactant molecule in solution. It was noticed that the CMC values of the prepared polymeric surfactants were lower than those of the prepared monomeric surfactants as shown in table (3). These results show that the polymeric surfactants have more hydrophilic and hydrophobic groups than those of the monomeric surfactants which enhance the solubility of the polymeric surfactants and lead to the decrease in CMC values.

Surfactant	T (⁰C)	CMC (mol/l)	Γ_{max} x 10^{-11} (mol.cm^{-2})	A_{min} nm^2	π_{CMC}
MC^{6}	25	0.0005	4.85828	3.39548	7.80
MC^{10}		0.00025	8.46044	1.94980	17.80
MC^{12}		0.00025	2.82015	5.84941	25.80
PC^{6}		0.00025	2.82015	5.84941	8.80
PC^{10}		0.0001	2.82015	5.84941	18.80
PC^{12}		0.0001	8.46044	1.94980	26.80
MC^{6}	35	0.0025	2.96452	5.56454	4.00
MC^{10}		0.001	7.41131	2.22582	12.00
MC^{12}		0.0005	4.44678	3.70969	20.00
PC^{6}		0.001	4.44678	3.70969	4.00
PC^{10}		0.0005	5.92905	2.78227	13.00
PC^{12}		0.00025	2.96452	5.56454	21.00
MC^{6}	50	0.001	3.47450	4.74779	1.00
MC^{10}		0.00025	8.06754	2.04476	6.00
MC^{12}		0.00025	2.68918	6.13429	16.00
PC^{6}		0.00025	2.68918	6.13429	1.00
PC^{10}		0.0001	2.68918	6.13429	9.00
PC^{12}		0.0001	8.06754	2.04476	17.00

Table 3. Critical micelle concentration (CMC), effectiveness π_{CMC}, Maximum surface Excess (Γ_{max}) and Minimum area (A_{min}) of monomeric surfactants (MC_6, MC_{10} and MC_{12}) and their analogs polymeric surfactants.

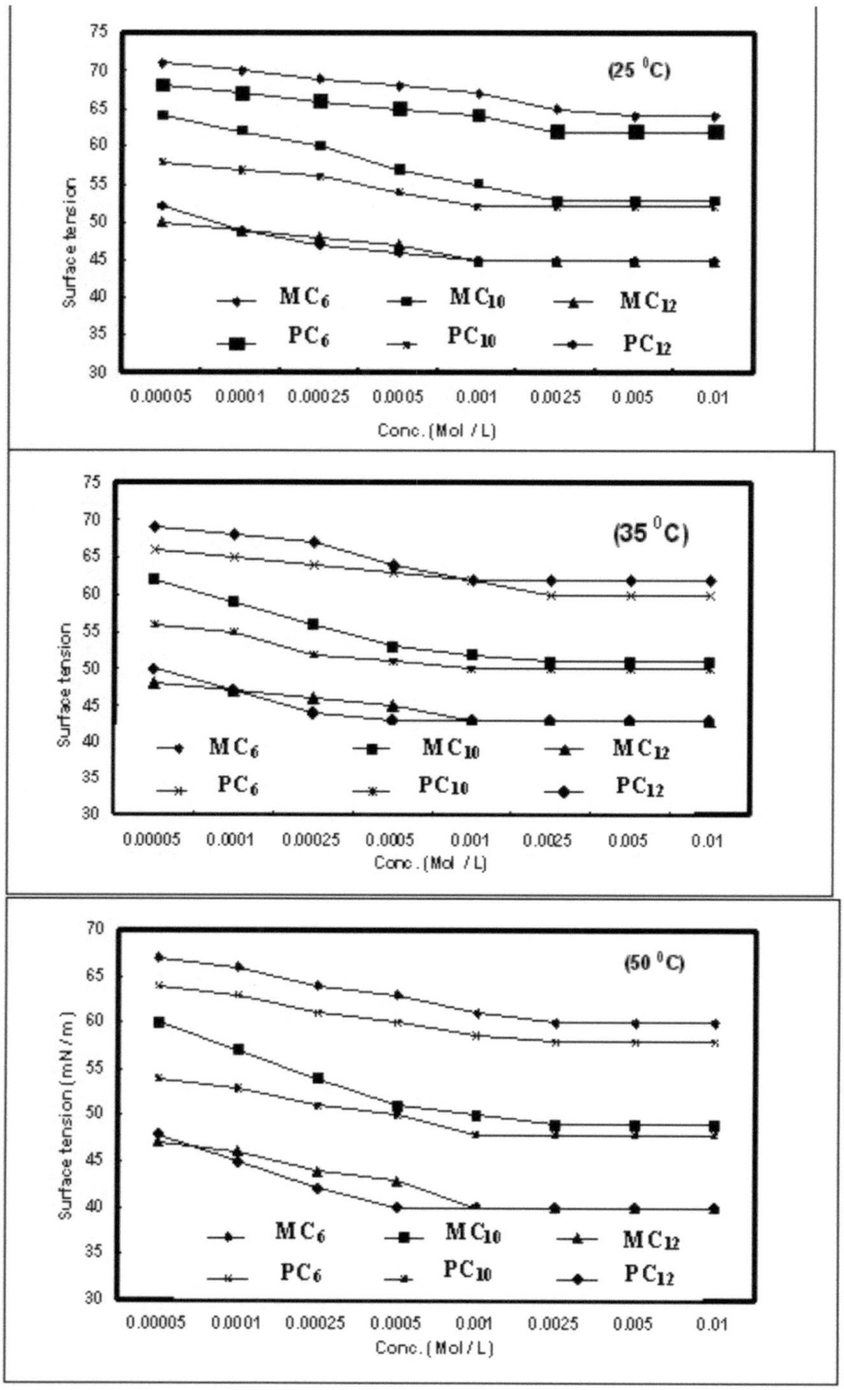

Figure 4. The relations between the surface tension (γ) and the different concentrations of the prepared monomeric and their analogues polymeric surfactants at 25, 35 and 50 °C

3.2.2. *Surface parameters of the prepared surfactants*

The surface parameters (effectiveness Π_{CMC}), maximum surface excess (Γ_{max}) and minimum area (A_{min}) of the surfactants were calculated according to Rosen et al. [90] and the data are summarized in table (3). These data show the good surface activity of the prepared surfactants and their ability for adsorption at air / water interface. The effectiveness values in table (3) show increasing with increasing of the alkyl chain moiety of the prepared monomeric and polymeric surfactants which give indication about the ability of these surfactants toward the adsorption at air / water interface and decrease the surface tension. Also, it is noticed that the effectiveness values of the prepared polymeric surfactants are higher than those of monomeric surfactants.

3.2.3. *Thermodynamic parameters of micellization*

Table (4) lists the free energies, ΔG_{mic}, enthalpies ΔH_{mic}, and entorpies, ΔS_{mic} of micellization for the prepared monomeric and polymeric surfactants in this study. These parameters were calculated using the following equations according to to Rosen et al. [90].

$$\Delta G_{mic} = 2RT \ln CMC \quad (1)$$

$$d\,[\Delta G_{mic}]/\Delta T = -\Delta S_{mic} \quad (2)$$

$$\Delta H_{mic} = \Delta G_{mic} + T\Delta S_{mic} \quad (3)$$

The ΔS_{mic} values in table (3) are all positive for the prepared surfactants, indicating increasing in randomness of the system upon transformation of the surfactant molecules into micelles. ΔH_{mic} in table (3) are significant factor in the process of micellization. The ΔH_{mic} values are positive for the prepared surfactants due to the endothermic solvation associated with micellization process. The data in table (4) show that the free energies ΔG_{mic} of micellization for the prepared monomeric and polymeric surfactants are always negative values, indicating that, micellization of such surfactants is a spontaneous process. It was noticed that the ΔG_{mic} values of the polymeric surfcatnts are higher than those of the monomeric surfactants which related to the more ability of the polymeric surfactants toward the micellization process than the monomeric surfactants.

3.2.4. *Thermodynamic parameters of adsorption*

The free energies ΔG_{ad}, enthalpies, ΔH_{ad}, and entropies, ΔS_{ad} values of adsorption listed in table (4) were calculated by use of the following equations according to Rosen et al. [90].

$$\Delta G_{ad} = \Delta G_{mic} - (6.023 \times 10^{-1} \pi_{cmc} A_{min}) \quad (4)$$

$$\Delta S_{ad} = d\,\Delta G_{ad} / \Delta T \qquad (5)$$

$$\Delta H_{ad} = \Delta G_{ad} + T\,\Delta S_{ad} \qquad (6)$$

According to the data given in table (5), the ΔG_{ad} values are negative, reflecting to the tendency of adsorption at liquid / air interface for the prepared surfactants. Comparing the data in table (5) it is noticed that the ΔG_{ad} values of the prepared monomeric and polymeric surfactants are higher than the G_{mic} values indicating that the prepared surfactants are favored the adsorption at air / water interface more than the micellization process. The ΔS_{ad} values are all positive this may reflect the greater freedom of motion of the hydrocarbon chains at the planar air / aqueous solution interface compared to that in the relatively cramped interior beneath the convex surface of the micelle. The most ΔH_{ad} values are less than ΔH_{mic} which indicates that less hydrogen bonds between the hydrophilic groups (SO_3Na) in the synthesized surfactants and water molecules are broken in the process of adsorption at the air/solution interface than in micellization process.

Name	T(°C)	$\Delta G^0_{mic.}$ kJmol^{-1}	ΔS^0_{mic} kJmol^{-1}.K^{-1}	ΔH^0_{mic} kJmol^{-1}	ΔG^0_{ad} kJmol^{-1}	ΔS^0_{ad} kJmol^{-1}.K^{-1}	ΔH^0_{ad} kJmol^{-1}
MC_6	25	-14.5788			-17.1930		
MC_{10}		-16.8084			-19.1947		
MC_{12}		-18.4950			-24.2596		
PC_6		-16.8084			-18.7746		
PC_{10}		-18.4950			-21.6455		
PC_{12}		-20.1816			-29.1637		
MC_6	35	-19.4419	0.06312		-20.2599		
MC_{10}		-21.2148	0.06888		-22.6241		
MC_{12}		-21.2148	0.06888		-28.9657		
PC_6		-21.2148	0.06888		-22.6241		
PC_{10}		-23.5586	0.07649		-28.1386		
PC_{12}		-23.5586	0.07649		-26.0247		
MC_6	50	-18.5294	0.13169	24.0051	-18.8154	0.05408	-1.3481
MC_{10}		-22.2480	0.25564	60.3235	-22.9870	0.19313	39.3944
MC_{12}		-22.2480	0.25564	60.3235	-28.1595	0.36555	89.9130
PC_6		-22.2480	0.25564	60.3235	-22.6175	0.18082	35.7859
PC_{10}		-24.7059	0.33757	84.3287	-28.0311	0.36127	88.6588
PC_{12}		-24.7059	0.33757	84.3287	-26.7995	0.32022	76.6306

Table 4. Thermodynamic parameters of micellization and adsorption for the prepared monomeric surfactantand their analogs polymeric surfactants.

http://dx.doi.org/10.5772/57350

3.3. Corrosion inhibition of aluminium using monomeric and polymeric surfactant in 0.5 M HCl solution

Corrosion inhibitors play a very important role in protecting metals and alloys. Recently many new corrosion inhibitors have been developed. Some authors [91, 92] indicate that polymeric surfactants constitute an important class of corrosion inhibitor for this reason we investigated the inhibition properties of the three monomeric surfactants ;viz, 3-(6-sodium sulphonate hexayloxy) aniline (MC_6), 3-(10-sodium sulphonate decyloxy) aniline (MC_{10}), 3-(12-sodium sulphonate dodecyloxy) aniline (MC_{12}) and their analogues polymeric surfactants poly 3-(hexayloxy sulphonic acid) aniline (PC_6), poly 3-(decyloxy sulphonic acid) aniline (PC_{10}) and poly 3-(dodecyloxy sulphonic acid) aniline (PC_{12}) for the corrosion of aluminium in 0.5 M HCl solution by using weight loss (chemical) and potentiodynamic polarization (electrochemical) techniques under different experimental conditions. The conditions included the influence of the immersion time, structure, concentrations of the surfactants and the solution temperature on the rate of aluminium corrosion in 0.5M HCl solution.

In acid medium, the corrosion of aluminium proceeds via two possible reactions; the partial anodic reaction which can be represented by the dissolution of the metal, according to the following reaction:

$$Al \longrightarrow Al^{+3} + 3e^- \quad (7)$$

and the partial cathodic reaction which can be represented by the overall hydrogen evolution reaction:

$$2H^+ + 2e^- \longrightarrow H_{2\,(g)} \quad (8)$$

The monomeric surfactants have the same structure but differ in the terminal of alkyl chain length. In the aqueous acidic solution, each monomer is ionized to an amphiphilic anion and Na^+ cation.

3.3.1. Weight loss technique

The weight loss (in mg cm^{-2}) of aluminium in 0.5 M HCl solution with and without different concentrations of the monomeric surfactants (MC_6), (MC_{10}) and (MC_{12}) and their polymers (PC_6), (PC_{10}) and (PC_{12}) was plotted as a function of immersion time at 30 ^{0}C. The plots are shown in figure (5). It is clear from this figure that, the weight loss of aluminium in 0.5 M HCl solution increases linearly with the immersion time. The linear increase in the weight loss with time in the absence of additives have greater value than in the presence of the prepared monomeric and polymeric surfactants. The slope of each line (mg.cm^{-2} min^{-1}) represents the corrosion rate of aluminium in these solutions. It is obvious that the rate of corrosion of aluminium decreases with increasing the immersion time under the prevailing conditions.

It follows from figure (6) that, the weight loss (i.e., the corrosion rate) is suppressed in the presence of the additives (the monomers and their corresponding polymers). These results reveal that the addition of each compound hinders the acid attach on the metal surface and prevent aluminium from dissolution, so these compounds act as corrosion inhibitors. The decrease in weight loss and hence the suppression in the rate of corrosion enhances with increasing the concentration of each additive. On the other hand, the weight loss and consequently the corrosion rate of aluminium in 0.5 M HCl solution in the absence and presence of surfactant increase with a rise in solution temperature.

The inhibition efficiency values (P %) of these surfactants at different inhibitor concentrations and temperatures were calculated using the following equation [93]:

$$P\% = 100 \times \left[1 - \left(W/W^{0}\right)\right] \tag{9}$$

Where W^0 and W are the weight loss per unit time in the absence and presence of definite inhibitor concentration, respectively. The calculated values of (P%) are listed in table (5).It is noticed from the obtained data that:

1. At a given temperature, the inhibition efficiency (P %) of each inhibitor increases with increasing inhibitor concentration and reaches a maximum at certain critical concentration. The critical concentration is 8.47 x 10^{-5} mol/L for MC_6, 1.43 x 10^{-5} mol/L for MC_{10}, 1.32 x 10^{-5} mol/L for MC_{12} and 4.11 x 10^{-6} mol/L for PC_6, 3.44x 10^{-6} mol/L for PC_{10}, 3.19 x 10^{-6} mol/L for PC_{12}). Beyond the critical concentration, the inhibition efficiency tends to decrease slightly and finally achieves steady state values [94].

At a given inhibitor concentration, the inhibition efficiency (P%) of each inhibitor decrease with the increasing of the temperature.

2. The inhibition efficiency (P %) increases with increasing the number of methylene group in the side terminal group, since the values of (P %) decrease in the following order: $MC_{12} < MC_{10} < MC_6$ for the monomers and also in case of polymers: $PC_{12} < PC_{10} < PC_6$. But, the inhibition efficiency of the monomers are lower than those of their analogues polymers.

The inhibition action of these surfactants is due to their adsorption and formation of barrier film on the metal surface which separates the metal from direct contact with corrosive medium and to protect the metal against corrosion. These monomers and their analogues polymers are characterized by the presence of several adsorption centers as N, S, and O atoms and aromatic rings in the molecule. Adsorption might arise via electrostatic interaction between the adsorption centers and the charged metal surface suggesting that the adsorption is of physical -type.

The increase in the inhibition efficiency observed with the increasing of each inhibitor concentration (before the critical concentration) indicating that more inhibitor species are adsorbed on the metal surface. This means the increase of the surface metal coverage.

It is found that under similar conditions the inhibition efficiency of the three monomers increases in the order: $MC_6 > MC_{10} > MC_{12}$. Since the three monomers have the same structure unit different number of carbon atoms in the alkyl chain, the increase of inhibition efficiency could be due to an increase in the number of carbon atoms in the alkyl chain. The obtained data are in good agreement with what found by Maitra. A., et.al [95] and Al-Sabagh et.al, [96]. The action of the long alkyl chain is to stabilize the adsorption of the ionic group cohesion on the metal surface thought van-der Waal forces which allow a more closely packed layer at the metal solution interface [97]. Higher efficiency of the polymers than those of their corresponding monomers could be due to the large size of the polymers which cover wide area of the metal surface and also to the presence of more adsorption centers in the polymers than their respective monomers [98].

In order to get more insight on the inhibition properties of the tested surfactants, the contact angles between the adsorbed inhibitors and aluminium surface in 0.5M HCl solution were measured. It is known that the contact angle (C.A) varies from 0° (perfectly wetting) to 180° (completely nonwetting). The measured contact angles in 0.5 M HCl solution containing different concentrations of monomeric surfactant and their analogues polymers at 30 °C are given in table (4). The obtained data proved the adsorption of these surfactants on the metal surface [99]. The data demonstrate that the inhibition efficiency of the inhibitors increases as their contact angles of the adsorption on aluminium surface decreases. Further more, it is clear that in all cases, the contact angles in the presence of polymers are lower than those in the presence of their respective monomers confirming the suggestion that the polymers used are more effective than their monomers for inhibition of aluminium corrosion in acid medium.

MC_6				MC_{10}				MC_{12}			
C(M)	P % (WT)	P % (Pot)	C.A	C(M)	P % (WT)	P % (Pot)	C.A	C(M)	P % (WT)	P % (Pot)	C.A
3.38×10^{-6}	20.4	18.7	98.6	2.84×10^{-6}	26.2	26.6	29.07	2.64×10^{-6}	28.1	25.8	83.16
1.02×10^{-5}	24.5	25.5	93.2	8.54×10^{-5}	42.2	44.0	25.97	7.91×10^{-6}	43.5	42.0	50.12
1.69×10^{-5}	27.3	28.9	73.4	1.42×10^{-5}	48.1	49.9	8.52	1.31×10^{-5}	55.0	56.5	6.82
3.38×10^{-5}	31.1	32.2	68.6	2.84×10^{-5}	47.0	46.4	10.33	2.64×10^{-5}	49.0	50.0	12.34
PC_6				PC_{10}				PC_{12}			
C(M)	P % (WT)	P % (Pot)	C.A	C(M)	P % (WT)	P % (Pot)	C.A	C(M)	P % (WT)	P % (Pot)	C.A
8.12×10^{-7}	30.4	25.8	91.5	6.87×10^{-7}	32.7	34.9	61.22	6.38×10^{-7}	35.5	36.2	76.91
2.43×10^{-6}	44.7	42.0	69.8	2.06×10^{-6}	46.7	50.8	48.01	1.91×10^{-6}	48.6	50.9	47.12
4.06×10^{-6}	56.1	56.5	0.0	3.43×10^{-6}	59.3	58.3	0.0	3.19×10^{-6}	71.3	69.0	10.78
8.12×10^{-6}	49.7	53.5	12.1	6.87×10^{-6}	53.0	54.0	22.9	6.38×10^{-6}	56.1	55.1	21.45
WT=Weight loss technique				Pot = Polarization technique				C.A= Contact angle			

Table 5. The variation of the inhibition efficiency(P%)and the contact angle (C.A) of aluminium with different inhibitor concentrations using weight loss and polarization measurements of the corrosion reaction at 30 °C.

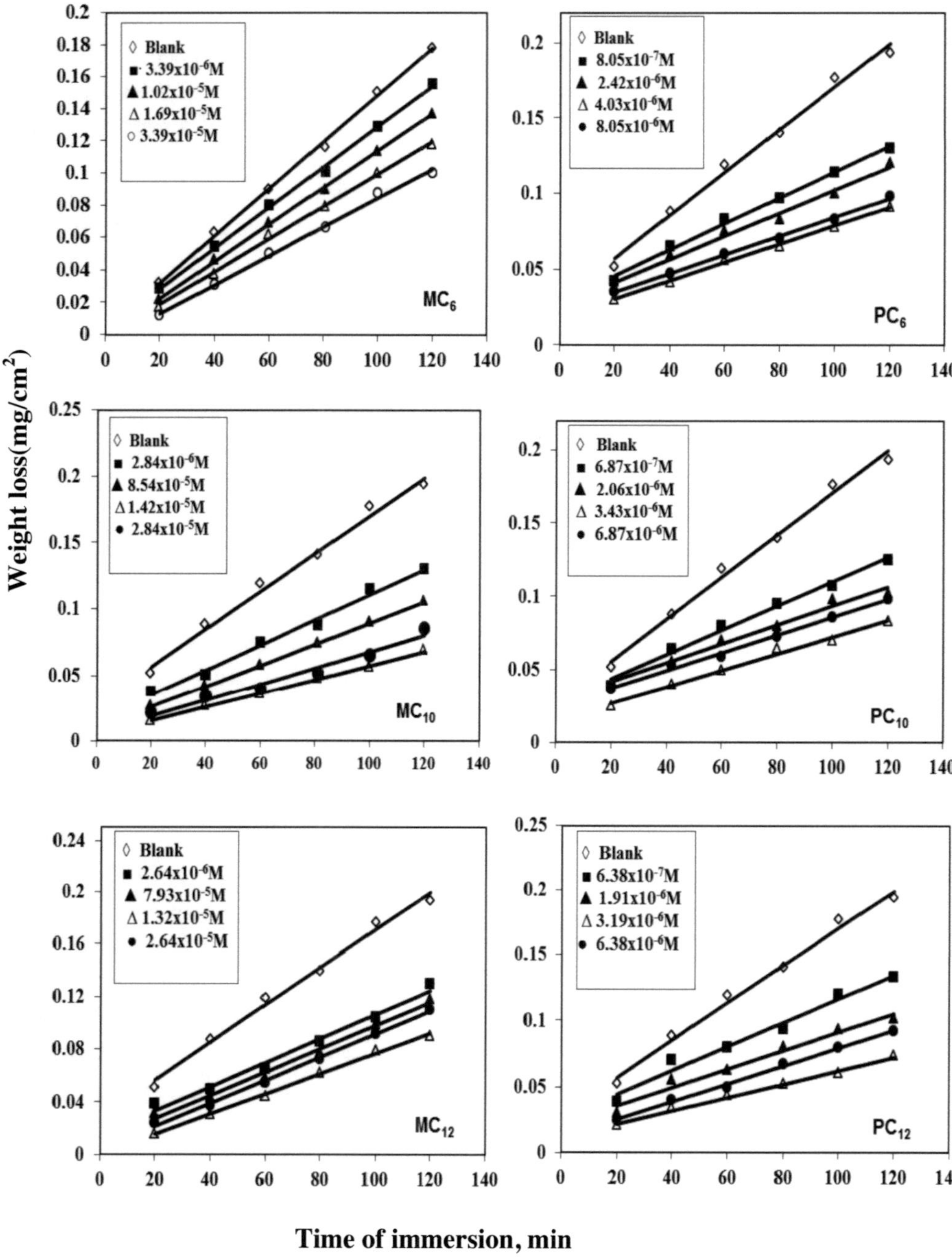

Figure 5. Weight loss vs. immersion time for Al in 0.5 M HCl solution with and without the addition of different concentrations of monomeric and polymeric surfactant at 30 ^{0}C.

http://dx.doi.org/10.5772/57350

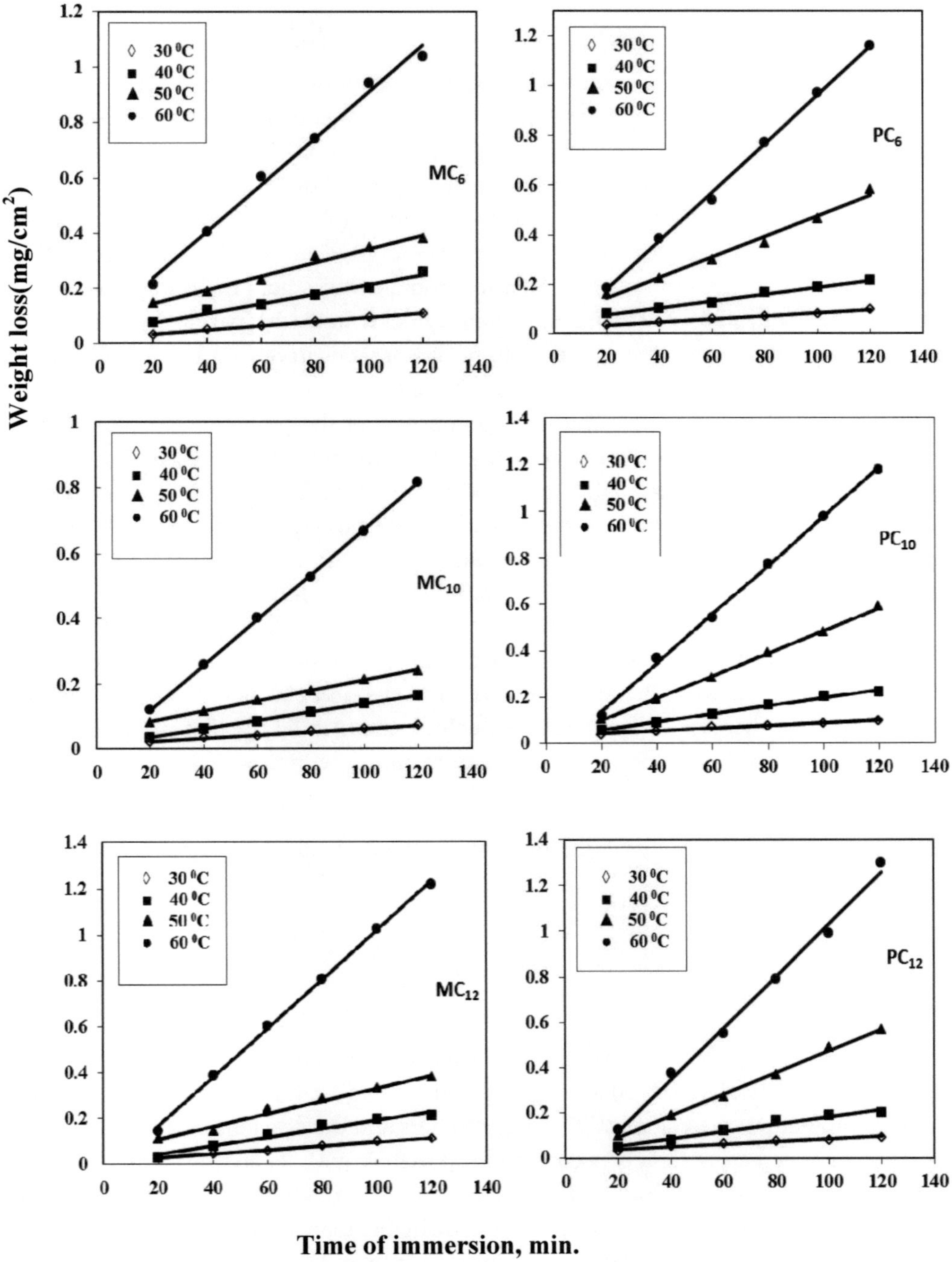

Figure 6. Weight loss vs. immersion time for Al in 0.5 M HCl solution containing 10 ppm of monomeric and polymeric surfactant at different temperatures.

3.3.2. Potentiodynamic polarization measurements

Potentiodynamic cathodic and anodic polarization curves for aluminium in 0.5 M HCl solution in the absence and presence of various concentrations of the prepared monomeric surfactants and their analogues polymers at 30^0 C are shown in figure (7). The curves were swept from -1.05 to 0.6 mV with scan rate 25 mVs^{-1}at 30^0C. On the other hand, figure (8) illustrate the influence of temperature on the polarization curves of aluminium in 0.5 M HCl containing $3.39x10^{-5}$ ML^{-1} MC_6, $2.84x10^{-5}$ ML^{-1} of MC_{10}, $2.64x10^{-5}$ ML^{-1} of MC_{12}, $8.05x10^{-6}$ ML^{-1} of PC_6, $6.87x10^{-6}$ ML^{-1} of PC_{10} and $6.38x10^{-6}$ ML^{-1} of PC_{12}. The figure display that the presence of inhibitor in 0.5 M HCl solution decreases both the cathodic and anodic overpotentials indicating that these surfactants inhibit both the partial hydrogen evolution and the partial dissolution of aluminium. The anodic curves are more polarized than the cathodic curves indicating that these inhibitors are considered as mixed-type inhibitors with anodic predominance. The inhibition action of the prepared monomeric surfactants and their analogues polymers are related to their physical adsorption on the metal surface. The adsorption of the inhibitors modifies the interface between the solution and metal surface, consequently both of the anodic, cathodic overpotential and the corresponding current densities decreases. These results reveal that the inhibitors act by simply blocking the available surface area [100]. In other words, the inhibitor decreases the surface area for corrosion without affecting the mechanism of the partial corrosion reactions and only causes inactivation of a part of surface area with respect to the corrosive medium. Moreover, it is observed that the value of (i_{corr}) in the absence and presence of inhibitor enhances with an increase of temperature. Such behavior indicates that the corrosion rate of aluminium in 0.5M HCl solution in all cases increases with raising temperature.

Since, the corrosion current density (i_{corr}) is directly proportional to the rate of corrosion, the inhibition efficiency (P %) of each inhibitor under the prevailing conditions was calculated using the following equation [101].

$$P\% = 100 \text{ x } \left[1- (i_{corr})I / (\, i_{corr})0\right] \tag{10}$$

Where $(i_{corr})_0$ and $(\, i_{corr})_I$ are the corrosion current densities in uninhibited and inhibited solution respectively. The calculated (P %) values are given in table (4). It follows from these data that at given temperature, the values of (P%)of monomers and their analogues polymers increase gradually with increasing their concentrations up to their critical concentrations. Beyond the critical concentrations, the values of (P%) decrease slightly and then tend to go to steady state values. The inhibition efficiency of the polymers are higher than those of respective monomers. Also, the inhibition efficiency increases with increasing the side chain length of the surfactants. By comparing the values of (P%)of the inhibitors obtained from weight loss measurements with those obtained from polarization measurements, it is seen that the data are comparable and in good agreements. The slight difference exist between these data may be attributed to the short time in case of polarization measurements.

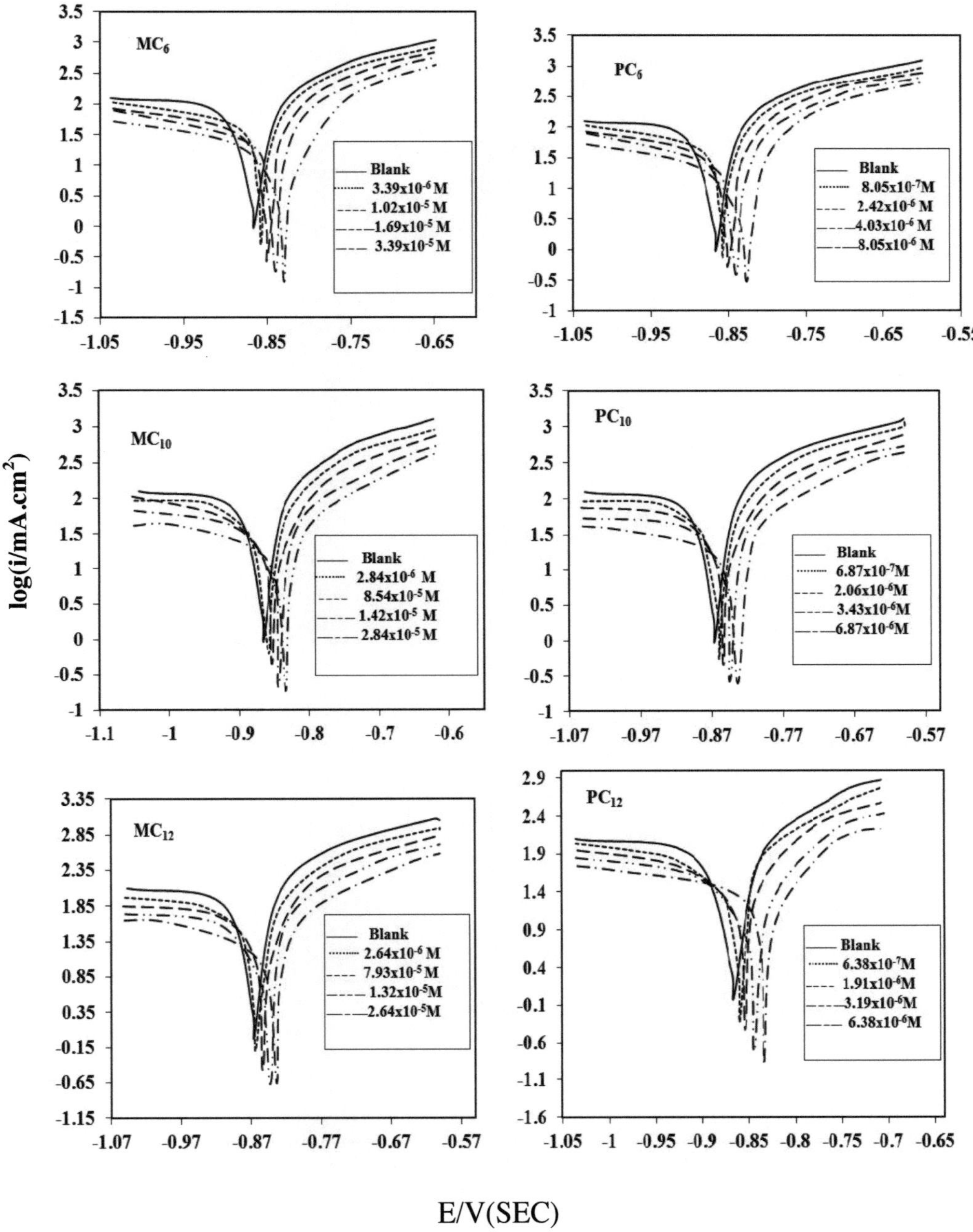

Figure 7. Effect of monomeric and polymeric surfactants concentrations on the cathodic and anodic polarization of Al in 0.5 M HCl solution with a scan rate of 25 mVs^{-1} at 30 ^{0}C.

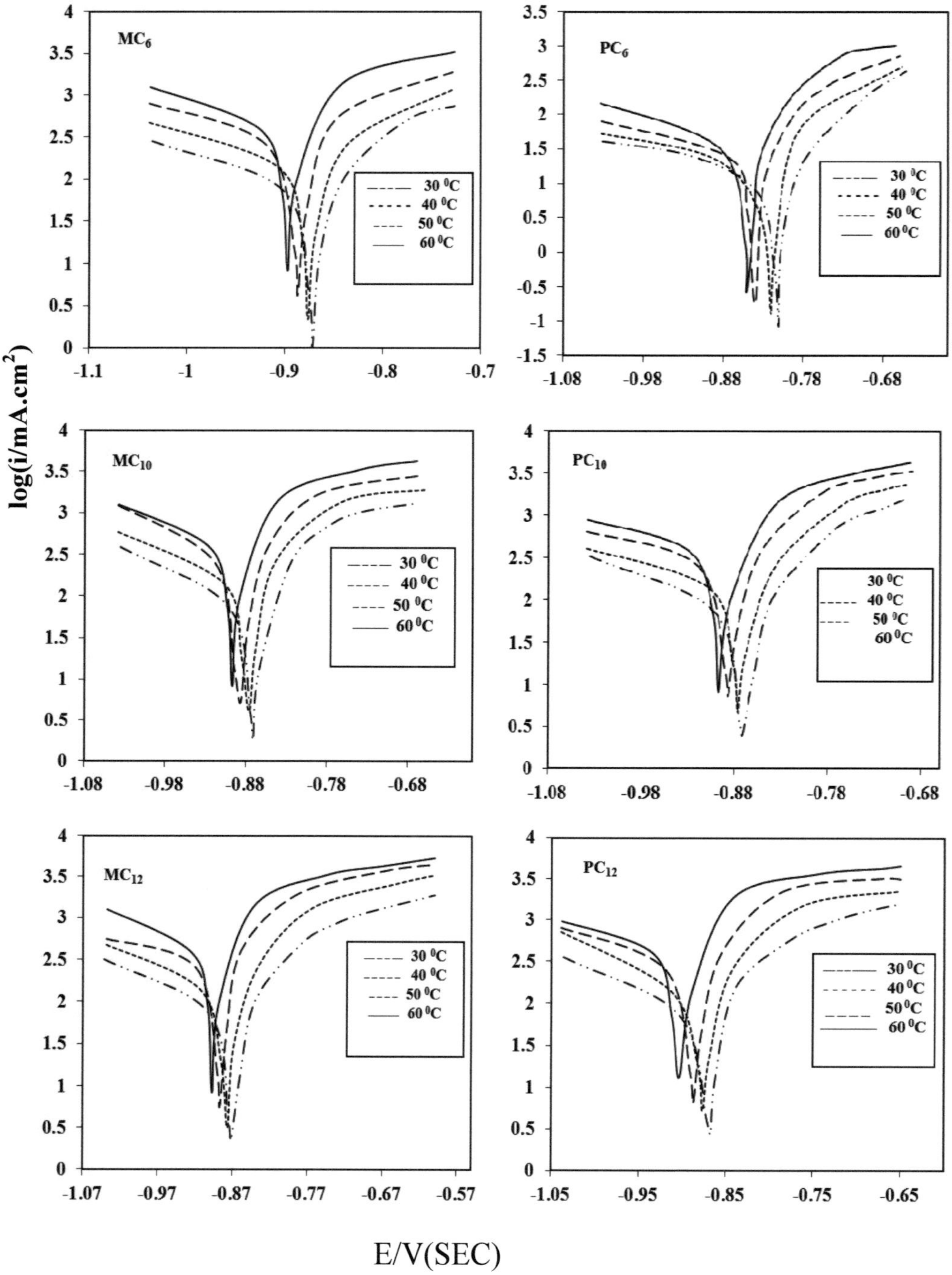

Figure 8. Effect of temperature on the cathodic and anodic polarization of Al in 0.5 M HCl in the presence of 10 ppm of monomeric surfactants and their polymers with a scan rate of 25 mVs^{-1}.

3.3.3. *Activation parameters*

The corrosion rate for each concentration of monomeric and polymeric surfactant was calculated at different temperatures, and the logarithm of the corrosion rate was plotted against the reciprocal of the absolute temperature (1/T) for each concentration according to Arrhenius equation [102]:

$$\text{Logarithm of the corrosion rate} = -E_a / 2.303\,RT + A \tag{11}$$

Where E_a is the apparent effective activation energy, R is the universal gas constant, and A is the Arrhenius pre-exponential factor. The values of E_a in the absence and presence of each concentration of both surfactants were calculated and are tabulated in tables (6,7).

An alternative formula of the Arrhenius equation is the transition state equation [102]:

$$\text{Rate} = RT/Nh \exp(\Delta S^0/R) \exp(-\Delta H^0/RT) \tag{12}$$

Where h is Planck's constant, N is Avogadro's number, ΔS^0 is the entropy of the activation, and ΔH^0 is the enthalpy of activation. The plot of the log(corrosion rate/T) versus 1/T gives a straight line with a slope of $(\_\Delta H^0/2.303R)$, from which the values of ΔH^0 were calculated; and are listed in tables (6, 7). The data in tables (5, 7)reveal that the addition of both monomeric and polymeric surfactants inhibitors leads to increase of the values of both E_a and ΔH^0. This enhancement may be due to the adsorption of surfactant inhibitors on the Al surface leading to an increase in the energy barrier of the corrosion reaction as the concentrations of the inhibitors increased. This means that the corrosion reactions will be further pushed to surface sites that are characterized by progressively higher values of E_a as the concentration of the inhibitor become larger [103]. Also the values of E_a in presence of the polymers are higher than those in presence of the monomers. The positive sign of the ΔH_a indicates that the corrosion process of aluminium in HCl solution is endothermic. Addition of inhibitor to the acid solution increases the values of ΔH_a to an extent depends on the type and concentration of inhibitor. It observed that the values of E_a and ΔH_a obtained from the two techniques are comparable.

	MC_6			**MC_{10}**			**MC_{12}**	
C(M)	**E_a kJmol^{-1}**	**ΔH_a kJmol^{-1}**	**C(M)**	**E_a kJmol^{-1}**	**ΔH_a kJmol^{-1}**	**C(M)**	**E_a kJmol^{-1}**	**ΔH_a kJmol^{-1}**
Blank	57.9	55.5	Blank	57.9	55.5	Blank	57.9	55.5
3.38×10^{-6}	61.6	57.9	2.84×10^{-6}	62.3	59.6	2.64×10^{-6}	64.5	62.3
1.02×10^{-5}	64.2	62.8	8.54×10^{-5}	67.9	64.6	7.91×10^{-6}	67.3	65.1
1.69×10^{-5}	67.2	64.5	1.42×10^{-5}	70.2	68.8	1.31×10^{-5}	72.8	70.0
3.38×10^{-5}	69.1	68.3	2.84×10^{-5}	69.3	67.0	2.64×10^{-5}	70.1	68.5
	PC_6			**PC_{10}**			**PC_{12}**	

C(M)	E_a kJmol^{-1}	ΔH_a kJmol^{-1}	C(M)	E_a kJmol^{-1}	ΔH_a kJmol^{-1}	C(M)	E_a kJmol^{-1}	ΔH_a kJmol^{-1}
Blank	57.9	55.5	Blank	57.9	55.5	Blank	57.9	55.5
8.12 x 10^{-7}	63.5	62.10	6.87 x 10^{-7}	66.7	63.51	6.38 x 10^{-7}	66.63	64.10
2.43 x 10^{-6}	69.4	69.30	2.06 x 10^{-6}	70.9	68.95	1.91 x 10^{-6}	73.73	70.50
4.06 x 10^{-6}	80.6	77.60	3.43 x 10^{-6}	81..4	78.90	3.19 x 10^{-6}	82.5	80.15
8.12 x 10^{-6}	77.7	75.80	6.87 x 10^{-6}	78.6	77.00	6.38 x 10^{-6}	80.11	78.60

Table 6. Effect of inhibitor concentrations on thermodynamic parameters for aluminium corrosion in 0.5 M HCl solution (obtained from Weight loss technique).

	MC_6			MC_{10}			MC_{12}	
C(M)	E_a kJmol^{-1}	ΔH_a kJmol^{-1}	C(M)	E_a kJmol^{-1}	ΔH_a kJmol^{-1}	C(M)	E_a kJmol^{-1}	ΔH_a kJmol^{-1}
Blank	57.3	54.7	Blank	57.3	54.7	Blank	57.3	54.7
3.38 x 10^{-6}	63.2	60.6	2.84 x 10^{-6}	61.7	59.2	2.64 x 10^{-6}	62.6	59.9
1.02 x 10^{-5}	65.7	63.2	8.54 x 10^{-5}	68.7	66.1	7.91 x 10^{-6}	66.4	64.3
1.69 x 10^{-5}	66.8	64.1	1.42 x 10^{-5}	70.1	65.4	1.31 x 10^{-5}	72.0	68.9
3.38 x 10^{-5}	68.2	65.5	2.84 x 10^{-5}	68.9	67.4	2.64 x 10^{-5}	69.8	67.6
	PC_6			PC_{10}			PC_{12}	
C(M)	E_a kJmol^{-1}	ΔH_a kJmol^{-1}	C(M)	E_a kJmol^{-1}	ΔH_a kJmol^{-1}	C(M)	E_a kJmol^{-1}	ΔH_a kJmol^{-1}
Blank	57.3	54.7	Blank	57.3	54.7	Blank	57.3	54.7
8.12 x 10^{-7}	64.0	61.3	6.87 x 10^{-7}	65.1	62.2	6.38 x 10^{-7}	65.3	62.6
2.43 x 10^{-6}	71.0	68.4	2.06 x 10^{-6}	71.3	68.7	1.91 x 10^{-6}	72.5	68.9
4.06 x 10^{-6}	78.2	75.6	3.43 x 10^{-6}	79.4	75.8	3.19 x 10^{-6}	80.4	76.8
8.12 x 10^{-6}	75.1	72.4	6.87 x 10^{-6}	74.8	71.4	6.38 x 10^{-6}	76.2	70.2

Table 7. Effect of inhibitor concentrations on thermodynamic parameters for aluminium corrosion in 0.5 M HCl solution (obtained from Polarization technique)

3.3.4. Adsorption isotherm

In order to gain more information about the mode of adsorption of the inhibitors on the metal surface at different concentrations and temperatures, the data obtained from weight loss and potentiodynamic polarization techniques have been tested with Frumkin adsorption isotherm [104]. Frumkin isotherm was found to fit well with our experimental data. The adsorption isotherm relationship of Frumkin isotherm is represented by the following equation:

$$\ln[\theta/C\,(1-\theta)] = \ln K_{ads} + 2a\theta \tag{13}$$

Where "a" the lateral interaction term describing the molecular interactions in the adsorption layer and the heterogeneity of the surface which reflect the steepness of the adsorption isotherm. Figure (9) represent curves fitting of monomeric and their analogues polymeric surfactants using weight loss, while figure (10) illustrate the data obtained from polarization measurements.

The average thermodynamic parameters viz; enthalpy and entropy of the adsorption process (ΔH_{ads} and ΔS_{ads}) of inhibitor adsorption on the aluminium surface in 0.5 M HCl solution at different temperatures were determined from the slopes and intercepts of the lines of log K_{ads} vs. 1/T according to the following equation [105]:

$$\text{Log } K_{ads} = -\Delta H_{ads}/2.303\,RT + \Delta S_{ads}/2.303\,R \tag{14}$$

The free energy change ΔG_{ads} of adsorption process was calculated from the equation: $\Delta G_{ads} = \Delta H_{ads} - T\Delta S_{ads}$.

Straight lines obtained when log K_{ads} were plotted against 1/T for weight loss and polarization data; with slopes of ($-\Delta H_{ads}/2.303R$) and intercept $\Delta S_{ads}/2.303$ R. The calculated values of ΔH_{ads}, ΔS_{ads} and ΔG_{ads} from weight loss and polarization measurements are listed in tables (8) and (9) respectively. Inspection of these data reveal that, the values of ΔG_{ads} in all cases are negative and less than – 40 kJmol^{-1} suggesting that the nature of inhibitor adsorption is mainly physisorption and spontaneous [106, 107]. The negative values of ΔH_{ads} imply that the adsorption process is an exothermic process [108]. The magnitude of the values of ΔH_{ads}, ΔS_{ads} are characteristic of the occurrence of replacement process during adsorption of surfactants on aluminium surface [109]. It is clear that the values of ΔH_{ads}, ΔS_{ads} and ΔG_{ads} obtained from weight loss measurements are in accordance with the respective values obtained from polarization measurements.

Surfactant	$-\Delta H_{ads}$ kJmol^{-1}	$-\Delta S_{ads}$ kJmol^{-1}K^{-1}	$-\Delta G_{ads}$ kJmol^{-1}
MC_6	109.6	0.232	39.30
MC_{10}	48.63	0.052	32.87
MC_{12}	65.95	0.1105	32.47
PC_6	45.11	0.0215	38.60
PC_{10}	51.95	0.0491	37.08
PC_{12}	46.32	0.0298	37.28

Table 8. The thermodynamic parameters of the adsorption process obtained by applying Frumkin isotherm from the weight loss data of Al in 0.5 M HCl.

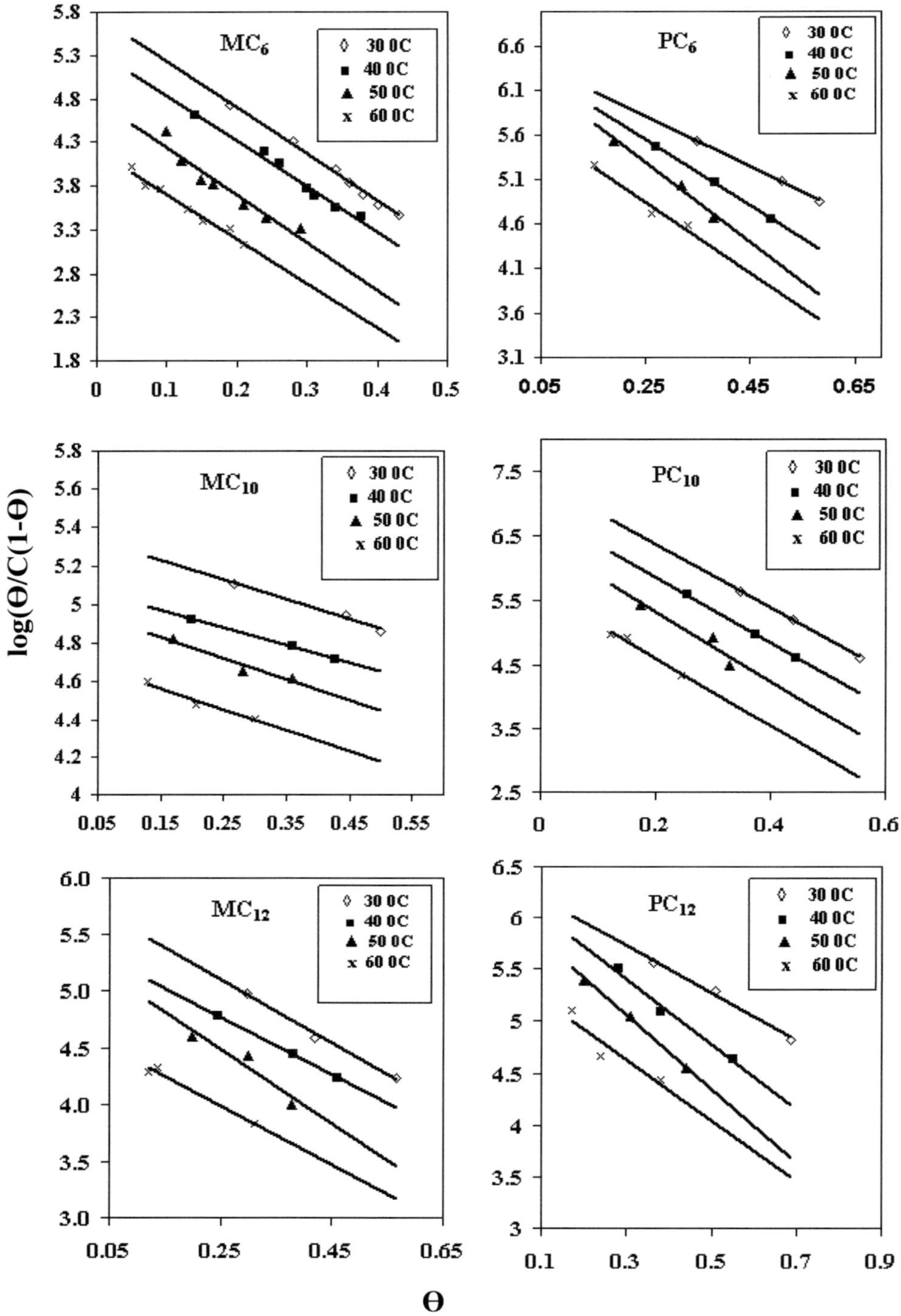

Figure 9. Curve fitting of weight loss data of Al in 0.5 M HCl solution containing various concentrations of monomeric and polymeric surfactant to Frumkin isotherm at different temperatures.

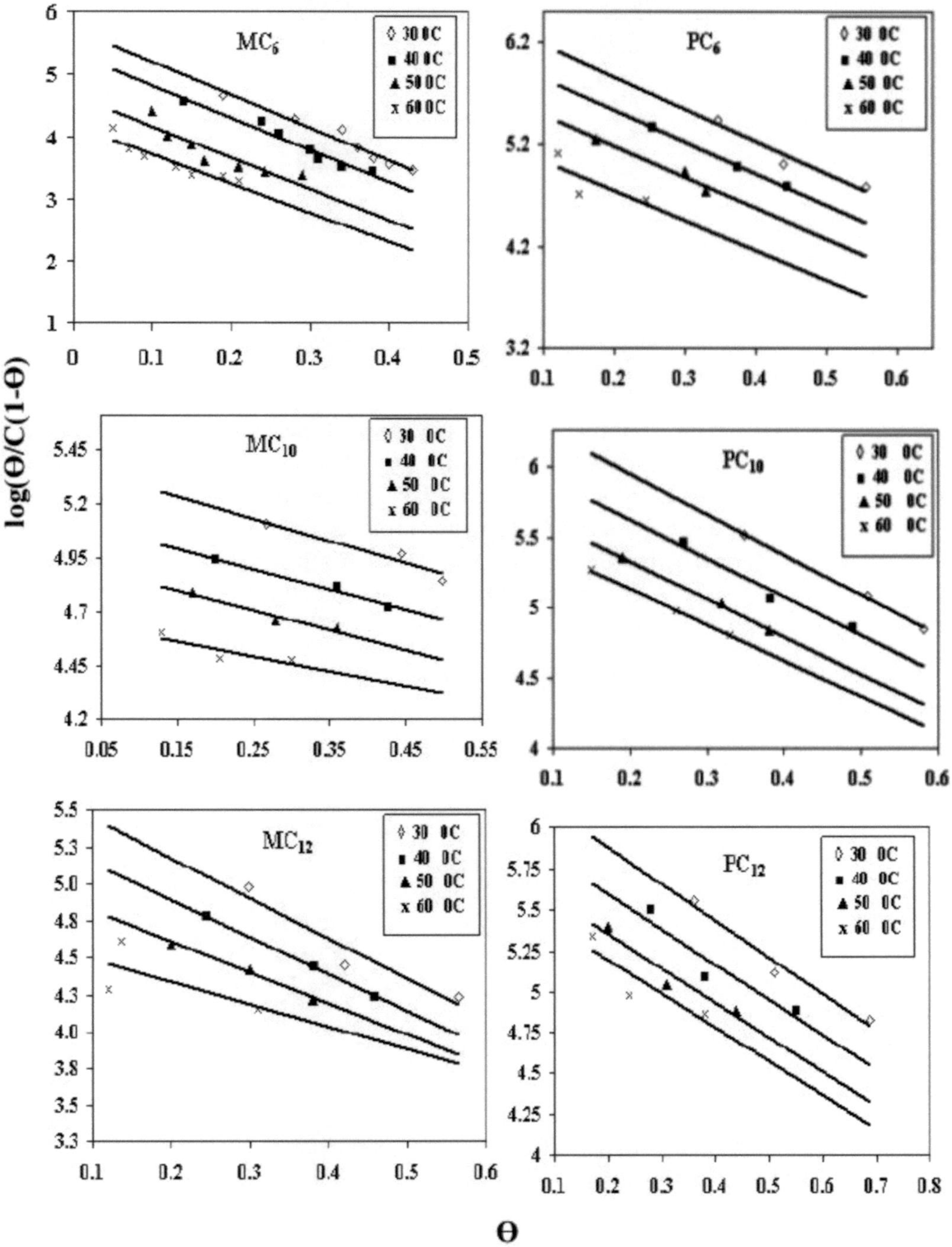

Figure 10. Curve fitting of polarization data of Al in 0.5 M HCl solution containing various concentrations of monomeric and polymeric surfactant to Frumkin isotherm at different temperatures.

Surfactant	$-\Delta H_{ads}$ kJmol^{-1}	$-\Delta S_{ads}$ kJmol^{-1}K^{-1}	$-\Delta G_{ads}$ kJmol^{-1}
MC_6	103.4	0.222	37.83
MC_{10}	45.76	0.048	31.258
MC_{12}	67.21	0.1128	33.01
PC_6	44.03	0.0216	37.48
PC_{10}	52.85	0.0504	37.59
PC_{12}	45.57	0.0301	36.45

Table 9. The thermodynamic parameters of the adsorption process obtained by applying Frumkin isotherm from the polarization data of aluminum in 0.5 M HCl.

3.3.5. Effect of terminal side chain of polymeric surfactant on the inhibition efficiency of aluminium corrosion in 0.5M HCl solution

The previous results show that, the inhibition efficiency of the monomeric and polymeric surfactants increases in the order: $MC_6 > MC_{10} > MC_{12} > PC_6 > PC_{10} > PC_{12}$. The inhibition efficiency of PC_{12} no more than (69%), however these surfactants are characterized by the presence of several adsorption centers as N, S, and O atoms and aromatic rings in the molecule. This may be due to the repulsion between the terminal side -SO_3H group and double layer exist on Al surface in 0.5 M HCl solution. To overcome this phenomenon, the -SO_3H group replaced by hydrogen atom as in poly 3-dodycyloxy aniline.

The suggested structure of poly 3-dodycyloxy aniline as mentioned by S. M. Sayyah et al [110], is given in scheme (4).

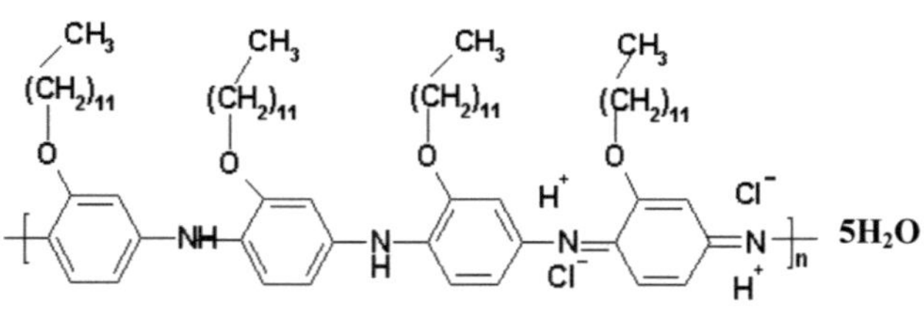

Scheme 4. Structure of poly 3-dodycyloxy aniline.

3.3.5.1. Potentiodynamic polarization measurements

Figure (11) shows the potentiodynamic cathodic and anodic polarization curves for Al in 0.5 M HCl in the absence and presence of different concentrations of $PC_{11}R$ at 30^{0}C.The curves were swept from -1.05 to 0.6 mV with scan rate 25 mVs^{-1}. The curves infer that the presence of $PC_{11}R$ in acid solution increases both the cathodic and anodic overpotentials (mainly the anodic overpotential). The obtained results indicate that, this polymer inhibit both the partial cathodic hydrogen evolution and the partial anodic dissolution of aluminium reaction. The anodic curves are more polarized than the cathodic curves indicating that, this inhibitor is considered as mixed-type inhibitor with anodic predominance.

http://dx.doi.org/10.5772/57350

Figure (12) illustrates the potentiodynamic polarization curves for Al in 0.5M HCl solution containing 8.04 x 10^{-6} Mole $PC_{11}R$ at different temperatures. It is observed that the current corrosion density (i_{corr}) in the absence and presence of the inhibitor increased with raising the temperature.

Figure (12) shows the variation of inhibition efficiency (P%) with different concentrations of PC_{12} and PC11R at different temperatures. Inspection of these data reveal that:

1. At a given temperature, the inhibition efficiency (P%) of this polymer increases with increasing its concentration as a result of increasing the surface coverage, thus the Al surface is efficiently separated from the corrosion medium. The results also reveal that when the concentration of this polymeric surfactant reaches a certain critical value, the inhibition efficiency slightly increases with concentration and tend to reach a steady state value. Such behavior denoted that the amount of adsorption and coverage does not change and the corrosion of Al tends to achieve equilibrium. At high concentrations (Close to its CMC, the surfactant molecules tend to form micelles in solution rather than adsorbing on metal surface).

2. At given, inhibitor concentration, the inhibition efficiency of $PC_{11}R$ decreases with an increase in temperature. This could be due to the increase in the corrosivity of medium and /or desorption of some inhibitor molecules from Al surface. The desorption process leads to a decrease in the surface coverage indicating that the adsorption is of physical nature.

3. Under identical condition, it is found that the inhibition efficiency of $PC_{11}R$ is higher than that of PC_{12}. This may be assigned to the elimination of the terminal -SO_3H groups in PC_{12}. The presence of terminal –CH_3 groups in polymer $PC_{11}R$ stabilizes the inhibition efficiency and may reduce repulsion between anionic head with similar charge as in PC_{12}. This allows a closed layer to form more easily and hence higher inhibition efficiency.

The apparent activation energy (E_a), enthalpy and entropy for corrosion of Al in the presence of $PC_{11}R$ was calculated using Arrhenius and alternative Arrhenius equations (11,12). The data of activation are calculated and tabulated in table (10), from which it is obvious that the values $E_{a,}$ ΔH_a and ΔS_a of the corrosion of Al in the presence of the inhibitor are higher than those of uninhibited solution and increase with an increase in the inhibitor concentration indicating that more energy barrier for the corrosion reaction in the presence of inhibitor is attained.

Concentration ppm	E_a kJ.mol^{-1}	ΔH kJmol^{-1}	ΔS kJmol^{-1}
Blank	57.30	54.70	0.381
1	71.61	68.97	0.267
3	71.99	69.36	0.268
5	99.37	96.88	0.039
10	109.31	106.65	0.016

Table 10. Effect of inhibitor ($PC_{11}CH_3$) concentrations on thermodynamic parameters for aluminium corrosion in 0.5 M HCl solution.

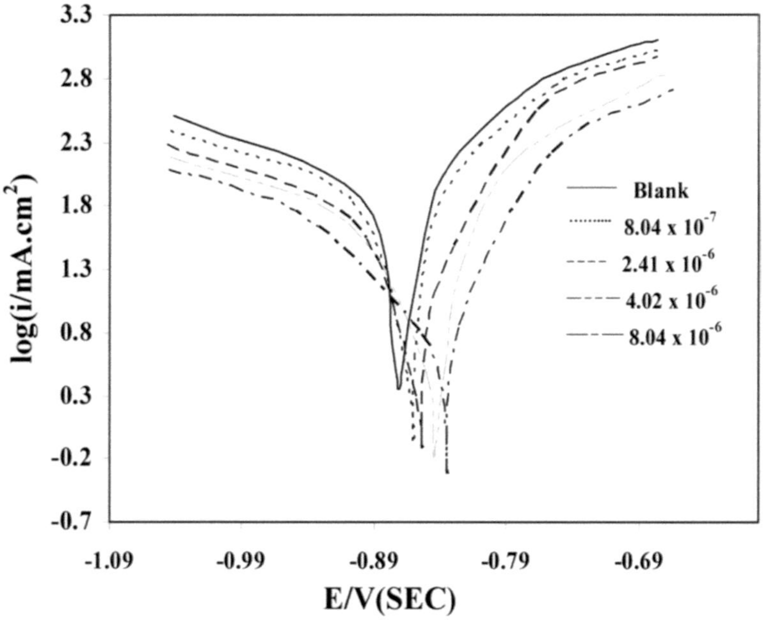

Figure 11. Effect of $PC_{11}R$ concentration on the anodic and cathodic polarization of Al in 0.5 M HCl with a scan rate of 25 mVs^{-1} at 30 ^0C

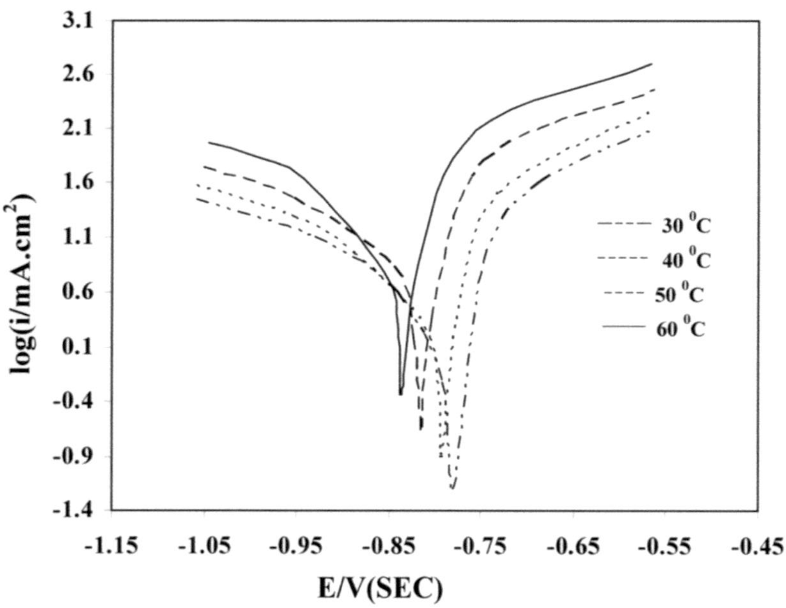

Figure 12. Effect of temperature on the anodic and cathodic polarization of aluminium in 0.5 M HCl in the absence and presence 8.04 x 10^{-6} Mole of $PC_{11}R$ with a scan rate of 25 mVs^{-1}.

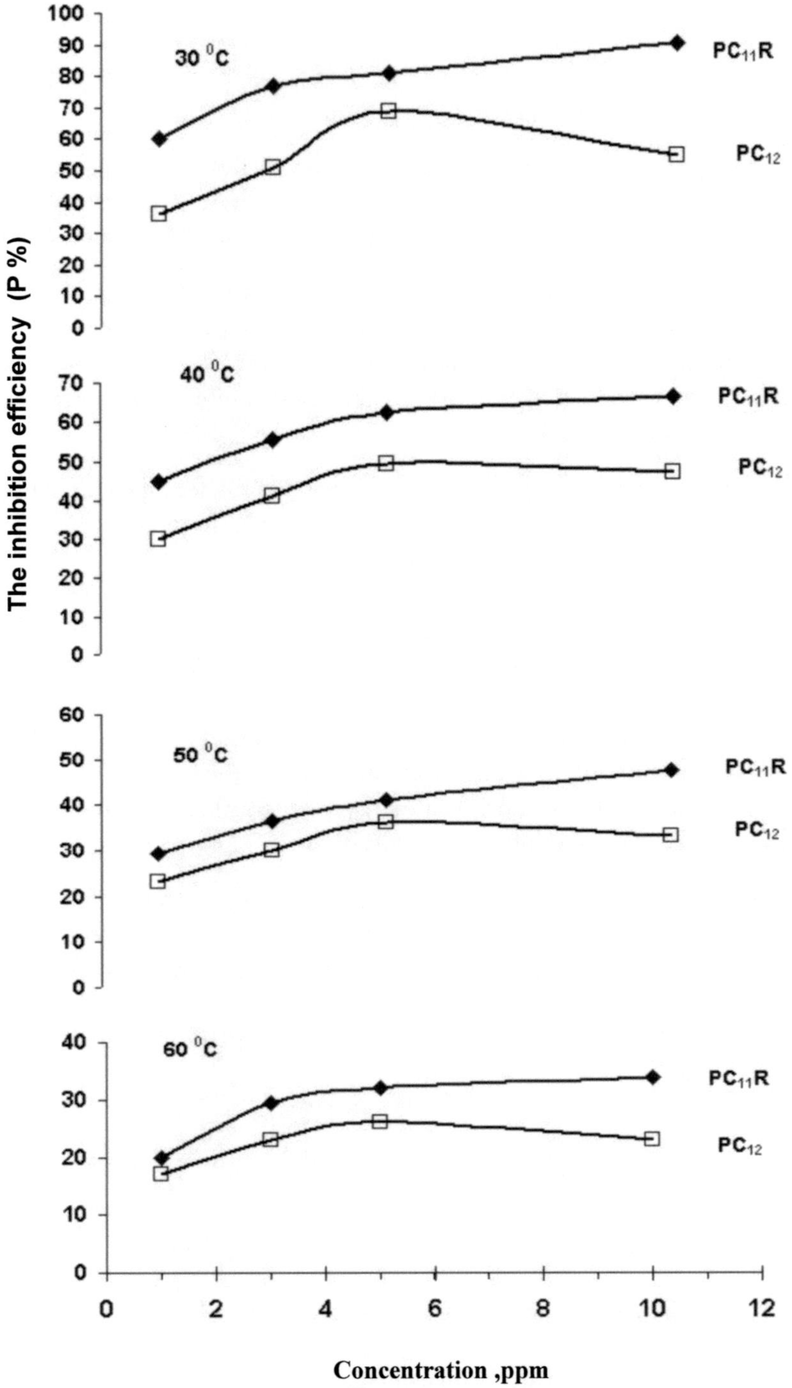

Figure 13. The variation of inhibition efficiency(P%) with different concentrations of PC_{12} and $PC_{11}R$ at different temperatures.

4. Conclusion

- Monomeric surfactants (MC_6),(MC_{10}) and (MC_{12}) and their analogues polymeric surfactants inhibit the corrosion of Al in a 0.5 M HCl solution.
- Weight loss measurements reveal that, the inhibition action of these surfactants is due to their adsorption and formation of barrier film on the Al surface which separates the Al from direct contact with corrosive medium and to protect the metal against corrosion.
- Corrosion tests display that, the presence of inhibitor in HCl solution decreases both the cathodic and anodic overpotentials indicating that, these surfactants inhibit both the partial hydrogen evolution and the partial dissolution of aluminium. The anodic curves are more polarized than the cathodic curves indicating that these inhibitors are considered as mixed-type inhibitors with anodic predominance.
- The inhibition efficiency of the polymeric surfactant was higher than that of the monomeric surfactant, and the inhibition efficiency increased with increasing inhibitor concentration but decreased with increasing temperature.
- Weight loss and potentiodynamic polarization techniques are in good agreements.
- The activation parameters for the corrosion of aluminium in 0.5M HCl solution of the monomers and their respective polymers were calculated from Arrhenius –type and alternatively, Arrhenius equations. The apparent activation energy (Ea), the enthalpy of activation(ΔHa) of aluminium corrosion and the extent of the increase is proportional to the inhibitor concentration. Such trend indicates that the energy barrier for the corrosion process enhances with increasing of inhibitor concentrations.
- The inhibition efficiency of $PC_{11}R$ is higher than that of PC_{12}. This may be assigned to the elimination of the terminal -SO_3H groups in PC_{12}. The presence of terminal –CH_3 groups stabilizes the inhibition efficiency. Also, the presence of terminal –CH_3 in the alkyl chain in $PC_{11}R$ molecules may reduce repulsion between anionic head with similar charge as in PC_{12}. This allows a closed layer to form more easily and hence higher inhibition efficiency.

Author details

S.M. Sayyah[1], S.S.Abd El-Rehim[2], M.M. El-Deeb[1] and S.M. Mohamed[1]

*Address all correspondence to: smsayyah@hotmail.com

1 Polymer Research Lab., Chemistry Department, Faculty of Science, Beni Suef, University Beni Suef, Egypt

2 Chemistry Department, Faculty of Science, Ain Shams University, Abasia, Egypt

References

[1] Umoren,S.A., (2009). "Polymers as corrosion inhibitors for metals in different media,"The Open Corros. J. 2, 175-188.

[2] Oguzie, E.E., (2007). "Corrosion inhibition of aluminium in acidic and alkaline media by Sansevieria trifasciata extract," Corros. Sci. 49, 1527–1539

[3] Sorkhabi,H. A., Ghasemi, Z., and Seifzadeh, D. (2005). "The inhibition effect of some amino acids towards the corrosion of aluminium in 1 M HCl + 1 M H2SO4 solution, " Appl. Suf. Sci. 247, 408-418.

[4] Roberge PR.(2006). "Corrosion Basics—An Introduction". 2nd ed. Houston, Tex: NACE International. P.364.

[5] Frankel,G. S., (1998). "Pitting Corrosion of Metals: A Review of the Critical Factors". J. Electrochem. Soc. 145(6): 2186-2198.

[6] Povroznik, V. S., and Shein,A. B., (2007) "Environmental and Inherent Factors That Affect Hydrogen Cathodic Evolution on Silicides of the Iron Family Metals"., Protection of Metals, 43(2), 216–221.

[7] Ammar, I. A. and Awad, S. A., (1956) "Effect of some corrosion inhibitors and activators on the hydrogen overpotential at Fe cathodes in NaOH solutions". J. of Phys. Chem., 60, 871-874.

[8] Bockris, J. O. and Conway, B. E. "Hydrogen overpotential and the partial inhibition of the corrosion of iron". J.Phys. & Colloid Chem., (1949) 53, 527-539.

[9] Zeng, X.X., Wang, J.M., Wang, Q.L., Kong, D.S., Shao, H.B., Zhang, J.Q., and Cao, C.N. (2010) "The effects of surface treatment and stannate as an electrolyte additive on the corrosion and electrochemical performances of pure aluminum in an alkaline methanol–water solution". Mater. Chem. & Phys. 121, (3),459-464.

[10] Kunitsugu, A., (2003). "Improvement in the self-healing ability of a protective film consisting of hydrated cerium(III) oxide and sodium phosphate layers on zinc" Corrs. Sci., 45 (2),451.

[11] Atashin, S., Toloei, A and Pakshir, M.,(2013)."Simultaneous Investigation of Marine Factors Effect on Corrosion Rate of SS 304 in Turbulent Condition". J.Mater. Eng.& Performance, 22 (7), 2038-2047.

[12] Kim, W. K., Koh, S. U., Yang, B. and Kim, K. Y., (2008)."Effect of environmental and metallurgical factors on hydrogen induced cracking of HSLA steels". Corrs. Sci., 50 (12), 3336-3342.

[13] Tomita, M., Yokoyama, K., and Sakai, J., (2008)."Effects of potential, temperature and pH on hydrogen absorption and thermal desorption behaviors of Ni–Ti superelastic alloy in 0.9% NaCl solution". Corrs. Sci., 50 (7).2061-2069.

[14] Adhikari, S., and Hebert, K., (2008)."Factors controlling the time evolution of the corrosion potential of aluminum in alkaline solutions". Corrs. Sci., 50 (5), 1414-1421.

[15] Zaid, B., Saidi, D., Benzaid, A. and Hadji, S., (2008)."Effects of pH and chloride concentration on pitting corrosion of AA6061 aluminum alloy". Corrs. Sci.; 50 (7), 1841-1847.

[16] Ha.,H.Y. and Kwon., H.S., (2012)."Effects of pH Levels on the Surface Charge and Pitting Corrosion Resistance of Fe". J. Electrochem. Soc., 159 (9) 416-421.

[17] Böhni, H. and Uhlig, H. H., (1969)."Environmental Factors Affecting the Critical Pitting Potential of Aluminum".J. Electrochem. Soc. 116(7): 906-910.

[18] Ahmad, S.,(2003), "Reinforcement corrosion in concrete structures, its monitoring and service life prediction—a review" Cement and Concrete Composites.,25,: 459–471.

[19] Leckie, H. P., and Uhlig, H. H. ., (1966). "Environmental Factors Affecting the Critical Potential for Pitting in 18–8 Stainless Steel" J. Electrochem. Soc. 113(12), 1262-1267.

[20] Xu, T., Senger, R., and Finsterle, S., (2008). "Corrosion-induced gas generation in a nuclear waste repository: Reactive geochemistry and multiphase flow effects" Applied Geochemistry 23(12), 3423–3433

[21] Wang, Y. and Bierwagen, G., (2009). " A new acceleration factor for the testing of corrosion protective coatings: flow-induced coating degradation". J. Coatings Technology & Research; 6 (4)., 429-436.

[22] Hassani, S., Roberts, K. R., Shirazi, S. A., Shadley, J. R., Rybicki, E. F., and Joia, C., (2012). "Flow Loop Study of NaCI Concentration Effect on Erosion, Corrosion, and Erosion-Corrosion of Carbon Steel in CO2-Saturated Systems. Corrosion; 68 (2)26001-26009.

[23] Stott, F. H., and Shih, C. Y., (2000)."High-Temperature Corrosion of Iron–Chromium Alloys in Oxidizing–Chloridizing Conditions". Oxidation of Metals., 54, 425-443.

[24] Tavakkolizadeh, M., and Saadatmanesh, H., (2001)."Galvanic Corrosion of Carbon and Steel in Aggressive Environments." J. Compos. Constr., 5(3), 200–210.

[25] Lichtenstein, J.,(2003)."Corrosion Inhibitors".Materials Performance; (12), 62-63.

[26] William C. Lyons., (19960 "Standard Handbook of Petroleum & Natural Gas Engineering". volume 1,p.6.

[27] Pryor, M. J., and Cohen, M., (1953) "The Inhibition of the Corrosion of Iron by Some Anodic Inhibitors". J. Electrochem. Soc 100 (5),3-215.

[28] Fontana, M. G., (1986). "Modern Theory—Principles, in Corrosion Engineering". McGraw-Hill, New York, pp.445-481.

[29] Miksic.BA and Vignetti.A. (2000) "Vapour Corrosion Inhibitors: successful field applications in electronics" NACE Corrosion Conference; Orlando; Paper # 7, 2000.

[30] Amin,M. A., Abd El-Rehim, S.S., El-Sherbini, S. E., Hazzazi,O. A., and Abbas, M.N. (2009). "Polyacrylic acid as a corrosion inhibitor for aluminium in weakly alkaline solutions. Part I: weight loss, polarization, impedance EFM and EDX studies." Corros. Sci. 51, 658-667.

[31] Breslin,C., Fenelon,A., and Conroy,K. (2005). "Surface engineering: corrosion protection using conducting polymers" Material and Design 26, 233-237.

[32] Wang,Y., Li,Y., and Wang, F. (2012). "Polymeric organo-silane coatings for aluminium alloy corrosion protection by self-assembled method." E-J. of Chem. 9(1), 435-442.

[33] Sultana, S., Mohammad,R., Khan,Z., Umar,K., and Muneer,M.(2013). "Electrical, thermal, photocatalytic and antibacterial studies of metallic oxide nanocomposite doped polyaniline." J. Mat. Sci. Tech. xxx-xxx.

[34] Rodriguez, J.G., Lucio-Garc´ıa. M.A., Nicho,M.E., Cruz-Silva, R., Casales, M., and Valenzuela, E. (2007)."Improvement on the corrosion protection of conductive polymers in pemfc environments by adhesives. " J. power sources 168, 184-190.

[35] Joseph, S., McClure,J.C., Sebastian, P.J., Moreira,J., and Valenzuela,E. (2008). "Polyaniline and polypyrrole coatings on aluminium for PEM fuel cell bipolar plates" J. power sources 177, 161-166.

[36] Lee,C., Lee,Y., Kim,K., Jeong,M., and Lim, D. (2013) "Electrically conductive polymer composite coating on aluminium for PEM fuel cells bipolar plate" Renewable Energy, 54, 46-50.

[37] Johansen,H. D., Brett,C. M., and Motheo,A. J. (2012). "Corrosion protection of aluminium alloy by cerium conversion and conducting polymer duplex coatings" Corros. Sci 63, 342- 350.

[38] Zubillaga,O., Cano,F.J., Azkarate,I., Molchan,I.S., Thompson,G.E., Cabral,A.M., Morais,P.J. (2008)."Corrosion performance of anodic films containing polyaniline and TiO2 nanoparticles on AA3105 aluminium alloy." Surface & Coatings technology, 202, 5936-5942.

[39] Gupta,K., Jana, P.C., Meikap,A.K. (2012). "Electrical transport and optical properties of the composite of polyaniline nanorod with gold" Solid State Sciences 14, 324-329.

[40] Shawn, E.B., Brock,A.W., Tito,V.(2012). "Electrical and thermal properties of graphite/polyaniline composites." J. Solid State Chem.196,309–313.

[41] Belabed,C., Abdi,A., Benabdelghani, Z., Rekhila,G., Etxeberria,A., and Trari, M. (2013). "Photoelectrochemical properties of doped polyaniline: Application to hydrogen photoproduction." Int. J. hydrogen energy 38, 6593 -6599.

[42] Sayyah,S.M., Azzam,E., Khalil,A., and Mohamed,S. M.(2013) "Improvement of the surface activity and the solubility for some synthesized polyaniline surfactants". Egy.J. Chem, xxx-xxx

[43] Kulkarni,M.V., Viswanath, A. K. (2004) "Comparative studies of chemically synthesized polyaniline and poly(o-toluidine) doped with p-toluene sulphonic acid" Eur. Polym. J. 40, 379–384.

[44] Gul,S., Shah,A., Bilal, Salma.,(2010) "Thermogravimetric study of polyaniline dodecylbenzene sulfonic acid" J.Pak. Mater.Soc. 4 (2),71-80.

[45] Ikkala, O.T., Lindholm,T.M., Ruohonen,H., Seläntaus,M., Väkiparta,K.(1995) "Phase behavior of polyaniline/dodecyl benzene sulphonic acid mixture." Synth. Met. 69, 135-136.

[46] Bıcak, N., Senkal,B., F., Sezer, E. (2005) "Preparation of organo-soluble polyanilines in ionic liquid." Synth. Met 155, 105–109.

[47] Lin,D.S.(2004). "Absorbance behavior of polyaniline-poly(styrenesulfonic acid)." J. Chin. Chem. Soc 51,1279-1286.

[48] Shi-Jian, S., Noriyuki K. (2000). "Synthesis of processable polyaniline complexed with anionic surfactant and its conducting blends in aqueous and organic system." Synth. Met. 108,121–126.

[49] (a) Watanabe, A., Mori, K., Iwabuchi, A., Iwasaki, Y., and Nakamura, Y., (1989).Macromolecules 22, 3521. (b) Wu, Y., Fuke, W. W., Wnik, G. E., Ray, A., and MacDiarmid, A. G., (1989) J. Phys. Chem. 93, 495. (c) Genies, E. M. and Noel, P., J. Electroanal. Chem. 310, 89 (1991). (d) Wang, L., Jing, X., and Wang, F., (1991) Synth. Met. 739. 41–43

[50] Sayyah,S.M., Abd El-Khalek,A.A., and Abd El-Salam,H. M. (2001). "Kinetic studies of the polymerization of substituted aniline in aqueous solutions and characterization of the polymer obtained."J. Polym. Mater 49,25-49.

[51] Sayyah,.S. M., Abd El-Salam,H.M., and Azzam,E. M. (2005). "Surface activity of monomeric and polymeric (3-alkyloxy aniline) surfactants." Int. J. Polym. Mater. 54, 541-554.

[52] (a) Matted, L. H., and Bulhoest, L. O. S., Synth Met 52, 471 (1992). (b) Gupta, M. C., and Umare, S. S., Macromolecules 25, 138 (1992). (c) Apramo, G. D. and LeClerc, M. J., Electroanal. Chem 351, 145 (1993). (d) Storrier, G. D., Colbran, S. B., and Hibbert, D. B., Synth. Met. 62, 179 (1994).

[53] Sayyah, S. M., El-Salam, H. M. A., and Bahgat, A. A. (2002)."Aqueous oxidative polymerization of 3-methoxyaniline and characterization of its polymer" J. Polym. Mater. 51,915-923.

[54] Sharma, S., and Kumar,D. (2010). "Study on salvatochromic behavior of polyaniline and alkyl substituted polyanilines. "Indian J.Engy. Mat. Sci. 17,231-237.

[55] Sayyah, S. M., Bahgat, A. A., and Abd El-Salam, H. M. (2002). "Kinetic studies of the aqueous oxidative polymerization of 3-hydroxyaniline and characterization of the polymer obtained." Int. J. Polym. Mater. 51,291-314.

[56] Sayyah, S. M., Abd El-Khalek, A. A., Bahgat, A. A., and Abd El-Salam, H. M. (2001). "kinetic study of the chemical polymerization of substituted aniline in aqueous solution and characterization of the polymer obtained,part1:3-chloroaniline." Int. J. Polym. Mater. 50, 197-208.

[57] Kumar,P.,and Ahmed.A.(2001)."Chemical synthesis, spectral characterization and stability of some electrically conducting polymers. " Chinese Journal of Polymer Science 28, 191-197.

[58] Dai,L., Xu,Y., He,Y., Gal.J., Wu., H. (2005)A comparative study on the electrochemical properties of ring-substituted polyaniline. Polym.Int.54, 1256-1261.

[59] Borkar, A. D., (2012). "Effect of copolymer composition on solubility and electrical conductivity of poly(aniline-co-o-chloroaniline). " Journal of Chemical and Pharmaceutical Research 4(7), 3526-3528

[60] Sayyah, S. M., and Abd El-Salam, H. M. (2003). "Aqueous oxidative polymerization of N-methyle aniline in acid medium and characterization of the obtained polymer." Int. J. Polym. Mater. 52, 1087-1111.

[61] Tadros, F., (ed.):The surfactants, academic Oress, London, 1984.

[62] Holmberg,K., Jonsson,B., Kronberg,B., and Lindman,B.(2003). "Surfactants and polymers solution 2nd edition, John Wiley & Sons, Chichester.

[63] Shaohua,L., and Jun, Wu., (2012)."Micellar evolution in mixed onionic/anionic surfactant systems". Journal of Colloid & nterface Science.,367 I(1), 272-279.

[64] Tsing-Hai Wang., Chi-Jung Hsieh., Shih-Min Lin., Ding-Chiang Wu., Ming-Hsu Li and Shi-Ping Teng., (2010)."Effect of Alkyl Properties and Head Groups of Cationic Surfactants on Retention of Cesium by Organoclays". Environmental Science & Technology. 44 (13),5142-5147.

[65] Liu, Jun.,Jiang, Yan.,Chen, Hong., Mao, Shi Zhen., Du, You Ru and Liu, Mai Li (2012). "Probing Dynamics and Mechanismof Exchange Processof Quaternary Ammonium Dimeric Surfactants, 14-s-14, in the Presence of Conventional Surfactants". Journal of Physical Chemistry., 116 (51), p14859-14868

[66] OsVan, N. M., Haak,J. R., Rupert,L. A. (1993)."Physico- chemical properties of selected anionic, cationic, and non-ionic surfactants." Elsevier, Amsterdam.

[67] Porter,M. R. (1994).Handbook of Surfactants, Blackie, London.

[68] Latief,F.H., and Sherif,E. S.M. (2012)."Effects of sintering temperature and graphite addition on the mechanical properties of aluminum" J.Ind. Eng.Chem 18, 2129-2134.

[69] Sherif,E. S.M., Almajid,A.A., Latif,F.H., and Junaedi,H. (2011)."Effects of Graphite on the Corrosion Behavior of Aluminum-Graphite Composite in Sodium Chloride Solutions." Int. J. Electro-chemical Sci 6(4), 1085-1099.

[70] Meng,G., Wei,L., Zhang, T., Shao,Y., Wang,F., Dong,C., and Li, X.(2009). "Effect of microcrystallization on pitting corrosion of pure aluminium" Corr. Sci.51, 2151-2157.

[71] Mousavifard,S.M., Nouri,P.M.M., Attar,M.M., and Ramezanzadeh,B. (2013). "The effects of zinc aluminum phosphate (ZPA) and zinc aluminum polyphosphate (ZAPP) mixtures on corrosion inhibition performance of epoxy/polyamide coating." J.Ind. Eng.Chem. 19, 1031-1039.

[72] Sherif,E.M., and Park,S.M. (2005) "Effects of 1,5-naphthalenediol on aluminum corrosion as a corrosion inhibitor in 0.5 M NaCl" J. Electr. Soci. 152, B205.

[73] Safak, S., Duran,B., Yurt,A., Türkoğlu,G. (2012). "Schiff bases as corrosion inhibitor for aluminium in HCl solution." Corros. Sci. 54, 251-259.

[74] El-Sayed, M. S., (2013) "Electrochemical investigations on the corrosion inhibition of aluminum by 3-amino-1,2,4-triazole-5-thiol in naturally aerated stagnant seawater" J. Ind.. Eng. Chem. xxx, xxx–xxx

[75] Awad, M.K., Metwally, M.S., Soliman, S.A., El-Zomrawy, A.A., and bedair, M.A., (2013) "Experimental and quantum chemical studies of the effect of poly 3 ethylene glycol as corrosion inhibitors of aluminum surface." J. Ind.. Eng. Chem. xxx, xxx–xxx

[76] Rethinnagiri,V., Jeyaprakash,P., Arunkumar,M., Maheswaran, V., and Madhiyalagan,A. (2012)."Investigation and inhibition of aluminium corrosion in hydrochloric acidsolutions by organic compound." Advances in Applied Science Research, 3 (3), 1718-1726.

[77] Muller, B., and Schemelich, T., (1994). "Corrosion and inhibition of corrosion of aluminum pigments in alkaline aqueous medium◎.Werkstoffe und Korrosion. 45 (5), 272-277.

[78] Muller, B., and Schemelich, T., (1994). "High-molecular weight styrene-maleic acid copolymers as corrosion inhibitors for aluminium pigments◎. Corros. Sci.37(6), 877-883.

[79] Ogurtsov, N.A., Pud,A.A., Kamarchik, P., and Shapoval, G.S., (2004). "Corrosion inhibition of aluminum alloy in chloride mediums by undoped and doped forms of polyaniline ◎. Synthetic Metals 143, 43–47.

[80] Abdallah, M., Megahed, H.E., El-Etre, A.Y., Obied, M.A., Mabrouk, E.M. (2004). "Polyamide compounds as inhibitors for corrosion of aluminum in oxalic acid solutions ◎.Bulletin of Electrochemistry, 20(6)277-281.

[81] Umoren, S.A., and Ebenso, E.E., (2008). "Blends of polyvinyl pyrrolidone and polyacrylamide as corrosion inhibitors for aluminium in acidic medium◎. Indian Journal of Chemical Technology, 15(4), 355-363.

[82] Amin, Mohammed A., El-Rehim, Sayed S. Abd., El-Sherbini, Essam E.F., Hazzazi, Omar A. and Abbas, Mohsen N.(2009). "Polyacrylic acid as a corrosion inhibitor for aluminium in weakly alkaline solutions. Part I: Weight loss, polarization, impedance EFM and EDX studies◎Corrosion Science. 51,(3), 658-667.

[83] Umoren, S.A., (2009). "Corrosion inhibition of aluminum alloy 3SR in HCl by polyvinylpyrrolidone and polyacrylamide: Effect of molecular structure on inhibition efficiency◎. Surface Review and Letters., 16 (6) 831-844.

[84] Doğru Mert, B. and Yazıcı, B., (2009). " The electrochemical synthesis of poly(pyrrole-co-o-anisidine) on 3102 aluminum alloy and its corrosion protection properties◎.Materials Chemistry and Physics., 125(3), 370–376.

[85] Oliveira,M.A.S., Moraes J.J., and Faez, R., (2011). " Impedance studies of poly(methylmethacrylate-co-acrylic acid) doped polyaniline films on aluminum alloy◎. Progress in Organic Coatings 65., 348–356.

[86] Zaafarany,I., (2012). " Corrosion Inhibition of Aluminum in Aqueous Alkaline Solutions by Alginate and Pectate Water-Soluble Natural Polymer Anionic Polyelectrolytes◎. Portugaliae Electrochimica Acta 2, 30(6), 419-426.

[87] Ghoreishi, S.M., Shabani-Nooshabadi, M., Behpour, M., and Jafari, Y., (2012). "Electrochemical synthesis of poly(o-anisidine) and its corrosion studies as a coating on aluminum alloy 3105◎. Progress in Organic Coatings 74, 502– 510.

[88] M.K. Awad, M.S. Metwallyb, S.A. Soliman, A.A. El-Zomrawy and M.A. bedair., (2013). "Experimental and quantum chemical studies of the effect of poly ethylene glycol as corrosion inhibitors of aluminum surface ◎. Journal of Industrial and Engineering Chemistry xxx (2013) xxx–xxx

[89] Hassan, R.M., Zaafarany, I.A. (2013). "Kinetics of corrosion inhibition of aluminum in acidic media by water-soluble natural polymeric pectates as anionic polyelectrolyte inhibitors◎.Materials, 6, (6),2436-2451.

[90] Rosen, M. J., and Hua, X. Y. (1982)."Surface concentrations and molecular interactions in binary mixtures of surfactants." J.Colloid Interface Sci. 86(1),164–172.

[91] El-Deeb, M. M., and Mohamed, S.M. (2011). "Corrosion inhibition of aluminum with a 3-(10-sodium sulfonate decyloxy) aniline monomeric surfactant and its analog polymer in a 0.5M hydrochloric acid solution." J. Appl. Polym. Sci. 122(5), 3030–3037.

[92] Gelling,V. J., Wiest,M. M., Tallman,D. E., Bierwagen,G. P., and Wallace, G.G. (2001). "Electroactive conducting polymers for Corrosion Control: 4. Studies of Poly(3-Octyl Pyrrole) and Poly(3-Octadecyl Pyrrole) on Aluminum 2024-T3 Alloy" Progress in Organic Coating., 43,149-157.

[93] Abd El Rehim, S.S., Hassan, H. H., and Amin, M. A. (2003) "The corrosion inhibition study of sodium dodecyl benzene sulphonate to aluminium and its alloys in 1.0 M HCl solution. " Mater. Chem. and Phys.,78,337- -348.

[94] Branzoi,V., Golgovici,F., and Branzoi,F. (2002)."Aluminium corrosion in hydrochloric acid solutions and the effect of some organic inhibitors." Mater. Chem. Phys. 78,122-131.

[95] Maitra, A., and Barua,S. (1974). "Dicyandiamide an inhibitor for acid corrosion of pure aluminium." Corro. Sci. 14(10), 587-590.

[96] Al-Sabagh, A.M., Abd-El-Bary, H.M., El-Ghazawy, R.A., and Mishrif, M.R. B.M. Hussein. (2011)."Corrosion inhibition efficiency of linear alkyl benzene derivatives for carbon steel pipelines in 1M HCl." Egy. J.Petro20, 33–45.

[97] Saleh,M., and Atia,A., (2006). "Effects of structure of the ionic head of cationic surfactant on its inhibition of acid corrosion of mild steel." J. Appl. Electrochem. 36,899-905.

[98] Morad,M. S., (2008). "Corrosion inhibition of mild steel in sulfamic acid solution by S-containing amino acids." J. Appl. Electrochem 38,1509-1519.

[99] Zhao,T., and Mu,G. (1999). "The adsorption and corrosion inhibition of anion surfactants on aluminium surface in hydrochloric acid." Corros. Sci. 41,1937-1944.

[100] Abd El Rehim,S. S., Amin,M. A., Moussa,S. O., and El lithy,A. S. (2008). "The corrosion inhibition of aluminum and its copper alloys in 1.0M H2SO4 solution using linear-sodium dodecyl benzene sulfonate as inhibitor." Mater. Chem. Phys 112, 898-906.

[101] Rafiquee,M. Z. A., Saxena,N., Khan,S., and Quraisshi,M. A. (2008)."Influence of surfactants on the corrosion inhibition behaviour of 2-aminophenyl-5-mercapto-1-oxa-3,4-diazole (AMOD) on mild steel." Mater. Chem. Phys 107,528-533.

[102] Abd El Rehim, S.S., Sayyah.,S.M. and Azooz, R.E. (2012). "Poly(p-Phenylenediamine) as an Inhibitor for Mild Steel in Hydrochloric Acid Medium." Portugaliae Electrochimica Acta. 30(1), 67-80.

[103] Amin,M. A. Abd El Rehim,S. S. El-Sherbini, E.E.F. and Bayoumi,R.S. (2008). Chemical and electrochemical (AC and DC) studies on the corrosion inhibition of low carbon steel in 1.0 M HCl solution by succinic acid –temperature effect, activation energies and thermodynamics of adsorptionInt. J. Electrochem. Sci.,3,199-210.

[104] Jevremović, I., and Stanković, V. M. (2012). "The inhibitive effect of ethanolamine on corrosion behavior of aluminium in NaCl solution saturated with CO2" Metall. Mater. Eng. 18, 241-248.

[105] El-Awady,Y. A. and Ahmed,A. I. (1985) "Effect of temperature and inhibitors on the corrosion of aluminium in 2N HCI solution, a kinetic study." J. Ind. Chem. 24A 601-610.

[106] Gomma,G.K., and Wahdan, M.H. (1995). "Schiff bases as corrosion inhibitors for Al in HCI acid solution." Chem.Phys.39, 209-2016.

[107] Marsh, J. (1988)."Advanced Organic Chemistry" 3rd edn; Wiely Eastern, New Delhy

[108] El-Sayed, A., Shaker,A. M., and Abd El-Lateef, H. M. (2010). "Corrosion inhibition of tin, indium and tin–indium alloys by adenine or adenosine in hydrochloric acid solution." Corro. Sci. 52, 72–81.

[109] Ampinen, M. J., and Fomino,M. (1993) "Analysis of free energy and entropy changes for half-cell Reactions" J. Electrochem.Soc, 140(12), 3537-3546.

[110] Sayyah, S. M., Abd El-Salam, H.M., and Azzam, E. S. (2006). "Oxidative Chemical Polymerization of Some 3-alkyloxyaniline Surfactants and Characterization of the Obtained Polymers".Inter. J. of Polym. Mate. 55(12), 1075-1093.

Chapter 21

Adsorption and Inhibitive Corrosion Properties of Some New Polymeric Compounds as Green Inhibitors on Carbon Steels in Cooling Water Systems

Florina Branzoi and Viorel Branzoi

Additional information is available at the end of the chapter

http://dx.doi.org/10.5772/57276

1. Introduction

Carbon steel, the most widely used engineering material, accounts for approximately 85% of the annual steel production world wide. Carbon steel is used in large tonnages in chemical processing, construction and metal –processing equipment, in marine applications and petroleum production and refining [1-4 and 4-7]. The use of organic inhibitors to decrease the rate of corrosion processes of carbon steels is quite varied [7-11 and 11-15]. Corrosion is a major problem in cooling water industrial systems, in oil and gas production systems. Therefore, the prevention from the corrosion of metals used in industrial applications is an important issue that the must be dealt with. Metal corrosion in water-conveying systems such as cooling water circuits is of major concern in industrial applications. It is well known that, in all the cases of cooling water systems at the metal/water interface contact appear frequent corrosion processes which determine deposition of corrosion products, like scales. Due to the scales formation the heat exchange becomes more difficult, that disturbs the normal function of industrial installation [15-18 and 18-22]. In this regard, aspects like cost of treatment versus cost of corrosion damages as well as process safety and the impact of corrosion and corrosion treatment on the environment have to be taken into account [22-26]. It is well known that, excessive corrosion does not only lead to serious damage of installations, it also causes considerable environmental pollution [26-31]. Most previous research into their usage has concentrated on relationships between chemical structure and inhibition performance [31-37 and 37-41]. Recent investigations have additionally emphasized the importance of the nature of the metal surface in inhibition performance [41-47 and 47-52]. In order to evaluate compounds as corrosion inhibitors and to design novel inhibitors, much more research works were concentrated on the studies of the relationship between structural characteristics of the organic compounds and their inhibiting effects [52-56]. It had been suggested that the most effective factors for the inhibiting effects are the electronegative atoms such as, N, S, P etc, the unsaturated bonds such as, double bonds or triple bonds etc., and the plane conjugated systems including all kinds of aromatic cycles, of which they can offer special active electrons or vacant orbital to donate or accept electrons [56-59 and 59-61]. A large number of organic molecules have at least one of the above-mentioned characteristics in their molecular structure according to the diversity of organic compounds [61-65 and 65-69]. Heterocyclic compounds, kind of effective inhibitors, have at least two factors within their structures, which is the reason of their structures, which is the reason of their effective inhibiting [69-73 and 73-75]. The ability of an inhibitor to provide corrosion protection therefore depends to a large extent upon the interaction between the inhibitor and the metal surface under corrosion conditions [75-81 and 81-84]. The adsorption of organic molecules at the metal/solution interface is of a great interest in surface science and can markedly change the corrosion resisting properties of metals [84-89, 89-91 and 91-96]. The protection of corroding surfaces prevents the waste of both resources and money during the industrial applications and it is vital for extension of the equipment and limiting the dissolution of the toxic metals from components into the environment [96-112]. Generally, it is assumed that strong adsorption of the inhibitors is a prerequisite [112-117]. The adsorption of inhibitors leads to the formation of a physical barrier that reduces the metal reactivity in the electrochemical reaction of corrosion [117-119]. Early studies considered the adsorption of inhibitors on metal surfaces to be primarily physical adsorption and/or chemisorptions [119-121]. New investigations have shown that adsorption could also occur through hydrogen bonding [121-123]. Most of these studies made use of sensitive surface analysis tools such as X-ray photoelectron spectroscopy (XPS) to resolve the nature of inhibitor adsorption [123-125]. Some earlier works has however been carried out under conditions that are not representative of on actual application. In this study, the inhibition of carbon steel corrosion in cooling waters by organic compounds was investigated by potentiodynamic polarizations, electrochemical impedance spectroscopy (EIS) measurements, FT-IR and metallography analysis. This study presents some attempts of analyzing of corrosive phenomena, which occur in cooling water systems, and relates to the protection of metallic surfaces from corrosion using these new polymers obtained in microwaves field. The use of inhibitors is one of the most practical methods for protecting against the corrosion and it is becoming increasingly popular.

The used organic inhibitors were six polymers, which were obtained by radicalic polymerization in presence of microwave field: PASAC-4 (maleic anhydride and urea molar ratio 1:0.8 at t=120°C), PASAC-5 (polyaspartic acid at t=120°C and reaction without catalyst), and PASAC-6 (polyaspartic acid and phosphoric acid molar ratio 1:0.04 at t=190°C), PASAC-7 (polyaspartic acid and phosphoric acid molar ratio 1:0.06 in ammonium salt at t=200°C), PASAC-8 (polyaspartic acid and H_3PO_4 molar ratio 1:0.025 at t=200°C in propylene carbonate), and PASAC-9 (maleic anhydride and urea molar ratio 1:0.6 at t=200°C) (These organic compounds are not commercial names-is abbreviation). The studied metals were the carbon steels type OL 37 and OLC 45.

Electrode	C%	Si%	Mn%	Fe%	P%	S%	Al%	Ni%	Cr%	Cu%	Sn%	As%
OLC 45	0.48	0.03	0.79	98.32	0.02	0.025	0.027	0.05	0.06	0.18	0.012	0.006
OL 37	0.15	0.09	0.4	99.293	0.023	0.02	0.022	0.001	0.001	-	-	-

Table 1. The chemical composition of the working electrodes

The working electrode was made from these metals materials and had a cylindrical shape. This shape is preferred, because it assures a greater surface and a reduce number of edges. Prior to each determination, the working electrode was mechanically ground and polished with emery paper of varied granulation up to mirror-luster, degreased in benzene at boiling temperature in order to remove all traced of fat and remained abrasive powder on the electrode surface after polishing. After that, the working electrode was washed with distilled water and inserted in the polarization cell, which was the usually three-electrode cell. All tests have been performed at 25°C under atmospheric oxygen without agitation.

The corrosion medium was industrial cooling water with the following chemical composition:

Indicators	UM	Water type SC_1, values of parameters
PH		8.42
Conductivity	µs/cm	1061
Alcalinity p	mval/L	0.1
Alcalinity m	mval/L	3.3
total Hardness	mval/L	8.3
calciu Hardness	mval/L	3.0
Chloride, Cl^-	mg/L	117.01
Sulfate	mg/L	155
Solid substances	mg/L	2.75
organic compounds	mg/L	11.37
Iron	mg/L	0.073
Aluminium,	mg/L	0.0175
Nitrite, NO_2^-	mg/L	<0.1
Nitrate, NO_3^-	mg/L	10
Phosphate, PO_4^{3-}	mg/L	0.046
Copper, Cu^{2+}	mg/L	<0.015
Zinc, Zn^{2+}	mg/L	<0.1

Table 2. The chemical composition of the cooling water type SC_1

Indicators	UM	Water type SC_2, values of parameters
PH		7.95
Conductivity	µs/cm	665
Alcalinity	mval/L	4.2
total Hardness	mval/L	4.18
calciu Hardness	mval/L	4.16
Chloride, Cl^-	mg/L	95.74
Sulfate	mg/L	73.15
Solid substances	mg/L	0.95
organic compounds	mg/L	3.47
Iron	mg/L	0.099
Aluminium,	mg/L	0.02
Nitrite, NO_2^-	mg/L	<0.1
Nitrate, NO_3^-	mg/L	35
Phosphate, PO_4^{3-}	mg/L	<0.04
Copper, Cu^{2+}	mg/L	<0.02
Zinc, Zn^{2+}	mg/L	<0.1

Table 3. The chemical composition of the cooling water type SC_2

In this study, we have been used as organic inhibitors the following polymers: PASAC-4, PASAC-5, PASAC-6 PASAC-7, PASAC-8 and PASAC-9. The inhibition activity analysis of these organic compounds was made by assuming that the mechanism of inhibition by organic molecules is chemisorptions and that the energetic of the corrosion process per se is unaffected by the addition of substituent on the parent compound.

2. Results and discussion

2.1. Potentiodynamic polarization

as, it is well known; less active metals are less accessible and more expensive in comparison with ordinary metals, which are too much exposed to corrosion process to be used. Protection of these metals can be achieved by multiple ways from which the treatment of the corrosive media is one of the most important. The treatment of the corrosive media can be achieved by removing the aggressive chemical agents or by using inhibitors, which control the corrosion of anodic or cathodic reaction of both. Here, the inhibition of the corrosion is the result of the adsorption of the organic compound on the metal surface forming an invisible film of a few molecular diameters thickness. The inhibition activity analysis of the organic compound was made by assuming that the mechanism of inhibition by organic molecules is chemisorption

The polarization behaviour of carbon steels mentioned above was studied through the plotting of the polarization curves obtained using the potentiodynamic method, finding the kinetic parameters of corrosion (especially the density of the corrosion current) from solutions without inhibitors and their comparison with the kinetic parameters from solutions with different concentrations of inhibitor. The corresponding Tafel parameters were obtained by Mansfeld′s method, employing polarization data near the corrosion potential. In the present study, when values of $E-E_{cor}$ are higher than about 70 mV, slight but significant changes in the anodic and cathodic Tafel slopes were found. Figures 1-4 show a series of potentiodynamic polarization curves of two-carbon steels electrode in industrial cooling water type SC_1 and SC_2 (it was not aerated) in absence and presence of different concentrations of inhibitors PASAC-4, PASAC-5, PASAC-6, PASAC-7, PASAC-8 and PASAC-9.

The corrosion parameters were calculated on the basis of potentiodynamic potential-current characteristics in the Tafel region ($E=E_{corr} \pm 150mV$) and the vicinity of the corrosion potential (($E=E_{corr} \pm 15mV$) according to Mansfield's theory.

$$\log i_a = \log i_{corr} + \left(E_i - E_{corr}\right) / b_a \tag{1}$$

$$\log i_c = \log i_{corr} + \left(E_{corr} - E_i\right) / b_c \tag{2}$$

This equation corresponded to linear anodic and cathodic Tafel lines. Current density i_{corr} was determined by extrapolating the Tafel lines to $E=E_{corr}$ or according to the Stern-Geary equation. This resulted in:

$$i_{corr} = b_a b_c / 2.303\left(b_a + b_c\right) R_p \tag{3}$$

Where R_p was the polarization resistance, defined as the tangent of a polarization curve at E_{corr}.

$$R_p = \left(\frac{dE}{di}\right)_{E=E_{corr}} \tag{4}$$

In the present study when the values of $E-E_{corr}$ are higher than 70mV, slight but significant changes in the anodic and cathodic Tafel slopes were found.

Inhibitor) (ppm)	i_{corr} (μA/cm²)	Rp KΩ/cm²	R_{mpy}	$P_{mm/year}$	K_g g/m²h	E(%)	E_{corr} (mV)	b_a (mV)	b_c (mV)	θ
0	13.01	1.78	6.07	0.15	0.137	-	-513	110	-215	-
50	4.011	6.56	1.8718	0.0475	0.0422	69.16	-366	108	-210	0.6916
100	4.319	4.71	2.0155	0.0511	0.0454	66.80	-467	108	-146	0.6680
300	0.942	17.57	0.4386	0.0111	0.0099	92.77	-326	80	-84	0.9277
500	0.992	17.86	0.462	0.0117	0.0104	92.39	-360	81	-76	0.9239
800	0.835	29.50	0.3896	0.0098	0.0087	93.58	-307	189	-109	0.9358
1000	1.001	21.91	0.4671	0.0118	0.0105	92.30	-332	153	-106	0.9230

Table 4. Kinetic corrosion parameters of carbon steel OL-37+ X ppm PASAC 4 $+S_1$ at 25°C

Inhibitor) (ppm)	i_{corr} (μA/cm²)	Rp KΩ/cm²	R_{mpy}	$P_{mm/year}$	K_g g/m²h	E(%)	E_{corr} (mV)	b_a (mV)	b_c (mV)	θ
0	13.01	1.78	6.07	0.15	0.137	-	-513	110	-215	-
50	7.62	2.93	3.556	0.0901	0.0802	41.42	-515	109	-1430	0.4142
100	7.08	2.70	3.304	0.0838	0.0745	45.58	-518	107	-150	0.4558
300	4.476	3.48	2.086	0.0529	0.0470	65.64	-709	81	-181	0.6564
500	2.64	4.84	1.232	0.0312	0.0278	79.70	-710	56	-54	0.7970
800	1.6	6.43	0.746	0.0189	0.0168	87.70	-760	85	-49	0.8770
1000	8.384	2.01	3.912	0.0993	0.0883	35.55	-694	78	-160	0.3555

Table 5. Kinetic corrosion parameters of carbon steel OL-37+ X ppm PASAC5$+S_1$ at 25°C

Inhibitor) (ppm)	i_{corr} (μA/cm²)	Rp KΩ/cm²	R_{mpy}	$P_{mm/year}$	K_g g/m²h	E(%)	E_{corr} (mV)	b_a (mV)	b_c (mV)	θ
0	13.01	1.78	6.07	0.15	0.137	-	-513	110	-215	-
100	3.34	3.11	1.558	0.0395	0.0351	74.32	-520	62	-76	0.7432
300	4.63	3.36	2.1606	0.0548	0.0487	64.41	-550	82	-124	0.6441
500	4.56	3.19	2.204	0.0559	0.04971	64.95	-533	99	-151	0.6495
800	5.04	2.82	2.436	0.05498	0.0741	62.18	-552	88	-133	0.6218

Table 6. Kinetic corrosion parameters of carbon steel OL-37+ X ppm PASAC 6 $+S_1$ at 25°C

The figures 1-5 show a series of potentiodynamic polarization curves for two carbon steel electrodes in a cooling water type S1, in absence and presence of different concentrations of inhibitor.

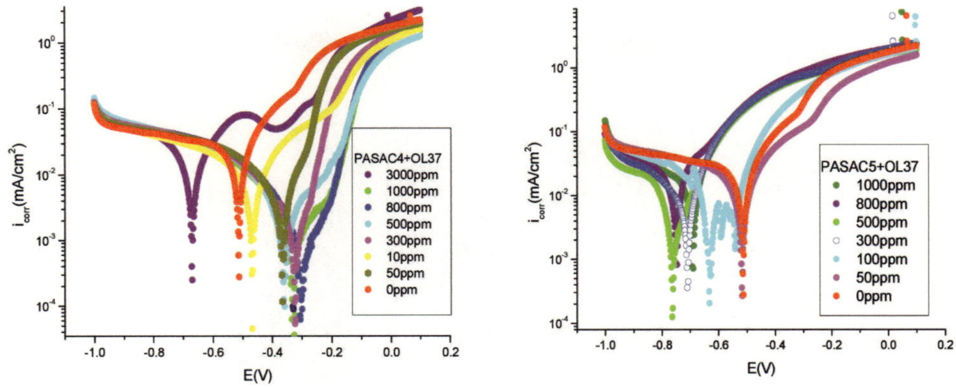

Figure 1. The polarization curves of OL-37 carbon steel in cooling water type S1 in presence of PASAC4 and PASAC5 at 25°C

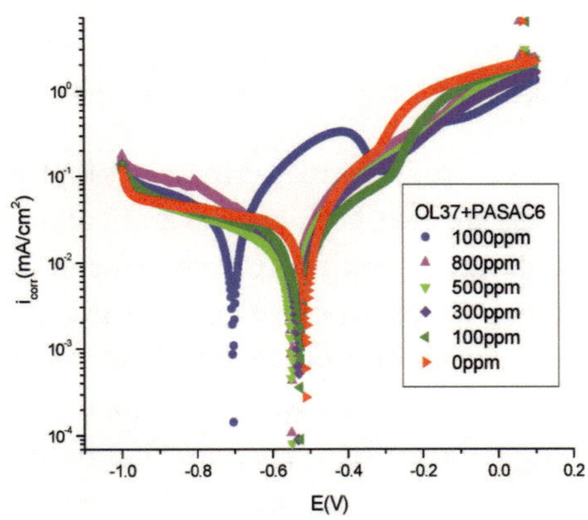

Figure 2. The polarization curves of OL-37 carbon steel in cooling water type S1 in presence of PASAC6 at 25°C

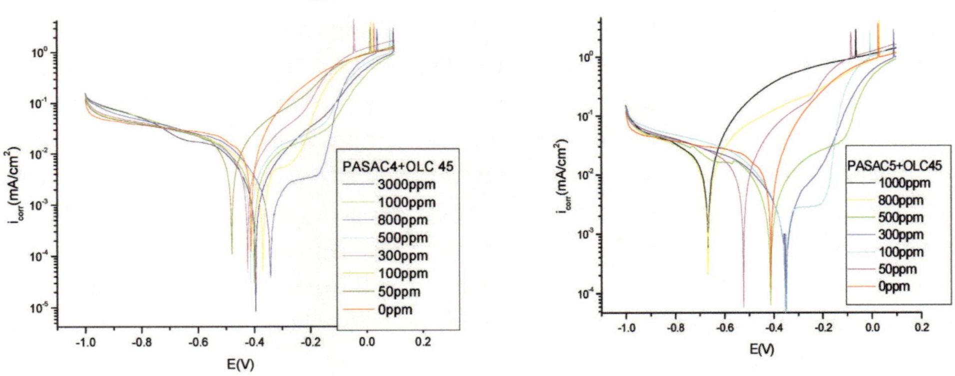

Figure 3. The polarization curves of OLC-45 carbon steel in cooling water type S1 in presence of PASAC4 and PASAC5 at 25°C

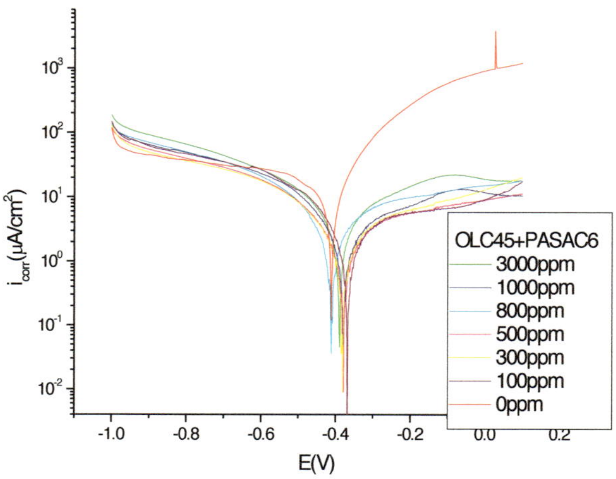

Figure 4. The polarization curves of OLC-45 carbon steel in cooling water type S1 in presence of PASAC6 at 25°C

Analysis of the polarization curves from figures 1-5 indicates that at low overvoltages, the Tafel relationship are followed, showing that both anodic and cathodic reactions are activation-controlled. At higher overvoltages a limiting current appears on the anodic and cathodic polarization curves showing that, the transport of ions towards the electrode surface becomes the rate-determining step (concentration polarization). Analyzing figures 1-5 it can be observed that on the anodic curves there is a low active range of potential where the relation Tafel is verified. After this range the current densities slightly increases and tends to a limit value. This behaviour points out that on this range the corrosion process is controlled by diffusion. Sometimes on the anodic curves appears oxidation peaks followed by the narrow passive range and a decrease of the current density (this behaviour can be explained due to formation of oxo-hydroxo-complexes of Fe). The maximum efficiency is obtained at the inhibitory concentration for PASAC 4 is 800ppm (OL37 and OLC 45), for PASAC5 is 500ppm (OLC45) and 800ppm (OL37) and for PASAC6 is 500ppm (OL37 and OLC 45). At the increasing inhibitor concentration over these concentrations (500 and 800 ppm) the inhibitor efficiency starts to decrease, respectively the corrosion current densities begin to decrease. Analyzing the cathodic polarization curves from figures 1-5 it can be observed that, on the large range of the potential the carbon steel electrodes behave very close to a passive behaviour. Practically, we can say that, in this potential range the electrode surface is passivated. We consider that, in this potential range, the cathodic reaction is hindered by the oxide film (passive film) from the electrode surface. In this potential range takes place the oxygen reduction cathodic reaction according to equation:

$$\frac{1}{2}O_2 + H_2O + 2e \rightarrow 2HO^- \qquad (5)$$

After this like passive range potential the cathodic current increases again and this increasing is due to the hydrogen evolution. From polarization curves obtained by potentiodynamic method were calculated all kinetic corrosion parameters which are given in tables 4-12.

Inhibitor(ppm)	i_{corr} (μA/cm²)	Rp KΩ/cm²	R_{mpy}	$P_{mm/year}$	K_g g/m²h	E(%)	E_{corr} (mV)	b_a (mV)	b_c (mV)	θ
0	7.12	2.62	3.32	0.084	0.075	-	-414	91	-188	-
50	5.59	3.55	2.608	0.066	0.0588	21.48	-482	113	-141	0.2148
100	3.263	5.62	1.5227	0.0386	0.0343	54.17	-424	106	-128	0.5417
300	2.1678	8.02	1.01164	0.0256	0.0228	69.55	-414	109	-106	0.6955
500	1.68	10.37	0.784	0.0198	0.0176	76.40	-372	94	-116	0.7640
800	1.027	20.43	0.479	0.0121	0.0108	85.57	-345	158	-106	0.8557
1000	1.17	13.43	0.546	0.0138	0.0123	83.56	-399	94	-94	0.8356

Table 7. Kinetic corrosion parameters of carbon steel OLC-45+ X ppm PASAC 4 +S1 at 25°C

Inhibitor) (ppm)	i_{corr} (μA/cm²)	Rp KΩ/cm²	R_{mpy}	$P_{mm/year}$	K_g g/m²h	E(%)	E_{corr} (mV)	b_a (mV)	b_c (mV)	θ
0	7.12	2.62	3.32	0.084	0.075	-	-414	91	-188	-
100	6.392	3.02	2.9829	0.0757	0.0673	10.22	-526	102	-166	0.1022
300	1.412	13.29	0.659	0.0167	0.0148	80.16	-357	145	-116	0.8016
500	1.234	14.14	0.5758	0.0146	0.0129	82.66	-355	101	-111	0.8266
800	2.356	8.75	1.0994	0.0279	0.0248	66.91	-416	121	-124	0.6691

Table 8. Kinetic corrosion parameters of carbon steel OLC-45+ X ppm PASAC5 +S1 at 25°C

Inhibitor (ppm)	i_{corr} (μA/cm²)	Rp KΩ/cm²	R_{mpy}	$P_{mm/year}$	K_g g/m²h	E(%)	E_{corr} (mV)	b_a (mV)	b_c (mV)	θ
0	7.12	2.62	3.32	0.084	0.075	-	-414	91	-188	-
100	1.34	17.12	0.6253	0.0158	0.0141	81.17	-371	169	-109	0.8117
300	1.08	19.86	0.4946	0.0125	0.0111	85.11	-385	95	-80	0.8511
500	1.18	20.77	0.5506	0.0139	0.0124	83.42	-384	146	-112	0.8342
800	1.305	15.89	0.609	0.0154	0.0137	81.67	-415	117	-100	0.8167
1000	1.45	14.11	0.6766	0.0171	0.0152	79.63	-390	130	-99	0.7963

Table 9. Kinetic corrosion parameters of carbon steel OLC-45+ X ppm PASAC 6 +S1 at 25°C

Analyzing these tables, it can be observed that, the addition of the organic inhibitor to the amounts shown in the tables 4-12 leads in all the cases to inhibition of the corrosion process.

Inhibitor ppm	i_{corr} μA/cm²	R_p kΩ/cm²	R_{mpy}	P mm/year	K_g g/m²h	E %	E_{corr} mV	ba mV	b_c mV	θ
0	13.01	1.78	6.07	0.15	0.137	-	-513	110	-215	-
50	4.46	4.53	2.081	0.0528	0.0469	65.71	-529	104	-160	0.6571
100	4.14	4.17	2.072	0.0525	0.0467	65.87	-563	89	-150	0.6587
300	3.01	5.85	1.404	0.0356	0.0316	76.86	-589	105	-94	0.7686
500	3.95	4.43	1.843	0.0467	0.0416	69.63	-680	80	-154	0.6963
800	4.44	3.94	2.072	0.0525	0.0467	65.87	-690	79	-148	0.6587

Table 10. Kinetic corrosion parameters of OL 37+ X ppm PASAC 7+S1 at 25°C

Inhibitor ppm	i_{corr} μA/cm²	R_p kΩ/cm²	R_{mpy}	P mm/year	K_g g/m²h	E %	E_{corr} mV	ba mV	b_c mV	θ
0	13.01	1.78	6.07	0.15	0.137	-	-513	110	-215	-
50	6.88	2.97	3.211	0.0814	0.0724	47.10	-550	100	-159	0.4710
100	6.55	2.96	3.056	0.0775	0.0689	49.65	-504	102	-158	0.4965
300	3.53	5.67	1.647	0.0418	0.0371	72.86	-446	111	-156	0.7286
500	2.67	7.88	1.248	0.0316	0.0281	79.43	-511	102	-208	0.7943
800	4.90	3.88	2.608	0.0661	0.0588	57.03	-633	349	-118	0.5703

Table 11. Kinetic corrosion parameters of OL 37+ X ppm PASAC 8+S1 at 25°C

Inhibitor ppm	i_{corr} μA/cm²	R_p kΩ/cm²	R_{mpy}	P mm/year	K_g g/m²h	E %	E_{corr} mV	ba mV	b_c mV	θ
0	13.01	1.78	6.07	0.15	0.137	-	-513	110	-215	-
50	6.36	2.87	2.968	0.0753	0.0669	51.11	-483	98	-134	0.511
100	6.27	3.29	2.661	0.0675	0.0600	56.16	-520	107	-167	0.561
300	5.703	3.24	2.926	0.0742	0.0660	51.80	-495	101	-137	0.518
500	3.683	5.11	1.718	0.0436	0.0387	71.69	-640	96	-143	0.717
800	3.55	5.12	1.656	0.0420	0.0373	72.71	-512	90	-133	0.7271

Table 12. Kinetic corrosion parameters of OL 37+ X ppm PASAC 9+S1 at 25°C

It can be observed from tables 9-12 that, the inhibitor PASAC8 has a higher efficiency for corrosion system OL37+S1, PASAC7 had a good efficiency for corrosion system OL 37 in S1 than for PASAC9 in same condition. The maximum efficiency is obtained at the inhibitor concentration for PASAC 7 is 300ppm, for PASAC8 is 500ppm and for PASAC9 is 800ppm. At the increasing inhibitor concentration over these concentrations (500 and 800 ppm) the inhibitor efficiency starts to decrease, respectively the corrosion current densities begin to increase again.

Analyzing in comparison the corrosion rate of organic inhibitors, in the same condition, one can see that, the PSAC-4 had a higher efficiency for corrosion system OL 37 in S1 and OLC 45 in S1, PASAC6, PASAC5, PASAC8, and PASAC7 had a good efficiency for corrosion system OL 37 in S1 than PASAC-9.

Analyzing these tables, it can be observed that, the addition of the organic inhibitor to the amounts shown in the tables 4-12 leads in all the cases to inhibition of the corrosion process.

This fact can be explained taking into account the effects of organic compounds on the electrochemical properties of the carbon steels in concordance with Donahue's theory [17]. According to this theory, the corrosion process of carbon steel could takes place thus:

$$Fe + OH^- \Leftrightarrow (FeOH)_{ads} + e^- \tag{6}$$

$$(FeOH)_{ads} + nI \Leftrightarrow [FeOH]In_{ads} \tag{7}$$

$$I + S_{ads} \Leftrightarrow Iads + S \tag{8}$$

$$(FeOH)_{ads} + nI_{\ ads} \Leftrightarrow [(FeOH)I_n]_{ads} \tag{9}$$

$$[(FeOH)I_n]_{ads} \rightarrow [(FeOH)I_n]^+ + e^- \tag{10}$$

$$(FeOH)_{ads} \rightarrow (FeOH)^+ + e^- \tag{11}$$

$$(FeOH)^{+} \rightarrow Fe^{2+} + OH^{-} \tag{12}$$

$$\left[(FeOH)I_n\right]^{+} \rightarrow \left[FeIn\right]^{2+} + OH^{-} \tag{13}$$

The symbol 'S' denotes the electrolyte or any specifically adsorbed ions and the symbol 'I' denotes the organic inhibitor. The step (9) would be the mode of inhibition, which has been designated as "blocking"or adsorption, while the steps (8) and (10) are the surface chelate modes. If the process described by (10) is the dominant mechanism for the production of chelate, then the formation of chelate would be enhanced by increased coverage of both adsorbed inhibitor and reaction intermediate. Since increased coverage of intermediate leads to product formation via step (12), the chelation process, i.e., step (10), must be fast or the surface coverage of adsorbed inhibitor must be high for inhibition to be maintained. Assuming the foregoing to be correct, i.e., chelate forms by (10) and that the equilibrium of (7) and (9) are shifted toward the right as adsorption sites are generated by dissolution, then importance of chelate ought to be increased with increasing immersion time [117]. The extent to which the steps (7) and (9) are shifted to the right relative to each other and the stoichiometry of (10), i.e., whether n ≥1, will determine whether inhibition will increase, decrease, or remain constant. Thus, until one may ascertain the critical parameters associated with (7), (9) and (10), only a qualitative discussion of the causes of the observed phenomena can be made. From the foregoing discussion, it should not be assumed that the mechanism given by (8) has been ruled out. On the contrary, such a mechanism is still attractive since organic compound is more readily available from the solution than it would be via some sort of surface diffusion mechanism. On the other hand, the fact that adsorbed organic compound is indeed present and the organic inhibitor and intermediate are assumed to gravitate toward the same surface sites suggests a preference for (10). We presume that both adsorption and surface chelate function in a more or less concerted manner to achieve inhibition (see tables 4-12 and figures 1-5).

The variation curves of the corrosion current density function of the inhibitor concentration are presented in figures 1-5. From figures, one can see much better the influence of these parameters on the polarization behaviour of the two carbon steels in cooling water system.

From polarization curves obtained by potentiodynamic method were calculated all kinetic corrosion parameters which are given in tables 4-12.

The variation curves of the corrosion current density and efficiency function of the inhibitor concentration are presented in figures 6-8. From figures, one can see much better the influence of these parameters on the polarization behaviour of the carbon steel OL 37 in cooling water system.

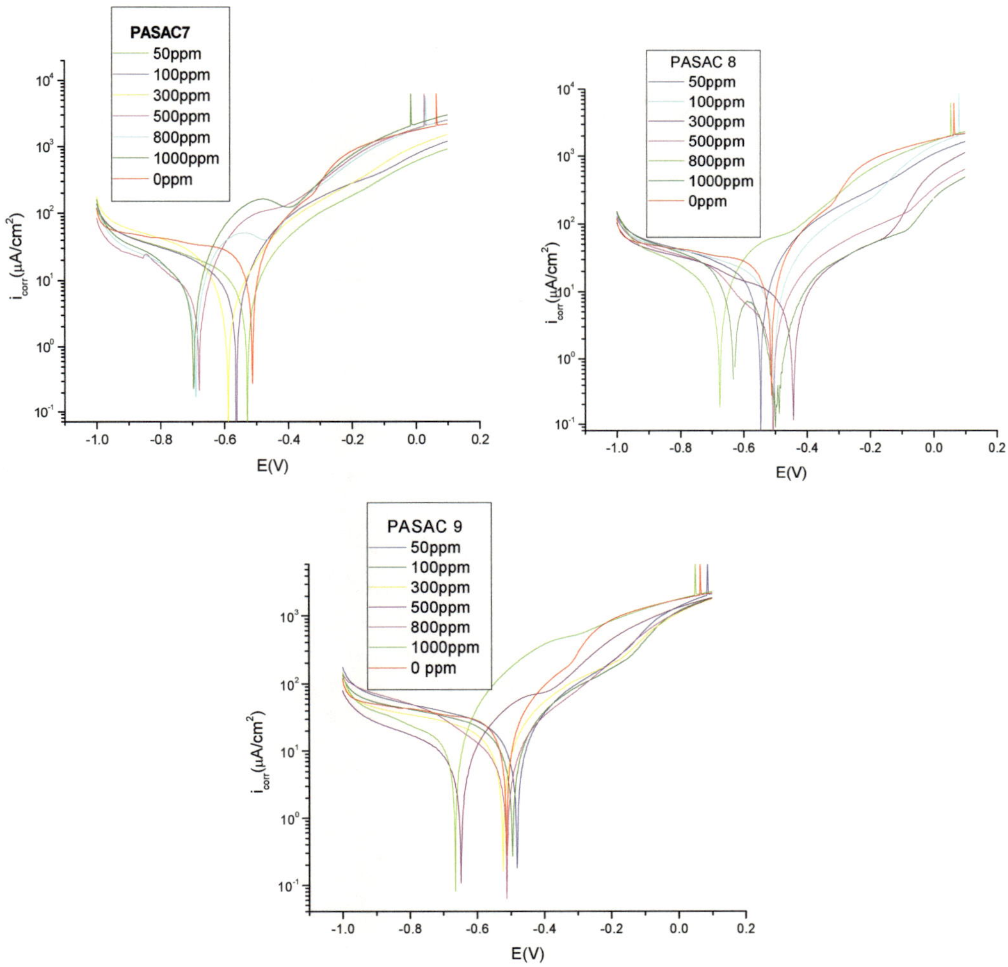

Figure 5. The polarization curves of OL 37 carbon steel in cooling water type S1 in presence of PASAC7, PASAC8 and PASAC9 at 25°C

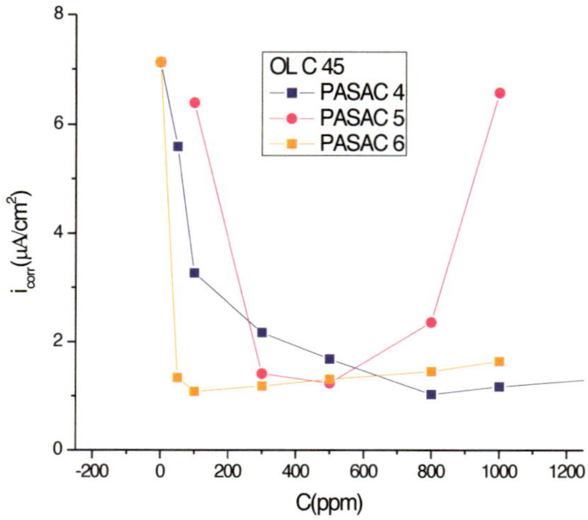

Figure 6. The influence of the inhibitor PASAC4, PASAC5 and PASAC6 concentration on the corrosion rate of carbon steel OLC 45 in cooling water type S_1 at 25^0

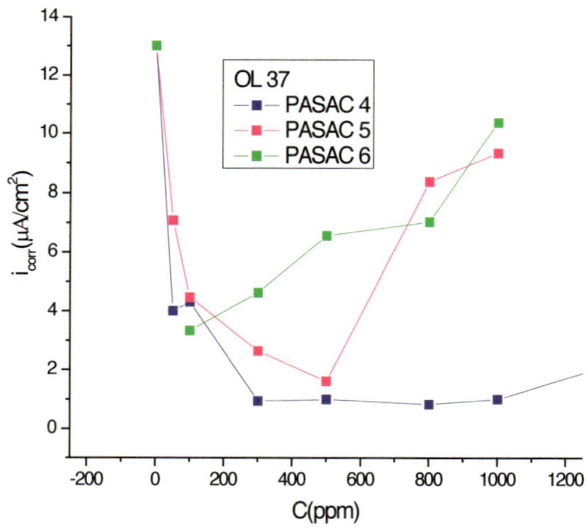

Figure 7. The influence of the inhibitor PASAC4, PASAC5 and PASAC6 concentration on the corrosion rate of carbon steel OL-37 in cooling water S_1 at 25^0

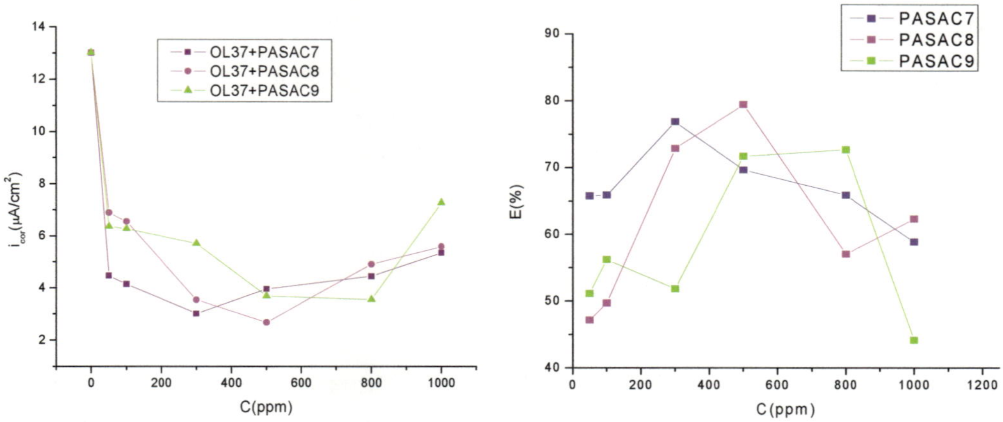

Figure 8. The influence of the inhibitor PASAC7, PASAC8 and PASAC9 concentration on the corrosion rate and efficiency of carbon steel OL-37 in cooling water type S_1 at 25⁰. The inhibition efficiency follows the order: $\mathbf{E_{PASAC4} > E_{PASAC6} > E_{PASAC5}\ E_{PASAC8} > E_{PASAC7} > E_{PASAC9}}$

2.2. The adsorptions of the organic compounds on the carbon steels surface obeyed Langmuir's isotherm

We presume that, the higher inhibitor efficiency is a consequence of the stronger adsorption process. The molecules of organic inhibitor are adsorbed on the metal surface and form a barriere film, which hindered the corrosion process. To quantify the effect of inhibitor concentration on the corrosion rate, it is common to fit the rate data to equilibrium adsorption expression, such as Langmuir equation:

$$\theta / \left(1-\theta\right) = Kc \tag{14}$$

Where θ is the fraction of surface coverage by the inhibitor and K is the equilibrium constant for the adsorption reaction. θ is given by:

$$\theta = \left(i_{corr} - i_{corr.inhib}\right) / \, i_{corr} \tag{15}$$

Where $i_{corr,\ inh}$ and i_{corr} are the corrosion rates in the industrial cooling water SC1 with and without inhibitor. Usage of the Langmuir treatment is often justified with the argument that inhibition must involve adsorption. In this study, the Langmuir isotherm is rearranged to give:

$$c / \theta = c + 1/K \tag{16}$$

c /θ is plotted against c, when a linear relationship is obtained for organic inhibitor and a slope of near unity which indicates an approximate Langmuir behaviour. The adsorption equilibrium constants (K) for our corrosion systems are given in table 13.

These values of K point out the adsorption process of organic inhibitors on the electrode surface and consequently the decrease of the corrosion rate. Further, we shell try to show what kind type of adsorption process takes place on the electrode surface. The adsorption equilibrium constant (K_{ads}) is related to the standard free energy of reaction by the equation:

$$\ln K_{ads} = -\left(\Delta G_T^{\circ} / RT\right) \quad (17)$$

The obtained values ΔG_{ads} up to -20Kjmol^{-1} are consistent with electrostatic interaction between the charged molecules (in our case, the inhibitor molecules) and the charged metal surface (physical adsorption), while those more negative than -40KJmol^{-1}, involve charge sharing or transfer from the inhibitor molecules to the metal surface to form a co-coordinative type of bond (chemisorptions see table 13) [119-121].

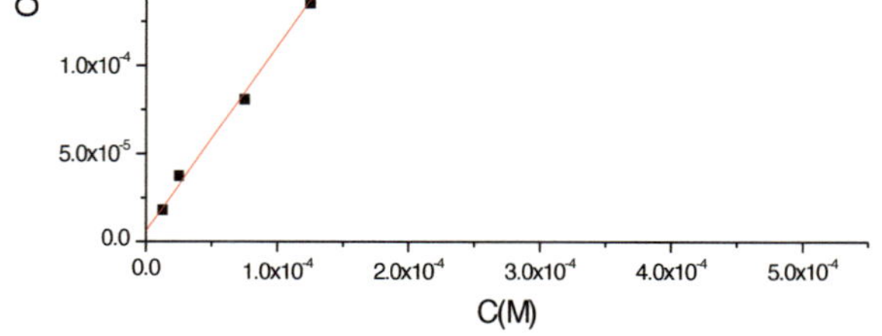

Figure 9. Langmuir plot for PASAC4+OL37 at different inhibitor concentrations

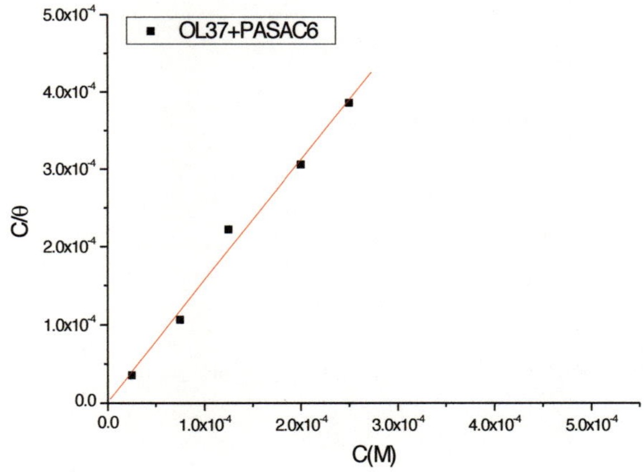

Figure 10. Langmuir plot for PASAC6+OL37 at different inhibitor concentrations

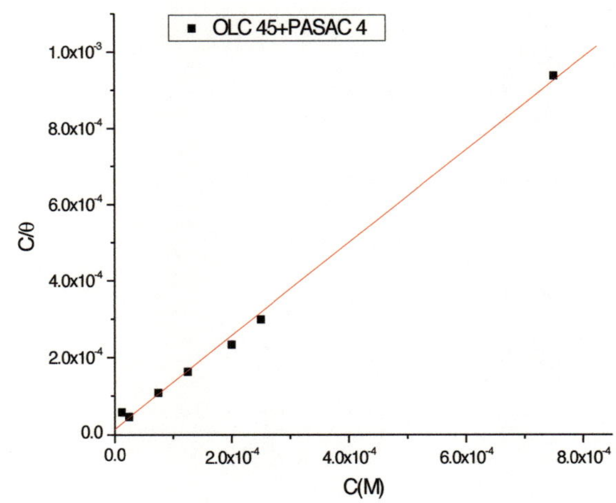

Figure 11. Langmuir plot for PASAC4+OLC-45 at different inhibitor concentrations

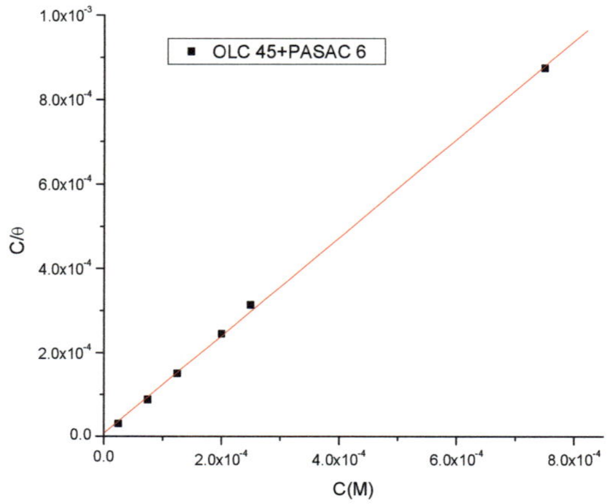

Figure 12. Langmuir plot for PASAC6+OLC-45 at different inhibitor concentrations

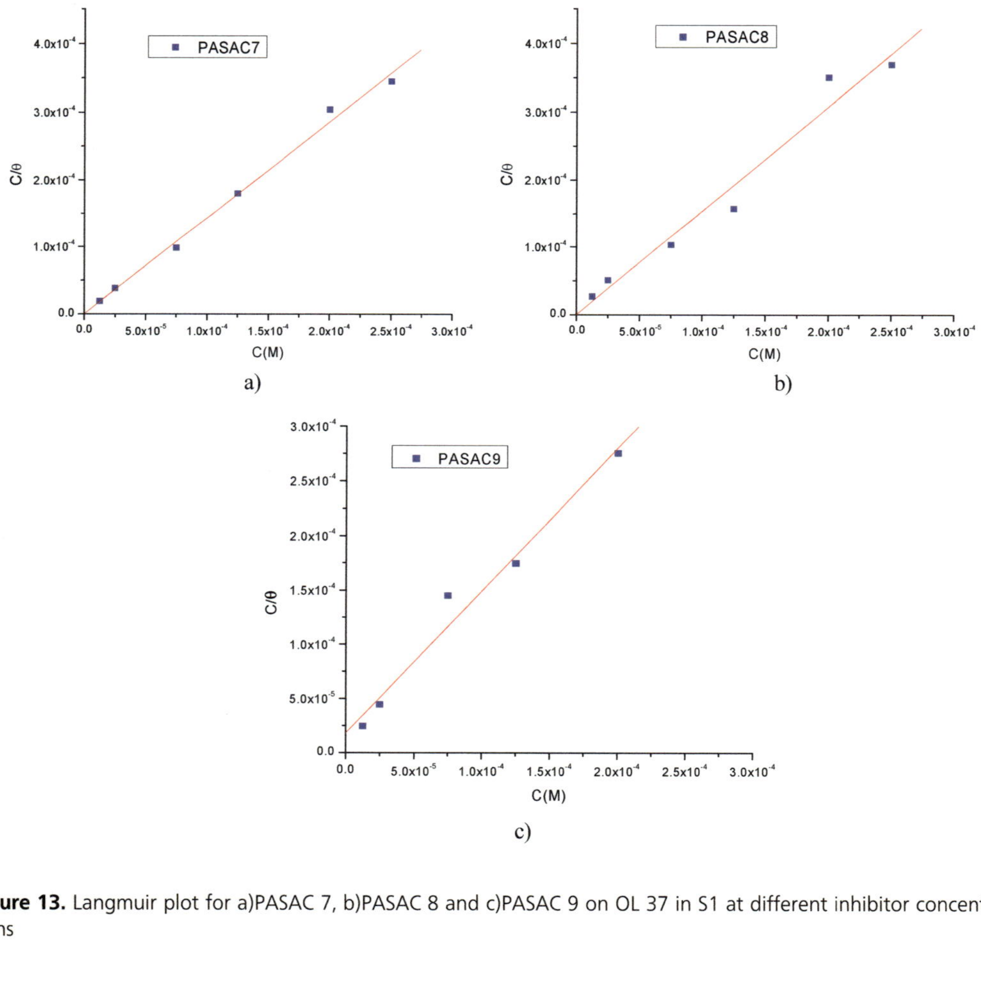

Figure 13. Langmuir plot for a)PASAC 7, b)PASAC 8 and c)PASAC 9 on OL 37 in S1 at different inhibitor concentrations

The system	Type of metallic material	Values of K_{ads}, M^{-1}	Values of ΔG°_{ads} KJM^{-1}	Type of adsorption
Cooling water type S_1+PASAC4	OL-37	1.675x10^5	-29.789	chemisorption and Physical adsorption
	OLC-45	6.959x10^4	-27.613	chemisorption and Physical adsorption
Cooling water S_1+PASAC6	OL-37	0. 5981x10^4	-21.5356	Physical adsorption and chemisorption
	OLC-45	1.33547x10^5	-29.226	chemisorption and Physical adsorption
Cooling water S_1+PASAC7	OL-37	3.7703x10^6	-37.499	chemisorption and Physical adsorption
Cooling water S_1+PASAC8	OL-37	6.1846x10^6	-38.724	chemisorption and Physical adsorption
Cooling water S_1+PASAC9	OL-37	5.4895x10^4	-27.025	chemisorption and Physical adsorption

Table 13. The values of K_{ads} and ΔG°_{ads} for studied systems

2.3. FT-IR studies (Fourier transform infrared spectroscopy)

In this study, FT-IR spectrometry was used to identify whether there was adsorption and to provide new bonding information on the steel surface after immersion in the cooling water system containing organic inhibitor [120-122]. All spectra in these experiments were obtained at a resolution 4cm^{-1} in the region 650- 4000 cm^{-1}.

The FT-IR spectrum of pure organic polymer PASAC 4 is depicted in figure 14a and the FT-IR spectrum obtained for the carbon steel specimens (ol 37, OLC 45) immersed in cooling water systems type S1 containing 800ppm PASAC4 inhibitor organic is presented in figure 14b. A broad peak at 3345 cm^{-1} indicates the presence of the C-H bond of the pASAC4 and the appearance of the peak in region 1636 cm^{-1} and 1570 cm^{-1} corresponds to the C=O and N-H symmetric and asymmetric stretching vibration of the carbonyl group. The presence of C-N stretching frequency is clearly manifest in the region 1315 to 1200 cm^{-1}.

The FT-IR spectra obtained for the carbon steel specimens (ol 37, OLC 45) immersed in cooling water systems type S1 containing 800ppm PASAC4 inhibitor organic is presented in figure 14b. This shows the characteristics the bands for the adsorbed pasac 4 on the metal surface. A weak band in the range from 3674 cm^{-1}is attributed to C-H, the appearance of the peak in region

1596cm^{-1} is assigned to N-H symmetric stretching vibration. The peaks for C-N stretching modes can be assigned in the region around 1459 cm^{-1}.The bands 1170 cm^{-1}and 1060 cm^{-1}are attributed to C-O and C-N. moreover, these FT-IR measurements indicated at 3840 cm-^{1} the direct bonding between Fe atoms and Pasac 4 molecules via O and N atoms, and the formation Fe-inhibitor complex and this reveal that there is only chemical adsorption occurred on the surface of the metal.

The FT-Ir spectrum of organic polymer PASAC6 and of carbon steel immersed in cooling water S1 containing 500ppm PASAC6 is show in figure 15a and 15b. In the spectra of PASAC6, the characteristics peaks at 1076 and 3291 cm^{-1} correspond to the C-O and O-H stretching of the COOH, peak at 1637 cm^{-1} indicates the presence of the C=O, the band at 1578 cm^{-1} is assigned to the bending of N-H. The band at 1393 cm^{-1} is ascribed to the stretching vibration C-N. The weak band at 1200-1000 cm^{-1} is attributed to C-H bending. The Ft-Ir spectra of adsorbed protective layer formed on the surface after immersion in S1 containing optimum concentration of inhibitor PASAC6 is shown in figure 15b. As can be seen all important peaks in pure compounds appeared in adsorption layer on the metal surface. The band around 3347 cm^{-1} is attributed to O-H stretching, which indicates that the protective film contains H_2O. The peak around 2978 cm^{-1} are assigned to C-H stretching vibration, the peaks at 1659 and 1433 cm^{-1} corresponds to the C=O and N-H.. The presence of C-N, C-O is indicated by their starching modes at 1078 and 820 cm^{-1}. The bands 3820 cm^{-1}and 3760 cm^{-1}are attributed to are assigned to Fe-O bending. This is already confirmed from the Langmuir adsorption isotherm studies

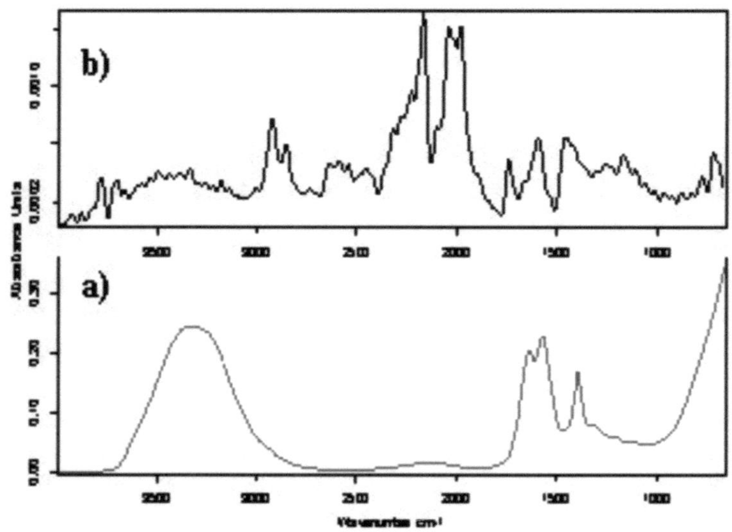

Figure 14. FT-IR spectra of (a) PASAC 4 and (b) OL 37+800 ppm PASAC 4

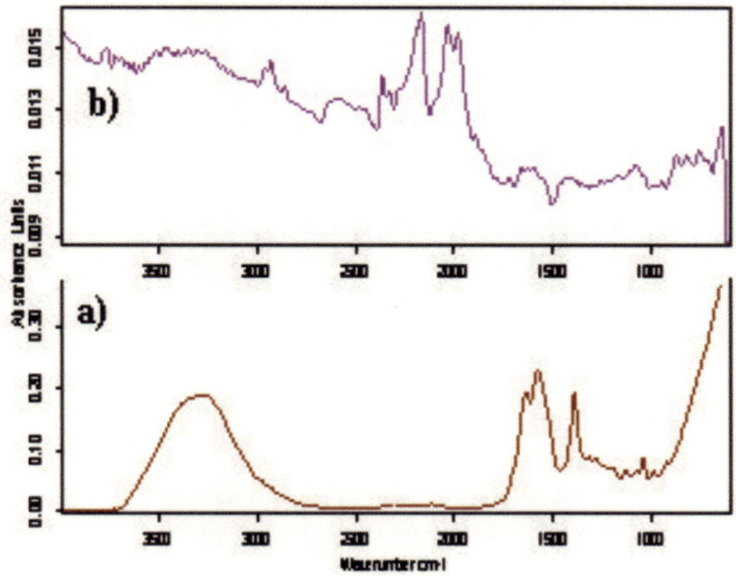

Figure 15. FT-IR spectra of (a) PASAC 6 and (b) OLC 45+500 ppm PASAC 6

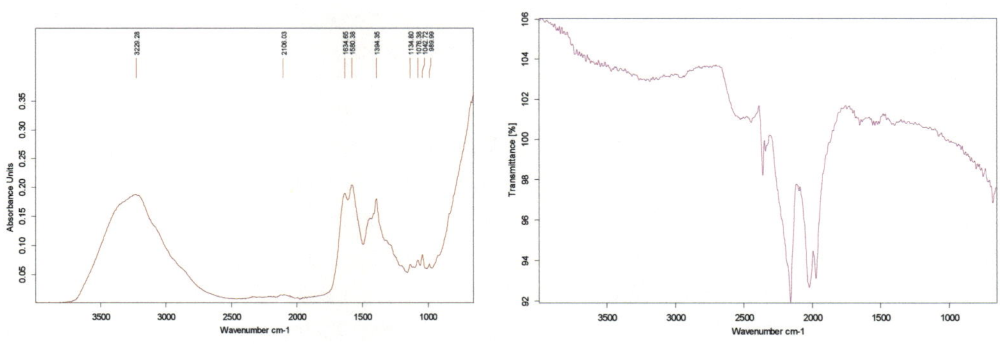

a) b)

Figure 16. FT-IR spectra of (a) PASAC7 and (b) OL 37+300 ppm PASAC7+S1

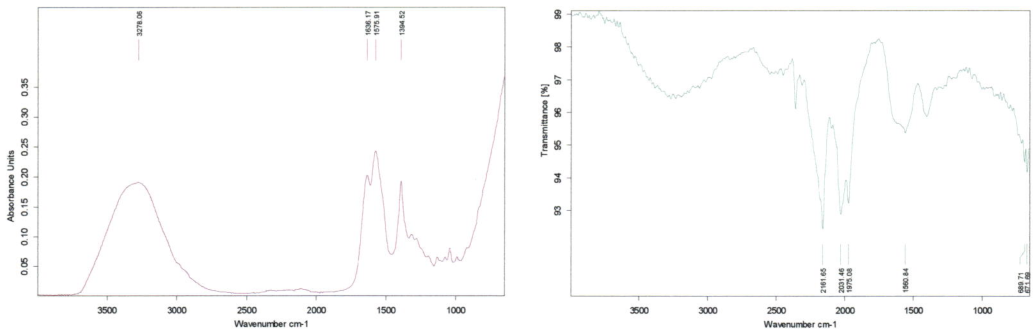

Figure 17. FT-IR spectra of (a) PASAC8 and (b) OL 37+500 ppm PASAC8+S1

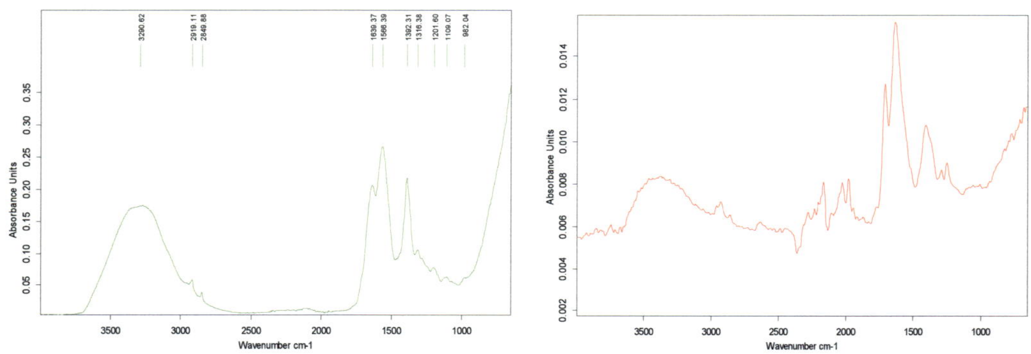

Figure 18. FT-IR spectra of (a) PASAC9 and (b) OL 37+800 ppm PASAC9+S1

The FT-Ir spectrum of organic polymer PASAC7 and 300ppm PASAC7+OL37+S1 is show in figure 16a and 16b. This shows the characteristic bands for the adsorbed PASAC 7 on the metal surface. A weak band in the range 3000-3500cm^{-1}, can be assigned to presence of O-H. The presence of C-H, C=O, N-H, C-N, and C-O is indicated by their stretching modes at 2992, 1642, 1443, 1062 and 810 cm^{-1} respectively.

The FT-IR spectrum of organic polymer PASAC8 and of carbon steel immersed in cooling water S1 containing 500ppm PASAC8 is show in figures 17a and 17b. The Ft-Ir spectra of adsorbed protective layer formed on the surface after immersion in S1 containing optimum concentration of inhibitor 500ppm PASAC8 is shown in figure 17b. As can be seen all important peaks in pure compounds appeared in adsorption layer on the metal surface. The band around 3247 cm^{-1} is attributed to O-H stretching, which indicates that the protective film contains H_2O. The peak around 2878 cm^{-1} are assigned to C-H stretching vibration, the peaks at 1560 and 1423 cm^{-1} corresponds to the C=O and N-H. The presence of C-N, C-O is indicated by their starching modes at 1078 and 789 cm^{-1}. The bands 3820 cm^{-1}and 3760 cm^{-1}are attributed to are assigned to Fe-O bending. This is already confirmed from the Langmuir adsorption isotherm studies.

The FT-Ir spectrum of pure organic polymer PASAC 9 is depicted in figure 18a and the FT-IR spectrum obtained for the ol 37 immersed in cooling water systems type S1 containing 800ppm PASAC9 inhibitor organic is presented in figure 18b. A broad peak at 3325 cm^{-1} indicates the presence of the C-H bond of the pASAC9 and the presence of C=O, N-H, C-H is indicated by their stretching modes at region 1632, 1581 1315 to 1200 cm^{-1} respectively. The FT-IR spectra obtained for the carbon steel specimen (ol 37) immersed in cooling water systems type S1 containing 800ppm PASAC9 inhibitor organic is presented in figure 18b. A broad band in the range from 3684 cm^{-1} is attributed to C-H, the appearance of the peak in region 1598cm^{-1} is assigned to N-H symmetric stretching vibration. The peaks for C-N stretching modes can be assigned in the region around 1459 cm^{-1}.The bands 1200 $cm^{-1}$and 1100 $cm^{-1}$are attributed to C-O and C-N. moreover, these FT-IR measurements indicated at 3800 cm^{-1} the direct bonding between Fe atoms and Pasac 9 molecules via O and N atoms, and the formation Fe-inhibitor complex and this reveal that there is only chemical adsorption occurred on the surface of the metal. This is already confirmed from the Langmuir adsorption isotherm studies.

2.4. EIS – (electrochemical impedance spectroscopy) studies

The corrosion of carbon steel in cooling water system S1and S2 in the absence and presence of PASAC-4, PASAC-5, PASAC-6, PASAC-7, PASAC-8 and PASAC-9 were investigated by EIS. Impedance measurements were performed at open circuit potential on the frequency range between 100 kHz and 40 mHz with an AC wave of ± 10 mV (peak-to-peak) and the impedance data were obtained at a rate of 10 points per decade change in frequency. Nyquist plots for carbon steel obtained at the interface in the presence of inhibitors at optimum concentration are given in figures 19-22. All impedance spectra exhibit one capacitive loop and the diameter of the semicircles increases on increasing the inhibitor concentration suggesting that the formed inhibitive film was strengthened by the addition of inhibitors. However, these diagrams are not perfect semicircles which are attributed to frequency dispersion. The semicircular appearance shows that the corrosion of steel is controlled by charge transfer and the presence of inhibitor does not change the mechanism of dissolution [121-125]. Figures 19-22 also indicates that the diameters of the capacitance loops in the presence of PASAC-4, PASAC-5, PASAC-6, PASAC-7, PASAC-8 and PASAC-9 are bigger than that in the absence of organic inhibitors, suggesting that PASAC-4, PASAC-5, PASAC-6, PASAC-7, PASAC-8 and PASAC-9 has good anticorrosion performance on the carbon steel in S1

Bode diagrams presented in figures 23-27- are in accordance with Nyquist diagrams. It can be observed that in absence of organic inhibitor the electrode presents one time constant corresponding to a phase angle of about 25° at medium and low frequencies, this fact indicates an inductive behaviour with low diffusive tendency. On the contrary, in the presence of the organic inhibitor, on the curve-phase angle versus log frequency appears a maximum very well defined corresponding to a phase angle of about 70° which means that in this case the electrode has a strong capacitive behaviour, according with the results obtained by electrochemical polarization and in concordance with the Nyquist diagrams. All the obtained plots show only one semicircle and they were fitted using one time constant equivalent mode with capacitance (C), the charge transfer resistance (R_{ct}) and Rs solution resistance. The lower

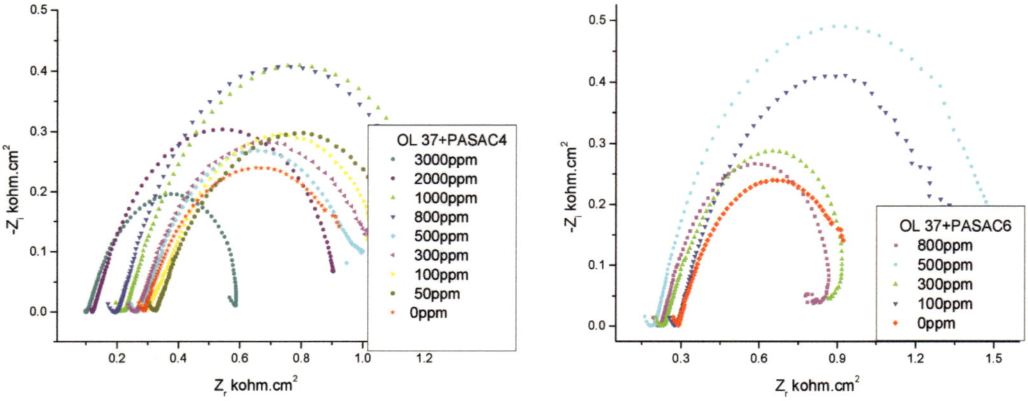

Figure 19. The Nyquist plot for OL 37+ PASAC4 and for OL 37+ PASAC6 in $S_1$with and without organic inhibitor at 25°C

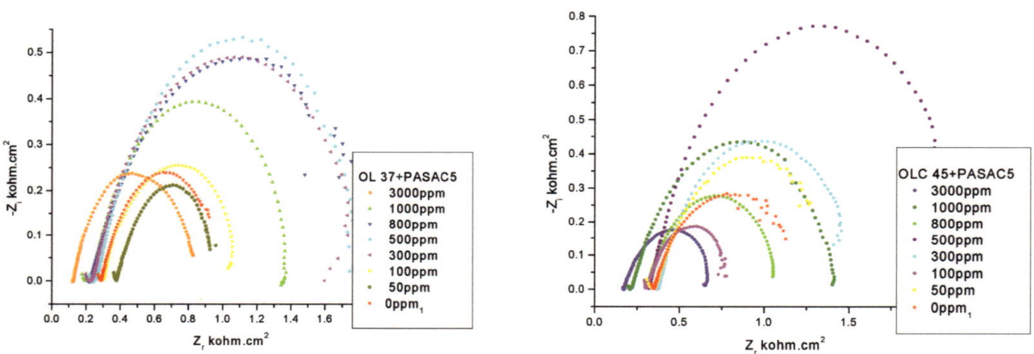

Figure 20. The Nyquist plot for OL 37+ PASAC5 and OLC 45+ PASAC5 in S1with and without organic inhibitor at 25°C

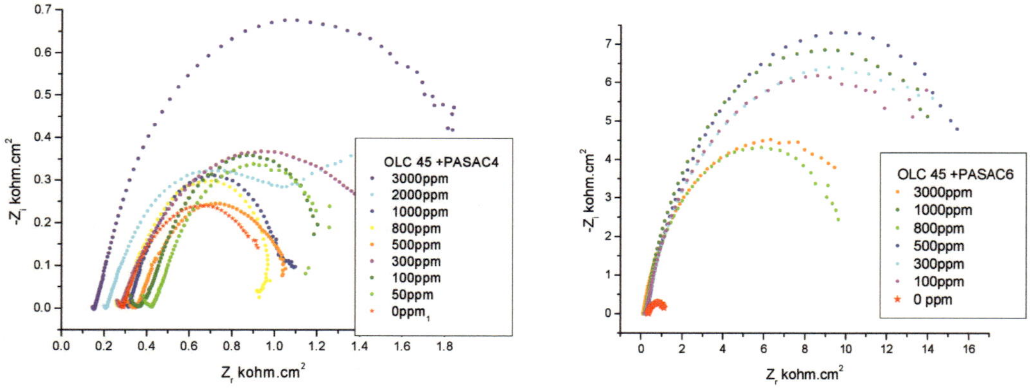

Figure 21. The Nyquist plot for OLC 45+ PASAC4 and OLC 45+ PASAC6 in $S_1$with and without organic inhibitor at 25°C

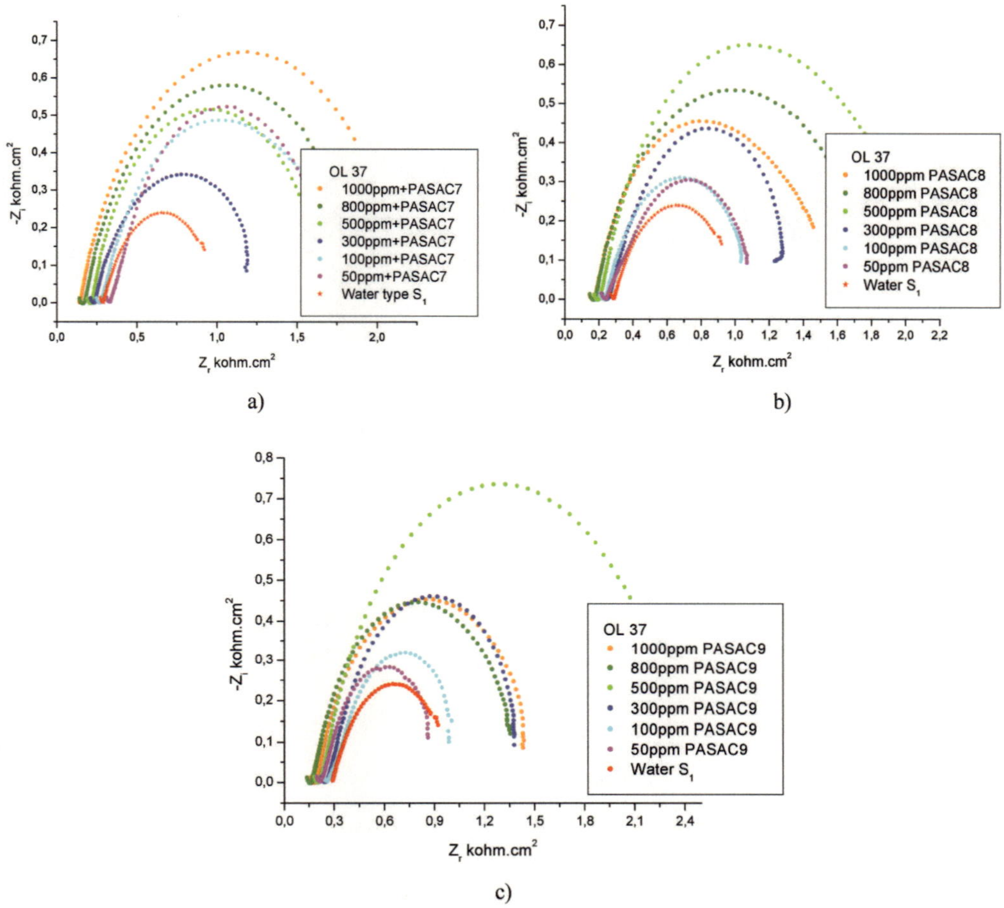

Figure 22. The Nyquist plot for OL 37 in S1 with and without organic inhibitor at 25°C a) PASAC7, b) PASAC8, c) PASAC9

capacitance value for S1+OLC 45 with PASAC 4 and PASAC6 medium indicates the inhomogeneity of surface of the metal roughened due to corrosion. The Cdl values decreases on the increasing the inhibitor concentration and reaches very low value for the optimum concentrations of all the studied systems indicating that the reduction of charged accumulated in the double layer due to the formation of adsorbed inhibitor layer.

The Bode plots are shown in figures 23-27 are in accordance with Nyquist diagrams. It can be observed that in absence of organic inhibitor the electrode presents one time constant corresponding to a phase angle of about 20° at medium and low frequencies, this fact indicates an inductive behaviour with low diffusive tendency. On the contrary, in the presence of the organic inhibitor, on the curve-phase angle versus log frequency appears a maximum very well defined corresponding to a phase angle of about 70° which means that in this case the electrode has a strong capacitive behaviour, according with the results obtained by electrochemical polarization and in concordance with the Nyquist diagrams.

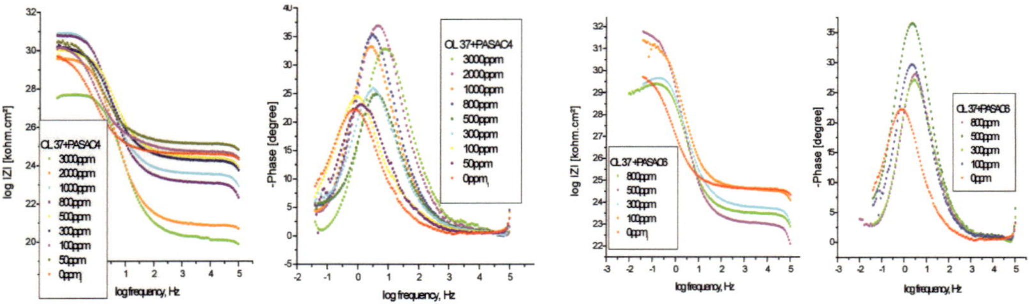

Figure 23. The Bode plot for OL 37+PASAC 4 and PASAC 6in S_1 at 25°C

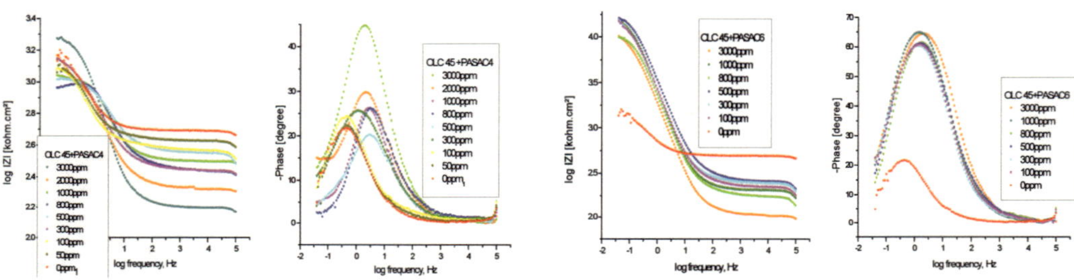

Figure 24. The Bode plot for OLC 45+PASAC4 and OLC 45+PASAC 6 in S_1 at 25°C

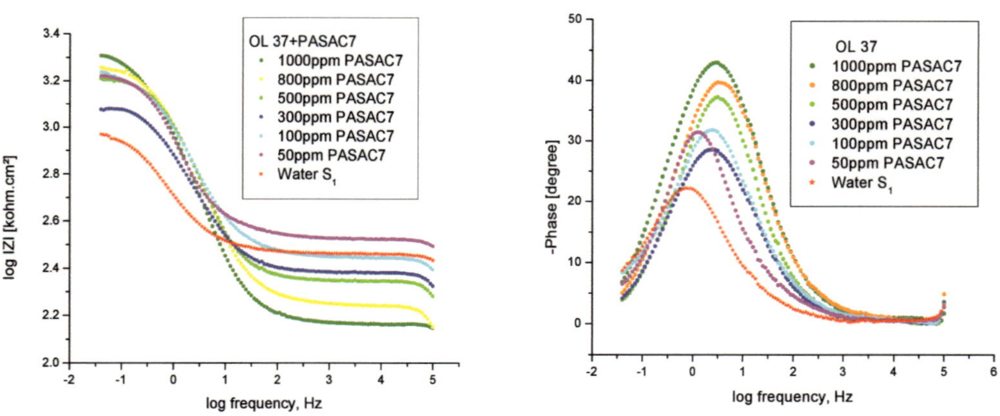

Figure 25. The Bode plot for OL 37+ PASAC7 in S_1 at 25°C

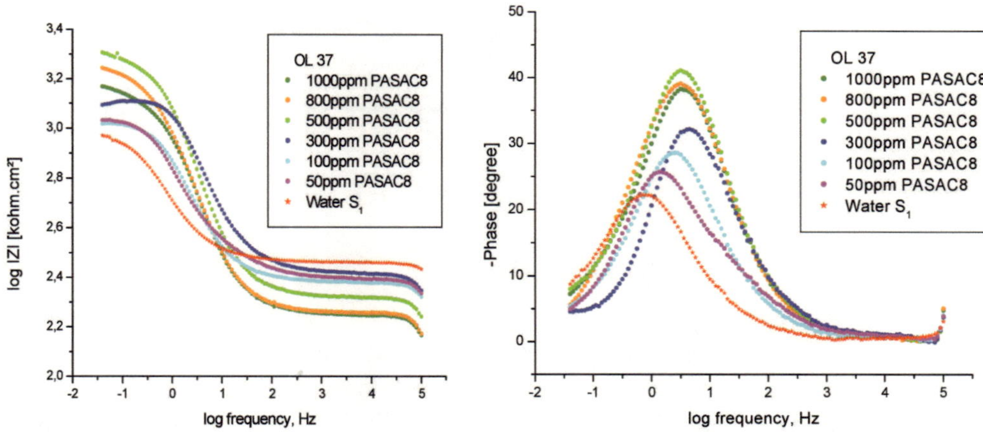

Figure 26. The Bode plot for OL 37+ PASAC8 in S_1 at 25°C

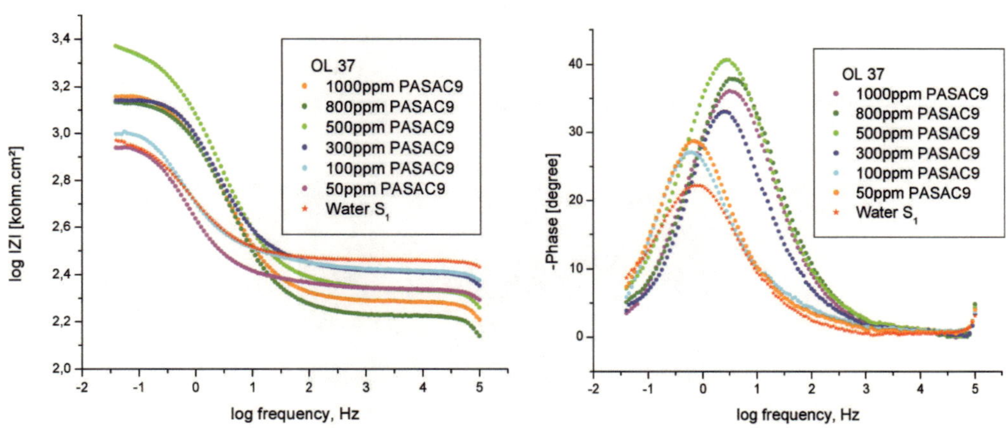

Figure 27. The Bode plot for OL 37+ PASAC9 in S_1 at 25°C

All the obtained plots show only one semicircle and they were fitted using one time constant equivalent mode (see figure 28) with capacitance (Cdl), values is affected by imperfections on the surface which is simulated via a constant phase element (CPE), the charge transfer resistance (R_2) and R_1 solution resistance. The lower capacitance value for S1+OL37 with PASAC 7 and PASAC8, medium indicates the inhomogeneity of surface of the metal roughened due to corrosion. The Cdl values decreases, the R_2 values increases on the increasing the inhibitor concentration and reaches very low value for the optimum concentrations of all the studied systems indicating that the reduction of charged accumulated in the double layer due to the formation of adsorbed inhibitor layer.

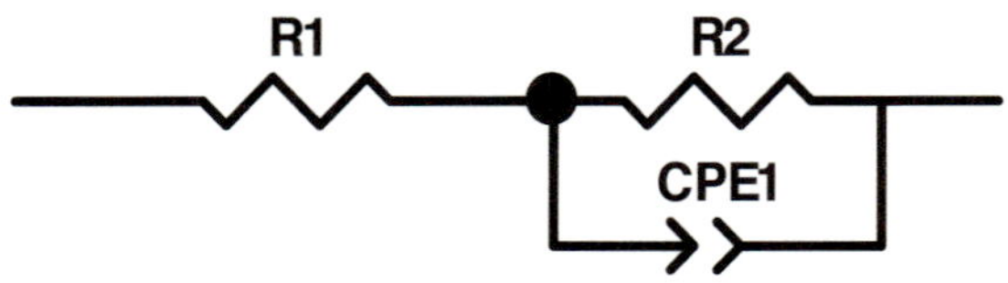

Figure 28. Equivalent circuit

2.5. Microscopic studies of the electrode surfaces

Further, using the metallographic microscope the electrode surfaces were analyzed before and after a certain immersion in cooling water type SC1and SC2. In figure 29-31 are given a few micrographies obtained for the following systems: carbon steel OL 37 and OLC-45 after a certain immersion in cooling water type SC1 and SC2 with and without organic inhibitor. As it can be observed from figures 29-31 the corrosive attack is more accentuated in the cooling water system where the organic inhibitor concentration is lower than in the cases for which the organic inhibitor concentration is higher (see in comparison the micrographies from figure 29-31) [125-127]. Analyzing in comparison the figures 29-31, it can be observed that, on the surface of micrographies there are the adsorbed films of inhibitor and corrosion products and that, these films are thicker if the inhibitor concentration are higher. These films behave like a barrier between corrosive medium and metal surface and as a consequence the corrosion process is inhibited - see in comparison the figures 29-31. Analyzing in comparison the figures

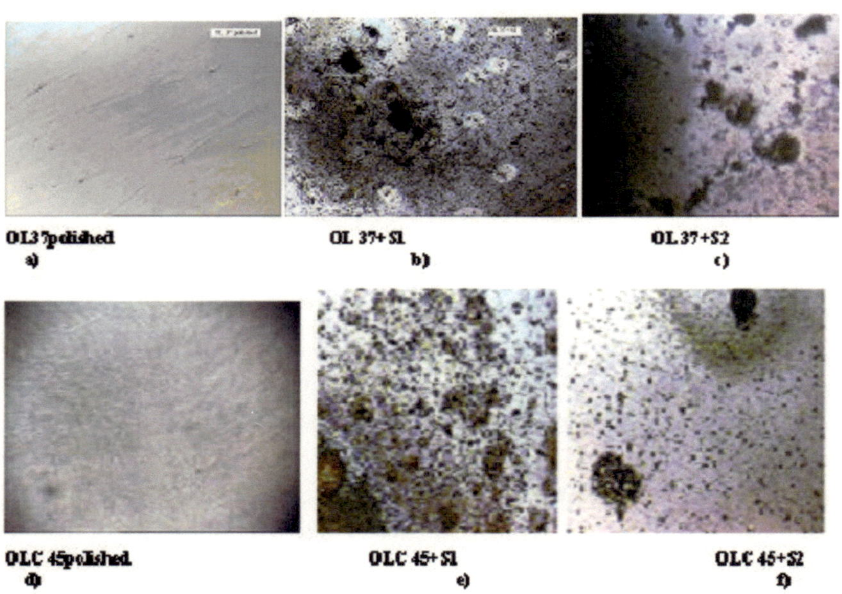

Figure 29. micrographies of the carbon steels in cooling water SC1 and SC2:a) OL 37 polished, b) Ol37+SC1, c) Ol37+SC2, d) OLC 45 polished, e), OlC 45+SC1 and f) OlC 45+SC2

29 a-d and figure 30-31 it can be observed that, the corrosive attack is much more accentuated in the case of OL 37+ water type S1 and OLC 45+ water type S1 system than in the case of OL-37+ water type S1 +800ppm PASAC4, 800ppm PASAC5, 800ppm/500ppm PASAC6, OL37+300ppm PASAC-7, OL37+500ppm PASAC-8 and OL37+800ppm PASAC-9 system. The same behaviour was observed for OLC 45 (see in comparison figures 30 e, f g, and h). This finding is in good concordance with the results obtained by electrochemical method (see tables 4 - 12) and the polarization curves from figures 1-5.

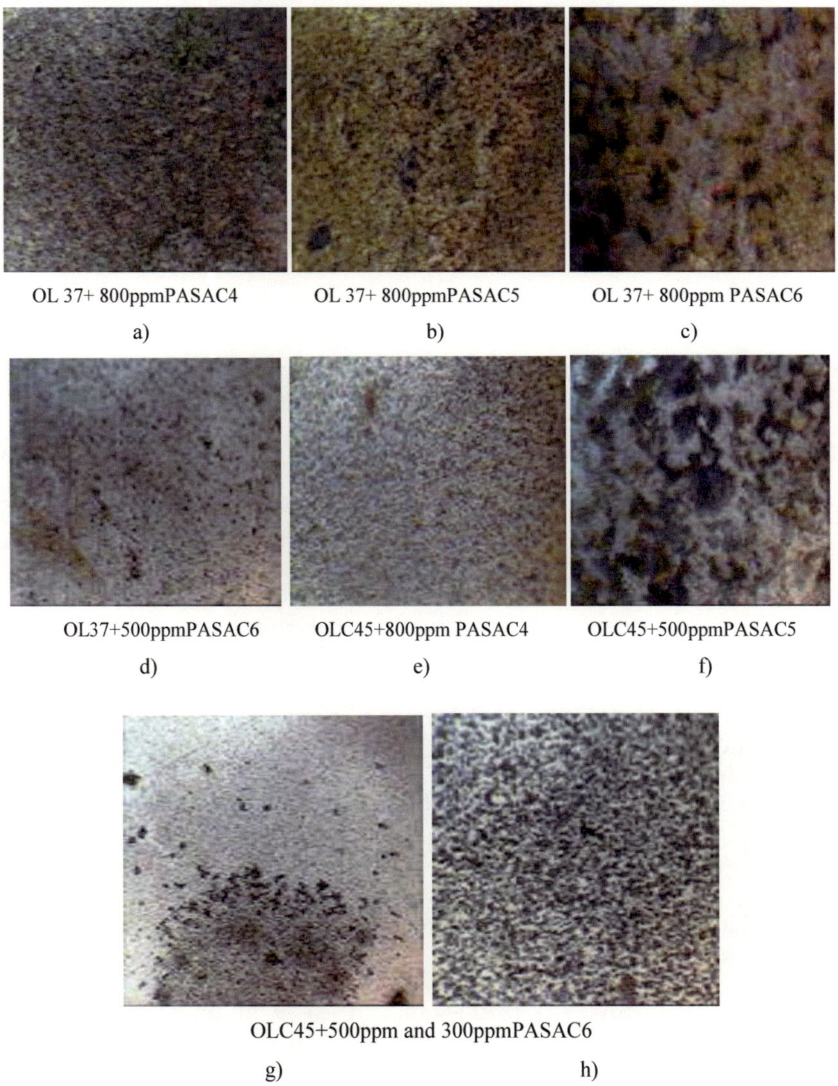

Figure 30. Micrographies of the carbon steel in cooling water SC1 with organic inhibitor; a) OL37+800ppm PASAC-4, b)OL37+800ppm PASAC-5, c) OL37+800ppm PASAC-6, d) OL37+500ppm PASAC-6, e) OLC45+800ppm PASAC-4, f) OLC 45+500ppm PASAC-5, g) OLC 45+500ppm PASAC-6 and h) OLC 45+300ppm PASAC-6

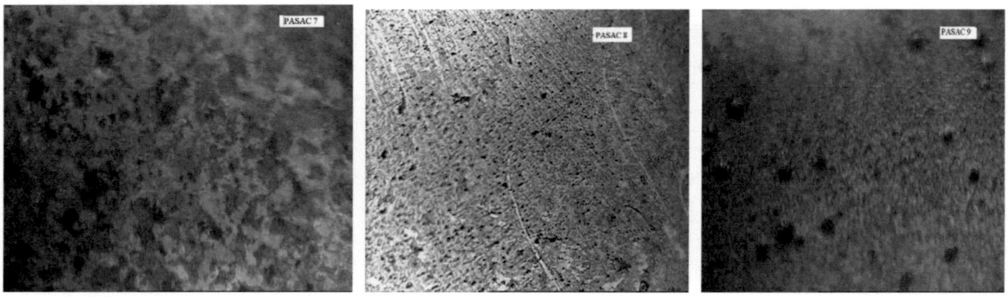

OL37+300ppm PASAC7+S1
a)

OL37+500ppm PASAC8+S1
b)

OL37+800ppm PASAC9+S1
c)

Figure 31. micrographies of the carbon steel in cooling water SC1 with organic inhibitor; a) OL37+300ppm PASAC-7 b)OL37+500ppm PASAC-8 and c) OL37+800ppm PASAC-9

3. Conclusions

Investigated corrosion systems at low overvoltages, the corrosion process are under activation control, while at high overvoltages is controlled by diffusion.

- The addition of organic inhibitors led in all cases to the inhibition of the corrosion process.
- The new organic polymer obtained by our team presented a good inhibitory action and a significant efficiency for decreasing of the rate corrosion of the studied carbon steels;
- The organic inhibitors were adsorbed on the carbon steel surface according to a Langmuir isotherm. The values of the adsorption constant determined from the plot of Langmuir isotherm pointed out that, there are both physical and chemical adsorptions.
- FT-IR spectra revealed very clear that the new organic inhibitors: PASAC-4, PASAC-5, PASAC-6, PASAC-7, PASAC-8 and PASAC-9 were adsorbed on the metal surface.
- The adsorption of investigated organic inhibitor follows the Langmuir isotherm and the FT-IR results, also reveals the adsorption of inhibitor molecule on the metal surface and blocking the active sites.
- EIS results are in very good concordance with results obtained by potentiodynamic and potentiostatic methods.
- PASAC-4, PASAC-6, PASAC-7, PASAC-8 and PASAC-9 inhibit both anodic and cathodic reactions by adsorption on the carbon steel surface and hence behave like mixed type inhibitor. PASAC5 acts preferentially as a cathodic inhibitor.
- In all of the cases, the organic inhibitor type PASAC-4 had a higher efficiency; PASAC-6 had a good efficiency than the organic inhibitor type PASAC-5. The inhibition efficiency follows the order: PASAC4>PASAC6>PASAC5

http://dx.doi.org/10.5772/57276

- In all of the cases, the organic inhibitor type PASAC-8 had a higher efficiency; PASAC-7 had a good efficiency than the organic inhibitor type PASAC-9. The inhibition efficiency follows the order: PASAC8>PASAC7>PASAC9.

The inhibition efficiency follows the order: $E_{PASAC4} > E_{PASAC6} > E_{PASAC5}$ $E_{PASAC8} > E_{PASAC7} > E_{PASAC9}$

Author details

Florina Branzoi[1] and Viorel Branzoi[2]

*Address all correspondence to: fbrinzoi@chimfiz.icf.ro

1 Institute of Physical Chemistry, Bucharest, Romania

2 University Politehnica of Bucharest, Bucharest, Romania

References

[1] H. H. Uhlig and R.W.Revie, "Corrosion and Corrosion Control", Whiley, New York, 3rd edn. 1985.

[2] D. Jones, "Principle and Prevention of Corrosion", MacMillan Publishing Company, New York, 1992.

[3] M.A.Quraishi, H.K.Sharma, *Mater.Chem.Phys*, 2002, 72,18

[4] W.H.Li, Q. He, C.L. Pei, B.R.Hou, Electrochim.Acta 52, 2007, 6386.

[5] L.Tang, X.Li,,G.Mu, G.Liu, *Mat.Chem.Phys*.97, 2006, 301

[6] Florina Branzoi, V. Branzoi, and I. Harabor, *Revue Roumaine de Chimie*, vol. 56 (2), 487-115-128, 2011.

[7] J. L. Rozenfeld, "Corrosion Inhibitors", McGraw-Hill, New York, 1981, p.109.

[8] V. Branzoi, and Florina Branzoi, *Rev Roum Chim*., 2002, *47*, 1193-1203

[9] V.Branzoi, F. Golgovici, Florina Branzoi, *Materials Chemistry and Physics*, 78, 2002, 122-131

[10] D. P. Schwensberg and V. Ashworth, *Corros. Sci*., 1988, *28*, 539.

[11] V. Branzoi, Florina Branzoi and L.Pilan, *Materials Chemistry and Physics*, 2009,*118*, 197.

[12] A. Stanca, V. Branzoi, Florina Branzoi, *International Journal of Chemical and Biomolecular Engineering*, vol.1, nr.2, pp.98-103, 2008

[13] T. Hirai, J. Yamaki, T. Okada and A. Yamagi, *Electrochim. Acta*, 1985, *30*, 61.

[14] D. G. Leaist, *J. Chem. Soc. Faraday Trans.*, 1990, *86*, 3487.

[15] E. Kamis, F. Bellucci, R. M. Latonision and E. S. H. El-Ashry, *Corrosion* 1991, *47*, 677.

[16] M. Beier and J. W. Schultze, *Electrochim. Acta* 1992, *37*, 2299.

[17] F. Donahue and K. Nobe, *J. Electrochemical. Soc.* 1965, *112*, 886.

[18] Y. A. Aleksanyan, I. I. Reformatskaya and A. N. Podobaev, *Protection of Metals*, 2007, *43*, 125-128.

[19] E.M. Sherif and Su-Moon Park, *Electrochimica Acta*, 2006, *51*, 4665-4673.

[20] V. Branzoi, Florina Branzoi, Luisa Pilan, Florentina Golgovici, *Molecular Crystal &Liquid Crystals*, vol.446, 2006

[21] M. J Incorvia and S. Contarini, *J.Electrochem.Soc.* 1989, *136*, 2493.

[22] F. Mansfeld, "Corrosion Mechanism", Marcel Dekker (ed.), New York, 1987, p.119.

[23] V. Branzoi, F. Branzoi and L. Pilan, *Molec. Crystal &Liquid Crystals*, 2006,*446*, 305.

[24] G. Banerjee and S. N Malhotra., *Corrosion*, 1992, *48*, 10.

[25] V. Branzoi, A. Pruna, Florina Branzoi, *Revista de Chimie*, vol.59, nr.5,pp.540-543, 2008

[26] V. Branzoi, A. Pruna and Florina Branzoi, *Rev Roum Chim.*, 2007, *52*, 587.

[27] P.Ramesh P, B.Babu, K.Thangavel, Anti-Corros Methods Mater, 2005, 52, 219.

[28] L Larabi, O Benali, Y Harek, Mater Lett, 2007, 61, 3287

[29] AS Fouda, HA Mostafa, GY Elewady, Corros Sci, 2005, 47, 1988

[30] R Yurchenko, L. Pogrebova, L Pilipenko, T Shubina, Russian J. Appl Chem, 2009, 79, 1100

[31] SA Hossain, AL Almarshad, Corros Eng Sci Technol, 2006, 41, 77.

[32] S Abd El-Wahaab, G.Gomma, HY El-Barradie, J Chem Technol Biotechnol, 2007, 36, 435

[33] S Muralidharan, SVK Iyer, Anti-Corros Methods Mater, 1997, 44, 100

[34] MA Quraishi, MAW Khan, M Ajmal, Anti-Corros Methods Mater, 1996, 43, 5

[35] F Bentis, M Traisnel, M Lagrenee, J Appl Electrochem, 2001, 31, 41

[36] M. Sahin, S Bilgic, Anti-Corros Methods Mater, 2003, 50, 34

[37] M Lebrini, F Bentis, F Bezin, M Lagrenee, Corros Sci, 2006, 48, 1279

[38] GK Goma, Mater Chem Chem, 1998, 34, 833

[39] L Larabi, Y Harek, m Traisnel, A Mansri, J. Appl Electrochem, 2004, 34, 833

[40] RS Chaudhary RS, H Kumar, Indian J Chem Tech, 2004, 11, 783-786

[41] E.E. Oguzie, Y. Li, F.H. Wang J.Colloid Interface Sci. 310 (2007) 90–98.

[42] P.C. Okafar, Y. Zheng, Corros. Sci. 51 (2009) 850–859.

[43] C. Jeyaprabha, S. Sathiyanarayanan, G. Venkatachari, J.Electroanal. Chem. 583 (2005) 232–240.

[44] S.S. Abdel Rehim, O.A. Hazzazi, M.A. Amin, K.F. Khaled, Corros. Sci. 50 (2008)

[45] 2258–2271.

[46] C. Jeyaprabha, S. Sathiyanarayanan, G. Venkatachari, Electrochim. Acta 51 (2006) 4080–4088.

[47] S.A. Umeron, O. Ogbobe, I.O. Igwe, E.E. Ebenso, Corros. Sci. 50 (2008) 1998–2006.

[48] G.K. Gomma, Mater. Chem. Phys. 55 (1998) 241–246.

[49] A.A. Atia, M.M. Saleh, J. Appl. Electrochem. 33 (2003) 171–177.

[50] M.M. Saleh,, Mat. Chem. Phys. 98 (2006) 83–89.

[51] M.M. Saleh, A.A. Atia, J. Appl. Electrochem.36 (2006) 899–905.

[52] F. Bentiss, M. Lagrenee, M. Traisnel, Corrosion 2000, 56, 733.

[53] P. Bothi Raja, M. G. Sethuraman, Mater. Corros. 2009, 60, 22.

[54] V. S. Sastry, Corrosion Inhibitors: Principles and Applications, John Wiley & Sons, New York, 1998.

[55] S. M. A. Hosseini, A. Azimi, Mater. Corros. 2008, 59, 41.

[56] M. Bouayed, H. Rabaa, A. Srhiri, J. Y. Saillard, A. Ben Bachir, Le Beuzed, Corros. Sci. 1999, 41, 501.

[57] L. B. Tang, X. M. Li, H. C. Liu, G. N. Mu, G. H. Liu, J. Mater.Sci. 2006, 41, 1991.

[58] Sekine, Y. Nakahata, H. Tanabe, Corros. Sci. 1988, 28, 987.

[59] M. G. Fontana, Corrosion Engineering, 3rd edition, McGraw-Hill, Singapore, 1986.

[60] H. Farooqi, A. Hussaiin, M. A. Quaraishi, P. A. Saini,Anti-Corrosion 1999, 46, 328.

[61] Z. Lu, Corros. Prot. 1999, 20, 201.

[62] G. Trabanelli, F. Mansfeld (Ed.), Corrosion Mechanisms, Marcel Dekker, Inc., New York, 1987.

[63] U.S. Congress, Office of Technology Assessment, Biopolymers: Making Materials Nature's Way, 1993.

[64] W. Hater, B. Mayer, M. Schweinsberg, Power Plant Chem.2000, 12, 721.

[65] C. S. Sikes, A. P. Wheeler, Chemtech 1988, 18, 620

[66] M. A. Amin, J. Appl. Electrochem. 2006, 36, 215.

[67] F. Mansfeld, M. W. Kending, S. Tsai, Corrosion 1982, 38, 570.

[68] C. B. Breslin, W. M. Carrol, Corros. Sci. 1993, 34, 327.

[69] M. G. A. Khedr, M. S. Lashien, Corros. Sci. 1992, 33, 137.

[70] S. S. A. Rehim, H. H. Hassan, M. A. Amin, Mater. Chem. Phys. 2001, 70, 64.

[71] M. R. Saleh, A. M. Shams El Din, Corros. Sci. 1981, 12, 688.

[72] K. Maayta, N. A. F. Al-Rawashdeh, Corros. Sci. 2004, 46,

[73] A. Mohammed, J. Appl. Electrochem. 2006, 36, 215.

[74] M. Abdallah, E. A. Helal, A. S. Fouda, Corros. Sci. 2006, 48,1639.

[75] A.Igual Muñoz, J. García Antón, J.L. Guiñón, V. Pérez Herranz, Electrochimica Acta 50(2004), 957

[76] P. Kern, D. Landolt, Electrochim. Act. 2001, 47, 589.

[77] P. Kern, D. Landolt, J. Electrochem. Soc. 2001, 148 (6), B228.

[78] P. Kern, D. Landolt, Electrochim. Act. 2002, 44, 1809.

[79] L. J. Vracar, D. M. Drazic, Corros. Sci. 2002, 44 1669.

[80] S. L. Granese, B. M. Rosales, C.Ovideo, J. O. Zerlino, Corros.Sci. 1992, 33, 1439.

[81] W. Pleith, Electrochim. Acta. 1992, 37, 2115.

[82] H. Akrout, L. Bousselmi, E. Triki, S. Maximovtch, F. Dalard, J. Mater. Sci. 2004, 39, 7341.

[83] E.M.Sherif, Su-Moon Park, Electrochimica Acta 51(2006) 6556

[84] El-Sayed M. Sherif, Applied surface science 252 (2006) 8615

[85] E.M.Sherif, Su-Moon Park, Corrosion science 48 (2006) 4065

[86] El-Sayed M.Sherif, A.M.Shamy, Mostafa M.Ramla, Ahmed O.H.El Nazhawy, Materials chemistry and physics 102 (2007) 231

[87] Gy. Vastag, E. Szöcs, A. Shaban, E. Kálmán, Pure Appl. Chem., 73 (2001) 1861

[88] F.Zucchi, G.Trabanelli, M.Fonsati, Corrosion science, 38 (1996) 2019

[89] E.Szõcs, Gy.Vastag, A.Shaban, E.Kálmán, Corrosion Science 47 (2005) 893

[90] X.R.Ye, X.Q.Xin, J.J.Zhu, Z.L.Xue, Applied Surface Science 135 (1998)307

[91] J.C.Marconato, L.O.Bulhões, M.L.Temperini, Electrochimica Acta 43(1998)771

[92] R.Subramanian, V. Lakshminarayanan, Corrosion Science 44 (2002) 535

[93] S. Ramesh, S.Rajeswari, Corrosion Science 47 (2005) 151

[94] El-Sayed M.Sherif, R.M.Erasmus, J.D.Comins, Journal of colloid and interface science 311 (2007), 144

[95] F. Bentiss, M. Lagrenee, M. Traisnel, Corrosion 56 (2000) 733–742.

[96] M. Bouayed, H. Rabaa, A. Srhiri, J.Y. Saillard, A. Ben Bachir, A. Le Beuze, Corros. Sci. 41 (1998) 501–517.

[97] L. Tang, X. Li, H. Liu, G. Mu, G. Liu, J. Mater. Sci. 41 (2006) 1991–1997.

[98] G. Moretti, G. Quartarone, A. Tassan, A. Zlngales, Electrochim. Acta 41, (1996) 1971–1980.

[99] Yıldırım, M. Cetin, Corros. Sci. 50 (2008) 155–165.

[100] N.M. Hashim, A.A. Rahim, H. Osman, P.B. Raja, Chem. Eng.Commun. 199 (2012) 751–766.

[101] P.B. Raja, M.G. Sethuraman, Mater. Lett. 62 (2008) 1602–1604.

[102] M.A. Amin, S.S. Abd El-Rehim, E.E.F. El-Sherbini, R.S. Bayoumi, I., Electrochim. Acta

[103] 52 (2007) 3588–3600.

[104] D. Kesavan, M. Gopiraman, N. Sulochana, Chem. Sci. Rev. Lett. 1 (2012) 1–8.

[105] A.Y. El-Etre, M. Abdallah, Z.E. El-Tantawy, Corros. Sci. 47 (2005) 385–395.

[106] O.K. Abiola, A.O. James, Corros. Sci. 47 (2010) 661–664.

[107] S. Deng, X. Li, Corros. Sci. 55 (2012) 407–415.

[108] S. Banerjee, V. Shrivatsava, M.M. Singh, Corros. Sci. 59 (2012) 35–41.

[109] S. Garai, S. Garai, P. Jaisankar, J.K. Singh, A. Elango, Corros. Sci. 60 (2012) 193–204.

[110] A.A. Rahim, E. Rocca, J. Steinmetz, M.J. Kassim, Corros. Sci. 50 (2008) 1546–1550.

[111] P.B. Raja, A.A. Rahim, H. Osman, K. Awang, Acta Phys. Chim. Sin. 26 (2010) (2010) 2171–2176.

[112] P.B. Raja, A.A. Rahim, H. Osman, K. Awang,Int. J. Miner. Metall. Mater. 18 (2011) 413–418.

[113] S. Nofrizal, A.A. Rahim, B. Saad, P.B. Raja, A.M. Shah, S. Yahya, Metall. Mater.

[114] K. Aramaki, J.Uehre and H. Nishihare, Proc. of the 11th International Corrosion Congress, Florence, Italy, 1990, 3, 331.

[115] G. B. Hunt and A. K. Holiday, "Organic Chemistry", London, United Kingdom, Buterworth, 1981, p. 229.

[116] R. D Braun, E.E Lopez and D.P. Vollmer, Corros. Sci. 1993, 34, 1251.

[117] Florina Branzoi, V. Branzoi, C. Licu, Materials and Corrosion, 2012, 63, No.9999, DOI10.1002/maco.201206579..

[118] Loupy, Chemistry Today, 2006, 24, 36.

[119] Q. Qu, S.Jiang, W.Bai and L.Li, Electrochim. Acta, 2007, 52,6811

[120] Q. Quing, L.Li, S.Jing and Z.Ding, J. Appl. Electrochem., 2009, 39,569

[121] M. F. L. Granero, P.H.L.S. Matai, I.V.Aoki and I.C.Guedes, J. Appl. Electrochem. 2009, 39, 1199.

[122] D. Gopi, K. M. Govindaraju and L. Kavitha, J. Appl. Electrochem. 2010, 40, 1349.

[123] Prună, V. Branzoi and F. Branzoi, Materials and Technologies, 2007, 23, 233.

[124] V.Branzoi, A. Pruna, Florina Branzoi, Revue Roumaine de Chimie, 52(6), 589-597, 2007

[125] Florina Branzoi, V. Branzoi and A. Stanca, Rev. Chim.,. 2008, 60, 797

[126] I.Harabor, V. Branzoi and Florina Branzoi, Revue Roumaine de Chimie,57 (6), pp. 735-745, 2012.

[127] Florina Branzoi, V. Branzoi, and I. Harabor, Revue Roumaine de Chimie, 55 (8), 487-500, 2010.

Section 4

Corrosion in Industry

Fractal Effect of Corrosion on Mechanical Behavior of Unprotected Structural Steel

Francisco Casanova del Angel

Additional information is available at the end of the chapter

http://dx.doi.org/10.5772/57061

1. Introduction

Currently, it is known that, most of the time, the approach to continuous media results are unsatisfactory for use in real materials. In man made structures, defects appear during the production stage at nano-, micro- and macro-scales, which evolve throughout its useful life, generating failures, some of which can be catastrophic. Fractography surveys show the non-Euclidean nature of fracture patterns [1]. Macro-scale breaking patterns are common in experimental laboratory research, while micro- and nano-scale processes, which govern the macroscopic behavior and deformed solids fracture, are not [2].

The fracturing of materials is an interesting problem from a scientific and a technological point of view, which has generated a lot of research regarding the formation and propagation of cracks. It is well established that fracture patterns may be treated from the point of view of fractal geometry [3]. Mandelbrot was one of the main researchers on fractal aspects of metal fracture surfaces. Results obtained on thermally treated steel samples fractured in impact tests suggested a strong correlation between macroscopic tenacity and the fractal dimension of the fracture surface [4]. Currently, it is known that fracture surfaces are fractal objects showing anisotropic scaling. Natural fractals are self-similar (remember that a self-similar object is exactly or approximately similar to a part of itself and in the case of natural fractals, they display self-similar structure over an extended -but finite- scale range), and fracture surfaces are self-similar from a statistical point of view. The self-affine case is described in terms of its parameters, regarding Hurst's exponent or rugosity exponent, ς, and the ξ correlation length, which is the higher limit for which the self-affine behavior is shown.

The fracturing of structures has been a problem since man began to model and build structures. This has become critical, due to complexity of modern structures and heavy operational loads. During the industrial revolution, there was a high increase of metals (iron and steel) usage for structural applications. Unfortunately, there were also many accidents leading to human and material losses due to failure of such structures. Some of them were due to defective designs, but it has also been discovered that the defects in materials like persistent cracks might have started cracking, thus leading to fracture. The surface of materials depends a lot on properties such as: adhesion, friction, wear, permeability, etc. It is true that surfaces exposed to the environment and interactions between both of them are responsible for the behavior of materials. Data that might be obtained from the surface may be related to microstructure and mechanical properties of material. It must also be remembered that when a metallic piece is under stress, corrosion concentrates causing transgranular or intergranular cracks.

In the specific case of surface fracture, its analysis is usually carried out through metallography, which may tell us about, among others things, the origin of the fracture, the direction of its propagation and the type of load that caused it. The surfaces of fractures are usually a set of repeated patterns, for example, fractures, fissures, ridges, holes and intergranular defects, among others. Quantitative fractography aims at translating such features to a parametrical shape, and that is where fractal geometry comes in [5].

Fractal geometry looks for regularity in the relationship of an object and its parts at various scales, that is, it studies invariable geometrical aspects, regardless of scale changes. In other words, fractal geometry studies non-differentiable or non-continuous geometrical shapes at any scale. From a mathematical point of view, a fractural is a subset of a metric space for which its Hausdorff-Besicovitch dimension is strictly higher than its topological dimension. In general, fractals have some type of similarity and it may be considered that they are made of small parts similar to the whole. This similarity may be strictly geometrical, or only approximate or statistical. Mandelbrot quantitatively explored for the first time, the fractal character of fracture surfaces subject to various thermal treatments and reported a correlation between fractal dimension and tenacity of the fracture, their property estimated through impact energy [4]. These results, though questioned in the beginning, started a new era in the fractography disciplines. From then, analysis of self-affinity of fracture surface is a very active research field, which has had the contribution of modern and sophisticated statistical and mathematical methods.

There are various methods for analyzing self-affinity of fracture surfaces, most of them use profiles extracted from surfaces through the use of experimental techniques. Highly refined experiments, focused on various materials, do not confirm correlation between the fractal dimension and mechanical properties [6]. It has been established that rugosity exponent ζ is the proper parameter by which to describe fracture surfaces in fast kinetic conditions analyzed mainly through scanning electron microscope (SEM). Bouchaud proposed a universal rugosity exponent $\zeta = 0.78$, regardless of microstructure and properties. Such universality was questioned by the discovery of another self-affine regime characterized by a rugosity exponent $\zeta = 0.5$ for fracture surfaces generated in slow crack propagation conditions or analyzed at manometric scales, using atomic force microscope (AFM). Coexistence of both regimes in various materials has also been reported [7]. Such regimes cross in a fracture length, which seems to be independent of kinetic conditions. Attempts to relate such fracture length with microstructural parameters of some materials continue to be carried out. [8]

The development of this research on behavior of unprotected A36 steel exposed to corrosion (the steel had no coating technique applied using zinc as the main protective element), resulted from observing occurrence of cracks on structures exposed to a marine environment, as well as the analysis of consequences implied by such fact, regarding marine corrosion. Research is shown from dimensioning of test specimens; laboratory tests with their chemical analysis; and, finally, a fractography of the corresponding test specimen, obtaining its fractal dimension based on the theoretical development shown.

2. Dimensioning of test specimens

In accordance with ASTM A-370 standard specifications [9] and complying with such parameters (geometrical relationships between length, width and height), Figure 1 shows the type of test specimen used. Table 1 shows proportions for dimensioning of the test specimen used in the laboratory, obtained in accordance with ASTM standards. Test specimens were made of A36 structural steel (C = 0.29 max., Mn = 1.20, Si = 0.40 max., P = 0.40 max., S = 0.05 max). In accordance with ASTM specifications, test specimen were made with A36 structural steel, chemical composition: C = 0.29 max., Mn = 1.20, Si = 0.40 max., P = 0.40 max., S = 0.05 max. Existing Mexican regulations were considered for tensile tests on steel products. [10]

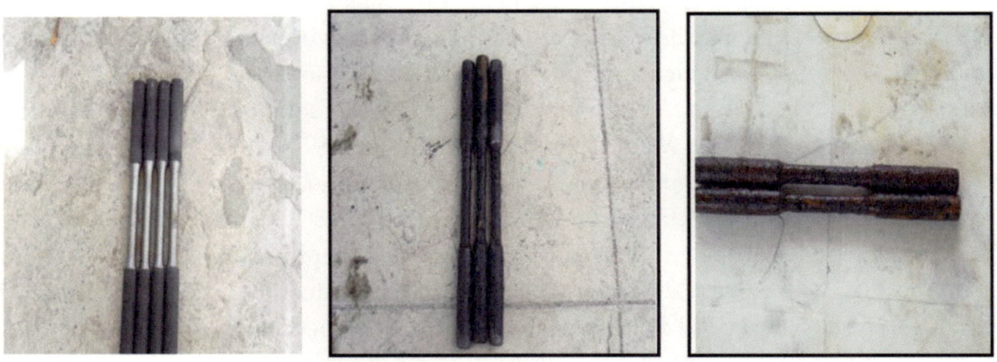

a) b) c)

Figure 1. a) Virgin A36 structural steel test specimen, b) after 6 months, and c) after 12 months

Diameter	Diameter in reduced section, D	Radio of transition area, R	Length of reduced section, A
12.0mm± 0.1mm	8.89mm ± 0.17mm	6.35mm± 0.1mm	58.0mm ± 0.1mm

Table 1. Dimension of test specimen to be used in accordance with ASTM A-370 Standard.

3. Laboratory tests and chemical analysis

Tensile tests were carried out in a 250 KN Universal Mechanical Testing machine. Conditions for the test were: testing speed: 0.60 mm/min.; sampling frequency: 5.0 points/s; downwards. Traction assays for unprotected A36 structural steel were carried out on three types of sample, taking into account the base material, and other two samples were applied under real marine corrosion, throughout 6-month and 1-year aging periods, Figures 1.a., 1.b and 1.c. Based on features and components of the metal used, the sample was roughed down and polished in order to remove scratches. The procedure used was the traditional one, consisting in roughing down the surface of the sample and sanding it with an ultimate approximation to a flat surface free from scratches with a moisturized spinning wheel. The sample was moisturized in a Petri dish with a solution of Nital dissolved in water for 10 seconds. Moisturizing time was carefully controlled. Action of moisturizing stopped when the sample was put under a water stream, cleaned with alcohol and a drier was used for finishing [11].

4. Results of mechanical properties of A36 structural steel

A fractographic analysis begins with a final observation of the fracture's surface features. Evidence of nucleation of fissure, mechanism and direction of propagation causes may be obtained, and an estimation of acting loads' magnitude may eventually be obtained. [12] and [13] Figure 2.a shows the results of the stress-strain graph for virgin material (i.e. not exposed to marine conditions), for which a maximum 369.18522 MPa load and a 255.1545 MPa yield strength were obtained. Figure 3.a shows its fractography. For structural steel aged more than 6 months at sea, a 351.50238 MPa maximum load and a 238.676 MPa yield strength were obtained, Figure 2.b. Fractography in Figure 3.b shows nucleated cracks and a behavior ductile to fracture. For structural steel aged more than 1 year, a 315.883 MPa maximum load and a 226.24 MPa yield strength were obtained, Figure 2.c. Fractography shown in Figure 3.c shows crack branching due to tensile testing, since material loses ductility due to corrosion. The second and third columns of Table 2 show the results of the mechanical properties of steel in accordance with the tensile test, and the fourth and fifth columns show specific values of steel regarding the Internet web page referred to [14]. It should be noticed that results obtained from tensile test for virgin A36 structural steel test specimen are very similar to those shown by that Web page.

Stress under which the material changes its behavior from elastic to plastic is not easily detected, so for this reason the yield strength has been determined drawing a line parallel to the starting section of the stress-strain curve, but displaced 0,002 mm/mm (0,2 % from the origin), crossing the curve of the diagram. The intersection point is the yield point.

To obtain fractal dimension of microcracks found in the material, the theoretical description of a fractal's behavior in unprotected A36 steel exposed to corrosion is shown [15].

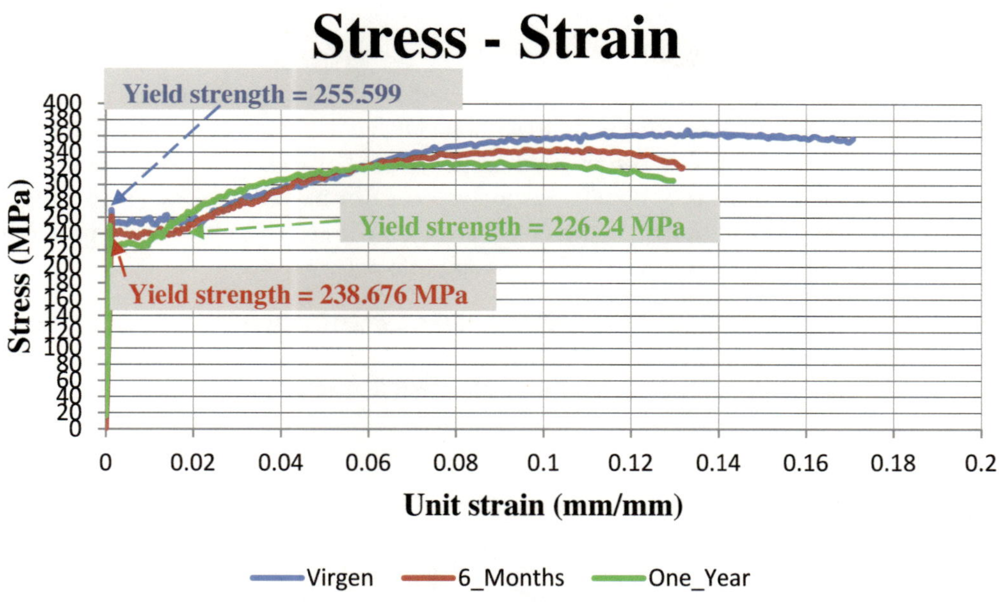

Figure 2. Yield strength for test specimen.

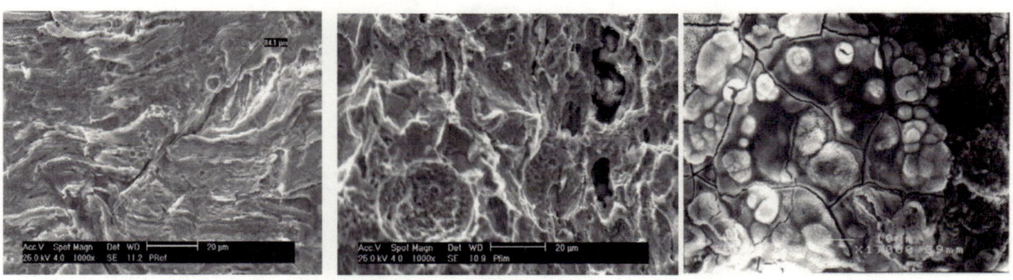

a) b) c)

Figure 3. a) Fractography of test specimen. b) Fractography of test specimen after 6 months in marine conditions, and c) Fractography of test specimen after 1 year in marine conditions.

	Tensile test			www.matweb.com	
Material. A36 structural steel	**Yield strength (f_y)**	**Elasticity module (E)**	**Maximum strength**	**Yield strength (f_y)**	**Elasticity module (E)**
Virgin	255.59 MPa	205317.72 MPa	369.185 MPa	250 MPa	200.000 MPa
Exposed 6 months	238.67 MPa	202474.23 MPa	351.502 MPa	X	X
Exposed 1 year	226.24 MPa	198716.21 MPa	315.883 MPa	X	X

Table 2. Mechanical properties of tensile test in accordance with reference [14].

5. A fractal theoretical desciption

Definition of generating curve. Let I_0 be a unitary lenght line segment, contained in a closed interval, that is, $I_0 \subset [a, b]$. Let I_1 be a set with sectioned behavior, consisting in three segments of a straight line which create, based on starting point a of I_0, two scalene triangles reflected regarding the middle point c of I_0, obtained as follows: the first half of segment I_1 is substituted or removed by the sides of triangle which create an angle with I_0. This process is repeated for the second half, but with the sides reflected from middle point c.

This process is known as *generator* which is called state 1. Construction of set I_2 is made applying the generator to every segment of I_1, which is called state 2. Thus, set I_k is created applying generator I_1 on every segment of I_{k-1}, which is called state k, Figure 4.a.

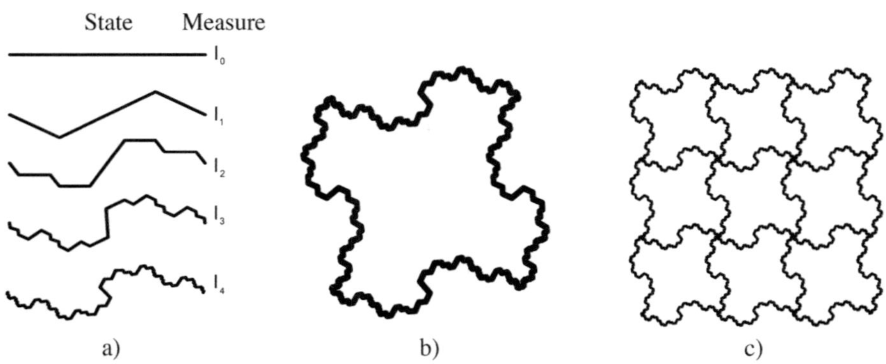

Figure 4. a) Construction of the fractal curve I. In every state I_k generator I_1 is applied on every segment of the curve. b) Fractal curve I on the plane. c) Fill the plane with the curve fractal.

Somer particular aspects presented in the construction of this type of fractal should be noticed when *generator* is created by combination or triangular layouts (in size and form):

- States I_{k-1} and I_k are different from each other in the sequence shown by poligonal curves $\forall$ $k \to \infty$,
- set I has a fine structure, that is, it contains every detail in every arbitrary small scale.
- although the *generator* is composed by two triangular figures corresponding to euclidean logic, the geometry of I_k $\forall$ $k \to \infty$ is too irregular to be described in classical geometric terms.

Figure 4.b shows the curve built on the plane through reflection of I_k on every side.

Curve I is characterized by being of a similar scale since, based on a transformation F: $\mathfrak{R}^n \to \mathfrak{R}^n$ with $\lambda_i > 0$ $\forall$ i $\exists_n$ $a, b \in \mathfrak{R}^n$ such that $| F_i(a)-F_i(b) | = \lambda_i | a-b |$. Similarity of scale is present for triangles created with I_0, by generator I_1. In a like manner, it also has the property of being affine, since, based on transformation F already defined, $F(a) = T(a) + \alpha$ with T a non-singular linear transformation and $\alpha \in \mathfrak{R}^n$. It must be remembered that affinity is conceived as a shear transformation or resistant to cutting, and is a contracting or expanding effect, not necessarily in the same direction. A curve type l_k satisfies the *scale principle* if all its relative figures are linked to each other by a scale law.

Let I be a Borel's set such that $I = \{I_1, I_2, I_3\}$, where I^j is a finite succesion of line segments creating the *generator* $\forall$ j = 1, 2, 3, in order that $I_i = \{U^3_{j=1} I^j_i\ \forall\ i = 1, 2, \dots\}$ is a countable sequence of sets. Thus, measurement μ, of segments I_i is defined as:

$$\mu(U^{\infty}_{j=1} I_i) = \sum^{\infty}_{i=1} \mu(U^{\infty}_{j=1} I^j_i) = \mu(I_i) = 3\mu(I_{i-1})\ \forall i=1, \tag{1}$$

When a geometrical discontinuity is of the fractal type, generated by a natural process, a uniform reticulate should be built. Let $(\chi, P(\chi), \mu)$ be a space with measures such that the sample space is $\chi = [0, 1]$ and $P(\chi)$ is a set of subsets of χ, where the measurement is μ. Since the system is dynamic, $\chi \subseteq \Re^p$ is the phase space. Let us consider a χ reticulate covered by p-dimensional boxes with radius δ_n, where $B_{\delta n}(t)$ is the neighbor box containing the segment of straight line or point t. Succession of neighbor boxes has radius $\delta_n \to 0$ as $n \to \infty$.

Let us suppose that there is subset $I \neq \phi$ from euclidian space n-dimensional, $\Re^n$, and that $|I| = \sup\{|a - b|$ such that $a, b \in I\}$. If $\{I_i\}$ is a countable set of neighbors with radius δ covering I, thus, there exists a subset II of $\Re^n$ such that $II \subset (\cup^{\infty}_{i=1} I_i)$ with $|I_i| \in [0, \delta]\ \forall\ i$. Therefore, $\{I_i\}$ is neighbor δ of I. If there is a $k > 0$, then for every neighbor $\delta > 0$ a function to minimze total covering II may be defined as follows:

$$Hk_\delta(II) = \inf\{\sum{}^{\infty}_{i=1} |\ I_i\ |\ k\} \qquad \forall\ \{I_i\} \text{ neighbor } \delta \text{ of } II \tag{2}$$

If in (2) we consider the limit, then lím $H^k_\delta(II) = H^k(II)$. It should be taken into account that $H^k(II)$ is known as Hausdorf's k-dimensional measurement.

6. Meshing and definition of fractal outline

Let $N(t^*, \Delta t^*)$ be the number of squares contained by reticulate, and $N(t, \Delta t)$ the number of squares intersected by the fractal curve. $D_{\theta°}(I)$ shall be the fractal dimension, L the total length of the object, and l the length of every segment. Therefore, L/l quotient defines the number of subdivisions contained by every side of the intersected reticulate. These scale properties correspond to a fragmented fractal, and the *multifractal* [16], pp. 45 concept is applied. Based on the above:

$$N(t^*, \Delta t^*) \cong \Delta t^* \varrho(t^*) t\ f(t^*) \tag{3}$$

where $t \in [t^*, t^*+\Delta t^*]$. $P(t^*)$ is the probability of distribution of intersection points $t \in [t^*, t^* +\Delta t^*]$ and $f(t^*)$ the fractal dimension of such points.

Considering a random generation of $f(t)$, the original curve is rotated to different angles, preferably constant, in order to calculate $D_{\theta°}$ for every case.

$$D_{\theta^\circ} = \frac{\sum\left\{D_{\theta^\circ} \ \forall\, i=1,\dots,n\right\}}{n} \tag{4}$$

In order to rotate the original curve a certain number of times, let us consider mapping M_n: $\chi \rightarrow \Re$, where $M_n(t) = -\log \mu[B_{\delta n}(t)]$, if $\mu[B_{\delta n}(t)] > 0$ then $C_n(t)$ is a re-scalated version of $M_n(t)$, that is:

$$C_n(t) = \frac{M_n(t)}{-\log \delta_n}$$

where C_n describes the local behavior of μ measurement.

7. Behavior pattern

The fractal behavior of a geometrial discontinuity takes us to the concept of diagonal self-affinity diagonal. In order to define the pattern of the fractal *generator*, let us begin drawing straight lines from its base to the points where such curves are all along its length, and thus obtain a fractioned curve. It is important to draw horizontal lines in case there is a change in its path behavior, in order to identify the affinity along such path.

Horizontal lines identifying the beginning of the generator, must show the feature of proportionality d such that:

$$d = \mathrm{L}\ sen\ \beta \quad \text{where} \quad \mathrm{sen}\,\beta = L\,/\,d \quad \text{and} \quad \cos\beta = k\,/\,L \tag{5}$$

L is the length of the generator, l the length of every segment, and L/l is the number of subdivisions contained by the *generator*. In order to observe the ideal behavior of the path, the scale relationship is defined as the average of the lengths of adjoining generators, that is, $(L_1 + L_2)/2$; Figure 2.a.

The scale factor of every rotation undergone by the generator or fractal curve, s, is defined as:

$$s = \log_{10}(N) \ \forall N \longrightarrow \infty \tag{6}$$

where the value of N is the average of the highest number of every rotation, that is:

$$N = \sum\nolimits_{\mathrm{i}} \left\{R_i \ \ \forall\, i = 1,\dots,n\right\} \tag{7}$$

Based on the above, we may get the scale factor of the generator, defined as the inversion of s which, applied, generates the geometrical structure.

8. Calculation of fractal dimension of the crack for 6-months test specimen

In order to calculate the fractal dimension for cracks generated on test specimen exposed to marine corrosion, both for 6 and 12 months, the theory presented above was applied to fragmented fractals, or fractals generated by natural processes [17]. The original crack was rotated 45°, 90°, 135° and 360° or 0° degrees in order to determine in a more reliable manner the fractal dimension, D_f. The process was to choose a picture for the 6-months test specimen and mesh it in order to calculate its fractal dimension, which is shown in Figures 5.a and 5.b.

In order to obtain fractal dimension D_f, an average of calculated dimensions for each of the four rotations applied to the fractal curve was obtained. Fractal dimension for a crack on structural steel exposed 6-months to marine conditions is:

$D_f = (D_{0°} + D_{45°} + D_{90°} + D_{135°})/4 = (1.34950+1.3429+1.3398+1.3432)/4 = 1.34385$

In order to verify that fractal dimension calculated under the theory developed above is valid, the commercial software *Benoit* was used. The fractal dimension obtained with such program is $D_f = 1.34029$ [18]. As may be seen, the results are similar.

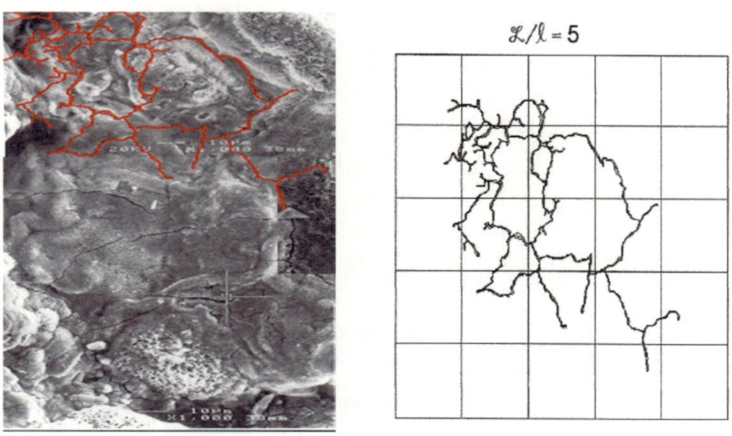

a) b)

Figure 5. a) Fractal behavior of the crack for 6-month test specimen. b) Meshing of fractal with 5 subdivisions and 0 degree rotation.

9. Calculation of fractal dimension of the crack for 1-year test specimen

Figure 6.a shows fractal behavior of the crack, and Figure 6.b shows its digitalization. To obtain fractal dimension of the 1-year test specimen in marine conditions, the same calculation procedure used for the 6-month test specimen was carried out.

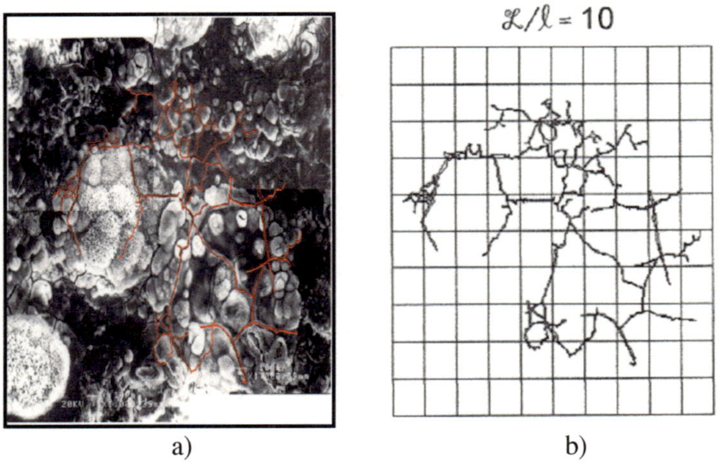

Figure 6. a) Fractal behavior of the crack for 1-year test specimen. b) Meshing of fractal with 10 subdivisions.

Fractal dimension for a crack in A36 structural steel exposed to marine conditions throughout a year is as follows:

$D_f = (D_{0°} + D_{45°} + D_{90°} + D_{135°})/4 = (1.3946+1.3943+1.4+1.3943)/4 = 1.3958$

That is, fractal dimension of the cracks in A36 structural steel exposed to marine conditions throughout a year has a 1.3958 value. In accordance with *Benoit* commercial software, its fractal dimension is 1.40040 [18].

10. Microanalysis

The X-ray microanalysis study on test specimen aged 12-months at sea shows presence of carbon, oxygen, iron, silica, sulphur, chloride, sodium, calcium and magnesium as the most representative elements, Figure 7. Chloride linked to sodium; basic constituents of sea water, are present as sodium chloride. Magnesium is present in constant relation with chloride which, when combined with other elements, creates magnesium chloride and magnesium sulfate. Sulphur is present in sulphates. All these elements have made possible acceleration of test specimen's corrosion.

11. Discussion of results

Virgin A36 structural steel showed a ductile fracture surface, which caused concentration of stresses and a local increase of plastic deformation. The elements showed uniform corrosion for both cases of marine exposition, which caused a generalized thinning on the entire surface. They also showed pitting corrosion and a fragile and porous surface.

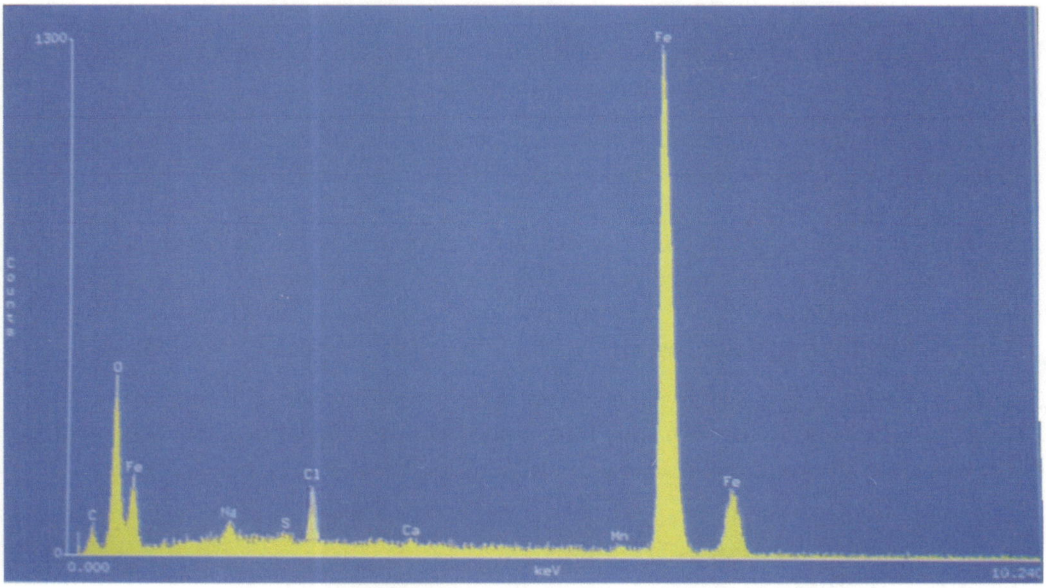

Figure 7. Analysis spectrum through electron microscopy.

A loss of ductility of 11.96 % (6 months) and 43.87 % (12 months) were recorded after marine exposition, regarding the virgin test specimen. These small values do not allow creation of a function for loss of ductility due to permanence of steel in Pacific Ocean coasts. This loss of ductility is mainly attributed to the presence of hydrogen, an element present in sea water and organic compounds on the site of marine exposition, causing it to be absorbed and transported to the inside, towards imperfections of the element, thus causing separation of grain borders.

In the case of tensile tests, breaking had to be achieved in order to calculate the fractal dimension of crack generated. The corresponding stress-strain graphical expression shows variation in the fluency strength in each exposition, due to the reduction of the area based on type of uniform corrosion found.

Once the crack was obtained, it was digitalized and the description format to obtain its fractal type was applied. Then, it was verified that such generator was applicable to fractal behavior of the crack. This being so, the crack was meshed, subdivided and rotated at various angles for calculation of its fractal dimension. Fractal dimension calculated shows a ± 0.0036 variation regarding commercial software, which may be due to the box counting algorithm. Fractal dimension obtained for the test specimen exposed 12 months at sea shows a fractal dimension higher than the 6-month test specimen, since the crack has more branches.

12. Conclusions

Regarding exposition of probes to the environment, it may be affirmed that, due to interaction of an alloy with the environment to which it was exposed, corrosion products were generated on it, which changed failure mechanism of probes subject to tensile assay. Probes exposed on both periods have significant stress concentrations on areas where corrosion developed. This was observed in laboratory assays, since failure appeared in such areas.

Regarding theoretical development of fractals obtained, it has been carried out from line segments creating a generating curve characterized by sets of Borel's theory, which is possible since generation of fractal curve is carried out through natural processes generating geometrical discontinuities of any typical fractal. Characterization of fractal is carried out on a reticulation and its value is based on count of segments for every box in such reticulation. The original crack was rotated 45°, 90°, 135° and 360° or 0° degrees in order to determine in a more reliable manner the fractal dimension, D_f. Self-affinity of the fractal is also tested.

Fractal dimension obtained after six months of exposition is lower regarding that obtained at 12 months, which shows that exposure time is directly proportional to cracks generated due to corrosion products and that crack ramification influences the corresponding value.

Acknowledgements

This document and its corresponding research was carried out, in part, with research Project IPN-SIP 20120585. Thanks to the Martín Armando Zamora García, M.Sc., for allowing the use of photographs used.

Author details

Francisco Casanova del Angel*

Address all correspondence to: fcasanova@ipn.mx; fcasanova49@prodigy.net.mx

SEPI of the ESIA, Unit ALM of the IPN, Mexico

References

[1] Cherepanov G. P and Balankin, A. S. 1997. Fractal and Fracture. A topical encyclopedia of current knowledge dedicated to Alan Arnold Griffith, G.P Cherepanov E.D., Krieger, Melbourne, pp. 104.

[2] Izotov, A. D, Balankin S and Lazarev V. 1993. Synergetics and Fractal Thermomechanics of Inorganic Materials. II. Fractal Geometry of Fracture in Solids. Inorganic Materials. Volume 29, No. 7, pp. 769-778.

[3] Caldarelli G., C. Castellano and Vespignani. 1994. Fractal and topological properties of directed fracture. Phys. Rev. E. E, 49 number 4, pp. 2673-2679. URL: http://

link.aps.org/doi/ 10.1103/PhysRevE.49.2673. Doi: 10.1103/PhysRevE.49.2673. PACS: 64.60.Ak, 62.20.Mk, 05.40.+j.

[4] Mandelbrot B. B. 1982. The Fractal Geometry of Nature. Freeman, New Cork, pp. 459. ISBN-13: 978-0716711865.

[5] Ipohorski, M. 1988. Fractografía-Aplicación al análisis de falla. Informe CNEA 490. Buenos Aires, Argentina.

[6] Bouchaud, E. 1997. Scaling Properties of Cracks. Journal of Physics: Condensed Matter. Volume 9, Number 21, pp. 4319-4344. ISSN: 1361-648X (Online). Doi: 10.1088/0953-8984/9/21/002

[7] Daguier P, Hénaux S, Bouchaud E and Creuzet F. 1996. Quantitative Analysis of a Fracture Surface by Atomic Force Microscopy. Phys. Rev. E, 53, 5637.

[8] Daguier P and Nghiem B. 1997. Pinning and Depinning of Crack Fronts in Heterogeneus Materials. Phys. Rev Left. Volume 78, Issue 6, pp. 1062. URL:http:// link.aps.org/doi/10.1103/ PhysRev Lett. 78.1062. Doi: 10.1103/ PhysRevLett.78.1062. PACS: 62.20.Mk, 05.40.+j,81. 40.Np.

[9] ASTM A-370. Standard Testy Method and Definition for Mechanical Testing of Steel Products. CFR Sections(s): 49 CFR 179.102-1(a)(1). American Society for Testing and Materials.

[10] NMX-B-310-1981. 1996. Métodos de prueba a la tensión para productos de acero. Secretaría de Comercio y Fomento Industrial. México. Norma Mexicana.

[11] Dowling, A. R and Townley, C. H. A. 1975. The effects of defects on structural failure: A two criteria Approch. International Journal of Pressure Vessel and Piping. Vol. 3 pp. 77-107. http://dx.doi.org/10.1016/0308-0161(75)90014-9

[12] Handbook ASM. 1987. Fractography. The Metals Information Society. Former Ninth Edition, Vol. 12.

[13] Roylance D. 2001. Introduction to Fracture Mechanics. Department of Materials Science and Engineering. Massachusetts Institute of Technology. Cambridge, MA. 02139.

[14] Web paged consulted and referred to in the text. http.//www.matweb.con/index. February 2012.

[15] Cortés P. R, Villanueva A. J, Ponce L. E, Rojas M. M and Rojas Z. E. 2004. Estudio de la soldabilidad y corrosión del acero inoxidable AISI 904L con los agentes utilizados en la lixiviación del cobre. Revista Facultad de Ingeniería de la Universidad de Tarapacá. Vol. 12 N°2, pp. 43-56. http://dx.doi.org/10.4067/S0718-13372 004000 200007.

[16] Mandelbrot B. B. 1997. Fractales, hasard et finance. Paris: Ed. Flammarion. ISBN: 2-08-08132-X.

[17] Bouchaud E, Lapasset G and Planès J. 1990. Fractal Dimension of Fracture Surfaces: A Universal Value?. EPL (Europhysics Letters). Vol. 3, No. 1. pp. 73. ISSN: 1286-4854 (online). Doi: 10.1209/0295-S075/13/1/013.

[18] Benoit – Fractal Analysis Software. 2012. www.trusoft.international.com/

[19] Arteaga A. J. 2005. Análisis de micro Fracture. Tesis de Maestría. SEPI-ESIA Zacatenco, IPN, México.

[20] Burdekin, F. M and Dawes, M. G. 1971. Practical Use of Linear Elastic and Yielding Fracture Mechanics with Particular Reference to Pressure Vessels. Proceedings of the Institute of Mechanical Engineer Conference, London, pp. 38-37.

[21] Budiman, H. T and Lagace, P. A. 1997. Non Dimensional Parameters for Geometric Nonlinear Effects in Pressurized Cylindrical with Axial Cracks. Transaction of ASME, Journal of Applied Mechanics, Vol.67.

[22] Chaaban, A and Ahaarani, A. 1991. A Proponed K1 Solution for Long Surface Cracks in Complex Geometries. Transaction of the ASME, Journal of Pressure Vessel Technology.

[23] Cherepanov G. P and Balankin, A. S. 1995. Fractal Fracture Mechanics. Eng. Fract. Mech., 51, pp. 997-1033.

[24] Folias, E. S. 1965. An Axial Cracks in Pressurized Cylindrical Shells. "International Journal of Fracture Mechanics", Vol. 1 N° 2, 104-113.

[25] González Velázquez J. L. 2000. Impacto en el análisis de integridad en el mantenimiento de ductos. Revistas Academia; pp, 7-13.

[26] González V. 1998. Mecánica de fractura bases y aplicación. Editorial Limusa.

[27] Harrison R. P, Loosemore. K. Milene, I and Dowling, A. R. 1980. Assessment of the integrity of structures Containing Defects. Central Electricity Generating Board Report R/H/R6-Rev2.

[28] Hinojosa M, Bouchaud E and Nghiem B. 1999. Materials Research Society Symposium Proceedings. Volume 539. Materials Research Society, Warrendale Pennsylvania. pp. 203-208.

[29] Irwin, G. R. 1948. Fracture Dynamics. Fracture of Metals. American Society for Metals. Cleveland, pp. 147-166.

[30] Newman, R. C. and Procter, R. P. M. 1990. Stress Corrosion Cracking. Journal British Corrosion Cracking. Vol. 25, N°. 4, pp. 259-269. Doi: http://dx.doi.org/ 10.11 79/00070599156 373.

[31] Rice, J. R. 1968. A Path Independent Integral and the Approximate Analysis of Straing Concentrations by Notches and Crack Problems. Journal of Applied Mechanics, Vol. 35, pp. 379-386. Doi: 10.1115/1.3601206

http://dx.doi.org/10.5772/57061

[32] Timoschenko Gere. 1986. Mecánica de materiales. Editorial Iberoamericana.

[33] Westergaard, H. M. 1939. Bearing Pressures and Cracks. Transaction, ASME, Journal of Applied Mechanics, A, 49, June.

[34] Wells, A. A. 1955. The Condition of Fast Fracture in Aluminum Alloys with Particular Reference to Comet Failures. British Welding Research Association Report.

[35] Wells, A. A. 1963. Application or Fracture Mechanics at and Beyond General Yielding. British Welding Journal. Vol. 10, pp. 563-570.

[36] Winne, D. H. and Wundt, B. M. 1958. Application of the Griffith Irwin Theory of Crack Propagation to the Bursting Behavior of Disk, Including Analytical and Experimental Studies. Transactions of the American Society of Mechanical Engineers, Vol. 20. pp. 1643-1655. ASME.

Chapter 23

Corrosion of Metals in Wood Products

Samuel L. Zelinka

Additional information is available at the end of the chapter

http://dx.doi.org/10.5772/57296

1. Introduction

The corrosion of metals in contact with wood has been studied for over 80 years, and in most situations wood is not corrosive [1]. Recently, however, the durability of fasteners in preservative-treated wood has become a concern. Changes in legislation and certification in the United States, the European Union, and Australasia have restricted the use of chromated copper arsenate (CCA), which was previously the most extensively used waterborne wood preservative [2, 3]. Following these changes, several different wood preservatives have come to the market, some of which are much more corrosive than CCA [4-7]. Although a lot of recent research has been conducted in this area, no attempt has been made to summarize all the recent advances, and confusion exists about the corrosiveness of alternatives to CCA and proper materials selection for use in treated wood. In this chapter, we summarize information on why metals corrode in wood, how fast this corrosion occurs, and techniques to minimize corrosion in wood products.

2. Wood as an electrolyte

Understanding the microstructure and chemistry of wood is helpful in understanding how and why metals corrode in wood. Wood is an anisotropic, cellular material; long, hollow, narrow cells are orientated along the root-to-crown direction of the living tree. The types of cells and the structure of the wood material depend on if the tree was a hardwood (angiosperm, deciduous leaves) or softwood (gymnosperm, evergreen). Many of the structural differences between hardwoods and softwoods are highlighted in Figure 1 [8].

Despite the diversity in wood microstructure across genera and species, the cell wall of all wood species is composed of the same three structural polymers: cellulose, hemicelluloses,

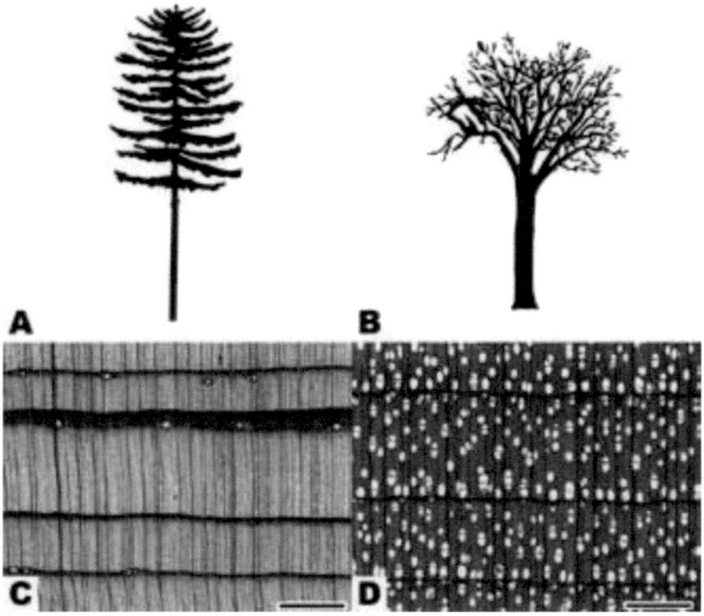

Figure 1. Generic shape of a softwood (A) and hardwood (B) and the corresponding microstructure (in cross section) of a softwood (C) and hardwood (D). Scale bar represents 780 μm. From [1].

and lignin, in roughly the same proportions [9]. In addition to the structural polymers, wood also contains a wide range of extractives, or chemicals that can be removed from the wood with various solvents. The types and amounts of extractives vary widely between wood species, and some of these extractives can affect the corrosion of embedded fasteners (see Section "Role of wood chemistry on corrosion").

Given the porous microstructure and relatively benign chemistry, wood may not seem like a challenging environment for corrosion. However, wood has complex interactions with water that greatly affect its physical, mechanical, and chemical properties, including corrosion. Wood is a hydrophilic material and can hold over 200% of its dry weight as water [10]. Moisture can exist in wood as free water (liquid water or water vapor in cell lumina and cavities) or as bound water (held by intermolecular forces within cell walls). The moisture content at which only the cell walls are completely saturated (all bound water) but no water exists in cell lumina is called the fiber saturation point. The fiber saturation point occurs at approximately 30% moisture content (MC, weight of water/weight of dry material) for many wood species [11].

In the absence of liquid water (and the resulting capillary water uptake), wood freely exchanges moisture with its environment until it reaches an equilibrium (that depends upon the relative humidity). The relationship between equilibrium moisture content and relative humidity at a given temperature is called a sorption isotherm. The sorption isotherm depends on the wood species and previous history of the wood, however the data in Figure 2 are frequently used for practical purposes to estimate moisture content for a given relative humidity and temperature in the absence of better data [10].

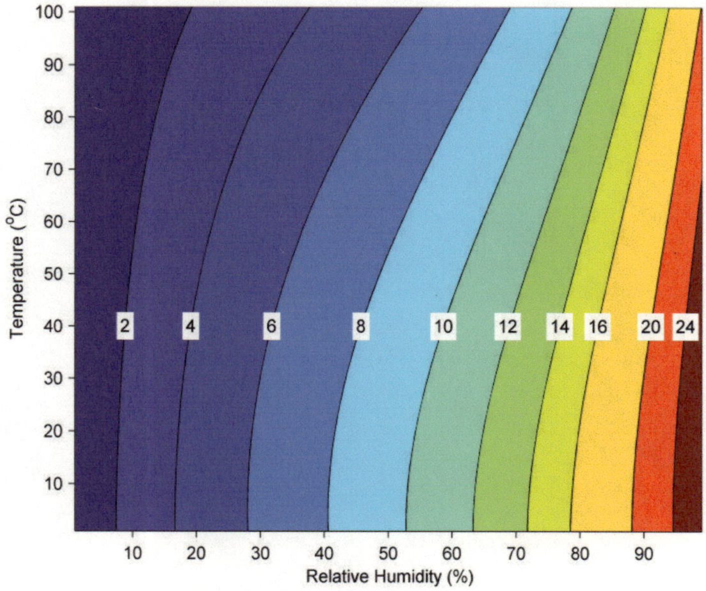

Figure 2. Moisture content (contours) as a function of relative humidity and temperature; data represent an average of adsorption/desorption and should be used for approximations only [3].

Even below fiber saturation, bound water is able to solvate and conduct ions. Zelinka et al. showed that ionic conduction in wood is a percolation phenomenon and that at approximately 16% MC, there is a continuous pathway for ion conduction in wood [12]. For reference, 16% MC corresponds with a relative humidity range of approximately 75%-85% [13]. The percolation threshold is related to a threshold moisture content for corrosion of metals in contact with wood (see Section "Role of wood moisture content on corrosion").

3. Wood preservatives

Wood preservatives are chemicals that are injected into the wood to help the wood resist attack by decay fungi, mold, and/or termites. Waterborne wood preservatives are commonly used when the wood may be in contact with humans or will be painted. Although many different formulations of waterborne preservative treatments have been developed, only a few of these have been used commercially. Most commercial treatments contain cupric ions, which give treated wood its characteristic greenish-brown coloration (Figure 3).

In 2004, regulation changes in the United States restricted the use of chromated copper arsenate (CCA), which had previously dominated the preservative market for many years. Similar regulation changes happened in the European Union and Australasia. Since the regulation change, alternatives to CCA have been introduced to the market. A brief summary of the commercially important alternative wood preservatives is provided below; more information is available in references [2, 3]. Although the exact formulations of the newer wood preserva-

Figure 3. (From left to right) Wood treated with MCQ, DDAC, and ACQ. Cupric ions from the wood preservative cause the dark coloration of the wood. Excess copper has deposited on the MCQ (green splotches) and the ACQ (along the end grain) samples. SCALE: The boards are approximately 38 × 140 mm.

tives are different from each other, they are similar in that they all have a higher percentage of copper than that of CCA. This is important because the corrosion mechanism involves the reduction of cupric ions from the preservative and the corrosion rate depends upon cupric ion concentration [14].

Alkaline copper quaternary (ACQ) is composed of copper oxide (67%) and a quaternary ammonium compound (DDAC- didecyldimethylammonium chloride or carbonate 33%). When it was first commercially available, the quaternary ammonium compound was made with a chloride formulation but was later almost exclusively replaced with a carbonate formulation. Several formulations of ACQ have been commercialized, and it can be treated with an amine or ammonia carrier.

Copper azole (CA) types B and C are composed of ammine copper (96%) and an azole (4%). In CA type B, the azole is entirely composed of tebuconazole. In type C, the azole is 50/50 mixture of tebuconazole and propiconazole. Although copper azole contains a higher percentage of copper than does ACQ, the retention required for aboveground use (Category U3 [15]) is lower and, therefore, the total amount of copper in the treated wood is less.

In addition to these preservatives (CCA, ACQ, CA) standardized by the American Wood Protection Association, several commercially important preservatives have been introduced

http://dx.doi.org/10.5772/57296

to the market by ICC-ES evaluation reports. These preservatives include "micronized" formulations of ACQ and CA, which have various trade names. In these formulations, soluble copper is not injected into the wood, rather solid copper, copper oxide, or copper carbonate is ground into submicron particles, or "micronized," and suspended in solution prior to injection. Several different formulations of these preservatives are covered by different ICC-ES evaluation reports. These formulations differ in listed uses and required retentions and have slight differences in formulations, but in general require less copper than the nonmicronized counterparts.

A summary of wood preservatives highlighting copper concentration is given in Table 1. Preservative composition is given in column 2; aboveground retention needed to meet Use Category 3b for aboveground use according to AWPA U1 is given in column 3 [15]. Column 4, calculated from columns 2 and 3, lists the amount of copper metal per volume of wood. Several studies have shown that as copper concentration in the wood is increased, corrosion of metal fasteners increases [4, 16].

Preservative	Composition	Aboveground retention (kg of preservative per m^3 of wood)	Copper concentration (g of copper per m^3 of wood)
CCA	47.5% chromium trioxide 34.0% arsenic pentoxide 18.5% copper as copper oxide	4	591
ACQ	67% copper as copper oxide 33%DDAC	4	2141
CA-B	96.1% amine copper as Cu 3.9% Tebuconazole	1.7	1634
CA-C	96.1% amine copper as Cu 1.95% Tebuconazole 1.95% Propiconazole	1.0	961
ESR-1721 (MCA-B)	96.1% amine copper as Cu 3.9% Tebuconazole	1.0	961
ESR-1721 (MCA-C)	96.1% amine copper as Cu 1.95% Tebuconazole 1.95% Propiconazole	0.8	769
ESR-1980	67% copper as copper oxide 33%DDAC	2.4	1285
ESR-2240	25/26 copper particles 1/26 Tebuconazole	1.0	961

Table 1. Summary of some waterborne wood preservatives and aboveground retentions highlighting the difference in copper concentration between preservatives. Data are combined from ([17, 18]).

4. Mechanism of corrosion in wood treated with copper-containing wood preservatives

Many of the concerns over corrosion of metals in treated wood involve wood treated with copper-based wood preservatives (e.g., those in Table 1). For wood treated with these preservatives, the corrosion mechanism involves the reduction of cupric ions available in the wood, which are reduced as the construction material, normally a steel or zinc galvanized connector, is oxidized. Figure 4 illustrates this mechanism of corrosion highlighting the transport of cupric ions to the metal surface, where there is a charge transfer reaction resulting in oxidation of the fastener and reduction of the cupric ions in the wood preservative.

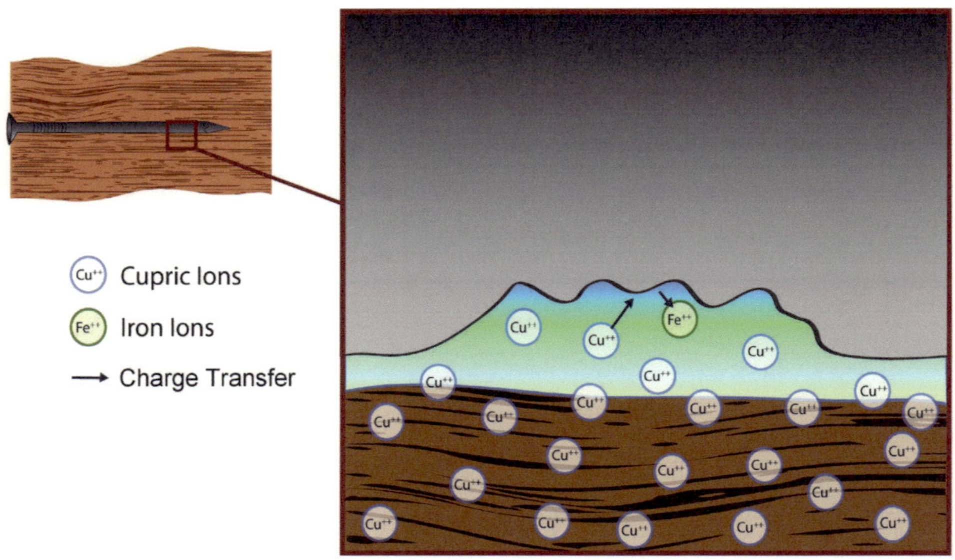

Figure 4. Schematic illustration of the mechanism of corrosion in treated wood (not to scale).

The role of cupric ions in the corrosion mechanism was first hypothesized by Baker [19] and later confirmed by work of Zelinka et al. [5, 14] and Kear et al. [4, 20] through energy dispersive x-ray analysis, Pourbaix diagrams, and examinations of the role of cupric ion concentration and acidity. Zelinka et al. performed electrochemical corrosion tests in water extracts of treated wood, which were later found to result in similar corrosion rates to those measured in solid wood [6, 14, 21]. Because the medium was aqueous and electrochemical techniques were used, ion concentrations, pH, and potentials could be measured and a Pourbaix diagram could be constructed. Figure 5 shows the Pourbaix diagram of copper with measured potentials in extracts of wood treated with different wood preservatives. The diagram shows that zinc-galvanized steel lies in a region where copper metal is the stable phase, and therefore cupric ions in the preservatives are thermodynamically unstable. Further work using energy dispersive x-ray spectroscopy (EDS) showed that fasteners placed in the extract had copper metal deposited on them.

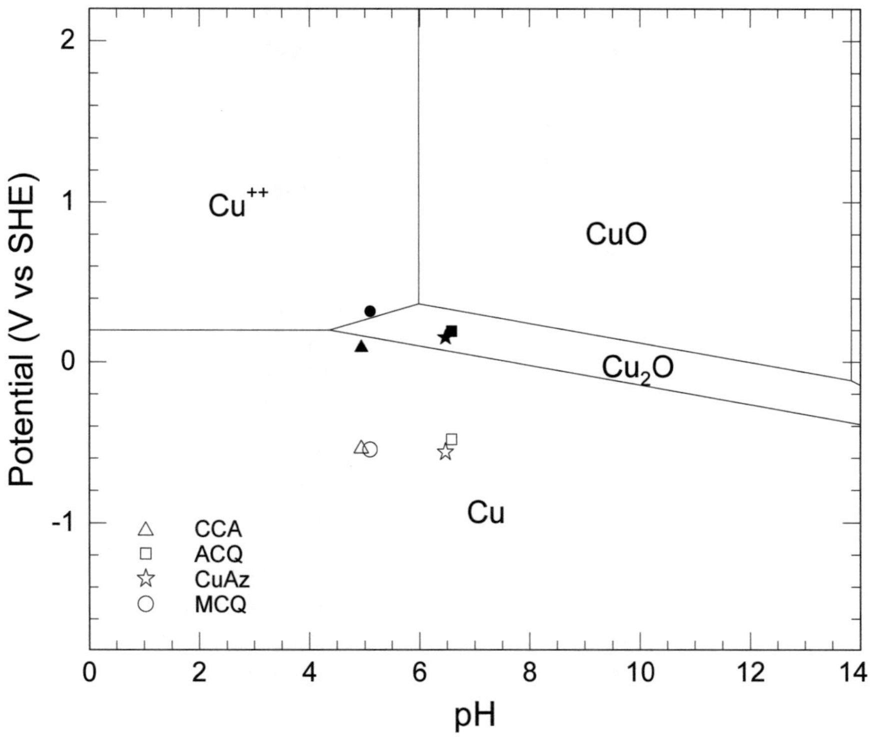

Figure 5. Pourbaix diagram with an assumed cupric ion concentration of 10^{-4} with the open circuit potentials of steel (filled symbols) and galvanized steel (open symbols) in water extracts of different wood preservatives.

Several researchers have shown that corrosion rate of metals in treated wood increases with increasing cupric ion concentration [4, 14, 16]. The clearest evidence of this was found in an unpublished report by the Forest Products Laboratory, the results of which were synthesized and republished by Zelinka and Rammer [16]. In these experiments, living trees were allowed to imbibe a copper-containing wood preservative before felling, and corrosion rate was measured in logs cut from these trees; corrosion rate quadrupled as copper concentration increased from 1 to 11 kg m^{-3}. Zelinka and Stone [14] examined differences in corrosion rates observed in different wood preservatives and showed that most differences could be explained by differences in cupric ion concentration and pH (MCQ, CCA were acidic, ACQ was alkaline). For example, at a given copper concentration, the preservative with the lower pH was more corrosive. Similarly, at a given pH, the preservative with the higher copper concentration was more corrosive.

Despite the fact that the corrosion mechanism clearly involves reduction of cupric ions in the preservative, several experiments that examined corrosion products of fasteners embedded in treated wood found no evidence of copper metal on the fasteners [5, 6]. This raises the question of what is happening to the reduced cupric ions: do they get bound to the wood near the metal surface, or do they get reoxidized or redissolved in a different reaction step? Further research is needed to explain this enigma.

5. Role of wood moisture content on corrosion

The most important environmental variable controlling corrosion of metals embedded in wood is moisture content of the wood. Below a threshold moisture content of 15%–18% MC, embedded metals do not corrode [12, 22-24]; above the threshold, corrosion rate increases with moisture content and plateaus at a maximum corrosion rate near or above fiber saturation point. Despite the importance of wood moisture content on corrosion rate, very little research has been conducted to examine the role of moisture on corrosion.

Two different long-term exposure tests examined the role of wood moisture content on corrosion by placing wood with embedded fasteners in chambers with a fixed relative humidity. Baechler [25, 26] used three chambers at 30%, 65%, and 90% RH. Kear et al. [4] examined 75% and 90% RH and "moisture saturated air." In these studies, corrosion rate was effectively zero at 30% and 65% RH and barely measureable at 75% RH. Corrosion rate then had noticeable increases between both the 75% and 90% RH and the 90% RH and moisture saturated air conditions.

The number of moisture contents at which a gravimetric corrosion rate can be determined is limited experimentally by time, expense, and difficulty maintaining the wood moisture content constant for long periods of time. To develop a more detailed curve, Dennis et al. used electrochemical tests to measure corrosion rate as a function of moisture content in solid wood treated with CCA [22]. In these tests, thin sheets of wood were pressed around a zinc working electrode and stainless steel counter electrode and corrosion rate was measured with the polarization resistance technique. A subset of the data are plotted in Figure 6. The data exhibit hysteresis with moisture content, which suggests that the wood specimens were not in equilibrium when measurements were taken. These data have been used to create a combined hygrothermal/corrosion model to predict the amount of corrosion one can expect in different environments [27]. The model calculates wood moisture content from hourly climatic data and calculates the hourly amount of corrosion from the wood moisture content.

6. Differences between atmospheric corrosion and corrosion of metals in wood products

Corrosion of metals embedded in wood has several differences from atmospheric corrosion, with implications for both materials selection and service life of metals in wood. The two most important differences are that (1) galvanized steel corrodes more rapidly than carbon steel when in contact with wood and (2) long-term corrosion kinetics are different in wood than in atmospheric conditions. Both of these differences can be attributed to the fact that corrosion products of metals in contact with wood are different than those that form in atmospheric conditions and therefore the passivation is different.

In atmospheric corrosion, zinc oxidizes to form hydrozincite $\{Zn_5(CO_3)_2(OH)_6\}$ and smithsonite ($ZnCO_3$), which passivate the zinc surface; that is, these oxidized species protect the metal from

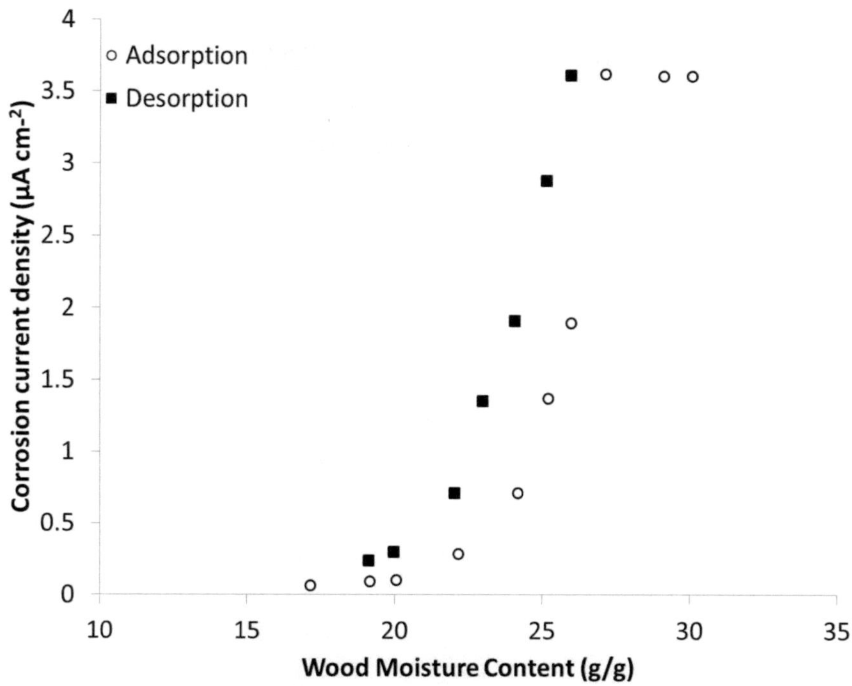

Figure 6. Selected data of Dennis et al. replotted to show the dependence of corrosion rate on wood moisture content. Data from [5].

further corrosion. Conversely, steel forms goethite (α-FeOOH), also called "red rust" in atmospheric conditions. Kinetically, hydrozincite and smithosnite are better at protecting the underlying metal than goethite, which is why zinc corrodes more slowly than steel in atmospheric conditions [28]. In certain environments, such as immersion in saltwater [29] or in environments with volatile acetic and formic acids [30], different corrosion products form and zinc corrodes more rapidly than steel. In addition to measuring corrosion rates, Zelinka et al. [5] examined corrosion products on fasteners removed from steel and galvanized steel fasteners in wood using X-ray diffraction and did not observe smithsonite on the zinc fasteners. Instead, they observed namuwite {$Zn_2(SO_4)(OH)_6 4H_2O$}, simonkolleite {$Zn_5(OH)_8Cl_2$ (H2O)}, and in some cases hydrozincite. The lack of a protecting passive layer explains why galvanized fasteners corrode more rapidly than steel fasteners in solid wood.

Passivation in atmospheric corrosion also causes a decrease in corrosion rate with time. Legault and Preban described corrosion kinetics under atmospheric conditions by

$$\Delta W = Kt^n \tag{1}$$

where ΔW is the change in weight, K is a constant (the 1-year corrosion rate), t is time in years, and n is an exponent that controls the kinetics and is less than or equal to unity [31]. Corrosion of metals in wood exhibits a constant rate with time (i.e., $n = 1$). This behavior was first observed by Baker [32], who found that the weight loss increased linearly with time in a 17-year exposure test. Zelinka et al. observed further evidence of activation control in measurements performed in water extracts of treated wood. They observed that corrosion rates measured in the extract were similar to those measured in solid wood, where the diffusion is slower [33].

7. Role of wood chemistry on corrosion

Even a single species of wood can have more than 700 different extractives, chemicals that can be solubilized in water or another solvent, and could, in theory, affect the corrosion of embedded metals [34]. Despite the large number of extractives that could potentially affect corrosion, very few of these compounds have been associated with corrosion of embedded metals. Zelinka and Stone [35] reviewed the literature on the role of extractives on corrosion and found that previous researchers had identified only three different compounds that affect corrosion of metals embedded in wood or exposed to the black liquors of wood pulp: small organic acids (acetic and formic acid) [30, 36-39], tannins (or more broadly, polyphenols) [34, 40-49], and phenols with two or three adjacent hydroxyl groups (e.g., catechol and pyrogallol) [45-50]. For solid wood, only organic acids and tannins have been mentioned in the literature; catechol (1,2-dihydroxybenzene) and pyrogallol (1,2,3-trihydroxybenze) are formed as lignin is destroyed in the pulping process [51, 52].

Zelinka and Stone further investigated the role of tannins and pH by making water extracts of different wood species known to vary in both tannin content and pH: pine (low pH, low tannin), oak (low pH, high tannin), elm (high pH, low tannin), and black locust (high pH, high tannin). A blue–black precipitate formed on the surface of steel in the extracts, suggesting the formation of a passive layer of iron-tannate (Figure 8). In further experiments they adjusted the extracts by changing the amount of tannins or the pH. In these experiments they observed that at a given pH, increasing the tannins decreased the corrosion rate; and at a given level of tannins, lowering the pH increased the corrosion rate. Based upon their measurements and simple kinetic models, Zelinka and Stone developed an isocorrosion map for the water extracts of different wood species (Figure 7). While absolute corrosion rates are higher than would be expected in solid wood, the same general trends are expected to apply.

8. Test methods to examine corrosion of metals in wood

One year after the 2004 regulation change in wood preservatives, Zelinka and Rammer [53] summarized 22 different methods that had been previously used to evaluate corrosion of metals in contact with wood. The review was divided into "exposure tests," where temperature and relative humidity were kept near realistic conditions, "accelerated tests," where temperature and/or a fog were used to accelerate corrosion, and "electrochemical tests," where the metal was polarized to evaluate corrosion rate. In general, many of the tests were for "com-

http://dx.doi.org/10.5772/57296

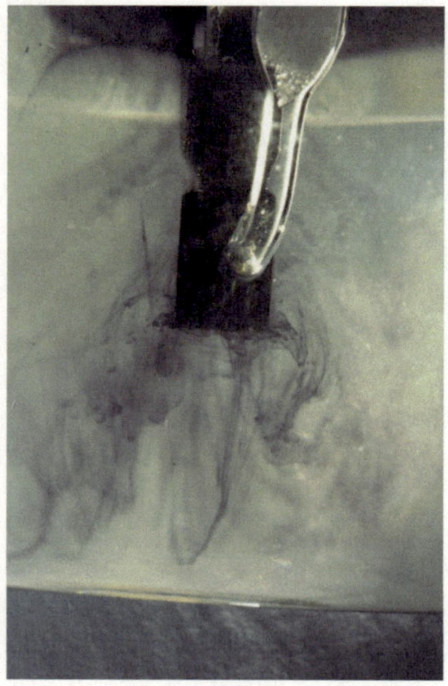

Figure 7. Iron-tannate precipitate forming on a steel plug in a solution with the same tannin concentration and pH as white oak. SCALE: The steel plug is approximately 9.5 mm in diameter.

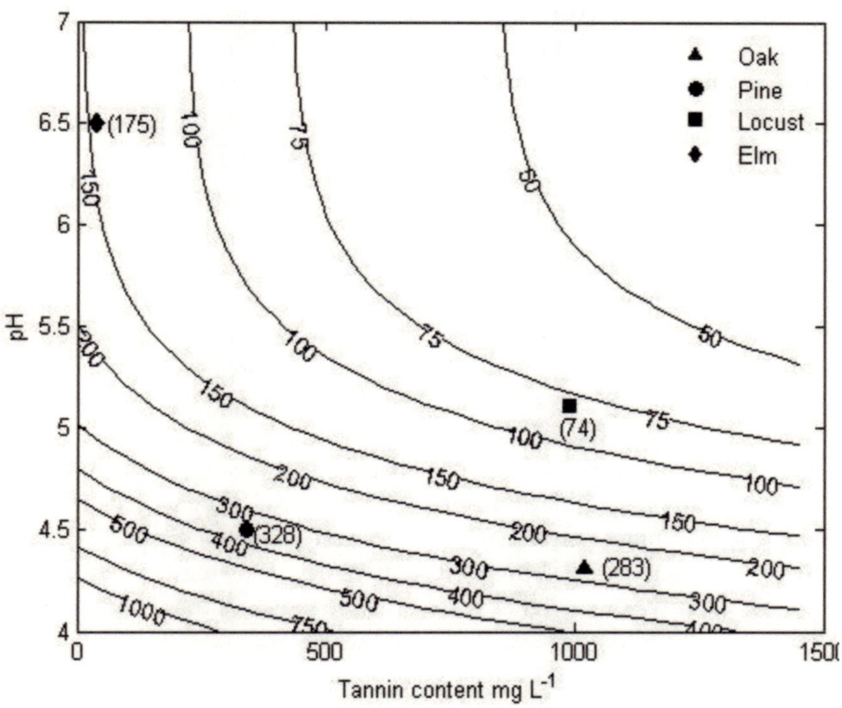

Figure 8. Showing the interplay between tannins (abscissa) and pH (ordinate). Contours represent combinations of tannins and pH that result in the same corrosion rate

parative purposes only," where conclusions could be drawn only between a treatment and a control group. In many cases, corrosion was reported as a percentage weight loss instead of a true corrosion rate because the surface area of threaded fasteners could not be calculated. It has been shown that this way of reporting corrosion rates distorted the performance of aluminum fasteners, which have a lower density than steel and galvanized steel fasteners [7].

Since the Zelinka and Rammer review of test methods, several new test methods have been developed. A new standard has been written to measure the corrosion rates of fasteners driven into treated wood [54], and the American Wood Protection Association (AWPA) E-12 standard has been greatly revised [55]. Furthermore, Rammer and Zelinka developed a method to calculate the surface area of threaded fasteners from digital images [56-58], which allows for true corrosion rates to be measured. The remainder of this section summarizes new developments in test methodologies.

AWPA standard E12, "Standard method of determining corrosion of metal in contact with treated wood," was the first standardized test method used to measure corrosion of metals in contact with wood [55]. The test method places metal coupons between two blocks of treated wood and exposes the "sandwich" to a high-temperature (50°C) high-humidity (90% RH) environment for 120 days. The blocks of wood are held together with nylon bolts. The E12 standard was revised and greatly improved in 2008. The new standard specifies an effective torque for the nylon bolts and gives much more detailed instructions for cleaning the corrosion products. Although the E12 test can be used to make a relative ranking between preservatives, information on the effect of temperature on the corrosion of embedded metals is lacking and it is impossible to relate corrosion rates measured in the E12 test to realistic temperatures for building enclosures or outdoor wood structures.

ASTM International standard G198, "Standard test method for determining the relative corrosion performance of driven fasteners in contact with treated wood," was published in 2011 [54]. The goal was to develop a test method to evaluate fasteners, which may have different metallurgical characteristics or coatings than the sheet metal used in the AWPA E12 test. Beyond the geometry of the metal sample, the biggest difference between the ASTM and AWPA standards is the exposure conditions. The AWPA test is conducted at a temperature of 50°C; the ASTM test is conducted at a lower and potentially more realistic temperature of 32°C. Relative humidity conditions are also different between the two standards. In the ASTM test, the wood and fasteners are exposed either to a "steady state" test of 95% RH or to a "cyclic fog" test where the fasteners are exposed to 48 hours of fog followed by a 72-hour dry cycle. Both tests are conducted for 120 days. The reasoning for including a cyclic fog test is that in accelerated testing conditions, galvanized fasteners perform better when exposed to alternating wet and dry cycles, which is similar to the environment they see in service [28, 29]. However, Zelinka recently showed that wood acts as a moisture buffer and does not dry out during the dry cycles; rather, it gains moisture throughout the entire test [59]. From this analysis it appears that the cyclic fog test may be more challenging than the steady-state moisture test.

Results of several exposure tests in newer wood preservatives have been reported as actual corrosion rates rather than percentage weight loss. Kear et al. exposed nails and metal coupons to treated wood exposed to three different relative humidity environments [4] (75% RH, 90%

RH, "moisture saturated air"). The surface area of the nails was calculated from the diameter and tip geometry. Zelinka et al. measured the corrosion rates of screws and nails at 100% relative humidity [5, 7]. Furthermore, using photographs from Baker's laboratory notebook, Zelinka and Rammer were able to convert Baker's seminal data from a 17 year exposure test in CCA- and ACA-treated wood from percentage weight loss to actual corrosion rates [7].

Additional work has further advanced electrochemical methods as a rapid technique to measure corrosion of metals in treated wood. Both Zelinka et al. [60] and Kear et al. [61, 62] have measured corrosion of metal fasteners in dilute solutions of wood preservatives. Zelinka et al. used the polarization resistance technique and concluded that these measurements could not be used to predict corrosion of metals embedded in treated wood. Kear et al. used impedance spectroscopy, large perturbation potentiodynamic polarization, and polarization resistance measurements to characterize the corrosion performance of mild steel, stainless steel, and galvanized steel in dilute solutions of wood preservatives. Kear et al. observed good correlation between trends found in these electrochemical measurements and gravimetric measurements in dilute solutions of wood preservatives as specified in the AWPA E17 test for measuring the corrosiveness of wood treatment solutions [20, 63], but not for metals embedded in treated wood.

Although these measurements in dilute solutions of wood preservatives were unable to predict the corrosion of embedded metals, a different method developed by Zelinka et al. [6] has good correlation to laboratory exposure tests of metals in treated wood. In this method, extracts of treated wood were made by grinding the treated wood, adding water (1:10 weight ratio), and extracting the water-soluble components of the treated wood for one week, after which the milled wood was filtered off. The rationale for the method was that corrosion of embedded metals is aqueous and that the water extract would have a similar chemistry to the micro-chemistry within the wood structure. Corrosion rates determined from polarization resistance measurements in the extract had excellent correlation to corrosion rates of fasteners embedded in wood conditioned at 100% RH for 1 year [7].

9. Review of corrosion rates measured in wood treated with different preservatives

Since the 2004 change in wood preservative regulation, several investigations of the corrosiveness of metals in contact with treated wood have been reported. The studies cover a wide range of preservatives, preservative retentions, wood moisture contents, and metals tested. Not surprisingly, a wide variation in corrosion rates has been reported (e.g., from 2–113 μm yr^{-1} for galvanized steel in ACQ-treated wood). Table 2 summarizes the corrosion data. Because of the wide variations of test conditions used, it is best to compare results only within a single study or across different studies with very similar conditions. The remainder of this section summarizes how the data in Table 2 were obtained.

Many design recommendations for materials selection in wood are based off the recommendations of Baker [32], who conducted a 17-year investigation of corrosion of metal fasteners embedded into chromated copper arsenate (CCA) and ammoniacal copper arsenate (ACA) treated wood exposed either underground or in a room maintained near 100% RH From these data, Baker concluded that at a minimum, hot-dip galvanized fasteners should be used in treated wood and cautioned against the use of aluminum fasteners. Importantly, Baker presented the corrosion data as percentage weight loss instead of corrosion rate because he could not calculate the surface area of threaded fasteners. Zelinka and Rammer reanalyzed Baker's data using data in Baker's laboratory notebook and an algorithm they had developed to measure surface area of threaded fasteners [7, 56, 57]. They found that when the corrosion rate was adjusted to a true corrosion rate, the corrosion rate of aluminum was in fact less than that of hot-dip galvanized steel.

Zelinka has published results of several different corrosion tests that were conducted at 27°C (80°F) and 100% RH [5, 7, 14, 64]. In one study [7], Zelinka and Rammer examined the corrosion of five different metal fasteners (carbon steel, hot-dip galvanized steel, electroplated galvanized steel, aluminum, and stainless steel) embedded in wood treated with ACQ-D to a retention of 4 kg m^{-3}. They found that the corrosion rate of hot-dip galvanized steel was the highest (62 μm yr^{-1}). The corrosion rate of stainless steel was statistically indistinguishable from zero. In another study [5], steel and hot-dip galvanized steel fasteners were exposed to six different wood treatments: chromated copper arsenate (CCA-C), alkaline copper quaternary (ACQ-D), copper azole (CA-B), micronized copper quaternary (MCQ), and didecyldimethylammonium carbonate (DDAC, or the "quat" in ACQ and MCQ). ACQ was found to be the most corrosive to both steel and galvanized steel.

Kear et al. examined the corrosion of three different metals (316 stainless steel, hot-dip galvanized steel, and plain carbon steel) in three different preservatives (CCA, ACQ, and CA) treated to three different retention levels with four different test methodologies [4]. Retention levels were specified in a New Zealand standard that specifies a mass basis (i.e., kg/kg) instead of a density basis and therefore cannot be directly compared to traditional U.S. retentions (in lbs ft^{-3} or kg m^{-3}). Differences in the corrosiveness of the preservatives were most apparent in a constant exposure to 90% RH.

Simpson Strong Tie corporation published a technical bulletin on the results of in-house AWPA-E12 [55] corrosion tests [65]. In this test method a metal plate is sandwiched between two blocks of wood and exposed to a high-temperature (50°C), high-humidity (90% RH) environment. They found that the corrosiveness of ACQ-D (carbonate) was roughly equivalent to that of CA, and both of these were more than twice as corrosive as CCA-C. They also found that ACZA was more than three times more corrosive than CCA and that borates were less corrosive than CCA. A footnote in the table mentioned that for micronized formulations like MCQ the "relative corrosiveness is somewhat lower than ACQ-D."

Copper napthenate is sometimes used as a preservative in timber bridges. Although no peer-reviewed data on the corrosiveness of copper napthenate have been reported, Anthony Forest Products company published a technical bulletin with the results of AWPA E12 tests that compares both waterborne and oilborne formulations of copper napthenate against ACQ-D and CCA-C for mild steel, hot-dip galvanized steel, aluminum, and red brass [66]. Although no data are reported on variability between replicates, the data clearly suggest that either formulation of copper napthenate is much less corrosive than ACQ and possibly less corrosive than CCA (Table 2). This bulletin was published by a company that supplies products treated with copper napthenate; nevertheless, the data suggest that copper napthenate is much less corrosive than ACQ and less than or equal in corrosiveness to CCA.

	Plain steel	Hot-dip galvanized steel	Stainless steel	Aluminum
17 years, 27 C (80 F), 100% RH [7, 32]		9 | CCA I 6 | CCA II	<1 | CCA I <1 | CCA II	3 | CCA I 4 | CCA II
1 year, 27 C (80 F), 100% RH [7]	34| ACQ	61| ACQ	1| ACQ	22| ACQ
1 year, 27 C (80 F), 100% RH [5]	9.4|CCA 16.8|ACQ 12.5|mCQ 11.1|CA-B 1.9|DDAC 0.7|Untreated	16.1|CCA 32.5|ACQ 19.4|mCQ 28.9|CA-B 5.5|DDAC 4.4|Untreated		
1 year, 27 C (80 F), "moisture saturated air" [4]	62|CCA 176|ACQ-B 84|CA-B	26|CCA 113|ACQ-B 87|CA-B	≤1|CCA ≤1|ACQ-B ≤1|CA-B	
1 year, 27 C (80 F), 90% RH [4]	4|CCA 45|ACQ-B 14|CA-B	5|CCA 26|ACQ-B 16|CA-B	≤1|CCA ≤1|ACQ-B ≤1|CA-B	
1 year, 27 C (80 F), 75% RH [4]	1|CCA 3|ACQ-B 2|CA-B	1|CCA 2|ACQ-B 2|CA-B	<1|CCA <1|ACQ-B <1|CA-B	
AWPA E12 [66]	4|CuN-W 1|CuN-O 124|ACQ-D 27|CCA-C	11|CuN-W 2|CuN-O 23|ACQ-D 10|CCA-C		(-1)|CuN-W 2|CuN-O 185|ACQ-D 7|CCA-C
AWPA E12 [67]	140| mCA 107|CA	23|mCA 51|CA		18|mCA (-3)|CA
AWPA E12 [67]	66|mCA 180|CA 41|mCQ 196|ACQ-D 41|CCA-C	69|mCA 76|CA 41|mCQ 81|ACQ-D 53|CCA-C		3|mCA 0|CA 0|mCQ 3|ACQ-D 3|CCA-C
AWPA E12 [67]	89|mCQ 307|ACQ-D 76|CCA-C	25|mCQ 28|ACQ-D 20|CCA-C		10|mCQ 10|ACQ-D 5|CCA-C

Table 2. Summary of published corrosion rates (in μm yr^{-1}) in treated wood. The table represents a wide variety of test methodologies—comparisons should be made only within a single row.

Freeman and McIntyre [67] summarized results of several unpublished corrosion tests of pressure-treated wood in contact with different metals that compared traditional (ACQ and

CA) against their micronized formulations (MCQ and mCA). Most of the data shown were from AWPA E-12 tests. The results were inconclusive. In one test, the micronized formulation of CA was less corrosive than the standard formulation for galvanized steel but more corrosive for plain carbon steel, and the micronized formulation was much more corrosive to aluminum. The remainder of the E-12 data have similar trends; in some cases the micronized formulations show lower corrosiveness, in other cases, the traditional formulations appear less corrosive. The only results that consistently show that the micronized formulations are less corrosive than the traditional formulations are from a test conducted using a protocol from the International Staple and Nail Tool Association (ISANTA). In this test, the micronized formulations appear slightly (approximately 30%–40%) less corrosive, with the notable exception of aluminum, in which case the micronized formulations are more corrosive.

Table 2 summarizes corrosion rates measured in a variety of test methods over a range of wood preservatives and metals. Each row in the table represents a paper or report, and the test method is briefly summarized in the first column. Corrosion rates for each metal are contained in columns followed by the type of wood preservative. The data can be used for observing general trends or relative comparison of a preservative or a metal type; however, it is dangerous to compare across rows because different test methods were used.

10. Effects of corrosion on the structural capacity of connections

The main effect of corrosion of fasteners embedded in wood is to reduce the strength of the connection. Recently, Zelinka and Rammer used the yield theory equations and measured corrosion rates to calculate how fastener corrosion reduces the structural capacity of connections [68]. Fastener connection performance can be calculated by the yield theory developed by Johansen, which treats both the wood and metal as elastic/perfectly plastic materials [69]. The yield theory equations are now the standard method for calculating design capacity for any dowel type connector (nail, screw, lag screw, bolt, etc.) in the United States through the National Design Specification for Wood Construction (NDS). Lateral design load, Z (in N), is determined by

$$Z = \min \begin{cases} \dfrac{D T_s F_{es}}{2.2} & (\text{Mode } \mathrm{I_s}) \\[2ex] \dfrac{k_2 D l_p F_{em}}{2.2(1+2R_e)} & (\text{Mode } \mathrm{III_m}) \\[2ex] \dfrac{k_3 D T_s F_{em}}{2.2(2+R_e)} & (\text{Mode } \mathrm{III_s}) \\[2ex] \dfrac{D^2}{2.2}\sqrt{\dfrac{2F_{em}F_{yb}}{3(1+R_e)}} & (\text{Mode IV}) \end{cases} \tag{2}$$

where

$$k_2 = -1 + \sqrt{2\left(1+R_e\right) + \frac{2F_{yb}\left(1+2R_e\right)D^2}{3F_{em}l_p^2}}$$

$$k_3 = -1 + \sqrt{\frac{2(1+R_e)}{R_e} + \frac{2F_{yb}(2+R_e)D^2}{3F_{em}T_s^2}} \tag{3}$$

and

$$R_e = F_{em}/F_{es}$$

and D is dowel diameter (mm), $F_{e\_}$ is dowel bearing stress of the main (m) or side (s) member (MPa), F_{yb} is bending yield stress of the nail (MPa), l_p is length of penetration into main member, and t_s is thickness of the side member.

Zelinka and Rammer took measured corrosion rates and converted them to reduction in fastener diameter. Because corrosion does not affect joint geometry, wood dowel bearing stress, or bending yield stress of the uncorroded fastener, design load becomes a function of fastener diameter only. Zelinka and Rammer observed that for most decking fasteners, a Mode IV failure would be observed. This is especially problematic since Mode IV failure is the most sensitive to corrosion rate; design load scales as the diameter squared in this case. Figure 9 illustrates reduction in relative capacity of a nailed joint with time for different corrosion rates. For the range of measured corrosion rates in wood (~1–100 μm yr^{-1}), relative capacity drops quickly with increasing corrosion rate.

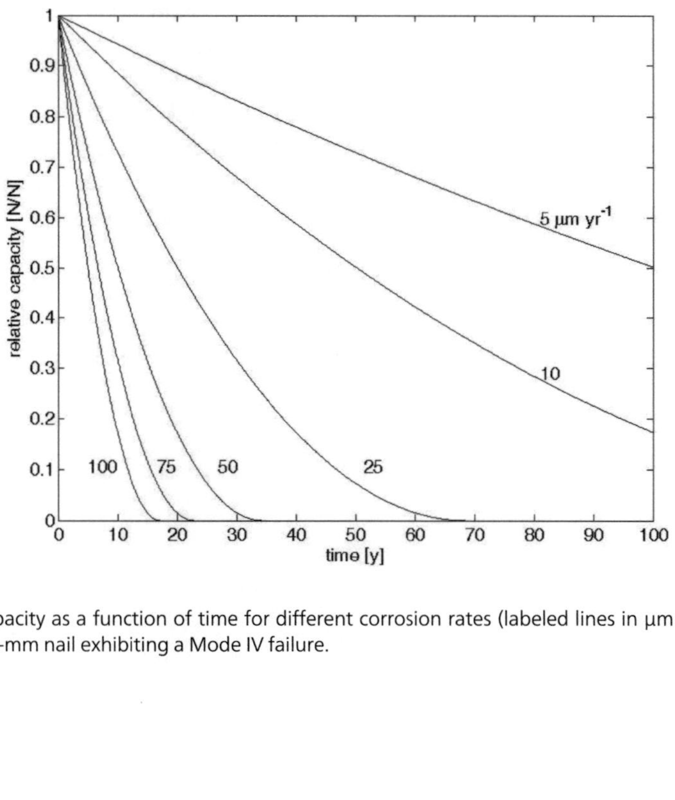

Figure 9. Loss of capacity as a function of time for different corrosion rates (labeled lines in μm yr^{-1}). The figure was constructed for a 3.4-mm nail exhibiting a Mode IV failure.

11. Design considerations for minimizing corrosion in wood

Metal fasteners are an essential part of wood construction, and their use in preservative-treated wood cannot be avoided. Therefore, it is important to design and utilize metals in wood so that they exhibit the least amount of corrosion. Here we discuss three design considerations that will help prolong the service life of metals in wood: proper moisture management, proper use of nonmetallic coatings and barriers, and avoiding mixed metals in construction.

11.1. Moisture management

Proper moisture management is the most important factor in reducing corrosion of metals in treated wood. If wood is kept dry, both the wood and the fasteners can last for centuries [70]. In most cases, preservative-treated wood is specified because the wood will be exposed in an outdoor environment where it will be exposed to moisture. However, sound design principles can maximize drainage of rainwater and minimize the amount of moisture that is absorbed by the wood.

Clausen and Glass recently published a design guide for minimizing wood decay in residential construction by keeping the wood dry [70]. Pertinent strategies that also apply to exterior structures involve (1) keeping rainwater from seeping in through the end grain and (2) designing roofs and overhangs so they do not drain onto lower structures.

11.1.1. End grain

Moisture transport in the longitudinal direction (along the grain) is more than 10 times faster than in the radial or tangential directions. Therefore, if the end grain of the wood member is allowed to come into contact with liquid water, the entire member will absorb significant amounts of moisture. It is important to design the structure so that end grain is not exposed to water. This frequently happens at joints between two members if the connection is improperly designed or installed and the joint traps water. Vertical members (i.e., posts) are especially prone to water uptake along the end grain. For these members, impermeable coatings or end caps should be used to prevent water uptake through the end grain.

11.1.2. Roof

For outdoor wooden structures with a roof, such as covered bridges, proper roof design and detailing can shield much of the structure from water and keep the wood moisture content low. The larger the overhangs of the roof, the less likely it is that drainage from the roof will strike the walls below (Figure 10).

11.2. Coated metals

Paints and other nonmetallic coatings are frequently used to protect metals from corrosion by isolating the metal from the corrosive environment. The annual cost of coatings used to prevent corrosion is estimated at $100 billion USD, which is 36% of the total cost of corrosion in the

Figure 10. Illustration of the importance of roof overhangs for protecting wood from biodeterioration and corrosion. The right side of the beam is protected by the large roof overhang, whereas the left side is exposed to rain. SCALE: The beam supporting the trellis is 0.66 m deep.

Unites States and 1.1% of the entire U.S. gross domestic product (GDP) [71]. Although many different formulations of organic coatings are available, the goal of nearly all coatings is to isolate the metal from the environment.

Coatings fail when the corrosive environment can reach the metal substrate. This can happen if the coating is mechanically damaged during installation (Figure 11). Zelinka et al. observed that screws with a nonmetallic coating had much worse corrosion performance after they had been driven into the wood or driven through a joist hanger into the wood [72] and attributed this decrease in performance to damage in the coating caused by inserting the fastener into wood. Extreme care should be taken if using coated metals in construction with wood so that the coating does not get damaged before or during installation.

11.3. Mixed metals

Special design consideration is needed if two different metals are used in contact with each other. Galvanic corrosion happens when three conditions are satisfied: (1) two dissimilar metals (2) are placed in electrical contact (3) in the presence of an electrolyte. Galvanic corrosion is a design concern in wood when a joist hanger or sign is attached to wood with a fastener. If the fastener is a different metal than the sheet metal, conditions for galvanic corrosion exist, wood acts as the electrolyte and the metals are in electrical contact between the head of the fastener and the sheet metal.

Figure 11. Corrosion underneath a coating failure. This lamp post was in service for less than 5 years before corrosion was visible at defects in the coating. SCALE: The lamp post is approximately 115 mm.

An example of galvanic corrosion with wood was observed by the Wisconsin Department of Transportation (WisDOT) [73]. In this case, aluminum signs were attached to sign posts made of treated wood using galvanized lag screws. The aluminum signs exhibited excessive corrosion around the lag screw, causing the sign to fall off in some cases. In this case, aluminum was acting as the anode; the galvanized lag screw, the cathode; and wood, the electrolyte. Because it would be difficult to change the sign material, and aluminum lag screws do not exist, it is impossible, from a design perspective to change one of the metals to avoid corrosion. A better design solution would be to electrically isolate the aluminum sign from the galvanized lag screw.

12. Conclusions

This chapter summarized recent research on the corrosion of metals in wood focusing on how metals corrode in wood, how fast this corrosion occurs, and techniques to minimize corrosion in wood products. Despite the complexity of the corrosion mechanism in wood, the most important considerations for materials selection and design can be summarized in twoshort phrases:

"Corrosion in wood is not atmospheric corrosion" Corrosion in wood is different from atmospheric corrosion. There are different thermodynamics, different kinetics, and different corrosion products form. It is not safe to assume that just because a solution works for atmospheric corrosion that it is a good idea to apply it to fastener corrosion in wood.

"Keep the wood moisture content below 18%" The corrosion of metals in wood is aqueous. The threshold moisture content at which moisture content occurs is between 15-18%, and the corrosion rate experiences a steep transition around 20% MC. The structural degradation of the joint depends strongly upon the corrosion rate, so even a few percent moisture content difference can make a large difference in the service life of the fasteners.

Author details

Samuel L. Zelinka*

Address all correspondence to: szelinka@fs.fed.us

Materials Research Engineer, Durability and Wood Protection, US Forest Service, Forest Products Laboratory, Madison, USA

References

[1] A.L. Heim, G.M. Davidson, A.R. Joyce, Report of sub-committee 4 - corrosoin of fastenings, in, American Wood Preservers' Association, New Orleans, LA, 1923, pp. 373-379.

[2] S. Lebow, Alternatives to Chromated Copper Arsenate for Residential Construction. Res. Pap. FPL-RP-618, in, U.S.D.A. Forest Service, Forest Products Laboratory, Madison, WI, 2004, pp. 9-9.

[3] S. Lebow, Wood Preservation, in: Wood handbook- Wood as an engineering material, U.S. Department of Agriculture, Forest Service, Forest Products Laboratory, Madison, WI, 2010, pp. 508 p.

[4] G. Kear, H.-Z. Wu, M.S. Jones, Weight loss studies of fastener materials corrosion in contact with timbers treated with copper azole and alkaline copper quaternary compounds, Corrosion Science, 51 (2009) 252-262.

[5] S.L. Zelinka, R.J. Sichel, D.S. Stone, Exposure testing of fasteners in preservative treated wood: gravimetric corrosion rates and corrosion product analyses, Corrosion Science, 52 (2010) 3943-3948.

[6] S.L. Zelinka, D.R. Rammer, D.S. Stone, Electrochemical corrosion testing of fasteners in wood extract, Corrosion Science, 50 (2008) 1251-1257.

[7] S.L. Zelinka, D.R. Rammer, Corrosion rates of fasteners in treated wood exposed to 100% relative humidity, ASCE Journal of Materials in Civil Engineering, 21 (2009) 758-763.

[8] A.C. Wiedenhoeft, Structure and Function of Wood, in: R.J. Ross (Ed.), U.S. Department of Agriculture, Forest Service, Forest Products Laboratory, 2010.

[9] J.F. Siau, V.P. Institute, S. University, Wood: Influence of moisture on physical properties, Department of Wood Science and Forest Products, Virginia Polytechnic Institute and State University Blacksburg, Virginia, 1995.

[10] S.V. Glass, S.L. Zelinka, Moisture Relations and Physical Properties of Wood, in: R.J. Ross (Ed.), U.S. Department of Agriculture, Forest Service, Forest Products Laboratory, 2010.

[11] S.L. Berry, M.L. Roderick, Plant–water relations and the fibre saturation point, New Phytologist, 168 (2005) 25-37.

[12] S. Zelinka, S. Glass, D. Stone, A Percolation Model for Electrical Conduction in Wood with Implications for Wood-Water Relations, Wood and Fiber Science, 40 (2008) 544-552.

[13] S.L. Zelinka, S.V. Glass, Water vapor sorption isotherms for southern pine treated with several waterborne preservatives, ASTM Journal of Testing and Evaluation, 38 (2010) 80-88.

[14] S.L. Zelinka, D.S. Stone, Corrosion of metals in wood: Comparing the results of a rapid test method with long-term exposure tests across six wood treatments, Corrosion Science, 53 (2011) 1708-1714.

[15] Anon, AWPA U1-06: Use category system, American Wood Preservers' Association, Granbury, TX, 2007.

[16] S.L. Zelinka, D.R. Rammer, Synthesis of Published and Unpublished Corrosion Data From Long Term Tests of Fasteners Embedded in Wood: Calculation of Corrosion Rates and the Effect of Corrosion on Lateral Joint Strength, in: CORROSION 2011, NACE International, Houston, TX, 2011, pp. Paper No. 11163.

[17] S.T. Lebow, Wood Preservation, in: R.J. Ross (Ed.), U.S. Department of Agriculture, Forest Service, Forest Products Laboratory, 2010.

[18] Anon, AWPA P5: Standard for waterborne preservatives, American Wood Preservers' Association, Granbury, TX, 2007.

[19] A.J. Baker, Corrosion of metals in preservative-treated wood, in: M. Hamel (Ed.) Wood Protection Techniques and the Use of Treated Wood in Construction, Forest Products Society, Madison, Wisconsin, 1988, pp. 99-101.

[20] G. Kear, H.Z. Wú, M.S. Jones, Corrosion of ferrous- and zinc-based materials in CCA, ACQ, and CuAz timber preservative solutions, Materials and Structures, 41 (2008) 1405-1417.

[21] S.L. Zelinka, D.R. Rammer, D.S. Stone, Evaluating the corrosiveness of southern pine treated with several wood preservatives using electrochemical techniques, in: CORROSION 2009, NACE International, Atlanta, GA, 2009, pp. Paper #09172.

[22] J.K. Dennis, C. Zou, N.R. Short, Corrosion behaviour of zinc and zinc alloy coated steel in preservative treated timber, Transactions of the Institute of Metal Finishing, 75 (1995) 96-101.

[23] N.R. Short, J.K. Dennis, Corrosion resistance of zinc-alloy coated steel in construction industry environments, Transactions of the Institute of Metal Finishing, 75 (1997) 47-52.

[24] A.J. Baker, Corrosion of metal in wood products, in: Durability of Building Materials and Components. ASTM STP 691., American Society for Testing and Materials, West Conshohocken, PA, 1980, pp. 981-993.

[25] R.H. Baechler, Corrosion of metal fastenings in zinc-chloride treated wood after ten years, in: Proceedings of the 35th Annual Meeting of the American Wood Preservers' Association, Washington, D.C., 1939, pp. 56-63.

[26] R.H. Baechler, Corrosion of metal fastenings in zinc-chloride treated wood after 20 years, in: Proceedings of the 45th Annual Meeting of the American Wood Preservers' Association, American Wood Preservers' Association, St. Louis, MO, 1949, pp. 381-397.

[27] S.L. Zelinka, D. Derome, S.V. Glass, Combining A Hygrothermal And Corrosion Model To Predict Corrosion Of Metal Fasteners Embedded In Wood, Building and Environment, 46 (2011) 2060-2068.

[28] X.G. Zhang, Corrosion of Zinc and Zinc Alloys, in: S.D. Cramer, B.S. Covino (Eds.) ASM Handbok Volume 13B, Corrosion:Materials, ASM International, Materials Park, OH, 2003, pp. 402-417.

[29] X.G. Zhang, J. Hwang, W.K. Wu, Corrosion testing of steel and zinc, in: Proceedings of the 4th International Conference on Zinc and Zinc Alloy Coated Steel Sheet (GALVATECH '98), Chiba, Japan, 1998.

[30] D. Knotkova-Cermakova, J. Vlckova, Corrosive effect of plastics, rubber and wood on metals in confined spaces, British Corrosion Journal, 6 (1971) 17-22.

[31] R.A. Legault, A.G. Preban, Kinetics of the atmospheric corrosion of low-alloy steels in an industrial environment, Corrosion, 31 (1975) 117-122.

[32] A.J. Baker, Corrosion of nails in CCA- and ACA-treated wood in two environments, Forest Products Journal, 42 (1992) 39-41.

[33] S.L. Zelinka, D.R. Rammer, D.S. Stone, Corrosion of metals in contact with treated wood: developing test methods, in: CORROSION 2008 NACE, New Orleans, LA, 2008, pp. Paper #09172.

[34] P.E. Hazlewood, P.M. Singh, J.S. Hsieh, Role of wood extractives in black liquor corrosiveness, Corrosion, 62 (2006) 911-917.

[35] S.L. Zelinka, D.S. Stone, The effect of tannins and pH on the corrosion of steel in wood extracts, Materials and Corrosion, 62 (2011) 739-744.

[36] D.F. Packman, The acidity of wood, Holzforschung, 14 (1960) 178-183.

[37] C.A. Smith, Corrosion of metals by wood, Anti-Corrosion Methods and Materials, 29 (1982) 16-17.

[38] R.H. Farmer, Corrosion of metals in association with wood. Part 2. Corrosion of metals in contact with wood, Wood, 27 (1962) 443-446.

[39] E.T. Bartel-Kornacka, Corrosion of iron by Ghana timbers, Wood, 32 (1967) 39-42.

[40] A. Krilov, R. Gref, Mechanism of sawblade corrosion by polyphenolic compounds, Wood Science and Technology, 20 (1986) 369-375.

[41] H. Winkelmann, E. Badisch, M. Roy, H. Danninger, Corrosion mechanisms in the wood industry, especially caused by tannins, Materials and Corrosion, 60 (2009) 40-48.

[42] H. Winkelmann, E. Badisch, S. Ilo, S. Eglsaer, Corrosion behaviour of tool steels in tannic acids, Materials and Corrosion, 60 (2009) 192-198.

[43] V.A. Pugsley, G. Korn, S. Luyckx, H.G. Sockel, On localised corrosive attack, stress corrosion cracking and corrosion fatigue effects in a hardmetal cutting-tool material, International Journal of Materials Research, 93 (2002) 745-749.

[44] V.A. Pugsley, G. Korn, S. Luyckx, H.G. Sockel, W. Heinrich, M. Wolf, H. Feld, R. Schulte, The influence of a corrosive wood-cutting environment on the mechanical properties of hardmetal tools, International Journal of Refractory Metals and Hard Materials, 19 (2001) 311-318.

[45] H. Maclean, J.A.F. Gardner, Heartwood extractives in digester corrosion, Pulp and Paper Magazine of Canada, 54 (1953) 125-130.

[46] S. Kannan, R.G. Kelly, The role of dihydroxybenzenes and oxygen on the corrosion of steel in black liquor, Corrosion Science, 38 (1996) 1051-1069.

[47] P.M. Singh, A. Anaya, Effect of wood species on corrosivity of black liquors, in, NACE International, Houston, Texas, 2001.

[48] P.M. Singh, A. Anaya, K. Frey, J. Mahmood, Corrosivity of black liquors- role of wood species pulped, in, VTT, Helsinki, Finland, 2001, pp. 409-425.

[49] P.M. Singh, A. Anaya, Effect of wood species on corrosion behavior of carbon steel and stainless steels in black liquors, Corrosion Science, 49 (2007) 497-509.

[50] J. Gust, J. Suwalski, Use of Mössbauer spectroscopy to study reaction products of polyphenols and iron compounds, Corrosion, 50 (1994) 355-365.

[51] I.T. Clark, J. Green, Production of phenols by cooking kraft lignin in alkaline solutions, Tappi Journal, 51 (1968) 44-48.

[52] K. Niemelä, GLC-MS Studies on Pine Kraft Black Liquors Part V. Identification of Catechol Compounds, Holzforschung, 43 (1989) 99-103.

[53] S.L. Zelinka, D.R. Rammer, Review of test methods used to determine the corrosion rate of metals in contact with treated wood. Gen. Tech. Rp. FPL-GTR-156, U.S. Department of Agriculture, Forest Service, Forest Products Laboratory, 2005.

[54] Anon, ASTM G198 - 11: Standard Test Method for Determining the Relative Corrosion Performance of Driven Fasteners in Contact with Treated Wood, ASTM International, West Conshohocken, PA, 2011.

[55] Anon, AWPA E-12: Standard method of determining corrosion of metal in contact with treated wood, American Wood Preservers' Association, Selma, AL, 2007.

[56] D.R. Rammer, S.L. Zelinka, Optical method for measuring the surface area of a threaded fastener, Experimental Techniques, 34 (2010) 36-39.

[57] D.R. Rammer, S.L. Zelinka, Analytical determination of the surface area of a threaded fastener, ASTM Journal of Testing and Evaluation, 36 (2008) 80-88.

[58] D.R. Rammer, S.L. Zelinka, Algorithm to calculate the surface area of a threaded fastener. United States Patent No. 8,041,150, 2011.

[59] S.L. Zelinka, Comparing the Methodologies in ASTM G198: Is There an Easy Way Out?, in: CORROSION 2013, NACE International, Orlando, FL, 2013, pp. Paper Number 2507.

[60] S.L. Zelinka, D.R. Rammer, D.S. Stone, J.T. Gilbertson, Direct current testing to measure corrosiveness of wood preservatives, Corrosion Science, 49 (2007) 1673-1685.

[61] G. Kear, H. Wú, M.S. Jones, F.C. Walsh, Direct-Current Methods for the Estimation of Corrosion Rates in Aqueous Timber Preservatives, Australian Journal of Chemistry, 61 (2008) 455-465.

[62] G. Kear, H.-Z. Wu, M.S. Jones, F. Walsh, Impedance spectroscopy studies of the dissolution of ferrous- and zinc-based materials in aqueous timber preservatives, Journal of Applied Electrochemistry, 38 (2008) 1599-1607.

[63] Anon, AWPA E-17: Standard Method for Determining Corrosion Rates of Metals in Contact with Treating Solutions, American Wood Protection Association, Selma, AL, 2013.

[64] S.L. Zelinka, Uncertainties in corrosion rate measurements of fasteners exposed to treated wood at 100% relative humidity, ASTM Journal of Testing and Evaluation, 35 (2007) 106-109.

[65] Anon, Preservative treated wood. Simpson Strong-Tie Technical Bulletin No. T-PTWOOD08-R, Simpson Strong Tie, Pleasanton, CA, 2008.

[66] Anon., Copper Naphthenate Corrosivity Data. Retrieved from http://www.anthony-forest.com/pdfs/cunap_corrosivity.pdf on April 1,2013, Anthony Forest Products n.d.

[67] M.A. Freeman, C.R. McIntyre, A comprehensive review of copper-based wood preservatives with a focus on new micronized or dispersed copper systems, Forest Products Journal, 58 (2008) 6-27.

[68] S.L. Zelinka, D.R. Rammer, Modeling the effect of nail corrosion on the lateral strength of joints Forest Products Journal, 62 (2012) 160-166.

[69] K.W. Johansen, Theory of timber connections, International Association of Bridge and Structural Engineering, Bern, Switzerland, 1949.

[70] C. Clausen, S.V. Glass, Build Green: Wood Can Last for Centuries, Forest Service, Forest Products Laboratory, General Technical Report, FPL-GTR-215, Madison, WI, 2012.

[71] K.B. Tator, Introduction to Coatings and Linings, in: S.D. Cramer, B.S. Covino (Eds.), ASM International, Materials Park, OH, 2003, pp. 814-816.

[72] S.L. Zelinka, L. Ortiz-Candelaria, Electrochemical impedance spectroscopy (EIS) as a tool for measuring corrosion of polymer-coated fasteners used in treated wood, Forest Products Journal, 59 (2009) 77-82.

[73] J. Wilson, Aluminum sign corrosion investigation. Final Report #WI-06-04 WisDOT Highway Research Study #WI-04-02, Wisconsin Department of Transportation Madison, WI, 2004.

Chapter 24

Effect of Alternative De-icers on the Corrosion Resistance of Reinforced Concrete Bridges and Highway Structures

S. O. Nwaubani and A. Katsanos

Additional information is available at the end of the chapter

http://dx.doi.org/10.5772/57551

1. Introduction

Prolonged periods of snowfall in countries with advanced infrastructure and transport systems have rendered the use of de-icing agents to a common occurrence on roads and highway structures. They are necessary in order to maintain a good level of service with respect to the transport systems, thus avoiding traffic jams and disruptions, but also to provide a high level of road safety. Today, chloride-based products, such as rock salt, are the most commonly encountered de-icers as they are easy to apply and store but mostly because they efficiently melt ice at an affordable price [1]. However, their widespread use over a long period has left the construction industry and the engineering community with a grave problem regarding the durability of highway reinforced concrete bridges and multi-storey parking structures [2], due mainly to the fact that they cause corrosion of the reinforcement and steel components [2, 3].

For reinforced concrete bridges, in particular, the chlorides reach the reinforcement through leaking joints and spray from passing traffic. The corrosion product is known to occupy 2 to 6 times the volume of the metal from which it originates and this results to spalling of the concrete cover [2]. This process has a multitude of implications for the durability of concrete structures as it may imply section loss of the reinforcement and compromise of their load bearing capacity. Loss of the protection provided by the concrete cover further promotes and accelerates corrosion, thus leading to structural deficiencies and reduction of the life cycle of the structure. In the US alone, there are 20-25 year-old bridges, which are already facing this type of problems due to heavy de-icer applications and inadequate maintenance [1]. Furthermore, chloride de-icers have also been reported to have a negative impact on the natural environment [4, 1], and accelerate the corrosion of vehicle parts and roadside objects [5, 1]. These factors have inevitably raised concerns regarding the sustainability of current de-icing practices.

2. Cost of corrosion in general

The annual cost of corrosion consists of both direct costs and indirect costs. The direct costs related to corrosion are made up of two main components:

a. The costs of design, manufacturing, and construction such as; use of new materials such as stainless steel to replace carbon steel; use of novel technologies to achieve increased wall thickness for corrosion allowance; innovations in the mitigation and/or prevention of corrosion, such as coatings, sealants, corrosion inhibitors, cathodic protection, and cost of labour and equipment.

b. The direct and indirect cost of management relating to the inspection, repair, maintenance, replacement of corroded parts, rehabilitation, and loss of productive time in corrosion-damaged structures.

Using highway bridges as an example, the optimized contribution of each of the contributing components is calculated through life cycle cost analysis and characterized by the annualized value. Relevant surveys carried out worldwide has highlighted the heavy cost of maintaining, repairing, and replacing concrete structures due to corrosion. In the United states a Corrosion Costs Study carried out by the National Association of Corrosion Engineers (NACE), revealed the direct cost of corrosion in the U.S. was $276 Billion in 1998, approximately 3.1 % of GDP. The indirect cost of corrosion was also estimated to be at least equal to the direct cost giving a total cost of 552 Billion. The direct and indirect costs increased to $468 Billion in 2011, giving a total cost due to corrosion of $936 Billion in that year. In fact, a significant milestone in the effect of corrosion on the U.S. economy was reached this year (2013), when **the total cost of corrosion in the US exceeded $1 trillion annually** for the first time (based on estimates of GDP from http://forecasts.org/gdp.htm). This represents 6.2% of the GDP.

	Year of Study		
	1998	2011	2013
Direct corrosion cost	$276Billion	$468 Billion	>$500 Billion
Indirect corrosion cost	$275 Billion	$468 Billion	>$500 Billion
Total Cost	$552 Billion	$936 Billion	>$1 Trillion

Table 1. Result from (NACE Corrosion Costs Study)

An earlier study by Koch, et al [6], established the cost of corrosion to the Government sector, the Production and Manufacturing sector, Utilities sector, Infrastructure and Transportation sector. Of these, the cost of corrosion in the infrastructure category was estimated at $22.6 billion, which was 16.4 % of the total cost of the sector categories examined in the study (see Figure 1).

The infrastructure category is divided into the following industry sectors: (a) highway bridges, (b) gas and liquid transmission pipelines, (c) waterways and ports, (e) hazardous materials storage, (f) airports, and (g) railroads.

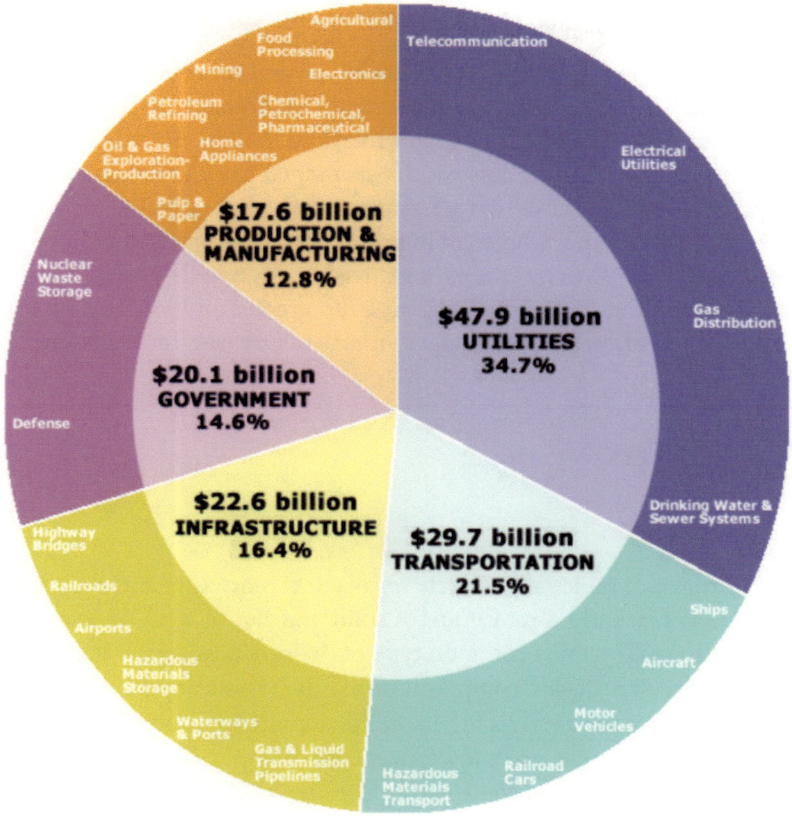

Figure 1. NACE Corrosion Costs Study [6]

2.1. Facts about reinforced concrete bridge corrosion

There are about three quarters of a million conventional reinforced concrete bridges in the USA. According to the latest Federal Highway Administration's National Bridge Inventory, at least 59% of these are reinforced concrete structures [6] and over 30% of the bridges are structurally deficient or functionally obsolete. The annual direct cost of corrosion for highway bridges is estimated to be $13.6 billion. In the United States, the direct annual cost of repair and replacement of deficient concrete bridges has been estimated to be around US$8.3 billion [7, 8], or 0.095% of the 1998 GDP of that country [9]. A similar survey conducted in Japan [10], estimated the corresponding expenditure to be ¥177.4 billion, or 0.059% of the 1999 GDP of that country [11]. In Britain the total estimated cost for the rehabilitation and repair of deficient concrete bridges was found to be £616.5 million spread over a period of ten years after 1989 when the report was published [12]. This would amount to about 0.013% of the 1989 GDP of the UK according to the Office for National Statistics and the figures does not include costs relating to damage on roadside objects, vehicles, the loss of floral and faunal species and surface water quality, the adverse effects on soil structure and the migration of sodium and chloride ions into water supplies [1]. It is also estimated that indirect costs, such as traffic delays and subsequent lost productivity, can amount to ten times the value of direct costs [1].

Cost of corrosion studies have been undertaken by several other countries, not cited above, including, Australia, Kuwait, Germany, Finland, Sweden, India, and China. The studies have ranged from formal and extensive efforts to informal and modest efforts. The common finding of these studies was that the annual corrosion costs ranged from approximately 1 to 5 percent of the Gross National Product (GNP) of each nation [5]. Furthermore, the cases cited above are mainly before the millennium, however an important point to note is that the cost of remediation of corrosion-damaged structures is ever increasing, even in this decade and millennium, with many developed countries spending an ever greater proportions of their GDP for maintenance and repair of corrosion-damaged infrastructure (see Table 1 below).

3. Facts about de-icers

3.1. Chloride-based de-icers

Due to extensive research on ice-melting chemicals in recent years and their many different applications, there is currently a multitude of solid and liquid road de-icing and anti-icing products on the market. The most common type of de-icers used today is the chloride-based de-icers and in particular rock salt, composed primarily by sodium chloride (NaCl). A figure typical of the extensiveness of its use is the 10 million tons of salt used in the US every year [13, 14]. It is used so widely because of its low price, ease of application, and documented effectiveness. Although chloride-based de-icers perform well with respect to ice melting, their aforementioned impact on the environment and infrastructure has led the industry to pursue different methods of winter operations, which could limit or eliminate their use [1]. In this context, the emerging, acetate-based de-icing chemicals have been chosen for investigation both on-site and in the laboratory by many transportation departments and research establishments especially in the US ([1], as there is evidence that they could provide a more sustainable solution towards winter operations, despite their, currently, high cost. In what follows, this chapter will explore in greater depth the properties of the two common Acetate-based de-icers: CMA (calcium magnesium acetate), and NAAC (sodium acetate)

3.2. Acetate-based de-icers

3.2.1. Calcium Magnesium Acetate (CMA)

Recognition of the negative effects of chloride-based de-icers has generated research into alternative de-icers, especially in the United States. Calcium magnesium acetate (CMA) is one of the chemicals considered for this purpose. CMA has been investigated for its ice-melting characteristics [15], its effects on concrete [16,17,18,19,20] and various metals including reinforcing steel [21,7,8], its environmental impact as well as other aspects. Almost all of these studies have found CMA to be effective and benign. The drawback however, is that its price currently ranges between 20 to 30 times the cost of rock salt [22, 1]. A report carried out

Country/Project	GDP	Annual Cost of Corrosion	Source of information
USA(1998)	$276 Billion	3.15% of GDP	Corrosion Costs and Preventive Strategies in the United States', Report FHWA-RD-01-156 - http://www.nace.org
USA(2007)	$429 Billion	$13,840 billion	NACE figures: http://events.nace.org/publicaffairs/cocorrindex.asp
USA(2013)	$16.535 Trillion	>1 trillion [6.2% GDP]	Bureau of Economic Analysis, Department of Commerce. July 31, 2013 Income and Product Accounts Gross Domestic Product, second quarter 2013 (advance estimate); http://www.bea.gov/newsreleases/national/gdp/2013/ . Retrieved August 1, 2013.
Canada(2005) Roads, Sidewalks, Bridges Toronto (Canada)	$992 Billion	$110 million spent in 2005 ... with a backlog of $235 million deferred due to budget constraints	K. McGran's article "On the road to ruin?", Toronto Star, February 5, 2005, pB4-B5.
Canada(2012)	$1.819 Trillion	2-4% of GDP	Bureau of Economic Analysis, Department of Commerce. July 31, 2013
Japan (1997)	$ 4.324 Trillion	0.8-1.0 % of GNP (1997 estimate of direct corrosion costs)	National Institute for Materials Science (Japan).
Japan(2012)	$5.964 Trillion	2-4% of GDP	Bureau of Economic Analysis, Department of Commerce. July 31, 2013
Australia(2008)		A$1-5 billion Australian dollars [~ 2% of GDP]	CSIRO-Re: Holistic Approach to Corrosion control, January 2008 web page.
Australia(2009)	$920 Billion	$70.6 billion	
Australia (2012)		$1,542 Trillion 2-4% of GDP	Bureau of Economic Analysis, Department of Commerce. July 31, 2013
United Kingdom (2007), Corrosion assessment, mitigation and repair of the old Severn Bridge.		£23 million	BBC News, March 7, 2007. (On-line Edition)
United Kingdom (2008)	$2.279 Trillion	$70.6 billion	GDP figures: http://www.economywatch.com/
United Kingdom (2012)	$2,441 Trillion	3-4% GDP	Structure and Infrastructure Engineering Maintenance, Management, Life-Cycle Design and Performance, Taylor & francis Volume 9, Issue 12, 2013, ISBN 1573-2479.

Table 2. Cost of Corrosion

following a request by the US congress [1], identified many benefits which could arise from the replacement of rock salt by CMA. Nevertheless, it highlighted the fact that it was doubtful whether a complete substitution would provide the expected financial benefits; a problem compounded by the fact that much of the data were not easy to quantify. Subsequent studies, by Katsanos and Nwaubani [23], along with data resulting from field experience [24, 25, 26, 27, 28] have all reinforced the view that the use of CMA is a safe and sustainable option.

3.2.2. Sodium Acetate (NAAC)

Another acetate de-icer, sodium acetate (NAAC), which has a similar cost to CMA, is currently used principally as an airport runway de-icer. It has been used been used for road de-icing in the towns of Aspen and Snowmass in the USA and has reportedly performed exceptionally well at very low temperatures [22]. NAAC is exothermic and this property makes it more effective in ice-melting as it stays active at lower temperatures than CMA [22]. Its potential use as a highway de-icer has not been fully researched, little or no few field experience has been documented regarding the effects of NAAC on corrosion behaviour of embedded steel reinforcement although it is commonly regarded as a non-corrosive de-icer [29, 30]. This is possibly due to concerns regarding its potential effects on surface waters. There are fears that, if used, it could increase the BOD of surface waters, more than other acetate de-icers, but less than urea [22]. If NAAC proves to be efficient in minimising reinforcement corrosion, and can be demonstrated to have an acceptable environmental impact, then it could provide a viable solution for use on corrosion prone structures in areas where temperatures fall below the active range of CMA.

De-icing chemicals are incapable of melting snow and ice in their dry (solid) state. They must first attract or encounter moisture to form brine (a chemical/water solution). The brine then penetrates down through the ice and snow until it reaches the pavement, then spreads outwards melting and undercutting the ice and snow for mechanical removal. Although the same chemicals behaviour are involved, there are remarkable differences in how de-icers work. A basic reason for these differences in de-icing is that some chemicals take longer to go into solution before they can begin penetrating ice and snow. Another reason for de-icing action is heat liberation. As they dissolve, some chemicals liberate heat due to exothermic reaction. In addition, pellet de-icers have been found to be more efficient than flake de-icers [24]. Generally, the performance of a de-icer is determined by its ability to melt, penetrate, undercut, and break the bond between the ice and snow. Two of the factors that determine the ability of a de-icer to perform these functions are the eutectic and effective temperatures of the de-icer. From basic physics, it is known that the eutectic temperature is the theoretical lowest temperature to which the de-icer can suppress the freezing point of water. However, this temperature can only be reached at the ideal percent concentration of the de-icer in water. All other concentrations will have higher freezing points [13]. Because de-icers become increasingly diluted as they melt ice, they remain at their ideal concentration only briefly. Consequently, the eutectic temperature has little bearing on real-life de-icing conditions. The effective temperature, which is the lowest temperature at which each de-icer remains effective, is a much more realistic measure of the de-icing capability of these chemicals [22]. Site experience gained by the use of acetate de-icers over recent years, has provided a clearer picture with respect to the way that acetate de-icers work and their effectiveness in comparison to chloride de-icers. Table 3 displays the eutectic and effective temperatures of the de-icers under consideration. It becomes obvious that the effective temperature of sodium acetate is much lower due to the fact that it possesses exothermic properties.

DE-ICER	EUTECTIC CONCENTRATION (%)	EUTECTIC TEMPERATURE (C°)	EFFECTIVE TEMPERATURE (C°)
CMA (anhydrous calcium magnesium acetate)	32.60	-27.78	-6.67
NAAC (anhydrous sodium acetate)	27.00	-21.67	-15.00
Sodium Chloride	23.00	-21.12	-9.44

Table 3. Eutectic and Effective Temperatures of selected de-icers [13,14]

In another study Shenk [31], observed that the slower rate of ice melting arising from the use of CMA was due primarily, to the tendency of the CMA crystals to be buoyed up in their solution. However, the study revealed that CMA, is in many ways superior to rock salt as a road de-icer, due to its ability to melt ice at lower temperatures, and a marked heat of solution of the anhydrous CMA.

3.3. Site experience on the effectiveness of CMA and sodium acetate deicers

Field evaluation of the performance of CMA was studied in two similar articles from the late 1980's [24, 25]. They discussed experience from CMA applications at different locations. A common finding from those field studies is that CMA works differently from Sodium chloride. CMA prevents snow pack when applied early in a storm, forming a fluffier, drier snow-pack easy to remove by ploughing or by passing traffic, while Sodium chloride produces a wetter and heavier snow-pack. Therefore, timing was crucial in the case of CMA application MI92 MI93.The downside to early application is that, if done before a snow storm, it runs the risk of being blown away by traffic due to its low density. If applied late it may be slower acting, taking around 20 minutes longer to act compared to rock salt [25].

CMA also seems to work at lower temperatures than rock salt, as tendencies to re-freeze after temperature drops seem less, according to the observations made at the Zilwaukee Bridge [27, 28]. It also lasts longer and it sometimes has to be applied once during a 12-hour period compared to two or three for rock salt. This could also be due to the residual effect that CMA seems to have on the road surface as liquefied CMA and pellets stick on the road surface minimising re-freezing and early morning deck icing [27]. Clearly, CMA acts in a different manner and this explains the initial tendency of early users and researchers to believe that a larger amount of CMA than rock salt would be required to achieve equal de-icing performance. A study conducted in 1993 [26] attributes the CMA-snow-pack interaction to the ability of the calcium acetate and the magnesium acetate to dissociate from the CMA compound and form independent hydrates. Similar studies have not, however been conducted for sodium acetate, and as a result we know little about its corrosiveness with respect to concrete reinforcement. Nonetheless, Sodium Acetate complies with SAE standard AMS 1431 regarding its suitability for use on runways, and is deemed non-corrosive to metal parts of aeroplanes.

3.4. Effects of acetate de-icers on corrosion of steel reinforcement

As the cost of maintaining, repairing and replacing concrete structures such as highway bridges or multi-storey parking buildings soared, academics and the industry were urged to investigate the corrosiveness of non-chloride-based de-icers. The chemical identified as the most likely substitute, as mentioned above, was CMA, and numerous projects have so far been undertaken in order to assess its effects on concrete and on steel reinforcement.

With respect to the corrosiveness of CMA, a study conducted by the Michigan transportation commission [7,8] deduced that prestressing steel strands immersed in NaCl solutions experienced 4 to 15 times more corrosion than identical specimens immersed in CMA solutions designed to have the same freezing point. The study also concluded that mixtures of the two materials produced results comparable to the ones produced by the CMA solutions down to at least a 0.46 CMA/NaCl weight ratio. However, this last claim is dismissed by a study which measured corrosion rates of steel wire embedded in mortar cylinders [32] by observing that corrosion can be inhibited by mixing CMA and NaCl, only for concentrations of about 0.05 M to 0.1 M NaCl, while CMA admixture does not prevent corrosion when the concentration of NaCl is more than 0.2 M. Similar results were obtained for sodium acetate additions. Another study conducted in Oklahoma, US [21], indicated that although CMA solutions ponded on chloride-contaminated reinforced concrete slabs were not able to passivate the reinforcement, the same solutions did not cause any corrosion on reinforcement embedded in uncontaminated concrete slabs. This result is supported by a 4-year long study by Chollar, B. H. & Virmani, Y. P[24], which reported that after four years of ponding on reinforced concrete slabs with NaCl and CMA solutions, no corrosion was produced by the CMA solutions in contrast to the heavy corrosion and subsequent cracking of the concrete produced by the NaCl solutions. Similar conclusions were drawn by two studies that compared the electrochemical behaviour of metals immersed in solutions of various concentrations of CMA and NaCl [33, 34]. In general, it can be said that CMA solutions do not induce corrosion of metals in almost any condition and in some cases even act as inhibitors.

3.5. Effects of acetate-based de-icers on concrete

Several studies have been carried out over the past two decades focussed on the effects of CMA deicers on the properties of concrete in contact with the applied deicer. In some cases, contradictory results have been recorded for similar kinds of tests by different researchers. This is mainly due to the diversity of conditions under which those experiments were undertaken. The effect of CMA can act very differently when in contact with concrete depending on

http://dx.doi.org/10.5772/57551

factors such as ambient temperature, Ca/Mg molar ratio of the CMA, concentration of the CMA solution, concrete mix design, cement composition, possible admixtures, type of aggregates etc. Consequently, the results from various studies cannot be directly compared. Nonetheless, they do offer an indication of the mechanism that governs the changes, which take place in concrete microstructure as well as in the composition of the CMA solution when these come in contact. Santagata & Collepardi [20] record severe deterioration of concrete after long-term, continuous immersion in concentrated CMA solutions (25% wt) at a temperature of 20°C. Another study by Lee et al [35]), involved tests at 58°C and found CMA to be particularly destructive due to the formation of expansive brucite and non-cementitious MSH. They also noted that the greater the Mg/Ca molar ratio, the greater the damage. This latter observation is also supported by a paper published in 1995 [18] which also documents that CMA solutions are more destructive in higher temperatures. However, this last study concluded that dilute solutions of CMA have a greater effect on compressive and flexural strength than concentrated solutions. The author went on to speculate that, this could be due to the fact that most of the magnesium in the solution is being consumed in the formation of $Mg(OH)_2$, and thus, most of the solution ends up comprising mostly calcium acetate, which he considers especially deleterious to concrete, although this is not supported by the tests involving this substance, conducted by Lee in his aforementioned study. However, most of the existing literature on the subject, support the concept of the formation of brucite and potentially the subsequent replacement of CSH in concrete with non-cementitious MSH as a result of the contact of concrete and CMA. Higher temperatures seem to have a detrimental effect on the cementitious matrix, and higher Mg/Ca molar ratios of the CMA appear to make it more destructive towards concrete. Higher concentrations of CMA solutions have a detrimental effect as documented in the literature, although some conflict exists on this matter, as one study has found dilute solutions to be more destructive.

3.6. Environmental effects of acetate de-icers

Modern concepts of sustainability and whole-life costing of structures cannot be complete without consideration for the environmental effect on the construction materials or how the use of structure during the period in service affects the environment. For highway structures, some of the basic environmental considerations, in the case of de-icers applied during the winter months, is whether they can alter the physicochemical properties of soil; their effect on water and air quality or on vegetation. Several studies have been conducted to this end and they have found CMA and NAAC to be environmentally low-risk de-icers in comparison with the chloride-based ones. More specifically, CMA has been found to be beneficial to soil [36, 37, 38] and harmless to vegetation [39, 4]. Certain concerns have been expressed, mainly regarding the possible future effects of the acetate-based de-icers on surface waters and treatment ponds[14, 38]. However, practical application of both CMA and NAAC has indicated no such effect [29]. A summary of the conclusions of some notable studies carried out are presented in Table 4 below.

DE-ICING AGENT	EFFECTS ON SOIL	EFFECTS ON VEGETATION	EFFECTS ON WATER QUALITY
NaCl	Potentially damaging. Exchangeable Sodium Percentage (ESP) more than 13 (toxic) [39].	Salt spray causes leaf burn. ESP higher than 4 kills some species. 0.5% concentration of chloride in plant tissue will result to plant death [39]. Browning, falling leaves, stem dieback, stunted or abnormal growth, premature death. Effect more apparent on grass rather than trees [14]..	Sodium and chloride concentrations in water supplies often higher than the standards set by health organisations [39].
CMA	Beneficial by increasing the pH and decreases solubility of trace metals. No upper limit defined [36, 37, 38]	No apparent adverse effect by spraying. Plants can withstand root exposure of up to 2500mg/l. [39, 4].	Biodegradable. No effects to water biota. It is better if the dilution rate is more than 100:1 [38]. It could cause eutrofication due to high BOD when re-aeration of waterways is limited by ice cover or in poorly flushed ponds [14]. It could significantly increase organic loading in treatment works when used in urban environments [40]. Despite concerns about BOD and DO levels of receiving waters practical experience has recorded no adverse effects.[29].
NAAC	No reports	No reports	Despite concerns about BOD and DO levels of receiving waters practical experience has recorded no adverse effects [29].

Table 4. Environmental effects of de-icers

4. Case study

4.1. Objectives

The main objectives of the laboratory study are:

- To monitor the corrosion rates, if any, of steel reinforcement in concrete exposed to CMA and Sodium Acetate solutions and to compare them with the corresponding effects on steel reinforcement in concrete exposed to NaCl solutions and water.

http://dx.doi.org/10.5772/57551

- To monitor the effects of varying the concentration of the solution of acetate-based de-icers in relation to their ability to prolong the time to corrosion initiation and to suppress corrosion rates.
- To investigate whether and how the addition of microsilica, varying the w/c ratio of concrete and the depth of cover can influence the time to corrosion initiation and corrosion rates if samples undergo the same testing regime.
- To determine whether the exposure of concrete to the de-icers under investigation could compromise the passivity of the rebars in the future, by causing carbonation of the concrete cover and thus alter the pH of the pore solution.
- To develop an understanding of the parameters controlling the behaviour of acetate-based de-icing agents in relation to the corrosion of steel reinforcement in concrete and their interaction with Portland cement and microsilica-containing paste by means of a micro-structural investigation.

4.2. Methods

4.2.1. Corrosion monitoring

The specimens used for this particular part of the study, had to be manufactured in such a way as to cover the needs of the proposed investigation. Thus, small-scale slab specimens (210mm x 330mm 100mm) were manufactured with three sets of three reinforcing bars each. The three sets of bars were placed at three different depths from the bottom of the ponding basin. Specifically, the depths chosen were 10mm, 25mm and 40mm, as shown in Figure 2.1(a) and 2b. The purpose of employing different cover depths was i) to establish the time to corrosion initiation induced by the different solutions at different depths and ii) to investigate the effect of the cover depth on the corrosion rates of the reinforcement which may, in turn, reveal information regarding the interaction between the de-icer and the concrete cover.

Steel electrodes were made from 8mm diameter mild steel bars. The bars were initially cut to a 310mm length each, and a 4mm hole was drilled on one side at a distance of 10mm from the edge of the bars. They were then turned to a bright steel finish and degreased with acetone in order to remove any manufacturing grease. Next, protective coating was applied on both ends of each bar, which served the purpose of ensuring that an equal area of steel was exposed in every bar, a detail essential as the anodic area has to be known for the calculation of the corrosion current. It also ensures that the protruding ends of the bars are protected from aggressive ions and are, therefore, not corroding. The protective coating was applied by dipping 95mm of either end of the bar thus, ensuring an exposed length of 120mm in the middle. The coating comprised initially two coats of binder paste with the same binder and the same water/binder ratio as the specimen in which the bars were to be imbedded in each case. Two coats of marine varnish were applied, also by dipping, to complete the coating process. The bars were then placed in the moulds as described above and the depth of cover checked.

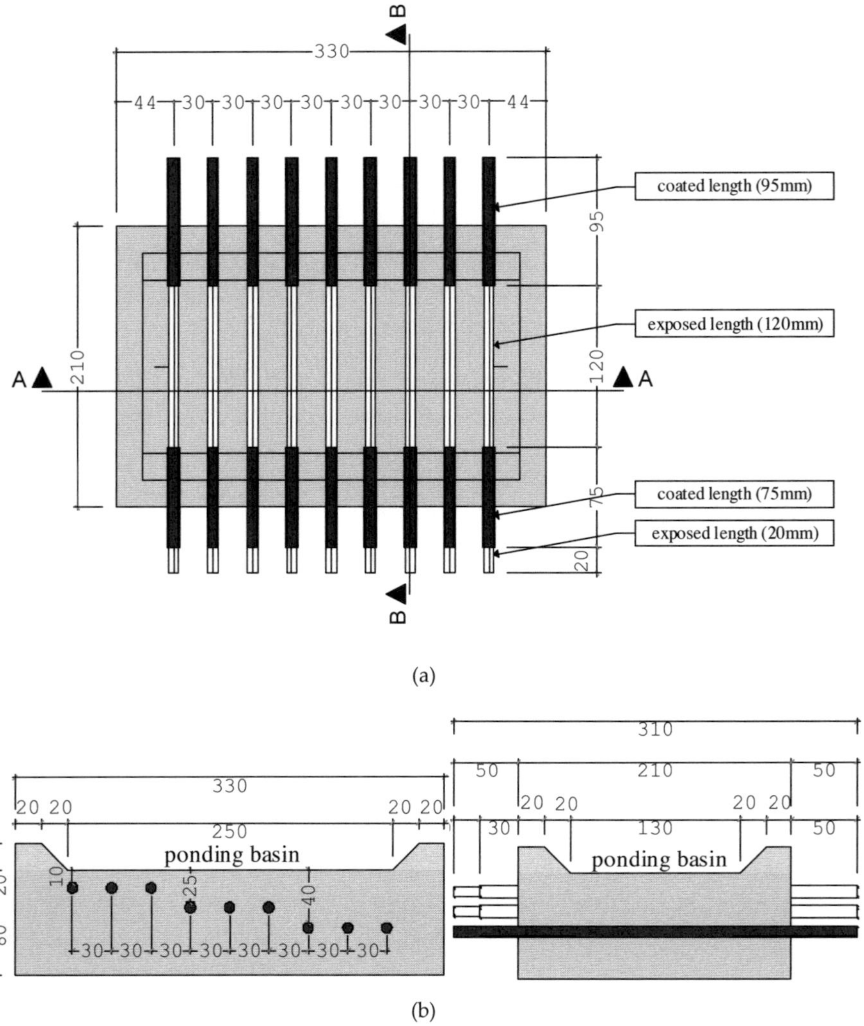

Figure 2. Constructional details of corrosion monitoring specimens, (a) Plan view, (b) Section A/A (c) Section B/B.

Standard batching and mixing techniques were adopted, and the same procedure was adopted for each mix. The mixer used for the preparation of all mixes was a Zyklos ZZ30. When filled and vibrated, the moulds were placed in the curing room (20 oC, 100% R.H.) where they were left to hydrate for 24 hours. Following that period, they were de-moulded and returned to the curing room for a further 27 days before testing. On the 27th and 28th days of curing two coats of anti-slip floor coating were applied on all sides of the specimens, including the protruding ends of the steel bars, except the ponding basin. After curing, the specimens were transferred to a room with constant temperature and relative humidity (20 oC and 70% R.H.), where they remained for the rest of the experimental process. Immediately after their removal from the curing room they were ponded with 300g of water saturated with calcium hydroxide. Forty-eight hours later the initial base-line measurements were taken as described below. Thereafter,

http://dx.doi.org/10.5772/57551

two-weekly dry-wet cycles (seven days dry, seven days wet) were followed whereby, the wet period of each cycle involved ponding with 300g of the corresponding solution for each specimen and sealing to prevent evaporation. Measurements were taken at the end of every wet half-cycle in order to ensure conductivity between the surface of the ponding basin and the steel bars.

Table 5., below outlines the specimens manufactured.

TYPE OF BINDER	W/B RATIO	NUMBER OF SLAB SAMPLES	NUMBER OF REINFORCEMENT BARS
OPC	0,40	6	54
	0,65	6	54
90%OPC / 10% Silica fume	0,40	6	54
	0,65	6	54
Total		24	216

Table 5. Slab specimens for corrosion measurements

It becomes obvious, therefore, that six samples were manufactured for each one of the four mixes designed for the project. This was done for comparison purposes as there were six different aqueous solutions chosen for ponding on the samples. The chosen solutions were as follows: i) Water, ii) 3.5% wt. NaCl solution, arbitrarily chosen to correspond to the average concentration of NaCl in sea water, ii)3.5% wt. CMA solution, chosen due to the indications from practical experience in the literature that equal amounts of CMA and NaCl produce equal de-icing performance, iv)6.181% wt. CMA solution, chosen to produce the same freezing point as 3.5% wt. NaCl, v)3.5% wt. NAAC solution, chosen due to the indications from practical experience in the literature that equal amounts of NAAC and NaCl produce equal de-icing performance, and vi)4.367% wt. NAAC solution, chosen to produce the same freezing point as 3.5% wt. NaCl.

Two well established types of corrosion measurement techniques were used so that their combined results could provide an insight into whether corrosion is taking place and what the corrosion rate is at any given time. The two techniques used were the half-cell potential measurement and Linear Polarisation Resistance technique. The corrosion measurements were conducted by the use of a Sycopel Scientific Superstat potentiostat that was operated manually throughout the duration of the experiments for most of the samples with the exception of eight samples. Namely, the two samples with pure OPC binder and a w/c ratio of 0.65 that were ponded with CMA solutions and all the six samples with a w/c ratio of 0.65 and a 10% silica fume admixture were tested using the Gamry PCI4-750 potentiostat after the completion of the 19th ponding cycle. A saturated KCl calomel electrode was used as a reference electrode. In both techniques, three measurements were taken at every depth for each sample. This provided an indication of the consistency of the results obtained.

4.2.2. pH shift measurements

A pH meter was used to measure the pH of the 300g of every ponding solution right before ponding and then right before the solutions were removed at the end of the wet cycles throughout the duration of the corrosion monitoring experiments. The negative and positive pH shifts were recorded and plotted over time. This was done to provide some means of observing the interaction between the solution and the specimen and to offer an indication of the exchange of ionic species and compounds between them.

4.2.3. Scanning electron microscope imaging

Small cement paste prisms (10 mm x 10 mm x 110 mm), were cast with a water/cement ratio of 0.4 and were immediately placed in the curing room at 100% RH. The following day they were de-moulded and placed back in the curing room for a further period of 27 days. After curing, the samples were transferred to a room maintained at constant temperature and relative humidity (20°C and 70% RH), where they remained for the rest of the experimental process. They were then coated with resin and left to dry overnight. The coating process was repeated the following day and they were, again, left to dry overnight. The following day the hardened resin along with the as-cast surface was removed from one of the two smaller faces of the sample, with a diamond saw. This way, it was ensured that the penetration of the de-icer solution from that contact face would be unidirectional so that the depth of penetration could be estimated with higher precision. Moreover, the removal of the as-cast surface of that particular face ensured that misinterpretations due to contamination would be avoided. The immersion regime, which the samples underwent, involved two-weekly cycles (one week immersed in solution and one week dry). Although wet-dry cycles may bear no particular significance in this case, the experiment was designed to correlate with the ponding regime performed on miniature slab samples designed for reinforcement corrosion testing, which forms the main core of this general project.

The effect of the immersion process was microscopically examined at predetermined times for a duration of 1 year after the end of the curing process, namely at 28 days, 90 days, 6 months and 1 year. The examination was performed with the help of a HITACHI 3200 scanning electron microscope. Specifically, the samples were cut in the middle with a diamond saw along their longitudinal axis, which, contrary to thin slicing, allows for visualisation of the effects of the solutions on the same sample, simply by moving the cut profile under the microscope. Samples were carbon coated before taken to the SEM for imaging. Images were taken at various locations, at different depths from the surface of the exposed face and at different magnifications, which varied from x300 to x9000. Element mapping was performed with the help of the microscope in order to help identify the elements and compound formations at locations of interest.

5. Results and discussion

In what follows, only selected parts of the results obtained from the different sets of experiments, which forms the key part of the data are presented.

5.1. Corrosion monitoring

The "Van Daveer" criteria for the interpretation of half-cell potential measurements taken with a copper/copper sulphate reference electrode specify that a potential more positive than -200mV indicate less than 10% probability of corrosion and a potential more negative than -350mV indicates more than 90% probability of corrosion [41]. The electrode used for the half-cell potential measurements was a saturated calomel electrode whose potential is 72mV more positive than the potential of a copper/copper sulphate electrode. Thus, a half-cell potential more negative than -278mV, indicates a 90% probability of corrosion. Similarly, potentials more positive than -128mV represent an indication of a less than 10% probability of corrosion. Values in between represent an uncertain corrosion activity [42]. Figures 3. and 4., illustrate the results produced so far by the measurement of the rest potential of the steel bars in OPC and 90% OPC / 10% Silica Fume specimens with a water/cement ratio of 0.65.

The results presented in figures 3. and 4.; concern only the half-cell potential measurements taken from the OPC and OPC/Silica Fume samples respectively, with a w/c ratio of 0.65. Clearly, no corrosion seems to be occurring in any of the corresponding samples with a w/c ratio of 0.4 after 25 wet-dry cycles.

In the OPC samples the only bars which seem to be exhibiting signs of a high probability of corrosion are the bars situated at 10mm depth in the OPC sample ponded with a 3.5% aqueous solution of NaCl (figure 3.a and 3.b). The bars at 25mm depth in the same specimen show a tendency to corrode but the half-cell potential values obtained are more positive than the value considered characteristic of a 90% probability of corrosion according to the aforementioned criteria. At the end of the 25th cycle these middle bars have been measured to have a potential of -187mV which is a sign of uncertain corrosion activity. The half-cell potential values of the bars at 40mm depth dwell at around -100mV, which is a value characteristic of a very low probability of corrosion. The difference in potential between bars situated at these three different depths probably results from differences in the concentration of free chloride ions around each bar. The free chloride ions can disturb the passive film on the surface of the steel bars an therefore act as catalysts towards the initiation of corrosion. In addition, the dissociation of iron chlorides can produce locally high levels of acidity at the surface of the steel leading to pit formation (Pullar-Strecker 2002). A careful observation of Figure 3. shows that, with the exception of the sample ponded with the NaCl solution, each one of the rest of the samples exhibits comparable values of rest potential over time at all three depths. The significance of this observation lies in the fact that it indicates the absence of diffusing aggressive ions, which could, directly cause corrosion after reaching a certain threshold, in the same way that chloride ions do. However, the absence of chlorides in a solution does not render it incapable of inducing rebar corrosion as its contact with concrete over a longer period of time could have other undesirable effects such as lowering of the pH of the concrete cover which would, in turn, render the reinforcement susceptible to corrosion.

It is also obvious in Figures 3.a and 3.b that, although the acetate solutions and the water do not seem to corrode the reinforcing bars after 25 ponding cycles, each one of those solutions produces a distinct pattern of potential values over time. The trend evident in the case of the sample ponded with the CMA solution closely resembles the trend produced by the potential of the bars in the sample ponded with water. This could mean that CMA does not contribute any ions more than water does, to the pore solution. Otherwise, this behaviour could indicate that CMA produces impervious deposits on the surface of the concrete, which block the capillary system of the cement paste and hinder the ingress of ions in the pore solution. In contrast, sodium acetate, at all three depths seems to produce the same behaviour as the graph that corresponds to the potentials measured at the lower bars, located at the 40mm depth, in the samples ponded with the NaCl solution. The chloride ions around these bars have not reached a threshold concentration, which could initiate depassivation and corrosion after 25 cycles of testing. In both cases of the samples ponded with NaCl and the ones ponded with NAAC, the pore solution has been influenced as indicated by the lower rest potentials of these bars compared to the rest potentials of the samples ponded with water. This similarity in patterns could be due to the ingress of sodium ions that are present both in NaCl and in NAAC, on the pore solution of the concrete.

The same patterns are evident in Figure 4., which illustrates the corresponding results for the samples containing Silica Fume. However, in these samples none of the bars has had the time to corrode within the 25 cycles of testing and the only bars which produce lower rest potentials and probably a tendency to initiate corrosion are the bars situated at a depth of 10mm from the surface in the sample ponded with NaCl solution. This is due to the documented effect of silica fume to alter the pore size distribution of the cement paste, reducing the size of the capillary pores, thus, making the concrete more impermeable and more resistant to attacks by chlorides and other aggressive ions [40]. However, in all the samples containing silica fume that were ponded with NAAC solutions the rest potential values tend to dwell consistently around 100 mV more positive than they do in the OPC samples. This could mean that the effects of NAAC on the pore solution are more evident in the samples with larger capillary pores, namely the OPC samples.

Finally, it should be noted that all samples were sectioned and sprayed with a phenolphthalein and alizarin pH colour indicators which showed no sign of reduced pH near the surface and therefore no carbonation taking place in any of them. The bars were extracted from all samples and were visually inspected; they all supported the results of the non-destructive corrosion monitoring, as corrosion was encountered only at the bars closest to the surface in the OPC samples with a w/c ratio of 0.65.

5.2. pH shift

Figures 5. and 6. illustrate the results for the pH shift monitoring described in section 3.2 above throughout the duration of the corrosion monitoring experiments. Although such a technique does not provide solid proof of any microstructural process, the results can be used as supporting evidence for the data derived from the widely used microstructural investigation techniques employed throughout this project. A simple observation of the information on figures 5. and 6. reveals a number of things worth noting.

The general trend observed is that all the plots tend to gradually decrease over time from more positive to more negative values and to eventually stabilise or at least decrease at a lower rate. This is possibly due to leaching of $Ca(OH)_2$ from the concrete and the subsequent dissociation

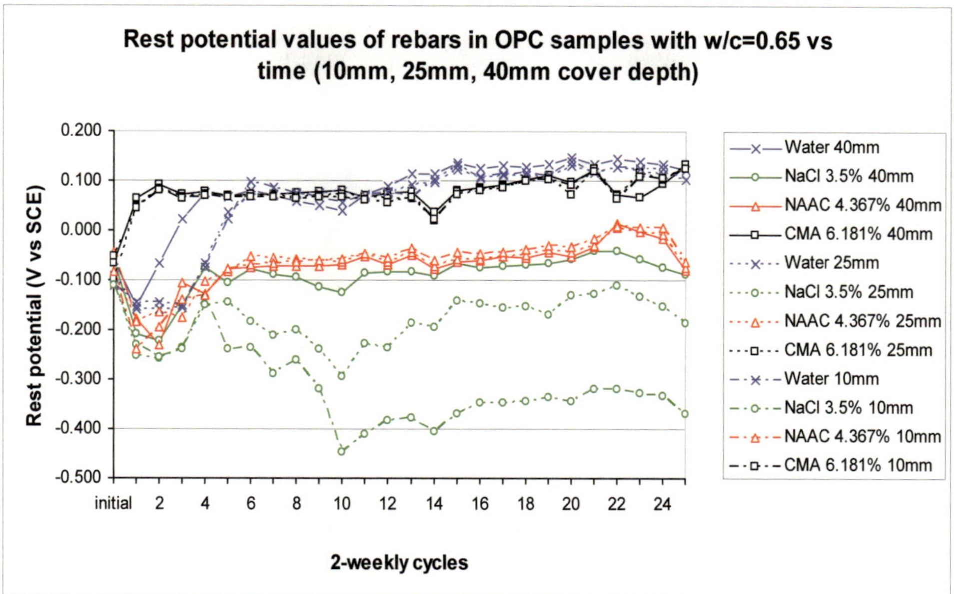

(a)

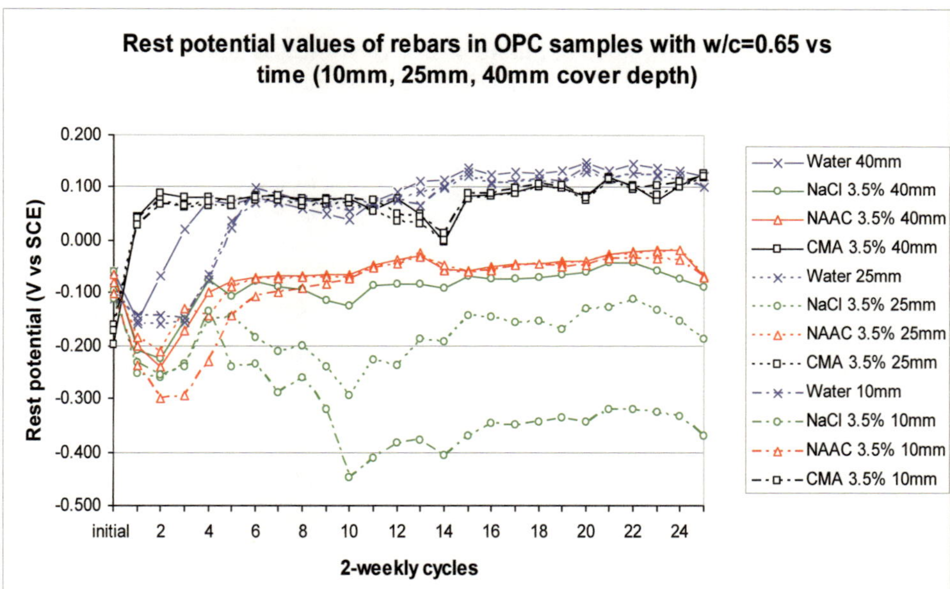

(b)

Figure 3. Half-cell potential values of rebars in OPC samples w/c=0.65 vs. time (10mm, 25mm, 40mm cover depth), a) Samples ponded with solutions with a freezing point equal to that of a 3.5% wt NaCl, b) Samples ponded with solutions of 3.5% wt

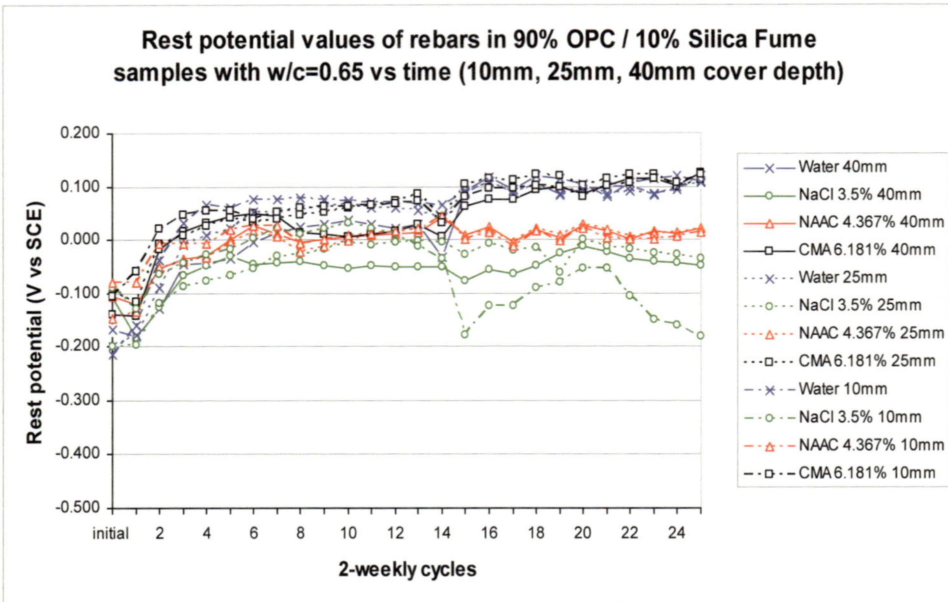

(a)

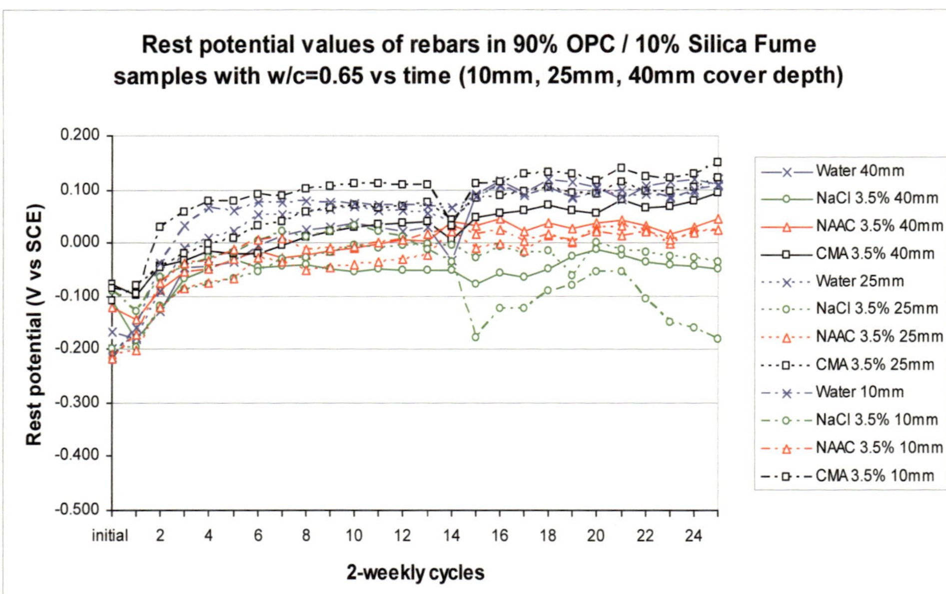

(b)

Figure 4. Half-cell potential values of rebars in 90% OPC/10% Silica Fume samples w/c=0.65 vs. time (10mm, 25mm, 40mm cover depth), a) Samples ponded with solutions with a freezing point equal to that of a 3.5% wt NaCl, b) Samples ponded with solutions of 3.5% wt.

http://dx.doi.org/10.5772/57551

of the calcium and hydroxyl ions, with the latter raising the pH value of the solution. The calcium hydroxide, however, becomes less available as the hydration process progresses and consumes it and the pores diminish in size. It also becomes less available due to the fact that much of it, previously located on and very close to the surface of the ponding basin of the samples has leached out during the initial stages of ponding. This latter phenomenon is probably what causes the abrupt drop of the pH shift graphs towards less positive (or more negative) values, within the first two to four cycles in most samples.

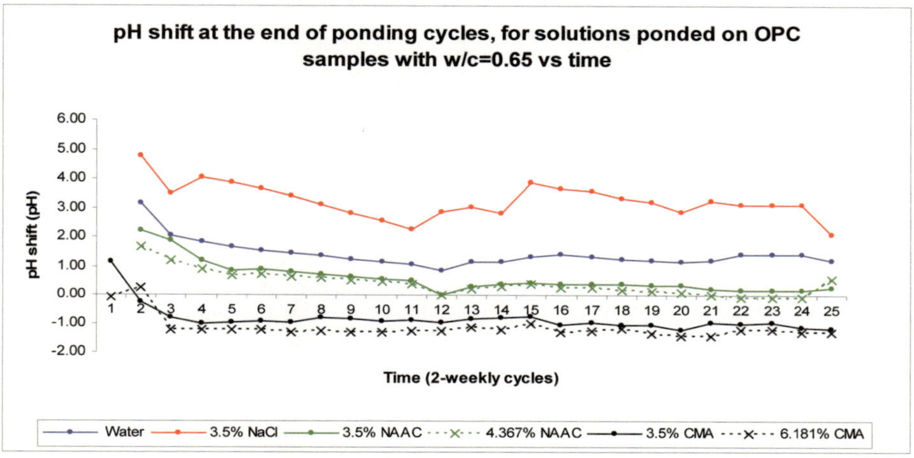

(a)

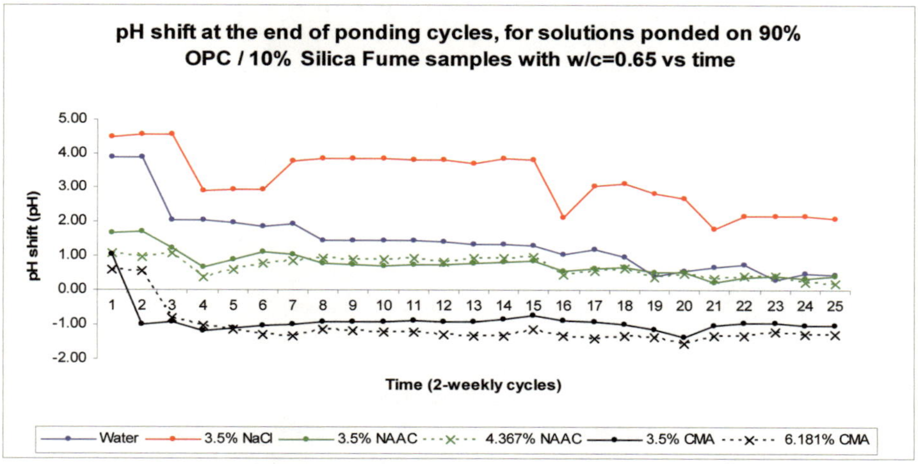

(b)

Figure 5. pH shift in ponding solutions at the end of ponding cycles for samples for samples with w/c=0.65, (a) OPC samples, (b)90% OPC / 10% Silica Fume samples

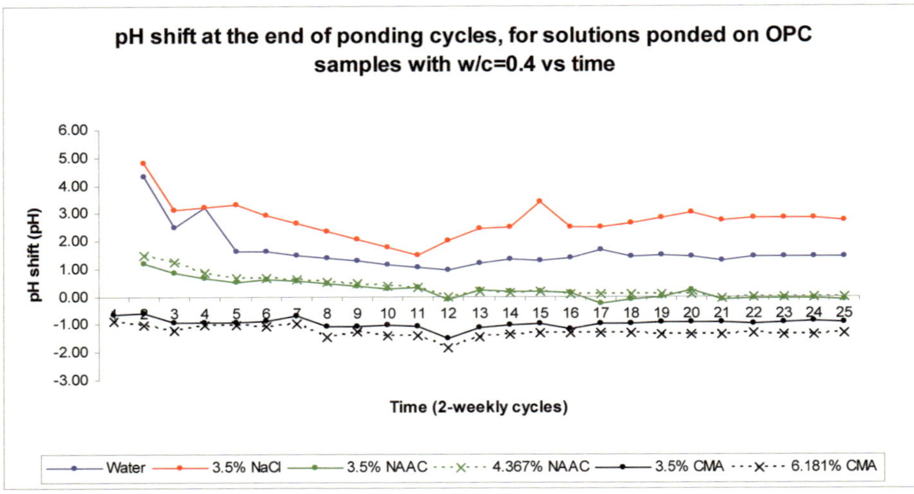

(a)

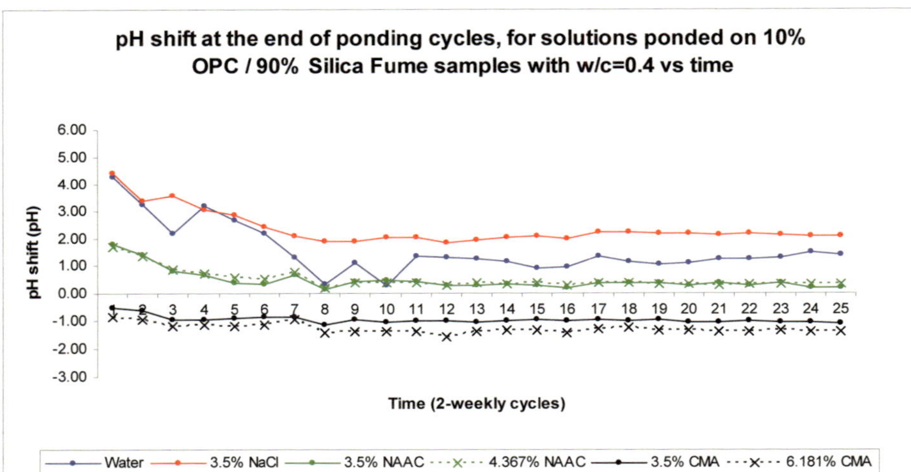

(b)

Figure 6. pH shift in ponding solutions at the end of ponding cycles for samples for samples with w/c=0.4, (a) OPC samples, (b)90% OPC / 10% Silica Fume samples

Varying the water cement ratio does not seem to have a major effect on the pH shift, as one might expect. This could be partly due to the fact that the samples with a 0.65 w/c ratio have larger pores while in the 0.4 w/c ratio samples, more calcium hydroxide is available. Therefore, leaching of calcium hydroxide could be facilitated in these two different ways, thus, producing seemingly comparable results. One, however, has to note that in the case of the CMA solutions on the 0.65 w/c ratio samples in figure 5., the graph decreases more abruptly producing positive pH shift initially within the first few cycles, than it does in the case of the corresponding solutions on samples with 0.4 w/c ratio as illustrated in figure 6. where the shift is negative from the onset of the measurements.

The 10% replacement by silica fume does not seem to have a significant effect on the pH shift for any of the solutions. Moreover, the variations in concentration of the acetate-based de-icer solutions seem to play no significant role either. This is possibly accounted for by the fact that the variations in concentration of the acetate de-icer solutions chosen for this project are relatively small.

It is interesting to note that the resulting pH shift for the CMA solutions is predominantly negative while for the NAAC solutions it is positive or in some cases marginally positive, throughout the graphs. Thus, despite the fact that NAAC and CMA solutions are both acetate solutions, equal or comparable concentrations of the two in this experiment, consistently produce a difference in pH shift at the end of ponding. This effect could be due to the strong presence of magnesium in the CMA. It is worth noting that the Ca/Mg molar ratio in the CMA used is 3/7. Therefore, when soluble lime leaches from the concrete into the CMA solution, and calcium and hydroxyl ions are released, dissociated magnesium is bound to form magnesium hydroxide otherwise known as brucite, a very insoluble and relatively impermeable compound. This possibly led to the decreased values of the pH at the end of the ponding cycles in CMA solutions.

6. Conclusions

This case study suggests that:

1. Acetate based de-icers, unlike NaCl, do not contain deleterious ions which cause the corrosion of the steel reinforcement bars.

2. Presence of Acetate de-icers neutralised the probability of corrosion.

3. The trend for corrosion probability evident in the case of the sample ponded with the CMA solution closely resembles the trend produced by the bars in the sample ponded with water. This could mean that CMA does not contribute any ions more than water does, to the pore solution.

4. NAAC and CMA solutions of equal or comparable concentrations, consistently produce a difference in pH shift at the end of ponding.

5. Varying the water cement ratio does not seem to have a significant effect on the pH shift, as one might expect but no corrosion seems to be occurring in any of the samples with a w/c ratio of 0.4 even after 25 wet-dry cycles.

6. The use of Acetate-based de-icers on reinforced concrete would lead to low maintenance and repair cost for concrete bridges.

7. Recommendation

For better **corrosion** prevention in highway structures, the following preventive strategies need to be considered :

1. The high cost of Acetate based de-icers could be solved by making a change in the de-icing tradition and protocols and on highways..
2. Reinforced concrete bridges may need to locally de-iced with the more expensive but non-corrosive Acetate-based de-icers.
3. There is need to increase the general awareness of the very high corrosion costs and potential savings.
4. There is need to change the misconception that nothing can be done about corrosion rather efforts should not be spared in the quest to reduce cost by effecting adequate change in existing policies, regulations, standards, and management practices to increase corrosion savings through sound corrosion management.

Author details

S. O. Nwaubani[1] and A. Katsanos[2]

1 Department of Engineering and the Built Environment, Anglia Ruskin University, Cambridge and Chelmsford, Essex, UK

2 Department of Civil Engineering, School of Engineering, University of Surrey, Guildford, Surrey, UK

References

[1] TRB, *Highway Deicing, Comparing salt and calcium magnesium acetate,* Transportation Research Board, National Research Council, Washington D.C., Special Report 235, 1991.

[2] Pullar-Strecker, P., *Concrete Reinforcement Corrosion, from assessment to repair decisions,* 1st edn, Thomas Telford Ltd, London, 2002.

[3] Richardson, M. G., *Fundamentals of Durable Reinforced Concrete,* 1st edn, F & EN Spon Press, London, 2002.

[4] Robidoux, P. Y. & Delisle, C. E., "Ecotoxicological evaluation of three deicers (NaCl, NaFo, CMA) - Effect on terrestrial organisms", *Ecotoxicology and Environmental Safety,* vol. 48, no. 2, pp. 128-139, 2001.

http://dx.doi.org/10.5772/57551

[5] Slick, D. S., *Effects of calcium magnesium acetate on pavements and motor vehicles*, 1988.

[6] G.H. Koch, M.P.H. Brongers, N.G. Thompson, Y.P. Virmani, and J.H. Payer, "Corrosion Costs and Preventive Strategies in the United States," U.S. Federal Highway Administration Report, FHWA-RD-01-156, March 2002.

[7] McCrum, R. L., *Corrosion Evaluation of Calcium Magnesium Acetate (CMA), Salt (NaCl), and CMA/Salt Solutions*, Michigan Department of Transportation, Research Laboratory Report R-1255, 1988.

[8] McCrum, R. L., "Calcium Magnesium Acetate and Sodium-Chloride As Highway Deicing Salts - A Comparative-Study", *Materials Performance*, vol. 28, no. 12, pp. 24-28, 1989.

[9] Louis D.Johnston & Samuel H.Williamson. The Annual Real and Nominal GDP for the United States, 1790 - Present.. Economic History Services. 2005.

[10] JACC, *Survey of Corrosion Cost in Japan*, Japan Association of Corrosion Control, 1999.

[11] Dori, J. & Fisher, R. D. Jr., *U.S And Asian Statistical Handbook 1999-2000* The Heritage Foundation, Washington, 2007.

[12] Wallbank, E. J., *The Performance of Concrete in Bridges, A survey of 200 Highway Bridges*, Her Majesty's stationery office, London, 1989.

[13] CRC Press, *Handbook of chemistry and physics : a ready-reference book of chemical and physical data*, 81st edn, Boca Raton, Fla., London, 2000.

[14] Mangold, T., *Road Salt Use for Winter Maintenance.A Review of Impacts, Alternatives, and Recommendations for the St. Paul Campus Stormwater Management Plan*, 2004.

[15] Mauritis, M., Mcgraw, J., & Jang, J. W., "Method for Determining Ice Undercut Temperature of Deicing Chemicals", *Journal of Materials in Civil Engineering*, vol. 7, no. 1, pp. 84-86, 1995.

[16] Bensted, J., "Chemical effects of cement mortar of calcium magnesium acetate as a deicing salt - Discussion", *Cement and Concrete Research*, vol. 26, no. 4, pp. 633-635, 1996.

[17] Fujii T. "Combined effect of acid and deicer solutions on fiber-reinforced hardened cement pastes", in *2nd International Conference on Concrete under Severe Conditions*, Sakai K., Banthia N., & Gjorv O.E., eds., E & FN Spon.

[18] Peterson, O., "Chemical Effects on Cement Mortar of Calcium-Magnesium Acetate As A Deicing Salt", *Cement and Concrete Research*, vol. 25, no. 3, pp. 617-626, 1995.

[19] Peterson, O., "Chemical effects of cement mortar of calcium magnesium acetate as a deicing salt - Reply", *Cement and Concrete Research*, vol. 26, no. 4, pp. 637-639.

[20] Santagata, M. C. & Collepardi, M. 2000, "The effect of CMA deicers on concrete properties", *Cement and Concrete Research*, vol. 30, no. 9, pp. 1389-1394, 1996.

[21] Locke, C. E., Kennelley, K. J., Boren, M. D., & Luster, V., *Study of corrosion properties of a new deicer, Calcium Magnesium Acetate*, 1987.

[22] Fischel, M., *Evaluation of selected deicers based on a review of the literature*, Colorado Department Of Transportation, CDOT-DTD-R-2001-15, 2001.

[23] Katsanos A. & Nwaubani S.O. "Viability of Alternative de-icing operations on European highway bridges", Global Construction: Ultimate Concrete Opportunities, ed..

[24] Chollar, B. H. & Virmani, Y. P., "Effects of calcium magnesium acetate on reinforced concrete steel", *Public Roads*, vol. 51, no. 4, pp. 113-115, 1988.

[25] Harrach, N. & Wyatt, J., "Fine tuning CMA for corrosion control", *Public Works*, vol. 121, no. 8, pp. 40-41, 1990.

[26] [Harris, G. P., Turner, R., & Nelson, R. J., "Field-Test Comparison of Calcium-Magnesium Acetate to Salt", *Journal of Transportation Engineering-Asce*, vol. 119, no. 6, pp. 889-904, 1993

[27] Miller, W. L. CMA Cuts Corrosion on the Zilwaukee Bridge. Better Roads [63], 18-19. 1-2-1993.

[28] Miller, W. L., "CMA Use on the Zilwaukee Bridge", *Chemical Deicers and the Environment* pp. 553-557, 1992.

[29] The City of Aspen. Water Quality. www.aspenpitkin.com/depts/44/waterquality.cfm. 2002b.

[30] The City of Aspen. Air Readings. www.aspenpitkin.com/depts/44/air_readings.cfm. 2002a. Ref Type: Electronic Citation

[31]]Schenk, R., *Ice Melting Characteristics of Calcium Magnesium Acetate*, FHWA, Washington D.C., FHWA/RD-86/005, 1985.

[32] Ushirode, W. M., Hinatsu, J. T., & Foulkes, F. R., "Voltammetric Behavior of Iron in Cement, Effect of Acetate and Urea Additions", *Journal of Applied Electrochemistry*, vol. 22, no. 3, pp. 224-229, 1992.

[33] .Kennelley, K. J. & Locke, C. E., "Electrochemical-Behavior of Steel in Calcium Magnesium Acetate", *Corrosion*, vol. 46, no. 11, pp. 888-895, 1990.

[34] Slick, D. S., *Effects of calcium magnesium acetate on pavements and motor vehicles*, 1988.

[35] Lee, Hyomin, Robert D. Cody, Anita M. Cody, and Paul G. Spry. "Effects of various deicing chemicals on pavement concrete deterioration." In Center for Transportation Research and Education (Ed.), Mid-Continent Transportation Symposium Proceedings, Iowa State University, Ames, USA, pp. 151-155. 2000.

[36] Amrhein, C., Mosher, P. A., Strong, J. E., & Pacheco, P. G., "Heavy-Metals in the Environment - Trace-Metal Solubility in Soils and Waters Receiving Deicing Salts", *Journal of Environmental Quality*, vol. 23, no. 2, pp. 219-227, 1994.

http://dx.doi.org/10.5772/57551

[37] Amrhein, C., Strong, J. E., & Mosher, P. A., "Effect of Deicing Salts on Metal and Organic-Matter Mobilization in Roadside Soils", *Environmental Science & Technology*, vol. 26, no. 4, pp. 703-709, 1992.

[38] Horner, R. R. & Brenner, M. V., "Environmental Evaluation of Calcium Magnesium Acetate for Highway Deicing Applications", *Resources Conservation and Recycling*, vol. 7, no. 1-3, pp. 213-237, 1992.

[39] Moran, V. M., Abron, L. A., & Weinberger, L. W., "A Comparison of Conventional and Alternative Deicers - An Environmental-Impact Perspective", *Chemical Deicers and the Environment* pp. 341-361, 1992.

[40] Rabideau Alan J.. & Matsumoto Mark R. 1987, "Impact of Calcium Magnesium Acetate Road Deicer on POWT Operation", *Journal of Water Resources Planning and Management*, vol. 113, no. 2, pp. 311-315.

[41] Pullar-Strecker, P., *Concrete Reinforcement Corrosion, from assessment to repair decisions*, 1st edn, Thomas Telford Ltd, London, 2002.

[42] Bjegovic, D., Mikulic D, & Sekulic D, "Non-destructive corrosion rate monitoring for reinforced concrete structures", in *15th World Conference on Non-Destructive Testing*, WCNDT, 200.

Chapter 25

Corrosion Detection for Automated Visual Inspection

Francisco Bonnin-Pascual and Alberto Ortiz

Additional information is available at the end of the chapter

http://dx.doi.org/10.5772/57209

1. Introduction

Vessels constitute one of the most cost effective forms of transporting bulk goods around the world. However, despite the efforts on reducing maritime accidents, they still occur and, from time to time, have catastrophic consequences both in personal, environmental and financial terms. Structural failure is the major cause of ships wreckages and, as such, Classification Societies impose extensive inspection schemes for assessing the structural integrity of vessels.

The external and internal parts of the hull can be affected by different kinds of defects typical of steel surfaces and structures, being corrosion of paramount importance. Nowadays, to detect these defects, visual hull inspections are carried out at a great cost [1]: the vessel has to be emptied and situated in a dockyard, where typically temporary staging, lifts, and/or movable platforms need to be installed to allow the workers for close-up inspection (and repair if needed) of all the different metallic surfaces and structures. Taking into account the huge dimensions of some vessels, this process can mean the visual assessment of more than 600,000 m^2 of steel. Besides, the surveys are on many occasions performed in hazardous environments for which the access is usually difficult, while the operational conditions turn out to be sometimes extreme for human operation. For large tonnage vessels, such as Ultra Large Crude Carriers (ULCC), total expenses can be as high as one million euros.

Corrosion is a clear indicator of the state of the hull metallic structures, and, thus, it is of great interest for the surveyor [2]. Different kinds of corrosion may arise: *general corrosion*, that appears as non-protective friable rust which can occur uniformly on uncoated surfaces; *pitting*, a localized process that is normally initiated due to local breakdown of coating and that derives, through corrosive attack, in deep and relatively small diameter pits that can in turn lead to hull penetration in isolated random places; *grooving*, again a localized process, but this time characterized by linear shaped corrosion which occurs at structural intersections where water collects and flows; and *weld metal corrosion*, which affects the weld deposits, mostly due to galvanic action with the base metal, and are likelier in manual welds than in machine welds.

The goal of the EU-funded FP7 MINOAS project [3] is to reengineer the inspection process through the incorporation of robotic technologies. Some of these robots are equipped with cameras which can provide images of the different areas of the vessel hull to be inspected [1]. This chapter revises related work in visual general defect detection, including corrosion, and then describes two pattern recognition approaches developed within the context of the MINOAS project to detect corrosion from digital images.

2. Related work

Talking about automated visual defect detection, the scientific literature contains an important number of proposals. These can be classified depending on the kind of surface in which they are looking for defects: some approaches face the defect detection on particular objects or surfaces (e.g. LCD displays [4], printed circuit boards [5], copper strips surfaces [6], ceramic tiles [7], etc) while other methods are intended for the detection of defects in generic surfaces (e.g. [8–11]).

A second classification can be established depending on the kind of defect that they try to detect. In this regard, many approaches intended for the inspection of generic surfaces look for general and unspecific defects, although there is an important amount of contributions that are dedicated to a specific type of defect. This is the case of [12], [13] and [14], which are part of a large collection of contributions to automated visual crack detection.

Regarding corrosion detection from images, just a few works can be found in the literature (see e.g. [15–18]). This reduced amount of contributions indicates that there is still much work to do. By way of example, the following sections introduce two novel algorithms for corrosion detection and assess their performance against a varied set of test images.

3. Weak-classifier Colour-based Corrosion Detector (WCCD)

3.1. Description of the algorithm

WCCD is a supervised classifier which has been built around a cascade scheme, although its two stages can be considered as weak classifiers. The idea is to chain different fast classifiers with poor performance in order to obtain a global classifier attaining a much better global performance. To this end, each weak classifier takes profit from different features of the items to classify, reducing the number of false positive detections at each stage. For a good global performance, the classifiers must present a false negative percentage close to zero.

The first stage of the cascade is based on the premise that a corroded area presents a rough texture. Roughness is then related to the energy of the symmetric *gray-level co-occurrence-matrix* (GLCM), calculated for downsampled intensity values between 0 and 31, for a given direction α and distance d [19]. The energy is obtained by means of Equation 1:

$$E = \sum_{i=0}^{31} \sum_{j=0}^{31} p(i,j)^2 , \tag{1}$$

http://dx.doi.org/10.5772/57209

where $p(i, j)$ is the probability of the occurrence of gray levels i and j at distance d and orientations α or $\alpha + \pi$. Patches with an energy lower than a given threshold τ_E, i.e. exhibit a rough texture, are finally candidates to be more deeply inspected.

The second stage filters the pixels of the patches that have passed the roughness stage. This stage makes use of the colour information that can be observed from corroded areas. More precisely, the classifier works over the Hue-Saturation-Value (HSV) space after the realization that HSV-values that can be observed in corroded areas are confined in a bounded subspace of the HS plane. Although the V component has been observed neither significant nor necessary to describe the color of corrosion, it is used to prevent the well-known instabilities in the computation of hue and saturation when color is close to white or black. In that case, the pixel is classified as non-corroded.

A training step is performed prior to the application of this second stage of the corrosion classifier. In this case, training consists of building a bi-dimensional histogram of HS values for image pixels known to be affected by corrosion in the training image set. The resulting histogram is subsequently filtered by zeroing entries whose value is below 10% the highest peak.

The classifier proceeds as follows for every 3-tuple (h, s, v):

1 pixels close to black, $v < mV$, or white, $v > MV \wedge s < mS$, are labeled as non-corroded, and

2 for the remaining pixels, the HS histogram is consulted and the pixel is labelled as corroded if $HS(h, s) > 0$,

for given thresholds mV, MV and mS.

Notice that the stages of the cascade cannot be reversed since they do not work with the same kind of entities: while the second stage works at the pixel level, the first stage operates over 15×15 -pixel image patches since it depends on texture, which necessarily involves a pixel neighborhood. Figure 1 shows the flow diagram of WCCD.

3.2. Performance of WCCD

The performance of WCCD depends on the performance of its different stages. To configure the parameters of the roughness stage, several experiments have been performed considering different values for d and α when calculating the GLCM and, consequently, its energy level. No significant differences have been observed among the output values, and so the parameter values are set to $d = 5$ (pixels) and $\alpha = 0$ (horizontal direction). As for the energy threshold τ_E, its value determines the algorithm performance in terms of computation time and number of false positives, since all patches with a high energy level are discarded and only those with a low value become input of the colour checking step.

The parameters of the colour-based step, mV, MV and mS, are set to prevent the instabilities of h and s values from affecting the pixel classification. Using 8-bit HSV values, the *minimum value* mV is set to 50, as well as the *minimum saturation* mS. The *maximum value* MV is set to 200.

Figure 2 provides classification outputs for the same input image using different energy thresholds τ_E. This parameter can be tuned to decrease false positives and just allow the

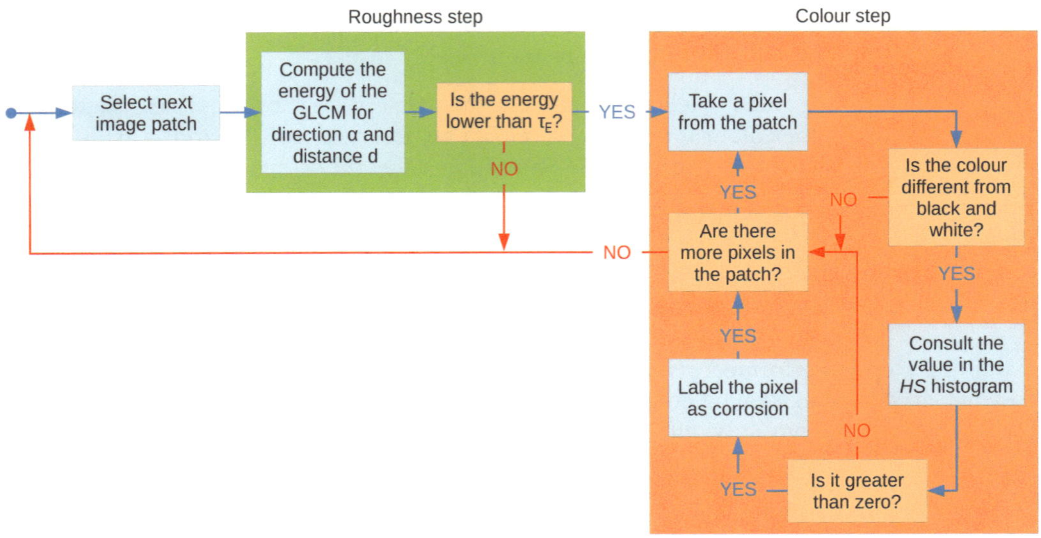

Figure 1. WCCD algorithm flowchart

detection of the most significant corroded areas. In the images, pixels labelled as corroded are colour-coded to indicate the probability of successful classification. To be more precise, the colour depends on the height of the corresponding histogram bin in the following way:

- red if $HS(h,s) \in [0.75\mathcal{HS}, 1.00\mathcal{HS}]$,
- orange if $HS(h,s) \in [0.50\mathcal{HS}, 0.75\mathcal{HS}]$,
- green if $HS(h,s) \in [0.25\mathcal{HS}, 0.50\mathcal{HS}]$ and
- blue if $HS(h,s) \in [0.10\mathcal{HS}, 0.25\mathcal{HS}]$,

where $\mathcal{HS} = \max\{HS(\cdot,\cdot)\}$.

The performance of the detector has been quantified by means of a varied set of test images and manually generated ground truth data (after the proper configuration of the different parameters, as explained above). False positive and false negative percentages, respectively *FP/no. pixels* and *FN/no. pixels*, have been used as the figures of merit. The process carried out to perform this assessment has entailed the implementation of different techniques to filter the *HS* histogram and a posterior comparison among the alternatives with regard to the generalization capability of the resulting classifier.

Downsampling the histogram to 32 levels for hue and saturation has been the first filter considered, which merely groups bins with similar hue-saturation values. As can be seen in the first and second rows of Table 1, this filter has resulted in a considerable improvement in comparison with the original 256 × 256 *HS* histogram, thus it has been considered as the reference for comparing with the other filtering strategies.

More specifically, two more attempts have been performed in order to reduce the false negative percentage while preserving the false positive percentage. On the one hand, the

http://dx.doi.org/10.5772/57209

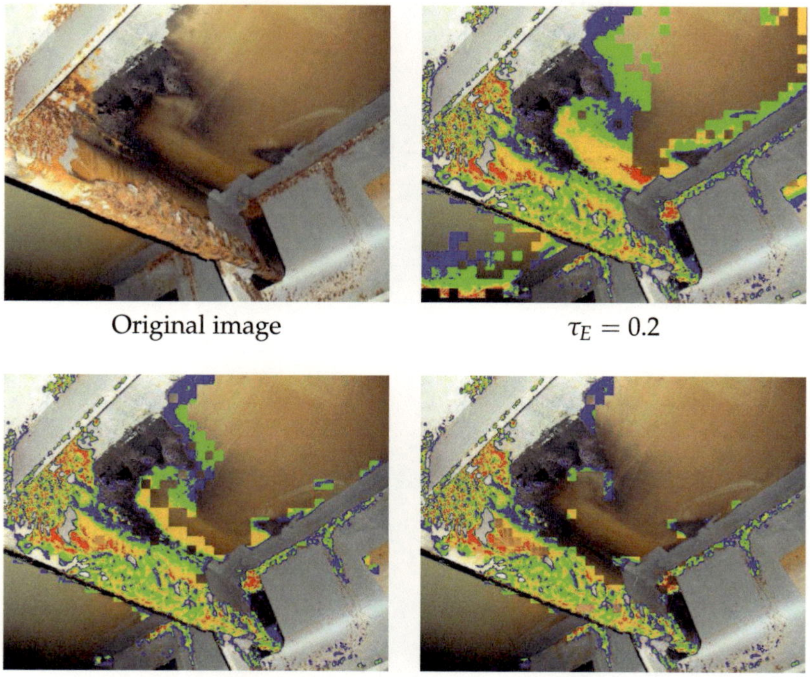

Figure 2. Corroded areas detected by WCCD for different energy threshold values

Parzen windows method [19] has been applied to the original 256 × 256 histogram using the two-dimensional Gaussian kernel shown in Equation 2:

$$G(x,y,\sigma) = \frac{1}{2\pi\,\sigma^2}\, e^{-\frac{1}{2}\left(\frac{(x-\mu_x)^2}{\sigma^2} + \frac{(y-\mu_y)^2}{\sigma^2}\right)}, \tag{2}$$

where μ_x and μ_y are the hue and saturation values for the neighbourhood center, x and y are the values for a nearby sample, and σ is the standard deviation. The algorithm performance has been assessed filtering the histogram using different values for σ. Best misclassification rates and percentages are shown in the third row of Table 1, which correspond to $\sigma = 12$.

On the other hand, a Bilateral filter [20] has been applied to the original 256 × 256 histogram, considering the bins height as the intensity values of an image. This approach filters the histogram using a kernel consisting of two Gaussians, one for the spatial domain and another for the range domain. After the different experiments carried out, the best performance has been obtained for $\sigma_{spatial} = 15$, $\sigma_{range} = 1$ and a kernel size of 30 pixels. The fourth row of Table 1 provides the resulting performance values. The resulting histograms for the different filters can be found in Figure 3. By way of conclusion, it seems the approach based on the bilateral filter is the one providing best results, although it is true the different strategies lead to a final similar performance. The bilateral filter has thus been selected for being part of WCCD.

	FP percentage	FN percentage
Original (256 bins)	0.80	13.56
Downsampled to 32 bins	9.78	6.10
Using Parzen windows	9.23	5.99
Using a Bilateral filter	9.80	5.86

Table 1. Misclassification measures for different HS histograms (WCCD)

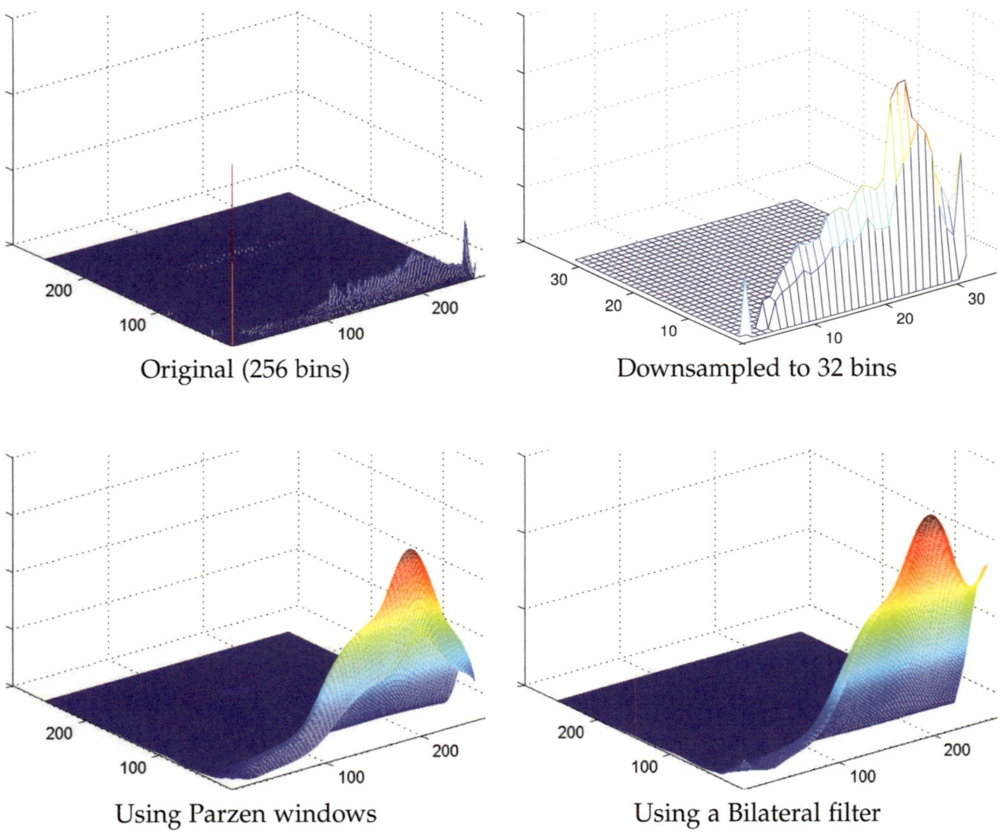

Figure 3. HS histograms resulting from the different filtering strategies (WCCD)

http://dx.doi.org/10.5772/57209

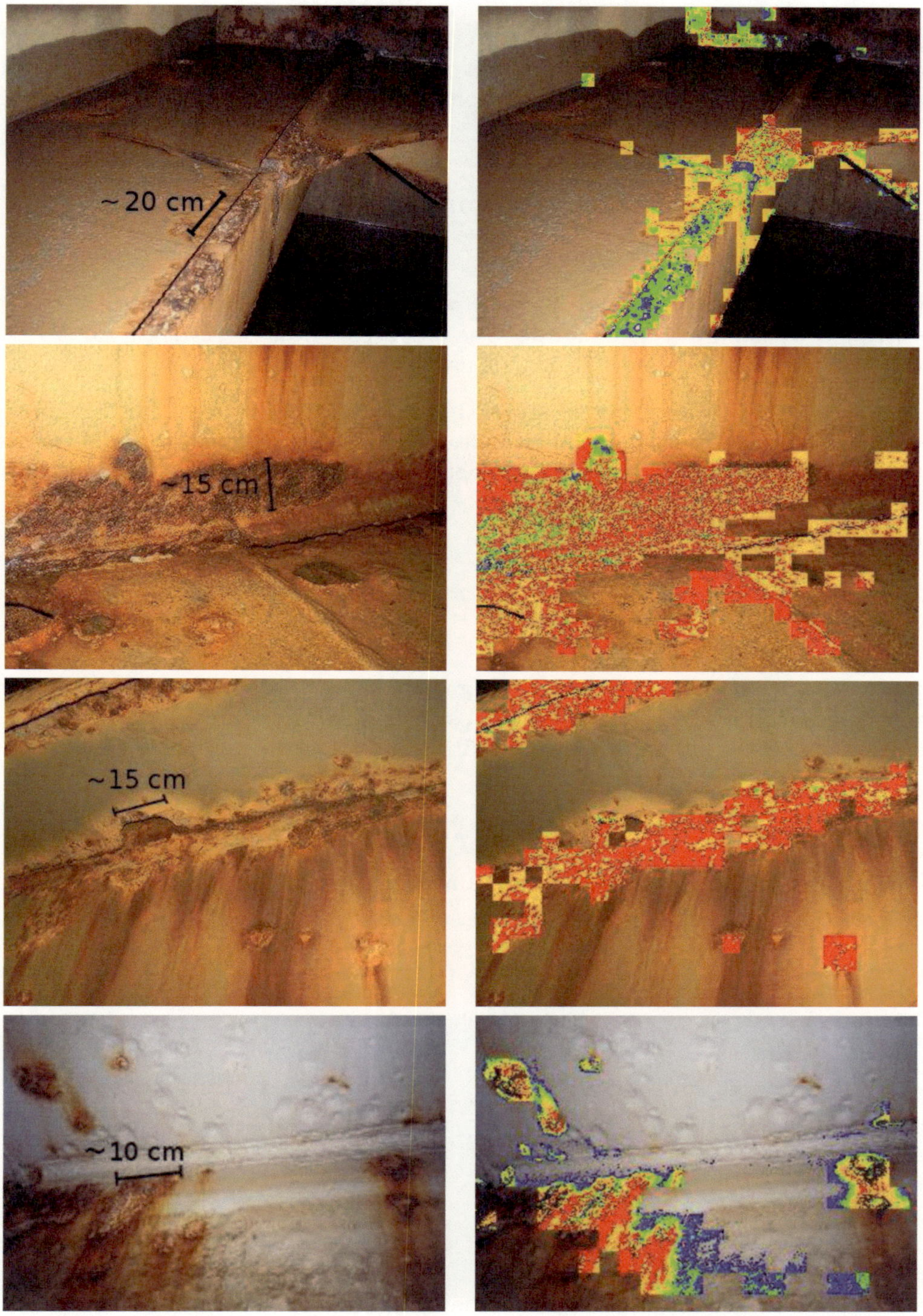

Figure 4. (1st column) Test images and (2nd column) corroded areas detected by WCCD

Examples of final classification outputs for WCCD are provided in Figure 4.

Regarding the execution times, WCCD provides corrosion-labelled images in 7-25 ms. These execution times correspond to images ranging from 120.000 to 172.000 pixels, and for a runtime environment comprising a laptop fitted with an Intel Core2 Duo processor (@2.20GHz, 4GB RAM).

4. AdaBoost based Corrosion Detector (ABCD)

4.1. Description of the algorithm

ABCD makes use of the *Adaptive Boosting* paradigm (AdaBoost) for both learning and classifying corroded areas. Decision trees, as produced by the *Classification and Regression Trees* (CART) learning technique [21, 22], are used as weak classifiers.

Laws' texture energy filter responses are used to feed the decision trees. This is so because these filters are able to enhance different features of every material texture. We use a filter bank with 48 different filters that are obtained after combining the following five 1D five-component basic filters:

level	L5 = [1 4 6 4 1]	edge	E5 = [-1 -2 0 2 1]
spot	S5 = [-1 0 2 0 -1]	wave	W5 = [-1 2 0 -2 1]
ripple	R5 = [1 -4 6 -4 1]		

To describe a texture, the corresponding gray-level patch is convolved with the set of energy filters ($T \otimes filter \rightarrow c$) and different statistical measures are taken over a 15×15 neighbourhood of the filter response, which finally constitute the texture descriptor:

$$\mu = \frac{\sum_{u=0,v=0}^{N,M} \mid c(u,v) \mid}{NM} \tag{3}$$

$$\sigma = \sqrt{\sum_{u=0,v=0}^{N,M} \frac{(c(u,v) - \mu)^2}{NM}} \tag{4}$$

$$\mu^+ = \frac{\sum_{u,v|c(u,v)>0}^{N,M} c(u,v)}{NM} \tag{5}$$

$$\mu^- = \frac{\sum_{u,v|c(u,v)<0}^{N,M} c(u,v)}{NM} \tag{6}$$

During the learning process, AdaBoost is fed with 192 statistical measures per image patch (4 statistical measures $\times$ 48 energy filters), together with the roughness of the patch (computed

http://dx.doi.org/10.5772/57209

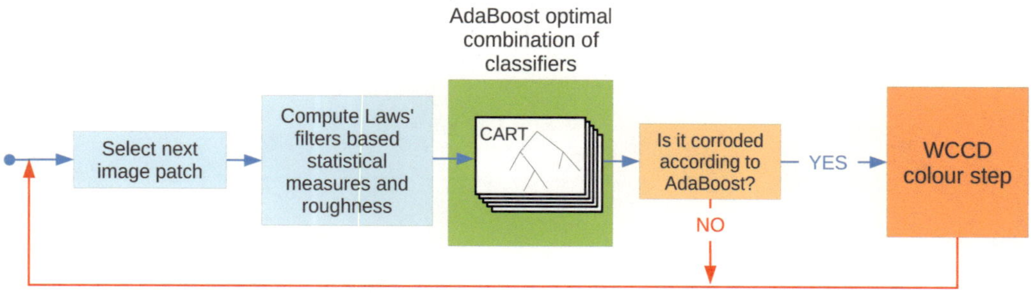

Figure 5. ABCD algorithm flowchart

as in the first stage of the WCCD detector) and its class label. The output is a set of weak classifiers together with their weights, what allows a correct discrimination of those patches which belong to defective areas from those which does not.

In addition, the same technique used in the second stage of the WCCD algorithm, based on a Hue-Saturation histogram, is also used in ABCD to filter the pixels from the patches that are classified as corroded by AdaBoost. The flow diagram for ABCD is shown in Figure 5.

4.2. Performance of ABCD

We have considered three versions of AdaBoost: *Real*, *Gentle* and *Modest*. *Real AdaBoost* is the generalization of a basic AdaBoost algorithm. *Gentle AdaBoost* [23] is a more robust and stable version of Real AdaBoost, used, for example, by the Viola-Jones object detector. Finally, *Modest AdaBoost* [24] is a version mostly aimed for better resistance to overfitting.

For training the algorithm, a total number of 39746 patches have been gathered from 25 different images. These patches have been labelled as *defective* (12952 patches) or *non-defective* (26794) by means of visual inspection. One half of the total amount of patches has been used to train the different versions of AdaBoost, while the other half has been used as control samples to assess the performance of the resulting classifiers.

The AdaBoost parameters have been configured as follows:

1. the maximum number of boosting iterations, i.e. the number of weak classifiers that make up the final classifier, has been set to 100.
2. the tree depth, that is, the depth of the CART, which determines how good the weak classifiers are, has been set to three levels.

Both parameters have been configured to improve the detection performance without prolonging the learning time unnecessarily.

After the execution of the three versions of AdaBoost, their performances have been assessed. Table 2 shows the error percentages obtained for the three versions.

As can be seen, best results are obtained for the Gentle version of AdaBoost. Some results obtained for this version, after adding the colour-based filter, are shown in Figure 6. Among

Real	Gentle	Modest
6.05	5.99	9.52

Table 2. Error percentage obtained for the different AdaBoost versions

FP %	FN %
17.16	3.39

Table 3. Misclassification measures for ABCD

the different images shown, special mention is done for the image in the first row, where corrosion in form of *pitting*, affecting a very reduced fraction of the image, is successfully detected.

The final performance of the complete algorithm has been analysed following the same procedure used for assessing the performance of WCCD. The false positive and false negative percentages have been again used as the figures of merit. The values obtained are shown in Table 3.

Regarding the execution times, ABCD provides corrosion-labelled images in 300-512 ms. These execution times correspond to the same images and runtime environment used for assessing WCCD.

5. Conclusions

In this chapter we have introduced two vision-based corrosion detection algorithms that have been developed within the context of the European project MINOAS. Both algorithms are based on the idea of combining weak classifiers for obtaining a good global performance.

After assessing their performance, the misclassification percentages obtained for both algorithms result to be not null. These results can be explained by analysing the kind of misclassifications and the areas where they appear. On the one hand, the FN percentages are not zero because the detectors tend to label as corrosion the center of the corroded area, while the borders are usually not totally labelled. On the other hand, the FP percentages are neither null due to the presence of different structures in the image that are misclassified as defects.

Nevertheless, if corroded areas are considered as entities and it is assumed that the labelling of a single pixel within a defective area is useful, then the ratio between the number of undetected defective areas and the true number of defective areas turns out to be zero for both algorithms, since all them are always detected. In this regard, it is important to remember that the detectors are intended to be used to facilitate visual inspections, and, thus, reporting about the existence of corroded areas in images is considered worth enough even if the areas are not completely labelled.

Taking into account the misclassification ratios, it is not clear whether WCCD performs better or worse than ABCD. However, based on the shorter execution times and the qualitative evaluation of the results, it seems that WCCD outperforms ABCD.

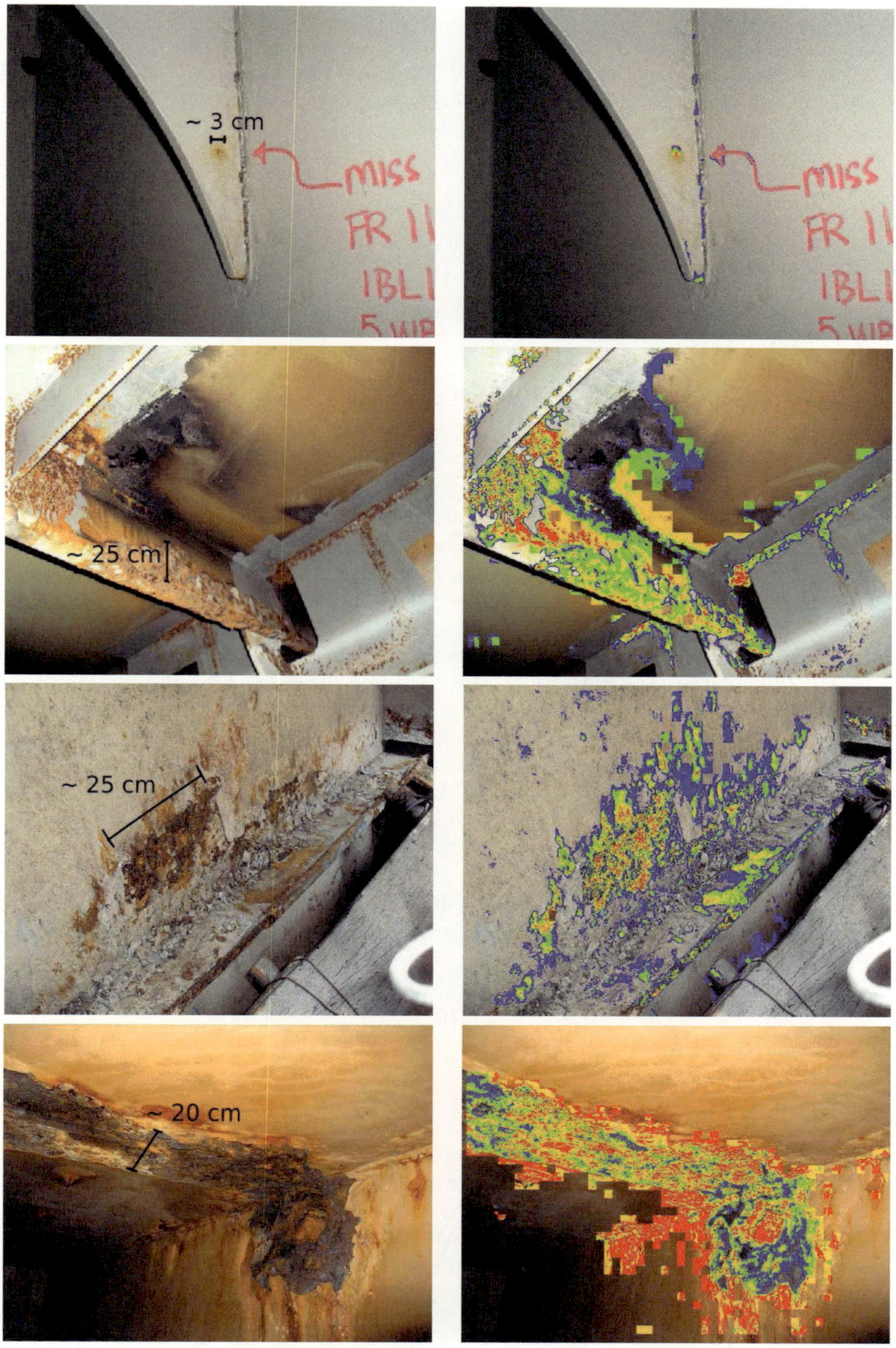

Figure 6. (1st column) Test images and (2nd column) corroded areas detected by ABCD

Acknowledgements

This work has been partially supported by FP7 projects MINOAS (GA 233715) and INCASS (GA 605200) and by the European Social Found through grant FPI10-43175042V (Conselleria d'Educacio, Cultura i Universitats, Govern de les Illes Balears).

Author details

Francisco Bonnin-Pascual*, Alberto Ortiz

Department of Mathematics and Computer Science, University of Balearic Islands, Palma de Mallorca, Spain

*Address all correspondence to xisco.bonnin@uib.es

References

[1] A. Ortiz, F. Bonnin-Pascual, A. Gibbins, P. Apostolopoulou, W. Bateman, M. Eich, F. Spadoni, M. Caccia, and L. Drikos. First steps towards a roboticized visual inspection system for vessels. In *Proc. of the IEEE Emerging Technologies and Factory Automation conference*, 2010.

[2] MINOAS. D1 - definition of the inspection plan / definition of acceptance criteria. In *MINOAS Public Deliverable*, 2011. http://minoasproject.eu/excibition/Publications/PDF/D1.pdf.

[3] M. Caccia, R. Robino, W. Bateman, M. Eich, A. Ortiz, L. Drikos, A. Todorova, I. Gaviotis, F. Spadoni, and V. Apostolopoulou. Minoas a marine inspection robotic assistant: system requirements and design. In *Proceedings of the IFAC Symposium on Intelligent Autonomous Vehicles*, 9 2010.

[4] C.-P. Hsu C.-L. Chang, H.-H. Chang. An intelligent defect inspection technique for color filter. In *Proceedings of the IEEE International Conference on Mechatronics*, pages 933–936, 2005.

[5] B. C. Jiang, C. C. Wang, and P. L. Chen. Logistic regression tree applied to classify pcb golden finger defects. *The International Journal of Advanced Manufacturing Technology*, 24(7-8):496–502, 2004.

[6] X. Zhang, R. Liang, Y. Ding, J. Chen, D. Duan, and G. Zong. The system of copper strips surface defects inspection based on intelligent fusion. In *Proceedings of the IEEE International Conference on Automation and Logistics*, pages 476–480, 2008.

[7] C. Boukouvalas, J. Kittler, R. Marik, M. Mirmehdi, and M. Petrou. Ceramic tile inspection for colour and structural defects. *Technical Report CS-EXT-1995-052*, 1995. Also: Proceedings of AMPT95, pp. 390-399, 1995.

[8] T. Amano. Correlation based image defect detection. In *Proceedings of the IAPR International Conference on Pattern Recognition*, pages 163–166, 2006.

[9] H. Castilho, J. Pinto, and A. Limas. An automated defect detection based on optimized thresholding. In *Proceedings of the International Conference on Image Analysis and Recognition*, volume 4142 of *Lecture Notes in Computer Science*, pages II:790–801, 2006.

[10] J. Hongbin, Y. Murphey, S. Jinajun, and C. Tzyy-Shuh. An intelligent real-time vision system for surface defect detection. In *Proceedings of the IAPR International Conference on Pattern Recognition*, pages III: 239–242, 2004.

[11] A. Kumar and H. Shen. Texture inspection for defects using neural networks and support vector machines. In *Proceedings of the IEEE International Conference on Image Processing*, pages III:353–356, 2002.

[12] Y. Fujita, Y. Mitani, and Y. Hamamoto. A method for crack detection on a concrete structure. In *Proceedings of the IAPR International Conference on Pattern Recognition*, pages III: 901–904, 2006.

[13] R. Oullette, M. Browne, and K. Hirasawa. Genetic algorithm optimization of a convolutional neural network for autonomous crack detection. In *Proceedings of the IEEE Congress on Evolutionary Computation*, pages I:516 – 521, 2004.

[14] T. Yamaguchi and S. Hashimoto. Fast crack detection method for large-size concrete surface images using percolation-based image processing. *Machine Vision and Applications*, 21(5):797–809, 2010.

[15] M.R. Jahanshahi and S.F. Masri. Effect of color space, color channels, and sub-image block size on the performance of wavelet-based texture analysis algorithms: An application to corrosion detection on steel structures. In *ASCE International Workshop on Computing in Civil Engineering*, 2013.

[16] G. Ji, Y. Zhu, and Y. Zhang. The corroded defect rating system of coating material based on computer vision. In *Transactions on Edutainment VIII*, volume 7220 of *Lecture Notes in Computer Science*, pages 210–220. Springer Berlin Heidelberg, 2012.

[17] M. Siegel and P. Gunatilake. Robotic enhanced visual inspection of aircraft skin, 1999.

[18] B. B. Zaidan, A. A. Zaidan, H. O. Alanazi, and R. Alnaqeib. Towards corrosion detection system. *International Journal of Computer Science Issues*, 7(1):33–35, 2010.

[19] S. Theodoridis and K. Koutroumbas. *Pattern Recognition, 3rd Edition*. Academic Press, 2006.

[20] C. Tomasi and R. Manduchi. Bilateral filtering for gray and color images. In *Proceedings of the IEEE International Conference on Computer Vision*, pages 839 – 846, 1998.

[21] L. Breiman, J. Friedman, R. Olshen, and C. Stone. *Classification and Regression Trees*. Wadsworth International, 1984.

[22] R.O. Duda, P.E. Hart, and D.G. Stork. *Pattern Classification*. Wiley, 2001.

[23] J. Friedman, T. Hastie, and R. Tibshirani. Additive logistic regression: A statistical view of boosting. *The Annals of Statistics*, 38(2):337 – 374, 2000.

[24] A. Vezhnevets and V. Vezhnevets. Modest adaboost-teaching adaboost to generalize better. Technical report, Graphicon, 2005.

Chapter 26

Corrosion of Biomaterials Used in Dental Reconstruction Dentistry

I. Patrascu, E. Vasilescu, E. Gatin and R.R. Cara-Ilici

Additional information is available at the end of the chapter

http://dx.doi.org/10.5772/57322

1. Introduction

Biocompatibility is a complex concept that takes into account all the processes occurring in the interaction between the biomaterial and a living organism. The biocompatibility means the property of a material to be compatible with living organism, that to be accepted a definitive manner by body without causing side effects and without chemical or mechanical damage [11, 47, 49].

Corrosion of biomaterials used in dentistry is the process of altering or destroying such materials in interaction with the oral environment [10]. Generally, all dental materials are subjected to the aggressiveness in the oral environment, in a certain period of time, longer or shorter, they shall chemically degrade. The term degradation of biomaterials in a biological environment combines metallic biomaterials corrosion or damage of ceramic and polymeric biomaterials with the host tissue reaction [11].

Oral environment is considered a highly chemical aggressive environment, characterized by frequent and important pH modifications due to various types of food or microbial flora. In this environment, dental materials can be dissolved in water or saliva or they can release constituents by the diffusion processes, they can be eroded in the presence of acids, they can change colour, or corrode.

Metallic biomaterials are a class of materials recommended for dental applications due to their very good mechanical properties and an acceptable biocompatibility. Metals and alloys commonly used as biomaterials are gold (Au), cobalt-chrome alloys (CoCr), austenitic stainless steel (316L), titanium and titanium alloys (TiNi, Ti-6Al-4V) and silver-mercury alloys (AgHg). Pure metals are seldom used, their alloys being mostly used due to the fact that by alloying, they enhance certain properties such as corrosion resistance and hardness. (e.g. pure gold, although biologically inert, it has poor mechanical properties, on the other hand, steels, which have excellent mechanical properties do not show a good corrosion resistance).

Titanium is a material reactive in water, air, or any other electrolyte where it covers spontaneously with a titanium oxide layer. It is considered an inert material, as in contact with the tissue, it is rapidly inactivated forming on its surface a thin, hard, and protective layer of oxide, in less than one second [9, 10, 11, 47]. The titanium oxide film formed spontaneously is continuously regenerating and it provides an in-depth protection for the metal towards chemical attack, including the aggressive attack produced by the liquids of the body. Titanium is still considered the ideal material in endosseous dental implant. It does not produce any magnetic effect, it does not produce any magnetic field to disturb the activity around the cells; oxides from the surface of the implant being very adherent and insoluble, and they prevent the release and direct contact between the potentially harmful metal ions and tissues (biological compatibility). Titanium does not produce organic-metallic compounds, which are toxic, or if they are produced, such organic-metallic are unstable. Surface of oxides, consisting of TiO, TiO_3, Ti_2O_3, and Ti_3O_4 attracts and binds biomolecules (Kasemo 1983). The major disadvantage of this metal is the difficulty to cast it. Today it is obtained by dissociation in vacuum at 1400°C having a purity of 99.85 to 99.95%. Titanium alloys are better tolerated than pure titanium because the oxide layer that forms is higher (of approx. 10-20 μm) [9, 10]. Recent researches have demonstrated that the oxide layer (TiO) considered so stable regenerates every nanosecond, and re-oxidation is a major advantage due to minimizing the risk of biodegradation. It has been proven that next to resistance to corrosion, biological compatibility, resistance and price, the alloys used in medicine are "conversion" alloys based on titanium. Resistance to corrosion can be increased by alloying with molybdenum, zirconium, rhenium, niobium, chromium, manganese. Biomedical titanium alloys are: Ti-Al-V, Ti-Al-Mo, Ti-Al-Cr, Ti-Al-Cr-Co. Frequent use of titanium alloy Ti-6Al-4V for implants is determined by a combination of the most numerous and more favourable characteristics, which include resistance to corrosion, durability, low elasticity module and the ability to adhere with bone and other tissues (osseointegration) [13, 14, 18, 20, 36]. However, there are a number of issues related to the effects the components of the alloy can have. Aluminium and vanadium are elements released into the tissue. Therefore, a number of titanium alloys have investigated (Ti-Al-Nb, Ti-Zr-Al) and it was demonstrated that the Ti-6Al-6Nb alloy has properties comparable to the Ti-6Al-4V alloy, but it shows a greater strength and resistance to corrosion [53, 55]. Analysis of possible reactions to prolonged contact of living tissues with the alignment elements of titanium alloys showed that the use of titanium alloys containing large amounts of vanadium, cobalt and nickel is not recommended. On the other hand, introducing the alloying elements into the titanium alloys such as molybdenum, niobium, zirconium, and tantalum is not limited quantitatively. They increase the anticorrosive resistance, facilitate the increase of strength and they are compatible with living tissues. The content of aluminium and vanadium must not exceed 6%, and the content of Fe, Cr, Mn, and Ni is of 1% [23, 46,47,48].

Release of vanadium ions in the body can produce serious damages to the respiratory system and of the blood plaquettes producing systems, but it is a long process. However, it is taken into account to replace V with Nb. In vitro studies have showed that the cells behave differently

http://dx.doi.org/10.5772/57322

Element	Ti6Al4V Wrought	Ti5Al2.5V Wrought	Ti6Al7Nb Wrought
Aluminium	5.5-6.75	4.5-5.5	5.5-6.5
Vanadium	3.5-4.5	-	Max.0.5 tantalum
Iron	Max.0.3	2-3	Max.0.25
Niobium	-	-	6.5-7.5
Oxygen	Max.0.2	Max.0.2	Max.0.2
Carbon	Max.0.08	Max.0.08	Max.0.8
Nitrogen	Max.0.05	Max.0.05	Max.0.05
Hydrogen	Max.0.015	Max.0.015	Max.0.009
Titanium	balance	balance	balance

Table 1. Chemical composition of titanium based alloys as implants for surgery [49]

in the presence of clutches generated by the wear of the two alloys. There is an increased release of prostaglandin E2 in response to contact with Ti-6Al-4V particles, and an increase in the release of other inflammatory cytokines compared to Ti-Al-Nb particles. These data suggest that Ti-6Al-4V stimulates phagocytic cells more than the Ti-Al-Nb and pure Ti. Exposure of bone marrow cells to Ti-6Al-4V particles induce a significant increase and release of proinflammatory and osteolytic mediators which are responsible for loss of dentures. Stainless steel alloys were widely used in the past because they were cheap, easy to process and with very good mechanical properties. However, stainless steel is very susceptible to corrosion in salty environments, as the tissue fluid is. By corrosion steel becomes a metal with low resistance to fatigue, the main cause of implants failure. The released corrosion products also determined inflammatory side-effects [14, 31]. In the endosseous implants, the inflammation prevents osseointegration and favourites fibrous capsule formation. Stainless steels are steels that contain more than 12% Cr. Chromium plays a protective role in steels, this metal having a high tendency to passivation. In this case too, passivation tendency occurs discountinually, i.d., at 1/8, 2/8, 3/8 chromium percents. The spectacular growth of the potential for positive values occurs in 12.5% chromium atoms as shown in Figure 2.4.a. The percentage of chromium required to achieve stability depends on the immediate work environment. Thus, in a solution of 33% HNO_3 7% Cr solution is enough, and for $FeSO_4$ solution 20% Cr is required. The first type of stainless steel used for implants was vanadium steel (18-8V), but its resistance to corrosion was not so good. To increase its resistance to corrosion, molybdenum was added (18-8Mo), which later became 316 stainless steel. In the 1950s, the carbon content of the 316 stainless steel was reduced from 0.08% to 0.03% in order to increase resistance to corrosion. Today it is known as the *316L stainless steel* and it has the following chemical composition: 0.03% carbon, 2% magnesium, 17-20% chromium, 12-14% nickel, 2-4% molybdenum and other elements in smaller quantities such as phosphorus, sulphur and silicon. The passive layer (resistant to corrosion) of these alloys is not as strong as in the case of titanium alloys. For this reason, stainless steels are used only for temporary medical implants such as screws and orthopaedic rods for fixation of fractures. In dentistry it is used for bolts, for dental coverage, to produce the nets in the dentures.

The problems raised by these steels in use as biocompatible materials are related to the loss of Ni ions, in particular as a result of the corrosion process. These ions are tolerated by the body only in small amounts.

Another large group of alloys used in prosthetic reconstruction is the Co-Cr alloys or stellits. They are cobalt based alloys with chromium as the main element and the alloying elements: molybdenum, nickel, titanium, tungsten added to improve the properties. These elements have a complex action, some of them dissolving producing solid solution hardening, but most form intermediary compounds that increase resistance to corrosion and mechanical properties. Chromium increases resistance to corrosion and oxidation forming an oxide film (Cr_2O_3) on the surface, thus ensuring continuous adherence and protection. On the other hand, chromium forms complex carbides with a role in increasing the mechanical properties. Nickel forms with cobalt a series of solid solution. It has an influence on mechanical properties and resistance to corrosion. Molybdenum lowers the allotropic processing temperature and improves mechanical properties by the densification of the solid solution, due to the formation of c compounds of $MoCo_3$, $MoCo_7$. Molybdenum contributes to a fine structure, resulting from the process of casting and forging. Tungsten increases the resistance to oxidation and density due to the formation of compounds of WCo_3 and carbides, if the alloy contains carbon. Four standard types of alloys are standardized (after ASTM): F62- Co-Cr-Ni-Mo (forged), F63- Co-Ni-Cr-Mo (forged), F76 - Co-Cr-Mo (cast), F90- Co-Cr-W-Ni (forged). Higher fatigue resistance and breaking of the CoNiCrMo alloy make it highly suitable for applications requiring long life, without cracking or material fatigue. Regarding the resistance to corrosion of Co alloys, they show characteristics similar to the stainless steels, having the advantage of practically zero toxicity.

Alloy	Metal converted into compound, ng/m^{2}h	Metal found in tissue, ng/m^{2}h
Stainless steel – mechanically polished	7, 8	0.274
(AISI 316L) – chemically polished	230	-
Vitallium – mechanically polished	150	0.249
(CoCrW-Ni alloy) – chemically polished	20	-
Ti – mechanically polished	4, 1	0.430
– chemically polished	3, 5	-

Table 2. Corrosion rates of biomaterials in Hank's solution [49]

In conclusion, choosing a metal should be based on the corrosive properties. Metals used nowadays as biomaterials include gold, Co-Cr alloys, 316 stainless steel, titanium, Ni-Ti alloy and Ag-Hg amalgam. Noble metals are resistant to corrosion and they would be the ideal materials if resistance to corrosion were the only condition. Gold is often used in dental reconstruction offering high performance and longevity. Titanium is a metal which forms a strong passivizing layer, remaining passive under physiological conditions. Corrosion currents in normal saline conditions are very weak: 10^{-8} A/cm^2. Titanium implants apparently remain unchanged. Ti provides superior resistance to corrosion, but it not as hard and resistant as steel. Co-Cr alloys, as the Ti too, are passive in the human body. Stainless steels contain sufficient chromium to have resistance to corrosion by passivation. The passive layer is not as strong as in the case of Ti or Co-Cr alloy. The most resistant to corrosion among the stainless steels are austenitic steels and are symbolized: 316, 316L, 317 (AISI) and 10TiMoNiCr175, 2MoNiCr175 (STAS) containing Mo.

Dental amalgam is an alloy of Hg, Ag, and Sn. Although the stages are passive to neutral pH, the transpassive potential for the $\gamma 2$ phase is exaggerated due to the interphase galvanic couples or their cells due to different aeration in the denture. Therefore, the amalgam corrodes and it is often the most active corrosive material used in dentistry.

2. Assessment of metal and dental alloys corrosion

Depending on the state of the environment in which it occurs and of its appearance, corrosion can be: dry corrosion caused by contact of the metal with the oxygen in the atmosphere and humid or galvanic corrosion, which occurs if the metal is in a humid environment, by the occurrence of electrolytic cells. Corrosion is uniform if it occurs on the whole surface of the metal or localized corrosion if it occurs only in specific points on the surface of the metal, being more dangerous.

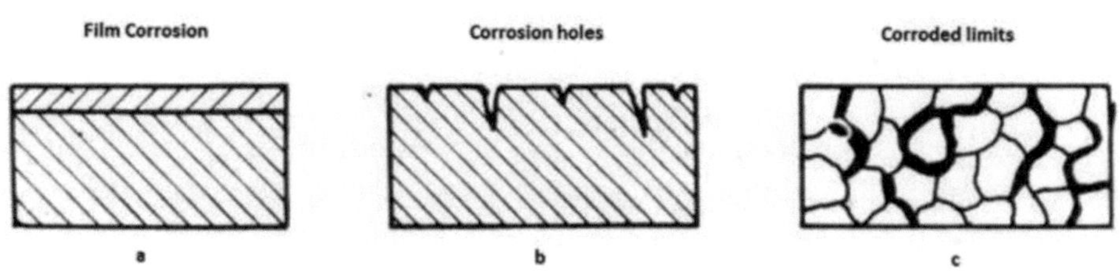

Figure 1. Types of corrosion: a– uniform; b– punctual; c– i ntercrystalline (at the crystal grains limit) [50, p.68]

The types of corrosion indicated are electrochemical corrosion, based on the formation of local galvanic elements at the contact of two metals with different electrode potential in the presence of an electrolyte. Electrochemical corrosion is the destruction of metals or alloys process in the presence of electrolyte solutions, by the electrochemical reactions that involve a transfer of ions and electrons under the influence of a difference of electric potential [10, 11, 37]. Metals release electrons by oxidation and its positive ions go into the solution. Formation of ions and electrons creates an electrical potential E (expressed in volts) to the material-solution interface called electrode potential which value depends on the nature of the metal on the one hand and on the other hand on the type of the solution. The electrode potential cannot be measured as such, but only the difference of potential at the terminals of a cell formed by a complete electrochemical chain. If the electrode inserted to form the chain is well-defined and stable, it is considered as a reference, making thus possible to compare different electrodes between them in relation to the reference adopted. The reference electrode is the hydrogen electrode, for which the oxidation process is described by the equation: $H_2 \rightarrow 2H^+ + 2e^-$.

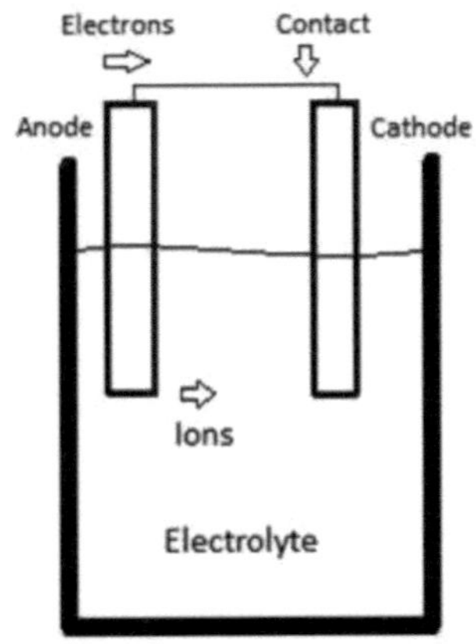

Figure 2. Electrochemical cell [49]

The electrolyte containing ions in solution which are also present in body fluids closes the electrical circuit. Chemical reactions that occur at each of the two electrodes of a galvanic cell (fig. 2) are called electrode reactions. Corrosion is a mixed process developing as follows:

- the anodic ionization reaction of metal and dissolution in the electrolyte of positively formed ions: $M \rightarrow M^{z+} + ze-$, $orM + zH_2O \rightarrow M(OH)_2 + zH^+ + ze-$
- takeover cathode reaction of electrons remained in the metal phase, by an existing electron acceptor existing in the solution; the agent is able to reduce itself has nobler potential equilibrium than the of the metal and is called depolarizing ($D + ze- \rightarrow Dze-$).

The tendency of metals to enter into the corrosion process is most simply expressed by standard electrochemical series of Nernst potentials (the table [47]). These potentials are obtained by electrochemical measurement in which an electrode is a standard hydrogen formed by a hydrogen bubble over a layer of platinum fine powder. The potential of this reference electrode is considered to be zero. Noble metals are those with a potential higher than the standard hydrogen electrode and base metals have lesser potential.

If two similar metals are present in the same environment, the one which is the most negative in the galvanic series becomes an anode and it shall corrode. The process is called bimetallic or galvanic corrosion and it can be much faster than the corrosion of a single metal.

Element	Electrode reaction	Electrode potential [V]
Lithium	$Li \rightarrow Li^{+} + e^{-}$	- 2.959
Rubidium	$Rb \rightarrow Rb^{+} + e^{-}$	- 2.925
Potassium	$K \rightarrow K^{+} + e^{-}$	- 2.294
Calcium	$Ca \rightarrow Ca^{2+} + 2e^{-}$	- 2.763
Sodium	$Na \rightarrow Na^{+} + e^{-}$	- 2.714
Magnesium	$Mg \rightarrow Mg^{2+} + 2e^{-}$	- 2.37
Beryllium	$Be \rightarrow Be^{2+} + 2e^{-}$	- 1.85
Aluminium	$Al \rightarrow Al^{3+} + 3e^{-}$	- 1.69
Titanium	$Ti \rightarrow Ti^{2+} + 2e^{-}$	- 1.63
Zinc	$Zn \rightarrow Zn^{2+} + 2e^{-}$	- 0.761
Chromium	$Cr \rightarrow Cr^{2+} + 2e^{-}$	- 0.71
Chromium	$Cr \rightarrow Cr^{3+} + 3e^{-}$	- 0.50
Iron	$Fe \rightarrow Fe^{2+} + 2e^{-}$	- 0.44
Cadmium	$Cd \rightarrow Cd^{2+} + 2e^{-}$	- 0.42
Nickel	$Ni \rightarrow Ni^{2+} + 2 e^{-}$	- 0.23
Tin	$Sn \rightarrow Sn^{2+} + 2e^{-}$	- 0.14
Tin	$Pb \rightarrow Pb^{2+} + 2e^{-}$	-0.13
Iron	$Fe \rightarrow Fe^{3+} + 3e^{-}$	- 0.045
Hydrogen	**$H_2(g) \rightarrow 1/2\ (H^{+} + e^{-})$**	**0.000(reference)**
Copper	$Cu \rightarrow Cu^{2+} + 2e^{-}$	+0.337
Oxygen	$O_2 + 2H_2O + 4e \rightarrow 4OH^{-}$	+0.401
Copper	$Cu \rightarrow Cu^{+} + e^{-}$	+0.522
Silver	$Ag \rightarrow Ag^{+} + e^{-}$	+0.797
Mercury	$Hg \rightarrow Hg^{2+} + 2e^{-}$	+0.798
Platinum	$Pt \rightarrow Pt^{2+} + 2e^{-}$	+1.20
Oxygen	$O_2 + 4H^{+} + 4e^{-} \rightarrow 2H_2O$	+1.229
Gold	$Au \rightarrow Au^{3+} + 3e^{-}$	+1.50

Table 3. Electrochemical series (normal electrode potential of hydrogen reduction) [47, p.92]

The potential difference E given by the concentration of metal ions in solution according to Nernst equation, is given by: $E = E_0 + (RT / nF) \ln [M^{n+}]$, where E_0 is the standard electrochemical potential, T is the absolute temperature, F is Farraday constant, (coulombs / mols), and n is the number of ions. The precedence of nobility observed in practice can be different as the one thermodynamically prescribed. Due to the phenomenon of passivation (metals cover with a passivation layer of reaction products, which protects the metal from further attack).

Localized corrosion can be, in its turn, crevicular or galvanic corrosion. Often, in the case of dry uniform corrosion, following the process of corrosion, a layer of corrosion products is formed on the surface of the metal, which are nothing else than the metal oxides. By thickening of the oxide layer, the metal surface shall be protected from exposure to the atmospheric oxygen and thus the corrosion process shall be self-limited in time. This type of corrosion occurs in metals where oxides formed are very stable. There are also oxides that are not as stable, so that the oxide layer can crack or run-out from the surface of metal and then the corrosion continues over time.

In the oral cavity, corrosion of metals and alloys is produced electrolytically in humid environment, so it is a galvanic corrosion. Galvanic corrosion occurs when an electrolytic cell is formed. For an electrolyte cell to form, an electrolyte (a liquid which can lead electricity, in our case saliva), an anode (electrode providing/giving up electrons) and a cathode (electrode receiving electrons) are needed. If two metals with different electrical potentials are found in wet environment, metal with a more negative electric potential (see table) shall oxidize, so it shall lose electrons becoming an anode, and the metal with a more positive electric potential shall become the cathode. There are significant differences between the oxidation potential of the various metals. The more susceptible to oxidation, the more reactive they are, with higher corrosion potential. The potential for corrosion of metals depends on their reactivity, i.e. their galvanic potentials. Galvanic corrosion can occur between two different metals or alloys, but galvanic corrosion is more insidious and difficult to detect occurring in the same alloy that is composed of different phases, with different oxidation rates.

Crevicular corrosion occurs if there is a crack in the metal surface, which can be filled with liquid saliva, meaning there are anaerobic conditions. However, the metal shall release the ions resulting from the corrosion process, but these released electrons shall not be able to react in the crack depth where there is no oxygen. Therefore they shall be forced to migrate to the surface of the crack, where the oxygen shall produce the oxidation reaction. This flow of electrons from the base to the surface of the crack, the crack base shall become the anode and the surface shall become the cathode, actually leading to the formation of an electrolytic cell, with loss of substance in the crack depth. As more corrosion products are formed, they tend to be deposited in the crack, further reducing the supply of oxygen at the base of the crack, thus increasing the potential difference between the core and the surface. Thus, the process is self-sustaining. This type of corrosion is more dangerous than others because by this mechanism, any microfissures in the metal surface is transformed over time, slowly, into deep fractures that shall lead to breaking the metal below its strength and often without any warning that this may happen.

In conclusion, the effects of corrosion on the dentures and organism in general are varied and consist mainly in the loss of metal ions, forming galvanic microcurrents (oral galvanism), metallic taste (due to the release of metal ions), opacity, adverse biological effects (rare).

If metal restorations are present in the oral cavity at a time from metals with very different electrical potentials in combination with oral fluid (which acts as the electrolyte), as we have seen, an electrolysis cell appears. The phenomenon is more intense if the two different metals are in contact, e.g. adjacent teeth. Under such circumstances, due to the difference of potential microcurrent galvanic occur, which results in pain in the pulp and/or metallic taste. This phenomenon is known as oral galvanism. With the occurrence of specific symptoms or signs of oral galvanism, it is necessary to replace one the metallic constructions with a non-metallic reconstruction.

Experimental studies have shown that in oral galvanism electrical currents occur with relatively high tensions. The threshold varies from patient to patient, most of them being sensitive to values between 20 and 50 μA. Oral Galvanism manifestations can be varied, with symptoms such as metallic taste, burning sensation, pain in teeth with metallic reconstruction (due to galvanic microcurrents), and trigeminal neuralgia. Objectively they are manifested by gingivitis and glossitis, hypertrophy and turgor of the lingual papilla, erosion and ulceration of the oral mucosa, late leukoplakia (4-5 years after application of dentures).

Galvanic corrosion in the oral cavity can be prevented by using the same type of alloy for all metal prosthetic reconstructions in the mouth, especially for those that come in direct contact. At the same time, the use of homogeneous alloys, which cannot produce potentially different phases, reduces the risk of intrinsic galvanic corrosion of the ally or perfect polishing of metal works (dentures or amalgam obturation) which reduce the risk of crevicular corrosion.

From the thermodynamic point of view [9, 46, 47] the tendency of metals to pass in the ionic state differ greatly from one metal to another and it can be energetically characterized by the variation in enthalpy (ΔG), which accompanies the process. Electrochemically, the enthalpy variation equals the electrical work performed by an equivalent of gram ions: $\Delta G = -Z F E$, where: E – electromotive tension of the cell in which the anode and cathode reversible reaction of the corrosion process is achieved; Z – number of electric charges involved in the reaction; F=96500 coulombs / equivalent gram; $E = E_C - E_A$, where: E_C - equilibrium potential of the cathode; and E_A - equilibrium potential of the anode. As it is well known a reaction of thermodynamically possible if it is accompanied by the decrease of the free enthalpy, namely $\Delta G < 0$. Correlating the relationships above it is obtained: $E_C < E_A$, stating that, the electro-chemical corrosion of a metal can occur if the equilibrium potential of the metal in the given solution is more electronegative than the equilibrium potential of a depolarizing in the solution.

Chemical stability of the metal and the type of the different corrosion products depend on the electrode potential and the pH* of the solution. Graphically, the equilibrium between metal and its various oxidizing species is represented by the diagram "potential – pH" in isothermal conditions or thermodynamic stability diagram called Pourbaix diagram. This provides thermodynamic data on the phenomenon of corrosion, indicating the equilibrium conditions of all reactions that can take place between the metal and the aggressive environment at a given temperature. Pourbaix diagram includes: the immune area where corrosion is energetically impossible, the conditions of corrosion area where the metal ionization occurs (corrosion), the passivity conditions area where the ionization of the metal is thermodynamically possible, but it does not occur due to the formation of a passivating film on the metal surface; in the passivation area, the stable solid constituent is an oxide, a hydroxide, a hydrate or a salt of the metal. In the case of biomaterials the significance of the Pourbaix diagram can be described as follows: different parts of the body have different pH and different oxygen concentrations. For example, a metal which behaves well (it is immune or passive) in a particular part of the body can have an enhanced corrosion elsewhere. Moreover, the pH may change its value in tissues that can be injured or infected. An ordinary liquid in the tissue has a pH of about 7.4, but in a wound it may drop to 3.5, and in infection can increase to 9.0 [24],

(*)-by definition pH= - log [H⁺], where [H⁺] is the concentration of H⁺ ions. The product [H⁺] [OH⁻] equals to 10^{-14}. For neutral solutions [OH⁻]= [H⁺] and pH=7. A pH<7 indicates an acidic solution (excess of H⁺ions) and a pH>7, an alkaline solution (OH⁻ ions excess⁻).

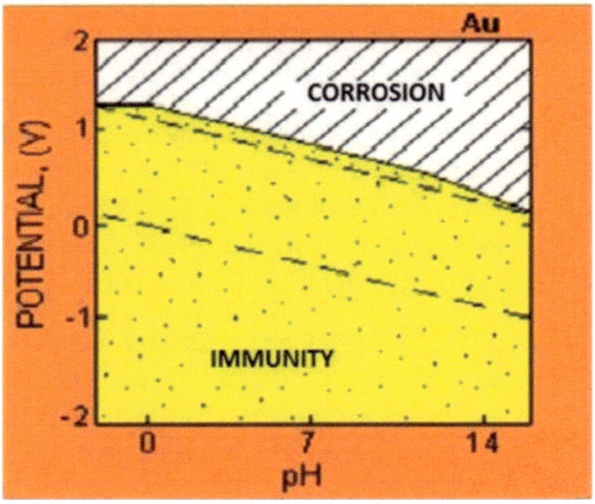

Figure 3. Pourbaix diagram for an immune metal (gold) (after M.Pourbaix, Atlas of Electrochemical Equilibria in Aqueous Solutions, NACE, Houston/CEBELCOR, Brussels, 1974) [47, pag.56]

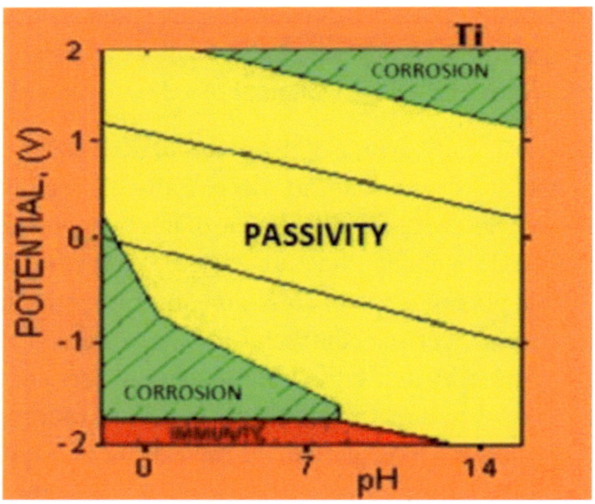

Figure 4. Pourbaix diagram for a passive metal (titanium) (after M.Pourbaix, Atlas of Electrochemical Equilibria in Aqueous Solutions, NACE, Houston/CEBELCOR, Brussels, 1974) [47, pag.56]

Pourbaix diagrams are useful, but are limited as they allow determining only the thermodynamics possibility of occurrence of a corrosion reaction. Completing them with kinetic data provides a real and useful guidance to assess/evaluate the level of metal's destruction in the specific environment.

http://dx.doi.org/10.5772/57322

Kinetics of electrochemical corrosion means the assessment of the *corrosion rate* which sets the rate of corrosion of a metal under the influence of the corrosive agent. The corrosion rate can be determined by direct methods and indirect methods. Direct methods determine the gravimetric index, or the index of penetration.

The gravimetric index or the corrosion rate(Vcor) is the change in sample mass (Δm), as a result of corrosion, per surface area unit *(S)* per time unit *(t)*: Vcor =Δm/(S t) which is usually considered as g/m^2 h. It is the most common way of expressing corrosion, being able to express the gain of weight of the sample, forming the corrosion products (metal oxidation) that remain adherent to metal or weight loss when corrosion products can be removed from the surface. The gravimetric index introduces errors in determination, as exact chemical composition of the corrosion products is not known. Therefore, the most used index is that corresponding to the weight loss, its precision depends on the removal of all corrosion products.

The penetration Index(I_p) is the depth by which the corrosion penetrated in the mass of metal, for one year. It is calculated from the gravimetric index V_{cor} and metal density ϱ [g/cm^3] as follows: I_p = (24 365 V_{cor})/(1000 ϱ) where 24 is the number of hours in a day; 365 – the number of days in a year; 1000 – the conversion factor of the measurement units.

Indirect methods to assess the rate of corrosion consist of electrochemical, electrical, acoustic, optical, etc. measurements. The electrochemical methods assess the quantity of corroded metal by measuring the current flowing in this process. If for each metal equivalent gram passed in solution, 96 500 coulombs are released, then for the electricity quantity the amount of electricity "It" flowing during corrosion, the amount of corroded metal "m" shall correspond. According to Faraday Law we can write: $m = Kit = \frac{A}{ZF} It$

The corrosion speed can be obtained comparing this quantity to the surface S and time t:

$$V = lg = \frac{m}{St} = \frac{A}{ZF} \cdot \frac{It}{St}; \; V = \frac{A}{ZF} \cdot \frac{I}{S} = \bar{K}.$$

Current density I, in A/cm²,

$I = \frac{\bar{K} ZF}{10^5 \cdot A}$, where: *K* – is the average corrosion speed; Z – valence of the ion that passes into the solution; F – 26.8 A/h.

Potentiodynamic methods provide useful information on the susceptibility of metals and alloys to generalized or pitting corrosion.

The evolution of the electrode potential in the open circuit is used as a corrosion behaviour criterion. This potential can vary over time as changes occur at the electrode surface (oxidation, formation of the passive layer or immunity). The physical and chemical reactions on the surface of the material change the solid-solution interface, which explains the development of the potential. After a period of immersion, it stabilizes around a stationary value.

Pitting corrosion and crevice corrosion is emphasized by cyclic polarization curves (CV/cyclic voltammetry). Cyclic polarization tests are commonly used to assess the susceptibility of

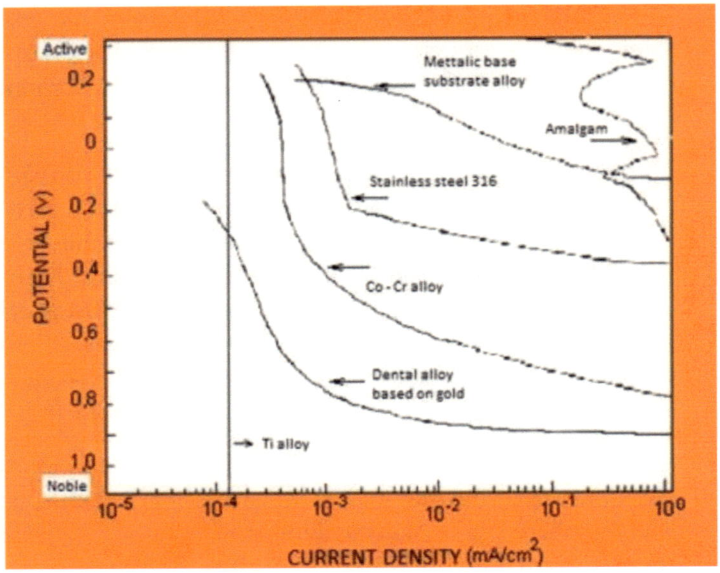

Figure 5. Curves of potential - current density for some biomaterials. (E.H.Greener, J.K. Harcourt, E.P. Lautenschlager, Materials Science in Dentistry, Williams and Wilkins, Baltimore, 1972) [47, pag.58]

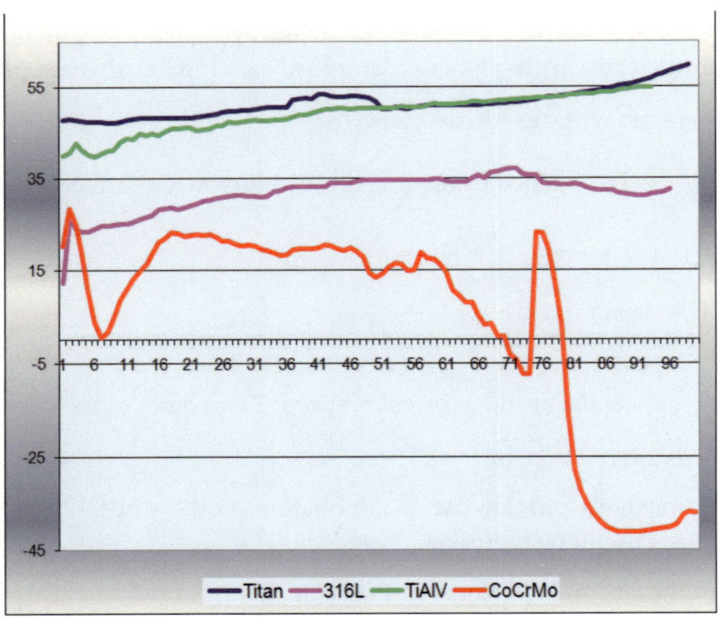

Figure 6. General diagram on the results of assessment tests of stationary potential variation [51]

metals and alloys to pitting corrosion (the hysteresis curve indicates pitting corrosion). The electrochemical impedance spectroscopy provides important information about the tendency of the alloy to the generalized corrosion process.

Conclusions

Pure titanium:

- the anodic curve has an active area between -0.19mA/cm^2 at a potential of -0.61V.
- Voltammetric curves show that the hysteresis area is small and negligible, which means that the Biomet not present and is not sensitive to the types of localized corrosion.

Alloyed Ti (Ti6Al4V):

- both the alloyed titanium and pure titanium indicate that the active corrosion area is very small or almost absent for pure titanium;
- experimental parameters: current density displays values from -0.15 la 0mA/cm^2; potential range is between 0.96 and 0.57V; the return voltammetry curve shows a small hysteresis compared to the previous biomaterial samples analyzed.
- the assessment of the voltametric cycle of the titanium alloy indicates that this biomaterial has a very good resistance to localized corrosion and general corrosion. These low values of corrosion are considered by the values of the maximum corrosion current, of the passivation current and the field of potential, when the current density is constant.

Voltammetric cycles (see figure6) - Comparative analysis indicates that in case of titanium and titanium alloy there is a surface unaffected by corrosion, result which is in correlation with diagrams obtained after the electrochemical tests. In case of 316L stainless a clouding of the submerged surface can be noticed and the occurrence of corrosion points.

3. Corrosive degradation-resin based composites

The introduction of resin-based composite dental materials around 1960s was a revolution in restorative dentistry. Resin-based composites are possibly the most used materials available in modern dentistry as they are used in a large variety of clinical applications, ranging from filling materials, luting agents, indirect restorations and metal facings to endodontic posts and cores. Composite restorative materials represent one of the many successes of modern biomaterials research, since they replace biological tissue in both esthetics and function. In anterior teeth composite is the clear material of choice among general dental practitioners. Direct composite restorations are increasingly employed also to restore posterior teeth due to their low cost and less need for the removal of sound tooth substance when compared to indirect restorations, as well as to their acceptable clinical performance [60].

Composite restorations must withstand an aggressive environment that is different from patient to patient. Mastication forces, occlusal habits, abrasive foods, chemically active foods and liquids, temperature fluctuations, humidity variation, bacterial by products, and salivary enzymes all contribute uncontrollable factors that affect composite restoration longevity [61].

To estimate how long posterior composite restorations last, the long-term studies are the ones to identify modes of failure and possible reasons for these failures. In the most recent review made by Demarco et al. 2012 [61] it has been searched the dental literature looking for clinical trials investigating posterior composite restorations over periods of at least 5 years of follow-up published between 1996 and 2011. As observed in their literature reviewed, a long survival rate for posterior composite restorations can be expected provided that patient, operator and materials factors are taken into account when the restorations are performed. The majority of the clinical studies indicated that annual failure rates between 1% and 3% can be achieved with Class I and II posterior composite restorations depending on several factors such as tooth type and location, operator, and socioeconomic, demographic, and behavioral elements. The material properties showed a minor effect on longevity. The results of this review reconfirmed that the main reasons for failure in the long term of composite restorations were secondary caries, related to the individual caries risk, and fracture, related to the presence of a lining or the strength of the material used as well as patient factors such as bruxism.

Taking into account the caries risk impact on the longevity of resin based composites materials is suggested that future composite research be directed toward developing materials that will suppress bacterial activity at the tooth–composite interface and counter the effects of caries [67,68,69,72]. Concern still exists when the composite materials are placed in high stress situations, especially in patients with bruxing or parafunctional habits or when placed in large preparations, perhaps on several teeth in a quadrant, and when used to replace cusps. The concern here is for fracture of the restoration as well as wear [60].

Degradation in dental composites may result in matrix and/or filler deterioration, due to mechanical and/or environmental loads, interfacial debonding, microcracking, and/or filler particle fracture. A continuous application of mechanical and environmental loads eventually leads to progressive degradation and crack initiation and growth, resulting in catastrophic failure of dental restorations [60-62].

The polymers used in resin composites are susceptible to absorption of solvents, especially water, and the loss of soluble components. The solvent molecule forces the polymer chains apart, causing swelling. As the strength of the bond decreases, the polymer becomes softer, the glass transition temperature is reduced and the strength may be lowered. Water sorption is a contributory factor to discoloration of the restorations and the hydrolytic degradation of the resin-filler interface. The second basic degradation process of the polymeric matrix involves the scission or breakdown of the covalent bonds. The scission of the polymer chain will reduce the molecular weight of the polymer, thus resulting on a significant loss of mechanical properties.

J.L.Drummond [62] has made a valuable review of the mechanisms and degradation effects due to aging of the resin based composites. During exposure to various environments, dental composites are subjected to material property changes due to degradation and aging. He concluded that these changes are due to: (a) chemical breakdown by hydrolysis; (b) chemical breakdown by stress-induced effects associated with swelling and applied stress; (c) chemical composition changes by leaching; (d) precipitation and swelling phenomena to produce voids and cracks, leaching the interface; and (e) loss of strength due to corrosion.

All of these degradation processes may lead to nucleation and the growth of microcracks. Over time, the leaching of the soluble components, the swelling and degradation of the cross-linked polymer matrix in the dental composite, and hydrolysis of the filler-matrix silane interfaces eventually lead to a decrease in mechanical properties. With respect to fracture toughness, water seems to lower the yield stress, release internal stress accumulated during polymer ization shrinkage, and increase the plastic zone ahead of the crack, which causes the increase in observed fracture toughness. Other theories as to the cause of the degradation of the dental resin include the formation of microcracks through repeated sorption/desorption cycles, leading to hydrolytic degradation of the polymer [62]. The extent of water uptake is dependent upon the monomer formulation and promising lowerwater uptake are observed for silorane-based systems already used by the practitioners as a filling alternative to the dymethacrylate-based composites. But as newer formulations of composites are designed to be self- adhesive, they will most likely be even more hydrophilic than current resins [59, 60].

Biotribocorrosion is a material degradation process due to the combined effect of corrosion and wear. Too little attention was given to the corrosive degradation of resin based composites in the dental literature so far. Most of research has focused on the clinical implication of the corrosive wear in the overall wear phenomenan of teeth and restorative materials [64-66]. Wear of teeth and restorative materials is the result of different complex processes that depend primarily on the abrasive nature of food, the properties of the antagonistic material, the thickness and hardness of enamel, the chewing behaviour along with parafunctional habits, and neuromuscular forces. Different wear phenomena may take place in the oral cavity. An overview of the types of wear, grouped as biotribocorrosion has been made by Lambrechts et al 2006 [64]. Wear as function of a tribological system is composed of three basic elements [64]: (a) the structure—the types of materials in contact and the contact geometry; (b) the interaction conditions— the loads, stresses, and duration of interaction; and (c) the environment and surface conditions—including the surface environment and chemistry, surface topography, and ambient temperature.

Weartribology and biotribocorrosion define wear as a complex phenomenon and an 'overall effect' of a number of interrelated processes. Depending on the parameters of the tribosystem the wear processes could be described with five terms [64]: two-body abrasion, three-body abrasion, fatigue wear, tribochemical wear (dental erosion, corrosion wear), adhesive wear. Abrasive wear describes the ploughing of hard asperities into softer surfaces, and may be further distinguished between abrasion and attrition. Abrasion occurs during mastication processes in the presence of food serving as a third body (three-body abrasion, whereas attrition is the result of direct contact of antagonistic teeth or restorations during mastication, swallowing, or occlusal movements as a two-body abrasion. Fatigue wear and corrosive wear are considered two important types of biotribocorrosion. Fatigue wear describes a process that is caused by subsurface cracks that proceed due to repeated load cycles, and tribochemical wear relates to a chemical reaction producing a surface layer that can be scraped away by antagonistic contact [62, 63].

Tribochemical wear [65] or corrosive wear [70] is caused when chemicals weaken the inter-molecular bonds of the surface and therefore potentiate the other wear processes. There is an interplay of erosion, attrition and three body abrasion in tooth wear. In the mouth this effect is normally caused by acids, which may be 'extrinsic' such as dietary acids or 'intrinsic' resulting from gastric reflux. On exposure to plaque acids, food-simulating constituents, and enzymes, resin composites have undergone softening and roughening [71, 78]. The most important thing to understand is that acids weaken only the surface molecules. In general, the corrosion is rapid initially, and tends to slow down, or it may even cease completely, after a cohesive film has been formed on the surface. However, when this pellicle is removed by the sliding of surfaces that meet over it, an unaltered surface is exposed, and the chemical attack continues. Mechanical tooth wear and chemical dissolution act simultaneously. Consequently, resin composites may show an increased abrasive wear rate.

Improvements in the properties of dental resin composites are constantly being sought. Given that secondary caries and fracture are the two primary reasons given for replacement of dental composites, it is warranted to continue to pursue improvements in wear resistance, as well as shrinkage and its accompanying stress. Shrinkage stress is often considered the most significant problem with current restoratives and a primary contributor to premature failure in composite restorations, since it is capable of deforming tooth structures and causing microcracks and adhesive failure [67]. This process is further assisted by voids introduced during material processing, imperfect interfaces, and residual stresses, making resistance to crack initiation and growth an important consideration for a reliable assessment of dental restorations. The gaps between dentin and adhesive system couldn't be attributed also to the shrinkage stress that accompanies the polymerisation process, but to the lower efficiency of the self-etch mechanism of adhesion [67] (Figure 7, 8). The marginal crevice caused between restoration and tooth by the polymerization shrinkage of the composite together with the voids between the adhesive layer and enamel/dentine where oxygen deficit can form can be considered sites prone to oxygen concentration cell attack or crevice corrosion and furher studies are needed to demonstrate this.

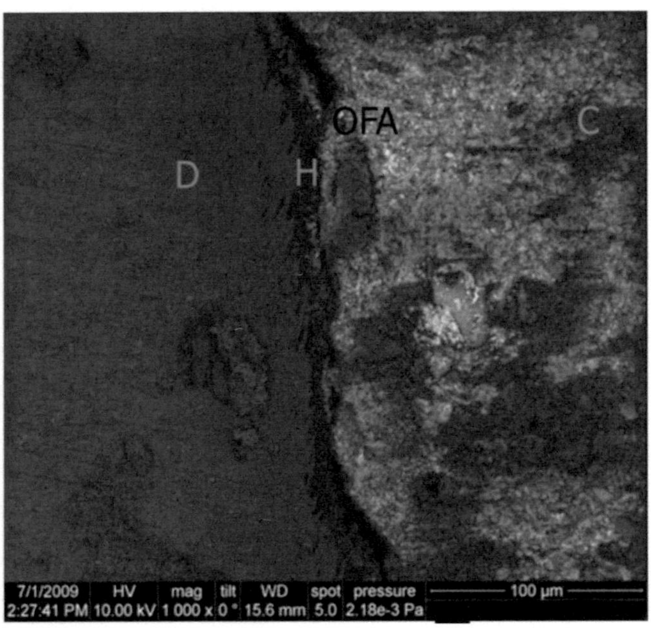

Figure 7. FE-SEM images of a sectioned restorations 1000X. Restoration/dentin interface of the cavity floor of dymethacrilate resin based composite. D-dentine, C-composite, OFA-adhesive OptiBond FL, H-Hybride layer, G-gap [79, 80]

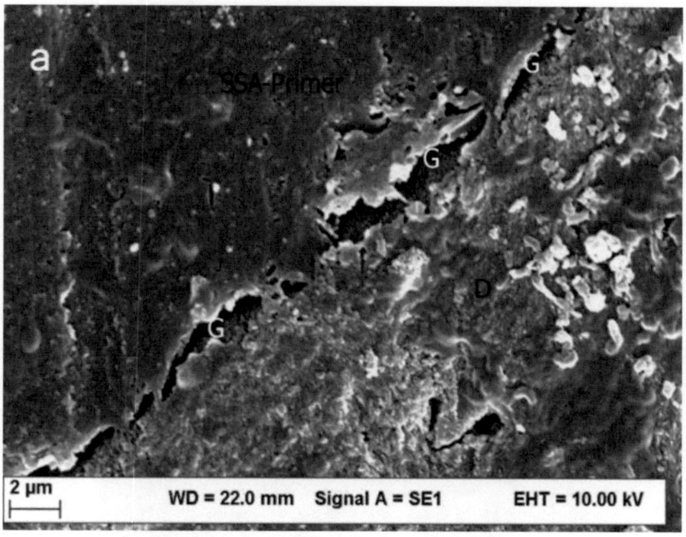

Figure 8. SEM images of sectioned restorations. Restoration/dentin interface of the cavity floor of a silorane based composite Filtek™ Silorane restoration (FS- Filtek™ Silorane, SSA -Primer-Silorane Adhesive System Primer, D- Dentin, G- gap [80].

The longer a restoration is in the clinical service life, the higher the failure rate, but the failure of resin composite restorative materials is far more complicated than just the material properties. As with all dental restorative materials, the proper technique, the appropriate materials, proper patient selection and monitorisation for repair vs. replacement usually ensure a successful clinical restoration.

4. Evaluation of corrosion for dental ceramics

The commonly and easiest method to investigate the corrosion decay of dental ceramics is to evaluate the weight loss of the samples after immersion in CH_3COOH solution 4%. Two different dental ceramics were investigated: alumina based ceramic (crystal structure) and zirconia based ceramic (Y-TZP, yttrium-stabilized tetragonal zirconia polycrystal structure). Both of them were sintered ceramics. The test sample specimens were rectangular shaped (l = 12 mm) as the blank ceramic shape with thickness d = 3 mm [80].

Samples were washed in distilled water and dried in a sterilized unit at 110±4°C for 2 hours. After determining the mass of the sample with the accuracy of $\pm 10^{-4}$ g (analytic scale, Precisa, 320XT), each sample was immersed in a recipient with CH_3COOH solution 4%. The recipients were placed in an usual thermostatic shaker at temperature t = 37°C for 4 hours. After the time has elapsed, the samples were washed with distilled water and dried in the sterilized unit at 110±5°C, for 2 hours time and weighed. The results obtained are depicted in Table4

Sample	Loss weigt (average value, µg/cm²)*
Alumina ceramic	14.30 ± 0.60
Zirconia ceramic (Y – TZP)	5.60 ± 0.60
*SD = ± 0.60	

Table 4. Loss weight (average value, µg/cm²) [80]

An important evaluation of the corrosion process is samples surface investigation by scanning electron microscopy (SEM). Information regarding all stages of corrosion process (galvanic, pitting, crevice and stress) can be obtained. Dental ceramic samples were examined before and after corrosion process according the above mentioned protocol. Results are depicted in Fig.9.

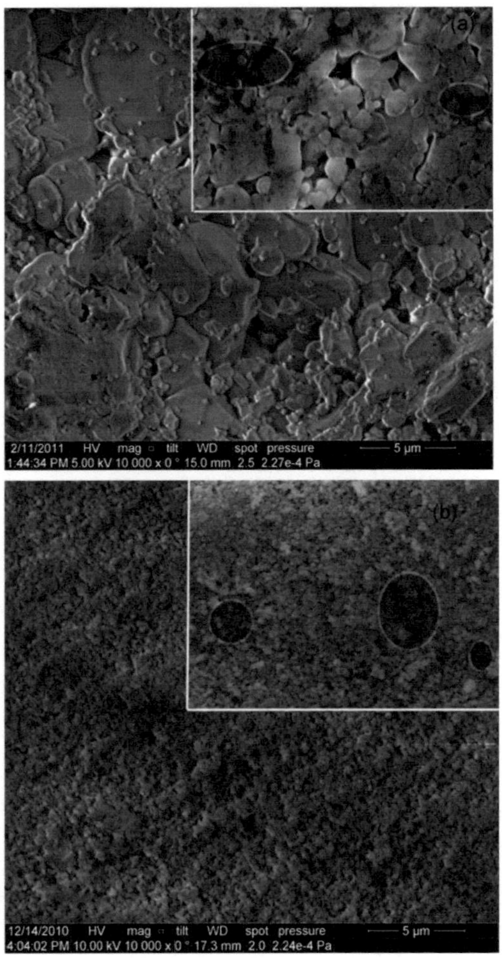

Figure 9. SEM micrographs of dental ceramic samples, before and after corrosion process (insets): (a) alumina ceramic, (b) zirconia ceramic. Details: for insets, shady areas are corresponding to the cooroded areas [79, 80, 81]

http://dx.doi.org/10.5772/57322

Examining the sample surfaces quality, we noticed that alumina ceramic sample is more corroded than zirconia sample. It can be observed small holes (gaps) on samples surface, larger for alumina than zirconia sapmple (Fig. 9). An important detail to be noticed, is that as morphological structure, zirconia is more compact than alumina ceramic [80, 82].

Figure 10. SEM micrographs of a dental restoration work with zirconia core after five years working time. Details: (a) interface zirconia core / dental luting cement, not corroded; (b) interface area affected by crack and crevice corrosion (shady areas) and surface deposits (black border shady areas); (c) surface deffects of zirconia dental ceramic core (shady areas) [79, 80]

Relevant for the study is the investigation of a dental restoration work, after a period of working time in the oral biotop. Hence a patient zirconia dental restoration work was investigated, with working time of aproximative 5 years. Using SEM technique, there were examined interface zirconia ceramic / dental lutting cement and the ceramic core surface (Fig.10).

It can be noticed that the corrosion process (corresponding to the crevice and crack corrosion) is quite aggresive at the interface ceramic / dental lutting cement (Fig. 10 (b), available for larger magnification), where small holes along the interface line are observed. Regarding the ceramic core of zirconia work (Fig. 10 (c)), the surface presents deffects and areas of stress corrosion and possible failure of the dental work. It is known that zirconia is sensitive to stress corrosion because of changing crystaline phases (T →M, tetragonal to monoclinic phase) [80, 82, 83]. That was the reason the dental work was replaced. Also bacteria deposits on dental ceramic work surface can be observed, those being improved by the surface roughness as is shown Fig. 10 (b), (c). Some areas are presented into good conditions, not affected by corrosion (Fig. 10 (a)) as a sign that a long lasting ressitance to corrosion is possible.

Conclusions

- The most efficient way for preventing corrosion effects, is that to minimize (to reduce) those factors conducting to the corrosion process (each type of corrosion).
- Pitting corrosion and crevice corrosion, although similar as mechanism, being initiated by differential aeration phenomena, are different as pitting corrosion is determined by submicroscopic defects, especially manganese sulphides oriented in the direction of deformation (sulphide being plastic), and crevice corrosion is determined by macroscopic defects of the surface oxide layer, these defects may be due to the degree of processing of the surface, respectively the broken pieces on the surface resulting in cells with differential aeration and initiates the crevice corrosion process. The softer the material, the more difficult its processing, and thus the possibility of developing these crevices is greater.
- Galvanic corrosion is the starting point of the corrosion process regarding the oral biotop. Some possibilities may be taken in account for reducing this corrosion effect:
- Reducing as much possible the number of materials used for dental restorations, same materials for the same pacient (each material has its own corroding potential);
- As much possible, using materials with similar corroding potential values; corroding potential being a criteria of material biocompatibility;
- Contact or open areas when using metal alloys, as much possible care should be taken to keep them electrically insulated;
- High quality smoothing surfaces (care must be taken during air abrasion process);
- In case of using alumina ceramic, using of zirconia as surface quality improver for given proper conditions of temperature and pressure during dental restoration manufacturing process (before applying veneer and glazer) [82, 83].

http://dx.doi.org/10.5772/57322

Author details

I. Patrascu[1], E. Vasilescu[3], E. Gatin[1,2] and R.R. Cara-Ilici[1]

1 University of Medicine and Pharmacy "Carol Davila", Faculty of Dentistry, Bucharest, Romania

2 University of Bucharest, Faculty of Physics, Bucharest, Romania

3 'Dunarea de Jos' University of Galati, Galati, Romania

References

[1] Al. Bucur (sub redacţia), C. Navarro Vila, J. Lowry, J. Acero — Compendiu de chirurgie oro-maxilo-facială, vol. I şi II, Editura Q Med Publishing, Bucureşti, 2009,

[2] Ardelean L., Materiale dentare, Curs litografiat, Timisoara, 2004,

[3] Anderson JM, et al. Host reactions to biomaterials and their evaluation. In: Ratner BD, et al., eds. Biomaterials Science: An Introduction to Materials in Medicine. London: Elsevier, 2004,

[4] ASM Handbook Committee, Metals Handbook-Corrosion of Metallic Implants and Prosthetic Devices, vol. 13, ninth ed., American Society for Metals, Metals Park, 1987.

[5] A.P.R. Alves*, F.A. Santana, L.A.A. Rosa, S.A. Cursino, E.N. Codaro, A study on corrosion resistance of the Ti-10Mo experimental alloy after different processing methods, Materials Science and Engineering Elsevier, p.693-696,

[6] Bratu D., Ciosescu D., Rominu M., Leretter M., Uram-Tuculescu S., Materiale dentare in cabinetul de stomatologie, ED. Helicon, Timisoara, 1998, ISBN 973-574-495,

[7] Bratu D., Materiale dentare Bazele fizico-chimice, Ed. Helicon Timisoara 1994, ISBN 975-9159-37-0,

[8] Brunski JB. Metals. In: Ratner BD, et al., eds. Biomaterial Science: An Introduction to Materials in Medicine. London: Elsevier Academic Press, 2004:137-53,

[9] Gadea, S., Petrescu, M., Metalurgie fizica si studiul metalelor, vol.3 EDP, Bucuresti 1983

[10] R.G. Craig, Restorative dental materials, Mosby, Chicago, 1993,

[11] Christoph Leyens, Manfred Peters, "Titanium and Titanium Alloys-Fundamentals and Applications", ISBN 3-527-30534-3, Ed. WILEY-VCH, pg. 4-16, 2004,

[12] Carew EO, Cooke FW, Lemons BD, et al. Properties of materials. In: Ratner BD, et al., eds. Biomaterials Science: An Introduction to Materials in Medicine. London: Elsevier Academic Press, 2004:23-65,

[13] Claudia Fleck a, *, Dietmar Eifler, Corrosion, fatigue and corrosion fatigue behaviour of., metal implant materials, especially titanium alloys, International Journal of Fatigue 32 (2010) 929–935,

[14] Claus Moseke Wolfgang Braun Andrea Ewald, Electrochemically Deposited Ca(OH)2 Coatings as a Bactericidal and Osteointegrative Modification of Ti Implants, Advanced Engineering Materials no.1/2009, p. 81-87,

[15] J.R. Davis, "Handbook of materials for medical devices", ISBN 0-87170-790-X, ASM International, pg. 38-50, 2003,

[16] Dumitriu H.T., Silvia Dumitriu, Anca Dumitriu — Parodontologie, Editura Viaţa Medicală Românească, Bucureşti, 2009,

[17] Dorobăţ Valentina, D. Stanciu — Ortodonţie şi ortopedie dento-facială, Editura Medicală, Bucureşti, 2003,

[18] E.Eisenbarth a, *, D.Velten a, M. M. ullera, R.Thull b, J.Breme Biocompatibility of b-stabilizing elements of titanium alloys, Biomaterials 25 (2004) p. 5705–5713,

[19] Forna N., C.De Baat, D. Bratu, V. Mercut, Al. Petre, S. Popşor, T.Trăistaru Protetică Dentară Vol. I, Editura Enciclopedică, Bucureşti, 2011,

[20] Gotman I. Characteristics of metals used in implants. J Endourol (1997), 11(6):383-9.

[21] Gerd Lütjering, James C. Williams, "Titanium-Engineering Materials and Processes", ISBN 3-540-42990-5, Springer-Verlag Berlin Heidelberg New York, pg. 21-26, 2003,

[22] J. Breme., Y. Zhou., L. Groh., Development of a titanium alloy suitable for an optimized coating with hydroxyapatite Biomaterials 16 (1995) p. 239– 244,

[23] Kenneth J. Anusavice, Phillips' Science of Dental Materials, Tenth Edition ISBN 81-7286-054-4, 1996,

[24] Karthega, M., Raman, V., Rajendran N., Influence of potential on the electrochemical behaviour of b titanium alloys in Hank's solution, Acta Biomaterialia 3 (2007) Elsevier, p. 1019–1023

[25] Liliane S. Morais., Glaucio G. Sera., Carlos A. Muller, ... Titanium alloy mini-implants for orthodontic anchorage: Immediate loading and metal ion release, Acta Biomaterialia 3 (2007) Elsevier, p. 331– 339

[26] Lipsa C-tin; Lipsa D., Biomateriale curs, Iasi 2009, ISBN 978-973-0- -7372-0

[27] M.A. Khan., R.L. Williams., D.F. Williams, In –vitro corrosion and wear of titanium alloys in the biological environment, Biomaterials 17 (1996), Elsevier, p. 2117-2126

[28] Nicola Codruta., Materiale dentare, Consideratii clinice si tehnologice, ISBN 978-973-133-467-7, Casa cartii de stiinta Cluj-Napoca, 2009

[29] Nicolas Schift., Brigitte Grosgogeat., Michele Lissac., Francis Dalard, Influence of floride content and pH on the corrosion resistance of titanium and its alloys, Biomaterials Elsevier 23 (2002) 1995–2002

[30] Yoshimitsu Okazaki, Effect of friction on anodic polarization properties of metallic biomaterials, Biomaterials Elsevier 23 (2002) 2071–2077

[31] Yoshimitsu Okazaki, Emiko Gotoh, Comparison of metal release from various metallic biomaterials in vitro, Biomaterials no.26/2005/11-12 Elsevier

[32] Park JB, et al. Biomaterials: Principles and Applications. Park JB, Bronzino JD, eds. Danvers, MA: CRC Press, 2003.

[33] Park JB. Biomaterials. In: Bronzino JD, ed. The Biomedical Engineering Handbook. Boca Raton, FL: CRC Press, 1995:537-719.

[34] I.Patrascu, Tehnologia aliajelor dentare, Materiale dentare- teste, ed. Libripress, 2000

[35] Ratner BD. A history of biomaterials. In: Ratner BD, et al., eds. Biomaterials Science: An Introduction to Materials in Medicine. San Diego, CA: Elsevier Academic Press, 2004:10-22

[36] H.J. Rack, J.I. Qazi, Titanium alloys for biomedical applications, Materials Science and Engineering C 26 (2006) Elsevier, p. 1269 – 1277

[37] E.E. Stanbury, R.A. Buchanan, Fundamentals of Electrochemical Corrosion, Materials Park, OH, ASM International, (2000), p. 219.

[38] S. Sathish., M. Geetha., N.D. Pandey., C. Richard., R.Asokamani, Studies on the corrosion and wear behavior of the laser nitride biomedical titanium and its alloys, Materials science and engineering C 30 (2010), p. 376 – 382.

[39] Sergio Luiz de Assis, Stephan Wolynec, Isolda Costa, Corrosion characterization of titanium alloys by electrochemical techniques, Electrochimica Acta 51 (2006) 1815–1819 Elsevier, p. 1815-1819

[40] Toru Nonami., Alkira Katsuyoshi, Naganuma, Tetsuya Kameyana, Implantation of hydroxyapatite granules into superplastic titanium alloy for biomaterials, Materials science and engineering C 6 (1998), p. 281 – 284.

[41] Vermeşan; H. Coroziune, Editura Risoprint Cluj-Napoca 2005, ISBN:973-656-881-4.

[42] D. Velten, V. Biehl, F.B. Valeske, W. Possart, J. Breme, J Biomed Mater Res (2002), 59: 18–28.

[43] Zhuo Caia, *, Ty Shafera, Ikuya Watanabea, Martha E. Nunnb, Toru Okabea, Electrochemical characterization of cast titanium alloys, Biomaterials 24 (2003) Elsevier p. 213–218

[44] Woo RK, Jenkins DD, Greco RS. Biomaterials: historical overview and current direction. In: Greco RS, Prinz FB, Smith RL, eds. Nanoscale Technology in Biological Systems. Boca Raton, FL: CRC Press, 2005:1-24

[45] William, R. Proffit, D.D.S., PHD....Henry W. Field, Jr., D.D.S., M.S., M.S.D. Contemporary Orthodontics Second Edition Mosby- Year Book the United State of America, 1993, ISBN 0-8016-6393-8

[46] www.ro.scribd.com/doc/55566517Biomateriale

[47] www.tuiasi.ro/user/112/V. Bulancea- Biomateriale

[48] www. scribd.com/doc/95682357/

[49] Ghiban, B., Metallic Biomaterials, Ed. Printech Bucuresti 1999

[50] Alexandru, I., s.a. Alegerea si utilizarea materialelor metalice, Ed. Didactica si Pedagogica Bucuresti 2000

[51] Mahmandiu Ch., Teza de doctorat UMF Bucuresti 2013

[52] C. Moranta, M.F. Lopezb, A. Gutierreza, J.A. Jimenezc AFM and SEM characterization of non-toxic vanadium-free Ti alloys used as biomaterials Applied Surface Science 220 (2003) 79–87

[53] Yoshimitsu Okazakia, Emiko Gotoh, Comparison of metal release from various metallic biomaterials in vitro Biomaterials 26 (2005) 11–21

[54] K. Bordji, J.Y. Jouzeau, D. Mainard, E. Payan, P. Netter, K. T. RieS, T. Stucky and M. Hage-Alis, Cytocompatibility of Ti-6Al4V and Ti-5Al-2.5Fe alloys according to three surface treatments, using human fibroblasts and osteoblasts Biomoterial 17 (1996) 929-940

[55] D.J. Wever, A.G. Veldhuizen, M.M. Sanders*, J.M. Schakenraad and J.R. van Horn, Cytotoxic, allergic and genotoxic activity of a nickel-titanium alloy Biomoterials 18 (1997) 1115-1120 Elsevier

[56] J.Breme, Y. Zhou and L. Groh, Development of a titanium alloy suitable for an optimized coating with hydroxyapatite, Biomaleriak 16 (1995) 239-244 Elsevier

[57] Oleg Prymaka, Denise Bogdanskib, Manfred Ko° llerb, Stefan A. Esenweinb, Gert Muhrb, Felix Beckmannc, Tilmann Donathc, Michel Assadd, Matthias Epplea, Morphological characterization and in vitro biocompatibility of a porous nickel–titanium alloy, Biomaterials 26 (2005) 5801–5807

[58] S. Sathish, M. Geetha, N.D. Pandey, C. Richard, R. Asokamani, Studies on the corrosion and wear behavior of the laser nitrided biomedical titanium and its alloys, Materials Science and Engineering C journal homepage: www.elsevier.com/locate/msec

[59] Sarrett DC. Clinical challenges and the relevance of materials testing for posterior composite restorations. Dent Mater 2005;21:9–20

[60] Ferracane JL. Resin composite-State of the art. Dent Mater 2011:27(1):29-38

[61] Demarco FF, Corrêa MB, Cenci MS, Moraes RR, Opdam NJM. Longevity of posterior composite restorations: Not only a matter of materials. Dent Mater 2012;28: 87-101

[62] Drummond JL. Degradation, Fatigue, and Failure of Resin Dental Composite Materials. J Dent Res 2008;87(8):710-719

[63] Correr GM, Bruschi Alonso RC, Correr Sobrinho L, Puppin-Rontani RM, Ferracane JL. In vitro wear of resin based materials—simultaneous corrosive and abrasive wear. J Biomed Mater Res B. Appl Biomater 2006;78: 105–114

[64] Lambrechts, P., Debels, E., Van Landuyt, K., Peumans, M., Van Meerbeek, B., 2006. How to simulate wear? Overview of existing methods. Dent. Mater. 22, 693–701

[65] Turssi CP, Purquerio BM, Serra MC. Wear of Dental Resin Composites: Insights into Underlying Processes and Assessment Methods—A Review. J Biomed Mater Res Part B: Appl Biomater 2003 65B: 280–285

[66] Cara Ilici RR, Gatin E, Matei E, Didilescu A, Nicola C, Patrascu I. Cuspal deflection and adhesive interface integrity of low-shrinking posterior composite restorations. Acta Stomatol Croat. 2010;44(3):142-51

[67] Komine T, Tomic M, Gerds T, Strub JR, Influence of different adhesive resin cements on the fracture strength of aluminium oxide ceramics posterior crowns. J. Prosthet. Dent, 2004, 92, 359 – 364;

[68] Burke FJ, Fleming GJ, Nathason D, Marquis PM, Are adhesive technologies needed to support ceramics? An assesment of current evidence. J. Adhes. Dent, 2002, 4, 7 – 22;

[69] Wolfart M, Lehmann F, Wolfart S, Kern M, Durability of the resin bond strength to zirconia ceramic after using different surface conditioning methods. Dent. Mater, 2007, 23, 45 – 50;

[70] Anusavice K J, Degradability of dental ceramics. Adv Dent Res, 1992, 6, 82-89;

[71] Jakovac M, Živko-Babić J, Ćurković L, Carek A, Chemical Durability of Dental Ceramic Material in Acid Medium, Acta Stomatol. Croat, 2006, 40, 65 – 71;

[72] Jakovac M, Živko-Babić J, Ćurković L, Aurer A, Measurement of ion elution from dental ceramics. J Eur Ceram Soc 2006, 26(9), 1695-1700;

[73] Milleding P, Wennerberg A, Alaeddin S, Karlsson S, Simon E. Surface corrosion of dental ceramics in vitro, Biomaterials 1999, 20(8), 733-746;

[74] Kukiattrakoon B, Junpoom P, Hengtrakool C, Vicker's microhardness and energy dispersive x-ray analysis of fluorapatite-leucite and fluorapatite ceramics cyclically immersed in acidic agents. J Oral Sci, 2009, 51(3), 443-450;

[75] Kukiattrakoon B, Hengtrakool C, Kedjarune-Leggat U, Effect of Acidic Agents on Surface Roughness of Dental Ceramics, J Dent Res, 2011, 8 (1), 6 – 15;

[76] Kennell GF, Heppner KL, Evitts RW, A Critical Crevice Solution and IR Drop Crevice Corrosion Model, Corrosion Science, 2008, 50, 1716 – 1725;

[77] ASM Handbook, "Corrosion", vol 13, ISBN 0-87170-007-7, ASM International, Ohio 1987;

[78] ASM International, Metals Handbook (Desk Edition) „Failure Analysis", vol. 11, American Society for Metals, (1997);

[79] Gatin E, Luculescu C, Birjega R, Ciobanu L, Patrascu I, Alumina versus zirconia comparative surveyof thermic influences during dental ceramic core manufacturating process, Part Scie Tech, 2013, 31 (2), 156 – 161;

[80] Gatin E, Luculescu C, Iordache S, Patrascu I, Complete morphological investigation by AFM technique applied to dental ceramics under thermal processing, 2013, JOAM, 15 (9 – 10), 1136 - 1141.

[81] Krell A, Blank P, Ma H, Hutzler T, Nebelung M, Processing of highdensity submicrometer Al2O3 for new applications, J. Am. Ceram. Soc, 2003, 86 (4), 546– 553;

[82] Wang X.H, Chen P.L, Chen I.W, Two-step sintering of ceramics with constant grain – size (I) Y2O3, J. Am. Ceram. Soc, 2006, 89 (2), 431– 437.

Chapter 27

Investigation of Al-Fe Aerospace Alloy Laser-Treated with Different Corrosion Characterization Techniques

Moisés Meza Pariona

Additional information is available at the end of the chapter

http://dx.doi.org/10.5772/57280

1. Introduction

One of the non-traditional surface engineering techniques, named laser surface melting (LSM) has attracted growing interest in the recent years for its ability to improving the corrosion performance of aluminum alloys. LSM is a versatile and promising technique that can be used for modifying the surface properties of the material without affecting its bulk properties.

An interesting study was conducted by authors Viejo et al. [1], in which they argue that in the recent years, the aluminum industry has been developed alloys with an increasing damage tolerance in order to finding the demands of the latest and upcoming generations of the commercial aircraft. Particularly the focus was on the Al–Cu–Li alloys (i.e. AA2050 or AA2198 alloys) that have been used in the recent past in the military and space sectors. Their favorable density, strength, toughness, fatigue behavior and thermal stability have made them attractive candidates in applications requiring both high specific strength and an excellent damage tolerance. And differently of the most conventional aerospace alloys, Al–Cu–Li alloys are fusion weldable, which opens up new opportunities in fuselage construction. Nevertheless, as with the others AA2xxx aluminum alloys, Al–Cu–Li alloys can be susceptible to localized and exfoliation corrosion, particularly in chloride-containing environments. Great efforts have been made for producing surface layers that are free from intermetallic precipitates in order to eliminating, or at least reducing, their detrimental effects. For example, it is generally accepted that laser surface melting (LSM) is an useful tool for improving the corrosion resistance of the aluminum alloys, as the result of the formation of the thin melted layers with refined microstructures that are virtually free from intermetallic precipitates and inclusions. Thus, LSM, using CO_2 or Nd:YAG laser irradiation, can improve the localized corrosion resistance by modifying the near-surface region through rapid melting and solidification processes.

The authors Yue et al. [2] have investigated the laser-treated surface using a KrF excimer laser, according to them it was found out that the laser-treated layer consists of the polycrystalline α-Al_2O_3 with some undetermined precipitates. They show that the size of α-Al_2O_3 crystalline is approximately 5–6 nm. In addition, they have verified by selected-area electron diffraction patterns (SAED) of the laser-melted zone that the structure was of crystalline form. These same authors have examined through the TEM the laser-treated specimen and they have concluded that it did not reveal any coarse second-phase particles, as found in the untreated specimen. However, a stable corrosion film had not been formed at the surface of the untreated specimen due to the presence of numerous second-phase particles, which readily have initiated pitting corrosion and also they have destroyed the integrity of the film. Being that the base the result of this study, the laser-treated layers mainly have consisted of nanocrystalline structures α-Al_2O_3, which is a chemically stable phase, serves as an effective barrier for protecting the matrix against corrosion attacks.

According to the study of Ryan and Pragnell [3] by the technique of pulsed laser surface melting (PLSM), they have showed that the increase in the corrosion performance of pulsed laser-treated alloys has been widely attributed to the formation of a surface layer that is much more chemically homogenous than to the bulk material. It has generally been assumed that the laser treatment removes second phase particles and partitionless re-solidification of the layer occurs. In order to preventing the formation of a cellular structure throughout the layer on re-solidification and ensure a reasonable chance of obtaining a chemically homogeneous layer, the melt depth is thus restricted to less than 10 μm thickness. The success of laser-treatment relies on the aluminum matrix and second phase particles being taken into the liquid phase and mixed with a combination of stirring and diffusion to form a uniform liquid. Ideally, the liquid film should then re-solidify sufficiently quickly to prevent interface instability, trapping the solute in solid-solution. However, due to the short time spent in the liquid phase, it is extremely difficult to achieve an uniform solute distribution. The study of the pulsed laser surface melting of aluminum has concluded that the improvements in corrosive properties are a result of the homogenization of the surface region.

Pariona et al. [4] reported a study of the laser treatment irradiated Al–1.5 wt. % Fe alloy with Yb-fiber laser beam. This laser treatment without an assisting gas jet was applied to augmenting the production of metal oxides on the laser-treated surface. The laser-treated samples were covered with several weld filets during the remelting process. The results reveal the formation of the weld filet structures with metastable phases and finely dispersed precipitates. The creation of a finely porous layer of the protective coating produced during the rapid remelting process contributed to increasing the corrosion resistance and homogeneous properties of the laser-treated samples when compared with untreated samples. The Yb-fiber laser beam technology applied to the surface treatment of aluminum alloys proved efficient for augmenting their corrosion resistance, thus deserving further investigation for aerospace and automotive applications.

Recently, Pariona et al. [5] have investigated AFM study of the effects of the laser surface remelting on the morphology of Al–Fe aerospace alloys. This work has been focused on the characterization of the surface roughness by AFM technique and cyclic voltammetry of the

Al-1.5 wt. % Fe alloy samples subjected to laser surface remelting (LSR). The AFM technique is a highly efficient tool for studying the topographies surface, providing details of the surface on a nanometric scale. This technique enables to the quantification of the peaks and valleys that characterize surface roughness. The analyses were performed on both laser-treated and untreated sanded surfaces, revealing significant differences. The region between weld filets in the laser-treated samples has showed by AFM technique the presence of the lamella-like morphology. The low-angle X-ray diffraction analysis has revealed the presence of the alumina, simple metals and metastable intermetallic phases, which considerably have improved the microhardness of the laser-remelted surfaces. The treated surfaces have showed passivity and stability characteristics by cyclic voltammetry in the electrolytic medium employed in this study. The morphology produced by the laser surface remelting enhanced the microstructure of the Al–Fe alloys by reducing their roughness and increasing their hardness.

Corrosion properties of the Al-1.5 wt. % Fe alloys as reviewed by Pariona et al. [6] have showed really interesting results, such as, the micrograph analyzed through SEM was found an homogenous microstructure as a result of the rapid solidification. The results obtained in this study have indicated a likely chemically stable phase, i.e. improved passive/oxide film after LSR treatment, which could serve as an effective barrier against corrosion attack in the aggressive sulfuric acidic environments. At an Open circuit potential (OCP) testing, the results have shown that the displacement for more anodic values for the laser-treated specimen when compared to the untreated specimen is attributed to the formation of an aluminum oxide, this oxide is chemically stable and passive that provides an active barrier against the corrosion. The potentiodynamic polarization results have shown that as a result of the laser-treatment, the corrosion current was reduced by as much as ten times and by the cyclic polarization a wide passive region was obtained. As a consequence of these tests the untreated sample is more susceptible to corrosion, while the laser-treated specimen is less susceptible to corrosion. In the cyclic polarization curves of the untreated sample it was observed greater area of hysteresis loop i.e. higher susceptibility to corrosion than to the laser-treated sample, it can therefore be concluded that LSR process indeed has an influence on surface film modification, which results in higher corrosion resistance.

In this paper the laser surface remelting (LSR) was applied in the Al-1.5 wt. % Fe alloy in order to investigating the treated layer with different corrosion characterization techniques and complementing with other techniques to elucidate the behavior of corrosion. For this purpose, multiple laser weld filets were generated on an entire surface of the samples by LSR technical. Thus, the morphological characteristic of cross-section of the LSR-treated surface was examined and also the existence of different phases in the treated sample was verified. In addition, corrosion testing was carried out, using different techniques to understand the performance of the samples and their stability in the aggressive environment of laser-treated surface in relation to the substrate material. The application possibilities of this technique may be in aeronautic, automotive, energy, electronic, biomedical implant applications, etc.

2. Experimental setup

2.1. Specimen preparation of Al-1.5 wt. % Fe

Usually cylindrical ingot of Al–1.5 wt. % Fe alloy was prepared from pure raw materials in our studies. Chemical composition of the Al–1.5 wt. % Fe alloy measured through the technique of florescence (Shimadzu, EDX-700) is shown in Table 1. The casting assembly of solidification experiments consists of water-cooled mold with heat being extracted only from the bottom, promoting vertical upward directional solidification.

Al	Fe	Cu	Ni
98.347	1.545	0.068	0.04

Table 1. Chemical composition of the Al-1.5 wt. % Fe alloy (wt. %).

2.2. Laser surface remelting of the Al-1.5 wt. % Fe alloy

The samples were cut, polished and sand-blasted to reduce their surface reflectance for the subsequent laser treatment, which was performed with a 2 kW Yb-fiber laser (IPG YLR-2000S) operating at wavelength of 1.06 μm, with an effective focal distance of 160 μm. The laser beam presents a near Gaussian intensity profile with a spot size of 50 μm and the sample surface was positioned 3 mm out of beam focus and the laser beam diameter was estimated at 560 μm on the sample surface. The laser scanning speed was kept at 40 mm/s^{-1}. The average power of the laser beam was fixed at 600 W and the power density on the surface of the sample was estimated as $4.8x10^5$ $W.cm^{-2}$. This laser treatment without an assisting gas jet was applied to augment the production of metal oxides on the laser-treated surface. The laser-treated samples were covered with several weld filets during the remelting process [4].

2.3. Phase, microstructure, and elements characterization

For the metallography characterization of the cross section, small specimens were cut and sanded with 600, 800, 1200 grit SiC sand paper, and polished with colloidal silica in a semi-automatic polishing machine (AROTEC Ind. and Com., Brazil). The micrographs were obtained by optical microscopy (OM, Olympus-BX51).

The low-angle X-ray diffraction (LA-XRD) technique was employed to identify the phases on the laser-treated surface. The profiles were recorded with a scan rate of 2°. min^{-1} and an inclination angle of 1.5°, using a Rigaku-ULTIMA IV X-ray diffractometer.

2.4. Surface roughness studies

The atomic force microscopy (AFM) analysis was performed with a Shimadzu SPM-9600 microscope in 400-nm minimum resolution, equipped with a 125x125x7 μm piezoelectric scanner operating in contact mode, operating point 1.5 V, integration gain 700, rate 0.8 Hz, resolution 512x512 pixels. Characteristic of cantilever, Olympus, COD OMCL-TR800PSA-1, resonance frequency 73 KHz, spring constant 0.57 N/m, rectangular shape, thickness 0.8 μm, probe shape pyramidal, height 2.9 μm, radius 15 nm, material SiN and coated with gold. Software used was SPM Manager versio 3.0, Shimatzu Corporation. For this study, a new cantilever was used every time. Analysis were performed on the material without laser treatment and sample sanded (SiC 1200#) and, as shown in the schema of Fig. 1, on the LSR-treated material on the region on the weld filet (I) and on the region between the weld filets (II).

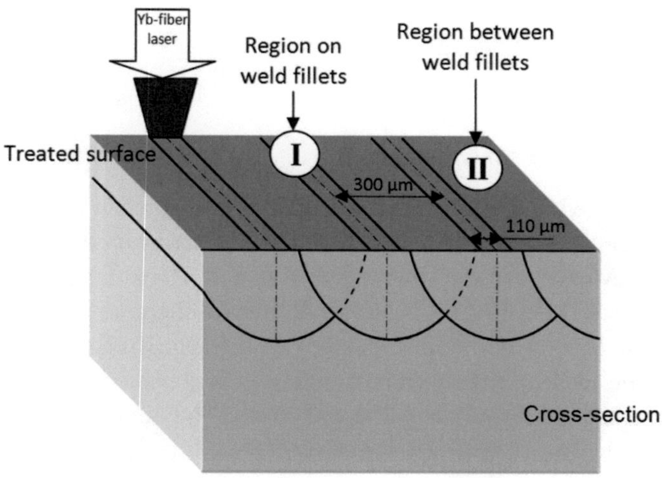

Figure 1. Laser Surface Remelting process schema showing the weld filets three-dimensional formed by the Yb-fiber laser. I – in the region on the weld filet and II – the region between weld filets.

2.5. Electrochemical measurements

Specimens of the untreated and laser-treated material were cut for appropriating the dimensions and then working electrodes were built using epoxy resin. Before the experiments, the untreated Al alloy was polished with 1200 grit SiC sand paper, washed with distilled water and dried with absorbent paper. The laser-treated specimens were only washed with distilled water and dried with absorbent paper.

The corrosion tests were performed in 0.1 $mol.L^{-1}$ of H_2SO_4 aqueous aerated solution using potentiostat Autolab – PGSTAT 30 system that was interfaced to a personal computer for controling the experiment and also for collecting data. The temperature was kept at 25 ± 1 °C during all the experiments using a Brookfield TC-501 thermostat. All the reagents used were analytically pure. Distilled water was used to prepare the aqueous solution.

A three-electrode glass corrosion cell was employed, with a saturated calomel reference electrode (SCE) and a platinum wire with 3.15 cm^2 as counter electrode (CE). The reference electrode was connected to a Luggin capillary for promoting the maximum approach of SCE with the working electrode.

The open circuit potential (OCP or E_{cor}) was determinated as a function of time, at zero current, after 55 minutes of immersion [7]. The micropolarization test has disturbed the system about ±10 mV of the E_{cor} while the macropolarization has been ±150 mV, with a scan rate of $1mV.s^{-1}$. Cyclic polarization tests have initial scan potentials close to the E_{cor}, where the electric currents associated to anodic and cathodic processes are equals. The maximum anodic applied potential was +3.0 V versus SCE, returning to potentials close to E_{cor}, with scan rate of 100 mV/s^{-1}.

3. Results and discussion

3.1. Microstructural and phase analysis

The surface of the Al–1.5 wt. % Fe specimen treated by LSR is shown by optical microscopy (OM) in Fig. 2. Fig. 2a shows the surface of the specimen with an arrangement of multiple weld filets covering the area. Micro-porosity was observed on the surface of the laser-treated sample [4] and more preferably on the region (I). In addition, the protuberances were observed on this surface, which corresponds to the region between weld filets (II). Fig. 2b shows the cross-section of the specimen, where three different regions are observed; the laser-melted zone (LMZ), the heat affect zone (HAZ) and the substrate. The first is characterized by an homogeneous microstructure; according to the study of Ryan and Pragnell [3] by the technique of pulsed laser surface melting (PLSM), the majority of the re-solidified material does not exhibit phase separation between the LMZ and HAZ, this feature was also confirmed by the authors [6,8,9], it is highlighted by dashed yellow lines and while on the substrate are observed grain boundaries.

The meltpool profile not only depends on the energy density but also on the laser beam profile used, which in this case was approximately Gaussian. This beam profile becomes more perceptible in the microstructure at low irradiance and residence times, which was observed in Fig. 2 and this study was confirmed by the authors Chikarakara et al. [10] and these authors also reported that, lower meltpool depths can be advantageous due to their corresponding lower induced thermal stress thus eliminating formation of cracks on the surface. Yue et al. [2] reported that in the solidification of the laser melted zone, a planar solid–liquid interface may prevail, although, this is different from the cellular dendritic structures obtained from the laser surface melting of Al-alloys using CO_2 lasers.

LMZ has the characteristic to be homogenous and many micro-porosities are observed on it [4,6,8]. Furthermore, LMZ has a higher level of homogeneity relative to the substrate material, meaning that the treatment could be considered an ideal tool for surface [3] and this result was coherent with this research. Wang and his collaborators [11] have shown a similar result (Fig. 2b), they reported that the microstructure of the cross-section is characterized by very fine granular grain in the coating and this result was confirmed in this research. In another research of laser treatment of Al alloys, given by Ryan and Pragnell [3], the behavior of homogenous chemical composition was attributed to the distribution of elements as a result of rapid solidification. In addition, low residence time leads to higher cooling rates, producing a more dynamic solidification and allowing the formation of novel phases and more homogeneous microstructures, meaning that apparently useful properties are achieved.

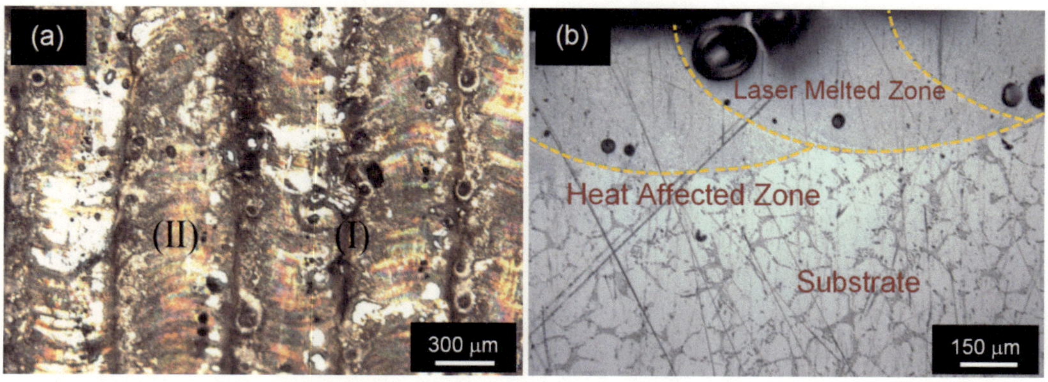

Figure 2. Specimen morphology of LSR by optical microscopy: (a) the LSR-treated surface of an as-received specimen showing in (I) the region on the weld filets, (II) the region between the weld filets and (b) cross-section of the specimen in relation to the weld filets.

3.2. X-ray diffraction analysis

The characterization by low-angle X-ray diffraction (LA-XRD) was made on the laser-treated and untreated specimens and the result is shown in Fig. 3. In this figure two intense peaks around 38° and 45° are observed.

In 2θ equal to 38° the intensity of the peak of the laser-treated specimen is very elevated, however the peak intensity of the untreated specimen is low. In both samples, the element Al is present; however, the phases alumina (Al_2O_3) and iron oxide (Fe_2O_3) are present in the laser-treated specimen, being both phases of the metastable characteristic. Nevertheless, for 2θ equal 45° the peak intensity of the laser-treated specimen is low but for the untreated specimen is high. Furthermore, in both samples in 45°, the phases Al, Fe and $Al_{13}Fe_4$ eutectic [12] are present and the metastable phases, Al_2O_3, AlFe and $FeAl_2$ are present only at the peak of the laser-treated specimen. In other angles the intensity of the peaks are despicable.

In the Al-Fe alloy the typical phase $Al_{13}Fe_4$ FCC [12] with eutectic structure occurs in solidification almost in equilibrium. When LSR is applied at the Al–1.5 wt. % Fe alloy the phase mentioned above turns into phases with metastable characteristics, due to the low residence time used in the experiment. That implies in high cooling rates forcing the formation of different phases, allowing very little time for the diffusion producing the equilibrium of the compositions phases [7].

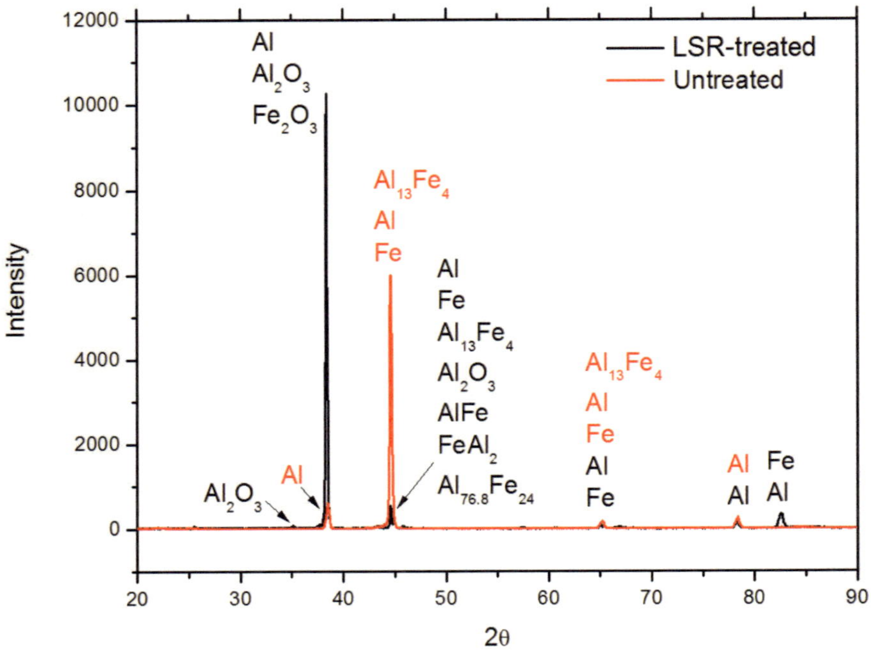

Figure 3. Low-angle X-ray diffraction (LA-XRD) technique for the untreated and laser-treated specimens.

3.3. Surface roughness

One of the advantages of the AFM technique is that it permits being operated in the air, compared to electron microscopy, where the specimens are examined under vacuum conditions causing drying of the specimens and structural alterations [13]. Our aim was to characterise the surface of Al–1.5 wt. % Fe microstructure that resulted from the LSR treatment in a nanoscale order, in the regions on weld filet (I) and between weld filets (II) and compare these results with the untreated specimen. One of the goals of this research is to enhance the understanding of the Yb-fiber laser beam treatment processing. The mathematical formulas that characterize the morphology of the studied surface by AFM is shown in Table 2. A summary of the image analysis results of the structures that were studied on the various surfaces are shown in Table 3. The measure of the roughness is given frequently by the parameter R_a. The size of the AFM area used for this study is shown in Figures 4 and 5.

Parameter	Description	Formula
R_a	Arithmetic average for absolute values **[13]**	$R_a=\frac{1}{N_xN_y}\sum_{i=1,j=1}^{N_x,Ny}\lvert z(i,j)-z_{mean}\rvert$ where $z_{mean}=\frac{1}{N_xN_y}\sum_{i=1,j=1}^{N_x,Ny} z_{ij}$
R_q, R_{RMS}	Root mean square [**13, 14**]	$R_q=\sqrt{\frac{1}{N_xN_y}\sum_{i=1,j=1}^{N_x,Ny}(z(i,j)-z_{mean})^2}$
R_V	Maximum valley depth **[14]**	$R_V=\min_i Z_i$
R_p	Maximum peak height **[14]**	$R_p=\max_i Z_i$
R_Z	Average distance between the highest peak and lowest valley in each sampling length **[15]**	$R_Z=R_p+R_V$
R_{Zjis}	Japanese Industrial Standard for R_z, based on the five highest peaks and lowest valleys over the entire sampling length. **[13]**	$R_{Zjis}=\frac{1}{5}\sum_{i=1}^{5} R_{pi}-R_{vi}$ where R_{pi} and R_{vi} are, respectively, the ith highest peak and lowest valley

Reprinted from Applied Surface Science 2013, 276,1, Pariona MM, et al., Influence of laser surface treated on the characterization and corrosion behavior of Al–Fe aerospace alloys, Copyright (2013), with permission from Elsevier [License Number:3176530612877] [6].

Table 2. Description of the parameters that characterize the morphology of the surface studied by AFM [13-15].

AFM parameter	On weld fillets (I)	Between weld fillets (II)	Untreated
Area [μm^2]	1600	1600	1600
R_a [nm]	178	211	142
R_z [nm]	1714	2097	1889
R_{zjis} [nm]	852	1025	917
R_q [nm]	236	266	190
R_p [nm]	936	1101	752
R_v [nm]	779	995	1137

Reprinted from Applied Surface Science 2013, 276,1, Pariona MM, et al., Influence of laser surface treated on the characterization and corrosion behavior of Al–Fe aerospace alloys, Copyright (2013), with permission from Elsevier [License Number:3176530612877] [6].

Table 3. Quantitative results of the morphology of the surface studied by AFM technique.

Fig. 4 corresponds to the result of the morphological characterization of the region II with R_a=178 nm (Table 2). Fig. 4a shows the result in 3-D, where it can be observed the surface topography that characterizes the different valleys and peaks. In Fig. 4b the rough surface in

2-D is observed, which displays a type of lamellae with different orientations covering the whole considered area.

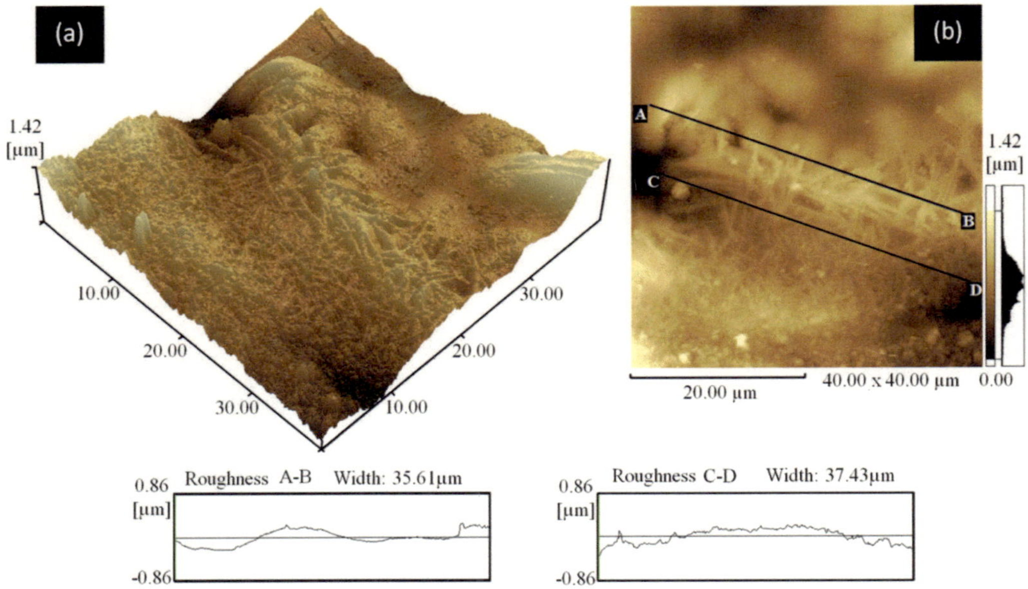

Figure 4. Image displays surface mapping for Al-1.5 wt. % Fe specimen around between weld filets (II): (a) visualization in 3D-AFM and (b) visualization in 2D-AFM.

Fig. 5 shows the result by AFM on region I with R_a= 211 nm. Fig. 5a shows the result in 3-D, where it can be noticed that all parameters corresponding to this area are greater than the ones corresponding to the region II. This fact may be justified due to the greater concentration of nano-porosity that was observed between weld filets (II) according to the micrograph of Fig. 2. In Fig. 5b, the 2-D surface, a not homogeneous feature is shown because of the increased presence of defects in the area, although, in this part it was not possible to observe lamellae, as in the previous case.

The morphology result for the untreated material is shown in Fig. 6 with R_a=142 nm. In Fig. 6a the morphological characteristic is shown in 3-D, and Fig. 6b in 2-D, where some nano-defects on the surface are noticed. According to the results in Table 3, the roughness parameter R_a was lower for the untreated sample than that of the laser-treated sample. This is associated with the presence of many nano-porosities, especially on region I and a certain vaulted shape of the specimen in the region II (Fig 2a). Also, the maximum peak (R_p) observed in the untreated sample was lower than on the treated specimen, but however, the depth of the maximum valley (R_v) was higher than the treated.

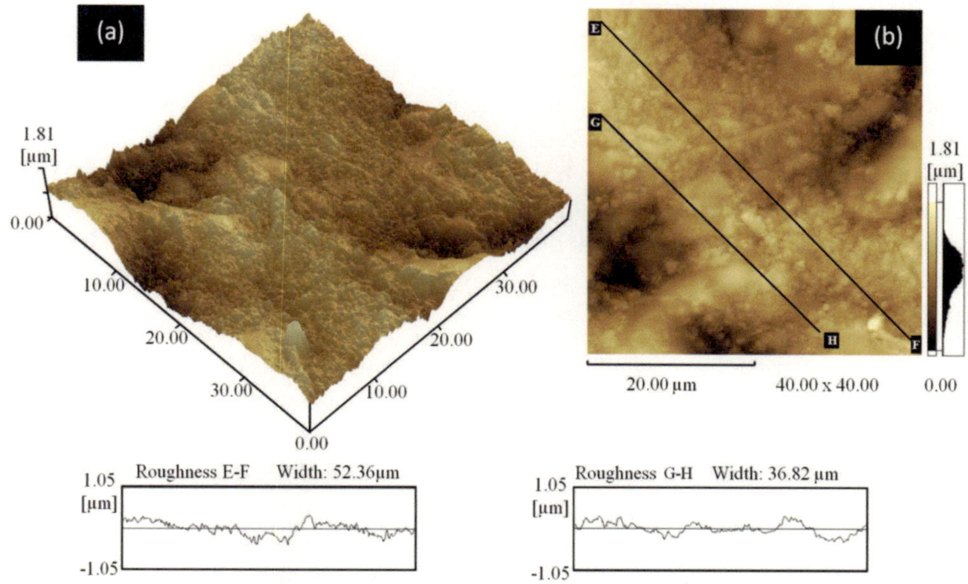

Figure 5. Images of the surfaces of Al-1.5 wt. % Fe around on weld filets (I): (a) visualization in 3D-AFM and (b) visualization in 2D-AFM.

Figure 6. AFM surface mapping of the untreated specimen, sanded with 1200#: (a) visualization in 3D and (b) visualization in 2D.

In this study it was verified that AFM is a suitable technique for examining the surface morphology of the laser-treated alloy [5]. Therefore, the AFM technique is related to the flatness requirement of the investigated area, because, through this characteristic it is possible to quantify the nano-roughness of the surface [5,6]. Thus in this work it was verified that the surface roughness of Al–1.5 wt. % Fe laser-treated is greater in relation to the untreated specimen. In the LSR-treated specimen the decrease in roughness in region II, it is mainly attributed to a more homogeneous morphology showing a lamella-like morphology covering the entire surface. Furthermore, the nano-roughness characteristics could be closely related to the melting characteristics and energy density. An increase in residence time of the laser beam resulted in an increased depth of processing [6,8].

In this study was varied the speed of the cantilever, as a result has been observed that for high speed of this device, the area of the samples analyzed by AFM showed little definition of the surface, however, for low speed cantilever the surface of the samples showed a better definition, therefore, at this work was used to the speed of 0.8 Hz, which is a relatively slow speed. In this work were done also tests for the cantilever in contact mode and non-contact, being that the best definition of the surface was for the contact mode, because this cantilever was more appropriate for material with characteristic surface rather uniform and with high hardness. Also, in this study were performed analysis by AFM in different areas and non-contact, being that the best definition of the surface was for the contact mode,, in this study were performed analysis by AFM in different areas of the surface, but, the characteristics of the lamellae between weld filets were always found [5,6]. Precisely, these characteristic and surface quality were assessed by different corrosion tests, the result is shown in the following section. The study done by the analysis using the AFM technique was necessary to complement the research done by corrosion.

3.4. Electrochemical studies

3.4.1. Cyclical polarization for untreated specimen

Cyclic voltammetry is an efficient and versatile electrochemical technique, used in the study of systems that involve oxidation-reduction reactions in electrolytic solutions, through the analysis of the behavior of anodic and/or cathodic current as a function of the applied potential. This technique allows obtaining the information on the electrode processes, such as, the reversibility of the electrochemical reactions and the determination of kinetic and thermodynamic parameters. Furthermore, the cyclic voltammetry was utilized for comprehension of the oxidation process of the laser-treated and as well as the untreated specimens in a sulphuric acid environment.

Fig. 7 shows two different cyclic polarizations for the untreated specimen in H_2SO_4 aerated solution. The first cycle starts in -0.50 V until the most anodic value +0.90 V, versus SCE, returning to -0.50V. During the anodic scan, there is a current increase near the -0.15 V, associated with the aluminum dissolution, followed by the formation of an aluminum oxide (Al_2O_3) film on the sanded substrate. The anodic current rapidly has increased until approximately +0.30 V, followed by a smaller growth rate until +0.90 V; it is characterized as a region of thickening of the oxide film. After the inversion potential, there is a sharp drop of the anodic current up to +0.60 V and, thereafter, reaching low values of currents in potential close to -0.50 V.

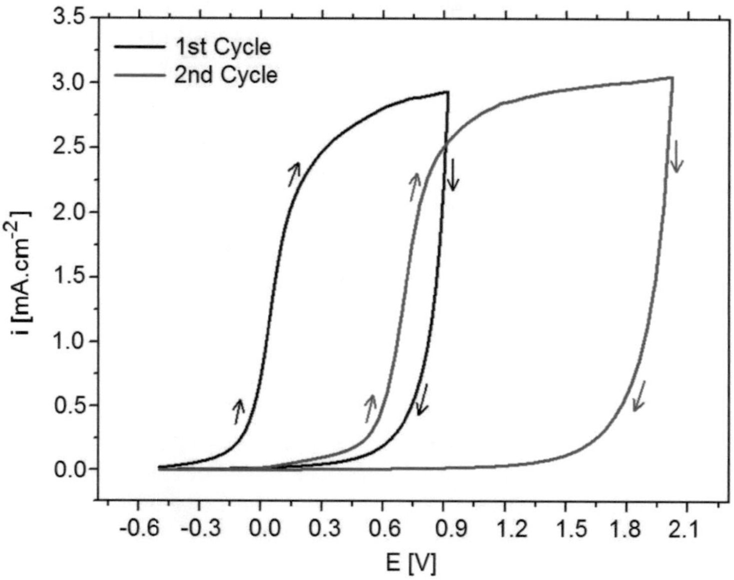

Figure 7. Cyclic polarization curves for untreated specimens, at 100 mV.s^{-1}, in 0.1 mol.L^{-1} aerated solution of H_2SO_4: (a) First scan cycle and (b) second scan cycle.

During the process, two chemical reactions are associated with the anodic potential scan. The first step is the electrochemical aluminum dissolution, followed by chemical reaction of Al^{+3} ions with the water molecules in the acid solution interface metal/solution, i.e. according to the reactions [17]:

$$Al_{(M)} \rightarrow Al^{+3}_{(M/sol)} + 3\ e^{-} \quad (1)$$

$$2\ Al^{+3}_{(M/sol)} + 3\ H_2O_{(M/sol)} \rightarrow Al_2O_{3(Ox)} + 6\ H^{+}_{(Ox/sol)} \quad (2)$$

Where M, M/sol, Ox e Ox/sol correspond to metal, interface metal/electrochemical solution, oxide layer and interface oxide/solution, respectively.

The authors Zankara et al. [18] and Santos et al. [19] showed that the aluminum oxides formed in acid electrolytes presents uniformly nanostruturated forms, like nanotubes with cylindrical pores, which the size orthe thickness depend on the electrical charge applied to the electrochemical system, and through this nanotubes there is Al^{+3}ions diffusion.

Finished the first cycle (Fig. 7), the electrode was maintained in the solution for 5 minutes to restore the equilibrium conditions on the interface electrode/solution, while keeping the system in open circuit potential (zero current). Right after, the second cycle began with -0.50 V to +2.0 V where the anodic current of aluminum dissolution starts increasing from +0.30 V, being around 600 mV more anodic when compared to the first cycle of potential scan; after de potential inversion, from +1.50 V, the values of anodic current density goes back to the same ones of the first cycle. The displacement for more positive potential values is associated with the growth of a second layer of an Al_2O_3 film over than the one obtained during the first cycle of potential scan. In the second cycle, the anodic current grows with lower rates from +1.20 V, due to the thickening of the Al_2O_3 film, and this current density is close to 3.0 $mA.cm^{-2}$ in that potential; although, in the inversion potential scan (+2.0 V), the anodic current diminished.

3.4.2. Cyclical polarization for untreated and laser-treated specimens

In this section a comparative study was carried out of the cyclic polarization to laser-treated and untreated samples with the purpose of studying the behavior of these materials in the same electrolyte solution. Fig. 8a shows the comparison between the cyclic polarizations curves for the laser-treated and untreated specimens, in aerated solution of H_2SO_4 0.1 $mol.L^{-1}$, at 25°C.

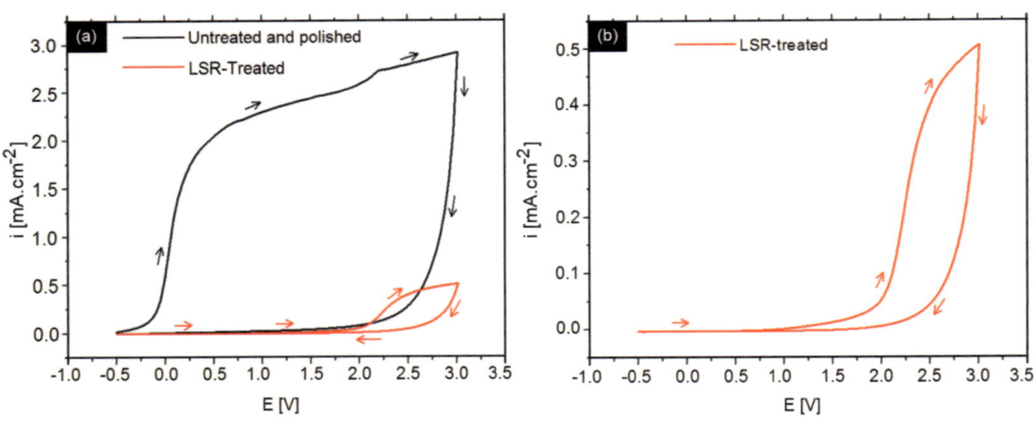

Figure 8. Cyclic polarization curves (a) untreated and laser-treated specimens and (b) magnification for the laser-treated specimen.

In Fig. 8a, the black line shows the cyclic polarization curve of the untreated specimen carried out in the range of potential between -0.50 V to +3.0 V versus SCE. It was noticed that the anodic current increases in the vicinity of -0.15 V and a change of curvature in the graphic occurs in +0.5 V, as it was also obtained in the first cycle of the cyclic polarization in Fig. 7. This change in the curvature is probably related to the reactions given by equations (1) and (2). Next to the potential of +2.10 V is observed, again, an increase in the current that can be associated to the thickening of the Al_2O_3 film, meanwhile, it can be related to the oxidation of the intermetallic phase Al_3Fe observed in Al–1.5 wt. % Fe alloy [4]. After the reversal potential, the dissolution current sharply decreases to +2.0 V, reaching low values of anodic current.

The cyclic polarization curve of Fig. 8b, is the same of Fig. 8a which was magnified forbettering the visualization. It is observed that the anodic current increases from the potential of +1.75 V, this can be associated with the growth of the aluminum oxide film and/or Al–Fe phases on the surface of laser-treated sample, which is constituted by different phases, such as Al_2O_3, Fe_2O_3, Al, AlFe, $Al_{13}Fe_4$ among others, this was confirmed by X-ray (Fig.3). A very similar point of view was reported in [4].

In the magnification Fig. 8b, the anodic potential that corresponds to the beginning of the oxidation on the surface of the LSR-treated was displaced to +1.90 V. In the anodic scan the elevation of the current starts at -0.15 V in the untreated sample, however, in the LSR-treated sample begins at +1.75 V, thus in the range of -0.5 V to +1.75 V the current is kept almost constant, showing the passivity of the LSR-treated sample. Thereby, during polarization as much forward as in return for this specimen, passivation had occurred, and it had also appeared in a spanning region over 2.25 V. Therefore, the surface of the LSR-treated has proved to be an effective way for achieving the passivation in this electrolytic environment, indicating the stability of the passive film against the dissolution. The authors Qu et al. [20] and Ng et al. [21] have observed a similar result for the Mg and Ti alloys, respectively. In addition, the authors Viejo et al. [1] have concluded that the change in the electrochemical behavior has been associated with the formation of a relatively uniform melted layer with an improved passivity properties in a magnesium alloy.

Nevertheless, the anodic current density in the inversion potential for the laser-treated alloy is 0.5 $mA.cm^{-2}$, as can be seen in Fig. 8b. This value is lower than the one observed for the untreated alloy, which presented a current of 2.92 $mA.cm^{-2}$. These results show that the laser-treated alloy surface has a different chemical composition and its own morphology, pre-oriented by surface, because of the formed Al_2O_3 film.

In Fig. 8 the hysteresis loops are observed, being greater for untreated sample. This characteristic is related to the susceptibility to attacked by the corrosion. Seah et al. [22] reported that the absence of any hysteresis loop demonstrates clearly the capability of the titanium to a quickly repassivate. Also Shih et al. [23] has pointed out that in general, the larger the area of the hysteresis loop, implies a greater susceptibility to pitting corrosion. Once again, it has been proved that the LSR-treatment was effectively improved corrosion resistance properties.

3.4.3. Open circuit potential (OCP) for laser-treated and untreated specimens

The nature of the oxide layer of Al_2O_3 formed during the LSR-treatment may have influence over the corrosion potential, as well as in polarization resistance (R_p) and in Tafels constant β_a and β_c, associated with the anodic and cathodic processes, respectively, which occurs at the interface understood by the laser-treated surface and the electrolyte solution. Moreover, the effect of laser-treatment in the Al–1.5 wt. % Fe alloy may directly affect the corrosion resistance of this material in the solution of H_2O_4 and these was verified in this study.

Fig. 9 shows the behavior of the OCP of laser-treated and untreated alloys as a function of exposure time in an acid medium. There was a shift in the initial potential of the untreated sample from -0.670 V to -0.720 V after approximately 750 seconds followed by stabilization of E_{cor} in -0.706 V, after 55 minutes in OCP. About the same period of time, the laser-treated alloy suffers a displacement of initial potential of -0.530 V until stabilization in -0.606 V.

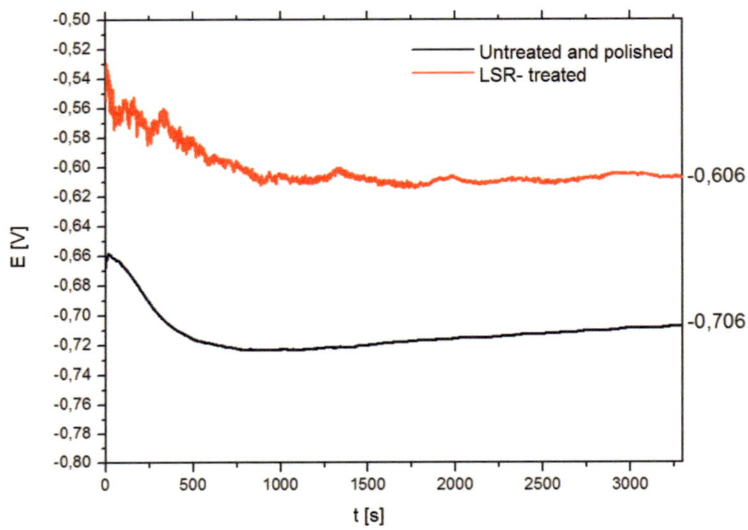

Figure 9. OCP versus saturated calomel electrode (SCE) in aerated solution of H_2SO_4 0.1 mol. L^{-1}.

The displacement of 100 mV for more anodic values in the corrosion potential for the laser-treated material is attributed to the formation of an aluminum oxide, this oxide is chemically stable that provides an active barrier against the corrosion of Al–1.5 wt. % Fe alloy. A similar result was attained by authors Viejo et al. [1] for aluminum alloy using the LSM technique.

3.4.4. The polarization resistance of laser-treated and untreated specimens

The micropolarization of ± 10 mV around corrosion potential promotes a perturbation in the equilibrium potential, giving the appearance of an anodic and cathodic current in the electrochemical cell circuit. This technique of the electric current versus applied potential was conducted to the laser-treated and untreated samples, and the result is shown in Fig. 10. The inverse slope of the line in Fig. 10 allows to determining the polarization resistance of the system that is associated with the charge transfer processes in the interface metal/oxide/

electrolyte solution. The values of 17.8 KΩ for the laser-treated alloy against 1.4 KΩ for the untreated alloy showed that the metal/solution interface is different and the polarization resistance has direct impact on corrosion rates of these surfaces in sulfuric acid. Therefore, the microstructure generated by the laser-treatment definitely influences the polarization resistance (R_p) of the alloy.

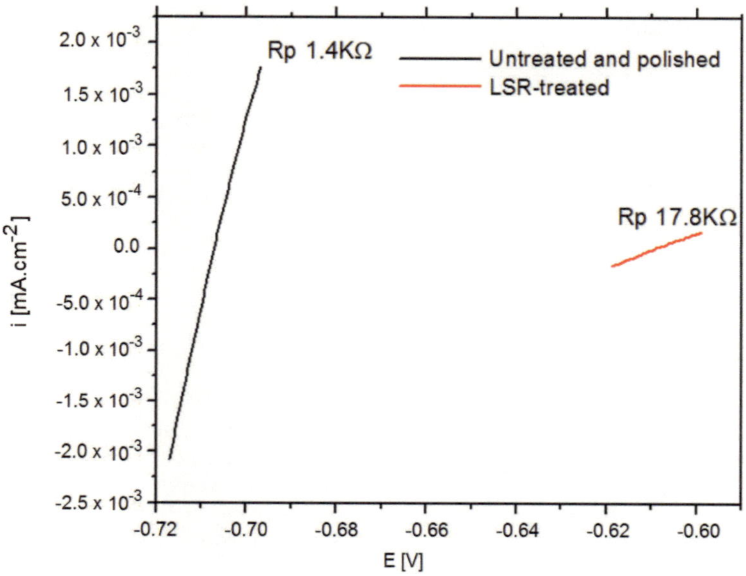

Figure 10. Micropolarization of ± 10 mV near of E_{cor} versus SCE in aerated solution of H_2SO_4 0.1 mol.L^{-1}, with scan rate of 1 mV.s^{-1}.

3.4.5. The corrosion current (i_{cor}) for laser-treated and untreated specimens

The potentiodynamic polarization curves for the Al–1.5 wt. % Fe alloy in 0.1 mol.L^{-1} of H_2SO_4, aerated at 25°C and with a scan rate of 1 mV.s^{-1} is shown in Fig. 11, varying from cathodic potentials at -150 mV toward E_{cor} and more positive to + 150 mV. According to this curves, the anodic (β_a) and cathodic (β_c) constants of Tafel are determinate.

Fig. 11 shows the behavior of Log (i) as a function of applied potentials over the alloy. At the point of inversion of cathodic to anodic current during the dynamic scanning, it represents approximately the corrosion potentials obtained after 55 minutes of immersion in OCP (Fig. 9). The different values of β_a and β_c are associated with the electrochemical reaction of oxidation and reduction over the surfaces of working electrodes.

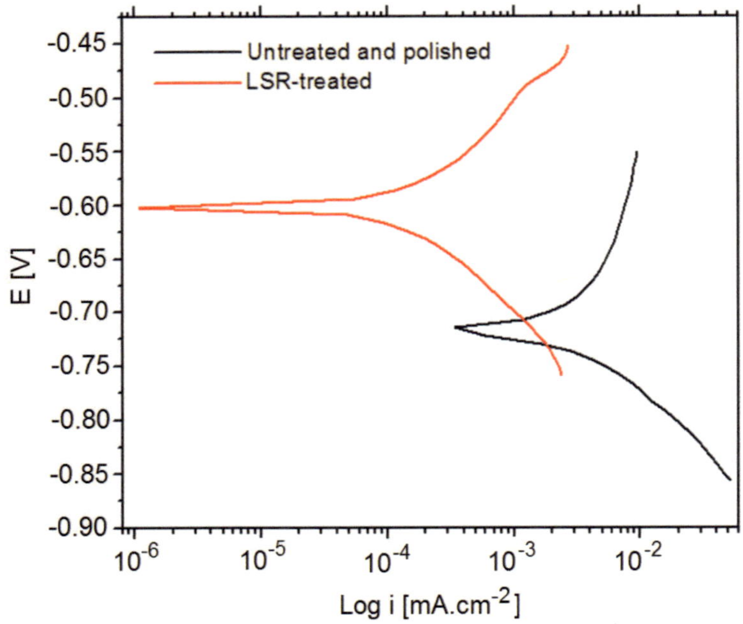

Figure 11. Potentiodynamic polarization for untreated and laser-treated alloys in the interval of ± 150 mV around E_{cor} versus SCE.

The corrosion current (i_{cor}) or the corrosion rate were calculated from the values of R_p, β_a and β_c appling Stern-Geary equation [20]. The main electrochemical parameters were calculated, and are shown in Table 4.

Condition	E_{cor} (V)	R_p (KΩ)	β_a ($V.dec^{-1}$)	β_c ($V.dec^{-1}$)	icor ($\mu A.cm^{-2}$)	Corrosion rate ($mm.year^{-1}$)
Laser-treated	0.606	7.8	0.161	0.110	11.7	0.1
Untreated	0.706	1.4	0.432	0.131	119.8	1.3

Table 4. Electrochemical parameters calculated for the Al-1.5 wt. % Fe alloy for laser-treated and untreated specimens in an aerated solution of H_2SO_4 0.1 $mol.L^{-1}$ at 25°C.

Table 4 shows the range of variation of the β_c for the untreated and laser-treated samples that is around 20 $mV.dec^{-1}$ which occurs in the linear region of the cathodic branch of Fig. 11. This process is associated with hydrogen reduction reaction over the working electrode. However, the β_a value had an expressive variation around 270 mV.dec^{-1}, it is referring to the linear region of anodic branch. This indicates that the oxidation process of the alloy is affected by the conditions of the electrodes.

Therefore, the value of the corrosion rate of laser-treated alloy is smaller than that of the untreated alloy in about 12 to 14 times. This occurs due to the presence of an oxide layer and because of the other phases produced by heat treatment with laser, turning this layer more resistant to the corrosion when compared to the oxide film formed on the surface of the untreated alloy during the interval of stabilization of the corrosion potential.

In this research, a great improvement in the performance of corrosion was confirmed on the laser-treated alloy, which showed a greater range of passivity and very high polarization resistance, a lower area of the hysteresis loop, and the corrosion rate was much smaller in relation to the untreated alloy. Through the analysis of the laser-treated sample by different techniques the following factors could have been contributed to a good performance of protection against corrosion: the micrograph presented an homogeneous microstructure; the surface roughness was lower by AFM; the presence of a lamellar structure were detected by AFM; the existence of several metastable phases by LA-XRD and non-existence of second-phase particles. Besides, in all the corrosion testing a reduction of current density was found in the laser-treated alloy; this result was confirmed by authors Viejo et al. [1]. They concluded that LSM-treated specimens exhibited a general reduction in current density over most of the examined potential range. They still have extended their research and reported that even after LSM; the alloy experienced a reduction in anodic and cathodic activities, mainly associated with the formation of a relatively uniform melted layer with passive properties and the diminished presence of intermetallic precipitates. They have also verified that LSM enhanced the intergranular corrosion resistance of the AA2050-T8 alloy.

The parameters that can have a strong influence on the increase in the corrosion performance are the homogenous microstructural and non-existence of the second-phase particles characteristics. To respect several authors have verified these characteristics. The authors Chikarakara et al. [10] used the energy dispersive X-ray spectroscopy (EDS) technique analysis and they demonstrated that laser surface modification produced a more homogenous chemical composition of the alloying elements compared to the untreated bulk metal. The authors Yue et al. [2] examined the laser-treated specimen with TEM and concluded that it did not reveal any coarse second-phase particles, as were found in the untreated specimen. However, a stable corrosion film had not been formed at the surface of the untreated specimen due to the presence of numerous second-phase particles, which readily have initiated pitting corrosion and have destroyed the integrity of the film. The laser-treated layers consisted mainly of the nanocrystalline structures α-Al_2O_3, which is a chemically stable phase and serves as an effective barrier to protect the matrix against corrosion attacks [2], possibly in this work these phases nanocrystalline structure facilitates the formation of lamellar structures identified by MFA, these lamellars provide in turn better behavior of surface treated against corrosion. According to the study of Ryan and Pragnell [3] by pulsed laser surface melting (PLSM) technique, they have showed that the increase in the corrosion performance of the pulsed laser-treated alloys has been widely attributed to the formation of a surface layer that is much more chemically homogenous than the bulk material. They assumed that the laser treatment removes second phase particles and partitionless re-solidification of the layer occurs. Redistribution of alloying elements within the treated titanium alloys is also known to improve pitting corrosion resistance due to preferential corrosion attack prevention [3]. In addition, Chikarakara and their collaborators [10] pointed out that a secular martensite structure enhances the materials wear and corrosion resistance.

4. Conclusions

Laser surface remelting (LSR) without protective coating with a 2 kW Yb-fiber laser (IPG YLR-2000S) was applied on Al–1.5 wt. % Fe alloy to investigate the corrosion behavior in an aerated H_2SO_4 solution (0.1 mol.L^{-1}). The results obtained permit the following conclusions:

1. An homogenous microstructure was verified as a result of the rapid solidification,
2. The LSR technique was successfully established to improve the surface properties of Al–1.5 wt. % Fe alloys in relation to the substrate alloy.
3. The results obtained in this study indicate a more likely chemically stable phase, i.e. improved passive/oxide film after LSR treatment, which could serve as an effective barrier against corrosion attack in an aggressive sulfuric acidic environments,
4. At an OCP testing, the results have shown that the displacement for more anodic values for the laser-treated specimen when compared to the untreated specimen is attributed to the formation of an aluminum oxide, this oxide is chemically stable and passive that provides an active barrier against the corrosion,
5. The potentiodynamic polarization results have shown that as a result of the laser-treatment, the corrosion current was reduced by as much as ten times and by the cyclic polarization a wide passive region was obtained. As a consequence of these tests the untreated sample is more susceptible to corrosion, while the laser-treated specimen is less susceptible to corrosion,
6. In the cyclic polarization curves of the untreated sample it was observed greater area of hysteresis loop i.e. higher susceptibility to corrosion than for the laser-treated sample,
7. It can therefore be concluded that LSR process indeed has an influence on surface film modification, which results in higher corrosion resistance.

Acknowledgements

The authors acknowledge the financial support of the Brazilian research funding agencies CNPq (National Council for Scientific, Technological Development), AF (Araucária Foundation), and FINEP (Research and Projects Financing Agency). We also thank CLABMU/ PROPESP UEPG for the analyses.

Author details

Moisés Meza Pariona

Address all correspondence to: mmpariona@uepg.br

Engineering and Materials Science, State University of Ponta Grossa (UEPG), Ponta Grossa, Brazil

References

[1] A.E. Coy, F. Viejo, F.J. Garcia-Garcia, Z. Liu, P. Skeldon, G.E. Thompson, Effect of excimer laser surface melting on the microstructure and corrosion performance of the die cast AZ91D magnesium alloy, Corros. Sci. 52 (2010) 387–397.

[2] T.M. Yue, L.J. Yan, C.P. Chan, C.F. Dong, H.C. Man, G.K.H. Pang, Excimer laser surface treatment of aluminum alloy AA7075 to improve corrosion resistance, Surf. Coat. Technol. 179 (2004) 158-164.

[3] P. Ryan, P.B. Prangnell, Grain structure and homogeneity of pulsed laser treated surfaces on Al-aerospace alloys and FSWs, Mater. Sci. Eng. A, 479 (2008) 65-75.

[4] M.M. Pariona, V. Teleginski, K. dos Santos, S. Machado, A.J. Zara, N.K. Zurba, R. Riva, Yb-fiber beam effects on the surface modification of Al-Fe aerospace alloy obtaining weld filet strutures, low fine porosity and corrosion resistence; Surf. Coat. Technol. 206 (2012) 2293-2301.

[5] M.M. Pariona, V. Teleginski, K. dos Santos, E. L. R. dos Santos, A. C. de Lima, A.J. Zara, R. Riva. AFM study of the effects of laser surface remelting on the morphology of Al–Fe aerospace alloys. Materials Characterization, (2012) 64-76.

[6] M.M. Pariona, V. Teleginski, K. dos Santos, A.A.O.C. de Lima, A. J. Zara, K. T. Micene,, R. Riva. Influence of laser surface treated on the characterization and corrosion behavior of Al–Fe aerospace alloys. Applied Surface Science. 276 (2013) 76-85.

[7] ASTM G59-97. Standard Test Method for Conducting Potentiodynamic Polarization Resistance Measurements, 2009.

[8] M. Ulmeanua, F. Jipa, C. Radu, M. Enculescu, M. Zamfirescu, Large scale microstructuring on silicon surface in air and liquid by femtosecond laser pulses, Appl. Surf. Sci. 258 (2012) 9314– 9317.

[9] F. Bertelli, E.S. Meza, P.R. Goulart, N. Cheung, R. Riva, A. Garcia, Laser remelting of Al–1.5 wt % Fe alloy surfaces: Numerical and experimental analyses, Opt. Laser Eng. 49 (2011) 490–497.

[10] E. Chikarakara, S. Naher, D. Brabazon, High speed laser surface modification of Ti–6Al–4V, Surf. Coat. Technol. 206 (2012) 3223–3229.

[11] A.H. Wang, H.B. Xia, W.Y. Wang, Z.K. Bai, X.C. Zhu, C.S. Xie, YAG laser cladding of homogenous coating onto magnesium alloy, Mater. Lett. 60 (2006) 850–853.

[12] Thermo Calc software, Stockholm, Sweden, 2011.

[13] A. Daskalova,_, S. Bashirb, W. Husinsky, Morphology of ablation craters generated by ultrashort laser pulses in dentin surfaces: AFM and ESEM evaluation, Appl. Surf. Sci. 257 (2010) 1119–1124.

[14] J. Podskocova, D. Chorvat–Jr, G. Kollarokova, I. Lacik, Characterization of polyelectrolyte microcapsules by confocal laser scanning microscopy and atomic force microscopy, Laser Phys. 15 (2005) 545–551.

[15] U. Khandey, Optimization of surface roughness, material removal rate and cutting tool flank wear in turning using extended taguchi approach, Master thesis, Mechanical Engineering, National Institute of Technology Rourkela, India, 2010.

[16] ISO 4287 – 1997 Geometrical Product Specifications (GPS) — Surface Texture: Profile Method—Terms, Definitions and Surface Texture Parameters.

[17] A. Santos, L. Vojkuvka, J. Pallanés, J. Ferré-Borrull, L.F. Marsal, In situ electrochemical dissolution of the oxide barrier of porous anodic alumina fabricated by hard anodization, J. Electroanal. Chem. 632 (2009) 139-142.

[18] L. Zaraka, G. D. Sulka, J. Szeremeta, M. Jaskula, Porous anodic alumina formed by anodization of aluminium alloy (AA1050) and high purity aluminium; Electrochim. Acta 55 (2010) 4377-4386.

[19] A. Santos, P. Formentin, J. Ferré-Borrull, J. Pallanés, L.F. Marsal, Nanoporous anodic alumina obtained without protetive oxide layer by hard anodization; Mater. Lett. 67 (2012) 296-299.

[20] Q. Qu, J. Ma, L. Wang, L. Li, W. Bai, Z. Ding, Corrosion behaviour of AZ31B magnesium alloy in NaCl solutions saturated with CO_2, Corros. Sci. 53 (2011) 1186–1193.

[21] K.W. Ng, H.C. Man, T.M. Yue, Characterization and corrosion study of NiTi laser surface alloyed with Nb or Co, Appl. Surf. Sci. 257 (2011) 3269–3274.

[22] K. H. W. Seah, R. Thampuran, S. H. Teoh, The influence of pore morphology on corrosion, Corros. Sci. 40 (1998) 547-556.

[23] C.-C Shih, C.-M. Shih, Y.-Y. Su, L. H. J. Su, M.-S. Chang, S.-J. Lin, Effect of surface oxide properties on corrosion resistance of 316L stainless steel for biomedical applications, Corros. Sci. 46 (2004) 427–441.

Chapter 28

Developments in Reliability-Based Assessment of Corrosion

Zahiraniza Mustaffa

Additional information is available at the end of the chapter

http://dx.doi.org/10.5772/57488

1. Introduction

The title of this book, *Developments in Corrosion Protection*, portrays one of the important issues addressed by the industries in the present world. This theme covers a wide range of topics with regards to corrosions which can be further expanded in different contexts. This chapter proposes the theme to be addressed based on the assessment aspects of managing corrosions, for which it is very much applicable to a structure that is already in operation. The discussion presented in this chapter is generally applicable to corrosions in general but special attention is given to corrosions in offshore pipelines. Thus this chapter attempts to brief readers on the development of the approaches used in assessing the reliability of corroded pipelines.

2. Disputes on the significance of corrosion parameters

A corrosion defect that forms in a pipeline is represented by several length scale parameters, namely *depth* (d), *longitudinal length* (l) and *circumferential width* (w), as shown in Table 1. Typically, it will have an irregular depth profile and extend in irregular pattern in both longitudinal and circumferential directions (Cosham et al., 2007). A defect will spread and develop in size with time. Its growth is described by the d, l and w dimensions. Therefore, one of the earliest concerns related to corrosion assessment in pipelines is about understanding the importance of these parameters with respect to the reliability of the pipe. Only the most governing parameters will be included in the design codes and standards.

Extensive experimental and numerical works have been dedicated to determine the best governing parameters to represent a corrosion shape. The current assessment practices use a single simple corrosion geometry and the corrosion circumferential width (w) is not considered

(Fu and Kirkwood, 1995). The longitudinal extent of a corroded area is the most important length parameter for the burst strength under internal pressure loading (Cosham et al., 2007). Defects in this orientation have been reported to be the most severe since it alters the hoop stress distribution and promotes bulging. Concurrently, Chouchaoui and Pick (1994), Fu and Kirkwood (1995) and Batte et al. (1997) have shown that the influence of corrosion circumferential width (w) to failures was not that significant.

There have been many arguments on the corrosion shapes assumed by the design standards. Corroded area has been argued from as simple shape as a rectangle to parabolic and average of rectangular and parabolic shapes, summary of which as shown in Table 1. The most conservative idealisation is a rectangular profile (Cosham et al., 2007). Even until today, one can never be too sure of the assumptions made in those standards.

Corrosion Shape	Calculation of Area, A	Corrosion Illustration
Rectangle	dl	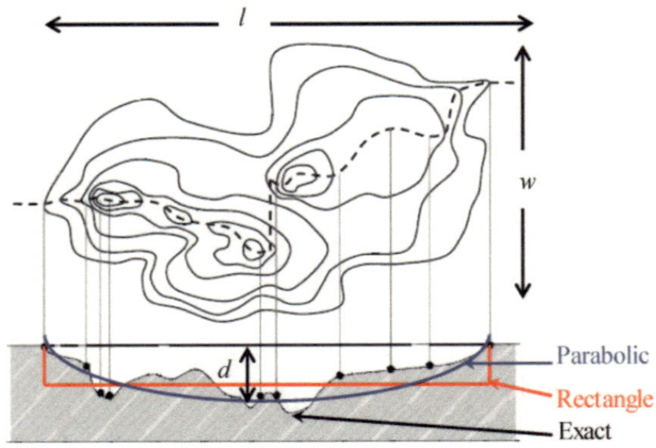
Parabolic	$2/3dl$	
Average of rectangle and parabolic	$0.85dl$	
'Exact' shape	'Exact' calculation	

Note = Corrosion parameters: depth (d), longitudinal length (l) and circumferential width (w)

Table 1. Disputes on corrosion shapes

3. Mechanics of corrosion models

The assessment models used to describe the reliability of corrosions in pipelines were originated from the mechanics of the circumferential stress or hoop stress (σ_h) acting on a pipeline. For this, consider a unit length (1 m long) pipeline containing fluids with external diameter (D_o), internal diameter (D_i), wall thickness (t), internal pressure (p_i), and external pressure (p_o), as shown in Fig.1(a). The idea is to determine the force that the internal pressure induces in the wall by considering the equilibrium of everything within the circumscribing rectangle drawn in Fig. 1(b). Half the pipe and half the contents are redrawn in Fig.1(b) as a free body diagram. The rectangle is bounded by the diameter, two tangents at the point where the diameter intersects the outside surface, and a tangent parallel to the diameter. The stress components that act across the boundaries of different parts of the rectangle are known as the hoop stress.

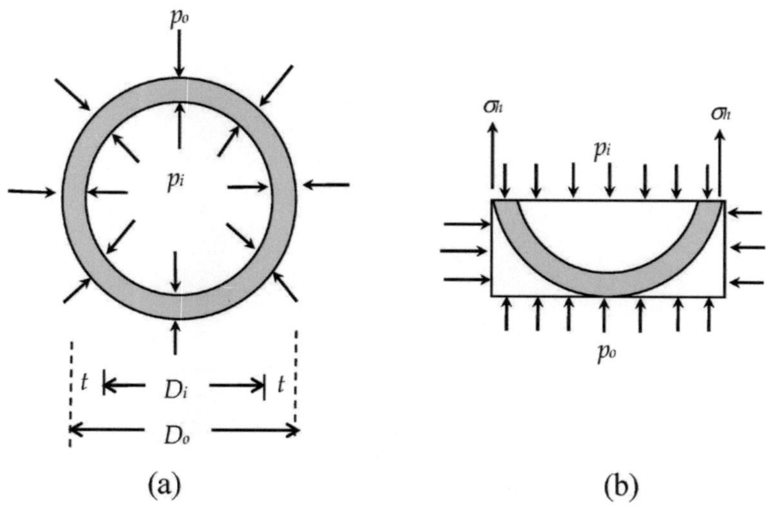

Figure 1. Circumferential stress in a pipeline pressurized internally and externally

The resultant force in the vertical direction must be zero, thus the equilibrium equation becomes,

$$p_o D_o + 2\sigma_h t = p_i D_i \tag{1}$$

Arranging Eq. (1),

$$\sigma_h = \frac{p_i D_i - p_o D_o}{2t} \tag{2}$$

Eq. (2) gives the mean circumferential stress exactly, whatever the diameter-to-thickness (*D*/*t*) ratio. There are various versions of Eq. (2) and the most widely used is the Barlow formula, given by,

$$\sigma_h = \frac{p_i D}{2t} \tag{3}$$

The above formula was derived by neglecting the external pressure term $p_o D_o$ in Eq. (2). Internal pressure from the contained fluid is the most important loading a pipeline has to carry (Palmer and King, 2008). *D* is normally taken as the outside diameter which is obviously larger than the inside parameter. This can be interpreted as a round-and-ready way of allowing for the small variation of hoop stress through the wall thickness (Palmer and King, 2008). Rearranging Eq. (3),

$$p_i = \frac{2\sigma_h t}{D} \tag{4}$$

It can be said that that,

$$p_i = f\left(\frac{t}{D}, \sigma_h\right) \tag{5}$$

The above equation implies that the internal pressure of an intact (no defect) pipe can withstand is a function of a wall thickness-to-diameter (t/D) ratio and its strength (or stress).

For the case of a pipeline with corrosion defects, Eq. (5) can be modified by incorporating the defect projected area (A) term into the equation. The same principle was applied when developing the failure pressure (PF) model; a model used for the assessment of remaining strength in a pipeline subjected to corrosions. Generally, the basic PF model can be expressed as,

$$p_i = PF = f\left(\frac{t}{D}, \sigma_h, A\right) \tag{6}$$

Batelle developed a semi-empirical equation for the remaining strength of corroded pipelines in early 1970 (Maxey et al., 1971; Kiefner and Duffy, 1971; Kiefner, 1974). The equation has been called the NG-18 equation and is given by,

$$PF = \frac{\sigma_{flow} . 2.t}{D}\left[\frac{1 - \frac{A}{A_o}}{1 - \frac{A}{A_o}\frac{1}{M}}\right] \tag{7}$$

where, A_o=dt, M is Folias bulging factor, σ_{flow} is flow stress, and d is maximum corrosion depth. Note that the σ_h term has been replaced by σ_{flow} here. Several modifications have been made to the above parameters depending on the available test data sets and study techniques. These includes (i) flow stress, σ_{flow}, (ii) defect profile or projected corrosion area, A, and (iii) geometry correction factor (also referred to as the Folias factor, or the bulging correction factor, M).

The flow stress (strength), σ_{flow} is a concept proposed in the 1960s to measure the strength of steel in the presence of a defect. The NG-18 equation here assumes that failure is due to a flow stress dependent mechanism and can, therefore be described by the tensile properties like yield strength or ultimate tensile strength (Cosham et al., 2007). The σ_{flow} has been proposed for several modifications, as listed below,

$\sigma_{\text{flow}} = 1.1$ SMYS

$\sigma_{\text{flow}} = 1.15$ SMYS

$\sigma_{\text{flow}} = 0.5$ (SMYS+SMTS)

$\sigma_{flow} = SMYS + 68.95$ MPa (or 10 ksi)

$\sigma_{flow} = x.SMYS$, where $x = 0.90$, 1.0 or 1.1

where, *SMYS* and *SMTS* is Specified Minimum Yield Stress and Specified Minimum Tensile Strength, respectively. The projected corrosion area, A has also undergone several propositions, namely,

$A = dl$ (rectangle)

$A = 2/3dl$ (parabolic)

$A = 0.85dl$ (approximate average of rectangle and parabolic)

$A =$ 'exact' calculation

with l as the defect longitudinal length (refer Table 1).

The geometry correction factor which is also referred to as the Folias factor, or the bulging correction factor, M developed by Folias (1964) to account for the stress concentration that is caused by radial deflection of the pipe surrounding a defect.

4. Reliability assessment of corrosion in pipelines

4.1. Deterministic vs. probabilistic method

The offshore technology is evolving and growing rapidly. As offshore knowledge continues to evolve, the recommended design practice gets revised accordingly as well. The basic approaches to structural reliability and design codes can be historically described based on three approaches, namely,

- Deterministic
- Semi-probabilistic
- Probabilistic

The *deterministic (traditional)* method originated from the knowledge of allowable stress design and plastic design while the *semi-probabilistic* method progresses from the ideology of partial factors design. The *probabilistic* method on the other hand, is based on the analytical methods and simulation. The difference between the early and recent practice in the design of structures can be summarized in Table 2.

Early Practice	Recent Practice
Design Margin	Probability of Failure
Factor of safety	Partial safety factors
Experience based	Probabilistic calculation
Natural scatter ignored	Statistical distributions used in practical engineering

Table 2. Comparison between the early and recent practice in the design concept of structures

In the deterministic method, the safety aspect of structures was expressed in terms of *safety margins* and *safety factors* to consider the effect for uncertainties in loading and material properties and inaccuracies in geometry and theory. The use of precisely defined point design (single) values represents not what an engineer needs to accomplish, but rather what is convenient to numerically solve, assuming inputs that are known precisely (Singh et al., 2007). It considers the *worse-case scenarios* to determine the load and capacity of a system. In most cases, such safety margins and factors are seldom based on any mathematical rigor or true knowledge of the underlying risk and results in an overdesign (Singh et al., 2007). Consequently, this leads to designs that are more 'heavy' and costly and even result in greater safety or reliability.

The probabilistic method deals with many uncertainties that are common to the data (random variables) employed. Both the strength (R) and load (S) can take on a wide range of values by explicitly incorporating uncertainty in system parameters. Note that the deterministic method does not give any idea of *probability of failure* (P_f) or *reliability*. The P_f is a rational measure of safety. The key to probabilistic method is the interference between load and strength to evaluate P_f. The approach treats both random variables in the form of probability density functions (with statistical parameters mean, μ and standard deviation, σ) rather than considering each input parameter as an average value, as what has been assumed in the deterministic approach. Fig. 2 illustrates the contradiction on these definitions. Measuring the safety of a structure by its reliability makes the reliability a useful decision parameter.

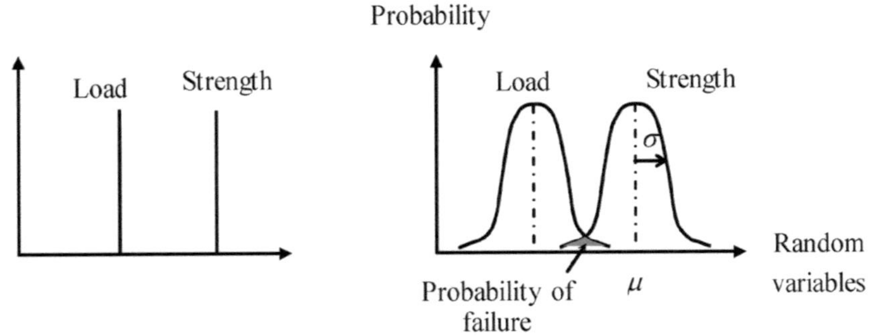

Figure 2. Comparison in load and strength from two different methods, (a) Deterministic (b) Probabilistic

http://dx.doi.org/10.5772/57488

The probability theory is considered to be more suitable since all design parameters used in engineering calculations have a degree of uncertainty. The uncertainties associated with these quantities arise because of many factors related to the measurement, calibration equipment, operator's error etc. Therefore, statistics and probability provides conventional means of reducing observed data to a form in which it can be interpreted, evaluated and effectively measured.

4.2. Design codes and standards

Several design codes and standards have been widely used to estimate the reliability assessment of corroded pipelines, namely ASME B31G, Modified ASME B31G, DNV 99, Shell 92, RSTRENG and PCORRC. Note that these codes take into account only the corrosion *depth* (*d*) and *longitudinal length* (*l*) parameters in their equations. The assessment in these codes is in the form of a *failure pressure* (PF) model, which is originated from the NG-18 criterion. Detailed discussions and comparison on the theories and development of the PF models have been carried out by Cosham et al. (2007), Bjørnøy and Marley (2001) and Cronin (2000), for instance while a summary of the equations are presented in Table 3.

Method	Basic equation	Flow stress	Defect shape	Bulging factor
NG-18	NG-18a	SMYS + 68.95 MPa	Rectangle (*dl*)	$\sqrt{1+0.6275\left(\frac{l}{\sqrt{Dt}}\right)^2-0.003375\left(\frac{l}{\sqrt{Dt}}\right)^4}$
ASME B31G	NG-18	1.1 SMYS	parabolic (2/3*dl*)	$\sqrt{1+0.8\left(\frac{l}{\sqrt{Dt}}\right)^2}$
Modified B31G	NG-18	SMYS + 68.95 MPa	arbitrary (0.85*dl*)	$\sqrt{1+0.6275\left(\frac{l}{\sqrt{Dt}}\right)^2-0.003375\left(\frac{l}{\sqrt{Dt}}\right)^4}$
RSTRENG	NG-18	SMYS + 68.95 MPa	river bottom profile	$\sqrt{1+0.6275\left(\frac{l}{\sqrt{Dt}}\right)^2-0.003375\left(\frac{l}{\sqrt{Dt}}\right)^4}$
SHELL 92	NG-18	SMTS	rectangle(*dl*)	$\sqrt{1+0.8\left(\frac{l}{\sqrt{Dt}}\right)^2}$
LPC	NG-18	SMTS	rectangle(*dl*)	$\sqrt{1+0.31\left(\frac{l}{\sqrt{Dt}}\right)^2}$
DNV-RP-F101	NG-18	SMTS	rectangle (*dl*) and river bottom profile	$\sqrt{1+0.31\left(\frac{l}{\sqrt{Dt}}\right)^2}$
PCORRC	Newb	SMTS	rectangle (*dl*)	c

Table 3. Design standards on the assessment of corrosion in pipelines (Adapted from Cosham et al., 2007)

[b]The basic equation of the PCORRC part-wall NG-18 failure criterion is,

$\sigma_\theta=\bar{\sigma}\left[\frac{1-\left(\frac{A}{A_0}\right)}{1-\left(\frac{A}{A_0}\right)^1/_M}\right]=\bar{\sigma}\left[\frac{1-\left(\frac{d}{t}\right)}{1-\left(\frac{d}{t}\right)^1/_M}\right]$, where, M is bulging factor and $\bar{\sigma}$ is flow stress.

[c] The basic equation for PCORRC failure criterion is,

$$\sigma_\theta = \bar{\sigma}\left[1-\left(\frac{d}{t}\right)\left(1-\exp\left[-0.16\left(\frac{l}{\sqrt{Rt}}\right)\left(1-\frac{d}{t}\right)^{-0.5}\right]\right)\right].$$

These design codes and standards have been critically discussed by Stephens and Francini (2000) and Cosham et al. (2007) for instance, and have further classified them into two categories, namely the 'old' and 'new' methods. Descriptions pertaining to these methods are summarized in Table 4.

Category	Description
'Old' Method	• Example: ASME B31G (or modified B31G and RSTRENG). • Predominantly developed and validated through full scale tests on older line pipe steels. • Empirically calibrated criteria that have been adjusted to be conservative for almost all corrosion defects, irrespective of the toughness of the line pipe (these criteria are variously based on the yield strength, the flow stress, or ultimate tensile strength).
'New' Method	• Example: DNV RP-F101 and PCORR. • Developed through tests on modern, high toughness, line pipe steels. • Developed based on plastic collapse criteria that are only appropriate for blunt defects in moderate to high toughness line pipe (these criteria are based on the ultimate tensile strength).

Table 4. Classification of the design standards on the assessment of corrosion in pipelines according to Stephens and Francini (2000) and Cosham et al. (2007)

The 'old' and 'new' methods were biased towards the type and toughness of the steels. Then, the difference between the behaviour of both categories can largely be attributed to the general increase in the toughness of line pipe, due to improvement in steel production and technological advances. Because of the 'old' methods demonstrate greater scatter than the 'new' methods when compared to the (relevant) published full-scale test data, the 'new' methods are more accurate (Cosham et al., 2007).

4.3. Limit state function models

With time, the *probabilistic* approaches have been identified as a more suitable approach to be used in the assessment of corrosions as compared to the deterministic method due to the conservatism of the latter (Mustaffa, 2011). The assessment of corroded pipelines has then been modified with the integration of probabilistic approaches into the existing failure pressure (PF) models. One of the common ways to represent probabilistic assessment is through the use of *limit state function* (LSF) equations. The LSF model is able to check the remaining strength of the pipe, for which its response towards operational loads can then be predicted.

A general limit state function, Z model can be written as,

$$Z = R - S \tag{8}$$

http://dx.doi.org/10.5772/57488

where R is the strength or more generally the resistance to failure and S is the load or that which is conducive to failure. The limit state is described by Z=0. Failures takes place when the failure surface falls in the region of $Z < 0$ while $Z > 0$ is a survival region. The probability of failure, P_f is then given by,

$$P_f = \mathrm{P_r}(Z \le 0) = \mathrm{P_r}(R \ge S) \tag{9}$$

The reliability is the probability $\mathrm{P_r}(Z \ge 0)$, and is therefore when described in term of probability of failure becomes,

$$\mathrm{P_r}(Z > 0) = 1 - P_f \tag{10}$$

The past limit state function models as given in Table 5 were mostly modified from the PF models originated from the NG-18 criterion. In these models, the strength/resistance (R) term of Eq. 8 has been aggressively studied by Ahammed and Melchers (1996), Pandey (1998), Ahammed (1998), De Leon and Macías (2005) and Teixeira et al. (2008), for instance. All models, however, assumed the same parameter for the load (S) term, represented by the *operational loading* (P_o) exerted by the transported hydrocarbon in the pipeline.

Limit State Function	Remarks
Ahammed and Melchers (1996): $Z = \left[2 m_f s_y \frac{t}{D} \frac{1-d/t}{1-d/(tM)}\right] - p_a$	A multiplying factor, m_f was introduced into the equation which is usually taken as between 1.10[10] and 1.15[10]
Pandey (1998): $Z = \left[2.3 s_y \frac{t}{D} \frac{1-d/t}{1-d/(tM)}\right] - p_a$	Flow stress, σ_{flow} coefficient of 1.15
Ahammed (1998) and De Leon and Macias (2005): $Z = \left\{2(s_y + 68.95)\frac{t}{D}\frac{1-[d_0 + R_d(T-T_0)]/t}{1-[d_0 + R_d(T-T_0)]/tM}\right\} - p_a$ with, $d=d_o+R_d(T-T_o)$ and $l=l_o+R_L(T-T_o)$	Revision in the σ_{flow} coefficient.
Teixeira et. al. (2008): $Z = \left(\frac{1.1 s_y 2t}{D}\right)[1-0.9435(d/t)^{1.6}(l/D)^{0.4}] - p_a$	Applied the Buckingham-π theorem and multivariate regression analysis techniques.

Table 5. Different limit state function models applied in the reliability assessment of corrosions in pipelines

For all expressions, the M term is given by,

$$M = \left(1 + 0.6275\frac{l^2}{Dt} - 0.003375\frac{l^4}{D^2 t^2}\right)^{1/2} \text{for } l^2/Dt \le 50 \text{ or}$$

$$M = 0.032\frac{l^2}{Dt} + 3.3 \text{for } l^2/Dt > 50$$

with s_y = yield stress
d = defect depth
l = defect longitudinal length
t = wall thickness
D = pipe outer diameter
M = Folias/bulging factor
p_a = applied/operating pressure
d_o = defect depth measured at time T_o
l_o = defect longitudinal length measured at time T_o
T = any future time
T_o = time of last inspection
R_d = radial corrosion rate ($=\Delta d/\Delta T$)
R_L = longitudinal corrosion rate ($=\Delta l/\Delta T$)

The Folias factor, M is a measure of stress concentration that is caused by radial deflection of the pipe surrounding a defect.

The performance of these models can be visualized in a plot as shown in Fig. 3. The figure illustrates the behaviour of P_f under varying pipeline operating pressures (P_o); a term that has been made dimensionless by dividing it with the specified minimum tensile strength ($SMTS$) term. The P_f increases as loads increases. Results from the figure showed the model developed by Ahammed and Melchers (1996) tends to fail first while those developed by Ahammed (1998) and De Leon and Macias (2005) seems to be the last to fail.

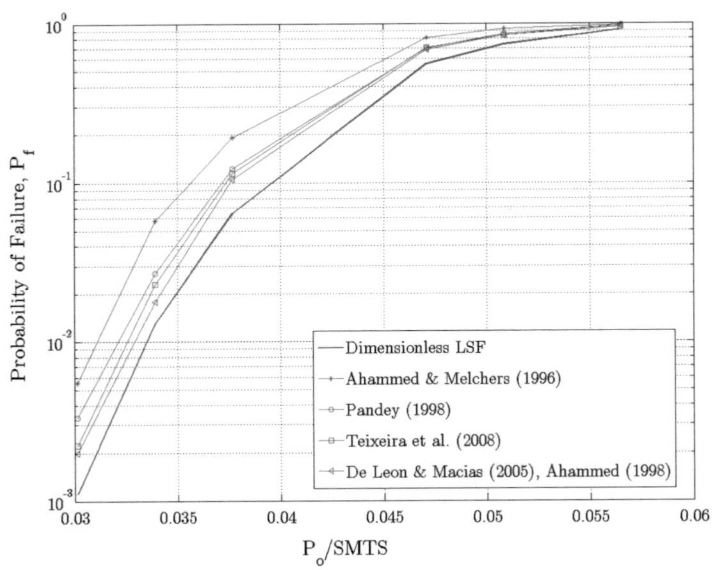

Figure 3. Probability of failure (P_f) computed for all limit state functions under varying operating pressures

This section has provided discussion on different methods and approaches used to estimate the reliability assessment of corrosion in pipelines, part of which can be summarized in Table 6 below.

Approach	Research	Related Literature
Deterministic approach: • Semi-empirical fracture-mechanical formula Finite element analysis	• Ambiguities in corrosion parameters and shapes. • Development of NG-18 criterion. • Development of *failure pressure* (PF) models design codes and standard criterion such as ASME B31G, Modified ASME B31G, DNV 99, Shell 92, RSTRENG and PCORRC.	*Critical summary prepared by:* Cronin (2000), BjØrnØy and Marley (2001) and Cosham et al. (2007) etc.
Probabilistic approach	• Corrosion assessment for *single pipeline*: Integration of *failure pressure* models into *limit state function* (LSF) models.	Ahammed and Melchers (1996), Pandey (1998), Ahammed (1998), De Leon and Macías (2005) and Teixeira et al. (2008) etc.
	• Corrosion assessment for *pipeline system*.	Li et al. (2009), De Leon and Macías (2005), Zhou (2010), and Mustaffa (2011).

Table 6. Brief summary on the development of reliability assessment of corrosions in pipelines

5. System reliability assessment of corrosions in pipelines

5.1. Reliability of pipeline in series

Several attempts have been made by previous researchers in understanding the response of corrosions towards the reliability of pipelines when acting as a *system*. A system can be defined as a group of elements connected either in *series* or *parallel,* having the same function or objective. Many structures are designed to be composed of multi components and the reliability/failure of a component may be triggered by other components as well. This is true as these components are sharing similar loads, thus their performance may be redundant.

A pipeline acts a system connected in series, as shown in Fig. 4. Therefore, the failure of any one or more of these components constitutes the failure of the system. In other words, the reliability or safety of the system requires that none of the components fail. From Fig. 4 for instance, the probability of failure, P_f of a corroded pipeline acting as a unit is not the same with those separated into different segments. Each small segment portrays smaller probability of failures (*i.e.* ~10^{-3}) as compared to the whole structure (*i.e.*~10^{-2}).

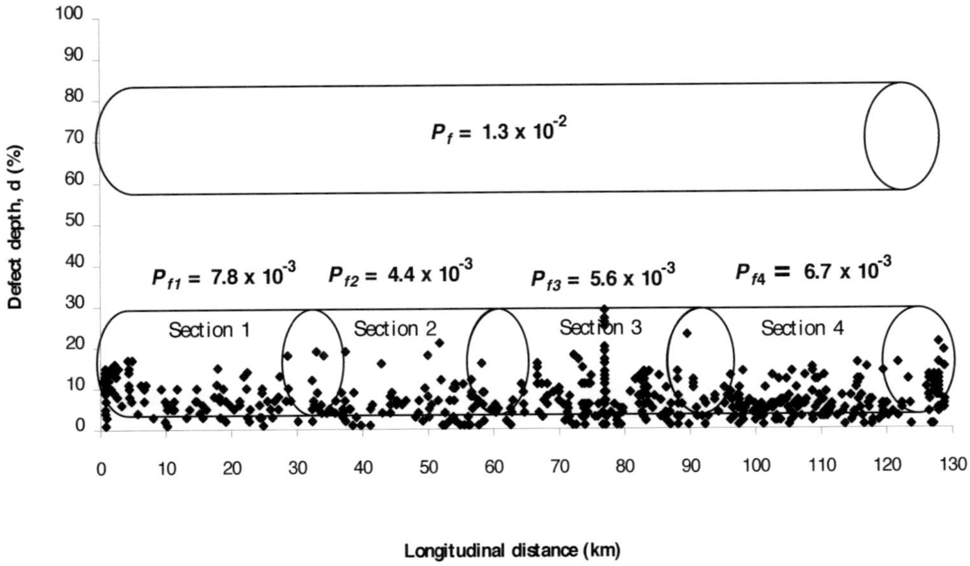

Figure 4. Comparison in probability of failure (P_f) between sectional and individual pipeline of corroded pipeline

Better understanding on pipelines acting as a system has paved the knowledge to interpret the correlation and characteristics among corrosion defects in the pipe. The interaction among multiple corrosion defects can best be assessed by assuming the pipeline operating as a series system. The interaction among defects is still not well defined. When it is conservative to assume that all of a cluster of adjacent defects interact (Cosham et al., 2007), Bjørnøy and Marley (2001) concluded that there are unlimited combinations of interaction of defects. Li et al. (2009) has studied the effect of correlation of corrosion defects and it was revealed that the assumption of independent corrosions defects lead to conservative results. Literatures on this, however, are still limited. To the author best knowledge, the reliability of corroded pipelines acting in a system has been reported by De Leon and Macías (2005), Zhou (2010) and Mustaffa (2011). A summary on their works are presented in Table 7.

Research on:	Related Literature
Spatial correlation between multiple defects on a pipeline system reliability under burst failure mode.	De Leon and Macías (2005)
Reliability of a corroding pipeline segment as a series system under three potential failure models (small leak, large leak and rupture).	Zhou (2010)
Length scale effect of a pipeline in series.	Mustaffa (2011)

Table 7. Literatures on corrosion assessment in pipeline systems

http://dx.doi.org/10.5772/57488

5.2. reliability-based maintenance approach

Most of the discussions on reliability assessment of pipelines presented in earlier sections of this chapter are those closely developed based on the mechanics of corrosions. The parameters addressed in the equations are very much depending on the design parameters of the pipe as well as the corrosion inspection data reported by the inspection tool *i.e.* intelligent pigging tool. Apparently, these are the important two factors required to carry out the assessment.

In reality, the development of corrosions in the pipe cannot be explicitly described. Not only corrosion growth is based on the science of corrosion, but other external factors are also believed to contribute to this development as well. These external factors can be further described in the aspects of operation and maintenance of the pipeline. Corrosions are mainly controlled by the use of *corrosion inhibitors*. A corrosion inhibitor is a chemical compound that, when added to a liquid or gas, decreases the corrosion rates of a material, typically a metal or an alloy. A pipeline is dosed continually with an inhibitor in order to mitigate against any corrosion that could occur. It acts as a film on a metal surface that provides physical protection against corrosive attack.

Describing the effect of corrosion inhibitors, however, is not a straightforward task (Nešić, 2007). The inhibitor is said to be only efficient when it could present in the water phase and reach the pipe wall. If the inhibitor residual concentration in the water phase exiting the pipeline is above a certain target level, then the whole pipeline is assumed to be protected by the inhibitor (Rippon, 2003). It is essential to keep the inhibitor concentration as close as to and above the minimum required at all time in order to effectively control corrosion in the pipeline in the most cost-effective ways. Longer exposure time may allow the inhibitor to perform better as suggested by Hong et al. (2002). In addition, Valor et al. (2010) have experimentally proven that corrosions do grow on a daily basis even though the increment is considered to be very small. The fact that corrosion grows every day, this then reflects the argument to also allow the presence of corrosion inhibitor in the pipe at the same time. Leaving a pipeline without inhibition within a period of time might cause the corrosion to grow rapidly. The inhibitor performance is the key for day to day assessment of the inhibitor system availability. Therefore the probability of an inhibitor to present and retain at the internal pipe circumference wall are something that should be addressed in a corrosion assessment model.

It has been assumed that the corrosion inhibitor will be injected into the system of pipeline at the correct dosage, without interruption during the lifetime of the system (Hedges et al., 2000), but experience has shown that these assumptions are not applicable for a variety of reasons for example pumps failure and interruption on inhibitor supplies. Also note that the practice of releasing inhibitor inside a pipe is mainly controlled by the pipeline operators. The inhibitor may not be injected accordingly to its dosage and schedule which can cause corrosion to happen rapidly in the pipelines. Consequently, the aspect of human intervention may then become an additional factor that indirectly contributes to the reliability assessment of corrosions in a pipeline (Mustaffa, 2011).

The practice of releasing inhibitor in a pipe is then considered as another source of uncertainty to the system. A reliability of a system that is exposed to such uncertainties can best be analysed using probabilistic approaches. This section illustrates an approach for which the effect of human intervention can be considered in a reliability model. This example portrays the inconsistency of releasing corrosion inhibitor in the pipeline, as shown in Fig. 5.

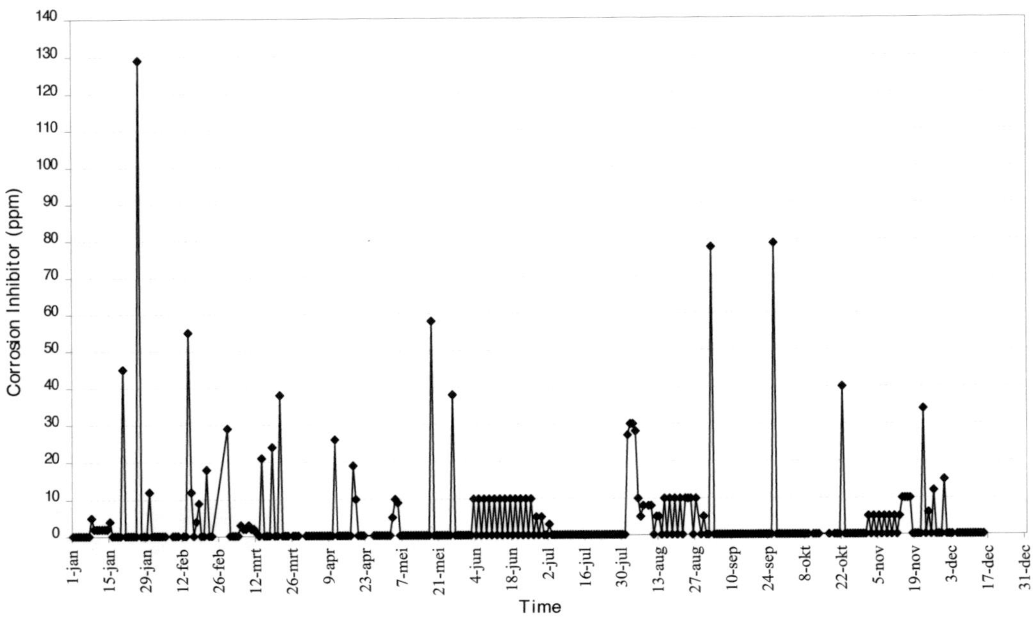

Figure 5. An illustration on the unavailability of corrosion inhibitor in pipeline in a year

In can be seen from Fig. 5 that the inhibitors were absent in the pipe for most of the days from a sample period of one year *i.e.* when inhibitor is 0 ppm. Such practice is believed to have direct impact towards corrosion growth in the pipe. It is proposed that the reliability for this scenario to be translated into the form of a limit state function model as well, part of which has been introduced in the earlier section. Therefore, the limit state function in Eq. 8 can be expanded to an equation described by,

$$g(x) = Z = d - (CR \times t_{abs}) \tag{11}$$

where d = allowable corrosion depth (mm), CR = corrosion rate (mm/yr), t_{abs}= time (days) when the corrosion inhibitor is absent in the pipeline. Eq. 11 translates the ideology of predicting corrosion in a particular pipeline given the amount of present corrosion, its expectation to evolve with time and the likelihood of having inhibitor in the pipeline. Eq. 11 also implicitly states that corrosions are allowed to grow based on the given rate, but their growth could be controlled by the practice of releasing inhibitor in the pipe.

In this model the availability of inhibitor is described as its time of absence per month. Herein, the corrosion parameters d and CR are identified based on the inspection and assessment reports while the time t_{abs} is entirely influenced by human intervention. The parameter t_{abs} on the other hand, is proposed to be treated as a variable so as to simulate its response towards the availability of inhibitor monitored in a month. The number of days of corrosion inhibitor absence in the pipeline decreased from 30 to 0 days and the effect on the pipeline integrity was investigated. Table 8 provides descriptive statistics of the random variables applied into the model.

Variables	Distribution	Mean, μ	Standard deviation, σ
Defect depth, *d*	Normal	15.69	0.76
Corrosion rate, *CR*	Normal	0.25	0.50
$t_{absence}$ @ 30 days	Normal	30.42	0.90
$t_{absence}$ @ 0 day	Normal	0.42	0.90

Table 8. Descriptive statistics of random variables

The probability of failures were computed using Eq. (11) based on the random tabulated in Table 8, with results as shown in Table 9.

t absence (day/month)	Probability of Failure (P_f)
30	0.312
28	0.278
26	0.247
24	0.231
22	0.188
20	0.149
18	0.109
16	0.087
14	0.081
12	0.032
10	0.009
8	0.003
6	0.002
4	0.000
2	0.000
0	0.000

Table 9. Probability of failure (P_f) reported according to $t_{absence}$

From the table, it can be seen that when the number of days of corrosion inhibitor absence in the pipeline decreases, the probability of failure also decreases. Results from the table portray the impact of corrosion failure as a result of the absence of corrosion inhibitor (in days per month) in a particular pipeline. It can be seen from the table that the probability of failure increases with the increase in the days of inhibitor absence. This is true because no continuous protection is given to the pipe giving more chance for the corrosions to grow rapidly. Thus it is important for the corrosion inhibitor to be present in the pipeline on a daily basis in order to keep the probability of failure lower.

6. Conclusions

This chapter discusses the development of reliability assessment models used for the offshore pipelines subjected to corrosion. It has been shown that the models were mostly developed based on the mechanics of corrosions acting on a pipe. Comparisons have also been made on different design codes and standards by the industry to estimate the reliability of corroded pipelines. Most of these standards, however, are deterministic in nature. Corrosions cannot be explicitly determined as there are so many uncertainties involved, thus probabilistic method seems to be more realistic to be applied instead. In the probabilistic approaches, literatures have integrated the failure pressure (PF) models into the limit state function (LSF) equation. Comparisons on the performance of these LSF models have been briefly discussed.

It is proposed that human intervention to be addressed when assessing the reliability of corroded pipelines. This is true as the growth of corrosions is very much depending on the effectiveness practice of releasing corrosion inhibitor in the pipe. Herein, a model has been proposed to the consider the aspect of human intervention in the assessment, and the simulation seems to provide favorable results.

Author details

Zahiraniza Mustaffa*

Civil Engineering Department, Universiti Teknologi PETRONAS, Tronoh, Perak, Malaysia

References

[1] Ahammed, M. and Melchers, R. E. (1996) Reliability Estimation of Pressurized Pipelines Subject to Localised Corrosion Defects, *International Journal of Pressure Vessels and Piping*, Vol 69, pp. 267-272.

[2] Ahammed, M. (1998) Probabilistic Estimation of Remaining Life of a Pipeline in the Presence of Active Corrosion Defects, *International Journal of Pressure Vessels and Piping*, Vol 75, pp. 321-329.

[3] Batte, A.D., Fu, B, Kirkwood, M.G. and Vu, D. (1997) New Methods for Determining the Remaining Strength of Corroded Pipelines, *the 16th International Conference on Ocean Mechanics and Arctic Engineering (OMAE).*

[4] Bjørnøy, O.H. and Marley, M.J. (2001) Assessment of Corroded Pipelines: Past, Present and Future, *International Journal Offshore and Polar Engineering Conference (ISOPE),* Vol 2, pp. 93-101.

[5] Chouchaoui, B.A. and Pick, R.J. (1994) Behaviour of Circumferentially Aligned Corrosion Pits, *International Journal of Pressure Vessels and Piping*, 57, pp. 187-200.

[6] Cosham, A, Hopkins, P. and Macdonald, K.A. (2007) Best Practice for the Assessment of Defects in Pipelines–Corrosions, *Engineering Failure Analysis*, No 14, pp. 1245-1265.

[7] Cronin, D.S. (2001) *Assessment of Corrosion Defects in Pipelines*, PhD, University of Waterloo.

[8] De Leon, D. and Macías, O.F. (2005) Effect of Spatial Correlation on the Failure Probability of Pipelines under Corrosion, *International Journal of Pressure Vessels and Piping*, No 82, pp. 123-128.

[9] Hedges, B., Paisley, D. and Woollam, R. (2000) Corrosion Inhibitor Availability Model, *Corrosion NACE International*, Paper No. 00034.

[10] Hong, T., Chen, Y., Gopal, M. and Jepson, W.P. (2000) EIS Study of a Corrosion Inhibitor Behavior Under Multiphase Flow Conditions, *Corrosion Science*, Vol 42, pp. 979-990.

[11] Fu, B. and Kirkwood, M.G. (1995) Predicting Failure Pressure of Internally Corroded Linepipe Using the Finite Element Method, *the 14th International Conference on Ocean Mechanics and Arctic Engineering(OMAE)*, Vol V, pp. 175-184.

[12] Kiefner, J.F. and Duffy, A.R. (1971) *Summary of Research to Determine the Strength of Corroded Areas in Line Pipe*, Presented to a Public Hearing at the U.S. Department of Transportation.

[13] Kiefner, J.F. (1974) Corroded Pipe: Strength and Repair Methods, *the 5th Symposium on Line Pipe Research*, Pipeline Research Committee of AGA.

[14] Maxey, W.A., Kiefner, J.F., Eiber, R.J. and Duffy A.R. (1971) Ductile Fracture Initiation, Propagation and Arrest in Cylindrical Vessels, Fracture Toughness, *Proc. of the 1971 National Symp. on Fracture Mechanics*, ASTM STP 514.

[15] Mustaffa, Z. and Van Gelder, P.H.A.J.M. (2010) A Review and Probabilistic Analysis of Limit State Functions of Corroded Pipelines, *the 20th International Offshore and Polar Engineering Conference (ISOPE*), Vol 4, pp. 625-632.

[16] Mustaffa, Z. (2011) *System Reliability Assessment of Offshore Pipeline.* Ph.D, Delft University of Technology, The Netherlands.

[17] Nešić, S. (2007) Key Issues Related to Modelling of Internal Corrosion of Oil and Gas Pipelines – A Review, *Corrosion Science*, Vol 49, pp. 4308-4338.

[18] Pandey, M. D. (1998) Probabilistic Models for Condition Assessment of Oil and Gas Pipelines, *NDT&E International*, Vol 31, No 5, pp. 349-358.

[19] Rippon, I. J. (2003) Corrosion Control System Availability Management for Reduced Cost and Extended Life, *Corrosion NACE International*, Paper No. 03313.

[20] Shitan, M. and Vazifedan, T. (2012) *Exploratory Data Analysis for Almost Anyone*, Universiti Putra Malaysia press.

[21] Singh, V.P., Jain, S.K. and Tyagi, A. (2007) *Risk and Reliability Analysis: A Handbook for Civil and Environmental Engineers*, United States of America, ASCE Press.

[22] Stephens, D.R., and Francini, R.B. (2000) A Review and Evaluation of Remaining Strength Criteria for Corrosion Defects in Transmission Pipelines, *Proc. of ETCE/ OMAE Joint Conference on Energy for the New Millennium.*

[23] Teixeira, A.P., Soares, C.G., Netto, T.A., and Estefen, S.F. (2008) Reliability of Pipelines With Corrosion Defects, *International Journal of Pressure Vessels and Piping*, Vol 85, pp. 228-237.

[24] Tukey, J.W. (1997) *Exploratory Data Analysis*, Addison-Wesley, Reading, Massachusetts.

[25] Valor, A., Caleyo, F., Rivas, D. and Hallen, J.M. (2010) Stochastic Approach to Pitting-Corrosion-Extreme Modelling in Low-Carbon Steel, *Corrosion Science*, 52, pp. 910-915.

[26] Zhou, W. (2010) System Reliability of Corroding Pipelines, *International Journal of Pressure Vessels and Piping*, Vol 87, pp. 587-595.